CREATING A VIEWING RECTANGLE

Each of the viewing rectangles shows portions of the graph of $y = 0.1x^4 - x^3 + 2x^2$.
The first viewing rectangle shows the most complete graph of the equation.

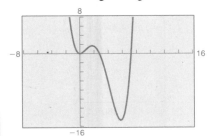
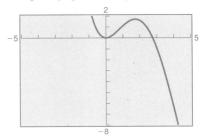
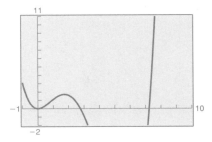

PARAMETRIC EQUATIONS OF A GRAPH

Circle
$$x = h + r \cos \theta$$
$$y = k + r \sin \theta$$

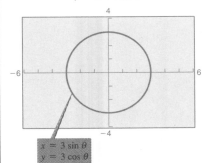

$x = 3 \sin \theta$
$y = 3 \cos \theta$

Ellipse
$$x = h + a \cos \theta$$
$$y = k + b \sin \theta$$

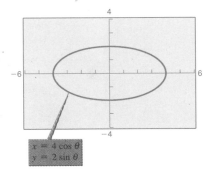

$x = 4 \cos \theta$
$y = 2 \sin \theta$

Hyperbola
$$x = h + a \sec \theta$$
$$y = k + b \tan \theta$$

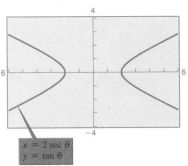

$x = 2 \sec \theta$
$y = \tan \theta$

GRAPHING POLAR EQUATIONS WITH A GRAPHING UTILITY

To graph the polar function $f(\theta)$, write the function in the parametric form $x = f(t) \cos t$ and $y = f(t) \sin t$.

Limaçon

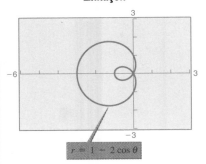

$r = 1 - 2 \cos \theta$

$x = (1 - 2 \cos t) \cos t$
$y = (1 - 2 \cos t) \sin t$
$0 \leq t \leq 2\pi$

Rose Curve

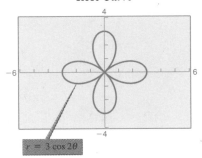

$r = 3 \cos 2\theta$

$x = (3 \cos 2t) \cos t$
$y = (3 \cos 2t) \sin t$
$0 \leq t \leq 2\pi$

Lemniscate

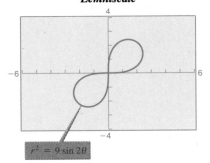

$r^2 = 9 \sin 2\theta$

$x = \left(3\sqrt{\sin(2t)}\right) \cos t$
$y = \left(3\sqrt{\sin(2t)}\right) \sin t$
$0 \leq t \leq 2\pi$

Trigonometry

A GRAPHING APPROACH
WITH TECHNOLOGY UPDATES

Trigonometry

A GRAPHING APPROACH
WITH
TECHNOLOGY UPDATES

Roland E. Larson
Robert P. Hostetler
THE PENNSYLVANIA STATE UNIVERSITY
THE BEHREND COLLEGE

Bruce H. Edwards
UNIVERSITY OF FLORIDA

WITH THE ASSISTANCE OF
David E. Heyd
THE PENNSYLVANIA STATE UNIVERSITY
THE BEHREND COLLEGE

Houghton Mifflin Company   Boston   New York

Sponsoring Editor: Christine B. Hoag

Senior Associate Editor: Maureen Brooks

Managing Editor: Catherine B. Cantin

Assistant Editor: Carolyn Johnson

Supervising Editor: Karen Carter

Associate Project Editor: Rachel D'Angelo Wimberly

Editorial Assistant: Caroline Lipscomb

Production Supervisor: Lisa Merrill

Art Supervisor: Gary Crespo

Marketing Manager: Charles Cavaliere

Marketing Associate: Ros Kane

Marketing Assistant: Kate Burden Thomas

Technical Art: Folium

Art and Photo Credits: p. 327, from *The Fractal Geometry of Nature* by Benoit B.
Mandelbrot, by permission; p. 328, from *Chaos and Fractals: The Mathematics
Behind the Computer Graphics,* edited by Robert L. Devaney and Linda Keen.
Copyright © 1989 by the American Mathematical Society. All rights reserved. Used
by permission of the publisher.

We have included examples and exercises that use real-life data. This would not have
been possible without the help of many people and organizations. Our wholehearted
thanks goes to all for their time and effort.

Printed in the U.S.A.

ISBN: 0-669-41760-2

123456789-Q-00 99 98 97 96

PREFACE

Trigonometry: A Graphing Approach with Technology Updates has two basic goals. The first is to help students develop a good understanding of trigonometry. The other goal is to show students how trigonometry can be used as a modeling language for real-life problems.

FEATURES

The text has several key features designed to help students develop their problem-solving skills, as well as acquire an understanding of mathematical concepts.

GRAPHICS

The ability to visualize a problem is a critical part of a student's ability to solve the problem. To encourage the development of this skill, the text has many figures in examples and exercise sets and in answers to odd-numbered exercises in the back of the text. Various types of graphics show geometric representations, including graphs of functions, geometric figures, symmetry, displays of statistical information, and numerous screen outputs from graphing technology. All graphs of functions, computer- or calculator-generated for accuracy, are designed to resemble students' actual screen outputs as closely as possible.

APPLICATIONS

Numerous pertinent applications are integrated throughout every section of the text, both as solved examples and as exercises. This encourages students to use and review their problem-solving skills. The text applications are current and often involve multiple parts. Students learn to apply the process of mathematical modeling to real-world situations in many areas, such as business, economics, biology, engineering, chemistry, and physics. Many applications in the text use real data, and source lines are included to help motivate student interest. We tried to use the data accurately—to give honest and unbiased portrayals of real-life situations. In the cases in which models were fit to data, we used the least-squares method. In all cases the square of the correlation coefficient r^2 was at least 0.95. In most cases it was 0.99 or greater.

EXAMPLES

Each example was carefully chosen to illustrate a particular concept or problem-solving technique. Examples are titled for quick reference, and many include color side comments to justify or explain the solution. We have included problems solved graphically, analytically, numerically, or by a combination of these strategies, and the text helps students choose appropriate approaches to the problems. Several examples also preview ideas from calculus, building an intuitive foundation for future study.

CONNECTIONS

When studying trigonometry, it is possible to see connections to areas outside mathematics as well as to other branches of mathematics. Both of these aspects of mathematical connections have been incorporated into this text. Connections between trigonometry and other disciplines are shown in the wide range of real-world applications in such areas as engineering, biology, and business. Similarly, connections between trigonometry and other areas of mathematics may be encountered throughout the text, including the connections between trigonometry and complex numbers (Chapter 5), polar coordinates (Chapter 7), and series (Appendix C). In addition, many examples are included that discuss techniques or graphically illustrate the concepts that are used in calculus. These help students develop a natural and intuitive foundation for later work.

DISCOVERY

Discovery boxes encourage students to strengthen their intuition and understanding by exploring the relationships between functions and the behaviors of functions. The powerful features of graphing utilities enhance the study of functions because a graph of each step in a solution can be generated quickly and easily for use in the problem-solving process.

EXERCISE SETS

Exercise sets, including warm-up exercises, appear at the end of each text section. Many sets include a group of exercises that provide the graphs of functions involved. Review exercises are included at the end of each chapter, and cumulative tests are included to review what students have learned from the preceding chapters. The opportunity to use calculators is provided with topics that allow students to see patterns, experiment, calculate, or create graphic models.

DISCUSSION PROBLEMS

The discussion problems offer students the opportunity to think, reason, and communicate about mathematics in different ways. Individually or in teams, for in-class discussion, writing assignments, or class presentations, students are encouraged to draw new conclusions about the concepts presented. The problem might ask for further explanation, synthesis, experimentation, or extension of the section concepts. Discussion problems appear at the end of each text section.

TECHNOLOGY NOTES

Technology notes to students appear in the margins throughout the text. These notes offer additional insights, help students avoid common errors, and provide opportunities for problem solving using technology.

WARM-UP EXERCISES AND CUMULATIVE TESTS

We have found that students can benefit greatly from reinforcement of previously learned concepts. Most sections in the text contain a set of ten warm-up exercises that efficiently give students practice using techniques studied earlier in the course that are necessary to master the new ideas presented in the section.

Cumulative tests are included after Chapters 2, 5, and 7. These tests help students assess their level of success and help them maintain the knowledge base they have been building throughout the text—preparing them for other exams and future courses.

These and other features of the text are described in greater detail on the following pages.

CHAPTER 2

TRIGONOMETRY

2.1   RADIAN AND DEGREE MEASURE

Angles / Radian Measure / Degree Measure / Applications

Angles

As derived from the Greek language, the word **trigonometry** means "measurement of triangles." Initially, trigonometry dealt with relationships among the sides and angles of triangles. As such, it was used in the development of astronomy, navigation, and surveying.

With the advent of calculus in the 17th century and a resulting expansion of knowledge in the physical sciences, a different perspective arose—one that viewed the classic trigonometric relationships as *functions* with the set of real numbers as their domains. Consequently, the applications of trigonometry expanded to include a vast number of physical phenomena involving rotations or vibrations. These include sound waves, light rays, planetary orbits, vibrating strings, pendulums, and orbits of atomic particles. Our approach to trigonometry incorporates *both* perspectives, starting with angles and their measure.

An **angle** is determined by rotating a ray (half-line) about its endpoint. The starting position of the ray is called the **initial side** of the angle, and the position after rotation is called the **terminal side,** as shown in Figure 2.1. The endpoint of the ray is called the **vertex** of the angle.

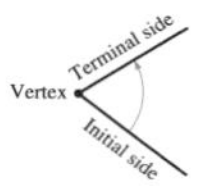

Vertex
Terminal side
Initial side

FIGURE 2.1

105

FEATURES OF THE TEXT

CHAPTER OPENER

Each chapter begins with a list of the topics to be covered. Each section begins with a list of important topics covered in that section.

DEFINITIONS

All of the important formulas and definitions are boxed for emphasis. Each is also titled for easy reference.

---

because (x, y) is on the unit circle, you know that $-1 \le y \le 1$ and $-1 \le x \le 1$, and it follows that the values of the sine and cosine also range between -1 and 1. That is,

$$-1 \le y \le 1 \qquad -1 \le x \le 1$$
$$-1 \le \sin t \le 1 \quad \text{and} \quad -1 \le \cos t \le 1.$$

Suppose you add 2π to each value of t in the interval $[0, 2\pi]$, thus completing a second revolution around the unit circle, as shown in Figure 2.22. The values of $\sin(t + 2\pi)$ and $\cos(t + 2\pi)$ correspond to those of $\sin t$ and $\cos t$. Similar results can be obtained for repeated revolutions (positive or negative) on the unit circle. This leads to the general result

$$\sin(t + 2\pi n) = \sin t \quad \text{and} \quad \cos(t + 2\pi n) = \cos t$$

for any integer n and real number t. Functions that behave in such a repetitive (or cyclic) manner are called **periodic.**

REMARK In Figure 2.22, *positive* multiples of 2π were added to the t-values. You could just as well have added *negative* multiples. For instance, $\pi/4 - 2\pi$ and $\pi/4 - 4\pi$ are also coterminal to $\pi/4$.

DISCOVERY

As the real number line is wrapped around the unit circle, each real number t will correspond with a point $(x, y) = (\cos t, \sin t)$ on the circle. You can visualize this graphically by setting your graphing utility to *simultaneous* mode. Using radian and parametric modes as well, let

$X_{1t} = \cos T$
$Y_{1t} = \sin T$
$X_{2t} = T$
$Y_{2t} = \sin T$.

Select the following ranges:
$Tmin = 0, Tmax = 6.3,$
$Tstep = .1, Xmin = -2,$
$Xmax = 7, Ymin = -3,$
$Ymax = 3$. Notice how the graphing utility traces out the unit circle and the sine function simultaneously. Try changing Y_{2t} to cos T or to tan T.

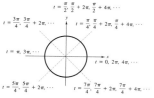

$t = \dfrac{\pi}{2}, \dfrac{\pi}{2} + 2\pi, \dfrac{\pi}{2} + 4\pi, \cdots$

$t = \dfrac{3\pi}{4}, \dfrac{3\pi}{4} + 2\pi, \cdots$

$t = \dfrac{\pi}{4}, \dfrac{\pi}{4} + 2\pi, \dfrac{\pi}{4} + 4\pi, \cdots$

$t = \pi, 3\pi, \cdots$

$t = 0, 2\pi, 4\pi, \cdots$

$t = \dfrac{5\pi}{4}, \dfrac{5\pi}{4} + 2\pi, \cdots$

$t = \dfrac{7\pi}{4}, \dfrac{7\pi}{4} + 2\pi, \dfrac{7\pi}{4} + 4\pi, \cdots$

Repeated Revolutions on the Unit Circle

FIGURE 2.22

DEFINITION OF A PERIODIC FUNCTION

A function f is **periodic** if there exists a positive real number c such that

$$f(t + c) = f(t)$$

for all t in the domain of f. The least number c for which f is periodic is called the **period** of f.

From this definition it follows that the sine and cosine functions are periodic and have a period of 2π. The other four trigonometric functions are also periodic, and we will say more about this in Section 2.6.

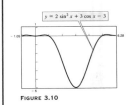

SECTION 3.3 SOLVING TRIGONOMETRIC EQUATIONS 223

SOLUTION

The graph of $y = 2 \sin^2 x + 3 \cos x - 3$, shown in Figure 3.10, indicates that there are three solutions in the interval $[0, 2\pi)$. Proceeding algebraically, you can write the following.

$$2 \sin^2 x + 3 \cos x - 3 = 0 \qquad \textit{Original equation}$$
$$2(1 - \cos^2 x) + 3 \cos x - 3 = 0 \qquad \textit{Pythagorean Identity}$$
$$2 \cos^2 x - 3 \cos x + 1 = 0 \qquad \textit{Multiply both sides by } -1$$
$$(2 \cos x - 1)(\cos x - 1) = 0 \qquad \textit{Factor}$$

By setting each factor equal to zero, you can find the solutions in the interval $[0, 2\pi)$ to be $x = 0$, $x = \pi/3$, and $x = 5\pi/3$. The general solution is therefore

$$x = 2n\pi, \qquad x = \frac{\pi}{3} + 2n\pi, \qquad x = \frac{5\pi}{3} + 2n\pi, \qquad \textit{General solution}$$

where n is an integer.

Squaring both sides of an equation can introduce extraneous solutions, as indicated in Example 7.

EXAMPLE 7 Squaring and Converting to Quadratic Type

$$\cos x + 1 = \sin x \qquad \textit{Original equation}$$
$$\cos^2 x + 2 \cos x + 1 = \sin^2 x \qquad \textit{Square both sides}$$
$$\cos^2 x + 2 \cos x + 1 = 1 - \cos^2 x \qquad \textit{Identity}$$
$$2 \cos^2 x + 2 \cos x = 0 \qquad \textit{Collect terms}$$
$$2 \cos x(\cos x + 1) = 0 \qquad \textit{Factor}$$

Setting each factor equal to zero produces the following.

$$2 \cos x = 0 \qquad \text{and} \qquad \cos x + 1 = 0$$
$$\cos x = 0 \qquad \qquad \cos x = -1$$
$$x = \frac{\pi}{2}, \frac{3\pi}{2} \qquad \qquad x = \pi$$

Because you squared both sides of the original equation, you must check for extraneous solutions. Of the three possible solutions, $x = 3\pi/2$ turns out to be extraneous. (Check this.) Thus, in the interval $[0, 2\pi)$, the only two solutions are $x = \pi/2$ and $x = \pi$. You can confirm this from the graph of $y = \cos x + 1 - \sin x$, shown in Figure 3.11. The general solution is $x = \pi/2 + 2n\pi$ and $x = \pi + 2n\pi$, where n is an integer.

FIGURE 3.10 $y = 2 \sin^2 x + 3 \cos x - 3$

FIGURE 3.11 $y = \cos x + 1 - \sin x$

CALCULATORS AND COMPUTER GRAPHING UTILITIES

To broaden the range of teaching and learning options, notes for working with calculators occur in many places. Students with access to graphics calculators or graphing utilities can solve exercises both graphically and analytically beginning with Chapter 1. Additionally, many exercises require a graphing utility.

R 2 TRIGONOMETRY

Translations of Sine and Cosine Curves

The constant c in the general equations

$$y = a \sin(bx - c) \quad \text{and} \quad y = a \cos(bx - c)$$

creates a horizontal shift of the basic sine and cosine curves. The graph of $y = a \sin(bx - c)$ completes one cycle from $bx - c = 0$ to $bx - c = 2\pi$. By solving for x in the inequality $0 \le bx - c \le 2\pi$, you can find the interval for one cycle to be

$$\underbrace{\frac{c}{b}}_{\text{Left endpoint}} \le x \le \underbrace{\frac{c}{b} + \frac{2\pi}{b}}_{\text{Right endpoint}}.$$

This implies that the period of $y = a \sin(bx - c)$ is $2\pi/b$, and the graph of $y = a \sin bx$ is shifted by an amount c/b. The number c/b is the **phase shift**.

DISCOVERY

Use a graphing utility to draw the graph of $y = \sin(x + c)$, where $c = -\pi/4$, 0, and $\pi/4$. (Use a viewing rectangle in which $0 \le X \le 6.3$ and $-2 \le Y \le 2$.) How does the value of c affect the graph?

DISCOVERY

Use your graphing utility to compare the graphs of $y_1 = \cos x$ and $y_2 = \cos(-x)$. Discuss the symmetry of the cosine function. Is $\cos x$ an even function, odd function, or neither? Repeat the experiment for $y_1 = \sin x$ and $y_2 = \sin(-x)$.

GRAPHS OF THE SINE AND COSINE FUNCTIONS

The graphs of $y = a \sin(bx - c)$ and $y = a \cos(bx - c)$ have the following characteristics. (Assume $b > 0$.)

$\quad$ Amplitude $= |a|$

$\quad$ Period $= 2\pi/b$

The left and right endpoints corresponding to a one-cycle interval of the graphs can be determined by solving the equations $bx - c = 0$ and $bx - c = 2\pi$.

EXAMPLES

The text contains over 350 examples. They are titled for easy reference, and many include side comments that explain or justify steps in the solution. Students are encouraged to check their solutions.

DISCOVERY

Throughout the text, the discovery boxes take advantage of the power of graphing utilities to explore and examine the behavior of complicated functions.

TECHNOLOGY NOTES

These notes appear in the margins. They provide additional insight and help students avoid common errors.

Technology Note

Some graphing utilities have the capability to perform vector operations, such as the cross product. For example, on the TI-85, you can use the VECTR menu to verify the cross product in Example 1 as follows:

cross ([1, 2, 1], [3, 1, 2])

EXAMPLE 1   **Finding Cross Products**

Given $\mathbf{u} = \mathbf{i} + 2\mathbf{j} + \mathbf{k}$ and $\mathbf{v} = 3\mathbf{i} + \mathbf{j} + 2\mathbf{k}$, find the following.

A. $\mathbf{u} \times \mathbf{v}$ **B.** $\mathbf{v} \times \mathbf{u}$ **C.** $\mathbf{v} \times \mathbf{v}$

SOLUTION

A. $\mathbf{u} \times \mathbf{v} = \begin{vmatrix} \mathbf{i} & \mathbf{j} & \mathbf{k} \\ 1 & 2 & 1 \\ 3 & 1 & 2 \end{vmatrix} = \begin{vmatrix} 2 & 1 \\ 1 & 2 \end{vmatrix} \mathbf{i} - \begin{vmatrix} 1 & 1 \\ 3 & 2 \end{vmatrix} \mathbf{j} + \begin{vmatrix} 1 & 2 \\ 3 & 1 \end{vmatrix} \mathbf{k}$

$= (4 - 1)\,\mathbf{i} - (2 - 3)\,\mathbf{j} + (1 - 6)\,\mathbf{k}$

$= 3\mathbf{i} + \mathbf{j} - 5\mathbf{k}$

B. $\mathbf{v} \times \mathbf{u} = \begin{vmatrix} \mathbf{i} & \mathbf{j} & \mathbf{k} \\ 3 & 1 & 2 \\ 1 & 2 & 1 \end{vmatrix} = \begin{vmatrix} 1 & 2 \\ 2 & 1 \end{vmatrix} \mathbf{i} - \begin{vmatrix} 3 & 2 \\ 1 & 1 \end{vmatrix} \mathbf{j} + \begin{vmatrix} 3 & 1 \\ 1 & 2 \end{vmatrix} \mathbf{k}$

$= (1 - 4)\,\mathbf{i} - (3 - 2)\,\mathbf{j} + (6 - 1)\,\mathbf{k}$

$= -3\mathbf{i} - \mathbf{j} + 5\mathbf{k}$

Note that this result is the negative of that in part **a**.

C. $\mathbf{v} \times \mathbf{v} = \begin{vmatrix} \mathbf{i} & \mathbf{j} & \mathbf{k} \\ 3 & 1 & 2 \\ 3 & 1 & 2 \end{vmatrix} = \mathbf{0}$

The results obtained in Example 1 suggest some interesting *algebraic* properties of the cross product. For instance

$$\mathbf{u} \times \mathbf{v} = -(\mathbf{v} \times \mathbf{u}) \quad \text{and} \quad \mathbf{v} \times \mathbf{v} = \mathbf{0}.$$

These properties, and several others, are summarized in the following list.

ALGEBRAIC PROPERTIES OF THE CROSS PRODUCT

Let $\mathbf{u}$, $\mathbf{v}$, and $\mathbf{w}$ be vectors in space and let c be a scalar.

1. $\mathbf{u} \times \mathbf{v} = -(\mathbf{v} \times \mathbf{u})$
2. $\mathbf{u} \times (\mathbf{v} + \mathbf{w}) = (\mathbf{u} \times \mathbf{v}) + (\mathbf{u} \times \mathbf{w})$
3. $c(\mathbf{u} \times \mathbf{v}) = (c\mathbf{u}) \times \mathbf{v} = \mathbf{u} \times (c\mathbf{v})$
4. $\mathbf{u} \times \mathbf{0} = \mathbf{0} \times \mathbf{u} = \mathbf{0}$
5. $\mathbf{u} \times \mathbf{u} = \mathbf{0}$
6. $\mathbf{u} \cdot (\mathbf{v} \times \mathbf{w}) = (\mathbf{u} \times \mathbf{v}) \cdot \mathbf{w}$

EXAMPLE 5   **Finding a Maximum *r*-Value of a Polar Graph**

Find the maximum r-value of the graph of $r = 2 \cos 3\theta$.

SOLUTION

Because $2 \cos 3\theta = 2 \cos(-3\theta)$, you can conclude that the graph is symmetric with respect to the polar axis. A set of parametric equations for the graph is given by

$$x = 2 \cos 3t \cos t$$
$$y = 2 \cos 3t \sin t.$$

REMARK The graph shown in Figure 7.63 is called a **rose curve**, and each of the loops on the graph is called a *petal* of the rose curve.

By letting t vary from 0 to π, you can use a graphing utility to display the graph shown in Figure 7.63. The maximum r-value for this graph is 2. This value occurs at *three* different points on the graph: $(r, \theta) = (2, 0)$, $(-2, \pi/3)$, and $(2, 2\pi/3)$.

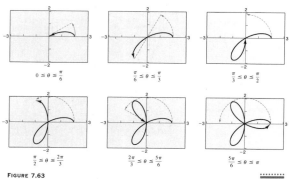

$0 \le \theta \le \dfrac{\pi}{6}$ $\dfrac{\pi}{6} \le \theta \le \dfrac{\pi}{3}$ $\dfrac{\pi}{3} \le \theta \le \dfrac{\pi}{2}$

$\dfrac{\pi}{2} \le \theta \le \dfrac{2\pi}{3}$ $\dfrac{2\pi}{3} \le \theta \le \dfrac{5\pi}{6}$ $\dfrac{5\pi}{6} \le \theta \le \pi$

FIGURE 7.63

Special Polar Graphs

Several important types of graphs have equations that are simpler in polar form than in rectangular form. For example, the circle $r = 4 \sin \theta$ in Example 1 has the more complicated rectangular equation $x^2 + (y - 2)^2 = 4$. The following list gives several other types of graphs that have simple polar equations.

INTUITIVE FOUNDATION FOR CALCULUS

Special emphasis has been given to skills that are needed in calculus. Many examples include algebraic techniques or graphically show concepts that are used in calculus, providing an intuitive foundation for future work.

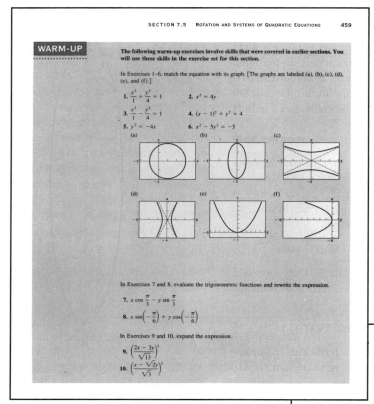

WARM-UPS

Each section contains a set of 10 warm-up exercises for students to review and practice the previously learned skills that are necessary to master the new skills and concepts presented in the section. All warm-up exercises are answered in the back of the text.

DISCUSSION PROBLEMS

A discussion problem appears at the end of each section. Each one encourages students to think, reason, and write about mathematics, individually or in groups. Presenting the mathematics in a different way from in the section, these problems emphasize synthesis and experimentation.

ADDITIONAL TOPICS IN TRIGONOMETRY

EXAMPLE 4   Using Heron's Area Formula

Find the area of the triangular region having sides of lengths $a = 47$ yards. $b = 58$ yards, and $c = 78.6$ yards.

SOLUTION

Because

$$s = \frac{1}{2}(a + b + c) = \frac{183.6}{2} = 91.8,$$

Heron's Formula yields

$$\text{Area} = \sqrt{s(s - a)(s - b)(s - c)}$$
$$= \sqrt{91.8(44.8)(33.8)(13.2)}$$
$$\approx 1354.58 \text{ square yards.}$$

DISCUSSION PROBLEM

THE AREA OF A TRIANGLE

We have now discussed three different formulas for the area of a triangle.

Standard formula: Area $= \frac{1}{2}bh$

Oblique triangle: Area $= \frac{1}{2}bc \sin A = \frac{1}{2}ab \sin C = \frac{1}{2}ac \sin B$

Heron's formula: Area $= \sqrt{s(s - a)(s - b)(s - c)}, \quad s = \frac{1}{2}(a + b + c)$

Use the most appropriate formula to find the area of each of the following triangles. Show your work and give your reason for choosing each formula.

(a)

(b)

(c)

APPLICATIONS

Real-world applications are integrated throughout the text in examples and exercises. This offers students insight about the usefulness of trigonometry, develops strategies for solving problems, and emphasizes the relevance of the mathematics. Titled for reference, many of the applications involve multiple parts, use current real data, and include source lines.

GRAPHICS

Students must be able to visualize problems in order to solve them. To develop this skill and reinforce concepts, the text has nearly 1000 figures.

---

272   CHAPTER 4   ADDITIONAL TOPICS IN TRIGONOMETRY

40. *Navigation* On a certain map, Minneapolis is 6.5 inches due west of Albany, Phoenix is 8.5 inches from Minneapolis, and Phoenix is 14.5 inches from Albany (see figure).
(a) Find the bearing of Minneapolis from Phoenix.
(b) Find the bearing of Albany from Phoenix.

FIGURE FOR 40

41. *Baseball* In a (square) baseball diamond with 90-foot sides, the pitcher's mound is 60 feet from home plate.
(a) How far is it from the pitcher's mound to third base?
(b) When a runner is halfway from second to third, how far is the runner from the pitcher's mound?

42. *Baseball* The baseball player in center field is playing approximately 330 feet from the television camera that is behind home plate. A batter hits a fly ball that goes to the wall 420 feet from the camera (see figure). Approximate the number of feet that the center fielder had to run to make the catch if the camera turned 9° in following the play.

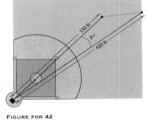

FIGURE FOR 42

43. *Awning Design* A retractable awning lowers at an angle of 50° from the top of a patio door that is 7 feet high (see figure). Find the length x of the awning if no direct sunlight is to enter the door when the angle of elevation of the sun is greater than 65°.

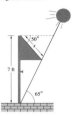

FIGURE FOR 43

44. *Circumscribed and Inscribed Circles* Let R and r be the radii of the circumscribed and inscribed circles of a triangle ABC, respectively, and let $s = (a + b + c)/2$ (see figure). Prove the following.
(a) $2R = \dfrac{a}{\sin A} = \dfrac{b}{\sin B} = \dfrac{c}{\sin C}$
(b) $r = \sqrt{\dfrac{(s - a)(s - b)(s - c)}{s}}$

$R =$ radius of the large circle

$r =$ radius of the small circle

FIGURE FOR 44

SECTION 2.5   GRAPHS OF SINE AND COSINE

In Exercises 59 and 60, find a and d for the function $f(x) = a \cos x + d$ so that the graph of f matches the figure.

59.   **60.**

In Exercises 61–64, find a, b, and c so that the graph of the function matches the graph in the figure.

61. $y = a \sin(bx - c)$ **62.** $y = a \sin(bx - c)$

63. $y = a \cos(bx - c)$ **64.** $y = a \sin(bx - c)$

65. Use a graphing utility to graph the functions $f(x) = 2e^x$ and $g(x) = 5 \cos x$. Approximate any points of intersection of the graphs in the interval $[-\pi, \pi]$.

66. Use a graphing utility to determine the smallest *integer* value of a such that the graphs of $f(x) = 2 \ln x$ and $g(x) = a \cos x$ intersect more than once.

67. *Respiratory Cycle* For a person at rest, the velocity v (in liters per second) of air flow during a respiratory cycle is

$v = 0.85 \sin \dfrac{\pi t}{3}$,

where t is the time in seconds. (Inhalation occurs when $v > 0$, and exhalation occurs when $v < 0$.)
(a) Find the time for one full respiratory cycle.
(b) Find the number of cycles per minute.
(c) Use a graphing utility to graph the velocity function.

68. *Respiratory Cycle* After exercising for a few minutes, a person has a respiratory cycle for which the velocity of air flow is approximated by

$v = 1.75 \sin \dfrac{\pi t}{2}$.

Use this model to repeat Exercise 67.

69. *Piano Tuning* When tuning a piano, tuning fork for the A above middle motion that can be approximated b

$y = 0.001 \sin 880\pi t$,

where t is the time in seconds.
(a) What is the period p of this fu
(b) The frequency f is given by f quency of this note?
(c) Use a graphing utility to graph this function.

70. *Blood Pressure* The function

$P = 100 - 20 \cos \dfrac{5\pi t}{3}$

approximates the blood pressure P in millimeters of mercury at time t in seconds for a person at rest.
(a) Find the period of the function.
(b) Find the number of heartbeats per minute.
(c) Use a graphing utility to graph the pressure function.

Sales In Exercises 71 and 72, use a graphing utility to graph the sales function over 1 year where S is sales in thousands of units and t is the time in months, with $t = 1$ corresponding to January. Use the graph to determine the month of maximum sales and the month of minimum sales.

71. $S = 22.3 - 3.4 \cos \dfrac{\pi t}{6}$

72. $S = 74.50 + 43.75 \sin \dfrac{\pi t}{6}$

In Exercises 73–76, describe the relationship between the graphs of the functions f and g.

73.   **74.**

75.   **76.**

GEOMETRY

Geometric formulas and concepts are reviewed throughout the text. For easy reference, common formulas are given inside the back cover.

EXERCISES

The approximately 3700 exercises—computational, conceptual, exploratory, and applied problems—are designed to build competence, skill, and understanding. Each exercise set is graded in difficulty to allow students to gain confidence as they progress. Many exercises require the use of a graphing utility. All odd-numbered exercises are solved in detail in the *Study and Solutions Guide,* with answers appearing in the back of the text.

40. $(3, 0, 0), (4, 1, 2), (3, -1, 4), (2, -2, 2), (-1, 5, 4), (0, 6, 6),$
$(-1, 4, 8), (-2, 3, 6)$

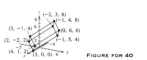

FIGURE FOR 40

In Exercises 41 and 42, prove the property of the cross product where $\mathbf{u} = \langle u_1, u_2, u_3 \rangle$ and $\mathbf{v} = \langle v_1, v_2, v_3 \rangle$.

41. $\mathbf{u} \times \mathbf{u} = \mathbf{0}$

42. $\mathbf{u} \times \mathbf{v}$ is orthogonal to both $\mathbf{u}$ and $\mathbf{v}$.

43. Consider the vectors $\mathbf{u} = \langle \cos \beta, \sin \beta, 0 \rangle$ and $\mathbf{v} = \langle \cos \alpha, \sin \alpha, 0 \rangle$ where $\alpha > \beta$. Find the cross product of the vectors and use the result to prove the identity
$$\sin(\alpha - \beta) = \sin \alpha \cos \beta - \cos \alpha \sin \beta.$$

CHAPTER 4 · REVIEW EXERCISES

In Exercises 1–16, solve the triangle (if possible). If two solutions exist, list both.

1. $a = 5, b = 8, c = 10$
2. $a = 6, b = 9, C = 45°$
3. $A = 12°, B = 58°, a = 5$
4. $B = 110°, C = 30°, c = 10.5$
5. $B = 110°, a = 4, c = 4$
6. $a = 80, b = 60, c = 100$
7. $A = 75° a = 2.5, b = 16.5$
8. $A = 130°, a = 50, b = 30$
9. $B = 115°, a = 7, b = 14.5$
10. $C = 50°, a = 25, c = 22$
11. $A = 15°, a = 5, b = 10$
12. $B = 150°, a = 64, b = 10$
13. $B = 150°, a = 10, c = 20$
14. $a = 2.5, b = 15.0, c = 4.5$
15. $B = 25°, a = 6.2, b = 4$
16. $B = 90°, a = 5, c = 12$

In Exercises 17–20, find the area of the triangle.

17. $a = 4, b = 5, c = 7$ **18.** $a = 15, b = 8, c = 10$
19. $A = 27°, b = 5, c = 8$ **20.** $B = 80°, a = 4, c = 8$

21. *Height of a Tree*   Find the height of a tree that stands on a hillside of slope 32° (from the horizontal) if from a point 75 feet down the hill from the tree the angle of elevation to the top of the tree is 48° (see figure)

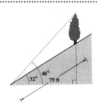

FIGURE FOR 21

22. *Surveying*   To approximate the length of a marsh, a surveyor walks 450 meters from point A to point B. Then the surveyor turns 65° and walks 325 meters to point C. Approximate the length AC of the marsh (see figure).

FIGURE FOR 22

REVIEW EXERCISES

A set of review exercises at the end of each chapter gives students an opportunity for additional practice. The review exercises include computational, conceptual, and applied problems covering a wide range of topics.

CHAPTERS 1 AND 2 · CUMULATIVE TEST

Take this test as you would take a test in class. After you are done, check your work against the answers in the back of the book.

In Exercises 1 and 2, solve the equation.

1. $1 - \frac{1}{2}(x - 3) = \frac{3}{4}$ **2.** $3x^2 + x - 10 = 0$

In Exercises 3–6, sketch the graph of the equation without the aid of a graphing utility.

3. $x - 3y + 12 = 0$ **4.** $y = x^2 - 4x + 1$
5. $y = \sqrt{4 - x}$ **6.** $y = 3 - |x - 2|$

7. A line passes through the point $(4, 9)$ with slope $m = -\frac{2}{3}$. Find the coordinates of three additional points on the line. (There are many correct answers.)

8. Find an equation of the line passing through the points $(-\frac{1}{2}, 1)$ and $(3, 8)$.

9. Evaluate (if possible) the function $f(x) = x/(x - 2)$ at the specified values of the independent variable.
  (a) $f(6)$ (b) $f(2)$ (c) $f(2t)$ (d) $f(s + 2)$

10. Express the area A of an equilateral triangle as a function of the length s of its sides.

11. Express the angle $4\pi/9$ in degree measure and sketch the angle in standard position.

12. Express the angle $-120°$ in radian measure as a multiple of π and sketch the angle in standard position.

13. The terminal side of an angle θ in standard position passes through the point $(12, 5)$. Evaluate the six trigonometric functions of the angle.

14. If $\cos t = -\frac{2}{3}$ where $\pi/2 < t < \pi$, find $\sin t$ and $\tan t$.

15. Use a calculator to approximate two values of $\theta(0° \le \theta < 360°)$ such that $\sec \theta = 2.125$. Round your answer to two decimal places.

16. Sketch a graph of each of the functions through two periods.
  (a) $y = -3 \sin 2x$ (b) $f(x) = 2 \cos\left(x - \frac{\pi}{2}\right)$
  (c) $g(x) = \tan \frac{\pi x}{2}$ (d) $h(t) = \sec t$

17. Write a sentence describing the relationship between the graphs of the functions $f(x) = \sin x$ and g.
  (a) $g(x) = 10 + \sin x$ (b) $g(x) = \sin \frac{\pi x}{2}$
  (c) $g(x) = \sin\left(x + \frac{\pi}{4}\right)$ (d) $g(x) = -\sin x$

18. Consider the function $f(x) = \sin 3x - 2 \cos x$.
  (a) Use a graphing utility to graph the function.
  (b) Approximate (accurate to one decimal place) the zero of the function in the interval $[0, 3]$.
  (c) Approximate (accurate to one decimal place) the maximum value of the function in the interval $[0, 3]$.

19. Evaluate the expression without the aid of a calculator.
  (a) $\arcsin \frac{1}{2}$ (b) $\arctan \sqrt{3}$

20. Write an expression that is equivalent to $\sin(\arccos 2x)$.

21. From a point on the ground 600 feet from the foot of a cliff, the angle of elevation to the top of the cliff is 32° 30'. How high is the cliff?

CUMULATIVE TESTS

Cumulative tests appear after Chapters 2, 5, and 7. These tests help students judge their mastery of previously covered concepts. They also help students maintain the knowledge base they have been building throughout the text, preparing them for other exams and future courses.

SUPPLEMENTS

This text is accompanied by a comprehensive supplements package for maximum teaching effectiveness and efficiency.

Instructor's Guide by Bruce H. Edwards, *University of Florida*, and David C. Falvo, *The Pennsylvania State University, The Behrend College*

Study and Solutions Guide by Bruce H. Edwards, *University of Florida*, and Dianna L. Zook, *Indiana University—Purdue University at Fort Wayne*

Test Item File and Resource Guide by David C. Falvo, *The Pennsylvania State University, The Behrend College*

Graphing Technology Guide by Benjamin N. Levy

Videotapes by Dana Mosely

Test-Generating Software (IBM, Macintosh)

BestGrapher Software (IBM, Macintosh) by George Best

Derive Software by Soft Warehouse—discounted for adopters

This complete supplements package offers ancillary materials for students, for instructors, and for classroom resources. Most items are keyed directly to the textbook for easy use. The components of this comprehensive teaching and learning package are outlined on the following pages.

PRINTED ANCILLARIES	SOFTWARE AND VIDEOS

INSTRUCTORS

Instructor's Guide
- Solutions to all even-numbered text exercises, all discussion problems, and all cumulative tests
- Project suggestions
- Chapter summaries with objectives and key term lists

Test Item File and Resource Guide
- Printed test bank
- Nearly 1850 test items
- Open-ended and multiple-choice test items
- Available as test-generating software
- Sample chapter tests
- Alternative assessment strategies
- Sample questions with standardized exam assessment formats

BestGrapher
- Function grapher
- Screen simultaneously displays equation, graph, and table of values
- Some features anticipate calculus
- Includes zoom and print features for use on assignments

Computerized Testing Software
- Test-generating software
- Nearly 1850 test items
- Also available as a printed test item file

Derive
- Computer algebra system
- Discount available to adopters

STUDENTS

Study and Solutions Guide
- Solutions to all odd-numbered text exercises
- Solutions match methods of text
- Summaries of key concepts in each text chapter
- Study strategies

Graphing Technology Guide
- Keystroke instructions for six different models of graphics calculators, representing Texas Instruments, Sharp, Casio, and Hewlett-Packard
- BestGrapher instructions for both IBM and Macintosh versions
- Step-by-step examples
- Numerous graphics screen output
- Technology tips

BestGrapher
- Function grapher
- Screen simultaneously displays equation, graph, and table of values
- Some features anticipate calculus
- Includes zoom and print features for use on assignments

Videotapes

Instructor's Guide

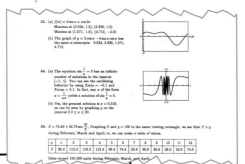

83. $g(x) = \dfrac{\sin x}{x}$

g approaches the value of 1 as x approaches 0.

85. $f(x) = \sin\left(\dfrac{1}{x}\right)$

f oscillates between -1 and 1 as x approaches 0.

87. $f(x) = \sin x + \cos\left(x + \dfrac{\pi}{2}\right)$

$g(x) = 0$
$f = g$
(The functions are identically 0.)

89. $f(x) = \sin^2 x$

$g(x) = \frac{1}{2}(1 - \cos 2x)$
$f = g$
(The functions are equal.)

Study and Solutions Guide

52. (a) $f(x) = 2\sin x + \cos 2x$

Maxima at (0.524, 1.5), (2.620, 1.5)
Minima at (1.571, 1.0), (4.712, −1.0)

(b) The graph of $y = 2\cos x - 4\sin x \cos x$ has the same x-intercepts: 0.524, 2.620, 1.571, 4.712.

54. (a) The equation $\sin\dfrac{1}{x} = 0$ has an infinite number of solutions in the interval $[-1, 1]$. You can see the oscillating behavior by using Xmin = −0.1 and Xmax = 0.1. In fact, any x of the form $x = \dfrac{1}{n\pi}$ yields a solution of $\sin\dfrac{1}{x} = 0$.

(b) Yes, the greatest solution is $x = 0.318$, as can be seen by graphing y on the interval $0.2 \le x \le 20$.

56. $S = 74.50 + 43.75\sin\dfrac{\pi t}{6}$. Graphing S and $y = 100$ in the same viewing rectangle, we see that $S > y$ during February, March and April; or, we can make a table of values.

y	1	2	3	4	5	6	7	8	9	10	11	12
S	96.4	112.4	118.3	112.4	96.4	74.5	52.6	36.6	30.8	36.6	52.6	74.5

Sales exceed 100,000 units during February, March, and April.

Test Item File and Resource Guide

6. The angle of depression from the top of one building to the foot of a building across the street is 63°. The angle of depression to the top of the same building is 33°. The two buildings are 40 feet apart. What is the height of the shorter building?

2—O—Answer: 52.5 feet

7. From a point 300 feet from a building, the angle of elevation to the base of an antenna on the roof is 26.6° and the angle of elevation to the top of the antenna is 31.5°. Determine the height, h, of the antenna.

(a) 42.0 feet   (b) 29.4 feet

(c) 33.6 feet   (d) 45.1 feet

(e) None of these

1—M—Answer: c

8. From a point on a cliff 75 feet above water level an observer can see a ship. The angle of depression to the ship is 4°. How far is the ship from the base of the cliff?

1—O—Answer: 1072.5 feet

9. The pilot of an airplane flying at an altitude of 3000 feet sights two ships traveling in the same direction as the plane. The angle of depression of the farther ship is 20° and the angle of depression of the other ship is 35°. Find the distance between the two ships.

(a) 470 feet   (b) 3541 feet   (c) 3358 feet   (d) 1009 feet   (e) None of these

2—M—Answer: c

10. The sun is 30° above the horizon. Find the length of a shadow cast by a person 6 feet tall.

(a) 7.9 feet   (b) 8 feet

(c) 9.6 feet   (d) 10.4 feet

(e) None of these

1—M—Answer: d

CHAPTER 2

2.1 RADIAN AND DEGREE MEASURE

2.2 THE TRIGONOMETRIC FUNCTIONS AND THE UNIT CIRCLE

2.3 TRIGONOMETRIC FUNCTIONS AND RIGHT TRIANGLES

2.4 TRIGONOMETRIC FUNCTIONS OF ANY ANGLE

2.5 GRAPHS OF SINE AND COSINE FUNCTIONS

2.6 OTHER TRIGONOMETRIC GRAPHS

2.7 INVERSE TRIGONOMETRIC FUNCTIONS

2.8 APPLICATIONS OF TRIGONOMETRY

TRIGONOMETRY

2.1 RADIAN AND DEGREE MEASURE

Angles / Radian Measure / Degree Measure / Applications

Angles

As derived from the Greek language, the word **trigonometry** means "measurement of triangles." Initially, trigonometry dealt with relationships among the sides and angles of triangles. As such, it was used in the development of astronomy, navigation, and surveying.

With the advent of calculus in the 17th century and a resulting expansion of knowledge in the physical sciences, a different perspective arose—one that viewed the classic trigonometric relationships as *functions* with the set of real numbers as their domains. Consequently, the applications of trigonometry expanded to include a vast number of physical phenomena involving rotations or vibrations. These include sound waves, light rays, planetary orbits, vibrating strings, pendulums, and orbits of atomic particles. Our approach to trigonometry incorporates *both* perspectives, starting with angles and their measure.

An **angle** is determined by rotating a ray (half-line) about its endpoint. The starting position of the ray is called the **initial side** of the angle, and the position after rotation is called the **terminal side**, as shown in Figure 2.1. The endpoint of the ray is called the **vertex** of the angle.

Vertex · Terminal side · Initial side

FIGURE 2.1

105

Test Item File and Resource Guide

BestGrapher

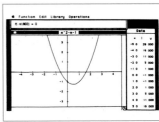

Videotapes

Computerized Testing Software

ACKNOWLEDGMENTS

We would like to thank the many people who have helped us prepare the text and supplements package. Their encouragement, criticisms, and suggestions have been invaluable to us.

Reviewers: Marilyn Carlson, University of Kansas; John Dersch, Grand Rapids Community College; Patricia J. Ernst, St. Cloud State University; Eunice F. Everett, Seminole Community College; James R. Fryxell, College of Lake County; Bernard Greenspan, University of Akron; Lynda Hollingsworth, Northwest Missouri State University; Spencer Hurd, The Citadel; Luella Johnson, State University of New York, College at Buffalo; Peter A. Lappan, Michigan State University; Marilyn McCollum, North Carolina State University; David R. Peterson, University of Central Arkansas; Antonio Quesada, University of Akron; Stephen Slack, Kenyon College; Judith Smalling, St. Petersburg Junior College; Howard L. Wilson, Oregon State University.

Focus Group: John Dersch, Grand Rapids Community College; Donald Shriner, Frostburg State University.

Telephone Focus Group: John Dersch, Grand Rapids Community College; Ruth Pruitt, Fort Hays State University; Sharon Sledge, San Jacinto College; Fredric W. Tufte, University of Wisconsin—Platteville; Darlene Whitkanack, Northern Illinois University.

Survey Respondents: William C. Allgyer, Mountain Empire Community College; Gloria Child, Rollins College; Allan C. Cochran, University of Arkansas at Fayetteville; Ronald Dalla, Eastern Washington University; Ann Dinkheller, Xavier University; Gloria Dion, Pennsylvania State University—Ogontz Campus; Iris B. Fetta, Clemson University; Spencer P. Hurd, The Citadel; Marvin L. Johnson, College of Lake County; Donald A. Josephson, Wheaton College (IL); Thomas J. Kearns, Northern Kentucky University; N. J. Kuenzi, University of Wisconsin—Oshkosh; Edward Laughbaum, Columbus State Community College; M. S. McCollum, North Carolina State University; Carolyn Meitler, Concordia University; Harold M. Ness, University of Wisconsin Centers—Fond du Lac; Michele Olson, College of the Redwoods; John Savige, St. Petersburg Junior College—Clearwater; Stephen Slack, Kenyon College; Marjie Vittum-Jones, South Seattle Community College; Charles Vander Embse, Central Michigan University.

Our thanks to the staffs of Texas Instruments Incorporated; Casio, Inc.; and Sharp Electronics Corporation for their help and cooperation with the programs.

A special thanks to all the people at Houghton Mifflin Company who worked with us in the development and production of the text, especially Chris Hoag, Sponsoring Editor; Cathy Cantin, Managing Editor; Maureen Brooks, Senior Associate Editor; Carolyn Johnson, Assistant Editor; Karen Carter, Supervising Editor; Rachel Wimberly, Associate Project Editor; Gary Crespo, Art Supervisor; Lisa Merrill, Production Supervisor; Ros Kane, Marketing Associate; and Carrie Lipscomb, Editorial Assistant.

We would also like to thank the staff at Larson Texts, Inc., who assisted with proofreading the manuscript, preparing and proofreading the art package, and checking and typesetting the supplements.

On a personal level, we are grateful to our wives, Deanna Gilbert Larson, Eloise Hostetler, and Consuelo Edwards, for their love, patience, and support. Also, a special thanks goes to R. Scott O'Neil.

If you have suggestions for improving the text, please feel free to write to us. Over the past two decades, we have received many useful comments from both instructors and students, and we value these very much.

Roland E. Larson
Robert P. Hostetler
Bruce H. Edwards

CONTENTS

Chapter 1   PREREQUISITES FOR TRIGONOMETRY   1

Chapter 2 TRIGONOMETRY   105

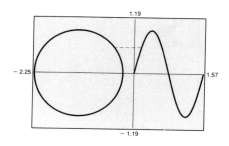

Chapter 3 ANALYTIC TRIGONOMETRY   203

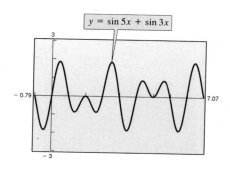

$y = \sin 5x + \sin 3x$

Chapter 4 ADDITIONAL TOPICS IN TRIGONOMETRY   253

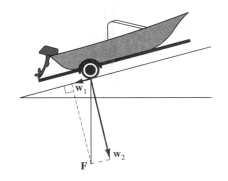

Chapter 5 COMPLEX NUMBERS   321

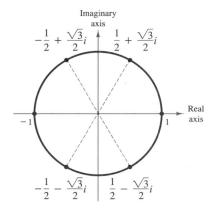

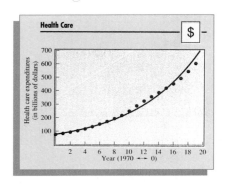
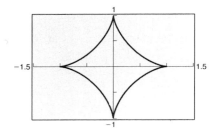

APPENDIXES

ANSWERS

INDEXES

INTRODUCTION TO CALCULATORS

Many examples and exercises in the text require the use of a calculator. Using this tool allows you to examine math in the context of real-life applications.

GRAPHING CALCULATORS

One of the basic differences in calculators is their order of operations. You should practice entering and evaluating expressions on your calculator to make sure you understand its order of operations. Some calculators use an order of operations called RPN (for Reverse Polish Notation). In this text, however, all calculator steps are given using algebraic order of operations. For example, the calculation

$$4.69[5 + 2(6.87 - 3.042)]$$

can be performed with the following steps using an algebraic order of operations.

4.69 ⊠ ⌈(⌉ 5 ⊞ 2 ⊠ ⌈(⌉ 6.87 ⊟ 3.042 ⌈)⌉ ⌈)⌉ ⌈ENTER⌉

This yields the value of 59.35664. Without parentheses, you could enter the expression from the inside out with the sequence

6.87 ⊟ 3.042 ⌈ENTER⌉ ⊠ 2 ⊞ 5 ⌈ENTER⌉ ⊠ 4.69 ⌈ENTER⌉

to obtain the same result.

Using a calculator with RPN, the calculation can be performed with the following steps.

6.87 ⌈ENTER⌉ 3.042 ⊟ 2 ⊠ 5 ⊞ 4.69 ⊠

You should also practice using your calculator's graphing features. To help you become familiar with the graphing features, we have included Appendix A,

Graphing Utilities. Be sure to read the examples in Appendix A and work the exercises. Answers to odd-numbered exercises in Appendix A are given in the back of the book.

You can program your calculator to perform various procedures. Try entering and using the selection of optional programs included in this text. For convenience, Appendix B, Programs, contains translations of the text programs for several types of graphing calculators.

ROUNDING NUMBERS

For all their usefulness, calculators do have a problem representing numbers because they are limited to a finite number of digits. For instance, what does your calculator display when you compute $2 \div 3$? Some calculators simply truncate (drop) the digits that exceed their display range and display .66666666. Others will round the number and display .66666667. Although the second display is more accurate, both of these decimal representations of $2/3$ contain a rounding error.

When rounding decimals, we suggest the following guidelines.

1. Determine the number of digits of accuracy you want to keep. The digit in the last position you keep is the **rounding digit,** and the digit in the first position you discard is the **decision digit.**

2. If the decision digit is 5 or greater, round up by adding 1 to the rounding digit.

3. If the decision digit is 4 or less, round down by leaving the rounding digit unchanged.

Here are some examples. Note that you round down in the first example because the decision digit is 4 or less, and you round up in the other two examples because the decision digit is 5 or greater.

Number	Rounded to Three Decimal Places	
a. $\sqrt{2} = 1.4142136$	1.414	*Round down*
b. $\pi = 3.1415927$	3.142	*Round up*
c. $\frac{7}{9} = 0.7777777$	0.778	*Round up*

One of the best ways to minimize error due to rounding is to leave numbers in your calculator until your calculations are complete. If you want to save a number for future use, store it in your calculator's memory.

Remember that once you (or your calculator) have rounded a number, a round-off error has been introduced. For instance, if a number is rounded to $x = 27.3$, then the actual value of x can lie anywhere between 27.25 and 27.35, or at 27.25 exactly. That is, $27.25 \leq x < 27.35$.

PROBLEM SOLVING USING A CALCULATOR

Here are some guidelines to consider when using any type of calculator in problem solving.

1. Be sure you understand the operation of your own calculator. You need to be skilled at entering expressions in a way that will guarantee that your calculator is performing the operations correctly.

2. Focus first on analyzing the problem. After you have developed a strategy, you may be able to use your calculator to help implement the strategy. Write down your steps in an organized way to clearly outline the strategy used and the results.

3. Most problems can be solved in a variety of ways. If you choose to solve a problem using a table, try checking the solution with an analytic (or algebraic) approach. If you choose to solve a problem using algebra, try checking the solution with a graphing approach. Or, if you choose to solve a problem using a graphing approach, try checking the solution with an algebraic approach.

4. After obtaining a solution with a calculator, be sure to ask yourself if the solution is reasonable (within the context of the problem).

5. To lessen the chance of errors, clear the calculator display (and check the settings) before beginning a new problem.

READING A MATHEMATICS TEXT

The following suggestions can help you make the most effective use of reading the text.

- Before class, read the portion of the text that is to be discussed during class. Read it literally, objectively, and interactively with pen or pencil and graphing calculator in hand. Try working the examples before reading further to see the solution, try the activities suggested in the Technology Notes and Discovery boxes on your graphing calculator, and try the end-of-section Discussion Problems to increase your depth of understanding.

- Because mathematics is a technical language with terminology and notation, reading a math text may take time and effort. It may help to take notes as you read, paying particular attention to terms, special symbols, and new ideas. This should allow you to look up unfamiliar terms or ideas easily as you encounter them in your reading.

- Identify the text organization. Notice that the major subdivisions of a section are listed at the beginning of each section. These are the key concepts and objectives for that section. Within a section or subsection, important terms are highlighted in **boldface** type. Make sure you understand this vocabulary. Study the side comments next to solution steps that indicate how to proceed from one step to the next.

- Especially when reading definitions, consider every sentence, word, and symbol carefully. It is helpful to read these more than once—first to get the gist of the statement and then a second time to absorb details, such as under what conditions the statement is true.

- Ask questions in class based on what you have read.

- After class, reread the text. Take your time. Have a pen or pencil and graphing calculator in hand. Make sure that you now understand any material that was unclear during your first reading.

- Be patient. The ability to read mathematics (or any technical material) is a skill that will be useful to you in this course, in other mathematics courses, and for any job with a technical component.

WRITING

Mathematics is a language—a way of communicating symbolically. As a mathematics student, you will communicate using the mathematics language to accomplish such tasks as showing the logical progression of a problem's solution steps. You will also communicate about mathematical concepts and ideas, such as explaining the real-world importance of a mathematical solution. Development of these skills will lead to the improvement of logical, critical, and organized thinking abilities. To improve your writing skills in mathematics, consider the following suggestions.

- When asked to discuss or explain a mathematical finding, you may find it helpful to start by sketching out a short outline of points to be covered. This will help you think through your argument logically and organize your thoughts.

- Use complete statements and explanations when writing mathematics for others to read.

- If you have difficulty starting a written solution, consider referring to the text solutions as models.

- For clarity, it can be useful to begin the solution to a problem by a partial restatement of the problem.

- In some cases you may find it helpful to construct your explanation or discussion around an example or counterexample.

- Be sure to use appropriate notation and math terminology in your writing. Take care to list any assumptions and define all variables used in your discussion or solution steps.

- When writing a solution to a problem, make an effort to explain or justify your reasoning for all intermediate steps as well as the final step.

- Remember that graphs are part of the language of mathematics. You may find it helpful to include and refer to graphs or a table of findings as appropriate in writing an explanation.

- When writing a mathematical argument or explanation, ask yourself if you have covered all the possible outcomes of the situation.

- Review your written responses for logical organization, detailed steps, and persuasiveness.

- Be neat.

- If research is involved, document your sources fully and accurately.

- The time you spend learning to write about mathematics is worth it. The ability to write a logical mathematical argument or explanation is a skill that will be useful in your trigonometry course, in other mathematics courses, and in the workplace.

CHAPTER 1

PREREQUISITES FOR TRIGONOMETRY

1.1 THE REAL NUMBER SYSTEM

The Real Number System / The Real Number Line / Ordering the Real Numbers / The Absolute Value of a Real Number / The Distance Between Two Real Numbers

The Real Number System

The algebraic techniques that are reviewed in this chapter will be used in the remaining chapters of the text. A clear understanding of the definitions, strategies, and concepts presented here will help you as you study trigonometry. **Real numbers** are used in everyday life to describe quantities such as age, miles per gallon, container size, population, and so on. To represent real numbers we use symbols such as

$$9, -5, 0, \frac{4}{3}, 0.6666\ldots, 28.21, \sqrt{2}, \pi, \quad \text{and} \quad \sqrt[3]{-32}.$$

The set of real numbers contains some important subsets with which you should be familiar:

$\{1, 2, 3, 4, \ldots\}$ *Set of **natural** numbers*

$\{0, 1, 2, 3, 4, \ldots\}$ *Set of **whole** numbers*

$\{\ldots, -3, -2, -1, 0, 1, 2, 3, \ldots\}$ *Set of **integers***

1

A real number is **rational** if it can be written as the ratio p/q of two integers, where $q \neq 0$. For instance, the numbers

$$\frac{1}{3} = 0.3333 \ldots, \quad \frac{1}{8} = 0.125, \quad \text{and} \quad \frac{125}{111} = 1.126126 \ldots$$

are rational. The decimal representation of a rational number either repeats (as in $3.1454545 \ldots$) or terminates (as in $\frac{1}{2} = 0.5$). A real number that cannot be written as the ratio of two integers is called **irrational.** Irrational numbers have infinite *nonrepeating* decimal representations. For instance, the numbers

$$\sqrt{2} \approx 1.4142135 \ldots \quad \text{and} \quad \pi \approx 3.1415926 \ldots$$

are irrational. (The symbol $\approx$ means "is approximately equal to.")

EXAMPLE 1   Identifying Real Numbers

Consider the following subset of real numbers:

$$\left\{ -8, -\sqrt{5}, 1, \frac{2}{3}, 0, -\frac{1}{7}, \sqrt{3}, \pi, 9 \right\}.$$

List the numbers in this set that are

A. Natural numbers   **B.** Integers
C. Rational numbers   **D.** Irrational numbers

SOLUTION

A. Natural numbers: 1, 9

B. Integers: -8, 1, 0, 9

C. Rational numbers: -8, 1, $\frac{2}{3}$, 0, $-\frac{1}{7}$, 9

D. Irrational numbers: $-\sqrt{5}$, $\sqrt{3}$, π

The Real Number Line

The model used to represent the real number system is called the **real number line.** It consists of a horizontal line with a point (the **origin**) labeled 0. Numbers to the right of 0 are positive, and numbers to the left of 0 are negative, as shown in Figure 1.1. We use the term **nonnegative** to describe a number that is either positive or zero.

The Real Number Line

FIGURE 1.1

One-to-One Correspondence

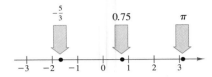

(a) Every real number corresponds to exactly one point on the real number line.

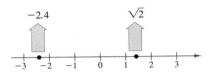

(b) Every point on the real number line corresponds to exactly one real number.

FIGURE 1.2

$a < b$ if and only if a lies to the left of b.

FIGURE 1.3

(a)

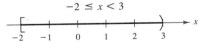

(b)

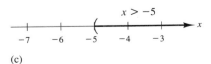

(c)

FIGURE 1.4

Each point on the real number line corresponds to one and only one real number and each real number corresponds to one and only one point on the real number line. This type of relationship is called a **one-to-one correspondence,** as shown in Figure 1.2.

The number associated with a point on the real number line is called the **coordinate** of the point. For example, in Figure 1.2(a), $-\frac{5}{3}$ is the coordinate of the leftmost point and π is the coordinate of the rightmost point.

Ordering the Real Numbers

One important property of real numbers is that they are **ordered.**

DEFINITION OF ORDER ON THE REAL NUMBER LINE

If a and b are real numbers, then a is **less than** b if $b - a$ is positive. We denote this order by the **inequality**

$a < b.$

This relationship can also be described by saying that b is **greater than** a and writing $b > a$. The inequality $a \leq b$ means that a is **less than or equal to** b and the inequality $b \geq a$ means that b is **greater than or equal to** a. The symbols $<, >, \leq,$ and $\geq$ are called **inequality symbols.**

Geometrically, this definition implies that $a < b$ if and only if a lies to the *left* of b on the real number line, as shown in Figure 1.3. For example, $1 < 2$ because 1 lies to the left of 2 on the real number line.

Inequalities are useful in denoting subsets of real numbers, as shown in Examples 2 and 3.

EXAMPLE 2   Interpreting Inequalities
. . . .

A. The inequality $x \leq 2$ denotes all real numbers less than or equal to 2.
B. The inequality $-2 \leq x < 3$ means that $x \geq -2$ *and* $x < 3$. This "double" inequality denotes all real numbers between -2 and 3, including -2 but *not* including 3.
C. The inequality $x > -5$ denotes all real numbers greater than -5.

The graphs of all three inequalities are shown in Figure 1.4. In the graphs, note that the brackets correspond to $\geq$ and $\leq$ and the parentheses correspond to $>$ and $<$.

Each of the subsets of real numbers in Example 2 is an **interval.**

BOUNDED INTERVALS ON THE REAL NUMBER LINE

Let a and b be real numbers such that $a < b$. The following intervals on the real number line are **bounded intervals.** The numbers a and b are the **endpoints** of each interval.

$[a, b]$	Closed	$a \leq x \leq b$	
(a, b)	Open	$a < x < b$	
$[a, b)$	Half-open	$a \leq x < b$	
$(a, b]$	Half-open	$a < x \leq b$	

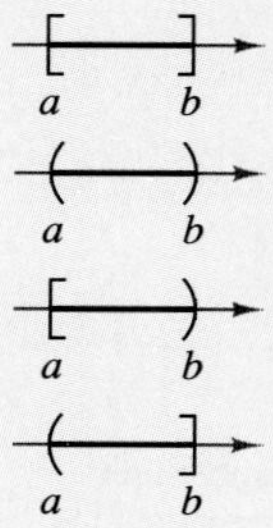

UNBOUNDED INTERVALS ON THE REAL NUMBER LINE

Let a and b be real numbers. The following intervals on the real number line are **unbounded intervals.**

Half-open	$x \geq a$	
Open	$x > a$	
Half-open	$x \leq b$	
Open	$x < b$	
Entire real line		

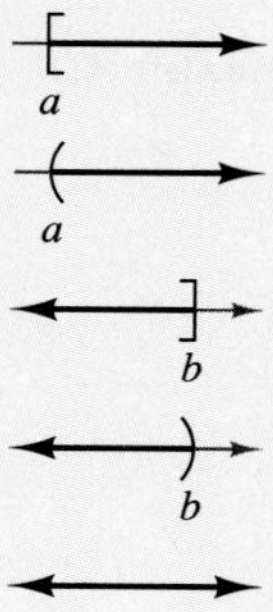

The symbols ∞ **(positive infinity)** and $-\infty$ **(negative infinity)** do not represent real numbers. They are simply convenient symbols used to describe the unboundedness of an interval such as $(1, \infty)$.

EXAMPLE 3   Intervals and Inequalities

Write an inequality to represent each of the intervals and state whether the interval is bounded or unbounded.

A. $(-3, 5]$ **B.** $(-3, \infty)$ **C.** $[0, 2]$

SOLUTION

A. $(-3, 5]$ corresponds to $-3 < x \leq 5$.   *Bounded*
B. $(-3, \infty)$ corresponds to $-3 < x$.   *Unbounded*
C. $[0, 2]$ corresponds to $0 \leq x \leq 2$.   *Bounded*

EXAMPLE 4   Using Inequalities to Represent Sets of Real Numbers

Use inequality and interval notation to describe each of the following.

A. c is nonnegative.   **B.** b is at most 5.
C. d is negative and greater than -3.
D. x is positive or x is less than -6.

SOLUTION

A. "c is nonnegative" means that c is greater than or equal to zero. This can be written as $c \geq 0$. The interval notation is $[0, \infty)$.
B. "b is at most 5" can be written as $b \leq 5$. The interval notation is $(-\infty, 5]$.
C. "d is negative" can be written as $d < 0$, and "d is greater than -3" can be written as $-3 < d$. Combining these two inequalities produces the double inequality $-3 < d < 0$. The interval notation is $(-3, 0)$.
D. "x is positive" can be written as $0 < x$, and "x is less than -6" can be written as $x < -6$. Combining these two inequalities produces $0 < x$ or $x < -6$. This yields *two* intervals: $(-\infty, -6), (0, \infty)$.

The Absolute Value of a Real Number

The **absolute value** of a real number is its *magnitude*.

DEFINITION OF ABSOLUTE VALUE

If a is a real number, then the **absolute value** of a is given by

$$|a| = \begin{cases} a, & \text{if } a \geq 0 \\ -a, & \text{if } a < 0. \end{cases}$$

Notice from this definition that the absolute value of a real number is never negative. For instance, if $a = -5$, then $|-5| = -(-5) = 5$. Zero is the only real number whose absolute value is zero. That is, $|0| = 0$.

The following list gives four useful properties of absolute value.

PROPERTIES OF ABSOLUTE VALUES

Let a and b be real numbers.

1. $|a| \geq 0$ 2. $|-a| = |a|$

3. $|ab| = |a||b|$ 4. $\left|\dfrac{a}{b}\right| = \dfrac{|a|}{|b|}, \quad b \neq 0$

The Distance Between Two Real Numbers

Absolute value is used to define the distance between two numbers on the real number line.

DISTANCE BETWEEN TWO POINTS ON THE REAL NUMBER LINE

Let a and b be real numbers. The **distance between a and b** is

$$d(a, b) = |b - a| = |a - b|.$$

A. The distance between $\sqrt{7}$ and 4 is given by

$$d\left(\sqrt{7}, 4\right) = \left|4 - \sqrt{7}\right| = 4 - \sqrt{7}.$$

B. The statement "the distance between c and -2 is at least 7" is written as

$$d(c, -2) = |c - (-2)| = |c + 2| \geq 7.$$

C. The distance between -4 and the origin is given by

$$d(-4, 0) = |-4 - 0| = |-4| = 4.$$

See Figure 1.5.

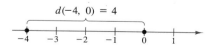

FIGURE 1.5

SECTION 1.1 · EXERCISES

In Exercises 1–6, determine which numbers in the set are
(a) natural numbers, (b) integers, (c) rational numbers, and
(d) irrational numbers.

*1. $-9, -\frac{7}{2}, 5, \frac{2}{3}, \sqrt{2}, 0, 1$

2. $\sqrt{5}, -7, -\frac{7}{3}, 0, 3.12, \frac{5}{4}$

3. $12, -13, 1, \sqrt{4}, \sqrt{6}, \frac{3}{2}$

4. $\frac{8}{2}, -\frac{8}{3}, \sqrt{10}, -4, 9, 14.2$

5. $-\pi, -\frac{1}{3}, \frac{6}{3}, \frac{1}{2}\sqrt{2}, -7.5$

6. $25, -17, -\frac{12}{5}, \sqrt{9}, 3.12, \frac{1}{2}\pi$

In Exercises 7–12, plot the two real numbers on the real number
line, and then write them with the appropriate inequality sign
($<$ or $>$) between them.

7. $\frac{3}{2}, 7$

8. $-3.5, 1$

9. $-4, -8$

10. $1, \frac{16}{3}$

11. $\frac{5}{6}, \frac{2}{3}$

12. $-\frac{8}{7}, -\frac{3}{7}$

In Exercises 13–16, write an inequality to represent the given
interval and state whether the interval is bounded or unbounded.

13. $[-1, 3]$

14. $(4, 10]$

15. $(10, \infty)$

16. $(-6, \infty)$

In Exercises 17–26, use interval notation to describe the subset
of real numbers that is represented by the inequality. Then
sketch the subset on the real number line.

17. $x \le 5$

18. $x \ge -2$

19. $x < 0$

20. $x > 3$

21. $x \ge 4$

22. $x < 2$

23. $-2 < x < 2$

24. $0 \le x \le 5$

25. $-1 \le x < 0$

26. $0 < x \le 6$

In Exercises 27–34, use inequality and interval notation to describe the set of real numbers.

27. x is negative.

28. z is at least 10.

29. y is no more than 25.

30. y is greater than 5 and less than or equal to 12.

31. The person's age A is at least 30.

32. The yield Y is no more than 45 bushels per acre.

33. The annual rate of inflation r is expected to be at least
3.5%, but no more than 6%.

34. The price p of unleaded gasoline is not expected to go
above $1.35 per gallon during the coming year.

In Exercises 35–40, write the expression without using absolute
value signs.

35. $\dfrac{-5}{|-5|}$

36. $|4 - \pi|$

37. $-3|-3|$

38. $|-1| - |-2|$

39. $-|16.25| + 20$

40. $2|33|$

In Exercises 41–46, fill in the blank with $<$, $>$, or $=$.

41. $|-3| \quad -|-3|$

42. $|-4| \quad |4|$

43. $-5 \quad -|5|$

44. $-|-6| \quad |-6|$

45. $-|-2| \quad -|2|$

46. $-(-2) \quad -2$

In Exercises 47–52, find the distance between a and b.

47.

48.

49. $a = 126, b = 75$

50. $a = -126, b = -75$

51. $a = 9.34, b = -5.65$

52. $a = \frac{16}{5}, b = \frac{112}{75}$

In Exercises 53–58, use absolute value notation to describe the
expression.

53. The distance between x and 5 is no more than 3.

54. The distance between x and -10 is at least 6.

55. The distance between z and $\frac{3}{2}$ is greater than 1.

56. The distance between z and 0 is less than 8.

57. y is at least 6 units from 0.

58. y is at most 2 units from a.

* Detailed solutions to all odd-numbered exercises can be found in the
Study and Solutions Guide.

In Exercises 59 and 60, use a calculator to order the numbers from smallest to largest.

59. $\frac{7071}{5000}, \frac{584}{413}, \sqrt{2}, \frac{47}{33}, \frac{127}{90}$

60. $\frac{26}{15}, \sqrt{3}, 1.7320, \frac{381}{220}, \sqrt{10} - \sqrt{2}$

In Exercises 61–64, use a calculator to find the decimal form of the rational number. If it is a nonterminating decimal, write the repeating pattern.

61. $\frac{5}{8}$ **62.** $\frac{1}{3}$

63. $\frac{41}{333}$ **64.** $\frac{6}{11}$

In Exercises 65–68, determine whether the statement is true or false. Explain your reasoning.

65. The reciprocal of every nonzero integer is an integer.
66. Every integer is a rational number.
67. Every real number is either rational or irrational.
68. The absolute value of every real number is positive.

1.2   SOLVING EQUATIONS

Equations and Solutions of Equations / Solving Linear Equations / Factorization Principle / Solving Quadratic Equations / Polynomial Equations of Higher Degree

Equations and Solutions of Equations

An **equation** is a statement that two algebraic expressions are equal. For example, $3x - 5 = 7$, $x^2 - x - 6 = 0$, and $\sqrt{2x} = 4$ are equations. To **solve** an equation in x means that you find all values of x for which the equation is true. Such values are **solutions.** For instance, $x = 4$ is a solution of the equation $3x - 5 = 7$ because $3(4) - 5 = 7$ is a true statement.

The solutions of an equation depend on the kinds of numbers being considered. For instance, in the set of rational numbers, the equation $x^2 = 10$ has no solution because there is no rational number whose square is 10. However, in the set of real numbers this equation has two solutions, $\sqrt{10}$ and $-\sqrt{10}$, because $\left(\sqrt{10}\right)^2 = 10$ and $\left(-\sqrt{10}\right)^2 = 10$.

A **polynomial equation** in x is of the form

$$a_n x^n + a_{n-1} x^{n-1} + \cdots + a_2 x^2 + a_1 x + a_0 = 0, \qquad a_n \neq 0,$$

where n is the **degree** of the equation. For instance, the degree of the polynomial equation $-x^3 + 2x - 1 = 0$ is 3.

An equation that is true of *every* real number in the domain of the variable is an **identity.** For example, $x^2 - 9 = (x + 3)(x - 3)$ is an identity because it is a true statement for any real value of x, and $x/3x^2 = 1/3x$, where $x \neq 0$, is an identity because it is true for any nonzero real value of x.

An equation that is true for just *some* (or even none) of the real numbers in the domain of the variable is a **conditional equation.** For example, the equation $x^2 - 9 = 0$ is conditional because $x = 3$ and $x = -3$ are the only values in the domain that satisfy the equation.

Solving Linear Equations

A **linear equation** in one variable x is an equation that can be written in the standard form

$$ax + b = 0,$$

where a and b are real numbers with $a \neq 0$. (Note that a linear equation in x is simply a first-degree polynomial equation in x.) To solve a linear equation in x, *isolate* the variable x by using operations that generate equivalent equations.

EXAMPLE 1   Solving a Linear Equation

$3x - 6 = 0$	*Given equation*
$3x = 6$	*Add 6 to both sides*
$x = 2$	*Divide both sides by 3*

After solving an equation, you should **check each solution** in the *original* equation. In Example 1, check that $x = 2$ is a solution by substituting in the original equation.

$$3(2) - 6 = 6 - 6 = 0 \qquad \text{\textit{Check solution}}$$

Some linear equations have no solution because all of the x-terms subtract out and a contradictory (false) statement such as $0 = 5$ or $12 = 7$ is obtained. Watch for this type of linear equation in the exercises.

To solve an equation involving fractional expressions, find the least common denominator of all terms in the equation and multiply every term by this LCD. This procedure clears the equation of fractions.

EXAMPLE 2   Solving an Equation Involving Fractional Expressions

$$\frac{x}{3} + \frac{3x}{4} = 2 \qquad \textit{Given equation}$$

$$(12)\frac{x}{3} + (12)\frac{3x}{4} = (12)2 \qquad \textit{Multiply by the LCD}$$

$$4x + 9x = 24 \qquad \textit{Reduce and multiply}$$

$$13x = 24 \qquad \textit{Combine like terms}$$

$$x = \frac{24}{13} \qquad \textit{Divide by 13}$$

Thus, the equation has one solution: $\frac{24}{13}$. Check this solution in the original equation.

When multiplying or dividing an equation by a *variable* quantity, it is possible to introduce an **extraneous** solution that does not satisfy the original equation.

EXAMPLE 3   An Equation with an Extraneous Solution

Solve the equation for x.

$$\frac{1}{x-2} = \frac{3}{x+2} - \frac{6x}{x^2 - 4}$$

SOLUTION

The LCD is $x^2 - 4 = (x+2)(x-2)$. Multiply every term by this LCD and reduce.

$$\frac{1}{x-2}(x+2)(x-2) = \frac{3}{x+2}(x+2)(x-2)$$

$$- \frac{6x}{x^2-4}(x+2)(x-2)$$

$$x + 2 = 3(x-2) - 6x, \qquad x \neq \pm 2$$

$$x + 2 = 3x - 6 - 6x$$

$$4x = -8$$

$$x = -2 \qquad \textit{Extraneous solution}$$

We know that in the original equation, $x = -2$ yields a denominator of zero. Therefore, $x = -2$ is an extraneous solution, and the equation has *no solution*.

Factorization Principle

To solve a polynomial equation of degree 2 or more, write the equation in standard form (with the polynomial on one side of the equation and 0 on the other side). Then we use the following Factorization Principle.

If $ab = 0$, then $a = 0$ or $b = 0$.   *Factorization Principle*

For instance, $25x^2 - 9$ factors as $(5x + 3)(5x - 3)$. Hence, the solutions of $25x^2 - 9 = 0$ can be obtained as follows.

$$25x^2 - 9 = 0 \qquad \textit{Standard form}$$
$$(5x + 3)(5x - 3) = 0 \qquad \textit{Factored form}$$
$$5x + 3 = 0 \longrightarrow x = -\tfrac{3}{5} \quad \textit{Set 1st factor equal to 0}$$
$$5x - 3 = 0 \longrightarrow x = \tfrac{3}{5} \quad \textit{Set 2nd factor equal to 0}$$

When factoring polynomials, the following special polynomial forms are useful.

FACTORING SPECIAL POLYNOMIAL FORMS

Factored Form	*Example*
Difference of Two Squares	
$u^2 - v^2 = (u + v)(u - v)$	$9x^2 - 4 = (3x)^2 - 2^2$
	$\quad\quad\quad = (3x + 2)(3x - 2)$
Perfect Square Trinomial	
$u^2 + 2uv + v^2 = (u + v)^2$	$x^2 + 6x + 9 = x^2 + 2(x)(3) + 3^2$
	$\quad\quad\quad = (x + 3)^2$
$u^2 - 2uv + v^2 = (u - v)^2$	$x^2 - 6x + 9 = x^2 - 2(x)(3) + 3^2$
	$\quad\quad\quad = (x - 3)^2$
Sum or Difference of Two Cubes	
$u^3 + v^3 = (u + v)(u^2 - uv + v^2)$	$x^3 + 8 = x^3 + 2^3$
	$\quad\quad\quad = (x + 2)(x^2 - 2x + 4)$
$u^3 - v^3 = (u - v)(u^2 + uv + v^2)$	$27x^3 - 1 = (3x)^3 - 1^3$
	$\quad\quad\quad = (3x - 1)(9x^2 + 3x + 1)$

The first step in factoring a polynomial is to check for a common factor. Once the common factor is removed, it is often possible to recognize patterns that were not immediately obvious. Watch for this in some of the examples that follow.

Solving Quadratic Equations

A **quadratic equation** in x is an equation that can be written in the standard form

$$ax^2 + bx + c = 0,$$

where a, b, and c are real numbers with $a \neq 0$. We can solve a quadratic equation either by factoring or by the Quadratic Formula. First, we demonstrate the solution by factoring.

EXAMPLE 4   A Quadratic Equation Involving the Difference of Squares

Solve the equation $3 - 12x^2 = 0$.

SOLUTION

First, factor out the common factor 3. Then use the difference-of-two-squares formula $u^2 - v^2 = (u + v)(u - v)$ with $u = 1$ and $v = 2x$.

$$3 - 12x^2 = 0 \qquad \textit{Standard form}$$
$$3(1 - 4x^2) = 0 \qquad \textit{Remove common factor}$$
$$3(1 + 2x)(1 - 2x) = 0 \qquad \textit{Factored form}$$
$$1 + 2x = 0 \quad \longrightarrow \quad x = -\frac{1}{2} \qquad \textit{Set 1st factor equal to 0}$$
$$1 - 2x = 0 \quad \longrightarrow \quad x = \frac{1}{2} \qquad \textit{Set 2nd factor equal to 0}$$

REMARK In Example 4, note that the common constant factor 3 does not affect the solution.

The Factorization Principle works *only* for equations written in standard form. Therefore, all terms must be collected to one side *before* factoring.

EXAMPLE 5   A Quadratic Equation Involving a General Trinomial

Solve the equation $2x^2 + x = 15$.

SOLUTION

First write the equation in standard form as follows.

$$2x^2 + x - 15 = 0$$

Next, attempt to factor the trinomial $2x^2 + x - 15$ as a product $(ax + b)(cx + d)$ of two binomials where a and c are factors of 2, and b and d are factors of -15. The eight possible factorizations are as follows.

$$(2x - 1)(x + 15) \qquad (2x + 1)(x - 15)$$
$$(2x - 3)(x + 5) \qquad (2x + 3)(x - 5)$$
$$(2x - 5)(x + 3) \qquad (2x + 5)(x - 3)$$
$$(2x - 15)(x + 1) \qquad (2x + 15)(x - 1)$$

By testing each of these possibilities, you will find the correct factorization to be

$$\overset{\text{Outer}}{2x^2 + x - 15 = (2x - 5)(x + 3).}$$

Inner

Therefore, the solutions are obtained as follows.

$$(2x - 5)(x + 3) = 0 \qquad \qquad \textit{Factored form}$$

$$2x - 5 = 0 \;\longrightarrow\; x = \frac{5}{2} \qquad \textit{Set 1st factor equal to 0}$$

$$x + 3 = 0 \;\longrightarrow\; x = -3 \qquad \textit{Set 2nd factor equal to 0}$$

Quadratic equations can also be solved by the **Quadratic Formula.**

THE QUADRATIC FORMULA

The solutions of the quadratic equation

$$ax^2 + bx + c = 0, \qquad a \neq 0$$

are given by the **Quadratic Formula**

$$x = \frac{-b \pm \sqrt{b^2 - 4ac}}{2a}.$$

The quantity under the radical sign, $b^2 - 4ac$, is the **discriminant** of the quadratic expression $ax^2 + bx + c$. It is used to determine the nature of the solutions of a quadratic equation.

1. If $b^2 - 4ac > 0$, then there are *two distinct real solutions.*
2. If $b^2 - 4ac = 0$, then there is *one repeated solution.*
3. If $b^2 - 4ac < 0$, then there are *no real solutions.*

The third case (no real solutions) will be studied in Section 5.2.

EXAMPLE 6   Using the Quadratic Formula: Two Distinct Solutions

Use the Quadratic Formula to solve $x^2 + 3x = 9$.

SOLUTION

$$x^2 + 3x = 9 \qquad \text{\textit{Given equation}}$$

$$x^2 + 3x - 9 = 0 \qquad \text{\textit{Standard form with}}$$
$$\text{\textit{a = 1, b = 3, c = −9}}$$

$$x = \frac{-b \pm \sqrt{b^2 - 4ac}}{2a} \qquad \text{\textit{Quadratic Formula}}$$

$$x = \frac{-3 \pm \sqrt{(3)^2 - 4(1)(-9)}}{2(1)} \qquad \text{\textit{Substitute}}$$

$$x = \frac{-3 \pm \sqrt{45}}{2}$$

$$x = \frac{-3 \pm 3\sqrt{5}}{2} \qquad \text{\textit{Solutions}}$$

Therefore, the equation has two solutions:

$$x = \frac{-3 + 3\sqrt{5}}{2} \quad \text{and} \quad x = \frac{-3 - 3\sqrt{5}}{2}.$$

Polynomial Equations of Higher Degree

A polynomial equation of degree greater than or equal to 2 can often be solved by factoring the polynomial into products of linear and quadratic factors. For instance, the third-degree equation $x^3 - 7x^2 + 12x = 0$ can be factored as

$$x(x^2 - 7x + 12) = x(x - 3)(x - 4) = 0.$$

By the Factorization Principle, the resulting solutions are $x = 0$, $x = 3$, and $x = 4$. In the next two examples, we show techniques commonly used to solve polynomial equations.

EXAMPLE 7   A Fourth-Degree Difference-of-Squares Equation

Solve the equation $x^4 - 16 = 0$.

SOLUTION

You can use the formula for the difference of two squares, as follows.

$$x^4 - 16 = 0 \qquad \text{\textit{Standard form}}$$
$$(x^2 + 4)(x^2 - 4) = 0 \qquad \text{\textit{Factor}}$$
$$(x^2 + 4)(x - 2)(x + 2) = 0 \qquad \text{\textit{Factor again}}$$

Setting the first factor equal to zero produces no real solutions. Therefore, the only real solutions are those obtained by setting the second and third factors equal to zero: $x = 2$ and $x = -2$.

EXAMPLE 8   A Fourth-Degree Trinomial Equation

Solve the equation $3x^4 - 6x^3 - 12x^2 = 0$.

SOLUTION

In this case, you first factor out the common monomial factor $3x^2$ and obtain $3x^2(x^2 - 2x - 4) = 0$. Setting the first factor equal to zero produces a repeated solution, $x = 0$. Setting the second factor equal to zero and using the Quadratic Formula produces the following.

$$x^2 - 2x - 4 = 0$$

$$x = \frac{-(-2) \pm \sqrt{4 - 4(1)(-4)}}{2(1)}$$

$$= \frac{2 \pm \sqrt{20}}{2}$$

$$= 1 \pm \sqrt{5}$$

Therefore, the four solutions are $x = 0$ (repeated), $x = 1 + \sqrt{5}$, and $x = 1 - \sqrt{5}$.

DISCUSSION PROBLEM

A MATHEMATICAL FALLACY

A mathematical **fallacy** is an argument that appears to prove something that we know is incorrect. For instance, the following argument appears to prove that $1 = 0$. Can you find the error in this argument?

$x = 1$	*Given equation*
$x - 1 = 0$	*Subtract 1 from both sides*
$x(x - 1) = 0$	*Multiply both sides by x*
$\dfrac{x(x - 1)}{x - 1} = \dfrac{0}{x - 1}$	*Divide both sides by x − 1*
$\dfrac{x(x - 1)}{x - 1} = 0$	*Reduce*
$x = 0$	*Solution*

WARM-UP

The following warm-up exercises involve skills that were covered in earlier sections. You will use these skills in the exercise set for this section.

In Exercises 1–10, perform the indicated operations and simplify your answer.

1. $-3(2x - 10)$

2. $8(-5x + 3)$

3. $2(x + 1) - (x + 2)$

4. $-3(2x - 4) + 7(x + 2)$

5. $(2x + 5)(2x - 5)$

6. $(3x - 4)^2$

7. $\dfrac{x}{3} + \dfrac{x}{5}$

8. $x - \dfrac{x}{4}$

9. $\dfrac{1}{x + 1} - \dfrac{1}{x}$

10. $\dfrac{2}{x} + \dfrac{3}{x}$

SECTION 1.2 · EXERCISES

In Exercises 1–6, determine whether the equation is an identity or a conditional equation.

1. $2(x - 1) = 2x - 2$

2. $3(x + 2) = 5x + 4$

3. $-6(x - 3) + 5 = -2x + 10$

4. $3(x + 2) - 5 = 3x + 1$

5. $3x^2 - 8x + 5 = (x - 4)^2 - 11$

6. $x^2 + 2(3x - 2) = x^2 + 6x - 4$

In Exercises 7–12, determine whether each given value of x is a solution of the equation.

Equation	Values	
7. $5x - 3 = 3x + 5$	(a) $x = 0$	(b) $x = -5$
	(c) $x = 4$	(d) $x = 10$
8. $7 - 3x = 5x - 17$	(a) $x = -3$	(b) $x = 0$
	(c) $x = 8$	(d) $x = 3$
9. $3x^2 + 2x - 5 = 2x^2 - 2$	(a) $x = -3$	(b) $x = 1$
	(c) $x = 4$	(d) $x = -5$
10. $5x^3 + 2x - 3 =$	(a) $x = 2$	(b) $x = -2$
$4x^3 + 2x - 11$	(c) $x = 0$	(d) $x = 10$
11. $\dfrac{5}{2x} - \dfrac{4}{x} = 3$	(a) $x = -\frac{1}{2}$	(b) $x = 4$
	(c) $x = 0$	(d) $x = \frac{1}{4}$
12. $x - \dfrac{15}{x} = 2$	(a) $x = -3$	(b) $x = 5$
	(c) $x = 1$	(d) $x = 0$

In Exercises 13–30, solve the equation (if possible) and check your answer.

13. $2(x + 5) - 7 = 3(x - 2)$

14. $2(13t - 15) + 3(t - 19) = 0$

15. $\dfrac{5x}{4} + \dfrac{1}{2} = x - \dfrac{1}{2}$

16. $0.60x + 0.40(100 - x) = 50$

17. $x + 8 = 2(x - 2) - x$

18. $3(x + 3) = 5(1 - x) - 1$

19. $\dfrac{100 - 4u}{3} = \dfrac{5u + 6}{4} + 6$

20. $\dfrac{17 + y}{y} + \dfrac{32 + y}{y} = 100$

21. $\dfrac{5x - 4}{5x + 4} = \dfrac{2}{3}$

22. $\dfrac{10x + 3}{5x + 6} = \dfrac{1}{2}$

23. $\dfrac{1}{x - 3} + \dfrac{1}{x + 3} = \dfrac{10}{x^2 - 9}$

24. $\dfrac{1}{x - 2} + \dfrac{3}{x + 3} = \dfrac{4}{x^2 + x - 6}$

25. $\dfrac{7}{2x + 1} - \dfrac{8x}{2x - 1} = -4$

26. $\dfrac{4}{u - 1} + \dfrac{6}{3u + 1} = \dfrac{15}{3u + 1}$

27. $(x + 2)^2 + 5 = (x + 3)^2$

28. $(x + 1)^2 + 2(x - 2) = (x + 1)(x - 2)$

29. $4 - 2(x - 2b) = ax + 3$

30. $5 + ax = 12 - bx$

In Exercises 31–38, solve the quadratic equation by factoring.

31. $6x^2 + 3x = 0$

32. $9x^2 - 1 = 0$

33. $x^2 - 2x - 8 = 0$

34. $x^2 + 10x + 25 = 0$

35. $16x^2 + 56x + 49 = 0$

36. $3 + 5x - 2x^2 = 0$

37. $2x^2 = 19x + 33$

38. $(x + a)^2 - b^2 = 0$

In Exercises 39–52, use the Quadratic Formula to solve the equation.

39. $2x^2 + x - 1 = 0$

40. $2x^2 - x - 1 = 0$

41. $16x^2 + 8x - 3 = 0$

42. $25x^2 - 20x + 3 = 0$

43. $2 + 2x - x^2 = 0$

44. $x^2 - 10x + 22 = 0$

45. $12x - 9x^2 = -3$

46. $16x^2 + 22 = 40x$

47. $4x^2 + 4x = 7$

48. $16x^2 - 40x + 5 = 0$

49. $(y - 5)^2 = 2y$

50. $(z + 6)^2 = -2z$

51. $\dfrac{1}{x} - \dfrac{1}{x + 1} = 3$

52. $\dfrac{x}{x^2 - 4} + \dfrac{1}{x + 2} = 3$

In Exercises 53–56, use a calculator to solve the quadratic equation. Round your answers to three decimal places.

53. $5.1x^2 - 1.7x - 3.2 = 0$

54. $10.4x^2 + 8.6x + 1.2 = 0$

55. $7.06x^2 - 4.85x + 0.50 = 0$

56. $-0.005x^2 + 0.101x - 0.193 = 0$

In Exercises 57–66, find all solutions of the equation. Check your answers in the original equation.

57. $4x^4 - 9x^2 = 0$

58. $20x^3 - 125x = 0$

59. $x^3 - 2x^2 - 3x = 0$

60. $2x^4 - 15x^3 + 18x^2 = 0$

61. $x^4 - 81 = 0$

62. $x^6 - 64 = 0$

63. $5x^3 + 30x^2 + 45x = 0$

64. $9x^4 - 24x^3 + 16x^2 = 0$

65. $x^6 + 7x^3 - 8 = 0$

66. $x^6 + 3x^3 + 2 = 0$

Negative Income Tax In Exercises 67 and 68, use the following information about a possible negative income tax for a family of two adults and two children. The plan would guarantee the poor a minimum income while encouraging families to increase their private incomes (see figure).

Family's earned income: $I = x$

Government payment: $G = 8000 - \frac{1}{2}x$, $\quad 0 \le x \le 16{,}000$

Spendable income: $S = I + G$

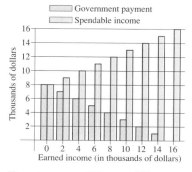

FIGURE FOR 67 AND 68

67. The spendable income is $11,800. Find the earned income x.

68. The spendable income is $10,500. Find the government payment G.

69. *Dimensions of a Corral*   A rancher has 200 feet of fencing to enclose two adjacent rectangular corrals (see figure). Find the dimensions that would create an enclosed area of 1400 square feet.

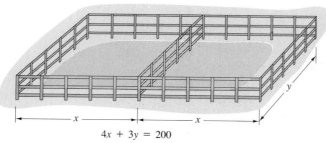

$$4x + 3y = 200$$

FIGURE FOR 69

70. *Dimensions of a Box*   An open box is to be made from a square piece of material by cutting 2-inch squares from each corner and turning up the sides (see figure). The volume of the finished box is to be 200 cubic inches. Find the size of the original piece of material.

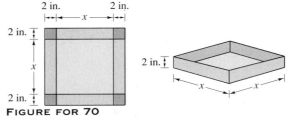

FIGURE FOR 70

71. *Sharing the Cost*   A college charters a bus for $1700 to take a group of students to a museum. When six more students join the trip, the cost per student drops by $7.50. How many students were in the original group?

72. *Airspeed*   An airline runs a commuter flight between two cities that are 720 miles apart. If the average speed of the planes is increased by 40 miles per hour, the travel time is decreased by 12 minutes. What speed is required to obtain this decrease in travel time?

1.3   THE CARTESIAN PLANE

The Cartesian Plane   /   The Distance Between Two Points in the Plane   /
The Midpoint Formula   /   The Equation of a Circle

The Cartesian Plane

Just as real numbers are represented by points on the real number line, ordered pairs of real numbers are represented by points in a plane. This plane is called a **rectangular coordinate system** or the **Cartesian plane,** after the French mathematician René Descartes (1596–1650).

The Cartesian plane is formed by two real number lines intersecting at right angles, as shown in Figure 1.6(a). The horizontal number line is usually called the ***x*-axis** and the vertical number line is usually called the ***y*-axis.** (The plural of axis is *axes.*) The point of intersection of the two axes is the **origin.** The axes separate the plane into four regions called **quadrants.**

REMARK   It is customary to use the notation (x, y) to denote both a point in the plane and an open interval on the real number line. The nature of a specific problem will show which of the two we are talking about.

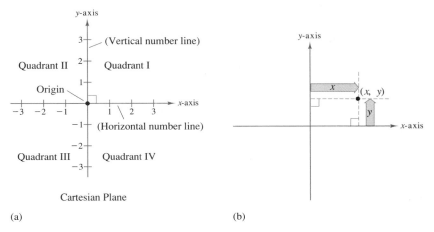

(a)   (b)

FIGURE 1.6

Each point in the plane corresponds to an **ordered pair** (x, y) of real numbers x and y, called **coordinates** of the point. The first number (***x*-coordinate**) tells how far to the left or right the point is from the vertical axis, and the second number (***y*-coordinate**) tells how far up or down the point is from the horizontal axis, as shown in Figure 1.6(b). Figure 1.7 shows several points that have been plotted in the plane.

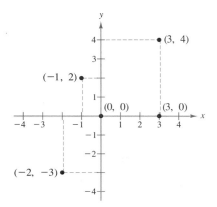

FIGURE 1.7

The rectangular coordinate system allows you to visualize relationships between variables x and y. Today, Descartes's ideas are commonly used in every scientific and business-related field.

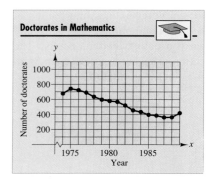

FIGURE 1.8

EXAMPLE 1   Number of Doctorates in Mathematics

The numbers of doctorates in mathematics granted to United States citizens by universities in the United States in the years 1974 to 1989 are given in Table 1.1. Plot these points on a rectangular coordinate system. (*Source*: American Mathematical Society).

TABLE 1.1

Year	1974	1975	1976	1977	1978	1979	1980	1981
Degrees	677	741	722	689	634	596	578	567
Year	1982	1983	1984	1985	1986	1987	1988	1989
Degrees	519	455	433	396	386	362	363	419

SOLUTION

The points are shown in Figure 1.8. Note that the break in the *x*-axis indicates that the numbers for years prior to 1974 have been omitted.

The Distance Between Two Points in the Plane

Let (x_1, y_1) and (x_2, y_2) represent two points in the plane that do not lie on the same horizontal or vertical line. With these two points, a right triangle can be formed, as shown in Figure 1.9. Note that the third vertex of the triangle is (x_1, y_2). Because (x_1, y_1) and (x_1, y_2) lie on the same vertical line, the length of the vertical side of the triangle is $|y_2 - y_1|$. Similarly, the length of the horizontal side is $|x_2 - x_1|$. Thus, by the Pythagorean Theorem, the square of the distance d between (x_1, y_1) and that (x_2, y_2) is

$$d^2 = |x_2 - x_1|^2 + |y_2 - y_1|^2.$$

Because the distance d must be positive, choose the positive square root and write

$$d = \sqrt{|x_2 - x_1|^2 + |y_2 - y_1|^2}.$$

Finally, replacing $|x_2 - x_1|^2$ and $|y_2 - y_1|^2$ by the equivalent expressions $(x_2 - x_1)^2$ and $(y_2 - y_1)^2$ gives the following formula for the distance between two points in a rectangular coordinate plane.

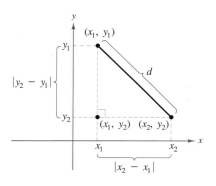

Distance Between Two Points

FIGURE 1.9

THE DISTANCE FORMULA

The distance d between the two points (x_1, y_1) and (x_2, y_2) in the coordinate plane is

$$d = \sqrt{(x_2 - x_1)^2 + (y_2 - y_1)^2}.$$

EXAMPLE 2   Finding the Distance Between Two Points

Find the distance between the points $(-2, 1)$ and $(3, 4)$.

SOLUTION

Letting $(x_1, y_1) = (-2, 1)$ and $(x_2, y_2) = (3, 4)$, apply the Distance Formula to obtain

$$d = \sqrt{[3 - (-2)]^2 + (4 - 1)^2}$$
$$= \sqrt{5^2 + 3^2}$$
$$= \sqrt{25 + 9}$$
$$= \sqrt{34}$$
$$\approx 5.83.$$

See Figure 1.10.

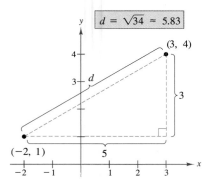

$d = \sqrt{34} \approx 5.83$

FIGURE 1.10

EXAMPLE 3   An Application of the Distance Formula

Show that the points $(2, 1)$, $(4, 0)$, and $(5, 7)$ are vertices of a right triangle.

SOLUTION

The three points are plotted in Figure 1.11. Using the Distance Formula, you can find the lengths of the three sides of the triangle:

$$d_1 = \sqrt{(5 - 2)^2 + (7 - 1)^2} = \sqrt{9 + 36} = \sqrt{45}$$
$$d_2 = \sqrt{(4 - 2)^2 + (0 - 1)^2} = \sqrt{4 + 1} = \sqrt{5}$$
$$d_3 = \sqrt{(5 - 4)^2 + (7 - 0)^2} = \sqrt{1 + 49} = \sqrt{50}.$$

Because $d_1^2 + d_2^2 = 45 + 5 = 50 = d_3^2$, you can conclude from the Pythagorean Theorem that the triangle is a right triangle.

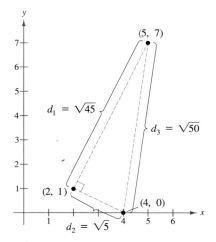

FIGURE 1.11

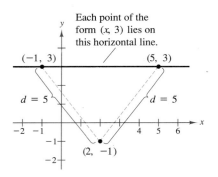

Each point of the form $(x, 3)$ lies on this horizontal line.

FIGURE 1.12

EXAMPLE 4   Finding Points at a Specified Distance from a Given Point

Find x so that the distance between $(x, 3)$ and $(2, -1)$ is 5.

SOLUTION

$$\sqrt{(x - 2)^2 + (3 + 1)^2} = 5 \qquad \textit{Distance Formula}$$
$$(x^2 - 4x + 4) + 16 = 25 \qquad \textit{Square both sides}$$
$$x^2 - 4x - 5 = 0 \qquad \textit{Standard form}$$
$$(x - 5)(x + 1) = 0 \qquad \textit{Factor}$$
$$x - 5 = 0 \;\longrightarrow\; x = 5 \qquad \textit{Set 1st factor equal to 0}$$
$$x + 1 = 0 \;\longrightarrow\; x = -1 \qquad \textit{Set 2nd factor equal to 0}$$

There are two solutions: $(5, 3)$ and $(-1, 3)$. Note that each of the points $(5, 3)$ and $(-1, 3)$ lies five units from the point $(2, -1)$, as shown in Figure 1.12.

The Midpoint Formula

The coordinates of the midpoint of a line segment are the average values of the corresponding coordinates of the two endpoints.

THE MIDPOINT FORMULA

The **midpoint** of the line segment joining the points (x_1, y_1) and (x_2, y_2) in the coordinate plane is

$$\left(\frac{x_1 + x_2}{2}, \frac{y_1 + y_2}{2} \right).$$

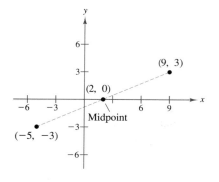

FIGURE 1.13

EXAMPLE 5   Finding the Midpoint of a Line Segment

Find the midpoint of the line segment joining the points $(-5, -3)$ and $(9, 3)$.

SOLUTION

Figure 1.13 shows the two given points and their midpoint. Using the Midpoint Formula, you can write

$$\text{Midpoint} = \left(\frac{-5 + 9}{2}, \frac{-3 + 3}{2} \right)$$
$$= (2, 0).$$

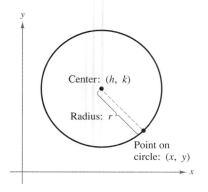

FIGURE 1.14

The Equation of a Circle

The Distance Formula provides a convenient way to define circles. A **circle of radius** r with **center** at the point (h, k) is shown in Figure 1.14. The point (x, y) is on this circle if and only if its distance from the center (h, k) is r. This means that a **circle** in the plane consists of all points (x, y) that are a given positive distance r from a fixed point (h, k). Using the Distance Formula, you can express this relationship by saying that the point (x, y) lies on the circle if an only if

$$\sqrt{(x - h)^2 + (y - k)^2} = r.$$

By squaring both sides of this equation, you can obtain the **standard form of the equation of a circle.**

STANDARD FORM OF THE EQUATION OF A CIRCLE

The **standard form of the equation of a circle** is

$$(x - h)^2 + (y - k)^2 = r^2.$$

The point (h, k) is the **center** of the circle, and the positive number r is the **radius** of the circle. The standard form of the equation of a circle whose center is the *origin* is $x^2 + y^2 = r^2$.

EXAMPLE 6   Finding an Equation of a Circle

The point $(3, 4)$ lies on a circle whose center is at $(-1, 2)$, as shown in Figure 1.15. Find an equation for the circle.

SOLUTION

The radius r of the circle is the distance between $(-1, 2)$ and $(3, 4)$.

$$r = \sqrt{[3 - (-1)]^2 + (4 - 2)^2}$$
$$= \sqrt{16 + 4}$$
$$= \sqrt{20}$$

Thus, the center of the circle is $(h, k) = (-1, 2)$ and the radius is $r = \sqrt{20}$, and you can write the standard form of the equation of the circle as follows.

$$(x - h)^2 + (y - k)^2 = r^2 \qquad \textit{Standard form}$$
$$[x - (-1)]^2 + (y - 2)^2 = (\sqrt{20})^2 \qquad \textit{Let } h = -1, k = 2, \text{ and } r = \sqrt{20}$$
$$(x + 1)^2 + (y - 2)^2 = 20 \qquad \textit{Equation of circle}$$

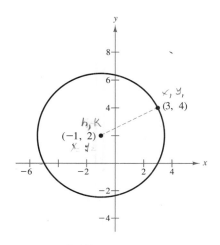

FIGURE 1.15

WARM-UP

The following warm-up exercises involve skills that were covered in earlier sections. You will use these skills in the exercise set for this section.

In Exercises 1–6, simplify the expression.

1. $\sqrt{(2 - 6)^2 + [-(-2)]^2}$

2. $\sqrt{(1 - 4)^2 + (-2 - 1)^2}$

3. $\dfrac{4 + (-2)}{2}$

4. $\dfrac{-1 + (-3)}{2}$

5. $\sqrt{18} + \sqrt{45}$

6. $\sqrt{12} + \sqrt{44}$

In Exercises 7–10, solve for x or y.

7. $\sqrt{(4 - x)^2 + (5 - 2)^2} = \sqrt{58}$

8. $\sqrt{(8 - 6)^2 + (y - 5)^2} = 2\sqrt{5}$

9. $\dfrac{x + 3}{2} = 7$

10. $\dfrac{-2 + y}{2} = 1$

SECTION 1.3 · EXERCISES

In Exercises 1–4, sketch the polygon with the indicated vertices.

1. Triangle: $(-1, 1)$, $(2, -1)$, $(3, 4)$

2. Triangle: $(0, 3)$, $(-1, -2)$, $(4, 8)$

3. Square: $(2, 4)$, $(5, 1)$, $(2, -2)$, $(-1, 1)$

4. Parallelogram: $(5, 2)$, $(7, 0)$, $(1, -2)$, $(-1, 0)$

In Exercises 5–8, find the distance between the points (*Note:* In each case, the two points lie on the same horizontal or vertical line.)

5. $(6, -3)$, $(6, 5)$

6. $(1, 4)$, $(8, 4)$

7. $(-3, -1)$, $(2, -1)$

8. $(-3, -4)$, $(-3, 6)$

In Exercises 9–12, (a) find the lengths of the two perpendicular sides of the right triangle and use the Pythagorean Theorem to find the length of the hypotenuse, and (b) use the Distance Formula to find the length of the hypotenuse of the triangle.

9.

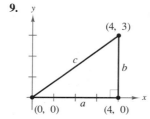

10.

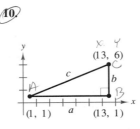

11.

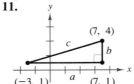

12.

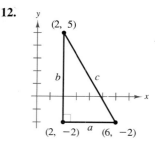

In Exercises 13–24, (a) plot the points, (b) find the distance between the points, and (c) find the midpoint of the line segment joining the points.

13. $(1, 1)$, $(9, 7)$

14. $(1, 12)$, $(6, 0)$

15. $(-4, 10)$, $(4, -5)$

16. $(-7, -4)$, $(2, 8)$

17. $(-1, 2)$, $(5, 4)$

18. $(2, 10)$, $(10, 2)$

19. $\left(\frac{1}{2}, 1\right)$, $\left(-\frac{5}{2}, \frac{4}{3}\right)$

20. $\left(-\frac{1}{3}, -\frac{1}{3}\right)$, $\left(-\frac{1}{6}, -\frac{1}{2}\right)$

21. $(6.2, 5.4)$, $(-3.7, 1.8)$

22. $(-16.8, 12.3)$, $(5.6, 4.9)$

23. $(-36, -18)$, $(48, -72)$

24. $(1.451, 3.051)$, $(5.906, 11.360)$

In Exercises 25 and 26, use the Midpoint Formula to estimate the sales of a company for 1991, given the sales in 1989 and 1993. Assume the sales followed a linear pattern.

25.

Year	1989	1993
Sales	$520,000	$740,000

26.

Year	1989	1993
Sales	$4,200,000	$5,650,000

In Exercises 27–30, show that the points form the vertices of the indicated polygon. (A rhombus is a parallelogram whose sides are all of the same length.)

27. Right triangle: $(4, 0), (2, 1), (-1, -5)$
28. Isosceles triangle: $(1, -3), (3, 2), (-2, 4)$
29. Rhombus: $(0, 0), (1, 2), (2, 1), (3, 3)$
30. Parallelogram: $(0, 1), (3, 7), (4, 4), (1, -2)$

In Exercises 31 and 32, find x so that the distance between the points is 13.

31. $(1, 2), (x, -10)$ **32.** $(-8, 0), (x, 5)$

In Exercises 33 and 34, find y so that the distance between the points is 17.

33. $(0, 0), (8, y)$ **34.** $(-8, 4), (7, y)$

In Exercises 35 and 36, find an equation that relates x and y so that (x, y) is equidistant from the two given points.

35. $(4, -1), (-2, 3)$ **36.** $(3, \frac{5}{2}), (-7, 1)$

In Exercises 37–48, determine the quadrant(s) in which (x, y) is located so that the given conditions are satisfied.

37. $x > 0$ and $y < 0$ **38.** $x < 0$ and $y < 0$
39. $x > 0$ and $y > 0$ **40.** $x < 0$ and $y > 0$
41. $x = -4$ and $y > 0$ **42.** $x > 2$ and $y = 3$
43. $y < -5$ **44.** $x > 4$
45. $xy > 0$ **46.** $xy < 0$
47. $(x, -y)$ is in the second quadrant.
48. $(-x, y)$ is in the fourth quadrant.

49. Plot the points $(2, 1), (-3, 5)$, and $(7, -3)$ on the rectangular coordinate system. Now plot the corresponding points when the sign of the x-coordinate is negated. What inference can you make about the result of the location of a point when the sign of the x-coordinate is changed?

50. Plot the points $(2, 1), (-3, 5)$, and $(7, -3)$ on the rectangular coordinate system. Now plot the corresponding points when the sign of the y-coordinate is negated. What inference can you make about the result of the location of a point when the sign of the y-coordinate is changed?

In Exercises 51–58, find the standard form of the equation of the specified circle.

51. Center: $(0, 0)$; radius: 3   **52.** Center: $(0, 0)$; radius: 5
53. Center: $(2, -1)$; radius: 4   **54.** Center: $(0, \frac{1}{3})$; radius: $\frac{1}{3}$
55. Center: $(-1, 2)$; solution point: $(0, 0)$
56. Center: $(3, -2)$; solution point: $(-1, 1)$
57. Endpoints of a diameter: $(0, 0), (6, 8)$
58. Endpoints of a diameter: $(-4, -1), (4, 1)$

1.4   GRAPHS AND GRAPHING UTILITIES

The Graph of an Equation / Using a Graphing Utility / Determining a Viewing Rectangle / Applications

The Graph of an Equation

News magazines often show graphs comparing the rate of inflation, the federal deficit, wholesale prices, or the unemployment rate to the time of year. Industrial firms and businesses use graphs to report their monthly production and sales statistics. Such graphs provide geometric pictures of the way one quantity changes with respect to another.

Frequently, the relationship between two quantities is expressed as an equation. This section introduces the basic procedure for determining the geometric picture associated with an equation.

For an equation in variables x and y, a point (a, b) is a **solution point** if the substitution of $x = a$ and $y = b$ satisfies the equation. Most equations have *infinitely* many solution points. For example, the equation $3x + y = 5$ has solution points $(0, 5)$, $(1, 2)$, $(2, -1)$, $(3, -4)$, and so on. The set of all solution points of an equation is the **graph** of the equation.

THE POINT-PLOTTING METHOD OF GRAPHING

To sketch the graph of an equation by point plotting, use the following steps.

1. If possible, rewrite the equation so that one of the variables is isolated on the left side of the equation.
2. Make up a table of several solution points.
3. Plot these points in the coordinate plane.
4. Connect the points with a smooth curve.

EXAMPLE 1   Sketching the Graph of an Equation by Point Plotting

Use point plotting and graph paper to sketch the graph of $3x + y = 6$.

SOLUTION

In this case you can isolate the variable y to obtain

$$y = 6 - 3x.$$

Using negative, zero, and positive values for x, you can obtain the following table of values (solution points).

x	-1	0	1	2	3
$y = 6 - 3x$	9	6	3	0	-3

Next, plot these points and connect them as shown in Figure 1.16. It appears that the graph is a straight line. (We will discuss lines extensively in Section 1.5.)

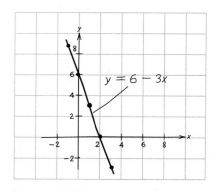

FIGURE 1.16

The points at which a graph touches or crosses an axis are the **intercepts** of the graph. For instance, in Example 1 the point $(0, 6)$ is the y-intercept of the graph because the graph crosses the y-axis at that point. The point $(2, 0)$ is the x-intercept of the graph because the graph crosses the x-axis at that point.

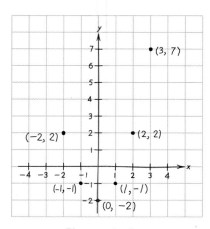

Plot several points.

FIGURE 1.17

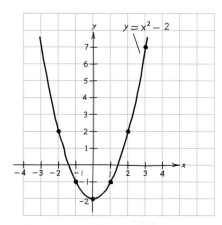

Connect points with a smooth curve.

FIGURE 1.18

EXAMPLE 2   Sketching the Graph of an Equation by Point Plotting

Use point plotting and graph paper to sketch the graph of $y = x^2 - 2$.

SOLUTION

First, make a table of values by choosing several convenient values of x and calculating the corresponding values of y.

x	-2	-1	0	1	2	3
$y = x^2 - 2$	2	-1	-2	-1	2	7

Next, plot the corresponding solution points, as shown in Figure 1.17. Finally, connect the points with a smooth curve, as shown in Figure 1.18. This graph is called a **parabola**.

Using a Graphing Utility

One of the disadvantages of the point-plotting method is that to get a good idea about the shape of a graph you need to plot *many* points. With only a few points, you could badly misrepresent the graph. For instance, consider the equation

$$y = \frac{1}{30}x(x^4 - 10x^2 + 39).$$

Suppose you plotted only five points: $(-3, -3)$, $(-1, -1)$, $(0, 0)$, $(1, 1)$ and $(3, 3)$, as shown in Figure 1.19. From these five points, a person might assume that the graph of the equation is a straight line. This, however, is not correct. By plotting several more points, you can see that the actual graph is not straight at all! (See Figure 1.20.)

Thus, the point-plotting method leaves us with a dilemma. On the one hand, the method can be very inaccurate if only a few points are plotted. But on the other hand, it is very time consuming to plot a dozen (or more) points. Technology can help solve this dilemma. Plotting several (even several hundred) points in a rectangular coordinate system is something that a computer or calculator can do easily.

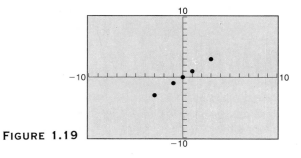

FIGURE 1.19

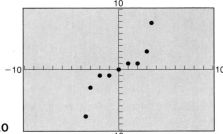

FIGURE 1.20

The point-plotting method is the method used by *all* graphing utilities. Each computer or calculator screen is made up of a grid of hundreds or thousands of small areas called **pixels.** Screens that have many pixels per inch are said to have a higher **resolution** than screens that don't have as many. Screens on most graphing calculators have 48 pixels per inch, whereas screens on computer monitors typically have between 32 and 100 pixels per inch.

With most graphing utilities, you can graph an equation by using the following steps.

USING A GRAPHING UTILITY TO GRAPH AN EQUATION

To graph an equation in x and y on a graphing utility, use the following steps.

1. Rewrite the equation so that y is isolated on the left side of the equation.
2. Enter the equation into a graphing utility.
3. Determine a **viewing rectangle** for the graph. For some graphing utilities, the standard viewing rectangle ranges between -10 and 10 for both the x and y values.
4. Activate the graphing utility.

Technology Note _____

There are often many equivalent ways to solve an equation for y in terms of x. Use your graphing utility to show that the following equations have the same graph as that of Example 3. Then explain how each equation was obtained from the equation $2y + x^3 = 4x$.

$$y = \frac{4x - x^3}{2}$$

$$y = 2x - \frac{x^3}{2}$$

$$y = -0.5x(x^2 - 4)$$

EXAMPLE 3   Using a Graphing Utility

Use a graphing utility to graph $2y + x^3 = 4x$.

SOLUTION

To begin, solve the equation for y in terms of x.

$2y + x^3 = 4x$	*Given equation*
$2y = -x^3 + 4x$	*Subtract x^3 from both sides*
$y = -\dfrac{1}{2}x^3 + 2x$	*Divide both sides by 2*

Now, by entering this equation into a graphing utility (using a standard viewing rectangle), you can obtain the graph shown in Figure 1.21.

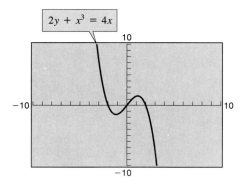

FIGURE 1.21

Determining a Viewing Rectangle

A **viewing rectangle** for a graph is a rectangular portion of the Cartesian plane. A viewing rectangle is determined by six values: the minimum x-value, the maximum x-value, the x-scale, the minimum y-value, the maximum y-value, and the y-scale. When you enter these six values into the graphing utility, you are setting the **range** of the viewing rectangle. The **standard** viewing rectangle for some graphing utilities is shown in Figure 1.22.

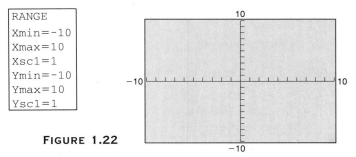

```
RANGE
Xmin=-10
Xmax=10
Xscl=1
Ymin=-10
Ymax=10
Yscl=1
```

FIGURE 1.22

By choosing different viewing rectangles for a graph, it is possible to obtain very different impressions of the graph's shape. For instance, Figure 1.23 shows four different viewing rectangles for the graph of

$$y = 0.1x^4 - x^3 + 2x^2.$$

Of these, the view shown in Figure 1.23(a) is the most complete because it shows more of the distinguishing portions of the graph.

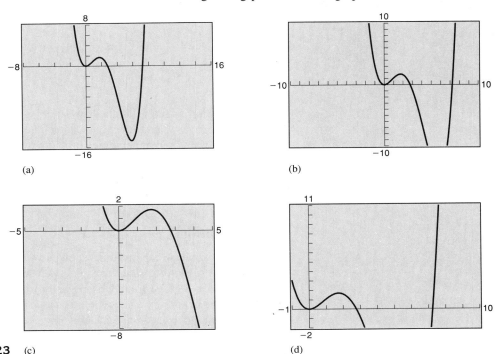

(a)

(b)

FIGURE 1.23   (c)

(d)

Technology Note _____

The standard viewing rectangle on many graphing utilities does not give a true geometric perspective. That is, perpendicular lines will not appear to be perpendicular and circles will not appear to be circular. To overcome this, you can use a square setting, as demonstrated in Example 4.

EXAMPLE 4   Sketching a Circle with a Graphing Utility

Use a graphing utility to graph $x^2 + y^2 = 9$.

SOLUTION

The graph of $x^2 + y^2 = 9$ is a circle whose center is the origin and whose radius is 3. To graph the equation, begin by solving the equation for y.

$$x^2 + y^2 = 9$$
$$y^2 = 9 - x^2$$
$$y = \pm\sqrt{9 - x^2}$$

The graph of $y = \sqrt{9 - x^2}$ is the upper semicircle. The graph of $y = -\sqrt{9 - x^2}$ is the lower semicircle. Enter *both* the equations in your

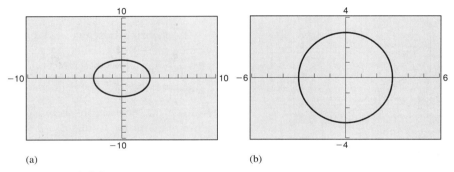

(a)   (b)

FIGURE 1.24

graphing utility and generate the resulting graphs. In Figure 1.24(a), note that if you use a standard viewing rectangle, the two graphs do not appear to form a circle. You can overcome this problem by using a **square setting,** in which the horizontal and vertical tick marks have equal spacing, as shown in Figure 1.24(b). On many graphing utilities, a square setting can be obtained by using the ratio $(Y_{max} - Y_{min})/(X_{max} - X_{min}) = 2/3$.   ■■■■■■■■■

Applications

The following two applications show how to develop mathematical models to represent real-world situations. You will see that both a graphing utility and algebra can be used to understand and resolve the problems posed.

Once an appropriate viewing rectangle is chosen for a particular graph, the *zoom* and *trace* features of a graphing utility are useful for approximating specific values from the graph.

EXAMPLE 5   Using the Zoom and Trace Features of a Graphing
················   Utility

A runner runs at a constant rate of 4.8 miles per hour. The verbal model and
algebraic equation relating distance run in terms of time are given by

VERBAL
MODEL   Distance = Rate · Time

EQUATION $d = 4.8t$

A. Use a graphing utility and an appropriate viewing rectangle to graph the
equation $d = 4.8t$. (Represent d by y and t by x.)
B. Use the zoom and trace features to estimate how far the runner can run in
3.2 hours.
C. Use the zoom and trace features to estimate how long it will take to run
a 26-mile marathon.

Technology Note

In applications, it is convenient to use
variable names that suggest real-life
quantities; d for distance, t for time,
and so on. Most graphing utilities,
however, require the variable names to
be x and y.

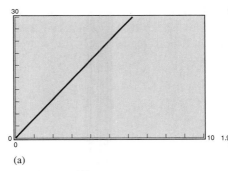
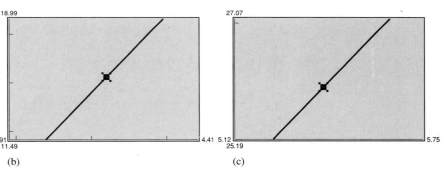

(a)   (b)   (c)

FIGURE 1.25

SOLUTION

A. An appropriate viewing rectangle and the associated graph are shown in
Figure 1.25(a).
B. Figure 1.25(b) shows the viewing rectangle after zooming in (near
$x = 3.2$) once by a factor of 4. Using the trace feature, you can determine
that for $x = 3.2$, the distance is $y \approx 15.36$ miles.
C. Figure 1.25(c) shows the viewing rectangle after zooming in (near $y = 26$)
twice by a factor of 4. Using the trace feature, you can determine that for
$y = 26$, the time is $x \approx 5.42$ hours.

REMARK   The viewing rectangle on
your graphing utility may differ from
those shown in parts (b) and (c) of
Figure 1.25.

EXAMPLE 6   An Application: Monthly Wages
··················

You receive a monthly salary of $2000 plus a commission of 10% of sales.

A. Find an equation expressing the monthly wages y in terms of sales x.
B. If sales are $x = 1480$ in August, what are your wages for that month?
C. If you receive $2225 in wages for September, what were your sales for that
month?

SOLUTION

A. The monthly wages are the sum of the fixed $2000 salary and the 10% commission on sales x.

VERBAL
MODEL   Wages = Salary + Commission on sales

EQUATION $y = 2000 + 0.1x$

B. If $x = 1480$, then the corresponding wages would be

$$y = 2000 + 0.1(1480) = 2000 + 148 = \$2148.$$

You can confirm this result using a graphing utility. Because $x \geq 0$ and the monthly wages are at least $2000, a reasonable viewing rectangle is shown below. [See Figure 1.26(a).] Using the zoom and trace features near $x = 1480$ shows that the wages are about $2148.

C. To answer the third question, you can use the graphing utility to find the value along the x-axis (sales) that corresponds to a y-value of 2225 (wages). Beginning with Figure 1.26(b) and using the zoom and trace features, you can estimate x to be about 2250. You can verify this answer by observing that

$$\text{Wages} = 2000 + 0.1(2250)$$
$$= 2000 + 225$$
$$= 2225.$$

```
RANGE
Xmin=0
Xmax=3000
Xscl=300
Ymin=2000
Ymax=2300
Yscl=30
```

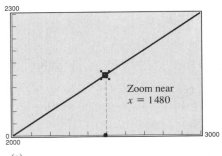

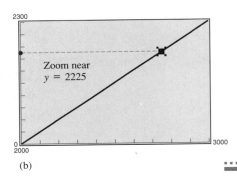

FIGURE 1.26   (a)   (b)

DISCUSSION PROBLEM

COMPARISON OF WAGES

Michael receives a monthly salary of $3100 plus commission of 7% of sales, whereas Janet receives a salary of $3400 plus a 6% commission. Discuss their relative wages depending on sales. Can there be a month in which they receive exactly the same amount of wages? Be sure to take advantage of your graphing utility.

WARM-UP

The following warm-up exercises involve skills that were covered in earlier sections. You will use these skills in the exercise set for this section.

In Exercises 1–6, simplify the expression.

1. $4(x - 3) - 5(6 - 2x)$ **2.** $-s(5s + 2) + 5(s^2 - 3s)$

3. $3y(-2y^2)^3$ **4.** $\dfrac{18a^2b^{-3}}{27ab^2}$

5. $\sqrt{150x^4}$ **6.** $8^{-1/3}$

In Exercises 7–10, completely factor the expression.

7. $2x^3 - 6x^2$ **8.** $2t(t + 3)^2 - 4(t + 3)$

9. $6z^2 + 5z - 50$ **10.** $4s^2 - 25$

SECTION 1.4 · EXERCISES

In Exercises 1–6, determine whether the points lie on the graph of the equation.

Equation	Points
1. $y = \sqrt{x} + 4$	(a) $(0, 2)$ (b) $(5, 3)$
2. $y = x^2 - 3x + 2$	(a) $(2, 0)$ (b) $(-2, 8)$
3. $2x - y - 3 = 0$	(a) $(1, 2)$ (b) $(1, -1)$
4. $x^2 + y^2 = 20$	(a) $(3, -2)$ (b) $(-4, 2)$
5. $x^2y - x^2 + 4y = 0$	(a) $(1, \frac{1}{5})$ (b) $(2, \frac{1}{2})$
6. $y = \dfrac{1}{x^2 + 1}$	(a) $(0, 0)$ (b) $(3, 0.1)$

In Exercises 7–10, find the constant C such that the ordered pair is a solution point of the equation.

7. $y = x^2 + C$, $(2, 6)$ **8.** $y = Cx^3$, $(-4, 8)$

9. $y = C\sqrt{x + 1}$, $(3, 8)$ **10.** $x + C(y + 2) = 0$, $(4, 3)$

In Exercises 11 and 12, complete the table and use the resulting solution points to graph the equation.

11. $2x + y = 3$

x	-4			2	4
y		7	3		
(x, y)					

12. $y = 4 - x^2$

x		-1		2	
y	0		4		-5
(x, y)					

In Exercises 13–18, match the equation with its graph and describe the given viewing rectangle. [The graphs are labeled (a), (b), (c), (d), (e), and (f).]

13. $y = 4 - x$ **14.** $y = x^2 + 2x$

15. $y = \sqrt{4 - x^2}$ **16.** $y = \sqrt{x}$

17. $y = x^3 - x$ **18.** $y = |x| - 2$

(a)
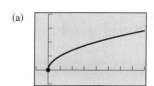

(b)
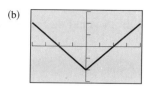

(c)
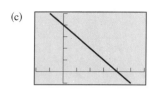

(d)
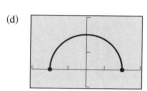

(e)
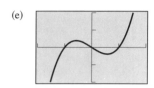

(f)
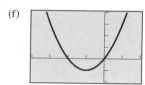

In Exercises 19–30, use the point-plotting method to graph the equation. Use a graphing utility to verify your graph. (Use the standard viewing rectangle.)

19. $y = -3x + 2$ **20.** $y = 2x - 3$

21. $y = 1 - x^2$ **22.** $y = x^2 - 1$

23. $y = x^3 + 2$ **24.** $y = x^3 - 1$

25. $y = (x - 3)(x + 2)$ **26.** $y = x(x - 5)$

27. $y = \sqrt{x - 3}$ **28.** $y = \sqrt{1 - x}$

29. $y = |x - 2|$ **30.** $y = 4 - |x|$

In Exercises 31–38, use a graphing utility to graph the equation. (Use the standard viewing rectangle.) Determine the number of times (if any) the graph intersects each coordinate axis.

31. $y = x^2 - 4x + 3$ **32.** $y = 4 - 4x - x^2$

33. $y = 3x^4 - 6x^2$ **34.** $y = \frac{1}{27}(x^4 + 4x^3)$

35. $y = x\sqrt{4 - x}$ **36.** $y = x\sqrt{4 - x^2}$

37. $y = \dfrac{10x}{x^2 + 1}$ **38.** $y = \dfrac{10}{x^2 + 1}$

In Exercises 39–44, use a graphing utility to graph the equation. Use the specified viewing rectangle. Determine the number of times the graph intersects each coordinate axis.

39. $y = x^4 - 4x^3 + 16x$ **40.** $y = 4x^3 - x^4$

RANGE
Xmin=-5
Xmax=5
Xscl=1
Ymin=-15
Ymax=30
Yscl=5

RANGE
Xmin=-2
Xmax=6
Xscl=1
Ymin=-2
Ymax=30
Yscl=2

41. $y = 100x\sqrt{25 - x}$ **42.** $y = 100x\sqrt{25 - x^2}$

RANGE
Xmin=-30
Xmax=30
Xscl=5
Ymin=-5000
Ymax=5000
Yscl=1000

RANGE
Xmin=-8
Xmax=8
Xscl=1
Ymin=-2000
Ymax=2000
Yscl=500

43. $x^2 - 100y - 1000 = 0$

RANGE
Xmin=-100
Xmax=100
Xscl=10
Ymin=-10
Ymax=10
Yscl=1

44. $2x^3 - 100x - 15,625 + 250y = 0$

RANGE
Xmin=-20
Xmax=25
Xscl=2
Ymin=-2
Ymax=100
Yscl=5

In Exercises 45–48, solve for y and use a graphing utility to graph each of the resulting equations on the same viewing rectangle. Adjust the viewing rectangle so a circle really does appear circular. (Your graphing utility may have a *square* setting that does this automatically.)

45. Circle: $x^2 + y^2 = 64$ **46.** Circle: $x^2 + y^2 = 49$

47. Ellipse: $6x^2 + y^2 = 72$ **48.** Ellipse: $x^2 + 9y^2 = 81$

In Exercises 49–52, describe the given viewing rectangle.

49. $9x + 27y - 1000 = 0$ **50.** $0.75x - 3y + 200 = 0$

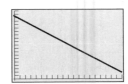 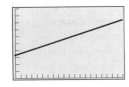

51. $y = -(x - 5)^2(x - 15)$ **52.** $y = \sqrt{x^3 + 8}$

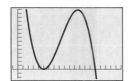 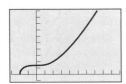

In Exercises 53 and 54, use a graphing utility to graph the equation using each of the suggested viewing rectangles. Assume that the equation gives the profit y when x units of a product are sold. Note that a graph can distort the information presented simply by changing the viewing rectangle. Which viewing rectangle would be selected by a person who wishes to argue that profits will increase dramatically with increased sales?

53. $y = 0.25x - 50$

```
RANGE
Xmin=-3
Xmax=800
Xscl=50
Ymin=-20
Ymax=100
Yscl=10
```

```
RANGE
Xmin=-3
Xmax=1000
Xscl=100
Ymin=-100
Ymax=500
Yscl=40
```

54. $y = 2.44x - \dfrac{x^2}{20,000} - 5000$

```
RANGE
Xmin=-5000
Xmax=22000
Xscl=5000
Ymin=-20000
Ymax=60000
Yscl=10000
```

```
RANGE
Xmin=-5000
Xmax=22000
Xscl=5000
Ymin=-5000
Ymax=24000
Yscl=5000
```

In Exercises 55 and 56, use the equation to complete each table. Note the importance of increasing the number of solution points to increase the accuracy of the graph.

(a) Use the table to sketch a graph of the equation.

x	-1	0	1
y			

(b) Use the table to sketch a graph of the equation.

x	-1	$-\frac{3}{4}$	$-\frac{1}{2}$	$-\frac{1}{4}$	0	$\frac{1}{4}$	$\frac{1}{2}$	$\frac{3}{4}$	1
y									

(c) Use a graphing utility to graph the equation for $-1 \leq x \leq 1$.

55. $y = \sqrt[3]{x}$ **56.** $y = x^3(3x + 4)$

In Exercises 57–60, use a graphing utility to graph the equation. Move the cursor along the curve to approximate the unknown coordinate(s) of the given solution point accurate to two decimal places. (*Hint:* You may need to use the zoom feature of the calculator to obtain the required accuracy.)

57. $y = \sqrt{5 - x}$ **58.** $y = x^3(x - 3)$
   (a) $(2, y)$ (a) $(2.25, y)$
   (b) $(x, 3)$ (b) $(x, 20)$

59. $y = x^5 - 5x$ **60.** $y = |x^2 - 6x + 5|$
   (a) $(-0.5, y)$ (a) $(2, y)$
   (b) $(x, -4)$ (b) $(x, 1.5)$

61. *Depreciation* A manufacturing plant purchases a new molding machine for $225,000. The depreciated value y after x years is given by $y = 225,000 - 20,000x$, $0 \leq x \leq 8$.
   (a) Use the constraints of the model to determine an appropriate viewing rectangle.
   (b) Graph the equation.

62. *Dimensions of a Rectangle* A rectangle of length x and width w has a perimeter of 12 meters.
   (a) Show that the area of the rectangle is $y = x(6 - x)$.
   (b) Use the physical constraints of the problem to determine the horizontal component of the viewing rectangle.
   (c) Use a graphing utility to graph the equation for the area. (*Note:* You may have to sketch the graph more than once in order to determine an appropriate vertical component of the viewing rectangle.)
   (d) From the graph in part (c), estimate the dimensions of the rectangle that yield a maximum area.

In Exercises 63 and 64, (a) graph the model and compare it with the data, and (b) use the model to predict y for the year 1994.

63. *Federal Debt*   The table gives the per capita federal debt for the United States for selected years from 1950 to 1990. (*Source:* U.S. Treasury Department)

Year	1950	1960	1970	1980	1985	1990
Per capita debt	$1688	$1572	$1807	$3981	$7614	$12,848

A model for the per capita debt during this period is

$$y = 0.40x^3 - 9.42x^2 + 1053.24,$$

where y represents the per capita debt and x is the year, with $x = 0$ corresponding to 1950.

64. *Life Expectancy*   The table gives the life expectancy of a child (at birth) for selected years from 1920 to 1989. (*Source:* Department of Health and Human Services)

Year	1920	1930	1940	1950
Life expectancy	54.1	59.7	62.9	68.2

Year	1960	1970	1980	1989
Life expectancy	69.7	70.8	73.7	75.2

A model for the life expectancy during this period is

$$y = \frac{x + 66.94}{0.01x + 1},$$

where y represents the life expectancy and x represents the year, with $x = 0$ corresponding to 1950.

65. *Earnings Per Share*   The earnings per share for Eli Lilly Corporation from 1980 to 1986 can be approximated by the model

$$y = 1.097x + 0.15, \qquad 0 \le x \le 6,$$

where y is the earnings and x represents the year, with $x = 0$ corresponding to 1980. Graph this equation. (*Source:* NYSE Stock Reports)

66. *Copper Wire*   The resistance y in ohms of 1000 feet of solid copper wire at 77 degrees Fahrenheit can be approximated by the model

$$y = \frac{10,770}{x^2} - 0.37, \qquad 5 \le x \le 100,$$

where x is the diameter of the wire in mils (0.001 in.). Use the model to estimate the resistance when $x = 50$.

67. Find a and b if the x-intercept of the graph of $y = \sqrt{ax + b}$ is (5, 0). (There is more than one correct answer.)

1.5   LINES IN THE PLANE AND SLOPE

The Slope of a Line   /   The Point-Slope Form of the Equation of a Line   /   Sketching Graphs of Lines   /   Changing the Viewing Rectangle   /   Parallel and Perpendicular Lines

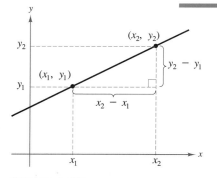

FIGURE 1.27

The Slope of a Line

In this section, you will study lines and their equations. Throughout this text, the term **line** always means a *straight* line.

The **slope** of a nonvertical line represents the number of units a line rises or falls vertically for each unit of horizontal change from left to right. For instance, consider the two points (x_1, y_1) and (x_2, y_2) on the line shown in Figure 1.27. As you move from left to right along this line, a change of $(y_2 - y_1)$ units in the vertical direction corresponds to a change of $(x_2 - x_1)$ units in the horizontal direction. That is,

D I S C O V E R Y

Use a graphing utility to compare the slopes of the lines given by $y = ax$ with $a = 0.5, 1, 2,$ and 4. What do you observe about the slopes of the lines? Repeat the experiment with $a = -0.5, -1, -2,$ and -4. What do you observe about the slopes of these lines? (*Hint:* Use a square setting to guarantee a true geometric perspective.)

$$y_2 - y_1 = \text{the change in } y$$

and

$$x_2 - x_1 = \text{the change in } x.$$

The slope of the line is given by the ratio of these two changes.

DEFINITION OF THE SLOPE OF A LINE

The **slope** m of the nonvertical line passing through the points (x_1, y_1) and (x_2, y_2) is

$$m = \frac{y_2 - y_1}{x_2 - x_1} = \frac{\text{change in } y}{\text{change in } x},$$

where $x_1 \neq x_2$.

When this formula is used, the *order of subtraction* is important. Given two points on a line, you are free to label either one of them (x_1, y_1), and the other as (x_2, y_2). However, once this is done, you must form the numerator and denominator using the same order of subtraction.

$$m = \frac{y_2 - y_1}{x_2 - x_1} \qquad m = \frac{y_1 - y_2}{x_1 - x_2} \qquad m = \frac{y_2 - y_1}{x_1 - x_2}$$

Correct   Correct   Incorrect

EXAMPLE 1 **Finding the Slope of a Line Passing Through Two Points**

Find the slopes of the lines passing through the following pairs of points.

A. $(-2, 0)$ and $(3, 1)$
B. $(-1, 2)$ and $(2, 2)$
C. $(0, 4)$ and $(1, -1)$

SOLUTION

Difference in y-values

A. $m = \dfrac{y_2 - y_1}{x_2 - x_1} = \dfrac{1 - 0}{3 - (-2)} = \dfrac{1}{3 + 2} = \dfrac{1}{5}$

Difference in x-values

B. $m = \dfrac{2 - 2}{2 - (-1)} = \dfrac{0}{3} = 0$

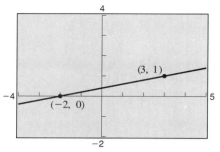

(a) The slope is $\frac{1}{5}$.

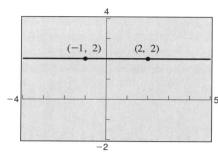

(b) The slope is zero.

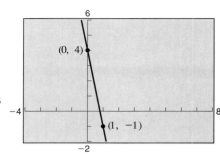

(c) The slope is −5.

FIGURE 1.28

c. $m = \dfrac{-1 - 4}{1 - 0} = \dfrac{-5}{1} = -5$

The graphs of the three lines are shown in Figure 1.28. In parts (a) and (c), note that the "square" setting gives the correct "steepness" of the line.

The definition of slope does not apply to vertical lines. For instance, consider the points $(3, 4)$ and $(3, 1)$ on the vertical line shown in Figure 1.29. Applying the formula for slope, you obtain

$$m = \frac{4 - 1}{3 - 3}.$$ *Undefined division by zero*

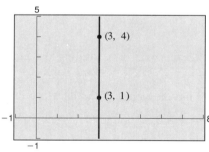

The slope is undefined.

FIGURE 1.29

Because division by zero is undefined, the slope of a vertical line is undefined.

From the slopes of the lines shown in Figures 1.28 and 1.29, you can make the following generalizations about the slope of a line.

1. A line with positive slope ($m > 0$) *rises* from left to right.
2. A line with negative slope ($m < 0$) *falls* from left to right.
3. A line with zero slope ($m = 0$) is *horizontal*.
4. A line with undefined slope is *vertical*.

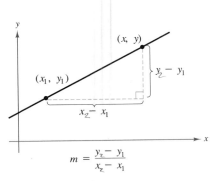

$$m = \frac{y_2 - y_1}{x_2 - x_1}$$

FIGURE 1.30

The Point-Slope Form of the Equation of a Line

If you know the slope of a line *and* you also know the coordinates of one point on the line, then you can find an equation for the line. For instance, in Figure 1.30, let (x_1, y_1) be a given point on the line whose slope is m. If (x, y) is any *other* point on the line, then it follows that

$$\frac{y - y_1}{x - x_1} = m.$$

This equation in the variables x and y can be rewritten in the form

$$y - y_1 = m(x - x_1),$$

which is the **point-slope form** of the equation of a line.

POINT-SLOPE FORM OF THE EQUATION OF A LINE

The **point-slope** form of the equation of the line that passes through the point (x_1, y_1) and has a slope of m is

$$y - y_1 = m(x - x_1).$$

EXAMPLE 2   The Point-Slope Form of the Equation of a Line

Find an equation of the line that passes through the point $(1, -2)$ and has a slope of 3.

SOLUTION

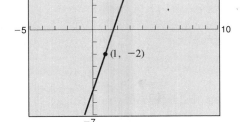

$$y - y_1 = m(x - x_1) \qquad \textit{Point-slope form}$$
$$y - (-2) = 3(x - 1) \qquad \textit{Substitute } y_1 = -2, x_1 = 1, \textit{ and } m = 3$$
$$y + 2 = 3x - 3$$
$$y = 3x - 5 \qquad \textit{Equation of line}$$

This line is shown in Figure 1.31.

FIGURE 1.31

The point-slope form can be used to find the equation of a nonvertical line passing through two points (x_1, y_1) and (x_2, y_2). First, use the formula for the slope of the line passing through two points.

$$m = \frac{y_2 - y_1}{x_2 - x_1}$$

Then, once you know the slope, use the point-slope form to obtain the equation

$$y - y_1 = \frac{y_2 - y_1}{x_2 - x_1}(x - x_1).$$

This is sometimes called the **two-point form** of the equation of a line.

EXAMPLE 3   A Linear Model for Sales Prediction

During the first two quarters of the year, a company had total sales of $3.4 million and $3.7 million, respectively.

A. Write a linear equation giving the total sales y in terms of the quarter x.
B. Use the equation to predict the total sales during the fourth quarter.

SOLUTION

A. In Figure 1.32, let (1, 3.4) and (2, 3.7) be two points on the line representing the total sales. The slope of the line passing through these two points is

$$m = \frac{3.7 - 3.4}{2 - 1} = 0.3.$$

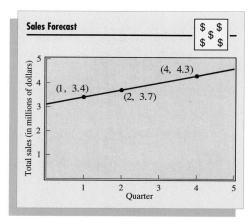

FIGURE 1.32

By the point-slope form, the equation of the line is as follows.

$$y - y_1 = m(x - x_1)$$
$$y - 3.4 = 0.3(x - 1)$$
$$y = 0.3x - 0.3 + 3.4$$
$$y = 0.3x + 3.1$$

B. Using the equation from part (A), estimate the fourth-quarter sales ($x = 4$) to be $y = 0.3(4) + 3.1 = 1.2 + 3.1 = \$4.3$ million.

The approximation method illustrated in Example 3 is **linear extrapolation.** Note in Figure 1.33 that for linear extrapolation, the estimated point lies outside of the given points. When the estimated point lies *between* two given points, the procedure is called **linear interpolation.**

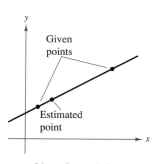

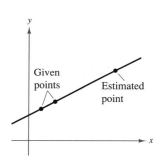

Linear Extrapolation

Linear Interpolation

FIGURE 1.33

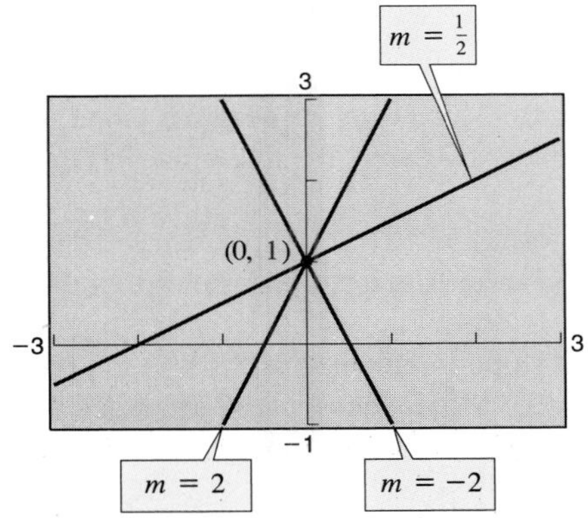

Sketching Graphs of Lines

Many problems in coordinate (or analytic) geometry can be classified in two basic categories.

1. Given a graph (or parts of it), find its equation.
2. Given an equation, find its graph.

For lines, the first problem is solved easily by using the point-slope form. This formula, however, is not particularly useful for solving the second type of problem. The form that is better suited to graphing linear equations is the **slope-intercept form** $y = mx + b$ of the equation of a line.

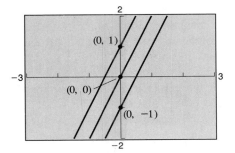

FIGURE 1.34

FIGURE 1.35

EXAMPLE 4   Determining the Slope and y-Intercepts

A. By graphing the equations $y = 2x + 1$, $y = \frac{1}{2}x + 1$, and $y = -2x + 1$ on the same coordinate axes, you can see that they each have the same y-intercept $(0, 1)$, but slopes of 2, $\frac{1}{2}$, and -2, respectively (see Figure 1.34).

B. By graphing the equations $y = 2x + 1$, $y = 2x$, and $y = 2x - 1$ on the some coordinate axes, you can see that they each have the same slope of 2, but y-intercepts of $(0, 1)$, $(0, 0)$, and $(0, -1)$, respectively (see Figure 1.35).

SLOPE-INTERCEPT FORM OF THE EQUATION OF A LINE

The graph of the equation

$$y = mx + b$$

is a line whose slope is m and whose y-intercept is $(0, b)$.

To derive algebraically the slope-intercept form, write the following.

$$y_2 - y_1 = m(x_2 - x_1) \qquad \text{\textit{Point-slope form}}$$
$$y_2 = mx_2 - mx_1 + y_1$$
$$y_2 = mx_2 + b \qquad \text{\textit{Slope-intercept form}}$$

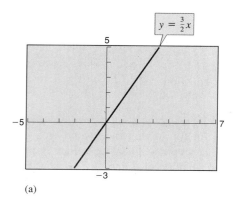

(a)

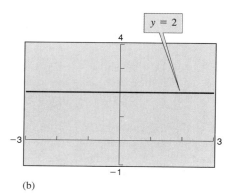

(b)

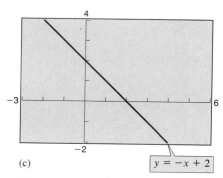

(c)

FIGURE 1.36

EXAMPLE 5   Using the Slope-Intercept Form

Describe the graphs of the linear equations.

A. $y = \dfrac{3}{2}x$

B. $y = 2$

C. $x + y = 2$

SOLUTION

A. Because $b = 0$, the y-intercept is $(0, 0)$. Moreover, because the slope is $m = \frac{3}{2}$, this line *rises* three units for every two units the line moves to the right, as shown in Figure 1.36(a).

B. By writing the equation $y = 2$ in the form $y = (0)x + 2$, you can see that the y-intercept is $(0, 2)$ and the slope is zero. A zero slope implies that the line is horizontal, as shown in Figure 1.36(b).

C. By writing the equation $x + y = 2$ in slope-intercept form, $y = -x + 2$, you can see that the y-intercept is $(0, 2)$. Moreover, because the slope is $m = -1$, this line *falls* one unit for each unit the line moves to the right, as shown in Figure 1.36(c).

From the slope-intercept form of the equation of a line, you can see that a horizontal line ($m = 0$) has an equation of the form $y = b$. This is consistent with the fact that each point on a horizontal line through $(0, b)$ has a y-coordinate of b.

Similarly, each point on a vertical line through $(a, 0)$ has an x-coordinate of a. Hence, a vertical line has an equation of the form $x = a$. This equation cannot be written in the slope-intercept form, because the slope of a vertical line is undefined. However, *every* line has an equation that can be written in the **general form** $Ax + By + C = 0$, where A and B are not *both* zero.

SUMMARY OF EQUATIONS OF LINES

1. General form: $Ax + By + C = 0$
2. Vertical line: $x = a$
3. Horizontal line: $y = b$
4. Slope-intercept form: $y = mx + b$
5. Point-slope form: $y - y_1 = m(x - x_1)$

Changing the Viewing Rectangle

When using a graphing utility to sketch a straight line, it is important to realize that the graph of the line may not visually appear to have the slope indicated by its equation. This occurs because of the viewing rectangle used for the graph. For instance, Figure 1.37 shows a graph of $y = 2x + 1$ using a graphing utility with three different viewing rectangles.

Notice that the slopes of the first two lines do not visually appear to be equal to 2. If you use the "square" viewing rectangle (right), then the slope will visually appear to be 2. In general, two graphs of the same equation can appear to be quite different depending upon the viewing rectangle selected.

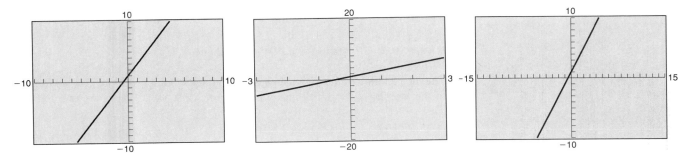

FIGURE 1.37

Effects of Different Viewing Rectangles on Graph of $y = 2x + 1$

EXAMPLE 6   **Different Viewing Rectangles**

The graphs of the two lines $y = -x - 1$ and $y = -10x - 1$ are shown in Figure 1.38. Even though the slopes of these lines are different (-1 and -10, respectively), the graphs seem similar because the viewing rectangles are different.

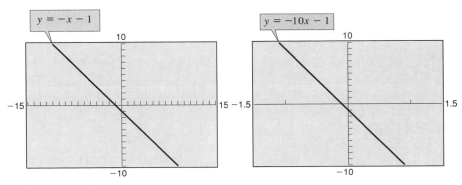

FIGURE 1.38

Parallel and Perpendicular Lines

The slope of a line is a convenient tool for determining whether two lines are parallel or perpendicular. Example 4(B) suggests the following property of parallel lines.

> ### PARALLEL LINES
>
> Two distinct nonvertical lines are **parallel** if and only if their slopes are equal.

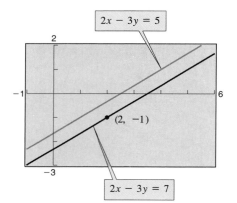

FIGURE 1.39

EXAMPLE 7   Equations of Parallel Lines

Find an equation of the line that passes through the point $(2, -1)$ and is parallel to the line $2x - 3y = 5$, as shown in Figure 1.39.

SOLUTION

Write the given equation in slope-intercept form.

$$2x - 3y = 5 \qquad \text{\textit{Given equation}}$$
$$3y = 2x - 5$$
$$y = \frac{2}{3}x - \frac{5}{3} \qquad \text{\textit{Slope-intercept form}}$$

Therefore, the given line has a slope of $m = \frac{2}{3}$. Because any line parallel to the given line must also have a slope of $\frac{2}{3}$, the required line through $(2, -1)$ has the following equation.

$$y - (-1) = \frac{2}{3}(x - 2) \qquad \text{\textit{Point-slope form}}$$
$$y = \frac{2}{3}x - \frac{4}{3} - 1$$
$$y = \frac{2}{3}x - \frac{7}{3} \qquad \text{\textit{Slope-intercept form}}$$

Notice the similarity between the slope-intercept form of the original equation and the slope-intercept form of the parallel equation.

Two nonvertical lines are *perpendicular* if and only if their slopes are negative reciprocals of each other. For instance, the lines $y = 2x$ and $y = -\frac{1}{2}x$ are perpendicular because one has a slope of 2 and the other has a slope of $-\frac{1}{2}$.

DISCOVERY

Use a graphing utility with the viewing rectangle $-15 \le x \le 15$ and $-10 \le y \le 10$ to graph the following linear equations.

$y_1 = -\frac{1}{3}x + 2$

$y_2 = 3x - 6$

$y_3 = 3x + 3$

What do you observe? How are the slopes of parallel lines related? How are the slopes of perpendicular lines related? Verify your conclusions using the lines $y_1 = -2x - 5$, $y_2 = -2x + 6$, and $y_3 = \frac{1}{2}x + 1$.

PERPENDICULAR LINES

Two nonvertical lines are **perpendicular** if and only if their slopes are negative reciprocals of each other. That is,

$$m_1 = -\frac{1}{m_2}.$$

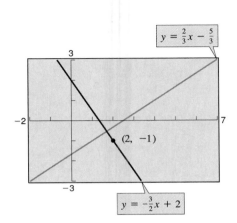

$y = \frac{2}{3}x - \frac{5}{3}$

(2, −1)

$y = -\frac{3}{2}x + 2$

FIGURE 1.40

EXAMPLE 8   Equations of Perpendicular Lines

Find an equation of the line that passes through the point $(2, -1)$ and is perpendicular to the line $2x - 3y = 5$.

SOLUTION

By writing the given line in the form $y = \frac{2}{3}x - \frac{5}{3}$ you can see that the line has a slope of $\frac{2}{3}$. Hence, any line that is perpendicular to this line must have a slope of $-\frac{3}{2}$ (because $-\frac{3}{2}$ is the negative reciprocal of $\frac{2}{3}$). Therefore, the required line through the point $(2, -1)$ has the following equation.

$$y - (-1) = -\frac{3}{2}(x - 2) \qquad \textit{Point-slope form}$$

$$y = -\frac{3}{2}x + 3 - 1$$

$$y = -\frac{3}{2}x + 2 \qquad \textit{Slope-intercept form}$$

The graphs of both equations are shown in Figure 1.40.

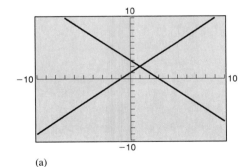

(a)

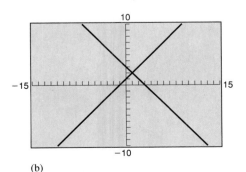

(b)

FIGURE 1.41

EXAMPLE 9   Graphs of Perpendicular Lines

Use a graphing utility to graph the lines given by $y = x + 1$ and $y = -x + 3$. Display *both* graphs on the same screen. The lines are supposed to be perpendicular (they have slopes of $m_1 = 1$ and $m_2 = -1$). Do they appear to be perpendicular on the display?

SOLUTION

If the viewing window has the *standard* range settings, then the tick marks on both the x-axis and the y-axis will vary between -10 and 10, as in Figure 1.41(a). However, because most display screens are not square, the lines might *not* appear to be perpendicular. That is, the graphs of two perpendicular lines will appear to be perpendicular only if the tick marks on the x-axis have the same spacing as the tick marks on the y-axis, as in Figure 1.41(b). This can be done by using a "square" setting for the viewing rectangle.

DISCUSSION PROBLEM

....

AN APPLICATION OF SLOPE

In 1982, a college had an enrollment of 5000 students. By 1992, the enrollment had increased to 7000 students.

(a) What was the average annual change in enrollment from 1982 to 1992?
(b) Use the average annual change in enrollment to estimate the enrollment in 1986, 1990, and 1994.

Year	1982	1986	1990	1992	1994
Enrollment	5000			7000	

(c) Graph the line represented by the data given in the table in part (b). What is the slope of this line?
(d) Write a short paragraph that compares the concepts of *slope* and *average rate of change*.

WARM-UP

....

The following warm-up exercises involve skills that were covered in earlier sections. You will use these skills in the exercise set for this section.

In Exercises 1 and 2, simplify the expression.

1. $\dfrac{4 - (-5)}{-3 - (-1)}$

2. $\dfrac{-5 - 8}{0 - (-3)}$

3. Find $-1/m$ for $m = 4/5$.

4. Find $-1/m$ for $m = -2$.

In Exercises 5–10, solve for y in terms of x.

5. $2x - 3y = 5$

6. $4x + 2y = 0$

7. $y - (-4) = 3[x - (-1)]$

8. $y - 7 = \frac{2}{3}(x - 3)$

9. $y - (-1) = \dfrac{3 - (-1)}{2 - 4}(x - 4)$

10. $y - 5 = \dfrac{3 - 5}{0 - 2}(x - 2)$

SECTION 1.5 · EXERCISES

In Exercises 1–6, estimate the slope of the line.

1.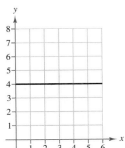

2.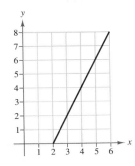

3.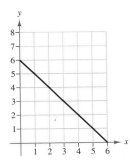

4.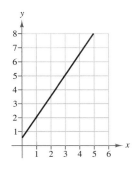

5.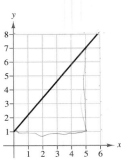

6.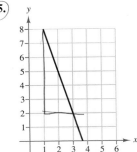

In Exercises 7 and 8, graph the lines through the given point with the indicated slopes. Graph the lines on the same set of coordinate axes.

Point	Slopes
7. (2, 3)	(a) 0 (b) 1 (c) 2 (d) −3
8. (−4, 1)	(a) 3 (b) −3 (c) $\frac{1}{2}$ (d) Undefined

In Exercises 9–14, plot the points and find the slope of the line passing through the points.

9. $(-3, -2), (1, 6)$ **10.** $(2, 4), (4, -4)$
11. $(-6, -1), (-6, 4)$ **12.** $(0, -10), (-4, 0)$
13. $(1, 2), (-2, -2)$ **14.** $\left(\frac{7}{8}, \frac{3}{4}\right), \left(\frac{5}{4}, -\frac{1}{4}\right)$

In Exercises 15–18, use the given point on the line and the slope of the line to find three additional points through which the line passes. (There are many correct answers.)

Point	Slope		Point	Slope
15. (2, 1)	$m = 0$		**16.** (−4, 1)	m is undefined
17. (5, −6)	$m = 1$		**18.** (10, −6)	$m = -1$

In Exercises 19–22, determine whether the lines L_1 and L_2 passing through the given pairs of points are parallel, perpendicular, or neither.

19. L_1: $(0, -1), (5, 9)$ **20.** L_1: $(-2, -1), (1, 5)$
L_2: $(0, 3), (4, 1)$ L_2: $(1, 3), (5, -5)$

21. L_1: $(3, 6), (-6, 0)$ **22.** L_1: $(4, 8), (-4, 2)$
L_2: $(0, -1), \left(5, \frac{7}{3}\right)$ L_2: $(3, -5), \left(-1, \frac{1}{3}\right)$

In Exercises 23 and 24, use the concept of slope to determine whether the three points are collinear.

23. $(0, -4), (2, 0), (3, 2)$ **24.** $(-2, 1), (-1, 0), (2, -2)$

25. *Mountain Driving* While driving down a mountain road, you notice a "12% grade" warning sign. This means that the slope of the road is $-\frac{12}{100}$. Determine the amount of horizontal change in your position if you note from elevation markers that you have descended 2000 feet vertically.

26. *Attic Height* The "rise to run" in determining the steepness of the roof on the house in the figure is 3 to 4. Determine the maximum height in the attic of the house if the house is 30 feet wide.

FIGURE FOR 26

In Exercises 27–30, find the slope and y-intercept (if possible) of the line specified by the equation. Graph the line.

27. $5x - y + 3 = 0$ **28.** $2x + 3y - 9 = 0$
29. $5x - 2 = 0$ **30.** $3y + 5 = 0$

In Exercises 31–36, find an equation for the line passing through the points.

31. $(5, -1), (-5, 5)$ **32.** $(4, 3), (-4, -4)$

33. $(2, \frac{1}{2}), (\frac{1}{2}, \frac{5}{4})$ **34.** $(-1, 4), (6, 4)$

35. $(-8, 1), (-8, 7)$ **36.** $(-8, 0.6), (2, -2.4)$

In Exercises 37–42, find an equation of the line that passes through the point and has the indicated slope.

Point	Slope		Point	Slope
37. $(0, -2)$	$m = 3$		**38.** $(0, 0)$	$m = 4$
39. $(4, \frac{5}{2})$	$m = -\frac{4}{3}$		**40.** $(-2, -5)$	$m = \frac{3}{4}$
41. $(6, -1)$	m is undefined		**42.** $(-10, 4)$	$m = 0$

In Exercises 43 and 44, use a graphing utility to graph the equation using each of the suggested viewing rectangles. Note that the viewing rectangle selected will alter the appearance of the slope.

43. $y = 0.5x - 3$

```
RANGE
Xmin=-5
Xmax=10
Xscl=1
Ymin=-10
Ymax=5
Yscl=1
```

```
RANGE
Xmin=-2
Xmax=10
Xscl=1
Ymin=-4
Ymax=1
Yscl=1
```

44. $y = -8x + 5$

```
RANGE
Xmin=-5
Xmax=5
Xscl=1
Ymin=-10
Ymax=10
Yscl=1
```

```
RANGE
Xmin=-5
Xmax=10
Xscl=1
Ymin=-80
Ymax=80
Yscl=20
```

In Exercises 45–48, use a graphing utility to graph the three equations on the same viewing rectangle. Adjust the viewing rectangle so the slope appears visually correct. (Your calculator may have a *square* setting that does this automatically.)

45. $y = 2x$ $y = -2x$ $y = \frac{1}{2}x$

46. $y = \frac{2}{3}x$ $y = -\frac{3}{2}x$ $y = \frac{2}{3}x + 2$

47. $y = -\frac{1}{2}x$ $y = -\frac{1}{2}x + 3$ $y = 2x - 4$

48. $y = x - 8$ $y = x + 1$ $y = -x + 3$

In Exercises 49 and 50, use the given values of a and b and a graphing utility to graph the equation of the line given by

$$\frac{x}{a} + \frac{y}{b} = 1, \qquad a \neq 0, b \neq 0.$$

Use the graphs to determine the meaning of the constants a and b.

49. $a = 5, \quad b = -3$ **50.** $a = -6, \quad b = 2$

In Exercises 51 and 52, use the results of Exercises 49 and 50 to write an equation of the line that passes through the given points.

51. x-intercept: $(2, 0)$ **52.** x-intercept: $(-\frac{1}{6}, 0)$
    y-intercept: $(0, 3)$ y-intercept: $(0, -\frac{2}{3})$

In Exercises 53–56, write an equation of the line through the point (a) parallel to the given line and (b) perpendicular to the given line.

Point	Line		Point	Line
53. $(2, 1)$	$4x - 2y = 3$		**54.** $(\frac{7}{8}, \frac{3}{4})$	$5x + 3y = 0$
55. $(-1, 0)$	$y = -3$		**56.** $(2, 5)$	$x = 4$

In Exercises 57–60, you are given the dollar value of a product in 1990 *and* the rate at which the value of the item is expected to increase during the next five years. Use this information to write a linear equation that gives the dollar value V of the product in terms of the year t. (Let $t = 0$ represent 1990.)

	1990 Value	Rate
57.	$2540	$125 per year
58.	$156	$4.50 per year
59.	$20,400	$2000 per year
60.	$245,000	$5600 per year

In Exercises 61–64, match the description with a graph. Determine the slope and how it is interpreted in the situation. [The graphs are labeled (a), (b), (c), and (d).]

61. A person is paying $10 per week to a friend to repay a $100 loan.

62. An employee is paid $12.50 per hour plus $1.50 for each unit produced per hour.

63. A sales representative receives $20 per day for food plus $0.25 for each mile traveled.

64. A typewriter purchased for $600 depreciates $100 per year.

(a)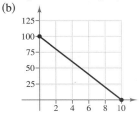

(b)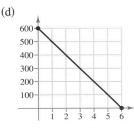

(c)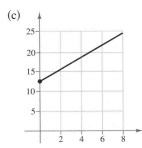

(d)

65. *Temperature*   Find an equation of the line giving the relationship between the temperature in degrees Celsius, C, and degrees Fahrenheit, F. Remember that water freezes at 0° Celsius (32° Fahrenheit) and boils at 100° Celsius (212° Fahrenheit).

66. *Temperature*   Use the result of Exercise 65 to complete the table.

C		$-10°$	$10°$			$177°$
F	$0°$			$68°$	$90°$	

67. *Annual Salary*   Suppose your salary was $28,500 in 1990 and $32,900 in 1992. Assume your salary follows a linear growth pattern.
   (a) Write a linear equation giving the salary S in terms of the year t where $t = 0$ corresponds to the year 1990.
   (b) Use the linear equation to predict your salary in 1995.

68. *College Enrollment*   A small college had 2546 students in 1990 and 2702 students in 1992. Assume the enrollment follows a linear growth pattern.
   (a) Write a linear equation giving the enrollment E in terms of the year t where $t = 0$ corresponds to the year 1990.
   (b) Use the linear equation to predict the enrollment in 1997.

69. *Straight-Line Depreciation*   A small business purchases a piece of equipment for $875. After 5 years the equipment will be outdated and have no value.
   (a) Write a linear equation giving the value y of the equipment in terms of the time x, $0 \le x \le 5$.
   (b) Use a graphing utility to graph the equation.
   (c) Move the cursor along the graph and estimate (to two decimal place accuracy) the value of the equipment when $x = 2$.
   (d) Move the cursor along the graph and estimate (to two decimal place accuracy) the time when the value of the equipment is $200.

70. *Hourly Wages*   A manufacturer pays its assembly line workers $11.50 per hour plus $0.75 per unit produced.
   (a) Write the linear equation for the hourly wages y in terms of the number of units x produced per hour.
   (b) Use a graphing utility to graph the equation.
   (c) Move the cursor along the graph and estimate (to two decimal place accuracy) the hourly wage when a worker produces $x = 6$ units per hour.
   (d) Move the cursor along the graph and determine the number of units that must be produced per hour to have an hourly wage of $18.25.

71. *Sales Commission*   A salesperson receives a monthly salary of $2500 plus a commission of 7% of sales. Write a linear equation for the salesperson's monthly wage W in terms of monthly sales S.

72. *Daily Cost*   A sales representative of a company receives $120 per day for lodging and meals plus $0.26 per mile driven. Write a linear equation giving the daily cost C to the company in terms of x, the number of miles driven.

73. *Contracting Purchase*   A contractor purchases a piece of equipment for $36,500. The equipment has an average expense of $5.25 per hour for fuel and maintenance, and the operator is paid $11.50 per hour.
   (a) Write a linear equation for the total cost C of operating the equipment for t hours. (Include the purchase cost.)
   (b) Customers are charged $37 per hour of machine use. Write an equation for the revenue R derived from t hours of use.
   (c) *Break-Even Point*   Use a graphing utility to graph the cost and revenue equations on the same viewing rectangle. Move the cursor to the point of intersection of the graphs to estimate (to the nearest hour) the number of hours the equipment must be used to break even.

74. *Real Estate*   A real estate office handles an apartment complex with 50 units. When the rent per unit is $380 per month, all 50 units are occupied. However, when the rent is $425 per month, the average number of occupied units drops to 47. Assume that the relationship between the monthly rent p and the demand x is linear.

(a) Write an equation of the line giving the demand x in terms of the rent p.

(b) Use a graphing utility to graph the demand equation.

(c) Move the cursor along the graph to predict the number of units occupied if the rent is raised to $455.

(d) Move the cursor along the graph to predict the number of units occupied if the rent is lowered to $395.

75.

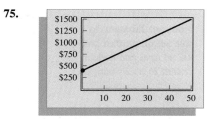

76.

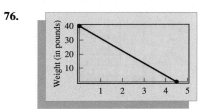

In Exercises 75 and 76, create a realistic problem that is modeled by the graph.

77. Use the theorem that the ratios of corresponding sides of similar triangles are equal to verify that any two points on a line can be used to calculate its slope.

1.6   FUNCTIONS

Introduction to Functions / Function Notation / The Domain of a Function / Function Keys on a Graphing Utility / Applications

Introduction to Functions

Many everyday phenomena involve two quantities that are related to each other by some rule of correspondence. Such a rule of correspondence is called a **function.** Here are two examples.

1. The simple interest I earned on $1000 for one year is related to the annual percentage rate r by the formula $I = 1000r$.
2. The area A of a circle is related to its radius r by the formula $A = \pi r^2$.

REMARK   Note that this use of the word *range* is not the same as the use of *range* relating to the viewing window for a graph.

DEFINITION OF A FUNCTION

A **function** f from a set A to a set B is a rule of correspondence that assigns to each element x in the set A exactly one element y in the set B. The set A is the **domain** (or set of inputs) of the function f, and the set B contains the **range** (or set of outputs).

To help understand this definition, look at the function illustrated in Figure 1.42. This function can be represented by the following set of ordered pairs.

$$\{(1, 9°), (2, 13°), (3, 15°), (4, 15°), (5, 12°), (6, 4°)\}$$

In each ordered pair, the first coordinate is the input and the second coordinate is the output.

The following characteristics are true of a function from a set A to a set B.

1. Each element in A must be matched with an element in B.
2. Some elements in B may not be matched with any element in A.
3. Two or more elements of A may be matched with the same element of B.

The converse of the third statement is not true. That is, an element of A (the domain) cannot be matched with two different elements of B.

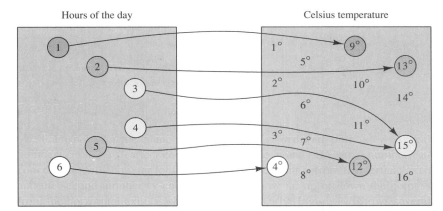

Set A is the domain.
Input: 1, 2, 3, 4, 5, 6

Set B contains the range.
Output: 4°, 9°, 12°, 13°, 15°

FIGURE 1.42   Function from Set A to Set B

EXAMPLE 1   Testing for Functions

Let $A = \{a, b, c\}$ and $B = \{1, 2, 3, 4, 5\}$. Does the set of ordered pairs or figures represent a function from set A to set B?

A. $\{(a, 2), (b, 3), (c, 4)\}$

B. $\{(a, 4), (b, 5)\}$

C.
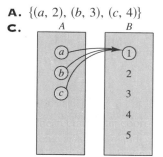

D.
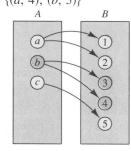

SOLUTION

A. Yes, because each element of A is matched with exactly one element of B.

B. No, because not all elements of A are matched with an element of B.

C. Yes. It does not matter that each element of A is matched with the same element of B.

D. No, because the element a in A is matched with *two* elements, 1 and 2, in B. This is also true of the element b.

Representing functions by sets of ordered pairs is a common practice in *discrete mathematics*. In algebra, however, it is more common to represent functions by equations or formulas involving two variables. For instance, the equation

$$y = x^2$$

represents the variable y as a function of the variable x. Here, x is the **independent variable** and y is the **dependent variable.** The domain of the function is the set of all values (inputs) taken on by the independent variable x, and the range of the function is the set of all values (outputs) taken on by the dependent variable y.

EXAMPLE 2   **Testing for Functions Represented by Equations**

Determine whether the equations represent y as a function of x.

A. $x^2 + y = 1$ **B.** $-x + y^2 = 1$

SOLUTION

In each case, to determine whether y is a function of x, it is helpful to solve for y in terms of x.

A. $x^2 + y = 1$ *Given equation*

$\qquad y = 1 - x^2$ *Solve for y*

To each value of x there corresponds one value of y. Therefore, y is a function of x.

B. $-x + y^2 = 1$ *Given equation*

$\qquad y^2 = 1 + x$ *Add x to both sides*

$\qquad y = \pm\sqrt{1 + x}$ *Solve for y*

The $\pm$ indicates that to a given value of x there correspond two values of y. Therefore, y is *not* a function of x. For instance, if $x = 3$, then y could be either 2 or -2.

D I S C O V E R Y

Use a graphing utility to graph $x^2 + y = 1$. Then use the graph to write a convincing argument that each x-value has at most one y-value.

Use a graphing utility to graph $-x + y^2 = 1$. (*Hint:* You will need to use two equations.) Then use the graph to find an x-value that corresponds to two y-values. Why does the graph not represent y as a function of x?

Function Notation

When using an equation to represent a function, it is convenient to name the function so that it can be referenced easily. For example, the equation $y = 1 - x^2$, Example 2(A), describes y as a function of x. Suppose you give this function the name "f." Then you can use the following **function notation.**

Input	Output	Equation
x	$f(x)$	$f(x) = 1 - x^2$

The symbol $f(x)$ is read as the **value of f at x** or simply "f of x." This corresponds to the y-value for a given x. Thus, you can write $y = f(x)$.

Keep in mind that f is the *name* of the function, whereas $f(x)$ is the *value* of the function at x. For instance, the function

$$f(x) = 3 - 2x$$

has *function values* denoted by $f(-1)$, $f(0)$, $f(2)$, and so on. To find these values, substitute the specified input values into the equation.

For $x = -1$, $f(-1) = 3 - 2(-1) = 3 + 2 = 5$.
For $x = 0$, $f(0) = 3 - 2(0) = 3 + 0 = 3$.
For $x = 2$, $f(2) = 3 - 2(2) = 3 - 4 = -1$.

Although it is convenient to use f as a function name and x as the independent variable, you can use other letters. For instance, $f(x) = x^2 - 4x + 7$, $f(t) = t^2 - 4t + 7$, and $g(s) = s^2 - 4s + 7$ all define the same function. In fact, the role of the independent variable in a function is simply that of a "placeholder." Consequently, the above function could be described by the form

$$f(\ ) = (\ )^2 - 4(\ ) + 7,$$

where the parentheses are used in place of a letter. To evaluate $f(-2)$, simply place -2 in each set of parentheses.

$f(\ ) = (\ )^2 - 4(\ ) + 7$
$f(-2) = (-2)^2 - 4(-2) + 7$ *Place -2 in each set of parentheses*
    $= 4 + 8 + 7$ *Evaluate each term*
    $= 19$ *Simplify*

Similarly, the value of $f(3x)$ is obtained as follows.

$f(\ ) = (\ )^2 - 4(\ ) + 7$
$f(3x) = (3x)^2 - 4(3x) + 7$ *Place $3x$ in each set of parentheses*
    $= 9x^2 - 12x + 7$ *Simplify*

EXAMPLE 3 Evaluating a Function

Let $g(x) = -x^2 + 4x + 1$ and find the following.

A. $g(2)$ **B.** $g(t)$ **C.** $g(x + 2)$

Technology Note

Most graphing utilities can be used to evaluate a function at a real value of x.

The technique depends on the graphing utility, but here is a sample program.

```
Prgml: EVALUATE
Lbl 1
:Disp "ENTER X"
:Input X
:Disp Y₁
:Goto 1
```

To use this program, enter a function in y_1. Then run the program—it will allow you to evaluate the function at several values of x.

REMARK Example 3 shows that $g(x + 2) \neq g(x) + g(2)$ because $-x^2 + 5 \neq (-x^2 + 4x + 1) + 5$. In general, $g(u + v) \neq g(u) + g(v)$.

SOLUTION

A. Replacing x with 2 in $g(x) = -x^2 + 4x + 1$ yields

$$g(2) = -(2)^2 + 4(2) + 1 = -4 + 8 + 1 = 5.$$

B. Replacing x with t yields

$$g(t) = -(t)^2 + 4(t) + 1 = -t^2 + 4t + 1.$$

C. Replacing x with $x + 2$ yields

$$\begin{aligned} g(x + 2) &= -(x + 2)^2 + 4(x + 2) + 1 \\ &= -(x^2 + 4x + 4) + 4x + 8 + 1 \\ &= -x^2 - 4x - 4 + 4x + 8 + 1 \\ &= -x^2 + 5. \end{aligned}$$

Sometimes a function is defined using more than one equation.

EXAMPLE 4 **A Function Defined by Two Equations**

Evaluate the function

$$f(x) = \begin{cases} x^2 + 1, & x < 0 \\ x - 1, & x \geq 0 \end{cases}$$

at $x = -1$, 0, and 1.

Technology Note

Most graphing utilities can graph functions that are defined using more than one equation. For example, on the TI-81, TI-82, or TI-85, you can obtain the graph of the function in Example 4 as follows:

$Y_1 = (x^2 + 1)(x < 0)$
$\qquad + (x - 1)(x \geq 0)$

SOLUTION

Because $x = -1$ is less than 0, use $f(x) = x^2 + 1$ to obtain

$$f(-1) = (-1)^2 + 1 = 2.$$

For $x = 0$, use $f(x) = x - 1$ to obtain

$$f(0) = (0) - 1 = -1.$$

For $x = 1$, use $f(x) = x - 1$ to obtain

$$f(1) = (1) - 1 = 0.$$

The Domain of a Function

The domain of a function may be explicitly described along with the function, or it may be *implied* by the expression used to define the function. The **implied domain** is the set of all real numbers for which the expression is defined. For instance, the function

$$f(x) = \frac{1}{x^2 - 4}$$

has an implied domain that consists of all real x other than $x = \pm 2$. These two values are excluded from the domain because division by zero is undefined. Another common type of implied domain is used to avoid even roots of negative numbers. For example, the function

$$f(x) = \sqrt{x}$$

is defined only for $x \geq 0$. Hence, its implied domain is the interval $[0, \infty)$. In general, the domain of a function *excludes* values that would cause division by zero *or* result in the even root of a negative number.

The *range* of a function is more difficult to find, and can best be obtained from the graph of the function (see Section 1.7).

EXAMPLE 5   **Finding the Domain of a Function**

Find the domain of each of the following functions.

A. f: $\{(-3, 0), (-1, 4), (0, 2), (2, 2), (4, -1)\}$

B. Volume of a sphere: $V = \dfrac{4}{3}\pi r^3$

C. $g(x) = \dfrac{1}{x + 5}$

D. $h(x) = \sqrt{4 - x}$

SOLUTION

A. The domain of f consists of all first coordinates in the set of ordered pairs, and is therefore the set

Domain = $\{-3, -1, 0, 2, 4\}$.

B. For the volume of a sphere you must choose nonnegative values for the radius r. Thus, the domain is the set of all real numbers r such that $r \geq 0$.

C. Excluding x-values that yield zero in the denominator, the domain of g is the set of all real numbers $x \neq -5$.

D. Choose x-values for which $4 - x \geq 0$. The domain is all real numbers that are less than or equal to 4.

REMARK   In Example 5(B), note that the domain of a function may be implied by the physical context. For instance, from the equation $V = \frac{4}{3}\pi r^3$, you would have no reason to restrict r to nonnegative values, but the physical context tells you that a sphere cannot have a negative radius.

Function Keys on a Graphing Utility

Calculators and computers have many built-in functions that can be evaluated with a simple keystroke. For instance, you can use the square root key to calculate square roots of nonnegative numbers. If you attempt to take the square root of a negative number, you will get an error message. Here are some more examples of built-in functions that you can verify on your calculator or computer.

Technology Note _____

You can evaluate the square root of a number in more than one way. For instance, the square root of 10 can be obtained in any of the following ways.

$\sqrt{10}$ $10 \wedge (\frac{1}{2})$ $10 \wedge .5$

Notice that you need parentheses around the fraction $\frac{1}{2}$.

EXAMPLE 6 **Function Keys**

A. $\sqrt{10} = 3.16227766$

B. $64^{1/3} = 4$

C. $|-3| = \text{abs}(-3) = 3$

D. $10^3 = 1000$

E. $\log 2 = 0.3010299957$ (You will study logarithms in Chapter 6).

Applications

EXAMPLE 7 The Dimensions of a Container

A standard soft-drink can has a height of about 4.75 inches and a radius of about 1.3 inches. For this standard can, the ratio of the height to the radius is about 3.65. Suppose you work in the marketing department of a soft-drink company, and are experimenting with a new soft-drink can that is slightly narrower and taller. For your experimental can, the ratio of the height to the radius is 4, as shown in Figure 1.43.

A. Express the volume of the can as a function of the radius r.

B. Express the volume of the can as a function of the height h.

$\dfrac{h}{r} = 4$

FIGURE 1.43

SOLUTION

The volume of a right circular cylinder is given by the formula

$$V = \pi(\text{radius})^2(\text{height}) = \pi r^2 h.$$

Because the ratio of the height to the radius is 4, you can write $h = 4r$.

A. To write the volume as a function of the radius, use the fact that $h = 4r$.

$$V = \pi r^2 h = \pi r^2(4r) = 4\pi r^3$$

B. To write the volume as a function of the height, use the fact that $r = h/4$.

$$V = \pi\left(\frac{h}{4}\right)^2 h = \frac{\pi h^3}{16}$$

EXAMPLE 8   The Path of a Baseball

A baseball is hit 3 feet above the ground at a velocity of 100 feet per second and at an angle of 45° with respect to the ground. The path of the baseball is given by the function

$$y = -0.0032x^2 + x + 3,$$

where the height y and the horizontal distance x are measured in feet, as shown in Figure 1.44. (From this equation, note that the height of the baseball is a function of the horizontal distance from home plate.) Will the baseball clear a 10-foot fence located 300 feet from home plate?

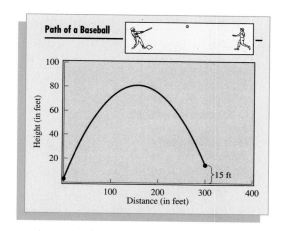

FIGURE 1.44

DISCOVERY

Use a graphing utility to graph $y = -0.0032x^2 + x + 3$ and $y = 10$ on the same viewing rectangle. (Use a viewing rectangle in which $0 \le x \le 300$ and $0 \le y \le 90$.) Explain how the graphs can be used to answer the question asked in Example 8.

Suppose the fence in Example 8 is raised to 20 feet and the distance from home plate is 295 feet. Will the ball clear this fence?

SOLUTION

When $x = 300$, the height of the baseball is

$$y = -0.0032(300^2) + 300 + 3 = 15 \text{ feet.}$$

Thus, the ball will clear the fence.

One of the basic definitions in calculus employs a ratio called a **difference quotient.**

$$\frac{f(x + h) - f(x)}{h}, \qquad h \neq 0$$

EXAMPLE 9   Evaluating a Difference Quotient

For the function given by $f(x) = x^2 - 4x + 7$, find

$$\frac{f(x + h) - f(x)}{h}.$$

SOLUTION

$$
\begin{aligned}
\frac{f(x + h) - f(x)}{h} &= \frac{[(x + h)^2 - 4(x + h) + 7] - [x^2 - 4x + 7]}{h} \\
&= \frac{x^2 + 2xh + h^2 - 4x - 4h + 7 - x^2 + 4x - 7}{h} \\
&= \frac{2xh + h^2 - 4h}{h} \\
&= \frac{h(2x + h - 4)}{h} \\
&= 2x + h - 4, \qquad (h \neq 0)
\end{aligned}
$$

SUMMARY OF FUNCTION TERMINOLOGY

Function
A **function** is a relationship between two variables such that to each value of the independent variable there corresponds exactly one value of the dependent variable.

Function Notation: $y = f(x)$

   f is the **name** of the function.
   y is the **dependent variable.**
   x is the **independent variable.**
   $f(x)$ is the **value of the function at x.**

Domain
The **domain** of a function is the set of all values (inputs) of the independent variable for which the function is defined. If x is in the domain of f, then f is **defined** at x. If x is not in the domain of f, then f is **undefined** at x.

Range
The **range** of a function is the set of all values (outputs) assumed by the dependent variable (that is, the set of all function values).

Implied Domain
If f is defined by an algebraic expression and the domain is not specified, then the **implied domain** consists of all real numbers for which the expression is defined.

DISCUSSION PROBLEM

·······················

DETERMINING RELATIONSHIPS THAT ARE FUNCTIONS

Write two statements describing relationships in everyday life that *are* functions and two that are *not* functions. Here are two examples.

(a) The statement "The sale price of an item is a function of the amount of sales tax on the item" is *not* a correct mathematical use of the word "function." The problem is that the sales tax does not determine the sale price. For instance, suppose the sales tax is 6%. Knowing that the sales tax on a particular item is $0.12 is not sufficient information to determine the sale price. (It could be any price between $1.92 and $2.08.)

(b) The statement "Your federal income tax is a function of your adjusted gross income" *is* a correct mathematical use of the word "function." Once you have determined your adjusted gross income, then your income tax can be determined.

WARM-UP

·················

The following warm-up exercises involve skills that were covered in earlier sections. You will use these skills in the exercise set for this section.

In Exercises 1–4, simplify the expression.

1. $2(-3)^3 + 4(-3) - 7$ **2.** $4(-1)^2 - 5(-1) + 4$

3. $(x + 1)^2 + 3(x + 1) - (x^2 + 3x)$ **4.** $(x - 2)^2 - 4(x - 2) - (x^2 - 4)$

In Exercises 5 and 6, solve for y in terms of x.

5. $2x + 5y - 7 = 0$ **6.** $y^2 = x^2$

In Exercises 7–10, simplify the expression. Assume the variables are positive.

7. $\sqrt{50x^3y^2}$ **8.** $\sqrt{18z^4 - 9z^2}$

9. $\dfrac{(x - y)^2}{x^2 - y^2}$ **10.** $\dfrac{x^2}{x + 2} \cdot \dfrac{x^2 - x - 6}{2x}$

SECTION 1.6 · EXERCISES

In Exercises 1 and 2, determine which of the sets of ordered pairs represents a function from A to B. Give reasons for your answers.

1. $A = \{0, 1, 2, 3\}$ and $B = \{-2, -1, 0, 1, 2\}$
   (a) $\{(0, 1), (1, -2), (2, 0), (3, 2)\}$
   (b) $\{(2, 2), (0, -2), (3, 0), (1, 1)\}$ *Not for*
   (c) $\{(0, 0), (1, 0), (2, 0), (3, 0)\}$
   (d) $\{(0, 2), (3, 0), (1, 1)\}$

2. $A = \{a, b, c\}$ and $B = \{0, 1, 2, 3\}$
   (a) $\{(a, 1), (c, 2), (c, 3), (b, 3)\}$
   (b) $\{(a, 1), (b, 2), (c, 3)\}$
   (c) $\{(1, a), (0, a), (2, c), (3, b)\}$
   (d) $\{(c, 0), (b, 0), (a, 3)\}$

In Exercises 3–10, state whether the equation determines y as a function of x.

3. $x^2 + y^2 = 4$

4. $x = y^2$

5. $x^2 + y = 4$

6. $x + y^2 = 4$

7. $x^2y - x^2 + 4y = 0$

8. $x^2 + y^2 - 2x - 4y + 1 = 0$

9. $y^2 = x^2 - 1$

10. $y = \sqrt{x + 5}$

In Exercises 11–14, fill in the blanks and simplify the result.

11. $f(x) = 6 - 4x$
  (a) $f(3) = 6 - 4(\qquad)$
  (b) $f(-7) = 6 - 4(\qquad)$
  (c) $f(t) = 6 - 4(\qquad)$
  (d) $f(c + 1) = 6 - 4(\qquad)$

12. $g(x) = x^2 - 2x$
  (a) $g(2) = (\qquad)^2 - 2(\qquad)$
  (b) $g(-3) = (\qquad)^2 - 2(\qquad)$
  (c) $g(t + 1) = (\qquad)^2 - 2(\qquad)$
  (d) $g(x + h) = (\qquad)^2 - 2(\qquad)$

13. $f(s) = \dfrac{1}{s + 1}$
  (a) $f(4) = \dfrac{1}{(\qquad) + 1}$
  (b) $f(0) = \dfrac{1}{(\qquad) + 1}$
  (c) $f(4x) = \dfrac{1}{(\qquad) + 1}$
  (d) $f(x + h) = \dfrac{1}{(\qquad) + 1}$

14. $f(t) = \sqrt{25 - t^2}$
  (a) $f(3) = \sqrt{25 - (\qquad)^2}$
  (b) $f(5) = \sqrt{25 - (\qquad)^2}$
  (c) $f(x + 5) = \sqrt{25 - (\qquad)^2}$
  (d) $f(2 + h) = \sqrt{25 - (\qquad)^2}$

In Exercises 15–22, evaluate (if possible) the function at the specified value of the independent variable and simplify the result.

15. $f(x) = 2x - 3$
  (a) $f(1)$
  (b) $f(-3)$
  (c) $f(x - 1)$
  (d) $f(\tfrac{1}{4})$

16. $V(r) = \tfrac{4}{3}\pi r^3$
  (a) $V(3)$
  (b) $V(0)$
  (c) $V(\tfrac{3}{2})$
  (d) $V(2r)$

17. $h(t) = t^2 - 2t$
  (a) $h(2)$
  (b) $h(-1)$
  (c) $h(x + 2)$
  (d) $h(1.5)$

18. $f(x) = \sqrt{x + 8} + 2$
  (a) $f(-8)$
  (b) $f(1)$
  (c) $f(x - 8)$
  (d) $f(h + 8)$

19. $f(x) = \dfrac{|x|}{x}$
  (a) $f(2)$
  (b) $f(-2)$
  (c) $f(x^2)$
  (d) $f(x - 1)$

20. $q(t) = \dfrac{2t^2 + 3}{t^2}$
  (a) $q(2)$
  (b) $q(0)$
  (c) $q(x)$
  (d) $q(-x)$

21. $f(x) = \begin{cases} 2x + 1, & x < 0 \\ 2x + 2, & x \ge 0 \end{cases}$
  (a) $f(-1)$
  (b) $f(0)$
  (c) $f(1)$
  (d) $f(2)$

22. $f(x) = \begin{cases} x^2 + 2, & x \le 1 \\ 2x^2 + 2, & x > 1 \end{cases}$
  (a) $f(-2)$
  (b) $f(0)$
  (c) $f(1)$
  (d) $f(2)$

In Exercises 23–30, find the domain of the function.

23. $f(x) = 5x^2 + 2x - 1$

24. $f(t) = \sqrt[3]{t + 4}$

25. $g(y) = \sqrt{y - 10}$

26. $f(x) = \sqrt[4]{1 - x^2}$

27. $h(t) = \dfrac{4}{t}$

28. $h(x) = \dfrac{10}{x^2 - 2x}$

29. $g(x) = \dfrac{1}{x} - \dfrac{3}{x + 2}$

30. $f(s) = \dfrac{\sqrt{s - 1}}{s - 4}$

In Exercises 31–34, assume that the domain of f is the set $A = \{-2, -1, 0, 1, 2\}$. Determine the set of ordered pairs representing the function f.

31. $f(x) = x^2$

32. $f(x) = \dfrac{2x}{x^2 + 1}$

33. $f(x) = \sqrt{x + 2}$

34. $f(x) = |x + 1|$

In Exercises 35–38, use a function key on a calculator to evaluate each expression. (Round the result to three decimal places.)

35. (a) 32.5^2 (b) 4.3^5

36. (a) $\sqrt{232}$ (b) $\sqrt[3]{2500}$

37. (a) $\dfrac{1}{8.5}$ (b) $\dfrac{1}{0.047}$

38. (a) $|-326.8|$ (b) $10^{3.8}$

In Exercises 39–42, select a function from (a) $f(x) = cx$, (b) $g(x) = cx^2$, (c) $h(x) = c\sqrt{|x|}$, or (d) $r(x) = c/x$ and determine the value of the constant c so that the function fits the data given in the table.

39.

x	-4	-1	0	1	4
y	-32	-2	0	-2	-32

40.

x	-4	-1	0	1	4
y	-1	$-\frac{1}{4}$	0	$\frac{1}{4}$	1

41.

x	-4	-1	0	1	4
y	-8	-32	undef.	32	8

42.

x	-4	-1	0	1	4
y	6	3	0	3	6

In Exercises 43–46, evaluate the difference quotient. Then simplify the result.

43. $f(x) = 2x$

$$\frac{f(x + h) - f(x)}{h}$$

44. $f(x) = 5x - x^2$

$$\frac{f(5 + h) - f(5)}{h}$$

45. $f(x) = x^3$

$$\frac{f(x + h) - f(x)}{h}$$

46. $f(t) = \dfrac{1}{t}$

$$\frac{f(t) - f(1)}{t - 1}$$

47. *Area of a Circle*   Express the area A of a circle as a function of its circumference C.

48. *Area of a Triangle*   Express the area A of an equilateral triangle as a function of the length s of its sides.

49. *Area of a Triangle*   A right triangle is formed in the first quadrant by the x- and y-axes and a line through the point $(1, 2)$ (see figure). Express the area of the triangle as a function of x, and determine the domain of the function.

50. *Area of a Rectangle*   A rectangle is bounded by the x-axis and the semicircle $y = \sqrt{25 - x^2}$ (see figure). Express the area of the rectangle as a function of x, and determine the domain of the function.

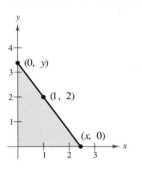

FIGURE FOR 49

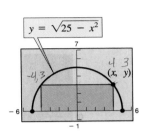

FIGURE FOR 50

51. *Volume of a Package*   A rectangular package to be sent by a postal service can have a maximum combined length and girth (perimeter of a cross section) of 108 inches (see figure). Express the volume of the package as a function of x. What is the domain of the function?

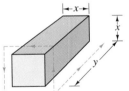

FIGURE FOR 51

52. *Volume of a Box*   An open box is to be made from a square piece of material 12 inches on a side by cutting equal squares from each corner and turning up the sides (see figure). Express the volume V of the box as a function of x. What is the domain of this function?

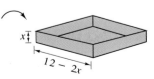

FIGURE FOR 52

53. *Height of a Balloon*   A transmitting balloon ascends vertically from a point 2000 feet from the receiving station (see figure). Let d be the distance between the balloon and the receiving station. Express the height of the balloon as a function of d. What is the domain of the function?

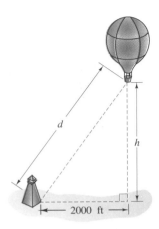

FIGURE FOR 53

54. *Price of Mobile Homes*   The average price p of a new mobile home in the United States from 1974 to 1988 can be modeled by the function

$$p(t) = \begin{cases} 19{,}503.6 + 1753.6t, & -6 \leq t \leq -1 \\ 19{,}838.8 + 81.11t^2, & 0 \leq t \leq 8 \end{cases}$$

where t is the year with $t = 0$ corresponding to 1980. Use this model to find the average price of a mobile home in 1978 and in 1988. (*Source:* U.S. Bureau of Census, Construction Reports)

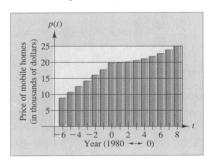

FIGURE FOR 54

55. *Cost, Revenue, and Profit*   A company invests $98,000 for equipment to produce a product. Each unit of the product costs $12.30 and is sold for $17.98. Let x be the number of units produced and sold.
(a) Write the total cost C as a function of x.
(b) Write the revenue R as a function of x.
(c) Write the profit P as a function of x.
   (*Note:* $P = R - C$)

56. *Charter Bus Fares*   For groups of 80 or more people, a charter bus company determines the rate per person according to the formula

$$\text{Rate} = 8 - 0.05(n - 80), \qquad n \geq 80,$$

where the rate is given in dollars and n is the number of people.
(a) Express the revenue R for the bus company as a function of n.
(b) Use the function from part (a) to complete the following table.

n	90	100	110	120	130	140	150
$R(n)$							

(c) Graph R and determine the number of people that will produce a maximum revenue.

1.7 GRAPHS OF FUNCTIONS

**The Graph of a Function / Increasing and Decreasing Functions /
Relative Minimum and Maximum Values / Step Functions /
Even and Odd Functions**

The Graph of a Function

In Section 1.6 you studied functions from an algebraic point of view. In this
section, you will study functions from a geometric perspective. The **graph of
a function** f is the collection of ordered pairs $(x, f(x))$ such that x is in the
domain of f. As you study this section, remember the following geometric
interpretation of x and $f(x)$.

$$x = \text{the directed distance from the } y\text{-axis} \quad \text{Domain}$$
$$f(x) = \text{the directed distance from the } x\text{-axis} \quad \text{RANGE}$$

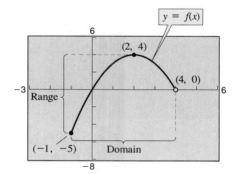

FIGURE 1.45

EXAMPLE 1 Finding Domain and Range from the Graph
of a Function

Use the graph of the function f, shown in Figure 1.45.

A. Find the domain of f.
B. Find the function values $f(-1)$ and $f(2)$.
C. Find the range of f.

SOLUTION

A. The closed dot (on the left) indicates that $x = -1$ is in the domain of f,
whereas the open dot (on the right) indicates $x = 4$ is not in the domain.
Thus, the domain of f is all x in the interval $[-1, 4)$.
B. Because $(-1, -5)$ is a point on the graph of f, it follows that $f(-1) = -5$.
Similarly, because $(2, 4)$ is a point on the graph of f, it follows that
$f(2) = 4$.
C. Because the graph does not extend below $f(-1) = -5$ or above
$f(2) = 4$, the range of f is the interval $[-5, 4]$.

REMARK In Figure 1.45, the solid dot
representing the point $(-1, -5)$ indicates that $(-1, -5)$ is a point on the
graph. The open dot indicates that
$(4, 0)$ is not a point on the graph.

By the definition of a function, at most one y-value corresponds to a given
x-value. It follows, then, that a vertical line can intersect the graph of a
function at most once. This observation provides a convenient visual test for
functions.

VERTICAL LINE TEST FOR FUNCTIONS

A set of points in a coordinate plane is the graph of y as a function of
x if and only if no vertical line intersects the graph at more than one
point.

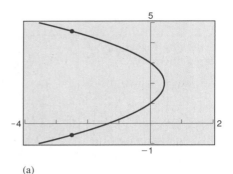

(a)

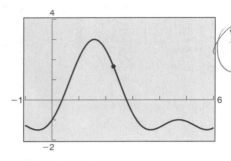

(b)

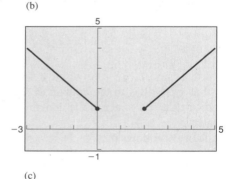

(c)

FIGURE 1.46

EXAMPLE 2   **Vertical Line Test for Functions**

Which of the graphs in Figure 1.46 represent y as a function of x?

SOLUTION

A. This is *not* a graph of y as a function of x because you can find a vertical line that intersects the graph twice.

B. This *is* a graph of y as a function of x because every vertical line intersects the graph at most once.

C. This *is* a graph of y as a function of x. (Note that if a vertical line does not intersect the graph, it simply means that the function is undefined for this particular value of x.)

Increasing and Decreasing Functions

The more you know about the graph of a function, the more you know about the function itself. Consider the graph shown in Figure 1.47. Moving from *left to right,* this graph falls from $x = -2$ to $x = 0$, is constant from $x = 0$ to $x = 2$, and rises from $x = 2$ to $x = 4$. These observations indicate that the function has the following characteristics.

1. The function is **decreasing** on the interval $(-2, 0)$.
2. The function is **constant** on the interval $(0, 2)$.
3. The function is **increasing** on the interval $(2, 4)$.

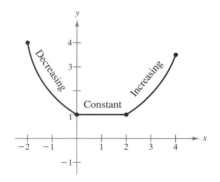

FIGURE 1.47

INCREASING, DECREASING, AND CONSTANT FUNCTIONS

A function f is **increasing** on an interval if, for any x_1 and x_2 in the interval, $x_1 < x_2$ implies $f(x_1) < f(x_2)$.

A function f is **decreasing** on an interval if, for any x_1 and x_2 in the interval, $x_1 < x_2$ implies $f(x_1) > f(x_2)$.

A function f is **constant** on an interval if, for any x_1 and x_2 in the interval, $f(x_1) = f(x_2)$.

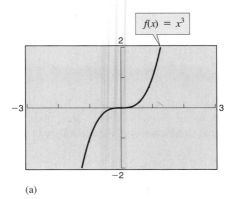

(a)

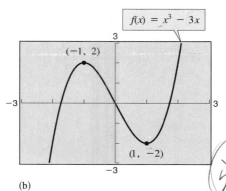

(b)

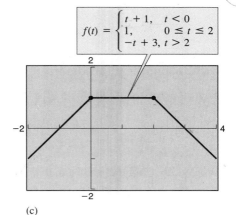

(c)

FIGURE 1.48

EXAMPLE 3 Increasing and Decreasing Functions

In Figure 1.48, determine the open intervals on which each function is increasing, decreasing, or constant.

SOLUTION

A. Although it might appear that there is an interval about zero over which this function is constant, you can see that if $x_1 < x_2$, then $f(x_1) = x_1^3 < x_2^3 = f(x_2)$. Thus, you can conclude that the function is increasing over the entire real line.

B. This function is increasing on the interval $(-\infty, -1)$, decreasing on the interval $(-1, 1)$, and increasing on the interval $(1, \infty)$.

C. This function is increasing on the interval $(-\infty, 0)$, constant on the interval $(0, 2)$, and decreasing on the interval $(2, \infty)$.

Relative Minimum and Maximum Values

The points at which a function changes its increasing, decreasing, or constant behavior are helpful in determining the relative maximum or relative minimum values of the function.

DEFINITION OF RELATIVE MINIMUM AND RELATIVE MAXIMUM

A function value $f(a)$ is called a **relative minimum** of f if there exists an interval (x_1, x_2) that contains a such that

$$x_1 < x < x_2 \quad \text{implies} \quad f(a) \leq f(x).$$

A function $f(a)$ is called a **relative maximum** of f if there exists an interval (x_1, x_2) that contains a such that

$$x_1 < x < x_2 \quad \text{implies} \quad f(a) \geq f(x).$$

Figure 1.49 shows two different examples of relative minimums and two of relative maximums.

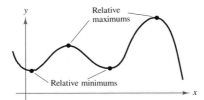

FIGURE 1.49

Technology Note _____

When you use a graphing utility to estimate the x- and y-values of a relative minimum or relative maximum, the automatic zoom feature will often produce graphs that are nearly flat. To overcome this problem, you can manually change the vertical setting of the viewing rectangle. You can stretch the graph vertically by making the values of Y_{min} and Y_{max} closer together.

Technology Note _____

Some graphing utilities, such as a TI-85, can automatically determine the maximum and minimum value of a function defined on a closed interval. If your graphing utility has this feature, use it to graph $y = -x^3 + x$ on the interval $-1 \le x \le 1$ and verify the results of Example 5. What happens if you use the interval $-10 \le x \le 10$?

EXAMPLE 4   Approximating a Relative Minimum

Use a graphing utility to approximate the relative minimum of the function $f(x) = 3x^2 - 4x - 2$.

SOLUTION

The graph of f is shown in Figure 1.50. By using the zoom and trace features of a graphing utility, you can estimate that the function has a relative minimum at the point

$(0.67, -3.33)$.   *Relative minimum*

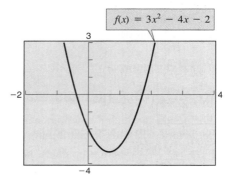

FIGURE 1.50

EXAMPLE 5   Approximating Relative Minimums and Maximums

Use a graphing utility to approximate the relative minimum and relative maximum of the function $f(x) = -x^3 + x$.

SOLUTION

A sketch of the graph of f is shown in Figure 1.51. By using the zoom and trace features of the graphing utility, you can estimate that the function has a relative minimum at the point

$(-0.58, -0.38)$ *Relative minimum*

and a relative maximum at the point

$(0.58, 0.38)$.   *Relative maximum*

If you go on to take a course in calculus, you will learn a technique for finding the exact points at which this function has a relative minimum and a relative maximum.

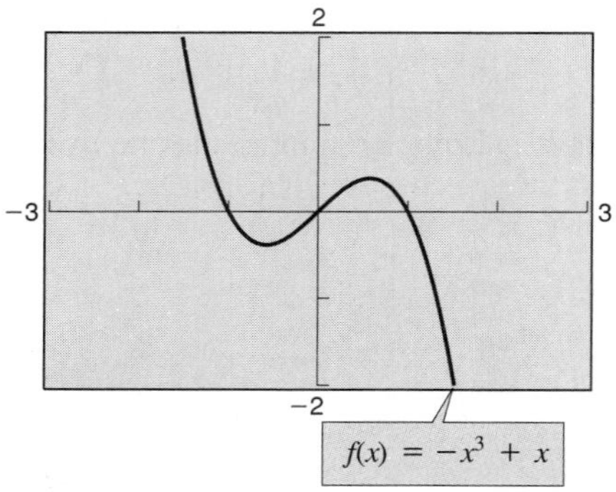

FIGURE 1.51

EXAMPLE 6   The Price of Diamonds

During the 1980s the average price of a 1-carat polished diamond decreased and then increased according to the model

$$C = -0.7t^3 + 16.25t^2 - 106t + 388, \qquad 2 \le t \le 10,$$

where C is the average price in dollars (on the Antwerp Index) and t represents the calendar year with $t = 2$ corresponding to January 1, 1982. (*Source: Diamond High Council*) According to this model, during which years was the price of diamonds decreasing? During which years was the price of diamonds increasing? Approximate the minimum price of a 1-carat diamond between 1982 and 1990.

SOLUTION

To solve this problem, sketch an accurate graph of the function, as shown in Figure 1.52. From the graph, you can see that the price of diamonds decreased from 1982 until late 1984. Then from late 1984 to 1990 the price increased. The minimum price during the 8-year period was approximately $175.

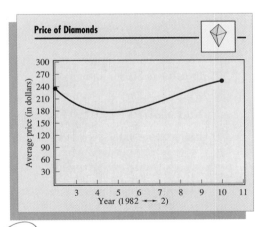

FIGURE 1.52

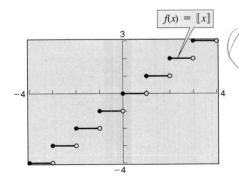

Greatest Integer Function

FIGURE 1.53

Step Functions

EXAMPLE 7   The Greatest Integer Function

The **greatest integer function** is denoted by $[\![x]\!]$ and is defined by

$$f(x) = [\![x]\!] = \text{the greatest integer less than or equal to } x.$$

The graph of this function is shown in Figure 1.53. Note that the graph of the greatest integer function jumps vertically one unit at each integer and is constant (a horizontal line segment) between each pair of consecutive integers.

Because of the jumps in its graph, the greatest integer function is an example of a **step function.** Some values of the greatest integer function are as follows.

$$[[-1]] = -1 \qquad [[-0.5]] = -1$$
$$[[0]] = 0 \qquad [[0.5]] = 0$$
$$[[1]] = 1 \qquad [[1.5]] = 1$$

The range of the greatest integer function is the set of all integers.

Technology Note _____

Most graphing utilities display graphs in what is called a *connected mode,* which means that the graph has no breaks. Thus, when you are sketching graphs that do have breaks, it is better to change the graphing utility to *dot mode.* Try sketching the graph of the greatest integer function [often called Int(x)] on a graphing utility in connected and dot modes, and compare the two results.

EXAMPLE 8   The Cost of a Telephone Call

Suppose the cost of a telephone call between Los Angeles and San Francisco is $0.50 for the first minute and $0.36 for each additional minute. The greatest integer function can be used to create a model for the cost of this call.

$$C = 0.50 + 0.36[[t]], \qquad 0 < t,$$

where C is the total cost of the call in dollars and t is the length of the call in minutes. Sketch the graph of this function.

SOLUTION

For calls up to 1 minute, the cost is $0.50. For calls between 1 and 2 minutes, the cost is $0.86, and so on.

Length of call: $0 < t < 1$ $1 \le t < 2$ $2 \le t < 3$ $3 \le t < 4$ $4 \le t < 5$
Cost of call: $0.50 $0.86 $1.22 $1.58 $1.94

Using these values, you can sketch the graph shown in Figure 1.54.

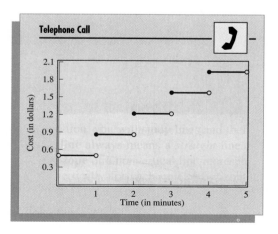

FIGURE 1.54

Even and Odd Functions

A graph has symmetry with respect to the y-axis if, whenever (x, y) is on the graph, so is the point $(-x, y)$. A function whose graph is symmetric with respect to the y-axis is an **even** function.

    A graph has symmetry with respect to the origin if, whenever (x, y) is on the graph, so is the point $(-x, -y)$. A function whose graph is symmetric with respect to the origin is an **odd** function.

    A graph has symmetry with respect to the x-axis if, whenever (x, y) is on the graph, so is the point $(x, -y)$. The graph of a (nonzero) function cannot be symmetric with respect to the x-axis.

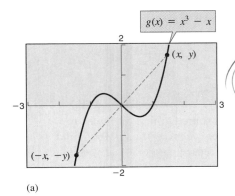

(a)

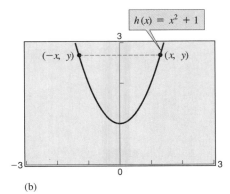

(b)

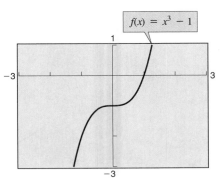

(c)

FIGURE 1.55

TEST FOR EVEN AND ODD FUNCTIONS

A function f is **even** if, for each x in the domain of f, $f(-x) = f(x)$.
A function f is **odd** if, for each x in the domain of f, $f(-x) = -f(x)$.

EXAMPLE 9 **Even and Odd Functions**

Determine whether the following functions are even, odd, or neither.

A. $g(x) = x^3 - x$
B. $h(x) = x^2 + 1$
C. $f(x) = x^3 - 1$

SOLUTION

A. This function is odd because

$$g(-x) = (-x)^3 - (-x) = -x^3 + x = -(x^3 - x) = -g(x).$$

B. This function is even because

$$h(-x) = (-x)^2 + 1 = x^2 + 1 = h(x).$$

C. Substituting $-x$ for x produces

$$f(-x) = (-x)^3 - 1 = -x^3 - 1.$$

Because $f(x) = x^3 - 1$ and $-f(x) = -x^3 + 1$, you can conclude that $f(-x) \neq f(x)$ and $f(-x) \neq -f(x)$. Hence, the function is neither even nor odd.

The graphs of the three functions are shown in Figure 1.55.

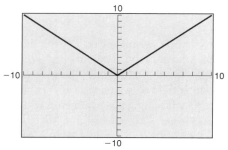

FIGURE 1.56

EXAMPLE 10   The Absolute Value Function

Figure 1.56 shows the V-shape of the absolute value function, $f(x) = |x|$. This was produced using the "abs" key and the standard viewing rectangle. Because the graph is symmetric about the y-axis, you can see that it is an even function. You can verify this by observing that $f(-x) = |-x| = |x| = f(x)$.

DISCUSSION PROBLEM

INCREASING AND DECREASING FUNCTIONS

Use your school's library or some other reference source to find examples of three different functions that represent quantities between 1980 and 1990. Find one that decreased during the decade, one that increased, and one that was constant. For instance, the value of the dollar decreased, the population of the United States increased, and the land size of the United States remained constant. Can you find three other examples? Present your results graphically.

WARM-UP

The following warm-up exercises involve skills that were covered in earlier sections. You will use these skills in the exercise set for this section.

1. Find $f(2)$ for $f(x) = -x^3 + 5x$.

2. Find $f(6)$ for $f(x) = x^2 - 6x$.

3. Find $f(-x)$ for $f(x) = \dfrac{3}{x}$.

4. Find $f(-x)$ for $f(x) = x^2 + 3$.

In Exercises 5 and 6, find the x- and y-intercepts of the graph of the line.

5. $5x - 2y + 12 = 0$

6. $-0.3x - 0.5y + 1.5 = 0$

In Exercises 7–10, find the domain of the function.

7. $g(x) = \dfrac{4}{x - 4}$

8. $f(x) = \dfrac{2x}{x^2 - 9x + 20}$

9. $h(t) = \sqrt[4]{5 - 3t}$

10. $f(t) = t^3 + 3t - 5$

SECTION 1.7 · EXERCISES

In Exercises 1–6, use a graphing utility to graph the function and find its domain and range.

1. $f(x) = \sqrt{x - 1}$ **2.** $f(x) = 4 - x^2$

3. $f(x) = \sqrt{x^2 - 4}$ **4.** $f(x) = |x - 2|$

5. $f(x) = \sqrt{25 - x^2}$ **6.** $f(x) = \dfrac{|x - 2|}{x - 2}$

In Exercises 7–12, use the vertical line test to determine whether y is a function of x. Describe how your graphing utility can be used to produce the given graph.

7. $y = x^2$ **8.** $y = x^3 - 1$

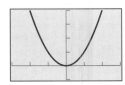

9. $x - y^2 = 0$ **10.** $x^2 + y^2 = 9$

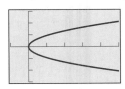 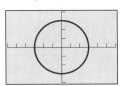

11. $x^2 = xy - 1$ **12.** $x = |y|$

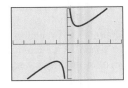 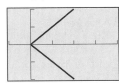

In Exercises 13–16, select the viewing rectangle on a graphing utility that shows the most complete graph of the function.

13. $f(x) = -0.2x^2 + 3x + 32$

(a)
RANGE
Xmin=-2
Xmax=20
Xscl=1
Ymin=-10
Ymax=30
Yscl=4

(b)
RANGE
Xmin=-10
Xmax=30
Xscl=5
Ymin=-5
Ymax=50
Yscl=5

(c)
RANGE
Xmin=0
Xmax=10
Xscl=0.5
Ymin=0
Ymax=200
Yscl=25

14. $f(x) = 6[x - (0.1x)^5]$

(a)
RANGE
Xmin=-500
Xmax=500
Xscl=50
Ymin=-500
Ymax=500
Yscl=50

(b)
RANGE
Xmin=-25
Xmax=25
Xscl=5
Ymin=-25
Ymax=25
Yscl=5

(c)
RANGE
Xmin=-20
Xmax=20
Xscl=5
Ymin=-100
Ymax=100
Yscl=20

15. $f(x) = 4x^3 - x^4$

(a)
RANGE
Xmin=-2
Xmax=6
Xscl=1
Ymin=-10
Ymax=30
Yscl=4

(b)
RANGE
Xmin=-50
Xmax=50
Xscl=5
Ymin=-50
Ymax=50
Yscl=5

(c)
RANGE
Xmin=0
Xmax=2
Xscl=0.2
Ymin=-2
Ymax=2
Yscl=0.5

16. $f(x) = 10x\sqrt{400 - x^2}$

(a)
RANGE
Xmin=-5
Xmax=50
Xscl=5
Ymin=-5000
Ymax=5000
Yscl=500

(b)
RANGE
Xmin=-20
Xmax=20
Xscl=2
Ymin=-500
Ymax=500
Yscl=50

(c)
RANGE
Xmin=-25
Xmax=25
Xscl=5
Ymin=-2000
Ymax=2000
Yscl=200

In Exercises 17–24, use a graphing utility to graph the function. Approximate the intervals over which the function is increasing, decreasing, or constant.

17. $f(x) = 2x$ **18.** $f(x) = x^2 - 2x$

19. $f(x) = x^3 - 3x^2$ **20.** $f(x) = \sqrt{x^2 - 4}$

21. $f(x) = 3x^4 - 6x^2$ **22.** $f(x) = x^{2/3} = (x^2)^{1/3}$

23. $f(x) = x\sqrt{x + 3}$ **24.** $f(x) = |x + 1| + |x - 1|$

In Exercises 25–30, use a graphing utility to approximate (to two decimal place accuracy) any relative minimum or maximum values of the function.

25. $f(x) = x^2 - 6x$ **26.** $f(x) = (x - 1)^2(x + 2)$

27. $g(x) = 2x^3 + 3x^2 - 12x$ **28.** $g(x) = x^3 - 6x^2 + 15$

29. $h(x) = (x - 1)\sqrt{x}$ **30.** $h(x) = x\sqrt{4 - x}$

31. *Maximum Area*   The perimeter of a rectangle is 100 feet (see figure).

(a) Show that the area of the rectangle is given by $A = x(50 - x)$, where x is its length.

(b) Use a graphing utility to graph the area function.

(c) Move the cursor along the graph to approximate (to two decimal place accuracy) the maximum of the area function. Approximate the dimensions of the rectangle that yield the maximum area.

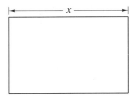

FIGURE FOR 31

32. *Minimum Cost*   A power station is on one side of a river that is $\frac{1}{2}$ mile wide, and a factory is 6 miles downstream on the other side (see figure). It costs $6 per foot to run power lines overland and $8 per foot to run them underwater.

(a) Show that the total cost for the power line is given by

$$T = 8(5280)\sqrt{x^2 + \tfrac{1}{4}} + 6(5280)(6 - x).$$

(b) Use a graphing utility to graph the cost function.

(c) To determine the most economical path for the power line, approximate (to two decimal place accuracy) the value of x that minimizes the cost function.

Factory

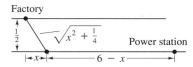

Power station

FIGURE FOR 32

33. *Maximum Profit*   The marketing department of a company estimates that the demand for a product is given by $p = 100 - 0.0001x$, where p is the price per unit and x is the number of units. The cost of producing x units is given by $C = 350,000 + 30x$, and the profit for producing and selling x units is given by

$$P = R - C = xp - C.$$

Graph the profit function and estimate the number of units that would produce a maximum profit.

34. *Maximum Revenue*   When a wholesaler sold a certain product at $25 per unit, sales were 800 units per week. After a price increase of $5, the average number of units sold dropped to 775 per week. Assume that the price p is a linear function of the demand x. Graph the revenue function $R = xp$ and estimate (to two decimal place accuracy) the price that will maximize total revenue.

In Exercises 35–38, use a graphing utility to obtain the graph of the compound function.

35. $f(x) = \begin{cases} 2x + 3, & x < 0 \\ 3 - x, & x \geq 0 \end{cases}$

36. $f(x) = \begin{cases} \sqrt{4 + x}, & x < 0 \\ \sqrt{4 - x}, & x \geq 0 \end{cases}$

37. $f(x) = \begin{cases} x^2 + 5, & x \leq 1 \\ -x^2 + 4x + 3, & x > 1 \end{cases}$

38. $f(x) = \begin{cases} 1 - (x - 1)^3, & x \leq 2 \\ \sqrt{x - 2}, & x > 2 \end{cases}$

In Exercises 39 and 40, use a graphing utility to graph the step function. State the domain and range of the function.

39. $s(x) = 2[[x - 1]]$ **40.** $g(x) = 6 - [[x]]$

41. *Price of a Telephone Call*   The cost of a telephone call between two cities is $0.65 for the first minute and $0.42 for each additional minute (or portion thereof).

(a) Use the greatest integer function to create a model for the cost C of a telephone call between the two cities lasting t minutes.

(b) Use a graphing utility to graph the cost model.

(c) Move the cursor along the graph to determine the maximum length of a call that cannot cost more than $6.

42. *Cost of Overnight Delivery*   Suppose the cost of sending an overnight package from New York to Atlanta is $9.80 for the first pound and $2.50 for each additional pound (or portion thereof).

(a) Use the greatest integer function to create a model for the cost C of overnight delivery of a package weighing x pounds.

(b) Use a graphing utility to graph the cost function.

(c) Move the cursor along the graph to determine the cost of overnight delivery of a package weighing 12 pounds.

In Exercises 43–46, find the coordinates of a second point on the graph of a function f if the given point is on the graph and the function is (a) even and (b) odd.

43. $(5, 6)$ **44.** $(-3, 7)$

45. $(-\frac{3}{2}, -2)$ **46.** $(\frac{3}{4}, -\frac{7}{8})$

In Exercises 47–56, use a graphing utility to graph the function and determine whether the function is even, odd, or neither.

47. $f(x) = 3$ **48.** $g(x) = x$

49. $f(x) = 5 - 3x$ **50.** $h(x) = x^2 - 4$

51. $g(s) = \dfrac{s^3}{4}$ **52.** $f(t) = -t^4$

53. $f(x) = \sqrt{1 - x}$ **54.** $f(x) = x^2\sqrt[3]{x}$

55. $g(x) = \dfrac{5x}{x^2 + 1}$ **56.** $f(x) = |x + 2|$

In Exercises 57–62, use an algebraic test to determine whether the function is even, odd, or neither. Verify your result graphically.

57. $f(x) = \frac{1}{3}x^6 - 2x^2$ **58.** $h(x) = x^3 - 5$

59. $g(x) = x^3 - 5x$ **60.** $f(x) = x\sqrt{1 - x^2}$

61. $f(t) = t^2 + 2t - 3$ **62.** $g(s) = 4s^{2/3}$

In Exercises 63–68, graph the function and determine the interval(s) (if any) on the real axis for which $f(x) \geq 0$.

63. $f(x) = 4 - x$ **64.** $f(x) = x^2 - 4x$

65. $f(x) = 1 - x^4$ **66.** $f(x) = \sqrt{x + 2}$

67. $f(x) = 5$ **68.** $f(x) = -(1 + |x|)$

In Exercises 69–72, express the height h of the rectangle as a function of x.

69.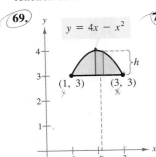

70.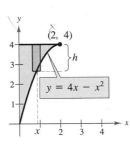

71.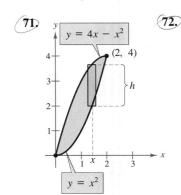

72.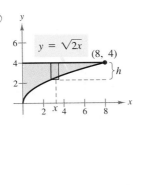

In Exercises 73 and 74, express the length L of the rectangle as a function of y.

73.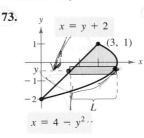

74.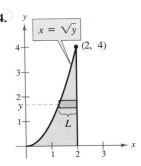

75. *Water Intake*   When the intake pipe of a 100-gallon tank is turned on, water flows into the tank at the rate of 10 gallons per minute. When either of the two drain pipes is turned on, water flows out of the tank at the rate of 5 gallons per minute. The graph below shows the volume of water in the tank over a 60-minute period. Which pipes are turned on during the various periods of time indicated by the graph? (There is more than one correct answer.)

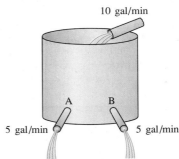

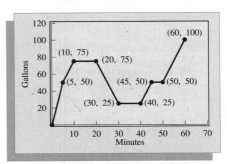

FIGURE FOR 75

76. Prove that the following function is odd.

$$f(x) = a_{2n+1}x^{2n+1} + a_{2n-1}x^{2n-1} + \cdots + a_3x^3 + a_1x$$

77. Prove that the following function is even.

$$f(x) = a_{2n}x^{2n} + a_{2n-2}x^{2n-2} + \cdots + a_2x^2 + a_0$$

1.8 SHIFTING, REFLECTING, AND STRETCHING GRAPHS

Summary of Graphs of Common Functions / Vertical and Horizontal / Shifts / Reflections / Nonrigid Transformations

Summary of Graphs of Common Functions

One of the goals of this text is to enable you to build your intuition for the basic shapes of the graphs of different types of functions. For instance, from your study of lines in Section 1.5, you can determine the basic shape of the graph of the linear function

$$f(x) = ax + b.$$

Specifically, you know that the graph of this function is a line whose slope is a and whose y-intercept is b.

The six graphs shown in Figure 1.57 represent the most commonly used functions in algebra. Familiarity with the basic characteristics of these simple graphs will help you analyze the shapes of more complicated graphs. Try using a graphing utility to verify these graphs.

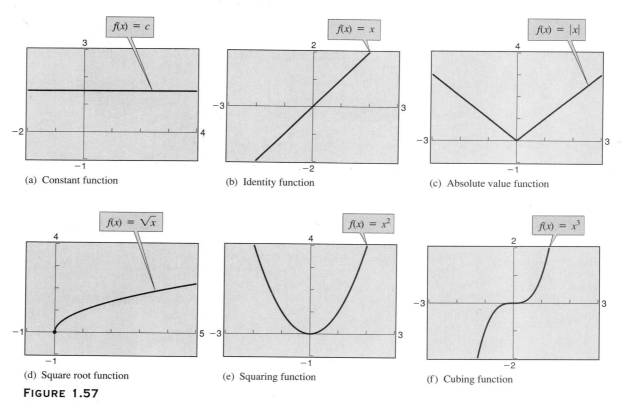

(a) Constant function

(b) Identity function

(c) Absolute value function

(d) Square root function

(e) Squaring function

(f) Cubing function

FIGURE 1.57

Vertical and Horizontal Shifts

Many functions have graphs that are simple transformations of the common graphs summarized in Figure 1.57. For example, you can obtain the graph of $h(x) = x^2 + 2$ by shifting the graph of $f(x) = x^2$ *up* two units, as shown in Figure 1.58. In function notation, h and f are related as follows.

$$h(x) = x^2 + 2 = f(x) + 2 \qquad \textit{Upward shift of 2}$$

Similarly, you can obtain the graph of $g(x) = (x - 2)^2$ by shifting the graph of $f(x) = x^2$ to the *right* two units, as shown in Figure 1.59. In this case, the functions g and f have the following relationship.

$$g(x) = (x - 2)^2 = f(x - 2) \qquad \textit{Right shift of 2}$$

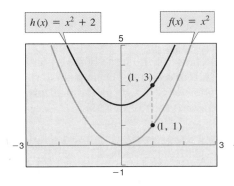

Vertical shift upward: 2 units

FIGURE 1.58

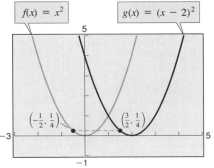

Horizontal shift to the right: 2 units

FIGURE 1.59

VERTICAL AND HORIZONTAL SHIFTS

Let c be a positive real number. **Vertical** and **horizontal shifts** in the graph of $y = f(x)$ are represented as follows.

1. Vertical shift c units **upward:** $h(x) = f(x) + c$
2. Vertical shift c units **downward:** $h(x) = f(x) - c$
3. Horizontal shift c units to the **right:** $h(x) = f(x - c)$
4. Horizontal shift c units to the **left:** $h(x) = f(x + c)$

EXAMPLE 1   **Shifts in the Graph of a Function**

Compare the graph of each function with the graph of $f(x) = x^3$. (See Figure 1.60.)

A. $g(x) = x^3 + 1$ **B.** $h(x) = (x - 1)^3$ **C.** $k(x) = (x + 2)^3 + 1$

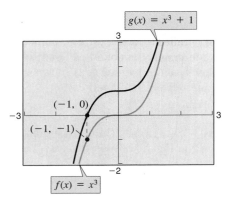

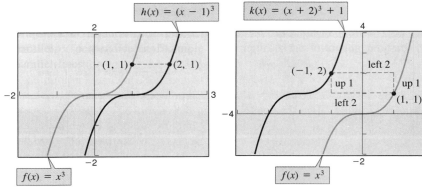

(a) Vertical shift: 1 unit up

(b) Horizontal shift: 1 unit right

(c) Horizontal shift: 2 units left
Vertical shift: 1 unit up

FIGURE 1.60

SOLUTION

Relative to the graph of $f(x) = x^3$, the graph of $g(x) = x^3 + 1$ is an upward shift of one unit. The graph of $h(x) = (x - 1)^3$ is a right shift of one unit. The graph of $k(x) = (x + 2)^3 + 1$ involves a left shift of two units *followed* by an upward shift of one unit.

EXAMPLE 2 **Finding Equations from Graphs**

Each of the graphs shown in Figure 1.61 is a vertical or horizontal shift of the graph of $f(x) = x^2$. Find an equation for each function.

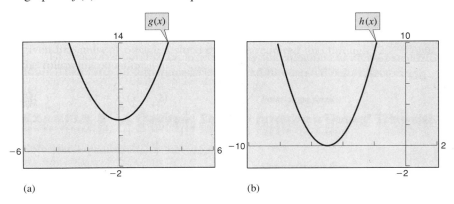

FIGURE 1.61 (a) (b)

SOLUTION

A. The graph of g is a vertical shift of four units upward of the graph of $f(x) = x^2$. Thus, the equation for g is $g(x) = x^2 + 4$.

B. The graph of h is a horizontal shift of five units to the left of the graph of $f(x) = x^2$. Thus, the equation for h is $h(x) = (x + 5)^2$.

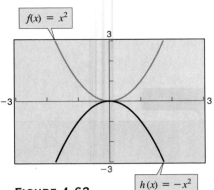

FIGURE 1.62

Reflections

The second common type of transformation is called a **reflection.** For instance, if you consider the x-axis to be a mirror, the graph of

$$h(x) = -x^2$$

is the mirror image (or reflection) of the graph of

$$f(x) = x^2,$$

as shown in Figure 1.62.

REFLECTIONS IN THE COORDINATE AXES

Reflections in the coordinate axes of the graph of $y = f(x)$ are represented as follows.

1. Reflection in the x-axis: $h(x) = -f(x)$
2. Reflection in the y-axis: $h(x) = f(-x)$

EXAMPLE 3   Finding Equations from Graphs

Each of the graphs shown in Figure 1.63(a) and 1.63(b) is a transformation of the graph of $f(x) = x^4$ (shown at left in Figure 1.63). Find an equation for each function.

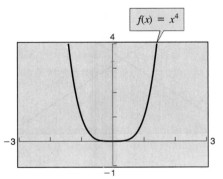

FIGURE 1.63

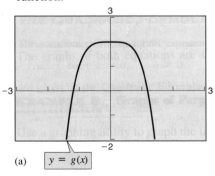

(a) $y = g(x)$

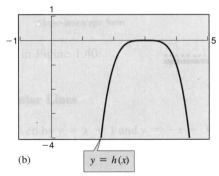

(b) $y = h(x)$

SOLUTION

A. The graph of g is a reflection in the x-axis *followed* by an upward shift of two units of the graph of $f(x) = x^4$. Thus, the equation for g is $g(x) = -x^4 + 2$.

B. The graph of h is a horizontal shift of three units to the right followed by a reflection in the x-axis of the graph of $f(x) = x^4$. Thus, the equation for h is $h(x) = -(x - 3)^4$.

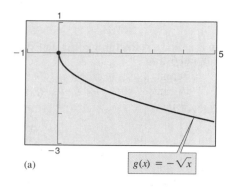

(a)

$g(x) = -\sqrt{x}$

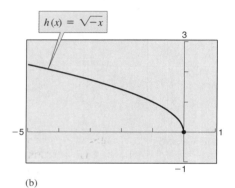

$h(x) = \sqrt{-x}$

(b)

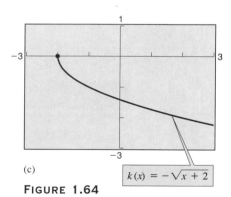

(c)

$k(x) = -\sqrt{x + 2}$

FIGURE 1.64

EXAMPLE 4   **Reflections and Shifts**

Compare the graph of each function with the graph of $f(x) = \sqrt{x}$.

A. $g(x) = -\sqrt{x}$

B. $h(x) = \sqrt{-x}$

C. $k(x) = -\sqrt{x + 2}$

SOLUTION

A. Relative to the graph of $f(x) = \sqrt{x}$, the graph of g is a reflection in the x-axis because

$$g(x) = -\sqrt{x} = -f(x).$$

B. The graph of h is a reflection of the graph of $f(x) = \sqrt{x}$ in the y-axis because

$$h(x) = \sqrt{-x} = f(-x).$$

C. From the equation

$$k(x) = -\sqrt{x + 2} = -f(x + 2)$$

you can conclude that the graph of k is a left shift of two units, followed by a reflection in the x-axis.

The graphs of all three functions are shown in Figure 1.64.

EXAMPLE 5   **A Program for Practice**

If you have a programmable calculator, try entering the following program. (These program steps are for a Texas Instruments TI-81. Programs for other calculators are listed in the appendix.)

```
:Rand→H                    :-6→Ymin
:-6+Int (12H)→H            :6→Ymax
:Rand→V                    :1→Yscl
:-3+ Int (6V)→V            :DispGraph
:Rand→R                    :Pause
:If R<.5                   :Disp "Y=R(X+H)²+V"
:-1→R                      :Disp "R="
:If R>.49                  :Disp R
:1→R                       :Disp "H="
:"R(X+H)²+V"→Y₁            :Disp H
:-9→Xmin                   :Disp "V="
:9→Xmax                    :Disp V
:1→Xscl                    :End
```

This program will sketch a graph of the function

$$y = R(x + H)^2 + V,$$

where $R = \pm 1$, H is an integer between -6 and 6, and V is an integer between -3 and 3. (Each time you run the program, different values of R, H, and V are possible.) From the graph, you should be able to determine the values of R, H, and V. After you have determined the values, press ENTER to see the answer. (To look at the graph again, press GRAPH.)

For example, in the graph shown in Figure 1.65, you can make the following conclusions. Because the graph of $f(x) = x^2$ has been reflected about the x-axis, you know that $R = -1$. Because the graph of $f(x) = x^2$ has been shifted four units to the left, you know that $H = 4$. Because the graph of $f(x) = x^2$ has been shifted three units up, you know that $V = 3$. Thus, the equation of the graph shown in Figure 1.65 must be

$$y = -(x + 4)^2 + 3.$$

Try running this program several times. It will give you practice in working with reflections, horizontal shifts, and vertical shifts.

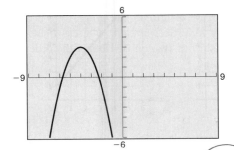

FIGURE 1.65

Nonrigid Transformations

Horizontal shifts, vertical shifts, and reflections are called **rigid** transformations because the basic shape of the graph is unchanged. These transformations change only the *position* of the graph in the xy-plane. **Nonrigid** transformations are those that cause a *distortion*—a change in the shape of the original graph. For instance, a nonrigid transformation of the graph of $y = f(x)$ is represented by $y = cf(x)$, where the transformation is a **vertical stretch** if $c > 1$ and a **vertical shrink** if $0 < c < 1$.

EXAMPLE 6   **Nonrigid Transformations**

Compare the graph of each function with the graph of $f(x) = |x|$.

A. $h(x) = 3|x|$ **B.** $g(x) = \dfrac{1}{3}|x|$

SOLUTION

A. Relative to the graph of $f(x) = |x|$, the graph of

$$h(x) = 3|x| = 3f(x)$$

is a vertical stretch (multiply each y-value by 3) of the graph of f.

B. Similarly, the equation

$$g(x) = \frac{1}{3}|x| = \frac{1}{3}f(x)$$

indicates that the graph of g is a vertical shrink of the graph of f.

The graphs of all three functions are shown in Figure 1.66.

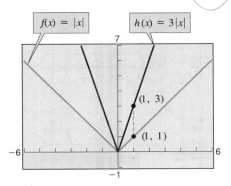

(a)

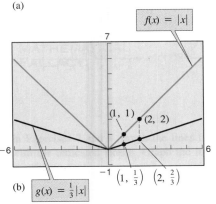

(b) $g(x) = \frac{1}{3}|x|$

FIGURE 1.66

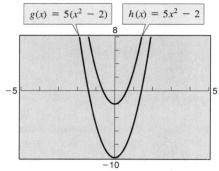

$g(x) = 5(x^2 - 2)$ $h(x) = 5x^2 - 2$

FIGURE 1.67

EXAMPLE 7   Sequences of Transformations

Use a graphing utility to graph the two functions $g(x) = 5(x^2 - 2)$ and $h(x) = 5x^2 - 2$ on the same screen. Describe how each function was obtained from $f(x) = x^2$ as a sequence of shifts and stretches.

SOLUTION

Notice that the two graphs in Figure 1.67 are different. The graph of g is a downward shift of two units followed by a vertical stretch, whereas h is a vertical stretch followed by a downward shift of two units. Hence, the order of applying the transformations is important.

DISCUSSION PROBLEM

COMPARING MATHEMATICAL MODELS

The following two second-degree polynomial functions are models of the United States population from 1800 to 1990. In the first model, $f(t)$ represents the population and t represents the year.

Model 1

$$f(t) = 6722.5t^2 - 24,201,000t + 21,787,790,000, \qquad 1800 \le t \le 1990$$

In the second model, $g(t)$ represents the population in millions and t represents the year, with $t = 0$ corresponding to 1800, $t = 1$ corresponding to 1810, and so on. (See figure.)

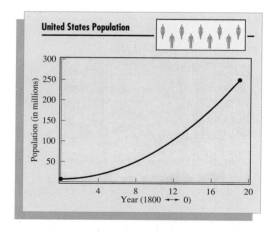

United States Population

Model 2

$$g(t) = 0.67225t^2 + 6.89, \qquad 0 \le t \le 19$$

These two models are related as follows.

$$f(t) = 1,000,000g\left(\frac{t - 1800}{10}\right)$$

Which model do you think is better? Write a short paragraph explaining your reasons.

WARM-UP

The following warm-up exercises involve skills that were covered in earlier sections. You will use these skills in the exercise set for this section.

In Exercises 1–6, graph the function. Determine its domain and range.

1. $f(x) = \sqrt{16 - x^2}$ **2.** $h(x) = \frac{1}{2}|x - 3|$

3. $g(x) = \frac{1}{4}x^2 - 1$ **4.** $f(x) = 1 - \frac{1}{3}x^3$

5. $f(s) = \dfrac{|s - 2|}{s - 2}$ **6.** $g(t) = \sqrt{t - 2}$

In Exercises 7–10, determine whether the function is even, odd, or neither.

7. $p(x) = -0.25x$ **8.** $h(x) = x^2 - 3$

9. $g(t) = \frac{3}{8}t^4 - t^2$ **10.** $f(s) = s^{4/3} + 3$

SECTION 1.8 · EXERCISES

In Exercises 1–8, sketch the graphs of the three functions *by hand* on the same coordinate system. Verify your result with a graphing utility.

1. $f(x) = x$
$g(x) = x - 4$
$h(x) = 3x$

2. $f(x) = \frac{1}{2}x$
$g(x) = \frac{1}{2}x + 2$
$h(x) = \frac{1}{2}(x - 2)$

3. $f(x) = x^2$
$g(x) = x^2 + 2$
$h(x) = (x - 2)^2$

4. $f(x) = x^2$
$g(x) = x^2 - 4$
$h(x) = (x + 2)^2 + 1$

5. $f(x) = -x^2$
$g(x) = -x^2 + 1$
$h(x) = -(x - 2)^2$

6. $f(x) = (x - 2)^2$
$g(x) = (x - 2)^2 + 2$
$h(x) = -(x - 2)^2 + 4$

7. $f(x) = x^2$
$g(x) = (\frac{1}{2}x)^2$
$h(x) = (2x)^2$

8. $f(x) = x^2$
$g(x) = (\frac{1}{4}x)^2 + 2$
$h(x) = -(\frac{1}{4}x)^2$

11.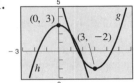

12.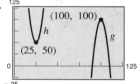

In Exercises 13–16, use a graphing utility to graph the three functions on the same set of coordinate axes. Describe the graphs of g and h relative to the graph of f.

13. $f(x) = x^3 - 3x^2$
$g(x) = f(x + 2)$
$h(x) = f(\frac{1}{2}x)$

14. $f(x) = x^3 - 3x^2 + 2$
$g(x) = f(x - 1)$
$h(x) = f(2x)$

15. $f(x) = x^3 - 3x^2$
$g(x) = -\frac{1}{3}f(x)$
$h(x) = f(-x)$

16. $f(x) = x^3 - 3x^2 + 2$
$g(x) = -f(x)$
$h(x) = f(-x)$

In Exercises 9–12, use the graph of $f(x) = x^2$ to write formulas for the functions g and h shown in the given graphs.

In Exercises 17 and 18, use the graph of $f(x) = x^3 - 3x^2$ (see Exercise 13) to write a formula for the function g shown in the given graph.

9.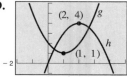

10.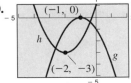

17.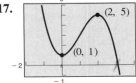

18.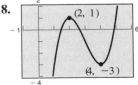

In Exercises 19–24, compare the graph of the given function with the graph of $f(x) = \sqrt{x}$.

19. $y = \sqrt{x} + 2$ **20.** $y = -\sqrt{x}$

21. $y = \sqrt{x} - 2$ **22.** $y = \sqrt{x + 3}$

23. $y = \sqrt{2x}$ **24.** $y = \sqrt{-x}$

In Exercises 25–30, compare the graph of the given function with the graph of $f(x) = \sqrt[3]{x}$.

25. $y = \sqrt[3]{x} - 1$ **26.** $y = \sqrt[3]{x + 1}$

27. $y = \sqrt[3]{x - 1}$ **28.** $y = -\sqrt[3]{x} - 2$

29. $y = \sqrt[3]{-x}$ **30.** $y = \frac{1}{2}\sqrt[3]{x}$

In Exercises 31–36, use the graph of $f(x)$ (see figure) to sketch the graph of the specified function.

31. $f(x - 4)$

32. $f(x + 2)$

33. $f(x) + 4$

34. $f(x) - 1$

35. $2f(x)$

36. $\frac{1}{2}f(x)$

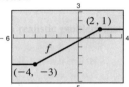
FIGURE FOR 31–36

In Exercises 37–42, specify a sequence of transformations that will yield the graph of the given function from the graph of the function $f(x) = x^3$.

37. $g(x) = 4 - x^3$ **38.** $g(x) = (x - 4)^3$

39. $h(x) = \frac{1}{4}(x + 2)^3$ **40.** $h(x) = -2(x - 1)^3 + 3$

41. $p(x) = (\frac{1}{3}x)^3 + 2$ **42.** $p(x) = [3(x - 2)]^3$

43. *Profit*   The profit P per week on a certain product is given by the model

$$P(x) = 80 + 20x - 0.5x^2, \qquad 0 \le x \le 20,$$

where x is the amount spent on advertising. In this model, x and P are both measured in hundreds of dollars.
(a) Use a graphing utility to graph the profit function.
(b) The business estimates that taxes and operating costs will increase by an average of $2500 per week during the next year. Rewrite the profit equation to reflect this expected decrease in profits. Identify the type of transformation made to the graph of the equation.
(c) Rewrite the original profit equation so that x measures advertising expenditures in dollars. [Find $P(x/100)$.] Describe the transformation made to the graph of the profit function.

44. *Automobile Aerodynamics*   The number of horsepower H required to overcome wind drag on a certain automobile is approximated by

$$H(x) = 0.002x^2 + 0.005x - 0.029, \qquad 10 \le x \le 100,$$

where x is the speed of the car in miles per hour.
(a) Use a graphing utility to graph the power function.
(b) Rewrite the power function so that x represents the speed in kilometers per hour. [Find $H(1.6x)$.] Describe the transformation made to the graph of the power function.

45. Use a graphing utility to graph $f(x) = x^2$, $g(x) = x^4$, and $h(x) = x^6$ in the same viewing rectangle. Describe any similarities and differences you observe among the graphs.

46. Use a graphing utility to graph $f(x) = x$, $g(x) = x^3$, and $h(x) = x^5$ in the same viewing rectangle. Describe any similarities and differences you observe among the graphs.

47. Use the results of Exercises 45 and 46 to sketch the graphs of $f(x) = x^{10}$ and $g(x) = x^{11}$, without the aid of a graphing utility.

48. Use the results of Exercises 45 and 46 to sketch the graphs of $f(x) = (x - 3)^3$, $g(x) = (x + 1)^2$, and $h(x) = (x - 4)^5$, without the aid of a graphing utility.

In Exercises 49–52, use the results of Exercises 45–48 and your imagination to guess the shape of the graph of the function. Then use a graphing utility to graph the function, and compare the result with the graph drawn by hand.

49. $f(x) = x^2(x - 6)^2$ **50.** $f(x) = x^3(x - 6)^2$

51. $f(x) = x^2(x - 6)^3$ **52.** $f(x) = x^3(x - 6)^3$

1.9   COMBINATIONS OF FUNCTIONS

Arithmetic Combinations of Functions / Compositions of Functions /
Applications

Arithmetic Combinations of Functions

Just as two real numbers can be combined by the operations of addition,
subtraction, multiplication, and division to form other real numbers, two *functions* can be combined to create new functions. For example, if

$$f(x) = 2x - 3 \quad \text{and} \quad g(x) = x^2 - 1,$$

you can form the sum, difference, product, and quotient of f and g as follows.

$$f(x) + g(x) = (2x - 3) + (x^2 - 1) = x^2 + 2x - 4 \qquad \textit{Sum}$$
$$f(x) - g(x) = (2x - 3) - (x^2 - 1) = -x^2 + 2x - 2 \qquad \textit{Difference}$$
$$f(x)g(x) = (2x - 3)(x^2 - 1) = 2x^3 - 3x^2 - 2x + 3 \qquad \textit{Product}$$
$$\frac{f(x)}{g(x)} = \frac{2x - 3}{x^2 - 1}, \qquad x \neq \pm 1 \qquad \textit{Quotient}$$

The domain of an arithmetic combination of functions f and g consists of all
real numbers that are common to the domains of f and g. In the case of the
quotient $f(x)/g(x)$, there is the further restriction that $g(x) \neq 0$.

**DEFINITIONS OF SUM, DIFFERENCE, PRODUCT,
AND QUOTIENT OF FUNCTIONS**

Let f and g be two functions with overlapping domains. Then, for all x
common to both domains, the **sum, difference, product,** and **quotient**
of f and g are defined as follows.

1. Sum: $(f + g)(x) = f(x) + g(x)$
2. Difference: $(f - g)(x) = f(x) - g(x)$
3. Product: $(fg)(x) = f(x) \cdot g(x)$
4. Quotient: $\left(\dfrac{f}{g}\right)(x) = \dfrac{f(x)}{g(x)}, \quad g(x) \neq 0$

EXAMPLE 1   Finding the Sum of Two Functions

Given $f(x) = 2x + 1$ and $g(x) = x^2 + 2x - 1$, find $(f + g)(x)$. Then evaluate this sum when $x = 2$.

SOLUTION

The sum of the functions f and g is given by

$$(f + g)(x) = f(x) + g(x)$$
$$= (2x + 1) + (x^2 + 2x - 1)$$
$$= x^2 + 4x.$$

When $x = 2$, the value of this sum is

$$(f + g)(2) = 2^2 + 4(2) = 12.$$

EXAMPLE 2   **Finding the Difference of Two Functions**

Given $f(x) = 2x + 1$ and $g(x) = x^2 + 2x - 1$, find $(f - g)(x)$. Then evaluate this difference when $x = 2$.

SOLUTION

The difference of the functions f and g is given by

$$(f - g)(x) = f(x) - g(x)$$
$$= (2x + 1) - (x^2 + 2x - 1)$$
$$= -x^2 + 2.$$

When $x = 2$, the value of this difference is

$$(f - g)(2) = -(2)^2 + 2 = -2.$$

In Examples 1 and 2, both f and g have domains that consist of all real numbers. Thus, the domain of both $(f + g)$ and $(f - g)$ is also the set of all real numbers. Remember that any restrictions on the domains of f or g must be taken into account when forming the sum, difference, product, or quotient of f and g. For instance, the domain of $f(x) = 1/x$ is all $x \neq 0$, and the domain of $g(x) = \sqrt{x}$ is $[0, \infty)$. This implies that the domain of $f + g$ is $(0, \infty)$.

EXAMPLE 3   **The Quotient of Two Functions**

Find $(f/g)(x)$ and $(g/f)(x)$ for the functions $f(x) = \sqrt{x}$ and $g(x) = \sqrt{4 - x^2}$. Then find the domains of f/g and g/f.

Technology Note _____

A graphing utility can be used to graph the sum of two functions f and g. For example, let $y_1 = 2x + 1$, $y_2 = x^2 + 2x - 1$, and $y_3 = y_1 + y_2$.
   Try graphing all three functions with your graphing utility set in simultaneous plotting mode. Notice, also, that you can evaluate y_3 at any value of x by simply storing the x-value, pressing y_3, and pressing ENTER.

SOLUTION

The quotient of f and g is given by

$$\left(\frac{f}{g}\right)(x) = \frac{f(x)}{g(x)} = \frac{\sqrt{x}}{\sqrt{4 - x^2}},$$

and the quotient of g and f is given by

$$\left(\frac{g}{f}\right)(x) = \frac{g(x)}{f(x)} = \frac{\sqrt{4 - x^2}}{\sqrt{x}}.$$

The domain of f is $[0, \infty)$ and the domain of g is $[-2, 2]$. The intersection of these two domains is $[0, 2]$. Thus, the domains for f/g and g/f are as follows.

$$\text{Domain of } \frac{f}{g}: [0, 2) \qquad \text{Domain of } \frac{g}{f}: (0, 2]$$

Can you see why these two domains differ slightly?

Compositions of Functions

Another way of combining two functions is to form the **composition** of one with the other. For instance, if $f(x) = x^2$ and $g(x) = x + 1$, then the composition of f with g is

$$f(g(x)) = f(x + 1) = (x + 1)^2.$$

This composition is denoted as $f \circ g$.

DEFINITION OF COMPOSITION OF TWO FUNCTIONS

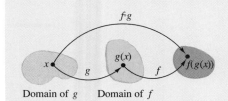

The **composition** of the functions f and g is

$$(f \circ g)(x) = f(g(x)).$$

The domain of $f \circ g$ is the set of all x in the domain of g such that $g(x)$ is in the domain of f. (See figure.)

EXAMPLE 4   Forming the Composition of f with g

Find $(f \circ g)(x)$ for $f(x) = \sqrt{x}$, $x \geq 0$, and $g(x) = x - 1$, $x \geq 1$. If possible, find $(f \circ g)(2)$ and $(f \circ g)(0)$.

SOLUTION

$$
\begin{aligned}
(f \circ g)(x) &= f(g(x)) && \textit{Definition of } f \circ g \\
&= f(x - 1) && \textit{Definition of } g(x) \\
&= \sqrt{x - 1}, \qquad x \geq 1 && \textit{Definition of } f(x)
\end{aligned}
$$

The domain of $f \circ g$ is $[1, \infty)$. Thus,

$$(f \circ g)(2) = \sqrt{2 - 1} = 1$$

is defined, but $(f \circ g)(0)$ is not defined because 0 is not in the domain of $f \circ g$.

The composition of f with g is generally *not* the same as the composition of g with f. This is illustrated in Example 5.

EXAMPLE 5   **Compositions of Functions**

Given $f(x) = x + 2$ and $g(x) = 4 - x^2$, find the following.

A. $(f \circ g)(x)$
B. $(g \circ f)(x)$

SOLUTION

A. $(f \circ g)(x) = f(g(x))$ *Definition of $f \circ g$*

$= f(4 - x^2)$ *Definition of $g(x)$*

$= (4 - x^2) + 2$ *Definition of $f(x)$*

$= -x^2 + 6$

B. $(g \circ f)(x) = g(f(x))$ *Definition of $g \circ f$*

$= g(x + 2)$ *Definition of $f(x)$*

$= 4 - (x + 2)^2$ *Definition of $g(x)$*

$= 4 - (x^2 + 4x + 4)$

$= -x^2 - 4x$

Note in this case that $(f \circ g)(x) \neq (g \circ f)(x)$.

EXAMPLE 6   **Finding the Domain of a Composite Function**

Given $f(x) = x^2 - 9$ and $g(x) = \sqrt{9 - x^2}$, find $(f \circ g)(x)$. Then find the domain of $f \circ g$.

SOLUTION

To begin, notice that the domain of g is $-3 \leq x \leq 3$.

$(f \circ g)(x) = f(g(x)),$ $-3 \leq x \leq 3$

$= f(\sqrt{9 - x^2}),$ $-3 \leq x \leq 3$

$= (\sqrt{9 - x^2})^2 - 9,$ $-3 \leq x \leq 3$

$= 9 - x^2 - 9,$ $-3 \leq x \leq 3$

$= -x^2,$ $-3 \leq x \leq 3$

Thus, the domain of $f \circ g$ is $-3 \leq x \leq 3$. To convince yourself of this, use a graphing utility to graph

$$y = (\sqrt{9 - x^2})^2 - 9,$$

as shown in Figure 1.68. Notice that the graphing utility does not extend the graph to the left of $x = -3$ or to the right of $x = 3$.

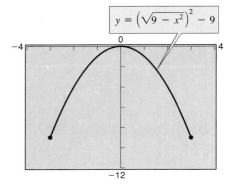

$$y = (\sqrt{9 - x^2})^2 - 9$$

FIGURE 1.68

EXAMPLE 7   A Case in Which $f \circ g = g \circ f$
·················

Given $f(x) = 2x + 3$ and $g(x) = \frac{1}{2}(x - 3)$, find the following.

A. $(f \circ g)(x)$
B. $(g \circ f)(x)$

SOLUTION

REMARK In Example 7, note that the two composite functions $f \circ g$ and $g \circ f$ are equal, and both represent the identity function. That is, $(f \circ g)(x) = (g \circ f)(x) = x$. We will discuss this special case in the next section.

A. $(f \circ g)(x) = f(g(x)) = f\left[\frac{1}{2}(x - 3)\right] = 2\left[\frac{1}{2}(x - 3)\right] + 3$

$\qquad = x - 3 + 3 = x$

B. $(g \circ f)(x) = g(f(x)) = g(2x + 3) = \frac{1}{2}[(2x + 3) - 3] = \frac{1}{2}(2x) = x$
·················

In Examples 5, 6, and 7 we formed the composition of two given functions. In calculus, it is also important to be able to identify two functions that *make up a given* composite function. For instance, the function h given by $h(x) = (3x - 5)^3$ is the composition of f with g, where $f(x) = x^3$ and $g(x) = 3x - 5$. That is,

$\qquad h(x) = (3x - 5)^3 = [g(x)]^3 = f(g(x)).$

Basically, to "decompose" a composite function, look for an "inner" and an "outer" function. In the function h above, $g(x) = 3x - 5$ is the inner function and $f(x) = x^3$ is the outer function.

EXAMPLE 8   Identifying a Composite Function
·················

Express the function

$\qquad h(x) = \dfrac{1}{(x - 2)^2}$

as a composition of two functions.

SOLUTION

One way to write h as a composite of two functions is to take the inner function to be $g(x) = x - 2$ and the outer function to be

$\qquad f(x) = \dfrac{1}{x^2} = x^{-2}.$

Then you can write

$\qquad h(x) = \dfrac{1}{(x - 2)^2} = (x - 2)^{-2} = f(x - 2) = f(g(x)).$

This function can be decomposed in other ways. For instance, let $g(x) = 1/(x - 2)$ and $f(x) = x^2$. Then $h(x) = f(g(x))$.
·················

Applications

EXAMPLE 9   Bacteria Count

The number of bacteria in a refrigerated food is given by

$$N(T) = 20T^2 - 80T + 500, \qquad 2 \le T \le 14,$$

where T is the Celsius temperature of the food. When the food is removed from refrigeration, the temperature is given by

$$T(t) = 4t + 2, \qquad 0 \le t \le 3,$$

where t is the time in hours. Find the following.

A. The composite function $N(T(t))$. What does this function represent?

B. The number of bacteria in the food when $t = 2$ hours

C. The time when the bacteria count reaches 2000

SOLUTION

A. $N(T(t)) = 20(4t + 2)^2 - 80(4t + 2) + 500$
$= 20(16t^2 + 16t + 4) - 320t - 160 + 500$
$= 320t^2 + 320t + 80 - 320t - 160 + 500$
$= 320t^2 + 420$

This composite function represents the number of bacteria as a function of the amount of time the food has been out of refrigeration.

B. When $t = 2$, the number of bacteria is

$$N = 320(2)^2 + 420 = 1280 + 420 = 1700.$$

C. The bacteria count will reach $N = 2000$ when $320t^2 + 420 = 2000$. You can solve this equation for t algebraically as follows.

$$320t^2 + 420 = 2000$$
$$320t^2 = 1580$$
$$t^2 = \frac{1580}{320} = \frac{79}{16}$$
$$t = \frac{\sqrt{79}}{4} \approx 2.2 \text{ hours}$$

Or you can use a graphing utility to approximate the solution, as shown in Figure 1.69.

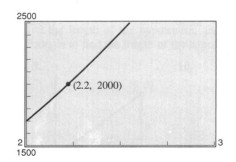

FIGURE 1.69

<table>
<tr><td>

DISCUSSION PROBLEM

THE COMPOSITION OF TWO FUNCTIONS

</td><td>

You are considering buying a new car for which the regular price is $15,800. The dealership has advertised a factory rebate of $1500 *and* a 12% discount. Using function notation, the rebate and the discount can be represented as

$$f(x) = x - 1500 \qquad \textit{Rebate of \$1500}$$

and

$$g(x) = 0.88x, \qquad \textit{Discount of 12\%}$$

where x is the price of the car, $f(x)$ is the price after subtracting the rebate, and $g(x)$ is the price after taking the 12% discount. Compare the sale price obtained by subtracting the rebate first and then taking the discount with the sale price obtained by taking the discount first and then subtracting the rebate.

</td></tr>
</table>

WARM-UP

The following warm-up exercises involve skills that were covered in earlier sections. You will use these skills in the exercise set for this section.

In Exercises 1–10, perform the indicated operations and simplify the result.

1. $\dfrac{1}{x} + \dfrac{1}{1-x}$

2. $\dfrac{2}{x+3} - \dfrac{2}{x-3}$

3. $\dfrac{3}{x-2} - \dfrac{2}{x(x-2)}$

4. $\dfrac{x}{x-5} + \dfrac{1}{3}$

5. $(x-1)\left(\dfrac{1}{\sqrt{x^2-1}}\right)$

6. $\left(\dfrac{x}{x^2-4}\right)\left(\dfrac{x^2-x-2}{x^2}\right)$

7. $(x^2-4) \div \left(\dfrac{x+2}{5}\right)$

8. $\left(\dfrac{x}{x^2+3x-10}\right) \div \left(\dfrac{x^2+3x}{x^2+6x+5}\right)$

9. $\dfrac{(1/x)+5}{3-(1/x)}$

10. $\dfrac{(x/4)-(4/x)}{x-4}$

SECTION 1.9 · EXERCISES

In Exercises 1–6, find (a) $(f+g)(x)$, (b) $(f-g)(x)$, (c) $(fg)(x)$, and (d) $(f/g)(x)$. What is the domain of f/g?

1. $f(x) = x + 1, \quad g(x) = x - 1$

2. $f(x) = 2x - 5, \quad g(x) = 5$

3. $f(x) = x^2 + 5, \quad g(x) = \sqrt{1-x}$

4. $f(x) = \sqrt{x^2-4}, \quad g(x) = \dfrac{x^2}{x^2+1}$

5. $f(x) = \dfrac{1}{x}, \quad g(x) = \dfrac{1}{x^2}$

6. $f(x) = \dfrac{x}{x+1}, \quad g(x) = x^3$

In Exercises 7–16, evaluate the indicated function for $f(x) = x^2 + 1$ and $g(x) = x - 4$.

7. $(f + g)(3)$ **8.** $(f - g)(-2)$

9. $(f - g)(2t)$ **10.** $(f + g)(t - 1)$

11. $(fg)(4)$ **12.** $(fg)(-6)$

13. $\left(\dfrac{f}{g}\right)(5)$ **14.** $\left(\dfrac{f}{g}\right)(0)$

15. $\left(\dfrac{f}{g}\right)(-1) - g(3)$ **16.** $(2f)(5)$

In Exercises 17–20, graph the functions f, g, and $f + g$ on the same set of coordinate axes.

17. $f(x) = \frac{1}{2}x$, $g(x) = x - 1$

18. $f(x) = \frac{1}{3}x$, $g(x) = -x + 4$

19. $f(x) = x^2$, $g(x) = -2x$

20. $f(x) = 4 - x^2$, $g(x) = x$

In Exercises 21 and 22, use a graphing utility to graph f, g, and $f + g$ in the same viewing rectangle. Which function contributes most to the magnitude of the sum when $0 \le x \le 2$? Which function contributes most to the sum when $x > 5$?

21. $f(x) = 3x$, $g(x) = -\dfrac{x^3}{10}$

22. $f(x) = \dfrac{x}{2}$, $g(x) = \sqrt{x}$

23. *Stopping Distance*   While traveling in a car at x miles per hour, you are required to stop quickly to avoid an accident. The distance the car travels during your reaction time is given by $R(x) = \frac{3}{4}x$. The distance traveled while braking is given by $B(x) = \frac{1}{15}x^2$. Find the function giving total stopping distance T. Graph the functions R, B, and T on the same set of coordinate axes for $0 \le x \le 60$.

24. *Comparing Sales*   Suppose you own two fast-food restaurants in town. From 1985 to 1990, the sales for one restaurant have been decreasing according to the function

$$R_1 = 500 - 0.8t^2, \qquad t = 5, 6, 7, 8, 9, 10,$$

where R_1 represents the sales for the first restaurant (in thousands of dollars) and t represents the calendar year with $t = 5$ corresponding to 1985. During the same 6-year period, the sales for the second restaurant have been increasing according to the function

$$R_2 = 250 + 0.78t, \qquad t = 5, 6, 7, 8, 9, 10.$$

Write a function that represents the total sales for the two restaurants. Use the *stacked bar graph* in the figure, which represents the total sales during the 6-year period, to determine whether the total sales have been increasing or decreasing.

FIGURE FOR 24

In Exercises 25–28, find (a) $f \circ g$, (b) $g \circ f$, and (c) $f \circ f$.

25. $f(x) = x^2$, $g(x) = x - 1$

26. $f(x) = \sqrt[3]{x - 1}$, $g(x) = x^3 + 1$

27. $f(x) = 3x + 5$, $g(x) = 5 - x$

28. $f(x) = x^3$, $g(x) = \dfrac{1}{x}$

In Exercises 29–34, find (a) $f \circ g$ and (b) $g \circ f$.

29. $f(x) = \sqrt{x + 4}$, $g(x) = x^2$

30. $f(x) = \sqrt[5]{x + 1}$, $g(x) = x^5 - 2$

31. $f(x) = \frac{1}{3}x - 3$, $g(x) = 3x + 1$

32. $f(x) = \sqrt{x}$, $g(x) = \sqrt{x}$

33. $f(x) = |x|$, $g(x) = x + 6$

34. $f(x) = x^{2/3}$, $g(x) = x^6$

In Exercises 35–38, use the graphs of f and g to evaluate the indicated functions.

35. (a) $(f + g)(3)$ (b) $\left(\dfrac{f}{g}\right)(2)$

36. (a) $(f - g)(1)$ (b) $(fg)(4)$

37. (a) $(f \circ g)(2)$ (b) $(g \circ f)(2)$

38. (a) $(f \circ g)(1)$ (b) $(g \circ f)(3)$

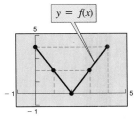

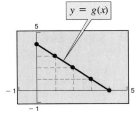

FIGURES FOR 35–38

In Exercises 39–46, find two functions f and g such that $(f \circ g)(x) = h(x)$. (There are many correct answers to these exercises.)

39. $h(x) = (2x + 1)^2$

40. $h(x) = (1 - x)^3$

41. $h(x) = \sqrt[3]{x^2 - 4}$

42. $h(x) = \sqrt{9 - x}$

43. $h(x) = \dfrac{1}{x + 2}$

44. $h(x) = \dfrac{4}{(5x + 2)^2}$

45. $h(x) = (x + 4)^2 + 2(x + 4)$

46. $h(x) = (x + 3)^{3/2}$

In Exercises 47–50, determine the domain of (a) f, (b) g, and (c) $f \circ g$.

47. $f(x) = \sqrt{x}$, $g(x) = x^2 + 1$

48. $f(x) = \dfrac{1}{x}$, $g(x) = x + 3$

49. $f(x) = \dfrac{3}{x^2 - 1}$, $g(x) = x + 1$

50. $f(x) = 2x + 3$, $g(x) = \dfrac{x}{2}$

51. *Ripples* A pebble is dropped into a calm pond, causing ripples in the form of concentric circles (see figure). The radius (in feet) of the outer ripple is given by $r(t) = 0.6t$, where t is the time in seconds after the pebble strikes the water. The area of the circle is given by the function $A(r) = \pi r^2$.
(a) Find and interpret $(A \circ r)(t)$.

(b) Use a graphing utility to graph the area as a function of t. Move the cursor along the graph to estimate (to two decimal place accuracy) the time required for the area enclosed by a ripple to increase to 20 square feet.

FIGURE FOR 51

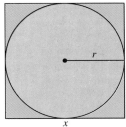

FIGURE FOR 52

52. *Area* A square concrete foundation was prepared as a base for a large cylindrical gasoline tank (see figure).
(a) Express the radius r of the tank as a function of the length x of the sides of the square.
(b) Express the area A of the circular base of the tank as a function of the radius r.
(c) Find and interpret $(A \circ r)(x)$.

53. *Cost* The weekly cost (in dollars) of producing x units in a manufacturing process is given by the function $C(x) = 60x + 750$. The number of units produced in t hours is given by $x(t) = 50t$.
(a) Find and interpret $(C \circ x)(t)$.
(b) Use a graphing utility to graph the cost as a function of time. Move the cursor along the curve to estimate (to two decimal place accuracy) the time that must elapse until the cost increases to $15,000.

54. *Air Traffic Control* An air traffic controller spots two planes at the same altitude flying toward the same point (see figure). Their flight paths form a right angle at point P. One plane is 150 miles from point P and is moving at 450 miles per hour. The second plane is 200 miles from point P and is moving at 450 miles per hour. Write the distance s between the planes as a function of time t.

FIGURE FOR 54

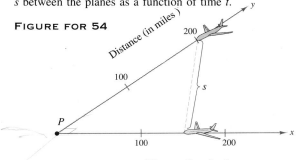

Distance (in miles)

55. *Rebate and Discount*   A car dealership is offering a rebate of $1200 *and* a discount of 15% on a car for which the regular price is $18,400. Write a function that represents each type of discount. Then use composition of functions to compare the sale price obtained by subtracting the rebate first with the sale price obtained by taking the discount first.

56. Prove that the product of two odd functions is an even function, and that the product of two even functions is an even function.

57. Use examples to hypothesize whether the product of an odd function and an even function is even or odd. Then prove your hypothesis.

58. Given a function f, prove that $g(x)$ is even and $h(x)$ is odd where

$$g(x) = \tfrac{1}{2}[f(x) + f(-x)] \quad \text{and} \quad h(x) = \tfrac{1}{2}[f(x) - f(-x)].$$

59. Use the result of Exercise 58 to prove that any function can be written as a sum of even and odd functions. (*Hint:* Add the two equations in Exercise 58.)

60. Use the result of Exercise 59 to write each of the following functions as a sum of even and odd functions.

(a) $f(x) = x^2 - 2x + 1$ (b) $f(x) = \dfrac{1}{x + 1}$

1.10   INVERSE FUNCTIONS

The Inverse of a Function   /   The Graph of the Inverse of a Function   /
The Existence of an Inverse Function   /   Finding the Inverse of a Function

The Inverse of a Function

In Section 1.6 you saw that a function can be represented by a set of ordered pairs. For instance, the function $f(x) = x + 4$ from the set $A = \{1, 2, 3, 4\}$ to the set $B = \{5, 6, 7, 8\}$ can be written as follows.

$$f(x) = x + 4: \{(1, 5), (2, 6), (3, 7), (4, 8)\}$$

By interchanging the first and second coordinates of each of these ordered pairs, you can form the **inverse function** of f, denoted by f^{-1}. This is a function from the set B to the set A, and can be written as follows.

$$f^{-1}(x) = x - 4: \{(5, 1), (6, 2), (7, 3), (8, 4)\}$$

Note that the domain of f is equal to the range of f^{-1}, and vice versa, as shown in Figure 1.70. Also note that the functions f and f^{-1} have the effect of "undoing" each other. In other words, when you form the composition of f with f^{-1} of the composition of f^{-1} with f, you obtain the identity function.

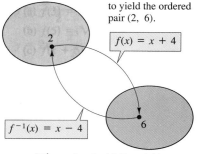

f matches 2 with 6 to yield the ordered pair (2, 6).

$f(x) = x + 4$

$f^{-1}(x) = x - 4$

f^{-1} matches 6 with 2 to yield the ordered pair (6, 2).

FIGURE 1.70

REMARK   Don't be confused by the use of -1 to denote the inverse function f^{-1}. In this book f^{-1} will *always* refer to the inverse of the function f and *not* to the reciprocal of $f(x)$.

Technology Note

On most graphing utilities you can graph $y = x^{1/3}$ in two ways:

$y = x \wedge (\frac{1}{3})$ or $y = \sqrt[3]{x}$. On most graphing utilities, however, you cannot obtain the complete graph of $y = x^{2/3}$ by entering $y = x \wedge (\frac{2}{3})$. Instead, you need to enter $y = (x \wedge (\frac{1}{3})) \wedge 2$ or $y = \sqrt[3]{x^2}$.

EXAMPLE 1   Finding Inverse Functions Informally

Find the inverse of the following.

A. $f(x) = 4x$

B. $f(x) = x - 6$

SOLUTION

A. The given function is a function that *multiplies* each input by 4. To "undo" this function, you need to *divide* each input by 4. Thus, the inverse function of $f(x) = 4x$ is

$$f^{-1}(x) = \frac{x}{4}.$$

B. The given function is a function that *subtracts* 6 from each input. To "undo" this function, you need to *add* 6 to each input. Thus, the inverse function of $f(x) = x - 6$ is

$$f^{-1}(x) = x + 6.$$

DEFINITION OF THE INVERSE OF A FUNCTION

Let f and g be two functions such that $f(g(x)) = x$ for every x in the domain of g and $g(f(x)) = x$ for every x in the domain of f. Then the function g is the **inverse** of the function f. The inverse function is denoted by f^{-1} (read "f-inverse"). Thus,

$$f(f^{-1}(x)) = x \quad \text{and} \quad f^{-1}(f(x)) = x.$$

The domain of f must be equal to the range of f^{-1}, and the range of f must be equal to the domain of f^{-1}.

Note from this definition that if the function g is the inverse of the function f, then it must also be true that the function f is the inverse of the function g.

EXAMPLE 2   Verifying Inverse Functions

Show that $f(x) = x^3 - 1$ and $g(x) = \sqrt[3]{x + 1}$ are inverses of each other.

DISCOVERY

Sketch the graphs of the functions given in Example 2, along with the graph of $y = x$.

$f(x) = x^3 - 1$

$g(x) = \sqrt[3]{x + 1}$

(Use the viewing rectangle $-6 \le x \le 6$ and $-4 \le y \le 4$.) Describe how the two graphs are related.

SOLUTION

Begin by noting that the domain (and the range) of both functions is the entire set of real numbers. To show that f and g are inverses of each other, you need to show that $f(g(x)) = x$ and $g(f(x)) = x$.

$$f(g(x)) = f\left(\sqrt[3]{x + 1}\right) = \left(\sqrt[3]{x + 1}\right)^3 - 1 = (x + 1) - 1 = x$$

$$g(f(x)) = g(x^3 - 1) = \sqrt[3]{(x^3 - 1) + 1} = \sqrt[3]{x^3} = x$$

You can see that the two functions f and g "undo" each other: the function f first cubes the input x and then subtracts 1, whereas the function g first adds 1 and then takes the cube root of the result.

EXAMPLE 3 Verifying Inverse Functions

Which of the functions

$$g(x) = \frac{x - 2}{5} \quad \text{and} \quad h(x) = \frac{5}{x} + 2$$

is the inverse of the function $f(x) = \dfrac{5}{x - 2}$?

SOLUTION

$$f(g(x)) = f\left(\frac{x - 2}{5}\right) = \frac{5}{[(x - 2)/5] - 2} = \frac{25}{(x - 2) - 10} = \frac{25}{x - 12} \neq x$$

Because this composition is not equal to the identity function x, you can conclude that g is *not* the inverse of f.

$$f(h(x)) = f\left(\frac{5}{x} + 2\right) = \frac{5}{(5/x) + 2 - 2} = \frac{5}{5/x} = x$$

Thus, it appears that h is the inverse of f. You can confirm this by showing that the composition of h with f is also equal to the identity function.

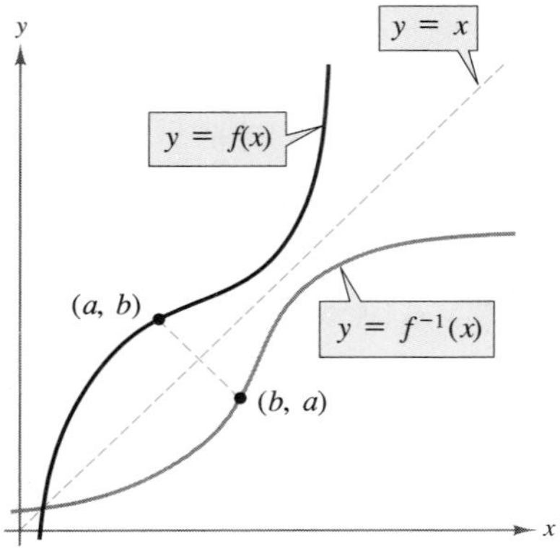

The graph of f^{-1} is a reflection of the graph of f in the line $y = x$.

FIGURE 1.71

The Graph of the Inverse of a Function

The graphs of f and f^{-1} are related to each other in the following way. If the point (a, b) lies on the graph of f, then the point (b, a) lies on the graph of f^{-1} and vice versa. This means that the graph of f^{-1} is a reflection of the graph of f in the line $y = x$, as shown in Figure 1.71.

In Examples 2 and 3, inverse functions were verified algebraically. A graphing utility can also be helpful in checking to see whether one function is the inverse of another function.

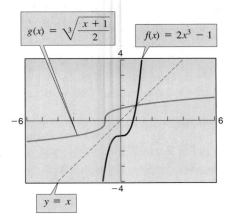

$g(x) = \sqrt[3]{\dfrac{x+1}{2}}$ $f(x) = 2x^3 - 1$

$y = x$

FIGURE 1.72

EXAMPLE 4   Graphical Check for Inverse Functions

Sketch the graphs of

$$f(x) = 2x^3 - 1 \quad \text{and} \quad g(x) = \sqrt[3]{\frac{x+1}{2}}$$

on the same coordinate plane and show that the graph of g is the reflection of the graph of f in the line $y = x$.

SOLUTION

Using a graphing utility, you can obtain the graphs shown in Figure 1.72. From this figure you can see that the graph of g is the reflection of the graph of f in the line $y = x$, which graphically confirms that g is the inverse of f.

EXAMPLE 5   A Graph Reflecting Program

The following program will graph a function f *and* its reflection in the line $y = x$. (The program steps listed are for a Texas Instruments TI-81. Program steps for other programmable calculators are given in the appendix.)

```
:2Xmin/3→Ymin          :Lbl 1
:2Xmax/3→Ymax          :PT-On(Y₁,X)
:Xscl→Yscl             :X+I→X
:"X"→Y₂                :If X>Xmax
:DispGraph             :End
:(Xmax-Xmin)/95→I      :Goto 1
:Xmin→X
```

Use this program to graph the inverse of $f(x) = x^3 + x + 1$.

SOLUTION

To run the program, enter the function f into the calculator. Then set a viewing rectangle. (For the program to run, the viewing rectangle must not contain x-values outside the domain of f.) The resulting display is shown in Figure 1.73.

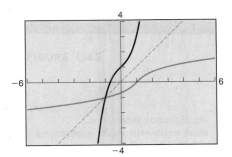

FIGURE 1.73

The Existence of an Inverse Function

A function need not have an inverse function. For instance, the function $f(x) = x^2$ has no inverse [assuming a domain of $(-\infty, \infty)$]. To have an inverse, a function must be **one-to-one,** which means that no two elements in the domain of f correspond to the same element in the range of f.

DEFINITION OF A ONE-TO-ONE FUNCTION

A function f is **one-to-one** if, for a and b in its domain,

$$f(a) = f(b) \quad \text{implies that} \quad a = b.$$

The function $f(x) = x + 1$ *is* one-to-one because $a + 1 = b + 1$ implies that a and b must be equal. However, the function $f(x) = x^2$ is *not* one-to-one because $a^2 = b^2$ does not imply that $a = b$. For instance, $(-1)^2 = 1^2$ and yet $-1 \neq 1$.

EXISTENCE OF AN INVERSE FUNCTION

A function f has an inverse function f^{-1} if and only if f is one-to-one.

EXAMPLE 6   Testing for One-to-One Functions

Which functions are one-to-one and have inverse functions?

A. $f(x) = x^3 + 1$ **B.** $g(x) = x^2 - x$ **C.** $h(x) = \sqrt{x}$

SOLUTION

A. Let a and b be real numbers with $f(a) = f(b)$.

$$a^3 + 1 = b^3 + 1 \qquad \textit{Set } f(a) = f(b)$$
$$a^3 = b^3$$
$$a = b$$

Therefore, $f(a) = f(b)$ implies that $a = b$. From this, it follows that f *is* one-to-one and has an inverse function.

B. Because $g(-1) = (-1)^2 - (-1) = 2$ and $g(2) = 2^2 - 2 = 2$, you have two distinct inputs matched with the same output. Thus, g *is not* a one-to-one function and has no inverse.

C. Let a and b be nonnegative real numbers with $h(a) = h(b)$.

$$\sqrt{a} = \sqrt{b} \qquad \textit{Set } h(a) = h(b)$$
$$a = b$$

Therefore, $h(a) = h(b)$ implies that $a = b$. Thus, h *is* one-to-one and has an inverse.

From its graph, it is easy to tell whether a function of x is one-to-one. Simply check to see that every *horizontal* line intersects the graph of the function at most once. For instance, Figure 1.74 shows the graphs of the three functions given in Example 6. On the graph of $g(x) = x^2 - x$, you can find a horizontal line that intersects the graph twice.

Two special types of functions that pass the **horizontal line test** are those that are increasing or decreasing on their entire domains.

1. If f is *increasing* on its entire domain, then f is one-to-one.
2. If f is *decreasing* on its entire domain, then f is one-to-one.

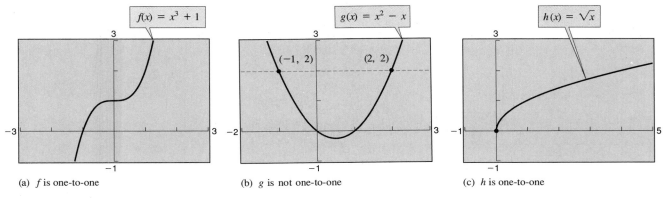

(a) f is one-to-one   (b) g is not one-to-one   (c) h is one-to-one

FIGURE 1.74

Finding the Inverse of a Function

For simple functions (such as the ones in Example 1) you can find inverse functions by inspection. For instance, the inverse of $f(x) = 8x$ is equal to $f^{-1}(x) = x/8$. For more complicated functions, however, it is best to use the following procedure for finding the inverse of a function.

Technology Note

Some graphing utilities can automatically draw the inverse of a function. For example, if $f(x) = 8x$ is entered into y1 on the TI-85, you can obtain the graph of f together with its inverse by using the DrInv command from the DRAW menu:

DrInv y1

   Try using this technique to verify the graphs in Examples 7 and 8 on the next page.

FINDING THE INVERSE OF A FUNCTION

To find the inverse of f, use the following steps.

1. In the equation for $f(x)$, replace $f(x)$ by y.
2. Interchange the roles of x and y.
3. If the new equation does not represent y as a function of x, then the function f does not have an inverse function. If the new equation does represent y as a function of x, then solve the new equation for y.
4. Replace y by $f^{-1}(x)$.
5. Verify that f and f^{-1} are inverses of each other by showing that $f(f^{-1}(x)) = x$ and $f^{-1}(f(x)) = x$.

EXAMPLE 7   Finding the Inverse of a Function

Find the inverse (if it exists) of

$$f(x) = \frac{5 - 3x}{2}.$$

SOLUTION

From Figure 1.75, you can see that f is one-to-one, and therefore has an inverse.

$$y = \frac{5 - 3x}{2}$$ *Write in form $y = f(x)$*

$$x = \frac{5 - 3y}{2}$$ *Interchange x and y*

$$2x = 5 - 3y$$

$$3y = 5 - 2x$$

$$y = \frac{5 - 2x}{3}$$ *Solve for y*

$$f^{-1}(x) = \frac{5 - 2x}{3}$$ *Replace y by $f^{-1}(x)$*

The domain and range of both f and f^{-1} consist of all real numbers.

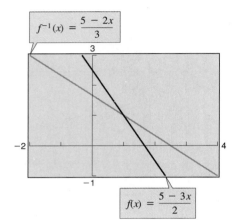

$$f^{-1}(x) = \frac{5 - 2x}{3}$$

$$f(x) = \frac{5 - 3x}{2}$$

FIGURE 1.75

EXAMPLE 8   Finding the Inverse of a Function

Find the inverse of the function $f(x) = \sqrt{2x - 3}$ and sketch the graphs of f and f^{-1}.

SOLUTION

$$y = \sqrt{2x - 3}$$

$$x = \sqrt{2y - 3}$$ *Interchange x and y*

$$x^2 = 2y - 3$$

$$2y = x^2 + 3$$ *Solve for y*

$$y = \frac{x^2 + 3}{2}$$ $x \geq 0$

$$f^{-1}(x) = \frac{x^2 + 3}{2}, \quad x \geq 0$$ *Replace y by $f^{-1}(x)$*

The graph of f^{-1} is the reflection of the graph of f in the line $y = x$, as shown in Figure 1.76. Note that the domain of f is the interval $[\frac{3}{2}, \infty)$ and the range of f if the interval $[0, \infty)$. Moreover, the domain of f^{-1} is the interval $[0, \infty)$ and the range of f^{-1} is the interval $[\frac{3}{2}, \infty)$.

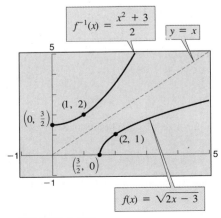

$$f^{-1}(x) = \frac{x^2 + 3}{2}$$ $y = x$

$(1, 2)$

$\left(0, \frac{3}{2}\right)$

$(2, 1)$

$\left(\frac{3}{2}, 0\right)$

$f(x) = \sqrt{2x - 3}$

FIGURE 1.76

The problem of finding the inverse of a function can be difficult (or even impossible) for two reasons. First, given $y = f(x)$, it may be algebraically difficult to solve for x in terms of y. Second, if f is not one-to-one, then f^{-1} does not exist.

DISCUSSION PROBLEM

································

THE EXISTENCE OF AN INVERSE FUNCTION

Write a short paragraph describing why the following functions do or do not have inverse functions.

(a) Let x represent the retail price of an item (in dollars), and let $f(x)$ represent the sales tax on the item. Assume that the sales tax is 7% of the retail price *and* that the sales tax is rounded to the nearest cent. Does this function have an inverse? (*Hint:* Can you undo this function? For instance, if you know that the sales tax is $0.14, can you determine *exactly* what the retail price is?)

(b) Let x represent the temperature in degrees Celsius, and let $f(x)$ represent the temperature in degrees Fahrenheit. Does this function have an inverse? (*Hint:* The formula for converting from degrees Celsius to degrees Fahrenheit is $F = \frac{9}{5}C + 32$.)

WARM-UP

················

The following warm-up exercises involve skills that were covered in earlier sections. You will use these skills in the exercise set for this section.

In Exercises 1–4, find the domain of the function.

1. $f(x) = \sqrt[3]{x + 1}$

2. $f(x) = \sqrt{x + 1}$

3. $g(x) = \dfrac{2}{x^2 - 2x}$

4. $h(x) = \dfrac{x}{3x + 5}$

In Exercises 5-8, simplify the expression.

5. $2\left(\dfrac{x + 5}{2}\right) - 5$

6. $7 - 10\left(\dfrac{7 - x}{10}\right)$

7. $\sqrt[3]{2\left(\dfrac{x^3}{2} - 2\right) + 4}$

8. $\left(\sqrt[5]{x + 2}\right)^5 - 2$

In Exercises 9 and 10, solve for x in terms of y.

9. $y = \dfrac{2x - 6}{3}$

10. $y = \sqrt[3]{2x - 4}$

SECTION 1.10 · EXERCISES

In Exercises 1–6, find the inverse f^{-1} of the function f informally. Verify that $f(f^{-1}(x))$ and $f^{-1}(f(x))$ are equal to the identity function.

1. $f(x) = 8x$

2. $f(x) = \frac{1}{5}x$

3. $f(x) = x + 10$

4. $f(x) = x - 5$

5. $f(x) = \sqrt[3]{x}$

6. $f(x) = x^5$

In Exercises 7–16, show that f and g are inverse functions (a) algebraically and (b) graphically.

7. $f(x) = 2x$, $\quad g(x) = \dfrac{x}{2}$

8. $f(x) = x - 5$, $\quad g(x) = x + 5$

9. $f(x) = 5x + 1$, $\quad g(x) = \dfrac{x - 1}{5}$

10. $f(x) = 3 - 4x$, $\quad g(x) = \dfrac{3 - x}{4}$

11. $f(x) = x^3$, $\quad g(x) = \sqrt[3]{x}$

12. $f(x) = \dfrac{1}{x}$, $\quad g(x) = \dfrac{1}{x}$

13. $f(x) = \sqrt{x - 4}$
$\quad g(x) = x^2 + 4$, $\quad x \geq 0$

14. $f(x) = 1 - x^3$
$\quad g(x) = \sqrt[3]{1 - x}$

15. $f(x) = 9 - x^2$, $\quad x \geq 0$
$\quad g(x) = \sqrt{9 - x}$, $\quad x \leq 9$

16. $f(x) = \dfrac{1}{1 + x}$, $\quad x \geq 0$
$\quad g(x) = \dfrac{1 - x}{x}$, $\quad 0 < x \leq 1$

In Exercises 17–22, use a graphing utility to graph the function and use the graph to determine whether the function is one-to-one.

17. $g(x) = \dfrac{4 - x}{6}$

18. $f(x) = 10$

19. $h(x) = |x + 4| - |x - 4|$

20. $g(x) = (x + 5)^3$

21. $f(x) = -2x\sqrt{16 - x^2}$

22. $f(x) = \frac{1}{8}(x + 2)^2 - 1$

In Exercises 23–32, find the inverse of the one-to-one function f. Then use a graphing utility to graph both f and f^{-1} in the same viewing rectangle.

23. $f(x) = 2x - 3$

24. $f(x) = 3x$

25. $f(x) = x^5$

26. $f(x) = x^3 + 1$

27. $f(x) = \sqrt{x}$

28. $f(x) = x^2$, $\quad x \geq 0$

29. $f(x) = \sqrt{4 - x^2}$, $\quad 0 \leq x \leq 2$

30. $f(x) = \dfrac{4}{x}$

31. $f(x) = \sqrt[3]{x - 1}$

32. $f(x) = x^{3/5}$

In Exercises 33–46, determine whether the given function is one-to-one. If it is, find its inverse.

33. $f(x) = x^4$

34. $f(x) = \dfrac{1}{x^2}$

35. $g(x) = \dfrac{x}{8}$

36. $f(x) = -3$

37. $f(x) = (x + 3)^2$, $\quad x \geq -3$

38. $q(x) = (x - 5)^2$

39. $h(x) = \dfrac{1}{x}$

40. $f(x) = |x - 2|$, $\quad x \leq 2$

41. $f(x) = \sqrt{2x + 3}$

42. $f(x) = \sqrt{x - 2}$

43. $g(x) = x^2 - x^4$

44. $f(x) = \dfrac{x^2}{x^2 + 1}$

45. $f(x) = 25 - x^2$, $\quad x \leq 0$

46. $f(x) = ax + b$, $\quad a \neq 0$

In Exercises 47–50, delete part of the graph of the function so that the part that remains is one-to-one. Find the inverse of the remaining part and give the domain of the inverse. (There is more than one correct answer.)

47. $f(x) = (x - 3)^2$

48. $f(x) = 16 - x^4$

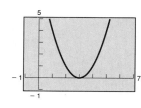

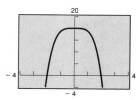

49. $f(x) = |x + 3|$ **50.** $f(x) = |x - 3|$

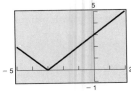

In Exercises 51 and 52, use the graph of the function f to complete the table and to graph f^{-1}.

51.

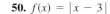

$(2,0)$ $(0,1)$ $(1,2)$ $(2,3)$ $(4,4)$

x	0	1	2	3	4
$f^{-1}(x)$					

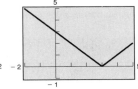

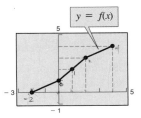

52.

x	0	2	4	6
$f^{-1}(x)$				

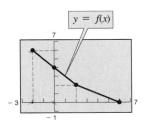

In Exercises 53–56, use the functions $f(x) = \frac{1}{8}x - 3$ and $g(x) = x^3$ to find the indicated value.

53. $(f^{-1} \circ g^{-1})(1)$ **54.** $(g^{-1} \circ f^{-1})(-3)$

55. $(f^{-1} \circ f^{-1})(6)$ **56.** $(g^{-1} \circ g^{-1})(-4)$

In Exercises 57–60, use the functions $f(x) = x + 4$ and $g(x) = 2x - 5$ to find the specified functions.

57. $g^{-1} \circ f^{-1}$ **58.** $f^{-1} \circ g^{-1}$

59. $(f \circ g)^{-1}$ **60.** $(g \circ f)^{-1}$

Diesel Engine   In Exercises 61 and 62, use the following function, which approximates the exhaust temperature y in degrees Fahrenheit

$$y = 0.03x^2 + 254.50, \quad 0 < x < 100,$$

where x is the percentage load for a diesel engine (see figure).

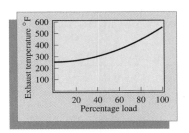

FIGURE

FOR 61 AND 62

61. (a) Find the inverse of the function and state what the variables x and y represent in the inverse function.
   (b) Graph the inverse function.

62. Determine the percentage load interval if the exhaust temperature of the engine must not exceed 500 degrees Fahrenheit.

In Exercises 63–66, determine whether the statement is true or false. Explain your reasoning.

63. If f is an even function, then f^{-1} exists.

64. If the inverse of f exists, then the y-intercept of f is an x-intercept of f^{-1}.

65. If $f(x) = x^n$ where n is odd, then f^{-1} exists.

66. There exists no function f such that $f = f^{-1}$.

CHAPTER 1 · REVIEW EXERCISES

In Exercises 1 and 2, determine which numbers in the given set are (a) natural numbers, (b) integers, (c) rational numbers, and (d) irrational numbers.

1. $11, -14, -\frac{8}{9}, \frac{5}{2}, \sqrt{6}, 0.4$

2. $\sqrt{15}, -22, -\frac{10}{3}, 0, 5.2, \frac{3}{7}$

In Exercises 3 and 4, plot the two real numbers on the real number line and place the appropriate sign ($<$ or $>$) between them.

3. $-4, -3$ **4.** $\frac{1}{2}, \frac{1}{3}$

In Exercises 5 and 6, give a verbal description of the subset of real numbers that is represented by the inequality, and sketch the subset on the real number line.

5. $x \le 7$ **6.** $x > 1$

In Exercises 7 and 8, write the expression without using absolute value signs.

7. $-|-14|$ **8.** $|-4 - 2|$

In Exercises 9 and 10, determine whether the value of x is a solution of the equation.

Equation	Values
9. $3x^2 + 7x + 5 = x^2 + 9$	(a) $x = 0$ (b) $x = -4$ (c) $x = \frac{1}{2}$ (d) $x = -1$
10. $6 + \dfrac{3}{x - 4} = 5$	(a) $x = 4$ (b) $x = 0$ (c) $x = -2$ (d) $x = 1$

In Exercises 11–22, solve the equation (if possible) and check your answer.

11. $3x - 2(x + 5) = 10$ **12.** $4x + 2(7 - x) = 5$

13. $3\left(1 - \dfrac{1}{5t}\right) = 0$ **14.** $\dfrac{1}{x - 2} = 3$

15. $6x = 3x^2$ **16.** $3x^2 + 1 = 0$

17. $x^2 - 12x + 30 = 0$ **18.** $x^2 + 6x - 3 = 0$

19. $5x^4 - 12x^3 = 0$ **20.** $4x^3 - 6x^2 = 0$

21. $4t^3 - 12t^2 + 8t = 0$ ✳**22.** $12t^3 - 84t^2 + 120t = 0$

In Exercises 23–26, find (a) the distance between the two points, (b) the coordinates of the midpoint of the line segment between the two points, and (c) an equation of the circle whose diameter is the line segment between the two points. *P.23*

23. $(0, 0)$, $(0, 10)$

24. $(-1, 4)$, $(2, 0)$

25. $(2, 1)$, $(14, 6)$

✳ **26.** $(-2, 2)$, $(3, -10)$

In Exercises 27 and 28, use the Midpoint Formula to estimate the sales of a company for 1991 given the sales in 1989 and 1993. Assume the sales followed a linear growth pattern.

27.

Year	1989	1993
Sales	640,000	810,000

28.

Year	1989	1993
Sales	3,250,000	5,690,000

In Exercises 29 and 30, plot the points and show that they are the vertices of the indicated polygon.

29. Parallelogram: $(1, 1)$, $(8, 2)$, $(9, 5)$, $(2, 4)$

30. Isosceles triangle: $(4, 5)$, $(1, 0)$, $(-1, 2)$

In Exercises 31–36, find the intercepts of the given graph and check for symmetry with respect to each of the coordinate axes and the origin.

31. $2y^2 = x^3$ **32.** $x^2 + (y + 2)^2 = 4$

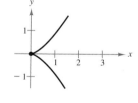
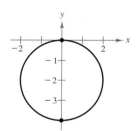

33. $y = \frac{1}{4}x^4 - 2x^2$ **34.** $y = \frac{1}{4}x^3 - 3x$

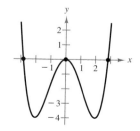
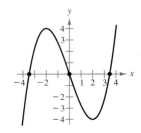

35. $y = x\sqrt{4 - x^2}$ **36.** $y = x\sqrt{x + 3}$

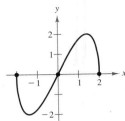

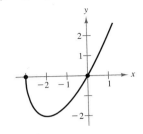

In Exercises 37–40, determine the center and radius of the circle, and sketch its graph.

37. $x^2 + y^2 - 12x - 8y + 43 = 0$

38. $x^2 + y^2 - 20x - 10y + 100 = 0$

39. $4x^2 + 4y^2 - 4x - 40y + 92 = 0$

40. $5x^2 + 5y^2 - 14y = 0$

In Exercises 41–54, sketch a graph of the equation *by hand*. Verify your result with a graphing utility.

41. $y - 2x - 3 = 0$ **42.** $3x + 2y + 6 = 0$

43. $x - 5 = 0$ **44.** $y = 8 - |x|$

45. $y = \sqrt{5 - x}$ **46.** $y = \sqrt{x + 2}$

47. $y + 2x^2 = 0$ **48.** $y = x^2 - 4x$

49. $y = \sqrt{25 - x^2}$ **50.** $y = -(x - 4)^2$

51. $y = \dfrac{2 + x}{1 - x}$ **52.** $y = \dfrac{2x}{x^2 + 4}$

53. $y = \dfrac{x}{x^2 - 1}$ **54.** $y = \dfrac{2x^2}{x^2 - 4}$

In Exercises 55–62, use a graphing utility to graph the equation. Determine the number of times (if any) the graph intersects the coordinate axes.

55. $y = \frac{1}{4}(x + 1)^3$ **56.** $y = 4 - (x - 4)^2$

57. $y = \frac{1}{4}x^4 - 2x^2$ **58.** $y = \frac{1}{4}x^3 - 3x$

59. $y = x\sqrt{9 - x^2}$ **60.** $y = x\sqrt{x + 3}$

61. $y = |x - 4| - 4$ **62.** $y = |x + 2| + |3 - x|$

In Exercises 63 and 64, solve for y and use a graphing utility to graph each of the resulting equations on the same viewing rectangle.

63. $2y^2 = x^3$ **64.** $(x + 2)^2 + y^2 = 16$

In Exercises 65–68, use the concept of slope to find t so that the three points are collinear.

65. $(-2, 5)$, $(0, t)$, $(1, 1)$ **66.** $(-6, 1)$, $(1, t)$, $(10, 5)$

67. $(1, -4)$, $(t, 3)$, $(5, 10)$ **68.** $(-3, 3)$, $(t, -1)$, $(8, 6)$

In Exercises 69–74, find an equation of the line through the two points.

69. $(0, 0)$, $(0, 10)$ **70.** $(-1, 4)$, $(2, 0)$

71. $(2, 1)$, $(14, 6)$ **72.** $(-2, 2)$, $(3, -10)$

73. $(-1, 0)$, $(6, 2)$ **74.** $(1, 6)$, $(4, 2)$

In Exercises 75–78, find an equation of the line that passes through the given point and has the specified slope. Sketch the graph of the line *by hand*.

Point	Slope		Point	Slope
75. $(0, -5)$	$m = \frac{3}{2}$		**76.** $(-2, 6)$	$m = 0$
77. $(3, 0)$	$m = -\frac{2}{3}$		**78.** $(5, 4)$	m is undefined

In Exercises 79 and 80, evaluate the function at the specified values of the independent variable. Simplify your answers.

79. $f(x) = x^2 + 1$
(a) $f(2)$ (b) $f(-4)$
(c) $f(t^2)$ (d) $-f(x)$

80. $h(x) = 6 - 5x^2$
(a) $h(2)$ (b) $h(x + 3)$
(c) $\dfrac{h(4) - h(2)}{4 - 2}$ (d) $h(x + 2) - h(x)$

81. *Fourth Quarter Sales*   During the second and third quarters of the year, a business had sales of $160,000 and $185,000, respectively. If the growth of sales follows a linear pattern, estimate sales during the fourth quarter.

82. *Dollar Value*   The dollar value of a product in 1990 is $85, and the item will increase in value at an expected rate of $3.75 per year.
(a) Write a linear equation that gives the dollar value V of the product in terms of the year t. (Let $t = 0$ represent 1990).
(b) Use a graphing utility to graph the sales equation.
(c) Move the cursor along the graph of the sales model to estimate the dollar value of the product in 1995.

In Exercises 83–88, determine the domain of the function. Verify your result with a graphing utility.

83. $f(x) = \sqrt{25 - x^2}$ **84.** $(x) = 3x + 4$

85. $g(s) = \dfrac{5}{3s - 9}$ ✳**86.** $f(x) = \sqrt{x^2 + 8x}$

87. $h(x) = \dfrac{x}{x^2 - x - 6}$ **88.** $h(t) = |t + 1|$

89. Sketch (on the same set of coordinate axes) a graph of f for $c = -2$, 0, and 2.
   (a) $f(x) = \frac{1}{2}x + c$ (b) $f(x) = \frac{1}{2}(x - c)$
   (c) $f(x) = \frac{1}{2}(cx)$

✳**90.** Sketch (on the same set of coordinate axes) a graph of f for $c = -2$, 0, and 2.
   (a) $f(x) = x^3 + c$ (b) $f(x) = (x - c)^3$
   (c) $f(x) = (x - 2)^3 + c$

In Exercises 91–96, (a) find f^{-1}, (b) use a graphing utility to graph both f and f^{-1} on the same viewing rectangle, and (c) verify that $f^{-1}(f(x)) = x$ and $f(f^{-1}(x)) = x$.

91. $f(x) = \frac{1}{2}x - 3$ **92.** $f(x) = 5x - 7$
93. $f(x) = \sqrt{x + 1}$ ✳**94.** $f(x) = x^3 + 2$
95. $f(x) = x^2 - 5,\ x \geq 0$
✳**96.** $f(x) = \sqrt[3]{x + 1}$

In Exercises 97–100, restrict the domain of the function f to an interval where the function is increasing and determine f^{-1} over that interval. Then use a graphing utility to graph both f and f^{-1} on the same viewing rectangle.

97. $f(x) = 2(x - 4)^2$ **98.** $f(x) = |x - 2|$
99. $f(x) = \sqrt{x^2 - 4}$ **100.** $f(x) = x^{4/3}$

In Exercises 101–108, let $f(x) = 3 - 2x$, $g(x) = \sqrt{x}$, and $h(x) = 3x^2 + 2$ and find the indicated value.

101. $(f - g)(4)$ **102.** $(f + h)(5)$

103. $(fh)(1)$ **104.** $\left(\dfrac{g}{h}\right)(1)$

105. $(h \circ g)(7)$ ✳**106.** $(g \circ f)(-2)$
107. $g^{-1}(3)$ ✳**108.** $(h \circ f^{-1})(1)$

109. *Monthly Profit*   The total profit for a company in October was 12% more than it was in September. The total profit for the two months was $689,000. Find the profit for each month.

110. *Starting Position*   A fitness center has two running tracks around a rectangular playing floor. The tracks are 3 feet wide and form semicircles at the narrow ends of the rectangular floor (see figure). Determine the distance between the starting positions if two runners must run the same distance to the finish line in one lap around the track.

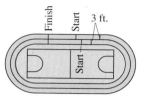

FIGURE FOR 110

111. *Cost Sharing*   A group of farmers agree to share equally in the cost of a $48,000 piece of machinery. If they could find two more farmers to join the group, each person's share of the cost would decrease by $4000. How many farmers are presently in the group?

112. *Average Speed*   Each week you must make a 180-mile trip to pick up supplies for your business. If you were to increase your average speed by 5 miles per hour, the trip would take 24 minutes less than usual. Find the usual average speed.

✳**113.** *Vertical Motion*   The velocity of a ball thrown vertically upward from ground level is given by

$$v(t) = -32t + 48,$$

where t is the time in seconds and v is the velocity in feet per second.
   (a) Find the velocity when $t = 1$.
   (b) Find the time when the ball reaches its maximum height. [*Hint:* Find the time when $v(t) = 0$.]
   (c) Find the velocity when $t = 2$.

114. *Cost and Profit*   A company produces a product for which the variable cost is $5.35 per unit and the fixed costs are $16,000. The company sells the product for $8.20, and can sell all that it produces.
   (a) Find the total cost as a function of x, the number of units produced.
   (b) Find the profit as a function of x.

115. *Dimensions of a Rectangle*   A wire 24 inches long is to be cut into four pieces to form a rectangle whose shortest side has a length of x. Express the area A of the rectangle as a function of x. Determine the domain of the function and sketch its graph over that domain.

CHAPTER 2

TRIGONOMETRY

2.1 RADIAN AND DEGREE MEASURE

Angles / Radian Measure / Degree Measure / Applications

Angles

As derived from the Greek language, the word **trigonometry** means "measurement of triangles." Initially, trigonometry dealt with relationships among the sides and angles of triangles. As such, it was used in the development of astronomy, navigation, and surveying.

With the advent of calculus in the 17th century and a resulting expansion of knowledge in the physical sciences, a different perspective arose—one that viewed the classic trigonometric relationships as *functions* with the set of real numbers as their domains. Consequently, the applications of trigonometry expanded to include a vast number of physical phenomena involving rotations or vibrations. These include sound waves, light rays, planetary orbits, vibrating strings, pendulums, and orbits of atomic particles. Our approach to trigonometry incorporates *both* perspectives, starting with angles and their measure.

An **angle** is determined by rotating a ray (half-line) about its endpoint. The starting position of the ray is called the **initial side** of the angle, and the position after rotation is called the **terminal side,** as shown in Figure 2.1. The endpoint of the ray is called the **vertex** of the angle.

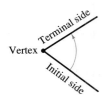

FIGURE 2.1

105

In a coordinate system, an angle is in **standard position** if its vertex is the origin and its initial side coincides with the positive *x*-axis, as shown in Figure 2.2. **Positive angles** are generated by counterclockwise rotation, and **negative angles** by clockwise rotation, as shown in Figures 2.3 and 2.4. To label angles in trigonometry, we use the Greek letters α (alpha), β (beta), and θ (theta), as well as uppercase letters *A*, *B*, and *C*. In Figure 2.4, note that the angles α and β have the same initial and terminal sides. Such angles are **coterminal.**

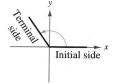

FIGURE 2.2

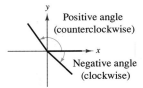

FIGURE 2.3

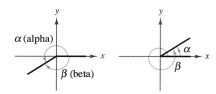

Coterminal Angles

FIGURE 2.4

Radian Measure

The **measure of an angle** is determined by the amount of rotation from the initial to the terminal side. One way to measure angles is in radians. This type of measure is needed in calculus. To define a radian we use a **central angle** of a circle, one whose vertex is the center of the circle, as shown in Figure 2.5.

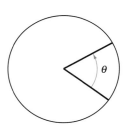

Central Angle θ

FIGURE 2.5

One **radian** is the measure of a central angle θ that subtends (intercepts) an arc *s* equal in length to the radius *r* of the circle (see figure).

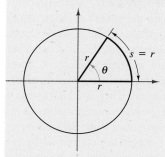

arc length = radius when θ = 1 radian

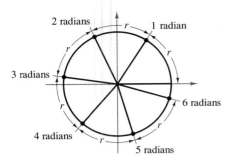

FIGURE 2.6

Because the circumference of a circle is $2\pi r$, it follows that a central angle of one full revolution (counterclockwise) corresponds to an arc length of $s = 2\pi r$. Moreover, because each radian intercepts an arc of length r, we conclude that one full revolution corresponds to an angle of

$$\frac{2\pi r}{r} = 2\pi \text{ radians.}$$

Note that since $2\pi \approx 6.28$, there are a little more than six radius lengths in a full circle, as shown in Figure 2.6.

In general, the radian measure of a central angle θ is obtained by dividing the arc length s by r. That is,

$$\frac{s}{r} = \theta, \qquad\qquad \textit{Radian measure}$$

where θ *is measured in radians*. Because the units of measure for s and r are the same, this ratio is unitless—it is simply a real number. The radian measures of several common angles are shown in Figure 2.7.

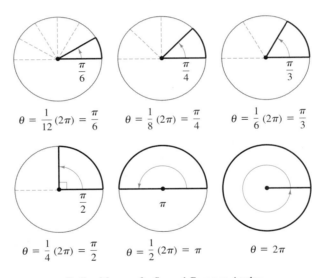

$$\theta = \frac{1}{12}(2\pi) = \frac{\pi}{6} \qquad \theta = \frac{1}{8}(2\pi) = \frac{\pi}{4} \qquad \theta = \frac{1}{6}(2\pi) = \frac{\pi}{3}$$

$$\theta = \frac{1}{4}(2\pi) = \frac{\pi}{2} \qquad \theta = \frac{1}{2}(2\pi) = \pi \qquad \theta = 2\pi$$

Radian Measure for Several Common Angles

FIGURE 2.7

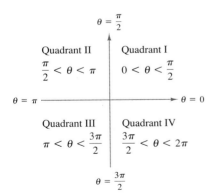

FIGURE 2.8

Recall that the four quadrants in a coordinate system are numbered counterclockwise as I, II, III, and IV. Figure 2.8 shows which angles between 0 and 2π lie in each of the four quadrants.

You can find an angle that is coterminal to a given angle θ by adding or subtracting 2π (one revolution), as demonstrated in Example 1. (Note that a given angle has many coterminal angles. For instance, $\theta = \pi/6$ is coterminal with both $13\pi/6$ and $-11\pi/6$.)

EXAMPLE 1   Sketching and Finding Coterminal Angles
· · · · · · · · · · · · · · · · · ·

A. To find an angle that is coterminal to the positive angle $\theta = 13\pi/6$, you can subtract 2π to obtain

$$\frac{13\pi}{6} - 2\pi = \frac{\pi}{6}.$$

Thus, the terminal side of θ lies in Quadrant I. Its sketch is shown in Figure 2.9(a).

B. To find an angle that is coterminal to the positive angle $\theta = 3\pi/4$, you can subtract 2π to obtain

$$\frac{3\pi}{4} - 2\pi = -\frac{5\pi}{4},$$

as shown in Figure 2.9(b).

C. To find an angle that is coterminal to the negative angle $\theta = -2\pi/3$, you can add 2π to obtain

$$\theta = -\frac{2\pi}{3} + 2\pi = \frac{4\pi}{3},$$

as shown in Figure 2.9(c).

REMARK   The phrase "the terminal side of θ lies in a quadrant" can be abbreviated by simply saying that "θ lies in a quadrant." The terminal sides of the quadrant angles 0, $\pi/2$, π, and $3\pi/2$ do not lie within any of the four quadrants.

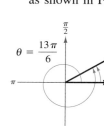
(a)

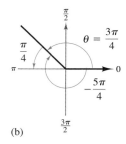
(b)

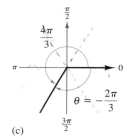
(c)

FIGURE 2.9

Figure 2.10 shows several common angles with their radian measures. Note that we classify angles between 0 and $\pi/2$ radians as **acute** and angles between $\pi/2$ and π as **obtuse.**

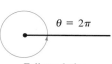

Acute angle:
between 0 and $\dfrac{\pi}{2}$

Right angle:
quarter revolution

$\theta = \dfrac{3\pi}{4}$

Obtuse angle:
between $\dfrac{\pi}{2}$ and π

$\theta = \pi$

Straight angle:
half revolution

$\theta = 2\pi$

Full revolution

FIGURE 2.10

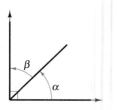

(a) Complementary angles

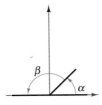

(b) Supplementary angles

FIGURE 2.11

Two *positive* angles α and β are said to be **complementary** (or complements of each other) if their sum is $\pi/2$, as shown in Figure 2.11(a). For example, $\pi/6$ and $\pi/3$ are complementary angles because

$$\frac{\pi}{6} + \frac{\pi}{3} = \frac{\pi}{2}.$$

Two positive angles are **supplementary** (or supplements of each other) if their sum is π, as shown in Figure 2.11(b). For example, $2\pi/3$ and $\pi/3$ are supplementary angles because

$$\frac{2\pi}{3} + \frac{\pi}{3} = \pi.$$

EXAMPLE 2   Complementary, Supplementary, and Coterminal Angles

A. The complement of $\theta = \pi/12$ is

$$\frac{\pi}{2} - \frac{\pi}{12} = \frac{6\pi}{12} - \frac{\pi}{12} = \frac{5\pi}{12},$$

as shown in Figure 2.12(a).

B. The supplement of $\theta = 5\pi/6$ is

$$\pi - \frac{5\pi}{6} = \frac{6\pi}{6} - \frac{5\pi}{6} = \frac{\pi}{6},$$

as shown in Figure 2.12(b)

C. In radian measure, a coterminal angle is found by adding or subtracting 2π. For $\theta = 17\pi/6$, you subtract 2π to obtain

$$\frac{17\pi}{6} - 2\pi = \frac{17\pi}{6} - \frac{12\pi}{6} = \frac{5\pi}{6},$$

as shown in Figure 2.12(c). Thus, $17\pi/6$ and $5\pi/6$ are coterminal.

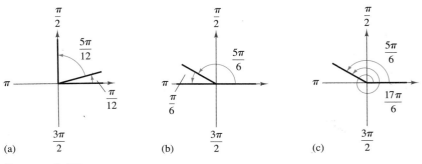

FIGURE 2.12

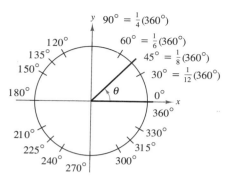

FIGURE 2.13

Degree Measure

A second way to measure angles is in terms of degrees. A measure of **one degree (1°)** is equivalent to $1/360$ of a complete revolution about the vertex. To measure angles in degrees, it is convenient to mark degrees on the circumference of a circle as shown in Figure 2.13. Thus, a full revolution (counter clockwise) corresponds to 360°, a half revolution to 180°, and a quarter revolution to 90°.

Because 2π radians is the measure of an angle of one complete revolution, degrees and radians are related by equations

$$360° = 2\pi \text{ rad} \quad \text{and} \quad 180° = \pi \text{ rad}.$$

From the latter equation, we obtain

$$1° = \frac{\pi}{180} \text{ rad} \quad \text{and} \quad 1 \text{ rad} = \left(\frac{180}{\pi}\right)°,$$

which lead to the following conversion rules.

REMARK Note that when no units of angle measure are specified, *radian measure is implied*. For instance, if we write $\theta = \pi$ or $\theta = 2$, we mean $\theta = \pi$ radians or $\theta = 2$ radians.

CONVERSIONS: DEGREES ↔ RADIANS

1. To convert degrees to radians, multiply degrees by $\dfrac{\pi \text{ rad}}{180°}$.

2. To convert radians to degrees, multiply radians by $\dfrac{180°}{\pi \text{ rad}}$.

To apply these two conversion rules, use the basic relationship π rad $= 180°$.

EXAMPLE 3　Converting from Degrees to Radians

A. $135° = (135 \text{ deg})\left(\dfrac{\pi \text{ rad}}{180 \text{ deg}}\right)$ *Multiply by $\pi/180$*

    $= \dfrac{3\pi}{4} \text{ rad}$

B. $540° = (540 \text{ deg})\left(\dfrac{\pi \text{ rad}}{180 \text{ deg}}\right)$ *Multiply by $\pi/180$*

    $= 3\pi \text{ rad}$

C. $-270° = (-270 \text{ deg})\left(\dfrac{\pi \text{ rad}}{180 \text{ deg}}\right)$ *Multiply by $\pi/180$*

    $= -\dfrac{3\pi}{2} \text{ rad}$

EXAMPLE 4   Converting from Radians to Degrees

A. $-\dfrac{\pi}{2} \text{ rad} = \left(-\dfrac{\pi}{2}\,\text{rad}\right)\left(\dfrac{180 \text{ deg}}{\pi \,\text{rad}}\right)$ *Multiply by* $180/\pi$

$\qquad\qquad = -90°$

B. $\dfrac{9\pi}{2} \text{ rad} = \left(\dfrac{9\pi}{2}\,\text{rad}\right)\left(\dfrac{180 \text{ deg}}{\pi \,\text{rad}}\right)$ *Multiply by* $180/\pi$

$\qquad\qquad = 810°$

C. $2 \text{ rad} = (2 \,\text{rad})\left(\dfrac{180 \text{ deg}}{\pi \,\text{rad}}\right) = \dfrac{360}{\pi} \text{ deg}$ *Multiply by* $180/\pi$

$\qquad\quad \approx 144.59°$

With calculators it is convenient to use *decimal* degrees to denote fractional parts of degrees. Historically, however, fractional parts of degrees were expressed in *minutes* and *seconds,* using the prime (′) and double prime (″) notations, respectively. That is,

$$1' = \text{one minute} = \frac{1}{60}(1°)$$

$$1'' = \text{one second} = \frac{1}{60}(1') = \frac{1}{3600}(1°).$$

Consequently, an angle of 64 degrees, 32 minutes, and 47 seconds is represented by $\theta = 64°\ 32'\ 47''$.

Many calculators have special keys for converting an angle in degrees, minutes, and seconds $(D°\ M'\ S'')$ into decimal degree form, and conversely. If your calculator does not have these special keys, you can use the techniques demonstrated in the next example to make the conversions.

EXAMPLE 5   Converting an Angle from $D°\ M'\ S''$ to Decimal Form

Convert $152°\ 15'\ 29''$ to decimal degree form.

SOLUTION

$$152°\ 15'\ 29'' = 152° + \left(\frac{15}{60}\right)° + \left(\frac{29}{3600}\right)°$$

$$\approx 152° + 0.25° + 0.00806°$$

$$= 152.25806°.$$

Applications

The *radian measure* formula, $\theta = s/r$, can be used to measure arc length along a circle. Specifically, for a circle of radius r, a central angle θ subtends an arc of length s given by

$$s = r\theta, \qquad\qquad \textit{Length of circular arc}$$

where θ is measured in radians.

EXAMPLE 6   Finding Arc Length

A circle has a radius of 4 inches. Find the length of the arc cut off (subtended) by a central angle of 240°, as shown in Figure 2.14.

SOLUTION

To use the formula $s = r\theta$, we must first convert 240° to radian measure.

$$240° = (240 \text{ deg})\left(\frac{\pi \text{ rad}}{180 \text{ deg}}\right) = \frac{4\pi}{3} \text{ rad}$$

Then, using a radius of $r = 4$ inches, we find the arc length to be

$$s = r\theta$$
$$= 4\left(\frac{4\pi}{3}\right)$$
$$= \frac{16\pi}{3}$$
$$\approx 16.76 \text{ inches.}$$

Note that the units for $r\theta$ are determined by the units for r because θ has no units.

The formula for the length of a circular arc can be used to analyze the motion of a particle moving at a *constant speed* along a circular path. Assume the particle is moving at a constant speed along a circular path (of radius r). If s is the length of the arc traveled in time t, then we say that the **speed** of the particle is

$$\text{Speed} = \frac{\text{distance}}{\text{time}} = \frac{s}{t}.$$

Moreover, if θ is the angle (in radian measure) corresponding to the arc length s, then the **angular speed** of the particle is

$$\text{Angular speed} = \frac{\theta}{t}.$$

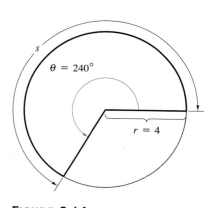

$\theta = 240°$

$r = 4$

FIGURE 2.14

EXAMPLE 7   Finding the Speed of an Object

The second hand on a clock is 4 inches long, as shown in Figure 2.15. Find the speed of the tip of this second hand.

SOLUTION

The time required for the second hand to make one full revolution is

$$t = 60 \text{ seconds} = 1 \text{ minute.}$$

The distance traveled by the tip of the second hand in one revolution is

$$s = 2\pi(\text{radius}) = 2\pi(4) = 8\pi \text{ inches.}$$

Therefore, the speed of the tip of the second hand is

$$\text{Speed} = \frac{s}{t} = \frac{8\pi \text{ inches}}{60 \text{ seconds}} \approx 0.419 \text{ in./sec.}$$

FIGURE 2.15

EXAMPLE 8   Finding Angular Speed and Linear Speed

A lawn roller that is 30 inches in diameter makes one revolution every $\frac{5}{6}$ second, as shown in Figure 2.16.

A. Find the angular speed of the roller in radians per second.
B. How fast is the roller moving across the lawn?

SOLUTION

A. Because there are 2π radians in one revolution, it follows that the angular speed is

$$\text{Angular speed} = \frac{\theta}{t}$$

$$= \frac{2\pi \text{ rad}}{\frac{5}{6} \text{ sec}}$$

$$= 2.4\pi \text{ rad/sec.}$$

B. Because the diameter is 30 inches, $r = 15$ and $s = 2\pi r = 30\pi$ inches. Thus,

$$\text{Speed} = \frac{s}{t}$$

$$= \frac{30\pi \text{ in.}}{\frac{5}{6} \text{ sec}}$$

$$= 36\pi \text{ in./sec}$$

$$\approx 113.1 \text{ in./sec.}$$

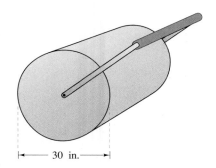

|←——— 30 in. ———→|

FIGURE 2.16

DISCUSSION PROBLEM

AN ANGLE-DRAWING PROGRAM

If you have a programmable calculator, try entering an "angle-drawing program." Then use the program to draw several different angles. The following program is for the TI-81. Programs for other calculators are given in Appendix B.

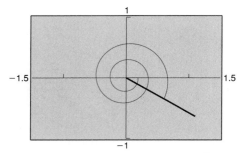

```
:Disp "ENTER,MODE"
:Disp "0,RADIAN"
:Disp "1,DEGREE"
:Input M
:Disp "ENTER,ANGLE"
:Input T
:If M=1
:πT/180→T
:Rad
:ClrDraw
:All-Off
:-1.5→Xmin
:1.5→Xmax
:1→Xscl
:-1→Ymin
:1→Ymax
:1→Yscl
```

```
:0→Tmin
:abs T→Tmax
:.15→Tstep
:cos T→A
:sin T→B
:Param
:1→S
:If T<0
:-1→S
:"(.25+.04T)cos T"→X₁ₜ
:"S(.25+.04T)sin T"→Y₁ₜ
:DispGraph
:Line(0,0,A,B)
:Pause
:Function
:End
```

To run this program, enter 0 for radian mode or 1 for degree mode. Then enter any angle (the angle can be negative or larger than 360°). After the angle is displayed, press ENTER to clear the display. The figure shows an angle of −750°, as drawn by this program.

WARM-UP

The following warm-up exercises involve skills that were covered in earlier sections. You will use these skills in the exercise set for this section.

In Exercises 1–10, solve for x.

1. $x + 135 = 180$

2. $790 = 720 + x$

3. $\pi = \dfrac{5\pi}{6} + x$

4. $2\pi - x = \dfrac{5\pi}{3}$

5. $\dfrac{45}{180} = \dfrac{x}{\pi}$

6. $\dfrac{240}{180} = \dfrac{x}{\pi}$

7. $\dfrac{\pi}{180} = \dfrac{x}{20}$

8. $\dfrac{180}{\pi} = \dfrac{330}{x}$

9. $\dfrac{x}{60} = \dfrac{3}{4}$

10. $\dfrac{x}{3600} = 0.0125$

SECTION 2.1 · EXERCISES

In Exercises 1–4, determine the quadrant in which the terminal side of the angle lies. (The angle is given in radians.)

1. (a) $\dfrac{\pi}{5}$ (b) $\dfrac{7\pi}{5}$

2. (a) $-\dfrac{\pi}{12}$ (b) $-\dfrac{11\pi}{9}$

3. (a) -1 (b) -2

4. (a) 5.63 (b) -2.25

In Exercises 5 and 6, determine the quadrant in which the terminal side of the angle lies.

5. (a) $130°$ (b) $285°$
6. (a) $-260°$ (b) $-3.4°$

In Exercises 7–10, sketch the angle in standard position.

7. (a) $\dfrac{5\pi}{4}$
(b) $\dfrac{2\pi}{3}$

8. (a) $-\dfrac{7\pi}{4}$
(b) $-\dfrac{5\pi}{2}$

9. (a) $30°$
(b) $150°$

10. (a) $405°$
(b) $-480°$

In Exercises 11 and 12, determine two coterminal angles (one positive and one negative) for the angle. Give your results in radians.

11. (a) $\theta = \dfrac{\pi}{9}$ (b) $\theta = \dfrac{4\pi}{3}$

 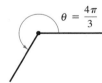

12. (a) $\theta = -\dfrac{9\pi}{4}$ (b) $\theta = -\dfrac{2\pi}{15}$

In Exercises 13–16, determine two coterminal angles (one positive and one negative) for the angle. Give your results in degrees.

13. (a) (b)

 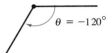

$\theta = 36°$ $\theta = -45°$

14. (a) $\theta = -120°$ (b) $\theta = 390°$

15. (a) $\theta = 300°$ (b) $\theta = 740°$

16. (a) $\theta = -420°$ (b) $\theta = 230°$

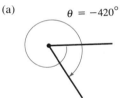 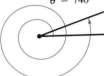

In Exercises 17–20, find (if possible) the positive angle complement and the positive angle supplement of the angle.

17. (a) $\dfrac{\pi}{3}$ (b) $\dfrac{3\pi}{4}$
18. (a) 1 (b) 2
19. (a) $18°$ (b) $115°$
20. (a) $79°$ (b) $150°$

In Exercises 21–24, express the angle in degree measure. (Do not use a calculator.)

21. (a) $\dfrac{3\pi}{2}$ (b) $\dfrac{7\pi}{6}$
22. (a) $-\dfrac{7\pi}{12}$ (b) $\dfrac{\pi}{9}$
23. (a) $\dfrac{7\pi}{3}$ (b) $-\dfrac{11\pi}{30}$
24. (a) $\dfrac{11\pi}{6}$ (b) $\dfrac{34\pi}{15}$

In Exercises 25–28, express the angle in radian measure as a multiple of π. (Do not use a calculator.)

25. (a) 30°   (b) 150°
26. (a) 315°   (b) 120°
27. (a) −20°   (b) −240°
28. (a) −270°   (b) 144°

In Exercises 29–32, convert the angle from degrees to radian measure. Express your result to three decimal places.

29. (a) 115°   (b) 87.4°
30. (a) −216.35°   (b) −48.27°
31. (a) 532°   (b) 0.54°
32. (a) −0.83°   (b) 345°

In Exercises 33–36, convert the angle from radian to degree measure. Express your result to three decimal places.

33. (a) $\dfrac{\pi}{7}$ (b) $\dfrac{5\pi}{11}$

34. (a) $\dfrac{15\pi}{8}$ (b) 6.5π

35. (a) -4.2π (b) 4.8
36. (a) −2   (b) −0.57

In Exercises 37 and 38, convert the angle measurement to decimal form.

37. (a) 245° 10′   (b) 2° 12′
38. (a) −135° 36″   (b) −408° 16′ 25″

In Exercises 39–42, convert the angle measurement to $D° M′ S″$ form.

39. (a) 240.6°   (b) −145.8°
40. (a) −345.12°   (b) 0.45
41. (a) 2.5   (b) −3.58
42. (a) −0.355   (b) 0.7865

In Exercises 43–46, find the radian measure of the central angle of a circle of radius r that intercepts an arc of length s.

Radius	Arc Length
43. 15 inches	4 inches
44. 16 feet	10 feet
45. 14.5 centimeters	25 centimeters
46. 80 kilometers	160 kilometers

In Exercises 47–50, on the circle of radius r find the length of the arc intercepted by the central angle θ.

Radius	Central Angle
47. 15 inches	180°
48. 9 feet	60°
49. 6 meters	2 radians
50. 40 centimeters	$\dfrac{3\pi}{4}$ radians

Distance Between Cities   In Exercises 51–54, find the distance between the two cities. Assume that the earth is a sphere of radius 4000 miles and that the cities are on the same meridian (one city is due north of the other).

City	Latitude
51. Dallas	32° 47′ 9″ N
Omaha	41° 15′ 42″ N
52. San Francisco	37° 46′ 39″ N
Seattle	47° 36′ 32″ N
53. Miami	25° 46′ 37″ N
Erie	42° 7′ 15″ N
54. Johannesburg, South Africa	26° 10′ S
Jerusalem, Israel	31° 47′ N

55. *Difference in Latitudes*   Assuming that the earth is a sphere of radius 4000 miles, what is the difference in latitude of two cities, one of which is 325 miles due north of the other?

56. *Difference in Latitudes*   Assuming that the earth is a sphere of radius 4000 miles, what is the difference in latitude of two cities, one of which is 500 miles due north of the other?

57. *Instrumentation* The pointer on a voltmeter is 2 inches long (see figure). Find the angle through which the pointer rotates when it moves $\frac{1}{2}$ inch on the scale.

FIGURE FOR 57

58. *Electric Hoist* An electric hoist is used to lift a piece of equipment (see figure). The diameter of the drum on the hoist is 8 inches and the equipment must be raised 1 foot. Find the number of degrees through which the drum must rotate.

FIGURE FOR 58

59. *Angular Speed* A car is moving at the rate of 50 miles per hour, and the diameter of each wheel is 2.5 feet. (a) Find the number of revolutions per minute of the rotating wheels. (b) Find the angular speed of the wheels in radians per minute.

60. *Angular Speed* A truck is moving at the rate of 50 miles per hour, and the diameter of each wheel is 3 feet. (a) Find the number of revolutions per minute the wheels are rotating. (b) Find the angular speed of the wheels in radians per minute.

61. *Angular Speed* A 2-inch-diameter pulley on an electric motor that runs at 1700 revolutions per minute is connected by a belt to a 4-inch-diameter pulley on a saw arbor. (a) Find the angular speed (in radians per minute) of each pulley. (b) Find the rotational speed (in rpm) of the saw.

62. *Angular Speed* How long will it take a pulley rotating at 12 radians per second to make 100 revolutions?

63. *Circular Saw Speed* The circular blade on a saw has a diameter of 7.5 inches and the blade rotates at 2400 revolutions per minute (see figure). (a) Find the angular speed in radians per second. (b) Find the speed of the saw teeth (in feet per second) as they contact the wood being cut.

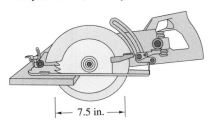

FIGURE FOR 63

64. *Speed of a Bicycle* The radii of the sprocket assemblies and the wheel of the bicycle in the figure are 4 inches, 2 inches, and 13 inches, respectively. If the cyclist is pedaling at the rate of 1 revolution per second, find the speed of the bicycle in (a) feet per second and (b) miles per hour.

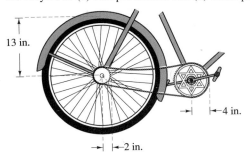

FIGURE FOR 64

2.2   THE TRIGONOMETRIC FUNCTIONS AND THE UNIT CIRCLE

....

The Unit Circle / The Trigonometric Functions / Domain and Period of Sine and Cosine / Evaluating Trig Functions with a Calculator

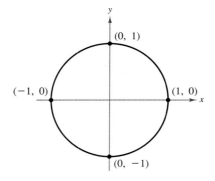

Unit Circle: $x^2 + y^2 = 1$

FIGURE 2.17

The Unit Circle

The two historical perspectives of trigonometry incorporate different methods for introducing the trigonometric functions. Our first introduction to these functions is based on the unit circle. Consider the **unit circle** given by

$$x^2 + y^2 = 1, \qquad \textit{Unit circle}$$

as shown in Figure 2.17. Imagine that the real number line is wrapped around this circle, with positive numbers corresponding to a counterclockwise wrapping and negative numbers corresponding to a clockwise wrapping, as shown in Figure 2.18.

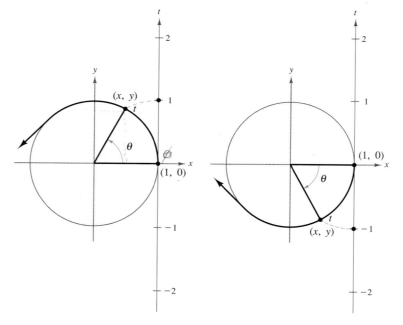

(a) Positive numbers   (b) Negative numbers

FIGURE 2.18

As the real number line is wrapped around the unit circle, each real number t will correspond with a point (x, y) on the circle. For example, the real number 0 corresponds to the point $(1, 0)$. Moreover, because the unit circle has a circumference of 2π, the real number 2π will also correspond to the point $(1, 0)$.

The Trigonometric Functions

From the preceding discussion, it follows that the coordinates x and y are two functions of the real variable t. These coordinates are used to define the six trigonometric functions of t.

sine	**cosecant**
cosine	**secant**
tangent	**cotangent**

These six functions are normally abbreviated as sin, csc, cos, sec, tan, and cot, respectively.

DEFINITION OF TRIGONOMETRIC FUNCTIONS

Let t be a real number at (x, y) the point on the unit circle corresponding to t.

$$\sin t = y \qquad \csc t = \frac{1}{y}, \quad y \neq 0$$

$$\cos t = x \qquad \sec t = \frac{1}{x}, \quad x \neq 0$$

$$\tan t = \frac{y}{x}, \quad x \neq 0 \qquad \cot t = \frac{x}{y}, \quad y \neq 0$$

REMARK   As an aid to memorizing these definitions, note that the functions in the second column are the *reciprocals* of the corresponding functions in the first column.

In the definition of the trigonometric functions note that the tangent or secant is not defined when $x = 0$. For instance, because $t = \pi/2$ corresponds to $(x, y) = (0, 1)$, it follows that $\tan(\pi/2)$ and $\sec(\pi/2)$ are *undefined*. Similarly, the cotangent or cosecant is not defined when $y = 0$. For instance, because $t = 0$ corrresponds to $(x, y) = (1, 0)$, cot 0 and csc 0 are *undefined*.

In Figure 2.19, the unit circle has been divided into eight equal arcs, corresponding to t-values of

$$0, \frac{\pi}{4}, \frac{\pi}{2}, \frac{3\pi}{4}, \pi, \frac{5\pi}{4}, \frac{3\pi}{2}, \frac{7\pi}{4}, \text{ and } 2\pi.$$

Similarly, in Figure 2.20, the unit circle has been divided into 12 equal arcs, corresponding to t-values of

$$0, \frac{\pi}{6}, \frac{\pi}{3}, \frac{\pi}{2}, \frac{2\pi}{3}, \frac{5\pi}{6}, \pi, \frac{7\pi}{6}, \frac{4\pi}{3}, \frac{3\pi}{2}, \frac{5\pi}{3}, \frac{11\pi}{6}, \text{ and } 2\pi.$$

Using the (x, y) coordinates in Figures 2.19 and 2.20, you can easily evaluate the trigonometric functions for common t-values. This procedure is demonstrated in Examples 1, 2, and 3.

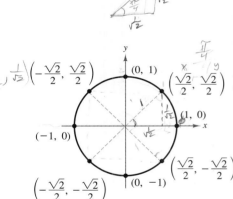

Unit Circle Divided into 8 Equal Arcs

FIGURE 2.19

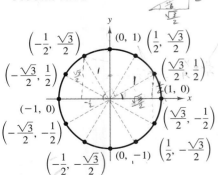

Unit Circle Divided into 12 Equal Arcs

FIGURE 2.20

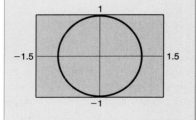
EXAMPLE 1 Evaluating Trigonometric Functions of Real Numbers

Evaluate the six trigonometric functions at the following real numbers.

A. $t = \dfrac{\pi}{6}$ **B.** $t = \dfrac{5\pi}{4}$

SOLUTION

A. Because $t = \pi/6$ corresponds to the first-quadrant point $(x, y) = \left(\sqrt{3}/2, 1/2\right)$, you can write the following.

$$\sin \frac{\pi}{6} = y = \frac{1}{2} \qquad\qquad \csc \frac{\pi}{6} = 2$$

$$\cos \frac{\pi}{6} = x = \frac{\sqrt{3}}{2} \qquad\qquad \sec \frac{\pi}{6} = \frac{2}{\sqrt{3}} = \frac{2\sqrt{3}}{3}$$

$$\tan \frac{\pi}{6} = \frac{y}{x} = \frac{1/2}{\sqrt{3}/2} = \frac{1}{\sqrt{3}} \qquad\qquad \cot \frac{\pi}{6} = \sqrt{3}$$

B. Because $t = 5\pi/4$ corresponds to the third-quadrant point $(x, y) = \left(-\sqrt{2}/2, -\sqrt{2}/2\right)$, you can write the following.

$$\sin \frac{5\pi}{4} = y = -\frac{\sqrt{2}}{2} \qquad\qquad \csc \frac{5\pi}{4} = -\frac{2}{\sqrt{2}} = -\sqrt{2}$$

$$\cos \frac{5\pi}{4} = x = -\frac{\sqrt{2}}{2} \qquad\qquad \sec \frac{5\pi}{4} = -\frac{2}{\sqrt{2}} = -\sqrt{2}$$

$$\tan \frac{5\pi}{4} = \frac{y}{x} = \frac{-\sqrt{2}/2}{-\sqrt{2}/2} = 1 \qquad\qquad \cot \frac{5\pi}{4} = 1$$

EXAMPLE 2 Evaluating Trigonometric Functions of Real Numbers

Evaluate the six trigonometric functions at the following real numbers.

A. $t = 0$ **B.** $t = \pi$

SOLUTION

A. $t = 0$ corresponds to the point $(x, y) = (1, 0)$ on the unit circle.

$$\sin 0 = y = 0 \qquad\quad \csc 0 \text{ is undefined}$$

$$\cos 0 = x = 1 \qquad\quad \sec 0 = 1$$

$$\tan 0 = \frac{y}{x} = 0 \qquad\quad \cot 0 \text{ is undefined}$$

B. $t = \pi$ corresponds to the point $(x, y) = (-1, 0)$ on the unit circle.

$$\sin \pi = y = 0 \qquad\qquad \csc \pi \text{ is undefined}$$

$$\cos \pi = x = -1 \qquad\qquad \sec \pi = -1$$

$$\tan \pi = \frac{y}{x} = \frac{0}{-1} = 0 \qquad \cot \pi \text{ is undefined}$$

EXAMPLE 3   Evaluating Trigonometric Functions of Real Numbers

Evaluate the six trigonometric functions at the following real numbers.

A. $t = -\dfrac{\pi}{3}$ **B.** $t = \dfrac{5\pi}{2}$

SOLUTION

A. Moving *clockwise* around the unit circle, you can see that $t = -\pi/3$ corresponds to the point $(x, y) = \left(1/2, -\sqrt{3}/2\right)$.

$$\sin\left(-\frac{\pi}{3}\right) = y = -\frac{\sqrt{3}}{2} \qquad \csc\left(-\frac{\pi}{3}\right) = -\frac{2}{\sqrt{3}}$$

$$\cos\left(-\frac{\pi}{3}\right) = x = \frac{1}{2} \qquad \sec\left(-\frac{\pi}{3}\right) = 2$$

$$\tan\left(-\frac{\pi}{3}\right) = \frac{y}{x} = -\sqrt{3} \qquad \cot\left(-\frac{\pi}{3}\right) = \frac{x}{y} = -\frac{1}{\sqrt{3}}$$

B. Moving *counterclockwise* around the unit circle one and a quarter revolutions, you can see that $t = 5\pi/2$ corresponds to the point $(x, y) = (0, 1)$.

$$\sin \frac{5\pi}{2} = y = 1 \qquad\qquad \csc \frac{5\pi}{2} = 1$$

$$\cos \frac{5\pi}{2} = x = 0 \qquad\qquad \sec \frac{5\pi}{2} \text{ is undefined}$$

$$\tan \frac{5\pi}{2} = \frac{y}{x} \text{ is undefined} \qquad \cot \frac{5\pi}{2} = \frac{x}{y} = 0$$

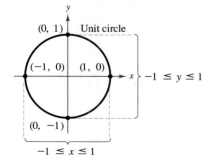

y

$(0, 1)$ Unit circle

$(-1, 0)$ $(1, 0)$

x $-1 \leq y \leq 1$

$(0, -1)$

$-1 \leq x \leq 1$

FIGURE 2.21

Domain and Period of Sine and Cosine

The *domain* of the sine and cosine functions is the set of all real numbers. To determine the *range* of these two functions, consider the unit circle shown in Figure 2.21. Because $r = 1$, it follows that $\sin t = y$ and $\cos t = x$. Moreover, because (x, y) is on the unit circle, you know that $-1 \le y \le 1$ and $-1 \le x \le 1$, and it follows that the values of the sine and cosine also range between -1 and 1. That is,

$$-1 \le y \le 1 \qquad\qquad -1 \le x \le 1$$
$$-1 \le \sin t \le 1 \quad \text{and} \quad -1 \le \cos t \le 1.$$

Suppose you add 2π to each value of t in the interval $[0, 2\pi]$, thus completing a second revolution around the unit circle, as shown in Figure 2.22. The values of $\sin(t + 2\pi)$ and $\cos(t + 2\pi)$ correspond to those of $\sin t$ and $\cos t$. Similar results can be obtained for repeated revolutions (positive or negative) on the unit circle. This leads to the general result

$$\sin(t + 2\pi n) = \sin t \quad \text{and} \quad \cos(t + 2\pi n) = \cos t$$

for any integer n and real number t. Functions that behave in such a repetitive (or cyclic) manner are called **periodic.**

REMARK In Figure 2.22, *positive* multiples of 2π were added to the t-values. You could just as well have added *negative* multiples. For instance, $\pi/4 - 2\pi$ and $\pi/4 - 4\pi$ are also coterminal to $\pi/4$.

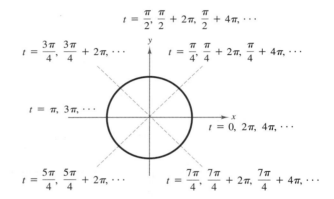

Repeated Revolutions on the Unit Circle

FIGURE 2.22

DISCOVERY

As the real number line is wrapped around the unit circle, each real number t will correspond with a point $(x, y) = (\cos t, \sin t)$ on the circle. You can visualize this graphically by setting your graphing utility to *simultaneous* mode. Using radian and parametric modes as well, let

$$X_{1t} = \cos T$$
$$Y_{1t} = \sin T$$
$$X_{2t} = T$$
$$Y_{2t} = \sin T.$$

Select the following ranges: *Tmin* = 0, *Tmax* = 6.3, *Tstep* = .1, *Xmin* = −2, *Xmax* = 7, *Ymin* = −3, *Ymax* = 3. Notice how the graphing utility traces out the unit circle and the sine function simultaneously. Try changing Y_{2t} to $\cos T$ or to $\tan T$.

DEFINITION OF A PERIODIC FUNCTION

A function f is **periodic** if there exists a positive real number c such that

$$f(t + c) = f(t)$$

for all t in the domain of f. The least number c for which f is periodic is called the **period** of f.

From this definition it follows that the sine and cosine functions are periodic and have a period of 2π. The other four trigonometric functions are also periodic, and we will say more about this in Section 2.6.

EXAMPLE 4   Using the Period to Evaluate Sine and Cosine

A. Because $\dfrac{13\pi}{6} = 2\pi + \dfrac{\pi}{6}$, you have

$$\sin \frac{13\pi}{6} = \sin\left(2\pi + \frac{\pi}{6}\right) = \sin \frac{\pi}{6} = \frac{1}{2}.$$

B. Because $-\dfrac{7\pi}{2} = -4\pi + \dfrac{\pi}{2}$, you have

$$\cos\left(-\frac{7\pi}{2}\right) = \cos\left(-4\pi + \frac{\pi}{2}\right) = \cos \frac{\pi}{2} = 0.$$

Recall from Section 1.7 that a function f is *even* if

$$f(-t) = f(t) \qquad\qquad \text{\textit{Even function}}$$

and is *odd* if

$$f(-t) = -f(t). \qquad\qquad \text{\textit{Odd function}}$$

Of the six trigonometric functions, two are even and four are odd, as stated in the following theorem.

EVEN AND ODD TRIGONOMETRIC FUNCTIONS

The cosine and secant functions are *even*.

$$\cos(-t) = \cos t \qquad \sec(-t) = \sec t$$

The sine, cosecant, tangent, and cotangent functions are *odd*.

$$\sin(-t) = -\sin t \qquad \csc(-t) = -\csc t$$
$$\tan(-t) = -\tan t \qquad \cot(-t) = -\cot t$$

Evaluating Trig Functions with a Calculator

From the arc length formula $s = r\theta$, with $r = 1$, you can see that each real number t measures a central angle (in radians). That is, $t = r\theta = 1(\theta) = \theta$ radians. Thus, when you are evaluating trigonometric functions, *it doesn't make any difference whether you consider t to be a real number or an angle given in radians.*

A scientific calculator can be used to obtain decimal approximations of the values of the trigonometric functions. Before doing this, however, you must be sure that the calculator is set to the correct mode: degrees or radians. Here are two examples.

Degree mode: cos 28   *Display 0.8829475929*

Radian mode: $\tan \dfrac{\pi}{12}$ *Display 0.2679491924*

Most calculators do not have keys for the cosecant, secant, and cotangent functions. To evaluate these functions, use the reciprocal key with the functions sine, cosine, and tangent. Here is an example.

$$\csc \frac{\pi}{8} = \frac{1}{\sin(\pi/8)} \approx 2.613.$$

EXAMPLE 5   **Using a Calculator to Evaluate Trigonometric Functions**

Use a calculator to evaluate the following. (Round to three decimal places.)

A. $\sin(-76.4)°$ **B.** cot 1.5   **C.** $\sec(5°\ 40'\ 12'')$

SOLUTION

Function	Mode	Display	Rounded to 3 Decimal Places
A. $\sin(-76.4)°$	Degree	−0.9719610006	−0.972
B. cot 1.5	Radian	0.0709148443	0.071
C. $\sec(5°\ 40'\ 12'')$	Degree	1.004916618	1.005

Note that $5°\ 40'\ 12'' = 5.67°$.

DISCUSSION PROBLEM

YOU BE THE INSTRUCTOR

Suppose you are tutoring a student who is just learning trigonometry. Your student was asked to evaluate the cosine of 2 radians and, using a calculator, obtained the following.

Keystrokes	Display
$\boxed{\text{COS}}$ 2	0.999390827

You know that 2 radians lies in the second quadrant. You also know that this implies that the cosine of 2 radians should be negative. What did your student do wrong?

WARM-UP

The following warm-up exercises involve skills that were covered in earlier sections. You will use these skills in the exercise set for this section.

In Exercises 1 and 2, simplify the expression.

1. $\dfrac{\frac{1}{2}}{\frac{-\sqrt{3}}{2}}$

2. $\dfrac{\frac{\sqrt{2}}{2}}{\frac{-\sqrt{2}}{2}}$

In Exercises 3 and 4, find an angle θ in the interval $[0, 2\pi]$ that is coterminal with the given angle.

3. $\dfrac{8\pi}{3}$

4. $-\dfrac{\pi}{4}$

In Exercises 5 and 6, convert the angle to radian measure.

5. $30°$

6. $135°$

In Exercises 7 and 8, convert the angle to degree measure.

7. $\dfrac{\pi}{3}$ radians

8. $-\dfrac{3\pi}{2}$ radians

9. Determine the circumference of a circle with radius 1.

10. Determine the arc length of a semicircle with radius 1.

SECTION 2.2 · EXERCISES

In Exercises 1–8, find the point (x, y) on the unit circle that corresponds to the real number t (see Figures 2.19 and 2.20).

1. $t = \dfrac{\pi}{4}$

2. $t = \dfrac{\pi}{3}$

3. $t = \dfrac{5\pi}{6}$

4. $t = \dfrac{5\pi}{4}$

5. $t = \dfrac{4\pi}{3}$

6. $t = \dfrac{11\pi}{6}$

7. $t = \dfrac{3\pi}{2}$

8. $t = \pi$

In Exercises 9–16, evaluate the sine, cosine, and tangent of the real number.

9. $t = \dfrac{\pi}{4}$

10. $t = -\dfrac{\pi}{4}$

11. $t = -\dfrac{5\pi}{4}$

12. $t = -\dfrac{5\pi}{6}$

13. $t = \dfrac{11\pi}{6}$

14. $t = \dfrac{2\pi}{3}$

15. $t = \dfrac{4\pi}{3}$

16. $t = \dfrac{7\pi}{4}$

In Exercises 17–22, evaluate (if possible) the six trigonometric functions of the real number.

17. $t = \dfrac{3\pi}{4}$

18. $t = -\dfrac{2\pi}{3}$

19. $t = \dfrac{\pi}{2}$

20. $t = \dfrac{3\pi}{2}$

21. $t = -\dfrac{4\pi}{3}$

22. $t = -\dfrac{11\pi}{6}$

In Exercises 23–30, evaluate the trigonometric function using its period as an aid.

23. $\sin 3\pi$

24. $\cos 3\pi$

25. $\cos \dfrac{8\pi}{3}$

26. $\sin \dfrac{9\pi}{4}$

27. $\cos \dfrac{19\pi}{6}$

28. $\sin\left(-\dfrac{13\pi}{6}\right)$

29. $\sin\left(-\dfrac{9\pi}{4}\right)$

30. $\cos\left(-\dfrac{8\pi}{4}\right)$

In Exercises 31–36, use the value of the trigonometric function to evaluate the indicated functions.

31. $\sin t = \frac{1}{3}$
  (a) $\sin(-t)$ (b) $\csc(-t)$

32. $\sin(-t) = \frac{2}{5}$
  (a) $\sin t$ (b) $\csc t$

33. $\cos(-t) = -\frac{7}{8}$
  (a) $\cos t$ (b) $\sec(-t)$

34. $\cos t = -\frac{3}{4}$
  (a) $\cos(-t)$ (b) $\sec(-t)$

35. $\sin t = \frac{4}{5}$
  (a) $\sin(\pi - t)$ (b) $\sin(t + \pi)$

36. $\cos t = \frac{4}{5}$
  (a) $\cos(\pi - t)$ (b) $\cos(t + \pi)$

In Exercises 37–44, use a calculator to evaluate the trigonometric function. (Set your calculator in the correct mode and round your answer to four decimal places.)

37. $\sin \dfrac{\pi}{4}$

38. $\tan \pi$

39. $\cos 34.2°$

40. $\cot 1$

41. $\tan 110.5°$

42. $\sec 54.9°$

43. $\csc 0.8$

44. $\sin(-0.9)$

In Exercises 45 and 46, use the accompanying figure and a straightedge to approximate the values of the trigonometric functions. In Exercises 47 and 48, approximate the solutions of the equations. Use $0 \le t \le 2\pi$.

45. (a) $\sin 5$ (b) $\cos 2$

46. (a) $\sin 0.75$ (b) $\cos 2.5$

47. (a) $\sin t = 0.25$ (b) $\cos t = -0.25$

48. (a) $\sin t = -0.75$ (b) $\cos t = 0.75$

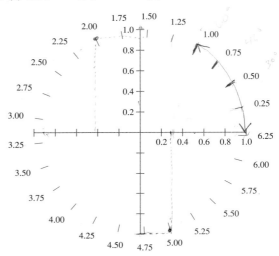

FIGURE FOR 45–48

49. *Harmonic Motion*   The displacement from equilibrium of an oscillating weight suspended by a spring is

$$y(t) = \tfrac{1}{4} \cos 6t,$$

where y is the displacement in feet and t is the time in seconds. Find the displacement when (a) $t = 0$, (b) $t = \frac{1}{4}$, and (c) $t = \frac{1}{2}$.

50. *Harmonic Motion*   The displacement from equilibrium of an oscillating weight suspended by a spring and subject to the damping effect of friction is

$$y(t) = \tfrac{1}{4}e^{-t} \cos 6t,$$

where y is the displacement in feet and t is the time in seconds. Find the displacement when (a) $t = 0$, (b) $t = \frac{1}{4}$, and (c) $t = \frac{1}{2}$.

51. *Electric Circuits*   The initial current and charge in the electrical circuit shown in the accompanying figure is zero. The current when 100 volts is applied to the circuit is given by

$$I = 5e^{-2t} \sin t$$

if the resistance, inductance, and capacitance are 80 ohms, 20 henrys, and 0.01 farads, respectively. Approximate the current $t = 0.7$ seconds after the voltage is applied.

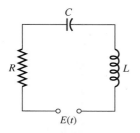

FIGURE FOR 51

52. Use the unit circle to verify that the cosine and secant functions are even and the sine, cosecant, tangent, and cotangent functions are odd.

2.3   TRIGONOMETRIC FUNCTIONS AND RIGHT TRIANGLES

Trigonometric Functions of an Acute Angle   /   Trigonometric Identities   /   Applications Involving Right Triangles

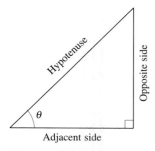

FIGURE 2.23

Trigonometric Functions of an Acute Angle

Our second look at the trigonometric functions is from a *right triangle* perspective. In Figure 2.23, the three sides of the triangle are the **hypotenuse,** the **opposite side** (the side opposite the angle θ), and the **adjacent side** (the side adjacent to the angle θ). Using the lengths of these three sides, you can form six ratios that define the six trigonometric functions of the acute angle θ.

In the following definition it is important to see that $0° < \theta < 90°$ and that for such angles the value of each of the six trigonometric functions is *positive.*

RIGHT TRIANGLE DEFINITION OF TRIGONOMETRIC FUNCTIONS

Let θ be an *acute* angle of a right triangle. Then the six trigonometric functions *of the angle* θ are defined as follows.

$$\sin \theta = \frac{\text{opp}}{\text{hyp}} \qquad \csc \theta = \frac{\text{hyp}}{\text{opp}}$$

$$\cos \theta = \frac{\text{adj}}{\text{hyp}} \qquad \sec \theta = \frac{\text{hyp}}{\text{adj}}$$

$$\tan \theta = \frac{\text{opp}}{\text{adj}} \qquad \cot \theta = \frac{\text{adj}}{\text{opp}}$$

The abbreviations opp, adj, and hyp represent the lengths of the three sides of a right triangle.

opp = the length of the side *opposite* θ
adj = the length of the side *adjacent* to θ
hyp = the length of the *hypotenuse*

EXAMPLE 1   Evaluating Trigonometric Functions

Find the values of the six trigonometric functions of θ in the right triangle shown in Figure 2.24.

SOLUTION

By the Pythagorean Theorem, $(\text{hyp})^2 = (\text{opp})^2 + (\text{adj})^2$, it follows that

$$\text{hyp} = \sqrt{3^2 + 4^2} = \sqrt{25} = 5.$$

Thus, adj = 3, opp = 4, and hyp = 5.

$$\sin \theta = \frac{\text{opp}}{\text{hyp}} = \frac{4}{5} \qquad \csc \theta = \frac{\text{hyp}}{\text{opp}} = \frac{5}{4}$$

$$\cos \theta = \frac{\text{adj}}{\text{hyp}} = \frac{3}{5} \qquad \sec \theta = \frac{\text{hyp}}{\text{adj}} = \frac{5}{3}$$

$$\tan \theta = \frac{\text{opp}}{\text{adj}} = \frac{4}{3} \qquad \cot \theta = \frac{\text{adj}}{\text{opp}} = \frac{3}{4}$$

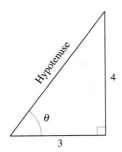

FIGURE 2.24

In Example 1, you were given the lengths of the sides of the right triangle, but not the angle θ. A much more common problem in trigonometry is to be asked to find the trigonometric functions for a *given* acute angle θ. To do this, construct a right triangle having θ as one of its angles.

EXAMPLE 2   Evaluating Trigonometric Functions of 45°

Find the values of sin 45°, cos 45°, and tan 45°.

SOLUTION

Construct a right triangle having 45° as one of its acute angles, as shown in Figure 2.25. Choose the length of the adjacent side to be 1. From geometry, you know that the other acute angle is also 45°. Hence, the triangle is isosceles and the length of the opposite side is also 1. Using the Pythagorean Theorem, you can find the length of the hypotenuse to be

$$\text{hyp} = \sqrt{1^2 + 1^2} = \sqrt{2}.$$

Finally, you can write the following.

$$\sin 45° = \frac{\text{opp}}{\text{hyp}} = \frac{1}{\sqrt{2}} = \frac{\sqrt{2}}{2}$$

$$\cos 45° = \frac{\text{adj}}{\text{hyp}} = \frac{1}{\sqrt{2}} = \frac{\sqrt{2}}{2}$$

$$\tan 45° = \frac{\text{opp}}{\text{adj}} = \frac{1}{1} = 1$$

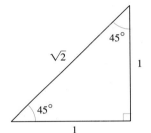

FIGURE 2.25

EXAMPLE 3   Evaluating Trigonometric Functions of 30° and 60°

Use the equilateral triangle shown in Figure 2.26 to find the values of sin 60°, cos 60°, sin 30°, and cos 30°.

SOLUTION

Try using the Pythagorean Theorem to verify the lengths of the sides given in Figure 2.26. For $\theta = 60°$, you have adj = 1, opp = $\sqrt{3}$, and hyp = 2, which implies that

$$\sin 60° = \frac{\text{opp}}{\text{hyp}} = \frac{\sqrt{3}}{2} \quad \text{and} \quad \cos 60° = \frac{\text{adj}}{\text{hyp}} = \frac{1}{2}.$$

For $\theta = 30°$, you have adj = $\sqrt{3}$, opp = 1, and hyp = 2, which implies that

$$\sin 30° = \frac{\text{opp}}{\text{hyp}} = \frac{1}{2} \quad \text{and} \quad \cos 30° = \frac{\text{adj}}{\text{hyp}} = \frac{\sqrt{3}}{2}.$$

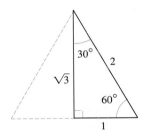

FIGURE 2.26

DISCOVERY

Set your graphing utility in degree mode and choose an angle x. Now evaluate cos(x) and sin(90 − x). What do you observe? Repeat this experiment with sin(x) and cos(90 − x).

Because the angles 30°, 45°, and 60° ($\pi/6$, $\pi/4$, and $\pi/3$) occur frequently in trigonometry, we suggest that you learn to construct the triangles shown in Figures 2.25 and 2.26.

SINE, COSINE, AND TANGENT OF SPECIAL ANGLES

$$\sin 30° = \sin \frac{\pi}{6} = \frac{1}{2}, \quad \cos 30° = \cos \frac{\pi}{6} = \frac{\sqrt{3}}{2}, \quad \tan 30° = \tan \frac{\pi}{6} = \frac{\sqrt{3}}{3}$$

$$\sin 45° = \sin \frac{\pi}{4} = \frac{\sqrt{2}}{2}, \quad \cos 45° = \cos \frac{\pi}{4} = \frac{\sqrt{2}}{2}, \quad \tan 45° = \tan \frac{\pi}{4} = 1$$

$$\sin 60° = \sin \frac{\pi}{3} = \frac{\sqrt{3}}{2}, \quad \cos 60° = \cos \frac{\pi}{3} = \frac{1}{2}, \quad \tan 60° = \tan \frac{\pi}{3} = \sqrt{3}$$

Trigonometric Identities

In the preceding list, note that $\sin 30° = \frac{1}{2} = \cos 60°$. This occurs because $30°$ and $60°$ are complementary angles. In general, it can be shown that *cofunctions of complementary angles are equal.* That is, if θ is an acute angle, then the following relationships are true.

$$\sin(90° - \theta) = \cos \theta \qquad \cos(90° - \theta) = \sin \theta$$
$$\tan(90° - \theta) = \cot \theta \qquad \cot(90° - \theta) = \tan \theta$$
$$\sec(90° - \theta) = \csc \theta \qquad \csc(90° - \theta) = \sec \theta$$

For instance, because $10°$ and $80°$ are complementary angles, it follows that $\sin 10° = \cos 80°$ and $\tan 10° = \cot 80°$.

FUNDAMENTAL TRIGONOMETRIC IDENTITIES

Reciprocal Identities

$$\sin \theta = \frac{1}{\csc \theta} \qquad \sec \theta = \frac{1}{\cos \theta} \qquad \tan \theta = \frac{1}{\cot \theta}$$

$$\csc \theta = \frac{1}{\sin \theta} \qquad \cos \theta = \frac{1}{\sec \theta} \qquad \cot \theta = \frac{1}{\tan \theta}$$

Quotient Identities

$$\tan \theta = \frac{\sin \theta}{\cos \theta} \qquad \cot \theta = \frac{\cos \theta}{\sin \theta}$$

Pythagorean Identities

$$\sin^2 \theta + \cos^2 \theta = 1 \quad 1 + \tan^2 \theta = \sec^2 \theta \quad 1 + \cot^2 \theta = \csc^2 \theta$$

Technology Note _____

Use your calculator to confirm several of the trigonometric identities at the right for various values of θ. For instance, calculate $(\sin 0.5)^2 + (\cos 0.5)^2$ and observe that the value is 1.

REMARK   We use $\sin^2 \theta$ to represent $(\sin \theta)^2$, $\cos^2 \theta$ to represent $(\cos \theta)^2$, and so on.

EXAMPLE 4   Applying Trigonometric Identities

Let θ be the acute angle such that $\sin \theta = 0.6$. Use trigonometric identities to find the values of the following.

A. $\cos \theta$ **B.** $\tan \theta$

SOLUTION

A. $\sin^2 \theta + \cos^2 \theta = 1$ *Pythagorean Identity*

$(0.6)^2 + \cos^2 \theta = 1$

$\cos^2 \theta = 1 - (0.6)^2 = 0.64$

$\cos \theta = \sqrt{0.64} = 0.8.$

B. Now, knowing the values of $\sin \theta$ and $\cos \theta$, you can find the value of $\tan \theta$.

$$\tan \theta = \frac{\sin \theta}{\cos \theta} = \frac{0.6}{0.8} = 0.75.$$

Try using the definitions of $\cos \theta$ and $\tan \theta$, and the triangle shown in Figure 2.27, to check these results.

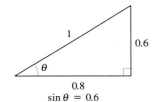

$\sin \theta = 0.6$

FIGURE 2.27

EXAMPLE 5   Applying Trigonometric Identities

Let θ be an acute angle such that $\tan \theta = 3$. Use trigonometric identities to find the values of the following.

A. $\cot \theta$ **B.** $\sec \theta$

SOLUTION

A. $\cot \theta = \dfrac{1}{\tan \theta}$ *Reciprocal Identity*

$= \dfrac{1}{3}$

B. $\sec^2 \theta = \tan^2 \theta + 1$ *Pythagorean Identity*

$\sec^2 \theta = 3^3 + 1$

$\sec^2 \theta = 10$

$\sec \theta = \sqrt{10}$

Try using the definitions of $\cot \theta$ and $\sec \theta$, and the triangle shown in Figure 2.28, to check these results.

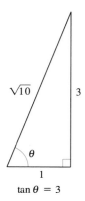

$\tan \theta = 3$

FIGURE 2.28

Applications Involving Right Triangles

Many applications of trigonometry involve a process called **solving right triangles.** In this type of application, you are usually given two sides of a right triangle and asked to find one of its acute angles, *or* you are given one side and one of the acute angles and asked to find one of the other sides.

EXAMPLE 6   Using Trigonometry to Solve a Right Triangle

A surveyor is standing 50 feet from the base of a large tree, as shown in Figure 2.29. The surveyor measures the angle of elevation to the top of the tree as 71.5°. How tall is the tree?

SOLUTION

From Figure 2.29, you can see that

$$\tan 71.5° = \frac{\text{opp}}{\text{adj}} = \frac{y}{x},$$

where $x = 50$ and y is the height of the tree. Thus, you can determine the height of the tree to be

$$y = x \tan 71.5° \approx 50(2.98868) \approx 149.4 \text{ feet.}$$

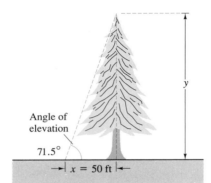

Angle of elevation

71.5°

$x = 50$ ft

FIGURE 2.29

EXAMPLE 7   Using Trigonometry to Solve a Right Triangle

A person is standing 200 yards from a river. Rather than walking directly to the river, the person walks 400 yards along a straight path to the river's edge. Find the acute angle θ between this path and the river's edge, as indicated in Figure 2.30.

SOLUTION

From Figure 2.30, you can see that the sine of the angle θ is

$$\sin \theta = \frac{\text{opp}}{\text{hyp}} = \frac{200}{400} = \frac{1}{2}.$$

Now, you can recognize that $\theta = 30°$.

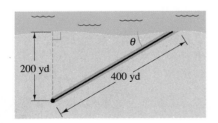

200 yd

θ

400 yd

FIGURE 2.30

In Example 7, you were able to recognize that the acute angle that satisfies the equation $\sin \theta = \frac{1}{2}$ is $\theta = 30°$. Suppose, however, that you were given the equation $\sin \theta = 0.6$ and asked to find the acute angle θ. Because

$$\sin 30° = \frac{1}{2} = 0.5000 \quad \text{and} \quad \sin 45° = \frac{1}{\sqrt{2}} \approx 0.7071,$$

you can figure that θ lies somewhere between $30°$ and $45°$. A more precise value of θ can be found using the inverse key on a calculator. (Consult your calculator manual to see how this key works on your own calculator.) For most calculators, one of the following keystroke sequences will work.

Degree mode: .6 **INV** **sin**   *Display 36.86989765*

Degree mode: **2nd** **sin** .6 **ENTER**   *Display 36.86989765*

Thus, you can conclude that if $\sin \theta = 0.6$, then $\theta \approx 36.87°$.

EXAMPLE 8   Using Trigonometry to Solve a Right Triangle

A 40-foot flagpole casts a 30-foot shadow, as shown in Figure 2.31. Find θ, the angle of elevation of the sun.

SOLUTION

Figure 2.31 shows that the *opposite* and *adjacent* sides are known. Thus,

$$\tan \theta = \frac{\text{opp}}{\text{adj}} = \frac{40}{30}.$$

With a calculator in degree mode you can obtain $\theta \approx 53.13°$.

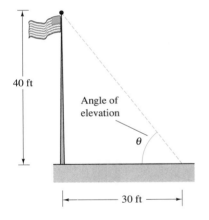

40 ft

Angle of elevation

θ

30 ft

FIGURE 2.31

DISCUSSION PROBLEM

COMPARING DEFINITIONS OF TRIGONOMETRIC FUNCTIONS

In Section 2.2 and in this section, we presented two different definitions of trigonometric functions. One was the "unit circle" definition and the other was the "right triangle" definition. Write a short paper that compares the two definitions. Then use both definitions to find the values of the six trigonometric functions at $\theta = 30°$. For this value of θ, which definition do you prefer? For $\theta = 3\pi$, which do you prefer and why?

WARM-UP

The following warm-up exercises involve skills that were covered in earlier sections. You will use these skills in the exercise set for this section.

In Exercises 1–4, find the distance between the points.

1. (3, 8), (1, 4)   **2.** (5, 2), (2, −7)

3. (−4, 0), (2, 8)   **4.** (−3, −3), (0, 0)

In Exercises 5–10, perform the indicated operation(s). (Round your result to two decimal places.)

5. 0.300×4.125 **6.** 7.30×43.50

7. $\dfrac{151.5}{2.40}$ **8.** $\dfrac{3740}{28.0}$

9. $\dfrac{19,500}{0.007}$ **10.** $\dfrac{(10.5)(3401)}{1240}$

SECTION 2.3 · EXERCISES

In Exercises 1–8, find the exact values of the six trigonometric functions of the angle θ. (Use the Pythagorean Theorem to find the third side of the triangle.)

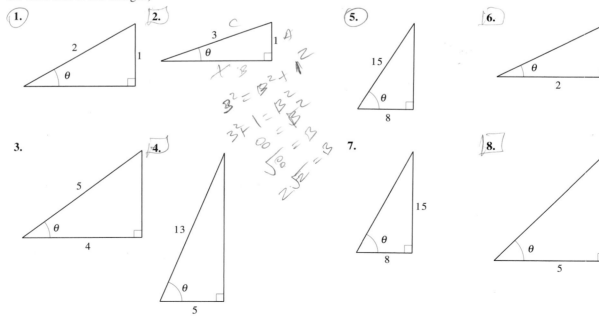

In Exercises 9–16, sketch a right triangle corresponding to the trigonometric function of the acute angle θ. Use the Pythagorean Theorem to determine the third side and then find the other five trigonometric functions of θ.

9. $\sin \theta = \frac{2}{3}$　　10. $\cot \theta = 5$

11. $\sec \theta = 2$　　12. $\cos \theta = \frac{5}{7}$

13. $\tan \theta = 3$　　14. $\csc \theta = \frac{17}{4}$

15. $\cot \theta = \frac{3}{2}$　　16. $\sin \theta = \frac{3}{8}$

In Exercises 17–20, use the given values and the trigonometric identities to evaluate the trigonometric functions.

17. $\sin 60° = \dfrac{\sqrt{3}}{2}$,　$\cos 60° = \dfrac{1}{2}$

　(a) $\tan 60°$　　　　(b) $\sin 30°$
　(c) $\cos 30°$　　　　(d) $\cot 60°$

18. $\sin 30° = \dfrac{1}{2}$,　$\tan 30° = \dfrac{\sqrt{3}}{3}$

　(a) $\csc 30°$　　　　(b) $\cot 60°$
　(c) $\cos 30°$　　　　(d) $\cot 30°$

19. $\csc \theta = 3$,　$\sec \theta = \dfrac{3\sqrt{2}}{4}$

　(a) $\sin \theta$　　　　(b) $\cos \theta$
　(c) $\tan \theta$　　　　(d) $\sec(90° - \theta)$

20. $\sec \theta = 5$,　$\tan \theta = 2\sqrt{6}$

　(a) $\cos \theta$　　　　(b) $\cot \theta$
　(c) $\cot(90° - \theta)$　　(d) $\sin \theta$

In Exercises 21–24, evaluate the trigonometric functions.

21. (a) $\cos 60°$　　　　(b) $\tan \dfrac{\pi}{4}$

22. (a) $\csc 30°$　　　　(b) $\sin \dfrac{\pi}{4}$

23. (a) $\cot 45°$　　　　(b) $\cos 45°$

24. (a) $\sin \dfrac{\pi}{3}$　　　　(b) $\csc 45°$

In Exercises 25–34, use a calculator to evaluate each function. Round your result to four decimal places. (Be sure the calculator is in the correct mode.)

25. (a) $\sin 10°$　　　　(b) $\cos 80°$

26. (a) $\tan 23.5°$　　　　(b) $\cot 66.5°$

27. (a) $\sin 16.35°$　　　　(b) $\csc 16.35°$

28. (a) $\cos 16° \, 18'$　　　　(b) $\sin 73° \, 56'$

29. (a) $\sec 42° \, 12'$　　　　(b) $\csc 48° \, 7'$

30. (a) $\cos 4° \, 50' \, 15''$　　　　(b) $\sec 4° \, 50' \, 15''$

31. (a) $\cot \dfrac{\pi}{16}$　　　　(b) $\tan \dfrac{\pi}{16}$

32. (a) $\sec 0.75$　　　　(b) $\cos 0.75$

33. (a) $\csc 1$　　　　(b) $\sec\left(\dfrac{\pi}{2} - 1\right)$

34. (a) $\tan \dfrac{1}{2}$　　　　(b) $\cot\left(\dfrac{\pi}{2} - \dfrac{1}{2}\right)$

In Exercises 35–40, find the value of θ in degrees $(0° < \theta < 90°)$ and radians $(0 < \theta < \pi/2)$ without using a calculator.

35. (a) $\sin \theta = \dfrac{1}{2}$　　36. (a) $\cos \theta = \dfrac{\sqrt{2}}{2}$

　(b) $\csc \theta = 2$　　　　(b) $\tan \theta = 1$

37. (a) $\sec \theta = 2$　　38. (a) $\tan \theta = \sqrt{3}$

　(b) $\cot \theta = 1$　　　　(b) $\cos \theta = \dfrac{1}{2}$

39. (a) $\csc \theta = \dfrac{2\sqrt{3}}{3}$　　40. (a) $\cot \theta = \dfrac{\sqrt{3}}{3}$

　(b) $\sin \theta = \dfrac{\sqrt{2}}{2}$　　　　(b) $\sec \theta = \sqrt{2}$

In Exercises 41–44, find the value of θ in degrees $(0° < \theta < 90°)$ and radians $(0 < \theta < \pi/2)$ by using the inverse key on a calculator.

41. (a) $\sin \theta = 0.8191$　　42. (a) $\cos \theta = 0.9848$
　(b) $\cos \theta = 0.0175$　　　　(b) $\cos \theta = 0.8746$

43. (a) $\tan \theta = 1.1920$　　44. (a) $\sin \theta = 0.3746$
　(b) $\tan \theta = 0.4663$　　　　(b) $\cos \theta = 0.3746$

45. Solve for y.

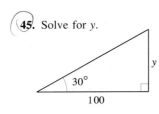

46. Solve for x.

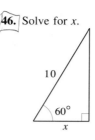

47. Solve for x.

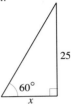

48. Solve for r.

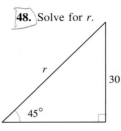

49. Solve for r.

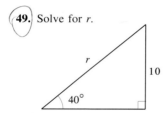

50. Solve for x.

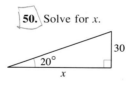

51. Solve for y.

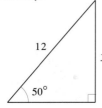

52. Solve for r.

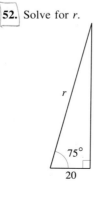

53. *Height*   A 6-foot person standing 12 feet from a streetlight casts an 8-foot shadow (see figure). What is the height of the streetlight?

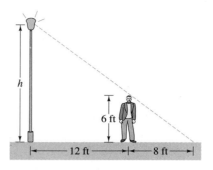

FIGURE FOR 53

54. *Height*   A 6-foot man walked from the base of a broadcasting tower directly toward the tip of the shadow cast by the tower. When he was 132 feet from the tower, his shadow started to appear beyond the tower's shadow. What is the height of the tower if the man's shadow is 3 feet long?

55. *Length*   A 20-foot ladder leaning against the side of a house makes a 75° angle with the ground (see figure). How far up the side of the house does the ladder reach?

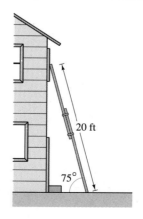

FIGURE FOR 55

56. *Width of a River*   A biologist wants to know the width w of a river in order to properly set instruments for studying the pollutants in the water. From point A, the biologist walks downstream 100 feet and sights to point C. From this sighting, it is determined that $\theta = 50°$ (see figure). How wide is the river?

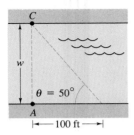

FIGURE FOR 56

57. *Distance*   From a 150-foot observation tower on the coast, a Coast Guard officer sights a boat in difficulty. The angle of depression of the boat is $4°$ (see figure). How far is the boat from the shoreline?

FIGURE FOR 57

58. *Angle of Elevation*   A ramp $17\frac{1}{2}$ feet in length rises to a loading platform that is $3\frac{1}{3}$ feet off the ground (see figure). Find the angle θ that the ramp makes with the ground.

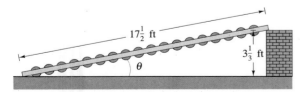

FIGURE FOR 58

59. *Machine Shop Calculations*   A steel plate has the form of a quarter circle with a radius of 24 inches. Two $\frac{3}{8}$-inch holes are to be drilled in the plate positioned as shown in the figure. Find the coordinates of the center of each hole.

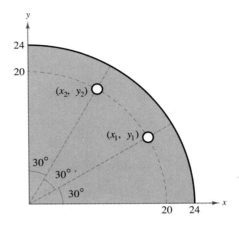

FIGURE FOR 59

60. *Machine Shop Calculations*   A tapered shaft has a diameter of 2 inches at the small end and is 6 inches long (see figure). If the taper is $3°$, find the diameter d of the large end of the shaft.

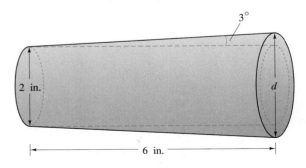

FIGURE FOR 60

61. *Trigonometric Functions by Actual Measurement* Use a compass to sketch a quarter of a circle of radius 10 centimeters. Using a protractor, construct an angle of 25° in standard position (see figure). Drop a perpendicular line from the point of intersection of the terminal side of the angle and the arc of the circle. By actual measurement, calculate the coordinates (x, y) of the point of intersection and use these measurements to approximate the six trigonometric functions of a 25° angle.

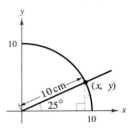

FIGURE FOR 61

62. *Trigonometric Functions by Actual Measurement* Repeat Exercise 61 using an angle of 75°.

In Exercises 63–68, determine whether the statement is true or false, and give a reason for your answer.

63. $\sin 60° \csc 60° = 1$

64. $\sec 30° = \csc 60°$

65. $\sin 45° + \cos 45° = 1$

66. $\cot^2 10° - \csc^2 10° = -1$

67. $\dfrac{\sin 60°}{\sin 30°} = \sin 2°$

68. $\tan[(0.8)^2] = \tan^2(0.8)$

69. A 30-foot ladder leaning against the side of a house is 4 feet from the house at the base (see figure).

(a) How far up the side of the house does the ladder reach? Express your answer accurate to two decimal places.

(b) Use the fact that $\cos \theta = 4/30$ to find the angle that the ladder makes with the ground. Express your answer in degrees accurate to two decimal places.

(c) Use the tangent of the angle found in part (b) to answer part (a) again.

(d) Why do your answers to parts (a) and (c) differ slightly?

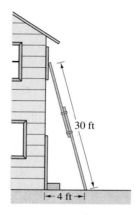

FIGURE FOR 69

2.4   TRIGONOMETRIC FUNCTIONS OF ANY ANGLE

Trigonometric Functions of Any Angle   /   Reference Angles

Trigonometric Functions of Any Angle

In Section 2.3, you learned to evaluate trigonometric functions of an acute angle. In this section you will learn to evaluate trigonometric functions of any angle.

DEFINITION OF TRIGONOMETRIC FUNCTIONS OF ANY ANGLE

Let θ be an angle in standard position with (x, y) any point (except the origin) on the terminal side of θ and $r = \sqrt{x^2 + y^2}$ (see figure).

$$\sin \theta = \frac{y}{r} \qquad\qquad \csc \theta = \frac{r}{y}, \quad y \neq 0$$

$$\cos \theta = \frac{x}{r} \qquad\qquad \sec \theta = \frac{r}{x}, \quad x \neq 0$$

$$\tan \theta = \frac{y}{x}, \quad x \neq 0 \qquad \cot \theta = \frac{x}{y}, \quad y \neq 0$$

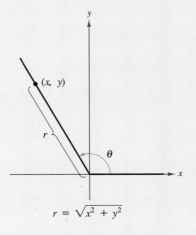

$$r = \sqrt{x^2 + y^2}$$

REMARK   Because $r = \sqrt{x^2 + y^2}$ *cannot* be zero, it follows that the sine and cosine functions are defined for any real value of θ. However, if $x = 0$, the tangent and secant of θ are undefined. For example, the tangent of $90°$ is undefined. Try calculating the tangent of $90°$ with your calculator. Similarly, if $y = 0$, the cotangent and cosecant of θ are undefined.

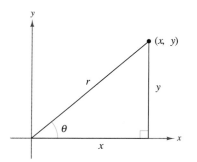

FIGURE 2.32

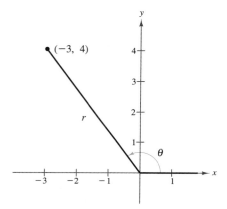

FIGURE 2.33

If θ is an *acute* angle, then these definitions coincide with those given in the previous section. To see this, note in Figure 2.32 that for an acute angle θ, $x = $ adj, $y = $ opp, and $r = $ hyp.

EXAMPLE 1 **Evaluating Trigonometric Functions**

Let $(-3, 4)$ be a point on the terminal side of θ. Find the sine, cosine, and tangent of θ.

SOLUTION

Referring to Figure 2.33, you can see that $x = -3$, $y = 4$, and

$$r = \sqrt{x^2 + y^2} = \sqrt{(-3)^2 + 4^2} = \sqrt{25} = 5.$$

Thus, you can write the following.

$$\sin \theta = \frac{y}{r} = \frac{4}{5}$$

$$\cos \theta = \frac{x}{r} = \frac{-3}{5} = -\frac{3}{5}$$

$$\tan \theta = \frac{y}{x} = \frac{4}{-3} = -\frac{4}{3}$$

The *signs* of trigonometric function values in the four quadrants can be determined easily from the definitions of the functions. For instance, because

$$\cos \theta = \frac{x}{r},$$

it follows that $\cos \theta$ is positive where $x > 0$, which is in Quadrants I and IV. (Remember, r is always positive.) In a similar manner you can verify the results shown in Figure 2.34.

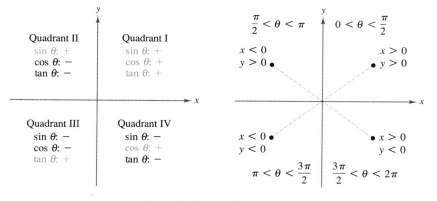

Signs of Trigonometric Functions

FIGURE 2.34

EXAMPLE 2   Evaluating Trigonometric Functions

Given $\tan \theta = -\frac{5}{4}$ and $\cos \theta > 0$, find $\sin \theta$ and $\sec \theta$.

SOLUTION

Note that θ lies in Quadrant IV because that is the only quadrant in which the tangent is negative and the cosine is positive. Moreover, using

$$\tan \theta = \frac{y}{x} = -\frac{5}{4}$$

and the fact that y is negative in Quadrant IV, you can let $y = -5$ and $x = 4$. Hence, $r = \sqrt{16 + 25} = \sqrt{41}$ and you have

$$\sin \theta = \frac{y}{r} = \frac{-5}{\sqrt{41}} \approx -0.7809 \quad \text{and} \quad \sec \theta = \frac{r}{x} = \frac{\sqrt{41}}{4} \approx 1.6008.$$

EXAMPLE 3   Trigonometric Functions of Quadrant Angles

Evaluate the sine function at the four quadrant angles 0, $\pi/2$, π, and $3\pi/2$.

SOLUTION

To begin, choose a point on the terminal side of each angle, as shown in Figure 2.35. For each of the four given points, $r = 1$, and you have

$$\sin 0 = \frac{y}{r} = \frac{0}{1} = 0 \qquad (x, y) = (1, 0)$$

$$\sin \frac{\pi}{2} = \frac{y}{r} = \frac{1}{1} = 1 \qquad (x, y) = (0, 1)$$

$$\sin \pi = \frac{y}{r} = \frac{0}{1} = 0 \qquad (x, y) = (-1, 0)$$

$$\sin \frac{3\pi}{2} = \frac{y}{r} = \frac{-1}{1} = -1. \qquad (x, y) = (0, -1)$$

Trying using Figure 2.35 to evaluate some of the other trigonometric functions at the four quadrant angles and check them on your calculator.

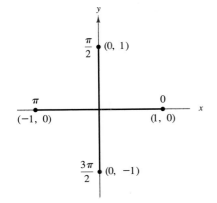

FIGURE 2.35

Reference Angles

The values of the trigonometric functions of angles greater than 90° (or less than 0°) can be determined from their values at corresponding acute angles called **reference angles.**

Quadrant II

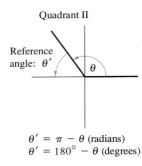

$\theta' = \pi - \theta$ (radians)
$\theta' = 180° - \theta$ (degrees)

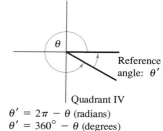

Reference angle: θ'

Quadrant III
$\theta' = \theta - \pi$ (radians)
$\theta' = \theta - 180°$ (degrees)

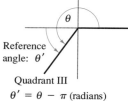

Reference angle: θ'

Quadrant IV
$\theta' = 2\pi - \theta$ (radians)
$\theta' = 360° - \theta$ (degrees)

FIGURE 2.36

DEFINITION OF REFERENCE ANGLES

Let θ be an angle in standard position. Its **reference angle** is the acute angle θ' formed by the terminal side of θ and the horizontal axis.

Figure 2.36 shows the reference angles for θ in Quadrants II, III, and IV.

EXAMPLE 4   Finding Reference Angles

Find the reference angle θ' for each of the following.

A. $\theta = 300°$ **B.** $\theta = 2.3$ **C.** $\theta = -135°$

SOLUTION

A. Because $\theta = 300°$ lies in Quadrant IV, the angle it makes with the x-axis is

$$\theta' = 360° - 300° = 60°. \qquad \textit{Degrees}$$

B. Because $\theta = 2.3$ lies between $\pi/2 \approx 1.5708$ and $\pi \approx 3.1416$, it follows that θ is in Quadrant II and its reference angle is

$$\theta' = \pi - 2.3 \approx 0.8416. \qquad \textit{Radians}$$

C. Because $\theta = -135°$ is coterminal with $225°$, it lies in Quadrant III. Hence, the reference angle is

$$\theta' = 225° - 180° = 45°. \qquad \textit{Degrees}$$

Figure 2.37 shows each angle θ and its reference angle θ'.

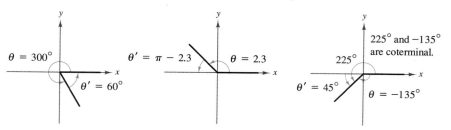

(a) θ in Quadrant IV   (b) θ in Quadrant II   (c) θ in Quadrant III

FIGURE 2.37

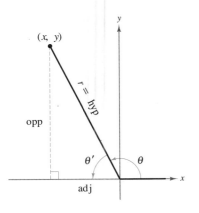

opp = |y|, adj = |x|

FIGURE 2.38

To see how a reference angle is used to evaluate a trigonometric function, consider the point (x, y) on the terminal side of θ, as shown in Figure 2.38. By definition, you know that

$$\sin \theta = \frac{y}{r} \quad \text{and} \quad \tan \theta = \frac{y}{x}.$$

For the right triangle with acute angle θ' and sides of lengths $|x|$ and $|y|$, you have

$$\sin \theta' = \frac{\text{opp}}{\text{hyp}} = \frac{|y|}{r} \quad \text{and} \quad \tan \theta' = \frac{\text{opp}}{\text{adj}} = \frac{|y|}{|x|}.$$

Thus, it follows that $\sin \theta$ and $\sin \theta'$ are equal, *except possibly in sign.* The same is true for $\tan \theta$ and $\tan \theta'$ *and* for the other four trigonometric functions. In all cases, the sign of the function value can be determined by the quadrant in which θ lies.

EVALUATING TRIGONOMETRIC FUNCTIONS OF ANY ANGLE

To find the value of a trigonometric function of any angle θ,

1. Determine the function value for the associated reference angle θ'.
2. Depending on the quadrant in which θ lies, prefix the appropriate sign to the function value.

By using reference angles and the special angles discussed in the previous section, you can greatly extend your scope of *exact* trigonometric values. Table 2.1 lists the function values for selected reference angles. For instance, knowing the function values of 30° means that you know the function values of all angles for which 30° is a reference angle.

TABLE 2.1

θ (degrees)	0°	30°	45°	60°	90°	180°	270°
θ (radians)	0	$\frac{\pi}{6}$	$\frac{\pi}{4}$	$\frac{\pi}{3}$	$\frac{\pi}{2}$	π	$\frac{3\pi}{2}$
$\sin \theta$	0	$\frac{1}{2}$	$\frac{\sqrt{2}}{2}$	$\frac{\sqrt{3}}{2}$	1	0	-1
$\cos \theta$	1	$\frac{\sqrt{3}}{2}$	$\frac{\sqrt{2}}{2}$	$\frac{1}{2}$	0	-1	0
$\tan \theta$	0	$\frac{\sqrt{3}}{3}$	1	$\sqrt{3}$	undef.	0	undef.

EXAMPLE 5   **Trigonometric Functions of Nonacute Angles**

Evaluate the following.

A. $\cos \dfrac{4\pi}{3}$ **B.** $\tan(-210°)$ **C.** $\csc \dfrac{11\pi}{4}$

SOLUTION

A. Because $\theta = 4\pi/3$ lies in Quadrant III, the reference angle is $\theta' = (4\pi/3) - \pi = \pi/3$, as shown in Figure 2.39(a). Moreover, the cosine is negative in Quadrant III, so that

$$\cos \frac{4\pi}{3} = (-)\cos \frac{\pi}{3} = -\frac{1}{2}.$$ *Reference angle, $\pi/3$*

B. Because $-210° + 360° = 150°$, it follows that $-210°$ is coterminal with the second-quadrant angle $150°$. Therefore, the reference angle is $\theta = 180° - 150° = 30°$, as shown in Figure 2.39(b). Finally, because the tangent is negative in Quadrant II, you have

$$\tan(-210°) = (-)\tan 30° = -\frac{\sqrt{3}}{3}.$$ *Reference angle, $30°$*

C. Because $(11\pi/4) - 2\pi = 3\pi/4$, it follows that $11\pi/4$ is coterminal with the second-quadrant angle $3\pi/4$. Therefore, the reference angle is $\theta' = \pi - (3\pi/4) = \pi/4$, as shown in Figure 2.39(c). Because the cosecant is positive in Quadrant II, you have

$$\csc \frac{11\pi}{4} = (+)\csc \frac{\pi}{4} = \frac{1}{\sin \pi/4} = \sqrt{2}.$$ *Reference angle, $\pi/4$*

Check the above values with your calculator.

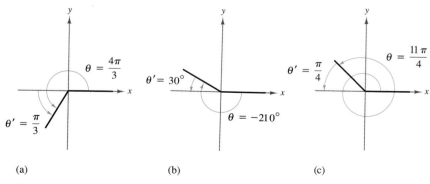

(a)   (b)   (c)

FIGURE 2.39

The fundamental trigonometric identities listed in the previous section (for an acute angle θ) are also valid when θ is any angle.

EXAMPLE 6   Using Identities to Evaluate Trigonometric Functions

Let θ be an angle in Quadrant II such that $\sin \theta = \frac{1}{3}$. Find the following.

A. $\cos \theta$ **B.** $\tan \theta$

SOLUTION

A. Because $\sin \theta = \frac{1}{3}$, use the Pythagorean Identity $\sin^2 \theta + \cos^2 \theta = 1$ to obtain

$$\left(\frac{1}{3}\right)^2 + \cos^2 \theta = 1$$

$$\cos^2 \theta = 1 - \frac{1}{9} = \frac{8}{9}.$$

Because $\cos \theta < 0$ in Quadrant II, use the negative root

$$\cos \theta = -\frac{\sqrt{8}}{\sqrt{9}} = -\frac{2\sqrt{2}}{3}.$$

B. Using the result from part A and the trigonometric identity $\tan \theta = \sin \theta / \cos \theta$, you obtain

$$\tan \theta = \frac{\frac{1}{3}}{-\frac{2\sqrt{2}}{3}} = -\frac{1}{2\sqrt{2}} = -\frac{\sqrt{2}}{4}.$$

Scientific calculators can be used to approximate the values of trigonometric functions of any angle, as demonstrated in Example 7.

EXAMPLE 7   Evaluating Trigonometric Functions with a Calculator

Use a calculator to approximate the following values. (Round your answers to three decimal places.)

A. $\cot 410°$ **B.** $\sin(-7)$ **C.** $\tan \dfrac{14\pi}{5}$

SOLUTION

Function	Mode	Display	Rounded to 3 Decimal Places
A. $\cot 410°$	Degree	0.8390996312	0.839
B. $\sin(-7)$	Radian	-0.6569865987	-0.657
C. $\tan \dfrac{14\pi}{5}$	Radian	-0.726542528	-0.727

Technology Note

Remember that most graphing utilities do not have keys for the cosecant, secant, and cotangent functions. To evaluate these functions, you need to use the reciprocal key with the sine, cosine and tangent. For instance, in Example 7(A), you can find the cotangent of 410° by evaluating 1/tan 410 (in degree mode.)

At this point, you have completed your introduction to basic trigonometry. You have measured angles in both degrees and radians. You have studied the definitions of the six trigonometric functions from a right triangle perspective and as functions of real numbers. In your remaining study of trigonometry, you will continue to rely on both perspectives.

For your convenience we have included in the endpapers of this text a summary of basic trigonometry.

DISCUSSION PROBLEM

....

MEMORIZATION AIDS

There are many different techniques that people use to memorize (or reconstruct) trigonometric formulas. Here is one that we like to use for the sines and cosines of common angles.

θ	$0°$	$30°$	$45°$	$60°$	$90°$
$\sin\theta$	$\dfrac{\sqrt{0}}{2}$	$\dfrac{\sqrt{1}}{2}$	$\dfrac{\sqrt{2}}{2}$	$\dfrac{\sqrt{3}}{2}$	$\dfrac{\sqrt{4}}{2}$
$\cos\theta$	$\dfrac{\sqrt{4}}{2}$	$\dfrac{\sqrt{3}}{2}$	$\dfrac{\sqrt{2}}{2}$	$\dfrac{\sqrt{1}}{2}$	$\dfrac{\sqrt{0}}{2}$

Write a paragraph describing the pattern indicated by this table. Discuss other memory aids for trigonometric formulas.

WARM-UP

....

The following warm-up exercises involve skills that were covered in earlier sections. You will use these skills in the exercise set for this section.

In Exercises 1–6, evaluate the trigonometric function from memory.

1. $\sin 30°$ **2.** $\tan 45°$

3. $\cos \dfrac{\pi}{4}$ **4.** $\cot \dfrac{\pi}{3}$

5. $\sec \dfrac{\pi}{6}$ **6.** $\csc \dfrac{\pi}{4}$

In Exercises 7–10, use the given trigonometric function of an acute angle θ to find the values of the remaining trigonometric functions.

7. $\tan \theta = \frac{3}{2}$ **8.** $\cos \theta = \frac{2}{3}$

9. $\sin \theta = \frac{1}{5}$ **10.** $\sec \theta = 3$

SECTION 2.4 · EXERCISES

In Exercises 1–4, determine the exact values of the six trigonometric functions of the angle θ.

1. (a)   (b)

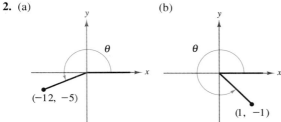

2. (a)   (b)

3. (a)   (b)

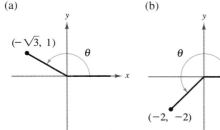

4. (a)   (b)

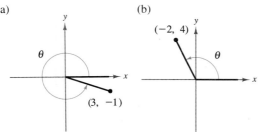

In Exercises 5–8, the point is on the terminal side of an angle in standard position. Determine the exact values of the six trigonometric functions of the angle.

5. (a) $(7, 24)$ (b) $(7, -24)$
6. (a) $(8, 15)$ (b) $(-9, -40)$
7. (a) $(-4, 10)$ (b) $(3, -5)$
8. (a) $(-5, -2)$ (b) $\left(-\frac{3}{2}, 3\right)$

In Exercises 9–12, use the two similar triangles in the figure to find (a) the unknown sides of the triangles and (b) the six trigonometric functions of the angles α_1 and α_2.

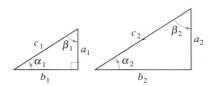

FIGURE FOR 9–12

9. $a_1 = 3$, $b_1 = 4$, $a_2 = 9$
10. $b_1 = 12$, $c_1 = 13$, $c_2 = 26$
11. $a_1 = 1$, $c_1 = 2$, $b_2 = 5$
12. $b_1 = 4$, $a_2 = 4$, $b_2 = 10$

In Exercises 13–16, determine the quadrant in which θ lies.

13. (a) $\sin \theta < 0$ and $\cos \theta < 0$
   (b) $\sin \theta > 0$ and $\cos \theta < 0$
14. (a) $\sin \theta > 0$ and $\cos \theta > 0$
   (b) $\sin \theta < 0$ and $\cos \theta > 0$
15. (a) $\sin \theta > 0$ and $\tan \theta < 0$
   (b) $\cos \theta > 0$ and $\tan \theta < 0$
16. (a) $\sec \theta > 0$ and $\cot \theta < 0$
   (b) $\csc \theta < 0$ and $\tan \theta > 0$

In Exercises 17–26, find the values (if possible) of the six trigonometric functions of θ using the given functional value and constraint.

Functional Value	*Constraint*
17. $\sin \theta = \frac{3}{5}$	θ lies in Quadrant II
18. $\cos \theta = -\frac{4}{5}$	θ lies in Quadrant III
19. $\tan \theta = -\frac{15}{8}$	$\sin \theta < 0$
20. $\cos \theta = \frac{8}{17}$	$\tan \theta < 0$
21. $\sec \theta = -2$	$\sin \theta > 0$
22. $\cot \theta$ is undefined	$\frac{\pi}{2} \le \theta \le \frac{3\pi}{2}$
23. $\sin \theta = 0$	$\sec \theta = -1$
24. $\tan \theta$ is undefined	$\pi \le \theta \le 2\pi$

25. The terminal side of θ is in Quadrant III and lies on the line $y = 2x$.

26. The terminal side of θ is in Quadrant IV and lies on the line $4x + 3y = 0$.

In Exercises 27–34, find the reference angle θ', and sketch θ and θ' in standard position.

27. (a) $\theta = 203°$ (b) $\theta = 127°$

28. (a) $\theta = 309°$ (b) $\theta = 226°$

29. (a) $\theta = -245°$ (b) $\theta = -72°$

30. (a) $\theta = -145°$ (b) $\theta = -239°$

31. (a) $\theta = \frac{2\pi}{3}$ (b) $\theta = \frac{7\pi}{6}$

32. (a) $\theta = \frac{7\pi}{4}$ (b) $\theta = \frac{8\pi}{9}$

33. (a) $\theta = 3.5$ (b) $\theta = 5.8$

34. (a) $\theta = \frac{11\pi}{3}$ (b) $\theta = -\frac{7\pi}{10}$

In Exercises 35–44, evaluate the sine, cosine, and tangent of each angle without using a calculator.

35. (a) $225°$ (b) $-225°$

36. (a) $300°$ (b) $330°$

37. (a) $750°$ (b) $510°$

38. (a) $-405°$ (b) $-120°$

39. (a) $\frac{4\pi}{3}$ (b) $\frac{2\pi}{3}$

40. (a) $\frac{\pi}{4}$ (b) $\frac{5\pi}{4}$

41. (a) $-\frac{\pi}{6}$ (b) $\frac{5\pi}{6}$

42. (a) $-\frac{\pi}{2}$ (b) $\frac{\pi}{2}$

43. (a) $\frac{11\pi}{4}$ (b) $-\frac{13\pi}{6}$

44. (a) $\frac{10\pi}{3}$ (b) $\frac{17\pi}{3}$

In Exercises 45–52, use a calculator to evaluate the trigonometric functions to four decimal places. (Be sure the calculator is set in the correct mode.)

45. (a) $\sin 10°$ (b) $\csc 10°$

46. (a) $\sec 225°$ (b) $\sec 135°$

47. (a) $\cos(-110°)$ (b) $\cos 250°$

48. (a) $\csc 330°$ (b) $\csc 150°$

49. (a) $\tan 240°$ (b) $\cot 210°$

50. (a) $\cot 1.35$ (b) $\tan 1.35$

51. (a) $\tan \frac{\pi}{9}$ (b) $\tan \frac{10\pi}{9}$

52. (a) $\sin(-0.65)$ (b) $\sin 5.63$

In Exercises 53–58, find two values of θ that satisfy the equation. Give your answers in degrees ($0° \le \theta < 360°$) and radians ($0 \le \theta < 2\pi$). Do not use a calculator.

53. (a) $\sin \theta = \frac{1}{2}$ (b) $\sin \theta = -\frac{1}{2}$

54. (a) $\cos \theta = \frac{\sqrt{2}}{2}$ (b) $\cos \theta = -\frac{\sqrt{2}}{2}$

55. (a) $\csc \theta = \frac{2\sqrt{3}}{3}$ (b) $\cot \theta = -1$

56. (a) $\sec \theta = 2$ (b) $\sec \theta = -2$

57. (a) $\tan \theta = 1$ (b) $\cot \theta = -\sqrt{3}$

58. (a) $\sin \theta = \frac{\sqrt{3}}{2}$ (b) $\sin \theta = -\frac{\sqrt{3}}{2}$

In Exercises 59 and 60, use a calculator to approximate two values of $\theta(0° \leq \theta < 360°)$ that satisfy the equation. Round to two decimal places.

59. (a) $\sin \theta = 0.8191$ (b) $\sin \theta = -0.2589$

60. (a) $\cos \theta = 0.8746$ (b) $\cos \theta = -0.2419$

In Exercises 61–64, use a calculator to approximate two values of $\theta(0 \leq \theta < 2\pi)$ that satisfy the equation. Round to three decimal places.

61. (a) $\cos \theta = 0.9848$ (b) $\cos \theta = -0.5890$

62. (a) $\sin \theta = 0.0175$ (b) $\sin \theta = -0.6691$

63. (a) $\tan \theta = 1.192$ (b) $\tan \theta = -8.144$

64. (a) $\cot \theta = 5.671$ (b) $\cot \theta = -1.280$

In Exercises 65–68, use the value of the given trigonometric function and trigonometric identities to find the required trigonometric function of the angle θ in the specified quadrant.

Given Function	Quadrant	Find
65. $\sin \theta = -\frac{3}{5}$	IV	$\cos \theta$
66. $\tan \theta = \frac{3}{2}$	III	$\sec \theta$
67. $\csc \theta = -2$	IV	$\cot \theta$
68. $\sec \theta = -\frac{9}{4}$	III	$\tan \theta$

In Exercises 69 and 70, evaluate the expression without using a calculator.

69. $\sin^2 2 + \cos^2 2$ **70.** $\tan^2 20° - \sec^2 20°$

71. *Average Temperature* The average daily temperature T (in degrees Fahrenheit) for a city is

$$T = 45 - 23 \cos\left[\frac{2\pi}{365}(t - 32)\right],$$

where t is the time in days with $t = 1$ corresponding to January 1. Find the average daily temperature on the following days.

(a) January 1   (b) July 4 ($t = 185$)

(c) October 18 ($t = 291$)

72. *Sales* A company that produces a seasonal product forecasts monthly sales over the next two years to be

$$S = 23.1 + 0.442t + 4.3 \sin \frac{\pi t}{6},$$

where S is measured in thousands of units and t is the time in months, with $t = 1$ representing January 1991. Predict sales for the following months.

(a) February 1991   (b) February 1992

(c) September 1991   (d) September 1992

73. *Distance* An airplane flying at an altitude of 5 miles is on a flight path that passes directly over an observer (see figure). If θ is the angle of elevation from the observer to the plane, find the distance from the observer to the plane when (a) $\theta = 30°$, (b) $\theta = 75°$, and (c) $\theta = 90°$.

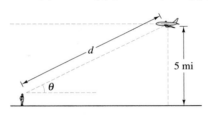

FIGURE FOR 73

74. Consider an angle in standard position with $r = 10$ cm, as shown in the figure. Write a short paragraph describing the changes in the magnitudes of x, y, $\sin \theta$, $\cos \theta$, and $\tan \theta$, as θ increases continuously from $0°$ to $90°$.

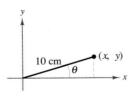

FIGURE FOR 74

2.5   GRAPHS OF SINE AND COSINE FUNCTIONS

Basic Sine and Cosine Curves / Key Points on Basic Sine and Cosine Curves / Amplitude and Period of Sine and Cosine Curves / Translations of Sine and Cosine Curves

Basic Sine and Cosine Curves

In this section you will study techniques for sketching the graphs of the sine and cosine functions. The graph of the sine function is called a **sine curve.** In Figure 2.40, the solid portion of the graph represents one period of the function and is called **one cycle** of the sine curve. The gray portion of the graph indicates that the basic sine wave repeats indefinitely to the right and left. The graph of the cosine function is shown in Figure 2.41. To produce these graphs with a graphing utility, make sure you have set the mode to *radians.*

Recall from Section 2.2 that the domain of the sine and cosine functions is the set of all real numbers. Moreover, the range of each function is the interval $[-1, 1]$, and each function has a period of 2π. Do you see how this information is consistent with the basic graphs given in Figures 2.40 and 2.41?

Note from Figures 2.40 and 2.41 that the sine graph is symmetric with respect to the *origin,* whereas the cosine graph is symmetric with respect to the *y*-axis. These properties of symmetry follow from the fact that the sine function is odd whereas the cosine function is even.

x	0	$\frac{\pi}{6}$	$\frac{\pi}{3}$	$\frac{\pi}{2}$	$\frac{3\pi}{4}$	π	$\frac{3\pi}{2}$	2π
$\sin x$	0	$\frac{1}{2}$	$\frac{\sqrt{3}}{2}$	1	$\frac{\sqrt{2}}{2}$	0	-1	0

Range: $-1 \le y \le 1$

FIGURE 2.40

Period: 2π

x	0	$\frac{\pi}{6}$	$\frac{\pi}{3}$	$\frac{\pi}{2}$	$\frac{3\pi}{4}$	π	$\frac{3\pi}{2}$	2π
$\cos x$	1	$\frac{\sqrt{3}}{2}$	$\frac{1}{2}$	0	$-\frac{\sqrt{2}}{2}$	-1	0	1

Range: $-1 \le y \le 1$

FIGURE 2.41

Period: 2π

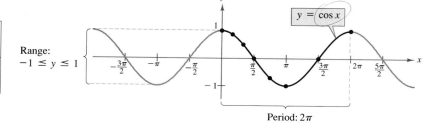

Key Points on Basic Sine and Cosine Curves

To construct the graphs of the basic sine and cosine functions *by hand,* it helps to note five **key points** in one period of each graph: the intercepts, maximum points, and minimum points. For the sine function, the key points are

SINE

Intercept	Maximum	Intercept	Minimum	Intercept

$$(0, 0), \quad \left(\frac{\pi}{2}, 1\right), \quad (\pi, 0), \quad \left(\frac{3\pi}{2}, -1\right), \quad (2\pi, 0).$$

For the cosine function, the key points are

COSINE

Maximum	Intercept	Minimum	Intercept	Maximum

$$(0, 1), \quad \left(\frac{\pi}{2}, 0\right), \quad (\pi, -1), \quad \left(\frac{3\pi}{2}, 0\right), \quad (2\pi, 1).$$

Note how the *x*-coordinates of these points divide the period of sin *x* and cos *x* into *four* equal parts, as indicated in Figure 2.42.

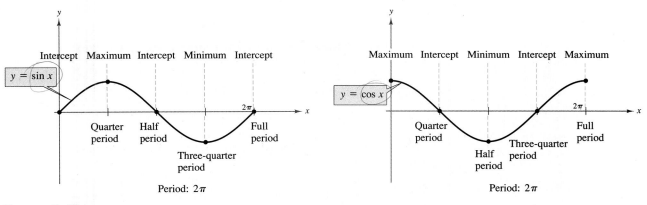

FIGURE 2.42

Technology Note _____

When using a graphing utility to graph trigonometric functions, pay special attention to the viewing rectangle you use. For instance, try graphing $y = [\sin(30x)]/30$ on the standard viewing rectangle. What do you observe? Use the zoom feature to find a viewing rectangle that displays a good view of the graph.

EXAMPLE 1
. . . .

Sketch the graph of $y = 2 \sin x$ on the interval $[-\pi, 4\pi]$.

SOLUTION

The *y*-values for the key points of $y = 2 \sin x$ have twice the magnitude of those of $y = \sin x$. Thus, the key points for $y = 2 \sin x$ are

$$(0, 0), \quad \left(\frac{\pi}{2}, 2\right), \quad (\pi, 0), \quad \left(\frac{3\pi}{2}, -2\right), \quad \text{and} \quad (2\pi, 0).$$

By connecting these key points with a smooth curve and extending the curve in both directions over the interval $[-\pi, 4\pi]$, you obtain the graph shown in Figure 2.43. Try using a graphing utility to check this result.

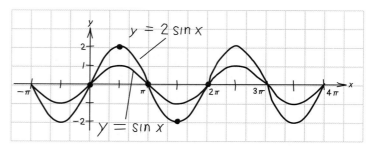

FIGURE 2.43

Amplitude and Period of Sine and Cosine Curves

In the rest of this section, you will study the graphic effect of each of the constants a, b, c, and d in equations of the forms

$$y = d + a \sin(bx - c) \quad \text{and} \quad y = d + a \cos(bx - c).$$

A quick review of the transformations studied in Section 1.8 should help in this investigation.

The constant factor a in $y = a \sin x$ acts as a *vertical stretch* or *vertical shrink* of the basic sine curve, as shown in Example 1. (If $|a| > 1$, the basic sine curve is stretched, and if $|a| < 1$, the basic sine curve is shrunk.) The result is that the graph of $y = a \sin x$ ranges between $-a$ and a instead of between -1 and 1. The absolute value of a is the **amplitude** of the function $y = a \sin x$.

DEFINITION OF AMPLITUDE OF SINE AND COSINE CURVES

The **amplitude** of $y = a \sin x$ and $y = a \cos x$ is the largest value of y and is given by

Amplitude $= |a|$.

EXAMPLE 2   Vertical Shrinking and Stretching

Sketch the graphs of $y = \frac{1}{2} \cos x$ and $y = 3 \cos x$.

SOLUTION

Because the amplitude of $y = \frac{1}{2} \cos x$ is $\frac{1}{2}$, the maximum value is $\frac{1}{2}$ and the minimum value is $-\frac{1}{2}$. For one cycle, $0 \le x \le 2\pi$, the key points are

$$\left(0, \frac{1}{2}\right), \quad \left(\frac{\pi}{2}, 0\right), \quad \left(\pi, -\frac{1}{2}\right), \quad \left(\frac{3\pi}{2}, 0\right), \quad \text{and} \quad \left(2\pi, \frac{1}{2}\right).$$

The amplitude of $y = 3 \cos x$ is 3, and the key points are

$$(0, 3), \quad \left(\frac{\pi}{2}, 0\right), \quad (\pi, -3), \quad \left(\frac{3\pi}{2}, 0\right), \quad \text{and} \quad (2\pi, 3).$$

The graphs of these two functions are shown in Figure 2.44. Try using a graphing utility to check this result.

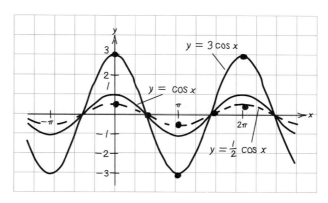

Amplitude Determines Vertical Stretch or Shrink

FIGURE 2.44

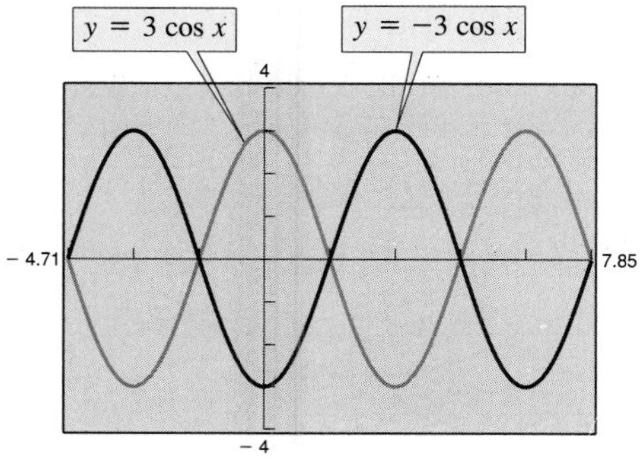

You know from Section 1.8 that the graph of $y = -f(x)$ is a **reflection** (in the x-axis) of the graph of $y = f(x)$. For instance, the graph of $y = -3 \cos x$ is a reflection of the graph of $y = 3 \cos x$, as shown in Figure 2.45.

Because $y = a \sin x$ completes one cycle from $x = 0$ to $x = 2\pi$, it follows that $y = a \sin bx$ completes one cycle from $bx = 0$ to $bx = 2\pi$. This implies that $y = a \sin bx$ completes one cycle from $x = 0$ to $x = 2\pi/b$.

$y = 3 \cos x$ $y = -3 \cos x$

Reflection in the x-Axis

FIGURE 2.45

DISCOVERY

Use a graphing utility to draw the graph of $y = \sin bx$, where $b = 0.5, 1,$ and 2. (Use a viewing rectangle in which $0 \le X \le 6.3$ and $-2 \le Y \le 2$.) How does the value of b affect the graph?

PERIOD OF SINE AND COSINE FUNCTIONS

Let b be a positive real number. The **period** of $y = a \sin bx$ and $y = a \cos bx$ is $2\pi/b$.

Note that if $0 < b < 1$, the period of $y = a \sin bx$ is greater than 2π and represents a *horizontal stretching* of the graph of $y = a \sin x$. Similarly, if $b > 1$, the period of $y = a \sin bx$ is less than 2π and represents a *horizontal shrinking* of the graph of $y = a \sin x$.

If b is negative, use the identities $\sin(-x) = -\sin x$ and $\cos(-x) = \cos x$ to rewrite the function.

EXAMPLE 3   Horizontal Stretching and Shrinking

Sketch the graph of

$$y = \sin \frac{x}{2}.$$

SOLUTION

The amplitude is 1. Moreover, because $b = \frac{1}{2}$, the period is $2\pi/(\frac{1}{2}) = 4\pi$. By dividing the period-interval $[0, 4\pi]$ into four equal parts with the values $\pi, 2\pi,$ and 3π, you obtain the following key points on the graph.

$$(0, 0), \quad (\pi, 1), \quad (2\pi, 0), \quad (3\pi, -1), \quad \text{and} \quad (4\pi, 0)$$

The graph is shown in Figure 2.46. Use a graphing utility to check this result.

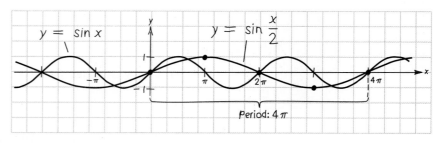

FIGURE 2.46

EXAMPLE 4   Horizontal Stretching and Shrinking

Sketch the graph of $y = \sin 3x$.

SOLUTION

The amplitude is 1, and because $b = 3$, the period is

$$\frac{2\pi}{b} = \frac{2\pi}{3}.$$

Dividing the period-interval $[0, 2\pi/3]$ into four equal parts, you obtain the following key points on the graph.

$$(0, 0), \quad \left(\frac{\pi}{6}, 1\right), \quad \left(\frac{\pi}{3}, 0\right), \quad \left(\frac{\pi}{2}, -1\right), \quad \text{and} \quad \left(\frac{2\pi}{3}, 0\right)$$

The graph is shown in Figure 2.47. Use a graphing utility to check this result.

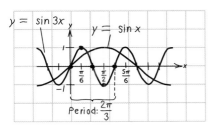

FIGURE 2.47

EXAMPLE 5   Using a Graphing Utility

Use a graphing utility to graph $y = 2 \cos 0.4x$ and $y = 2 \cos 4x$ on the same screen. Determine the period of each function from its graph.

SOLUTION

The standard viewing rectangle is not appropriate for graphing most sine and cosine functions. You should choose a viewing rectangle that accommodates the amplitudes of the functions as well as any stretching or shrinking that may occur. For these graphs, a y-scale ranging between -3 and 3 will accommodate the amplitude of 2. Figure 2.48 shows the two graphs and an appropriate viewing rectangle. From the graph, you can estimate the period of $y = 2 \cos 0.4x$ to be about 16 and the period of $y = 2 \cos 4x$ to be about 1.5. Algebraically, the period of $y = 2 \cos 0.4x$ is

$$\frac{2\pi}{0.4} = 5\pi \approx 15.71,$$

and the period of $y = 2 \cos 4x$ is

$$\frac{2\pi}{4} = \frac{\pi}{2} \approx 1.57.$$

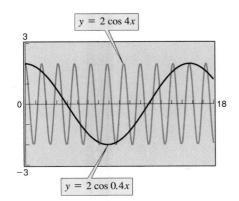

FIGURE 2.48

Translations of Sine and Cosine Curves

The constant c in the general equations

$$y = a \sin(bx - c) \quad \text{and} \quad y = a \cos(bx - c)$$

creates a horizontal shift of the basic sine and cosine curves. The graph of $y = a \sin(bx - c)$ completes one cycle from $bx - c = 0$ to $bx - c = 2\pi$. By solving for x in the inequality $0 \le bx - c \le 2\pi$, you can find the interval for one cycle to be

Left endpoint   Right endpoint

$$\frac{c}{b} \quad \le x \le \quad \frac{c}{b} + \frac{2\pi}{b}.$$

Period

This implies that the period of $y = a \sin(bx - c)$ is $2\pi/b$, and the graph of $y = a \sin bx$ is shifted by an amount c/b. The number c/b is the **phase shift.**

GRAPHS OF THE SINE AND COSINE FUNCTIONS

The graphs of $y = a \sin(bx - c)$ and $y = a \cos(bx - c)$ have the following characteristics. (Assume $b > 0$.)

Amplitude $= |a|$

Period $= 2\pi/b$

The left and right endpoints corresponding to a one-cycle interval of the graphs can be determined by solving the equations $bx - c = 0$ and $bx - c = 2\pi$.

EXAMPLE 6   Horizontal Shift

Describe the graph of

$$y = \frac{1}{2} \sin\left(x - \frac{\pi}{3}\right).$$

SOLUTION

Because $a = \frac{1}{2}$ and $b = 1$, the amplitude is $\frac{1}{2}$ and the period is 2π. By solving the inequality

$$0 \leq x - \frac{\pi}{3} \leq 2\pi$$

$$\frac{\pi}{3} \leq x \qquad \leq \frac{7\pi}{3},$$

you can see that the interval $[\pi/3, 7\pi/3]$ corresponds to one cycle of the graph. Dividing this interval into four equal parts produces the following key points.

$$\left(\frac{\pi}{3}, 0\right), \quad \left(\frac{5\pi}{6}, \frac{1}{2}\right), \quad \left(\frac{4\pi}{3}, 0\right), \quad \left(\frac{11\pi}{6}, -\frac{1}{2}\right), \quad \text{and} \quad \left(\frac{7\pi}{3}, 0\right)$$

The graph is shown in Figure 2.49.

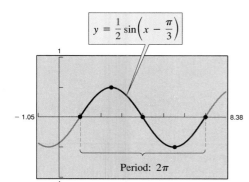

$y = \frac{1}{2} \sin\left(x - \frac{\pi}{3}\right)$

Period: 2π

FIGURE 2.49

The final type of transformation is the *vertical shift* caused by the constant d in the equations

$$y = d + a \sin(bx - c) \quad \text{and} \quad y = d + a \cos(bx - c).$$

The shift is d units upward for $d > 0$ and downward for $d < 0$. In other words, the graph oscillates about the horizontal line $y = d$ instead of the x-axis.

EXAMPLE 7   Vertical Shift

Describe the graph of $y = 2 + 3 \sin 2x$.

SOLUTION

The amplitude is 3 and the period is π. The key points over the interval $[0, \pi]$ are

$$(0, 2), \quad \left(\frac{\pi}{4}, 5\right), \quad \left(\frac{\pi}{2}, 2\right), \quad \left(\frac{3\pi}{4}, -1\right), \quad \text{and} \quad (\pi, 2).$$

The graph is shown in Figure 2.50.

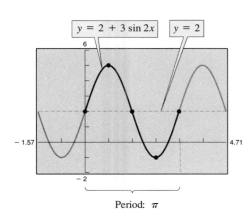

$y = 2 + 3 \sin 2x$ $y = 2$

Period: π

FIGURE 2.50

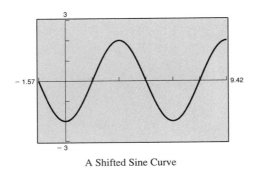

A Shifted Sine Curve

FIGURE 2.51

EXAMPLE 8   Finding an Equation for a Given Graph

....

Find the amplitude, period, and phase shift for the sine function whose graph is shown in Figure 2.51. Write an equation for this graph.

SOLUTION

The amplitude for this sine curve is 2. The period is 2π, and there is a right phase shift of $\pi/2$. Thus, you can write the following equation.

$$y = 2 \sin\left(x - \frac{\pi}{2}\right)$$

Try finding a cosine function with the same graph.

DISCUSSION PROBLEM

....

A SINE SHOW

If you have a programmable calculator, try running the following "sine-show program." This program simultaneously draws the unit circle and the corresponding points on the sine curve. After the circle and sine curve are drawn, you can connect points on the unit circle with their corresponding points on the sine curve by repeatedly pressing ENTER. After the program is complete, the screen should look like that shown in the figure. (This program is for the TI-81. Program steps for other calculators are given in Appendix B.)

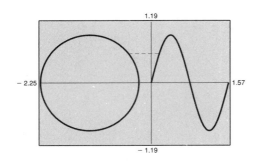

```
:Rad
:ClrDraw
:All-Off
:Param
:Simul
:-2.25→Xmin
:π/2→Xmax
:3→Xscl
:-1.19→Ymin
:1.19→Ymax
:1→Yscl
:0→Tmin
:6.3→Tmax
:.15→Tstep
:"-1.25+cos T"→X₁ₜ
:"sin T"→Y₁ₜ
:"T/4"→X₂ₜ
:"sin T"→Y₂ₜ
```

```
:Line(-1.25,1.19,-1.25,1.19)
:DispGraph
:0→N
:Lbl 1
:N+1→N
:Nπ/6.5→T
:-1.25+cos T→A
:sin T→B
:T/4→C
:Line(A,B,C,B)
:Pause
:If N=12
:Goto 2
:Goto 1
:Lbl 2
:Function
:Sequence
:End
```

WARM-UP

The following warm-up exercises involve skills that were covered in earlier sections. You will use these skills in the exercise set for this section.

In Exercises 1 and 2, simplify the expression.

1. $\dfrac{2\pi}{\frac{1}{3}}$

2. $\dfrac{2\pi}{4\pi}$

In Exercises 3–6, solve for x.

3. $2x - \dfrac{\pi}{3} = 0$

4. $2x - \dfrac{\pi}{3} = 2\pi$

5. $3\pi x + 6\pi = 0$

6. $3\pi x + 6\pi = 2\pi$

In Exercises 7–10, evaluate the trigonometric function without using a calculator.

7. $\sin \dfrac{\pi}{2}$

8. $\sin \pi$

9. $\cos 0$

10. $\cos \dfrac{\pi}{2}$

SECTION 2.5 · EXERCISES

In Exercises 1–10, determine the period and amplitude of the function. Then describe the viewing rectangle for Exercises 1–6.

1. $y = 2 \sin 2x$

2. $y = 3 \cos 3x$

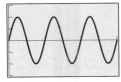

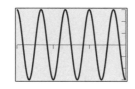

3. $y = \dfrac{3}{2} \cos \dfrac{x}{2}$

4. $y = -2 \sin \dfrac{x}{3}$

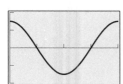

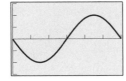

5. $y = \frac{1}{2} \sin \pi x$

6. $y = \dfrac{5}{2} \cos \dfrac{\pi x}{2}$

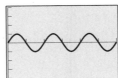

7. $y = -3 \sin 10x$

8. $u = -\dfrac{5}{2} \cos \dfrac{x}{4}$

9. $y = 3 \sin 4\pi x$

10. $y = \dfrac{2}{3} \cos \dfrac{\pi x}{10}$

In Exercises 11–16, describe the relationship between the graphs of f and g.

11. $f(x) = \sin x$
   $g(x) = \sin(x - \pi)$

12. $f(x) = \cos x$
   $g(x) = \cos(x + \pi)$

13. $f(x) = \cos 2x$
   $g(x) = -\cos 2x$

14. $f(x) = \sin 3x$
   $g(x) = \sin(-3x)$

15. $f(x) = \sin x$
   $g(x) = 2 + \sin x$

16. $f(x) = \cos 4x$
   $g(x) = -2 + \cos 4x$

In Exercises 17–24, sketch the graphs of the two functions on the same coordinate plane. (Include two full periods, and use a graphing utility to verify your result.)

17. $f(x) = -2 \sin x$
   $g(x) = 4 \sin x$

18. $f(x) = \sin x$
   $g(x) = \sin \dfrac{x}{3}$

19. $f(x) = \cos x$
   $g(x) = 1 + \cos x$

20. $f(x) = 2 \cos 2x$
   $g(x) = -\cos 4x$

21. $f(x) = -\dfrac{1}{2} \sin \dfrac{x}{2}$
   $g(x) = 3 - \dfrac{1}{2} \sin \dfrac{x}{2}$

22. $f(x) = 4 \sin \pi x$
   $g(x) = 4 \sin \pi x - 3$

23. $f(x) = 2 \cos x$
   $g(x) = 2 \cos(x + \pi)$

24. $f(x) = -\cos x$
   $g(x) = -\cos(x - \pi)$

In Exercises 25–28, sketch the graphs of f and g on the same coordinate axes and show that $f(x) = g(x)$ for all x. (Include two full periods, and use a graphing utility to verify your result.)

25. $f(x) = \sin x$
   $g(x) = \cos\left(x - \dfrac{\pi}{2}\right)$

26. $f(x) = \sin x$
   $g(x) = -\cos\left(x + \dfrac{\pi}{2}\right)$

27. $f(x) = \cos x$
   $g(x) = -\sin\left(x - \dfrac{\pi}{2}\right)$

28. $f(x) = \cos x$
   $g(x) = -\cos(x - \pi)$

In Exercises 29–42, sketch the graph of the function. (Include two full periods, and use a graphing utility to verify your result.)

29. $y = -2 \sin 6x$

30. $y = -3 \cos 4x$

31. $y = \cos 2\pi x$

32. $y = \dfrac{3}{2} \sin \dfrac{\pi x}{4}$

33. $y = -\sin \dfrac{2\pi x}{3}$

34. $y = 10 \cos \dfrac{\pi x}{6}$

35. $y = 2 - \sin \dfrac{2\pi x}{3}$

36. $y = 2 \cos x - 3$

37. $y = \sin\left(x - \dfrac{\pi}{4}\right)$

38. $y = \dfrac{1}{2} \sin(x - \pi)$

39. $y = 3 \cos(x + \pi)$

40. $y = 4 \cos\left(x + \dfrac{\pi}{4}\right)$

41. $y = \dfrac{1}{10} \cos 60\pi x$

42. $y = -3 + 5 \cos \dfrac{\pi t}{12}$

In Exercises 43–54, use a graphing utility to graph the function. (Include two full periods.)

43. $y = 3 \cos(x + \pi) - 3$

44. $y = 4 \cos\left(x + \dfrac{\pi}{4}\right) + 4$

45. $y = \dfrac{2}{3} \cos\left(\dfrac{x}{2} - \dfrac{\pi}{4}\right)$

46. $y = -3 \cos(6x + \pi)$

47. $y = -2 \sin(4x + \pi)$

48. $y = -4 \sin\left(\dfrac{2}{3}x - \dfrac{\pi}{3}\right)$

49. $y = \cos\left(2\pi x - \dfrac{\pi}{2}\right) + 1$

50. $y = 3 \cos\left(\dfrac{\pi x}{2} + \dfrac{\pi}{2}\right) - 2$

51. $y = -0.1 \sin\left(\dfrac{\pi x}{10} + \pi\right)$

52. $y = 5 \sin(\pi - 2x) + 10$

53. $y = 5 \cos(\pi - 2x) + 2$

54. $y = \dfrac{1}{100} \sin 120\pi t$

In Exercises 55–58, use the graph of the trigonometric function to find all real numbers x in the interval $[-2\pi, 2\pi]$ that give the specified function value.

Function	Function Value
55. $\sin x$	$-\dfrac{1}{2}$
56. $\cos x$	-1
57. $\cos x$	$\dfrac{\sqrt{2}}{2}$
58. $\sin x$	$\dfrac{\sqrt{3}}{2}$

In Exercises 59 and 60, find a and d for the function $f(x) = a \cos x + d$ so that the graph of f matches the figure.

59.   **60.**

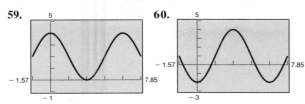

In Exercises 61–64, find a, b, and c so that the graph of the function matches the graph in the figure.

61. $y = a \sin(bx - c)$ **62.** $y = a \sin(bx - c)$

63. $y = a \cos(bx - c)$ **64.** $y = a \sin(bx - c)$

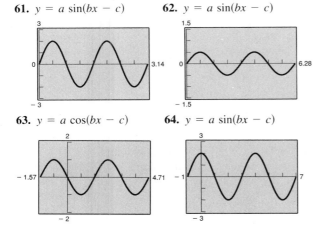

65. Use a graphing utility to graph the functions $f(x) = 2e^x$ and $g(x) = 5 \cos x$. Approximate any points of intersection of the graphs in the interval $[-\pi, \pi]$.

66. Use a graphing utility to determine the smallest *integer* value of a such that the graphs of $f(x) = 2 \ln x$ and $g(x) = a \cos x$ intersect more than once.

67. *Respiratory Cycle*   For a person at rest, the velocity v (in liters per second) of air flow during a respiratory cycle is

$$v = 0.85 \sin \frac{\pi t}{3},$$

where t is the time in seconds. (Inhalation occurs when $v > 0$, and exhalation occurs when $v < 0$.)
   (a) Find the time for one full respiratory cycle.
   (b) Find the number of cycles per minute.
   (c) Use a graphing utility to graph the velocity function.

68. *Respiratory Cycle*   After exercising for a few minutes, a person has a respiratory cycle for which the velocity of air flow is approximated by

$$v = 1.75 \sin \frac{\pi t}{2}.$$

Use this model to repeat Exercise 67.

69. *Piano Tuning*   When tuning a piano, a technician strikes a tuning fork for the A above middle C and sets up wave motion that can be approximated by

$$y = 0.001 \sin 880 \pi t,$$

where t is the time in seconds.
   (a) What is the period p of this function?
   (b) The frequency f is given by $f = 1/p$. What is the frequency of this note?
   (c) Use a graphing utility to graph this function.

70. *Blood Pressure*   The function

$$P = 100 - 20 \cos \frac{5 \pi t}{3}$$

approximates the blood pressure P in millimeters of mercury at time t in seconds for a person at rest.
   (a) Find the period of the function.
   (b) Find the number of heartbeats per minute.
   (c) Use a graphing utility to graph the pressure function.

Sales   In Exercises 71 and 72, use a graphing utility to graph the sales function over 1 year where S is sales in thousands of units and t is the time in months, with $t = 1$ corresponding to January. Use the graph to determine the month of maximum sales and the month of minimum sales.

71. $S = 22.3 - 3.4 \cos \dfrac{\pi t}{6}$

72. $S = 74.50 + 43.75 \sin \dfrac{\pi t}{6}$

In Exercises 73–76, describe the relationship between the graphs of the functions f and g.

73.   **74.**

75.   **76.**

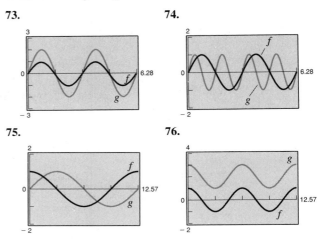

2.6   OTHER TRIGONOMETRIC GRAPHS

Graphs of Tangent and Cotangent Functions / Graphs of the Reciprocal
Functions / graphs of Combinations of Trigonometric Functions /
Combinations of Algebraic and Trigonometric Functions /
Damped Trigonometric Graphs

Graphs of Tangent and Cotangent Functions

Recall from Section 2.2 that the tangent functions is odd. That is, $\tan(-x) = -\tan x$. Consequently, the graph of $y = \tan x$ is symmetric with respect to the origin. From the identity $\tan x = \sin x/\cos x$, you know that the tangent is undefined when $\cos x = 0$. Two such values are $x = \pm\pi/2 \approx \pm 1.5708$.

x	$-\dfrac{\pi}{2}$	-1.57	-1.5	-1	0	1	1.5	1.57	$\dfrac{\pi}{2}$
$\tan x$	undef.	-1255.8	-14.1	-1.56	0	1.56	14.1	1255.8	undef.

$\tan x$ approaches $-\infty$ as x approaches $-\pi/2$ from the right

$\tan x$ approaches ∞ as x approaches $\pi/2$ from the left

As indicated in the table, $\tan x$ increases without bound as x approaches $\pi/2$ from the left, and decreases without bound as x approaches $-\pi/2$ from the right. Thus, the graph of $y = \tan x$ has *vertical asymptotes* at $x = \pi/2$ and $-\pi/2$, as shown in Figure 2.52. Moreover, because the period of the tangent function is π, vertical asymptotes also occur when $x = \pi/2 \pm n\pi$. The domain of the tangent function is the set of all real numbers other than $x = \pi/2 \pm n\pi$, and the range is the set of all real numbers.

Sketching the graph of a function with the form $y = a\tan(bx - c)$ is similar to sketching the graph of $y = a\sin(bx - c)$ in that you locate key points which identify the intercepts and asymptotes. Two consecutive asymptotes can be found by solving the equations

$$bx - c = -\frac{\pi}{2} \quad \text{and} \quad bx - c = \frac{\pi}{2}.$$

The midpoint between two consecutive asymptotes is an x-intercept of the graph. After plotting the asymptotes and the x-intercept, plot a few additional points between the two asymptotes and sketch one cycle. Finally, sketch one or two additional cycles to the left and right.

Period: π
Domain: all $x \neq \dfrac{\pi}{2} \pm n\pi$

Range: $(-\infty, \infty)$
Vertical asymptotes: $x = \dfrac{\pi}{2} \pm n\pi$

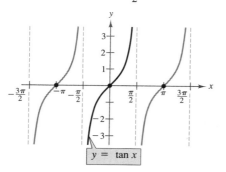

$y = \tan x$

FIGURE 2.52

REMARK   The period of the function $y = a\tan(bx - c)$ is the distance between two consecutive asymptotes. The amplitude of a tangent function is not defined.

EXAMPLE 1   Sketching the Graph of a Tangent Function

Sketch the graph of $y = \tan \dfrac{x}{2}$.

SOLUTION

By solving the inequality

$$-\frac{\pi}{2} < \frac{x}{2} < \frac{\pi}{2}$$

$$-\pi < x < \pi$$

you can see that two consecutive asymptotes occur at $x = -\pi$ and $x = \pi$. Between these two asymptotes, plot a few points including the x-intercept, as shown in the table. Three cycles of the graph are shown in Figure 2.53. Use a graphing utility to confirm this result.

x	$-\dfrac{\pi}{2}$	0	$\dfrac{\pi}{2}$
$\tan \dfrac{x}{2}$	-1	0	1

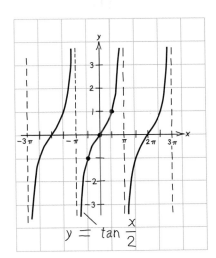

FIGURE 2.53

EXAMPLE 2   Sketching the Graph of a Tangent Function

Sketch the graph of $y = -3 \tan 2x$.

SOLUTION

By solving the inequality

$$-\frac{\pi}{2} < 2x < \frac{\pi}{2}$$

$$-\frac{\pi}{4} < x < \frac{\pi}{4}$$

you can see that two consecutive asymptotes occur at $x = -\pi/4$ and $x = \pi/4$. Between these two asymptotes, plot a few points as shown in the table, and complete one cycle. Four cycles of the graph are shown in Figure 2.54. Use a graphing utility to confirm this result.

x	$-\dfrac{\pi}{8}$	0	$\dfrac{\pi}{8}$
$-3 \tan 2x$	3	0	-3

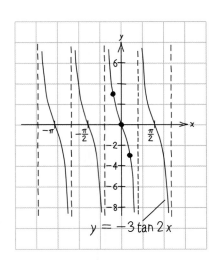

FIGURE 2.54

Period: π
Domain: all $x \neq n\pi$
Range: $(-\infty, \infty)$
Vertical asymptotes: $x = n\pi$

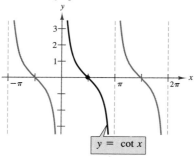

FIGURE 2.55

Technology Note _____

You can use the tangent function on your graphing utility to obtain the graph of the cotangent function. For example, to graph the function $y = 2 \cot(x/3)$ from Example 3, let

$y_1 = 2/\tan(x/3)$.

If you select the viewing rectangle $-9 \leq x \leq 18$ and $-6 \leq y \leq 6$, you should obtain a graph similar to that of Figure 2.56.

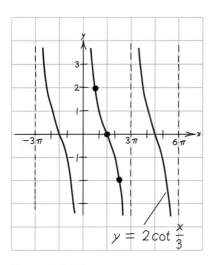

FIGURE 2.56

By comparing the graphs in Examples 1 and 2, you can see that the graph of

$$y = a \tan(bx + c)$$

is increasing between consecutive vertical asymptotes if $a > 0$ and decreasing between consecutive vertical asymptotes if $a < 0$. In other words, the graph for $a < 0$ is a reflection of the graph for $a > 0$.

The graph of the cotangent function is similar to the graph of the tangent function. It also has a period of π. However, from the identity

$$y = \cot x = \frac{\cos x}{\sin x}$$

you can see that the cotangent function has vertical asymptotes at $x = n\pi$ (because $\sin x$ is zero at these x-values). The graph of the cotangent function is shown in Figure 2.55.

EXAMPLE 3 **Sketching the Graph of a Cotangent Function**

Sketch the graph of $y = 2 \cot \dfrac{x}{3}$.

SOLUTION

To locate two consecutive vertical asymptotes of the graph, you can solve the following inequality.

$$0 < \frac{x}{3} < \pi$$

$$0 < x < 3\pi$$

Then, between these two asymptotes, plot the points shown in the following table, and complete one cycle of the graph. (Note that the period is 3π, the distance between consecutive asymptotes.) Three cycles of the graph are shown in Figure 2.56.

x	$\dfrac{3\pi}{4}$	$\dfrac{3\pi}{2}$	$\dfrac{9\pi}{4}$
$2 \cot \dfrac{x}{3}$	2	0	-2

Graphs of the Reciprocal Functions

The graphs of the two remaining trigonometric functions can be obtained from the graphs of the sine and cosine functions using the reciprocal identities

$$\csc x = \frac{1}{\sin x} \quad \text{and} \quad \sec x = \frac{1}{\cos x}.$$

For instance, at a given value for x, the y-coordinate for sec x is the reciprocal of the y-coordinate for cos x. Of course, when cos $x = 0$, the reciprocal does not exist. Near such values for x, the behavior of the secant function is similar to that of the tangent function. In other words, the graphs of tan $x = (\sin x)/(\cos x)$ and sec $x = 1/(\cos x)$ have vertical asymptotes at $x = (\pi/2) + n\pi$, because the cosine is zero at these x-values. Similarly, cot $x = (\cos x)/(\sin x)$ and csc $x = 1/(\sin x)$ have vertical asymptotes where sin $x = 0$, that is, at $x = n\pi$.

To sketch the graph of a secant or cosecant function, we suggest that you first make a sketch of its reciprocal function. For instance, to sketch the graph of $y = \csc x$, first sketch the graph of $y = \sin x$. Then take reciprocals of the y-coordinates to obtain points on the graph of $y = \csc x$. You can use this procedure to obtain the graphs shown in Figure 2.57.

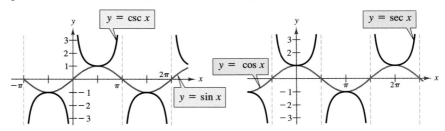

Period: 2π
Domain: all $x \neq n\pi$
Range: all y not in $(-1, 1)$
Vertical asymptotes: $x = n\pi$
Symmetry: origin

FIGURE 2.57

Period: 2π
Domain: all $x \neq \dfrac{\pi}{2} + n\pi$
Range: all y not in $(-1, 1)$

Vertical asymptotes: $x = \dfrac{\pi}{2} + n\pi$

Symmetry: y-axis

In comparing the graphs of the secant and cosecant functions with those of the sine and cosine functions, note that the "hills" and "valleys" are interchanged. For example, a hill (or maximum point) on the sine curve corresponds to a valley (a local minimum) on the cosecant curve. Similarly, a valley (or minimum point) on the sine curve corresponds to a hill (a local maximum) on the cosecant curve, as shown in Figure 2.58.

EXAMPLE 4   *Graphing a Cosecant Function*

Use a graphing utility to graph

$$y = 2 \sin\left(x + \frac{\pi}{4}\right) \quad \text{and} \quad y = 2 \csc\left(x + \frac{\pi}{4}\right).$$

D I S C O V E R Y

Use a graphing utility to graph the functions $y_1 = \sin x$ and $y_2 = \csc x = 1/\sin x$ on the same viewing rectangle. How are the graphs related? What happens to the graph of the cosecant function as x approaches the zeros of the sine function? Similarly, graph $y_1 = \cos x$ and $y_2 = \sec x = 1/\cos x$ on the same viewing rectangle. How are these functions related?

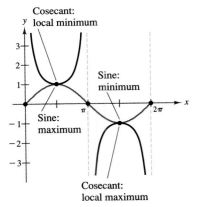

FIGURE 2.58

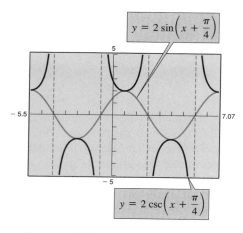

FIGURE 2.59

SOLUTION

The two graphs are shown in Figure 2.59. Note how the "hills" and "valleys" of each graph are related. For the function $y = 2 \sin[x + (\pi/4)]$, the amplitude is 2 and the period is 2π. One cycle of the sine function corresponds to the interval from $x = -\pi/4$ to $x = 7\pi/4$. Because the sine function is zero at the endpoints of this interval, the corresponding cosecant function

$$y = 2 \csc\left(x + \frac{\pi}{4}\right) = 2\left[\frac{1}{\sin\left(x + \frac{\pi}{4}\right)}\right]$$

has vertical asymptotes at $x = -\pi/4$, $x = 3\pi/4$, and $7\pi/4$.

In Figure 2.60, we summarize the graphs, domains, ranges, and periods of the six basic trigonometric functions.

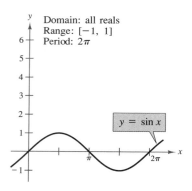

Domain: all reals
Range: $[-1, 1]$
Period: 2π

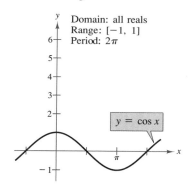

Domain: all reals
Range: $[-1, 1]$
Period: 2π

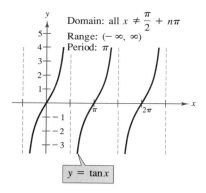

Domain: all $x \neq \dfrac{\pi}{2} + n\pi$
Range: $(-\infty, \infty)$
Period: π

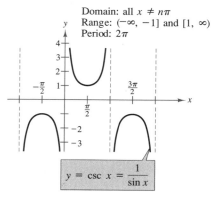

Domain: all $x \neq n\pi$
Range: $(-\infty, -1]$ and $[1, \infty)$
Period: 2π

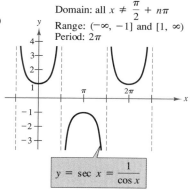

Domain: all $x \neq \dfrac{\pi}{2} + n\pi$
Range: $(-\infty, -1]$ and $[1, \infty)$
Period: 2π

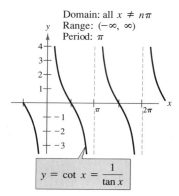

Domain: all $x \neq n\pi$
Range: $(-\infty, \infty)$
Period: π

Graphs of the Six Trigonometric Functions

FIGURE 2.60

Graphs of Combinations of Trigonometric Functions

Sums, differences, products, and quotients of periodic functions are also periodic. The period of the combined function is the least common multiple of the periods of the component functions. This period is important for determining an appropriate viewing rectangle for a graphing utility.

EXAMPLE 5   Finding the Period and Relative Extrema of
a Function

Graph $y = \sin x - \cos 2x$. Find the period and the relative minimums and maximums of this function.

SOLUTION

The period of $\sin x$ is 2π and the period of $\cos 2x$ is π. Thus, the period of the given function is 2π, because 2π is the least common multiple of 2π and π. This conclusion is further reinforced by graphing the function, as shown in Figure 2.61. From the graph, it appears that the function has two relative maximums and two relative minimums in each complete cycle. For instance, between 0 and 2π, the graph appears to have relative maximums when $x = \pi/2 \approx 1.57$ and when $x = 3\pi/2 \approx 4.71$. Using the zoom and trace features of the graphing utility, you can find that the relative minimums occur when $x \approx 3.40$ and when $x \approx 6.03$.

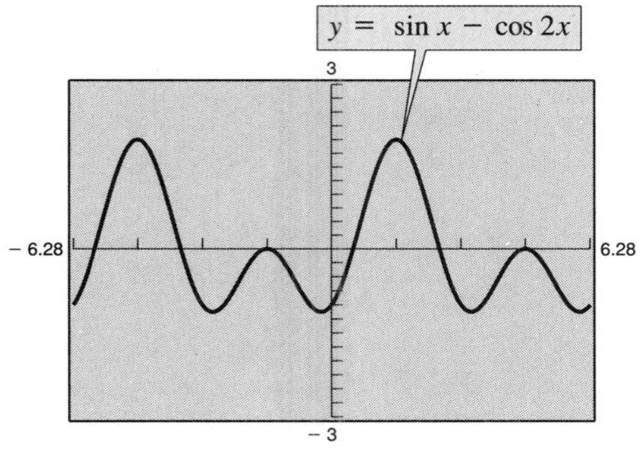

$\quad$ *Relative maximums:* $\quad$ (1.57, 2.00) $\qquad$ (4.71, 0.00)
$\quad$ *Relative minimums:* $\quad$ (3.40, -1.13) $\qquad$ (6.03, -1.13)

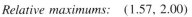

FIGURE 2.61

$y = \sin x - \cos 2x$

EXAMPLE 6   Finding the Period and Range of a Function

Graph $y = 2 \sin 6x + \sin 4x$. Find the period and range of this function.

SOLUTION

The period of $2 \sin 6x$ is $\pi/3$ and the period of $\sin 4x$ is $\pi/2$. Thus, the period of the given function is π, because π is the least common multiple of $\pi/3$ and $\pi/2$. This conclusion is further reinforced by graphing the function, as shown in Figure 2.62. In the interval from 0 to π, the maximum y-value occurs when $x \approx 0.29$ and $y \approx 2.89$. The minimum value in this interval occurs when $x \approx 2.86$ and $y \approx -2.89$. Thus, the range of the function is approximated by

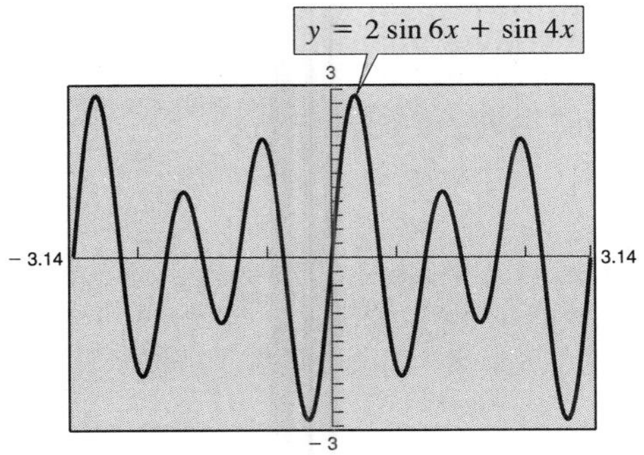

$\qquad -2.89 \leq y \leq 2.89$. $\qquad$ *Range*

FIGURE 2.62

$y = 2 \sin 6x + \sin 4x$

Combinations of Algebraic and Trigonometric Functions

Functions that are combinations of algebraic and trigonometric functions are not, in general, periodic.

EXAMPLE 7   The Graph of a Nonperiodic Function

Graph $y = x + \cos x$. Find the domain and range of the function. Approximate any zeros of the graph.

SOLUTION

The graph of the function is shown in Figure 2.63. Notice that even though the function is not periodic, it does have a pattern that repeats. Notice that the graph of $y = x + \cos x$ oscillates about the line $y = x$. Both the domain and range of the function are the set of all real numbers. Using the zoom feature, you can find that the zero of $y = x + \cos x$ is approximately $x = -0.739$.

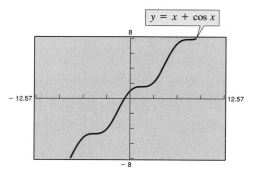

FIGURE 2.63

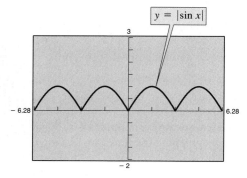

FIGURE 2.64

EXAMPLE 8   A Function Involving Absolute Value

Graph $y = |\sin x|$, and find the domain and range of the function.

SOLUTION

The domain of the function is the set of all real numbers. The graph of the function is shown in Figure 2.64. Notice that the minimum value of the function is 0 and the maximum value is 1. Thus, the range of the function is given by $0 \le y \le 1$.

Damped Trigonometric Graphs

A *product* of two functions can be graphed using properties of the individual functions involved. For instance, consider the function

$$f(x) = x \sin x$$

as the product of the functions $y = x$ and $y = \sin x$. Using properties of absolute value and the fact that $|\sin x| \leq 1$, you obtain $0 \leq |x||\sin x| \leq |x|$. Consequently,

$$-|x| \leq x \sin x \leq |x|,$$

which means that the graph of $f(x) = x \sin x$ lies between the lines $y = -x$ and $y = x$. Furthermore, since

$$f(x) = x \sin x = \pm x \quad \text{at} \quad x = \frac{\pi}{2} + n\pi$$

$$f(x) = x \sin x = 0 \quad \text{at} \quad x = n\pi,$$

the graph of f touches the line $y = -x$ or the line $y = x$ at $x = (\pi/2) + n\pi$ and has x-intercepts at $x = n\pi$. The graph of f, together with $y = x$ and $y = -x$, is shown in Figure 2.65.

In the function $f(x) = x \sin x$, the factor x is called the **damping factor.** By changing the damping factor, you can change the graph significantly. For example, look in Figure 2.66 at the graphs of

$$y = \frac{1}{x} \sin x$$

and

$$y = e^{-x} \sin 3x.$$

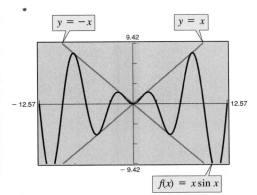

FIGURE 2.65

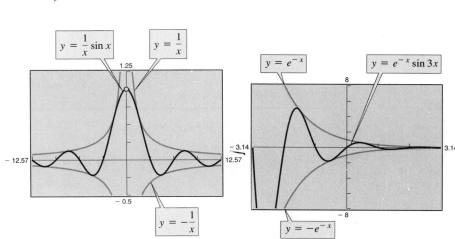

FIGURE 2.66

EXAMPLE 9   Damped Cosine Wave

Graph $f(x) = 2^{-x/2} \cos x$.

SOLUTION

A graph of $f(x) = 2^{-x/2} \cos x$ is shown in Figure 2.67. To analyze this fuction further, consider $f(x)$ as the product of the two functions

$$y = 2^{-x/2} \quad \text{and} \quad y = \cos x,$$

each of which has the set of real numbers as its domain. For any real number x, you know that $2^{-x/2} \geq 0$ and $|\cos x| \leq 1$. Therefore, $|2^{-x/2}||\cos x| \leq 2^{-x/2}$, which means that

$$-2^{-x/2} \leq 2^{-x/2} \cos x \leq 2^{-x/2}.$$

Furthermore, since

$$f(x) = 2^{-x/2} \cos x = \pm 2^{-x/2} \quad \text{at} \quad x = n\pi$$

and

$$f(x) = 2^{-x/2} \cos x = 0 \quad \text{at} \quad x = \frac{\pi}{2} + n\pi,$$

the graph of f touches the curve $y = -2^{-x/2}$ or the curve $y = 2^{-x/2}$ at $x = n\pi$ and has intercepts at $x = (\pi/2) + n\pi$.

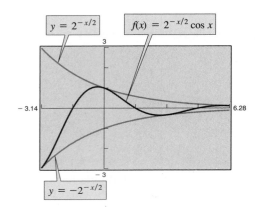

FIGURE 2.67

DISCUSSION PROBLEM

GRAPHING UTILITIES

Use a graphing utility to graph the following functions for varying values of a, b, c, and d.

$$y = d + a \sin(bx + c) \qquad y = d + a \cos(bx + c)$$
$$y = d + a \tan(bx + c) \qquad y = d + a \cot(bx + c)$$
$$y = d + a \sec(bx + c) \qquad y = d + a \csc(bx + c)$$

In a paper or a discussion, summarize the effects of the constants a, b, c, and d in these graphs.

WARM-UP

The following warm-up exercises involve skills that were covered in earlier sections. You will use these skills in the exercise set for this section.

In Exercises 1–4, find the x-values in the interval $[0, 2\pi]$ for which $f(x)$ is $-1, 0$, or 1.

1. $f(x) = \sin x$ **2.** $f(x) = \cos x$ **3.** $f(x) = \sin 2x$ **4.** $f(x) = \cos \dfrac{x}{2}$

In Exercises 5–8, graph the function.

5. $y = |x|$ **6.** $y = e^{-x}$ **7.** $y = \sin \pi x$ **8.** $y = \cos 2x$

In Exercises 9 and 10, evaluate $f(x)$ when $x = 0, \pi/6, \pi/4, \pi/3$, and $\pi/2$.

9. $f(x) = x \cos x$ **10.** $f(x) = x + \sin x$

SECTION 2.6 · EXERCISES

In Exercises 1–8, match the trigonometric function with its graph and describe the viewing rectangle. [The graphs are labeled (a), (b), (c), (d), (e), (f), (g), and (h).]

1. $y = \sec 2x$

2. $y = \tan 3x$

3. $y = \tan \dfrac{x}{2}$

4. $y = 2 \csc \dfrac{x}{2}$

5. $y = \cot \pi x$

6. $y = \frac{1}{2} \sec \pi x$

7. $y = -\sec x$

8. $y = -2 \csc 2\pi x$

(c)   (d)

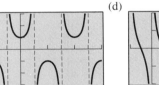

(e)   (f)

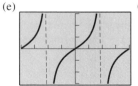

(a)   (b)

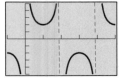

(g)   (h)

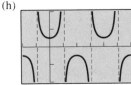

In Exercises 9–30, sketch the graph of the function. (Include two full periods, and use a graphing utility to verify your result.)

9. $y = \frac{1}{3} \tan x$ **10.** $y = \frac{1}{4} \tan x$

11. $y = \tan 2x$ **12.** $y = -3 \tan \pi x$

13. $y = -\frac{1}{2} \sec x$ **14.** $y = \frac{1}{4} \sec x$

15. $y = -\sec \pi x$ **16.** $y = 2 \sec 4x$

17. $y = \sec \pi x - 1$ **18.** $y = -2 \sec 4x + 2$

19. $y = \csc \frac{x}{2}$ **20.** $y = \csc \frac{x}{3}$

21. $y = \cot \frac{x}{2}$ **22.** $y = 3 \cot \frac{\pi x}{2}$

23. $y = \frac{1}{2} \sec 2x$ **24.** $y = -\frac{1}{2} \tan x$

25. $y = \tan \frac{\pi x}{4}$ **26.** $y = \sec(x + \pi)$

27. $y = \csc(\pi - x)$ **28.** $y = \sec(\pi - x)$

29. $y = \frac{1}{4} \csc\left(x + \frac{\pi}{4}\right)$ **30.** $y = 2 \cot\left(x + \frac{\pi}{2}\right)$

In Exercises 31–40, use a graphing utility to graph the function. (Include two full periods.)

31. $y = \tan \frac{x}{3}$ **32.** $y = -\tan 2x$

33. $y = -2 \sec 4x$ **34.** $y = \sec \pi x$

35. $y = \tan\left(x - \frac{\pi}{4}\right)$ **36.** $y = -\csc(4x - \pi)$

37. $y = \frac{1}{4} \cot\left(x - \frac{\pi}{2}\right)$ **38.** $y = \frac{1}{3} \sec\left(\frac{\pi x}{2} + \frac{\pi}{2}\right)$

39. $y = 2 \sec(2x - \pi)$ **40.** $y = 0.1 \tan\left(\frac{\pi x}{4} + \frac{\pi}{4}\right)$

In Exercises 41–44, use the graph of the trigonometric function to find all real numbers x in the interval $[-2\pi, 2\pi]$ that give the specified function value.

Function	*Function Value*
41. $\tan x$	1
42. $\cot x$	$-\sqrt{3}$
43. $\sec x$	-2
44. $\csc x$	$\sqrt{2}$

In Exercises 45 and 46, use the graph of the function to determine whether the function is even, odd, or neither.

45. $f(x) = \sec x$ **46.** $f(x) = \tan x$

47. Consider the functions $f(x) = 2 \sin x$ and $g(x) = \frac{1}{2} \csc x$ over the interval $(0, \pi)$.
  (a) Use a graphing utility to graph f and g in the same viewing rectangle.
  (b) Approximate the interval where $f > g$.
  (c) Describe the behavior of each of the functions as x approaches π. How is the behavior of g related to the behavior of f as x approaches π?

48. Consider the functions $f(x) = \tan(\pi x/2)$ and $g(x) = \frac{1}{2} \sec(\pi x/2)$ over the interval $(-1, 1)$.
  (a) Use a graphing utility to graph f and g in the same viewing rectangle.
  (b) Approximate the interval where $f < g$.

In Exercises 49–54, sketch the graph of the function. (Use a graphing utility to verify your result.)

49. $y = 2 - 2 \sin \frac{x}{2}$ **50.** $y = -3 + \cos x$

51. $y = 4 - 2 \cos \pi x$ **52.** $y = 5 - \frac{1}{2} \sin 2\pi x$

53. $y = 1 + \csc x$ **54.** $y = 2 + \tan \pi x$

In Exercises 55–60, use a graphing utility to graph the function. Determine the period of the function and approximate any relative minimums and maximums of the function through one period.

55. $y = \sin x + \cos x$ **56.** $y = \cos x + \cos 2x$

57. $f(x) = 2 \sin x + \sin 2x$ **58.** $f(x) = 2 \sin x + \cos 2x$

59. $g(x) = \cos x - \cos \frac{x}{2}$ **60.** $g(x) = \sin x - \frac{1}{2} \sin \frac{x}{2}$

In Exercises 61–64, use a graphing utility to graph the function through three periods.

61. $h(x) = \sin x + \frac{1}{3} \sin 5x$

62. $h(x) = \cos x - \frac{1}{4} \cos 2x$

63. $y = -3 + \cos x + 2 \sin 2x$

64. $y = \sin \pi x + \sin \frac{\pi x}{2}$

In Exercises 65–72, use a graphing utility to graph the given function and the algebraic component of the function in the same viewing rectangle. Notice that the graph of the function oscillates about the graph of the algebraic component.

65. $y = x + \sin x$

66. $y = x + \cos x$

67. $f(x) = \frac{1}{2}x - 2\cos x$

68. $f(x) = 2x - \sin x$

69. $g(t) = -t + \sin \dfrac{\pi t}{2}$

70. $h(s) = -\dfrac{s}{2} - \dfrac{1}{2}\sin \dfrac{\pi s}{4}$

71. $y = \dfrac{x^2}{8} + \sin \dfrac{\pi x}{2}$

72. $y = 4 - \dfrac{x^2}{16} + 4\cos \pi x$

In Exercises 73–76, use a graphing utility to graph the function and the damping factor of the function in the same viewing rectangle. Describe the behavior of the function as x increases without bound.

73. $f(x) = 2^{-x/4} \cos \pi x$

74. $f(t) = e^{-t} \cos t$

75. $g(x) = e^{-x^2/2} \sin x$

76. $h(t) = 2^{-t^2/4} \sin t$

In Exercises 77–80, use a graphing utility to graph the function and the equations $y = x$ and $y = -x$ in the same viewing rectangle. Describe the behavior of the given function as x approaches zero.

77. $f(x) = x \cos x$

78. $f(x) = |x \sin x|$

79. $g(x) = |x| \sin x$

80. $g(x) = |x| \cos x$

In Exercises 81–86, use a graphing utility to graph the function. Describe the behavior of the function as x approaches zero.

81. $y = \dfrac{6}{x} + \cos x, \quad x > 0$

82. $y = \dfrac{4}{x} + \sin 2x, \quad x > 0$

83. $g(x) = \dfrac{\sin x}{x}$

84. $f(x) = \dfrac{1 - \cos x}{x}$

85. $f(x) = \sin \dfrac{1}{x}$

86. $h(x) = x \sin \dfrac{1}{x}$

In Exercises 87–90, use a graphing utility to graph the functions f and g in the same viewing rectangle. Use the graphs to determine the relationship between the functions.

87. $f(x) = \sin x + \cos\left(x + \dfrac{\pi}{2}\right), \quad g(x) = 0$

88. $f(x) = \sin x - \cos\left(x + \dfrac{\pi}{2}\right), \quad g(x) = 2\sin x$

89. $f(x) = \sin^2 x, \quad g(x) = \frac{1}{2}(1 - \cos 2x)$

90. $f(x) = \cos^2 \dfrac{\pi x}{2}, \quad g(x) = \dfrac{1}{2}(1 + \cos \pi x)$

In Exercises 91–94, use a graphing utility to graph the function over the specified interval.

Function	Interval
91. $f(t) = t^2 \sin t$	$[0, 2\pi]$
92. $f(x) = \sqrt{2x} \sin x$	$[0, 4\pi]$
93. $f(x) = \sin x - \frac{1}{3}\sin 3x + \frac{1}{5}\sin 5x$	$[0, \pi]$
94. $f(x) = \dfrac{1}{2} - \dfrac{4}{\pi^2}\left(\cos \pi x + \dfrac{1}{9}\cos 3\pi x\right)$	$[0, 2]$

95. *Distance*   A plane flying at an altitude of 6 miles over level ground will pass directly over a radar antenna (see figure). Let d be the ground distance from the antenna to the point directly under the plane and let x be the angle of elevation to the plane from the antenna. Write d as a function of x, and graph the function over the interval $0 < x < \pi$.

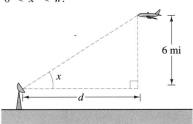

FIGURE FOR 95

96. *Television Coverage*   A television camera is on a reviewing platform 100 feet from the street on which a parade will be passing from left to right (see figure). Express the distance d from the camera to a particular unit in the parade as a function of the angle x, and graph the function over the interval $-\pi/2 < x < \pi/2$. (Consider x as negative when a unit in the parade approaches from the left.)

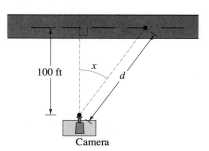

Camera

FIGURE FOR 96

97. *Sales* The projected monthly sales S (in thousands of units) of a seasonal product is modeled by

$$S = 74 + 3t + 40 \sin \frac{\pi t}{6},$$

where t is the time in months, with $t = 1$ corresponding to January. Graph this sales function over 1 year.

98. *Sales* The projected monthly sales S (in thousands of units) of a seasonal product is modeled by

$$S = 25 + 2t + 20 \sin \frac{\pi t}{6},$$

where t is the time in months, with $t = 1$ corresponding to January. Graph this sales function over 1 year.

99. *Predator-Prey Problem* Suppose the population of a certain predator at time t (in months) in a given region is estimated to be

$$P = 10{,}000 + 3000 \sin \frac{2\pi t}{24},$$

and the population of its primary food source (its prey) is estimated to be

$$p = 15{,}000 + 5000 \cos \frac{2\pi t}{24}.$$

Graph both of these functions in the same viewing rectangle, and explain the oscillations in the size of each population.

100. *Normal Temperatures* The normal monthly high temperature for Erie, Pennsylvania, is approximated by

$$H(t) = 54.33 - 20.38 \cos \frac{\pi t}{6} - 15.69 \sin \frac{\pi t}{6},$$

and the normal monthly low temperature is approximated by

$$L(t) = 39.36 - 15.70 \cos \frac{\pi t}{6} - 14.16 \sin \frac{\pi t}{6},$$

where t is the time in months, with $t = 1$ corresponding to January. (*Source:* NOAA) Use a graphing utility to graph the functions over a period of 1 year, and use the graphs to answer the following questions.

(a) During what part of the year is the difference between the normal high and low temperatures greatest? When is it smallest?

(b) The Sun is farthest north in the sky around June 21, but the graph shows the warmest temperatures at a later date. Approximate the lag time of the temperatures relative to the position of the Sun.

101. *Harmonic Motion* An object weighing W pounds is suspended from the ceiling by a steel spring (see figure). The weight is pulled downward (positive direction) from its equilibrium position and released. The resulting motion of the weight is described by the function

$$y = \tfrac{1}{2}e^{-t/4} \cos 4t, \qquad t > 0,$$

where y is the distance in feet and t is the time in seconds. Graph the function.

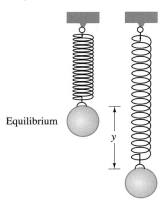

Equilibrium

FIGURE FOR 101

2.7   INVERSE TRIGONOMETRIC FUNCTIONS
Inverse Sine Function   /   Other Inverse Trigonometric Functions   /
Compositions of Trigonometric and Inverse Trigonometric Functions

Inverse Sine Function

Recall that for a function to have an inverse, it must be one-to-one. From
Figure 2.68 it is clear that $y = \sin x$ is not one-to-one because different values
of x yield the same y-value. If, however, you restrict the domain to the interval
$-\pi/2 \leq x \leq \pi/2$ (corresponding to the darker portion of the graph in Figure 2.68), the following properties hold.

1. On the interval $[-\pi/2, \pi/2]$, the function $y = \sin x$ is increasing.
2. On the interval $[-\pi/2, \pi/2]$, $y = \sin x$ takes on its full range of values,
   $-1 \leq \sin x \leq 1$.
3. On the interval $[-\pi/2, \pi/2]$, $y = \sin x$ is a one-to-one function.

Thus, on the restricted domain $-\pi/2 \leq x \leq \pi/2$, $y = \sin x$ has a unique
inverse called the **inverse sine function.** It is denoted by

$$y = \arcsin x \quad \text{or} \quad y = \sin^{-1} x.$$

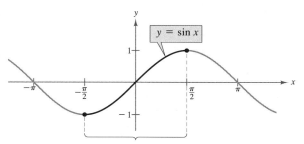

Sin x is one-to-one on this interval.

FIGURE 2.68

The notation $\sin^{-1} x$ is consistent with the inverse function notation $f^{-1}(x)$,
which is used in Section 1.10. The arcsin x notation (read as "the arcsine of x")
comes from the association of a central angle with its subtended *arc length* on
a unit circle. Thus, arcsin x means the angle (or arc) whose sine is x. Both
notations, arcsin x and $\sin^{-1} x$, are commonly used in mathematics, so remember that $\sin^{-1} x$ denotes the *inverse* sine function rather than $1/\sin x$.

The values of arcsin x lie in the interval $[-\pi/2, \pi/2]$. The graph of $y =$
arcsin x is shown in Example 2. Try producing this graph with a graphing
utility. Note that the domain of $y = \arcsin x$ is $-1 \leq x \leq 1$.

DEFINITION OF INVERSE SINE FUNCTION

The **inverse sine function** is defined by

$$y = \arcsin x \quad \text{if and only if} \quad \sin y = x,$$

where $-1 \leq x \leq 1$ and $-\pi/2 \leq y \leq \pi/2$. The domain of $y = \arcsin x$ is $[-1, 1]$, and the range is $[-\pi/2, \pi/2]$.

As with trigonometric functions, much of the work with inverse trigonometric functions can be done by *exact* calculations rather than by calculator approximations. Exact calculations help to increase your understanding of the inverse functions by relating them to the triangle definitions of the trigonometric functions.

EXAMPLE 1   Evaluating the Inverse Sine Function

Find the values of the following (if possible).

A. $\arcsin\left(-\dfrac{1}{2}\right)$ **B.** $\sin^{-1} \dfrac{\sqrt{3}}{2}$ **C.** $\sin^{-1} 2$

SOLUTION

A. By definition, $y = \arcsin(-\frac{1}{2})$ implies that

$$\sin y = -\frac{1}{2}, \quad \text{for} \quad -\frac{\pi}{2} \leq y \leq \frac{\pi}{2}.$$

Because $\sin(-\pi/6) = -\frac{1}{2}$, you can conclude that $y = -\pi/6$ and

$$\arcsin\left(-\frac{1}{2}\right) = -\frac{\pi}{6}.$$

B. By definition, $y = \sin^{-1}\left(\sqrt{3}/2\right)$ implies that

$$\sin y = \frac{\sqrt{3}}{2}, \quad \text{for} \quad -\frac{\pi}{2} \leq y \leq \frac{\pi}{2}.$$

Because $\sin(\pi/3) = \sqrt{3}/2$, you can conclude that $y = \pi/3$ and

$$\sin^{-1} \frac{\sqrt{3}}{2} = \frac{\pi}{3}.$$

C. It is not possible to evaluate $y = \sin^{-1} x$ when $x = 2$, because there is no angle whose sine is 2. Remember that the domain of the inverse sine function is $[-1, 1]$.

From Section 1.10, you know that graphs of inverse functions are reflections of each other in the line $y = x$.

EXAMPLE 2   **Sketching the Graph of the Arcsine Function**

Sketch a graph of $y = \arcsin x$ *by hand.*

SOLUTION

By definition, the equations

$$y = \arcsin x \quad \text{and} \quad \sin y = x$$

are equivalent for $-\pi/2 \le y \le \pi/2$. Hence, their graphs are the same. By assigning values to y in the second equation, you can construct the following table of values.

y	$-\dfrac{\pi}{2}$	$-\dfrac{\pi}{4}$	$-\dfrac{\pi}{6}$	0	$\dfrac{\pi}{6}$	$\dfrac{\pi}{4}$	$\dfrac{\pi}{2}$
$x = \sin y$	-1	$-\dfrac{\sqrt{2}}{2}$	$-\dfrac{1}{2}$	0	$\dfrac{1}{2}$	$\dfrac{\sqrt{2}}{2}$	1

The resulting graph of $y = \arcsin x$ is shown in Figure 2.69. Note that it is the reflection (in the line $y = x$) of the darker part of Figure 2.68. Be sure you see that Figure 2.69 shows the *entire* graph of the inverse sine function. Remember that the range of $y = \arcsin x$ is the closed interval $[-\pi/2, \pi/2]$. You can verify this graph with a graphing utility for the function $y = \arcsin x$.

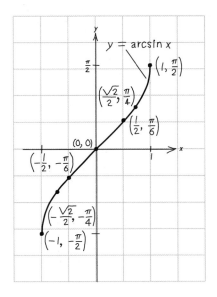

FIGURE 2.69

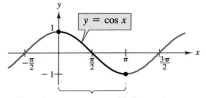

Cos *x* is one-to-one on this interval.

FIGURE 2.70

Other Inverse Trigonometric Functions

The cosine function is decreasing on the interval $0 \leq x \leq \pi$, as shown in Figure 2.70. Consequently, on this interval the cosine has an inverse function, which is called the **inverse cosine function** and is denoted by

$$y = \arccos x \quad \text{or} \quad y = \cos^{-1} x.$$

Similarly, you can define an **inverse tangent function** by restricting the domain of $y = \tan x$ to the interval $(-\pi/2, \pi/2)$. The following list summarizes the definitions of the three most common inverse trigonometric functions. The remaining three are discussed in the exercise set.

DEFINITION OF THE INVERSE TRIGONOMETRIC FUNCTIONS

Function	Domain	Range
$y = \arcsin x$ if and only if $\sin y = x$	$-1 \leq x \leq 1$	$-\dfrac{\pi}{2} \leq y \leq \dfrac{\pi}{2}$
$y = \arccos x$ if and only if $\cos y = x$	$-1 \leq x \leq 1$	$0 \leq y \leq \pi$
$y = \arctan x$ if and only if $\tan y = x$	$-\infty < x < \infty$	$-\dfrac{\pi}{2} < y < \dfrac{\pi}{2}$

The graphs of these three inverse trigonometric functions are shown in Figure 2.71.

Domain: $[-1, 1]$
Range: $\left[-\dfrac{\pi}{2}, \dfrac{\pi}{2}\right]$

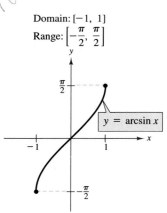

Domain: $[-1, 1]$
Range: $[0, \pi]$

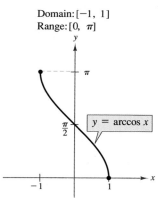

Domain: $(-\infty, \infty)$
Range: $\left(-\dfrac{\pi}{2}, \dfrac{\pi}{2}\right)$

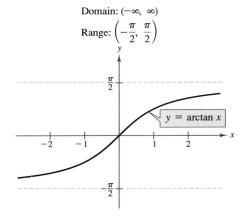

FIGURE 2.71

EXAMPLE 3   Evaluating Inverse Trigonometric Functions

Find the exact values of the following.

A. $\arccos \dfrac{\sqrt{2}}{2}$ **B.** $\arccos(-1)$ **C.** $\arctan 0$

SOLUTION

A. Because $\cos(\pi/4) = \sqrt{2}/2$ and $\pi/4$ lies in $[0, \pi]$, it follows that

$$\arccos \frac{\sqrt{2}}{2} = \frac{\pi}{4}.$$

B. Because $\cos \pi = -1$ and π lies in $[0, \pi]$, it follows that

$$\arccos(-1) = \pi.$$

C. Because $\tan 0 = 0$ and 0 lies in $(-\pi/2, \pi/2)$, it follows that

$$\arctan 0 = 0.$$

REMARK In Example 4, had you set the calculator to the degree mode, the display would have been in degrees rather than radians. This convention is peculiar to calculators. By definition, the values of inverse trigonometric functions are always *in radians*.

In Example 3, you were able to find the *exact* values of the given inverse trigonometric functions without a calculator. In the next example, a calculator is necessary to approximate the function values.

Technology Note _____

Most graphing utilities do not have keys for evaluating the inverse cotangent function, the inverse secant function, or the inverse cosecant function. Is it still possible to calculate arcsec 3.4? If you let $x = \operatorname{arcsec} 3.4$, then $\sec x = 3.4$ and $\cos x = 1/\sec x = 1/3.4$. Hence, using the inverse cosine function key, $x \approx 1.272$. Similarly, to evaluate the other inverse trigonometric functions, use the following identities.

$$\operatorname{arccsc} x = \arcsin \frac{1}{x}$$

$$\operatorname{arccot} x = \frac{\pi}{2} + \arctan(-x)$$

EXAMPLE 4   Using a Calculator to Evaluate Inverse Trigonometric Functions

Use a calculator to approximate the values (if possible).

A. $\arctan(-8.45)$ **B.** $\arcsin 0.2447$ **C.** $\arccos 2$

SOLUTION

Function	Mode	Display	Rounded to 3 Decimal Places
A. $\arctan(-8.45)$	Radian	-1.453001005	-1.453
B. $\arcsin 0.2447$	Radian	0.2472102741	0.247
C. $\arccos 2$	Radian	ERROR	

Note that the *error* in part (C) occurs because the domain of the inverse cosine function is $[-1, 1]$.

Compositions of Trigonometric and Inverse Trigonometric Functions

Recall from Section 1.10 that inverse functions possess the properties

$$f(f^{-1}(x)) = x \quad \text{and} \quad f^{-1}(f(x)) = x.$$

The inverse trigonometric versions of these properties are as follows.

REMARK   Keep in mind that these inverse properties do not apply for arbitrary values of x and y. For instance,

$$\arcsin\left(\sin\frac{3\pi}{2}\right) = \arcsin(-1)$$

$$= -\frac{\pi}{2} \neq \frac{3\pi}{2}.$$

In other words, the property $\arcsin(\sin y) = y$ is not valid for values of y outside the interval $[-\pi/2, \pi/2]$.

INVERSE PROPERTIES

If $-1 \le x \le 1$ and $-\pi/2 \le y \le \pi/2$, then

$$\sin(\arcsin x) = x \quad \text{and} \quad \arcsin(\sin y) = y.$$

If $-1 \le x \le 1$ and $0 \le y \le \pi$, then

$$\cos(\arccos x) = x \quad \text{and} \quad \arccos(\cos y) = y.$$

If $-\pi/2 < y < \pi/2$, then

$$\tan(\arctan x) = x \quad \text{and} \quad \arctan(\tan y) = y.$$

DISCOVERY

(a) Use a graphing utility to graph $y = \arcsin(\sin x)$. What are the domain and range of the function? Explain why $\arcsin(\sin 4)$ does not equal 4.

(b) Use a graphing utility to graph $y = \sin(\arcsin x)$. What are the domain and range of the function? Explain why $\sin(\arcsin 4)$ is not defined.

EXAMPLE 5   Using Inverse Properties

If possible, find the exact values.

A. $\tan[\arctan(-5)]$ **B.** $\arcsin\left(\sin\dfrac{5\pi}{3}\right)$ **C.** $\cos(\cos^{-1}\pi)$

SOLUTION

A. Because -5 lies in the domain of arctan x, the inverse property applies, and you have $\tan[\arctan(-5)] = -5$.

B. In this case, $5\pi/3$ does not lie within the range of the arcsine function, $-\pi/2 \le x \le \pi/2$. However, $5\pi/3$ is coterminal with

$$\frac{5\pi}{3} - 2\pi = -\frac{\pi}{3},$$

which does lie in the range of the arcsine function, and you have

$$\arcsin\left(\sin\frac{5\pi}{3}\right) = \arcsin\left[\sin\left(-\frac{\pi}{3}\right)\right] = -\frac{\pi}{3}.$$

C. The expression $\cos(\cos^{-1}\pi)$ is not defined, because $\cos^{-1}\pi$ is not defined. Remember that the domain of the inverse cosine function is $[-1, 1]$.

REMARK   Try verifying these results with your calculator.

Example 6 shows how to use right triangles to find exact values of functions of inverse functions. Example 7 shows how to use triangles to convert a trigonometric expression into an algebraic one. This conversion technique is used frequently in calculus.

EXAMPLE 6   Evaluating Functions of Inverse Trigonometric Functions

Find the exact values of the following.

A. $\tan\left(\arccos\dfrac{2}{3}\right)$ **B.** $\cos\left[\arcsin\left(-\dfrac{3}{5}\right)\right]$

SOLUTION

A. If you let $u = \arccos\frac{2}{3}$, then $\cos u = \frac{2}{3}$. Because $\cos u$ is positive, u is a *first*-quadrant angle. You can sketch and label angle u as shown in Figure 2.72. Consequently,

$$\tan\left(\arccos\frac{2}{3}\right) = \tan u = \frac{\text{opp}}{\text{adj}} = \frac{\sqrt{5}}{2}.$$

B. If you let $u = \arcsin(-\frac{3}{5})$, then $\sin u = -\frac{3}{5}$. Because $\sin u$ is negative, u is a *fourth*-quadrant angle. You can sketch and label u as shown in Figure 2.73. Consequently,

$$\cos\left[\arcsin\left(-\frac{3}{5}\right)\right] = \cos u = \frac{\text{adj}}{\text{hyp}} = \frac{4}{5}.$$

EXAMPLE 7   Some Problems from Calculus

Write each of the following as an algebraic expression in x.

A. $\sin(\arccos 3x)$, $0 \le x \le \dfrac{1}{3}$ **B.** $\cot(\arccos 3x)$, $0 \le x < \dfrac{1}{3}$

SOLUTION

If you let $u = \arccos 3x$, then $\cos u = 3x$. Because

$$\cos u = \frac{3x}{1} = \frac{\text{adj}}{\text{hyp}},$$

you can sketch a right triangle with acute angle u, as shown in Figure 2.74. From this triangle, you can convert each expression to algebraic form.

A. $\sin(\arccos 3x) = \sin u = \dfrac{\text{opp}}{\text{hyp}} = \sqrt{1 - 9x^2},$ $0 \le x \le \dfrac{1}{3}$

B. $\cot(\arccos 3x) = \cot u = \dfrac{\text{adj}}{\text{opp}} = \dfrac{3x}{\sqrt{1 - 9x^2}},$ $0 \le x < \dfrac{1}{3}$

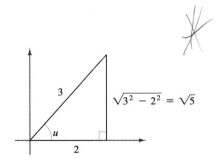

FIGURE 2.72

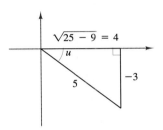

FIGURE 2.73

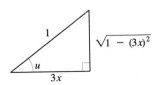

FIGURE 2.74

REMARK   In Example 7, a similar argument can be made for x-values lying in the interval $[-1/3, 0]$. Why do we restrict $x < \frac{1}{3}$ in part (B)?

DISCUSSION PROBLEM

•••••••••••••••••••

INVERSE FUNCTIONS

You have studied inverses for several types of functions. Match each of the functions in the left column with its inverse function in the right column.

1. $f(x) = x$
2. $f(x) = x^2, \quad 0 \le x$
3. $f(x) = x^3$
4. $f(x) = e^x$
5. $f(x) = \ln x$
6. $f(x) = \sin x, \quad -\dfrac{\pi}{2} \le x \le \dfrac{\pi}{2}$
7. $f(x) = \cos x, \quad 0 \le x \le \pi$
8. $f(x) = \tan x, \quad -\dfrac{\pi}{2} \le x \le \dfrac{\pi}{2}$

(a) $f^{-1}(x) = \arcsin x$
(b) $f^{-1}(x) = \ln x$
(c) $f^{-1}(x) = \sqrt{x}$
(d) $f^{-1}(x) = \arctan x$
(e) $f^{-1}(x) = \arccos x$
(f) $f^{-1}(x) = \sqrt[3]{x}$
(g) $f^{-1}(x) = e^x$
(h) $f^{-1}(x) = x$

Provide reasons for your answers.

WARM-UP

•••••••••••••••••

The following warm-up exercises involve skills that were covered in earlier sections. You will use these skills in the exercise set for this section.

In Exercises 1–4, evaluate the trigonometric function without using a calculator.

1. $\sin\left(-\dfrac{\pi}{2}\right)$ 2. $\cos \pi$

3. $\tan\left(-\dfrac{\pi}{4}\right)$ 4. $\sin \dfrac{\pi}{4}$

In Exercises 5 and 6, find a real number x in the interval $[-\pi/2, \pi/2]$ that has the same sine value as the given value.

5. $\sin 2\pi$ 6. $\sin \dfrac{5\pi}{6}$

In Exercises 7 and 8, find a real number x in the interval $[0, \pi]$ that has the same cosine value as the given value.

7. $\cos 3\pi$ 8. $\cos\left(-\dfrac{\pi}{4}\right)$

In Exercises 9 and 10, find a real number x in the interval $(-\pi/2, \pi/2)$ that has the same tangent value as the given value.

9. $\tan 4\pi$ 10. $\tan \dfrac{3\pi}{4}$

SECTION 2.7 · EXERCISES

In Exercises 1–16, evaluate the expression without using a calculator.

1. $\arcsin \frac{1}{2}$

2. $\arcsin 0$

3. $\arccos \frac{1}{2}$

4. $\arccos 0$

5. $\arctan \dfrac{\sqrt{3}}{3}$

6. $\arctan(-1)$

7. $\arccos\left(-\dfrac{\sqrt{3}}{2}\right)$

8. $\arcsin\left(-\dfrac{\sqrt{2}}{2}\right)$

9. $\arctan\left(-\sqrt{3}\right)$

10. $\arctan\left(\sqrt{3}\right)$

11. $\arccos\left(-\dfrac{1}{2}\right)$

12. $\arcsin \dfrac{\sqrt{2}}{2}$

13. $\arcsin \dfrac{\sqrt{3}}{2}$

14. $\arctan\left(-\dfrac{\sqrt{3}}{3}\right)$

15. $\arctan 0$

16. $\arccos 1$

In Exercises 17–28, use a calculator to approximate the given value. (Round your result to two decimal places.)

17. $\arccos 0.28$

18. $\arcsin 0.45$

19. $\arcsin(-0.75)$

20. $\arccos(-0.8)$

21. $\arctan(-2)$

22. $\arctan 15$

23. $\arcsin 0.31$

24. $\arccos 0.26$

25. $\arccos(-0.41)$

26. $\arcsin(-0.125)$

27. $\arctan 0.92$

28. $\arctan 2.8$

In Exercises 29 and 30, use a graphing utility to graph f, g, and $y = x$ in the same viewing rectangle to verify geometrically that g is the inverse of f. (Be sure to properly restrict the domain of f.)

29. $f(x) = \tan x$, $g(x) = \arctan x$

30. $f(x) = \sin x$, $g(x) = \arcsin x$

In Exercises 31–36, use the properties of inverse trigonometric functions to evaluate the expression.

31. $\sin(\arcsin 0.3)$

32. $\tan(\arctan 25)$

33. $\cos[\arccos(-0.1)]$

34. $\sin[\arcsin(-0.2)]$

35. $\arcsin(\sin 3\pi)$

36. $\arccos\left(\cos \dfrac{7\pi}{2}\right)$

In Exercises 37–44, find the exact value of the expression without using a calculator. (*Hint:* Make a sketch of a right triangle, as illustrated in Example 6.)

37. $\sin(\arctan \frac{3}{4})$

38. $\sec(\arcsin \frac{4}{5})$

39. $\cos(\arctan 2)$

40. $\sin\left(\arccos \dfrac{\sqrt{5}}{5}\right)$

41. $\cos(\arcsin \frac{5}{13})$

42. $\csc[\arctan(-\frac{5}{12})]$

43. $\sec[\arctan(-\frac{3}{5})]$

44. $\tan[\arcsin(-\frac{3}{4})]$

In Exercises 45 and 46, use a graphing utility to graph f and g in the same viewing rectangle to verify that the two are equal. Identify any asymptotes of the graphs.

45. $f(x) = \sin(\arctan 2x)$, $g(x) = \dfrac{2x}{\sqrt{1 + 4x^2}}$

46. $f(x) = \tan\left(\arccos \dfrac{x}{2}\right)$, $g(x) = \dfrac{\sqrt{4 - x^2}}{x}$

In Exercises 47–56, write an algebraic expression that is equivalent to the given expression. (*Hint:* Sketch a right triangle, as demonstrated in Example 7.)

47. $\cot(\arctan x)$

48. $\sin(\arctan x)$

49. $\cos(\arcsin 2x)$

50. $\sec(\arctan 3x)$

51. $\sin(\arccos x)$

52. $\cot\left(\arctan \dfrac{1}{x}\right)$

53. $\tan\left(\arccos \dfrac{x}{3}\right)$

54. $\sec[\arcsin(x - 1)]$

55. $\csc\left(\arccos \dfrac{x}{\sqrt{2}}\right)$

56. $\cos\left(\arcsin \dfrac{x - h}{r}\right)$

In Exercises 57–60, fill in the blanks.

57. $\arctan \dfrac{9}{x} = \arcsin\left(\rule{1cm}{0.4pt}\right)$, $x \neq 0$

58. $\arcsin \dfrac{\sqrt{36 - x^2}}{6} = \arccos\left(\rule{1cm}{0.4pt}\right)$, $0 \leq x \leq 6$

59. $\arccos \dfrac{3}{\sqrt{x^2 - 2x + 10}} = \arcsin\left(\rule{1cm}{0.4pt}\right)$

60. $\arccos \dfrac{x - 2}{2} = \arctan\left(\rule{1cm}{0.4pt}\right)$, $|x - 2| \leq 2$

In Exercises 61–64, graph the function.

61. $f(x) = \arcsin(x - 1)$

62. $f(x) = \dfrac{\pi}{2} + \arctan x$

63. $f(x) = \arctan 2x$

64. $f(x) = \arccos \dfrac{x}{4}$

In Exercises 65 and 66, write the given function in terms of the sine function by using the identity

$$A \cos \omega t + B \sin \omega t = \sqrt{A^2 + B^2} \sin\left(\omega t + \arctan \dfrac{B}{A}\right).$$

Verify your result by using a graphing utility to graph both forms of the function.

65. $f(t) = 3 \cos 2t + 3 \sin 2t$

66. $f(t) = 4 \cos \pi t + 3 \sin \pi t$

67. *Photography*   A photographer is taking a picture of a 4-foot-square painting hung in an art gallery. The camera lens is 1 foot below the lower edge of the painting (see figure). The angle β subtended by the camera lens x feet from the painting is given by

$$\beta = \arctan \dfrac{4x}{x^2 + 5}, \qquad x > 0.$$

(a) Use a graphing utility to graph β as a function of x.

(b) Move the cursor along the graph to approximate the distance from the picture when β is maximum.

(c) Identify any asymptote of the graph and discuss its meaning in the context of the problem.

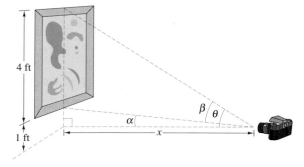

FIGURE FOR 67

68. *Photography*   A television camera at ground level is filming the lift-off of the space shuttle at a point 2000 feet from the launch pad (see figure). If θ is the angle of elevation to the shuttle and s is the height of the shuttle in feet, write θ as a function of s. Find θ when (a) $s = 1000$ and (b) $s = 4000$.

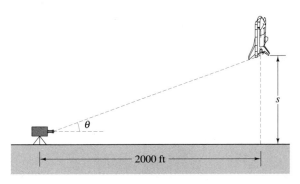

FIGURE FOR 68

69. *Docking a Boat*   A boat is pulled in by means of a winch located on a dock 12 feet above the deck of the boat (see figure). If θ is the angle of elevation from the boat to the winch and s is the length of the rope from the winch to the boat, write θ as a function of s. Find θ when (a) $s = 48$ feet and (b) $s = 24$ feet.

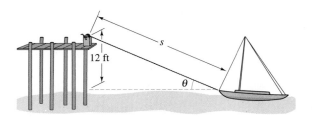

FIGURE FOR 69

70. *Area*   In calculus, it is shown that the area of the region bounded by the graphs of $y = 0$, $y = 1/(x^2 + 1)$, $x = a$, and $x = b$ is given by

Area $= \arctan b - \arctan a$

(see figure). Find the area for the following values of a and b.

(a) $a = 0$, $b = 1$ (b) $a = -1$, $b = 1$
(c) $a = 0$, $b = 3$ (d) $a = -1$, $b = 3$

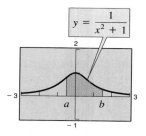

$$y = \frac{1}{x^2 + 1}$$

FIGURE FOR 70

71. Define the inverse cotangent function by restricting the domain of the cotangent to the interval $(0, \pi)$, and sketch its graph.

72. Define the inverse secant function by restricting the domain of the secant to the intervals $[0, \pi/2)$ and $(\pi/2, \pi]$, and sketch its graph.

73. Define the inverse cosecant function by restricting the domain of the cosecant to the intervals $[-\pi/2, 0)$ and $(0, \pi/2]$, and sketch its graph.

74. Use the results of Exercises 71–73 to evaluate the following without using a calculator.

(a) arcsec $\sqrt{2}$ (b) arcsec 1
(c) arccot $-\sqrt{3}$ (d) arcsec 2

In Exercises 75–79, verify the identity with a calculator, then prove the identity.

75. $\arcsin(-x) = -\arcsin x$

76. $\arctan(-x) = -\arctan x$

77. $\arccos(-x) = \pi - \arccos x$

78. $\arctan x + \arctan \dfrac{1}{x} = \dfrac{\pi}{2}$, $x > 0$

79. $\arcsin x + \arccos x = \dfrac{\pi}{2}$

80. The Chebyshev polynomial of degree n is defined by the formula $T_n(x) = \cos(n \arccos x)$ for $-1 \le x \le 1$ and $n = 1, 2, 3, \ldots$.

(a) Show that $T_0(x) = 1$.
(b) Show that $T_1(x) = x$.
(c) Find the quadratic polynomial $T_2(x)$.
(d) Graph $T_3(x)$ and $4x^3 - 3x$ on the same viewing rectangle. What do you observe?

2.8   APPLICATIONS OF TRIGONOMETRY

Applications Involving Right Triangles / Trigonometry and Bearings /
Harmonic Motion

Applications Involving Right Triangles

In keeping with our twofold perspective of trigonometry, this section includes
both right triangle applications and applications that emphasize the periodic
nature of the trigonometric functions.

In this section, the three angles of a right triangle are denoted by the letters
A, B, and C (where C is the right angle), and the lengths of the sides opposite
these angles are denoted by the letters a, b, and c (where c is the hypotenuse).

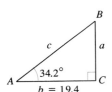

FIGURE 2.75

EXAMPLE 1   Solving a Right Triangle, Given One Acute Angle and
One Side

Solve the right triangle having $A = 34.2°$ and $b = 19.4$, as shown in Figure 2.75.

SOLUTION

Because $C = 90°$, it follows that $A + B = 90°$ and

$$B = 90° - 34.2° = 55.8°.$$

To solve for a, use the fact that

$$\tan A = \frac{\text{opp}}{\text{adj}} = \frac{a}{b} \longrightarrow a = b \tan A.$$

Thus,

$$a = 19.4 \tan 34.2° \approx 13.18.$$

Similarly, to solve for c, use the fact that

$$\cos A = \frac{\text{adj}}{\text{hyp}} = \frac{b}{c} \longrightarrow c = \frac{b}{\cos A}.$$

Thus,

$$c = \frac{19.4}{\cos 34.2°} \approx 23.46.$$

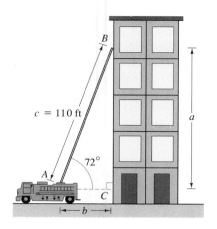

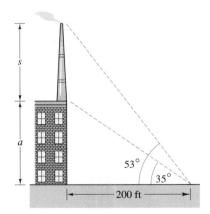

FIGURE 2.76

EXAMPLE 2   Finding a Side of a Right Triangle

A safety regulation states that the maximum angle of elevation for a rescue ladder is 72°. If a fire department's longest ladder is 110 feet, what is the maximum safe rescue height?

SOLUTION

A sketch is shown in Figure 2.76. From the equation

$$\sin A = \frac{a}{c},$$

it follows that

$$a = c \sin A = 110(\sin 72°) \approx 104.6 \text{ feet.}$$

EXAMPLE 3   Finding a Side of a Right Triangle

At a point 200 feet from the base of a building, the angle of elevation to the *bottom* of a smokestack is 35°, whereas the angle of elevation to the *top* is 53°, as shown in Figure 2.77. Find the height s of the smokestack alone.

SOLUTION

Note from Figure 2.77 that this problem involves two right triangles. In the smaller right triangle, use the fact that $\tan 35° = a/200$ to conclude that the height of the building is

$$a = 200 \tan 35°.$$

Now, from the larger right triangle, use the equation

$$\tan 53° = \frac{a + s}{200}$$

to conclude that $a + s = 200 \tan 53°$. Hence, the height of the smokestack is

$$s = 200 \tan 53° - a = 200 \tan 53° - 200 \tan 35° \approx 125.4 \text{ feet.}$$

FIGURE 2.77

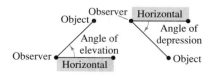

FIGURE 2.78

Examples 2 and 3 used the term **angle of elevation** to represent the angle from the horizontal upward to an object. For objects that lie below the horizontal, it is common to use the term **angle of depression,** as shown in Figure 2.78.

In Examples 1 through 3, you found the lengths of the sides of a right triangle, given an acute angle and the length of one of the sides. You can also find the angles of a right triangle given only the lengths of two sides, as demonstrated in Example 4.

EXAMPLE 4   Finding an Acute Angle of a Right Triangle

A swimming pool is 20 meters long and 12 meters wide. The bottom of the pool is slanted so that the water depth is 1.3 meters at the shallow end and 4 meters at the deep end, as shown in Figure 2.79. Find the angle of depression of the bottom of the pool.

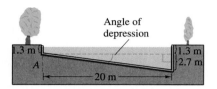

Angle of
depression

1.3 m

A

1.3 m
2.7 m

20 m

FIGURE 2.79

REMARK   Note that the width of the pool, 12 meters, is irrelevant to the problem.

SOLUTION

Using the tangent function, you can see that

$$\tan A = \frac{\text{opp}}{\text{adj}} = \frac{2.7}{20} = 0.135.$$

Thus, the angle of depression is given by

$$A = \arctan 0.135 \approx 0.13419 \text{ radians} \approx 7.69°.$$

Trigonometry and Bearings

In surveying and navigation, directions are generally given in terms of **bearings.** A bearing measures the acute angle a path or line of sight makes with a fixed north-south line. For instance, in Figure 2.80(a), the bearing is S 35° E, meaning 35 *degrees east of south*. Similarly, the bearings in Figure 2.80(b) and (c) are N 80° W and N 45° E, respectively.

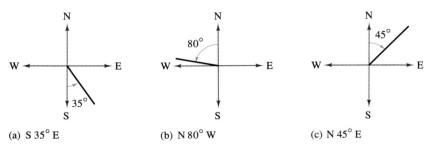

(a) S 35° E   (b) N 80° W   (c) N 45° E

FIGURE 2.80

EXAMPLE 5 Finding Directions in Terms of Bearings

A ship leaves port at noon and heads due west at 20 knots (nautical miles per hour). At 2 P.M., to avoid a storm, the ship changes course to N 54° W, as shown in Figure 2.81. Find the ship's bearing and distance from the port of departure at 3 P.M.

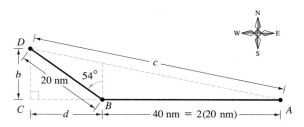

FIGURE 2.81

SOLUTION

Because the ship travels west at 20 knots for two hours, the length of *AB* is 40. Similarly, *BD* is 20. In triangle *BCD*, you have $B = 90° − 54° = 36°$. The two sides of this triangle are determined as follows.

$$\sin B = \frac{b}{20} \qquad \text{and} \quad \cos B = \frac{d}{20}$$

$$b = 20 \sin 36° \qquad\qquad d = 20 \cos 36°$$

Now, in triangle *ACD*, you can determine angle *A* as follows.

$$\tan A = \frac{b}{d + 40} = \frac{20 \sin 36°}{20 \cos 36° + 40} \approx 0.2092494$$

$$A \approx \arctan 0.2092494 \approx 0.2062732 \text{ radians} \approx 11.82°$$

The angle with the north-south line is $90° − 11.82° = 78.18°$. Therefore, the bearing of the ship is

N 78.18° W.   *Bearing*

Finally, from triangle *ACD*, you have $\sin A = b/c$, which yields

$$c = \frac{b}{\sin A} = \frac{20 \sin 36°}{\sin(11.82)}$$

$$c \approx 57.4 \text{ nautical miles.} \qquad \textit{Distance from port}$$

Harmonic Motion

The periodic nature of the trigonometric functions is useful for describing the motion of a point on an object that vibrates, oscillates, rotates, or is moved by wave motion.

For example, consider a ball that is bobbing up and down on the end of a spring, as shown in Figure 2.82. Suppose that 10 centimeters is the maximum distance the ball moves vertically upward or downward from its equilibrium (at rest) position. Suppose further that the time it takes for the ball to move from its maximum displacement above zero to its maximum displacement below zero and back again is $t = 4$ seconds. Assuming the ideal conditions of perfect elasticity and no friction or air resistance, the ball would continue to move up and down in a uniform and regular manner.

From this spring you can conclude that the period (time for one complete cycle) of the motion is

Period = 4 seconds

and that its amplitude (maximum displacement from equilibrium) is

Amplitude = 10 centimeters.

Motion of this nature can be described by a sine or cosine function, and is called **simple harmonic motion.**

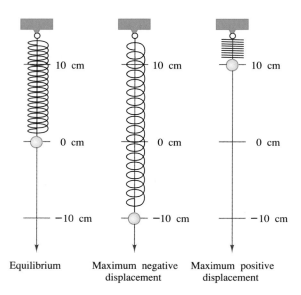

Simple Harmonic Motion

FIGURE 2.82

DEFINITION OF SIMPLE HARMONIC MOTION

A point that moves on a coordinate line is in **simple harmonic motion** if its distance d from the origin at time t is given by either

$$d = a \sin \omega t \quad \text{or} \quad d = a \cos \omega t,$$

where a and ω are real numbers such that $\omega > 0$. The motion has **amplitude** $|a|$, **period** $2\pi/\omega$, and **frequency** $\omega/2\pi$.

EXAMPLE 6   Simple Harmonic Motion

Write the equation for the simple harmonic motion of the ball described in Figure 2.82, where the period is 4 seconds. What is the frequency of this motion?

SOLUTION

Because the spring is at equilibrium ($d = 0$) when $t = 0$, use the equation

$$d = a \sin \omega t.$$

Moreover, because the maximum displacement from zero is 10 and the period is 4, you have

Amplitude $= |a| = 10$

$$\text{Period} = \frac{2\pi}{\omega} = 4 \quad \longrightarrow \quad \omega = \frac{\pi}{2}.$$

Consequently, the equation of motion is

$$d = 10 \sin \frac{\pi}{2} t.$$

Note that the choice of $a = 10$ or $a = -10$ depends on whether the ball initially moves up or down. The frequency is given by

$$\text{Frequency} = \frac{\omega}{2\pi} = \frac{\pi/2}{2\pi} = \frac{1}{4} \text{ cycle per second.}$$

FIGURE 2.83

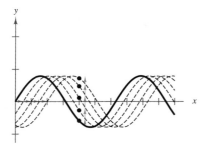

A fishing bob moves in a vertical
direction as waves move to the right.

FIGURE 2.84

One illustration of the relation between sine waves and harmonic motion is seen in the wave motion resulting from dropping a stone into a calm pool of water. The waves move outward in roughly the shape of sine (or cosine) waves, as shown in Figure 2.83. As an example, suppose you are fishing and your fishing bob is attached so that it does not move horizontally. As the waves move outward from the dropped stone, your fishing bob will move up and down in simple harmonic motion, as shown in Figure 2.84.

EXAMPLE 7   Simple Harmonic Motion

Given the equation for simple harmonic motion,

$$d = 6 \cos \frac{3\pi}{4}t,$$

find the following.

A. The maximum displacement
B. The frequency
C. The value of d when $t = 4$
D. The least positive value of t for which $d = 0$

SOLUTION

The given equation has the form $d = a \cos \omega t$, with $a = 6$ and $\omega = 3\pi/4$.

A. The maximum displacement (from the point of equilibrium) is given by the amplitude. Thus, the maximum displacement is 6.

B. Frequency $= \dfrac{\omega}{2\pi} = \dfrac{3\pi/4}{2\pi} = \dfrac{3}{8}$ cycle per unit of time

C. $d = 6 \cos\left[\dfrac{3\pi}{4}(4)\right] = 6 \cos 3\pi = 6(-1) = -6$

D. To find the least positive value of t for which $d = 0$, solve the equation

$$d = 6 \cos \frac{3\pi}{4}t = 0$$

to obtain

$$\frac{3\pi}{4}t = \frac{\pi}{2}, \frac{3\pi}{2}, \frac{5\pi}{2}, \ldots$$

$$t = \frac{2}{3}, 2, \frac{10}{3}, \ldots.$$

Thus, the least positive value of t is $t = \frac{2}{3}$.

Technology Note

You can use your graphing utility to solve part D as follows. Graph the function $y_1 = 6 \cos(\frac{3\pi}{4})$ on the viewing rectangle $0 \le x \le 6$, $-7 \le y \le 7$, and observe that the first x-intercept is approximately $x = 0.7$. Using the zoom and trace features, you will see that $x \approx 0.667$.

Many other physical phenomena can be characterized by wave motion. These include electromagnetic waves such as radio waves, television waves, and microwaves. Radio waves transmit sound in two different ways. For an AM station, the *amplitude* of the wave is modified to carry sound (AM stands for **amplitude modulation**). See Figure 2.85(a). An FM radio signal has its *frequency* modified in order to carry sound, hence the term **frequency modulation.** See Figure 2.85(b).

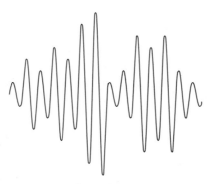

(a) AM: Amplitude modulation

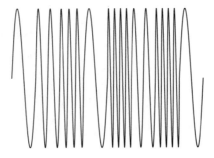

(b) FM: Frequency modulation

FIGURE 2.85

Radio Waves

DISCUSSION PROBLEM

....

YOU BE THE INSTRUCTOR

Suppose you are teaching a class in trigonometry. Write two "right triangle problems" that you think would be reasonable to ask your students to solve. (Assume that your students have 5 minutes to solve each problem.)

WARM-UP

The following warm-up exercises involve skills that were covered in earlier sections. You will use these skills in the exercise set for this section.

In Exercises 1–4, evaluate the expression and round to two decimal places.

1. $20 \sin 25°$

2. $42 \tan 62°$

3. arcsin 0.8723

4. arctan 2.8703

In Exercises 5 and 6, solve for x and round to two decimal places.

5. $\cos 22° = \dfrac{x + 13 \sin 22°}{13 \sin 54°}$

6. $\tan 36° = \dfrac{x + 85 \tan 18°}{85}$

In Exercises 7–10, find the amplitude and period of the function.

7. $f(x) = -4 \sin 2x$

8. $f(x) = \frac{1}{2} \sin \pi x$

9. $g(x) = 3 \cos 3\pi x$

10. $g(x) = 0.2 \cot \dfrac{x}{4}$

SECTION 2.8 · EXERCISES

In Exercises 1–10, solve the right triangle shown in the figure. (Round your result to two decimal places.)

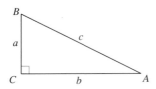

FIGURE FOR 1–10

1. $A = 20°,\quad b = 10$

2. $B = 54°,\quad c = 15$

3. $B = 71°,\quad b = 24$

4. $A = 8.4°,\quad a = 40.5$

5. $A = 12° \, 15',\quad c = 430.5$

6. $B = 65° \, 12',\quad a = 14.2$

7. $a = 6,\quad b = 10$

8. $a = 25,\quad c = 35$

9. $b = 16,\quad c = 52$

10. $b = 1.32,\quad c = 9.45$

In Exercises 11 and 12, find the altitude of the isosceles triangle shown in the figure. (Round your result to two decimal places.)

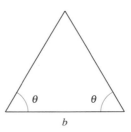

FIGURE FOR 11 AND 12

11. $\theta = 52°,\quad b = 4$ inches

12. $\theta = 18°,\quad b = 10$ meters

13. *Length of a Shadow*   If the sun is 30° above the horizon, find the length of a shadow cast by a silo that is 70 feet high (see figure).

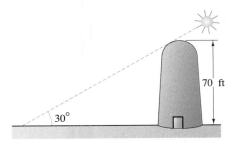

FIGURE FOR 13

14. *Length of a Shadow*   The sun is 20° above the horizon. Find the length of a shadow cast by a building that is 600 feet high (see figure).

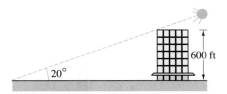

FIGURE FOR 14

15. *Height*   A ladder of length 16 feet leans against the side of a house (see figure). Find the height h of the top of the ladder if the angle of elevation of the ladder is 74°.

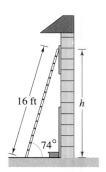

FIGURE FOR 15

16. *Height*   The length of a shadow of a tree is 125 feet when the angle of elevation of the sun is 33° (see figure). Approximate the height h of the tree.

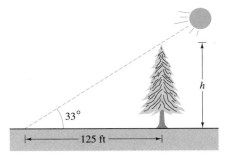

FIGURE FOR 16

17. *Angle of Elevation*   An amateur radio operator erects a 75-foot vertical tower for his antenna (see figure). Find the angle of elevation to the top of the tower at a point on level ground 50 feet from the base.

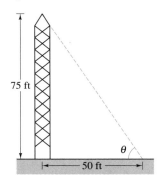

FIGURE FOR 17

18. *Angle of Elevation*   The height of an outdoor basketball backboard is $12\frac{1}{2}$ feet, and the backboard casts a shadow $17\frac{1}{3}$ feet long (see figure). Find the angle of elevation of the sun.

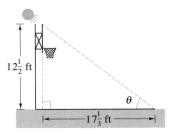

FIGURE FOR 18

19. *Angle of Depression*   A spacecraft is traveling in a circular orbit 100 miles above the surface of the earth (see figure). Find the angle of depression from the spacecraft to the horizon. Assume that the radius of the earth is 4000 miles.

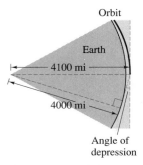

Orbit

Earth

4100 mi

4000 mi

Angle of depression

FIGURE FOR 19

20. *Angle of Depression*   Find the angle of depression from the top of a lighthouse 250 feet above water level to the water line of a ship 2 miles offshore.

21. *Airplane Ascent*   When an airplane leaves the runway (see figure), its angle of climb is 18° and its speed is 275 feet per second. Find the altitude of the plane after 1 minute.

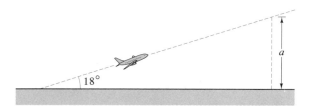

18°

a

FIGURE FOR 21

22. *Mountain Descent*   A sign on the roadway at the top of a mountain indicates that for the next 4 miles the grade is 12.5° (see figure). Find the change in elevation for a car descending the mountain.

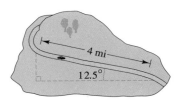

4 mi

12.5°

FIGURE FOR 22

23. *Height*   From a point 50 feet in front of a church, the angles of elevation to the base of the steeple and to the top of the steeple are 35° and 47° 40′, respectively (see figure). Find the height of the steeple.

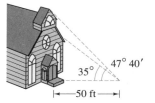

35°   47° 40′

50 ft

FIGURE FOR 23

24. *Height*   From a point 100 feet in front of a public library, the angles of elevation to the base of the flagpole and to the top of the pole are 28° and 39° 45′, respectively. The flagpole is mounted on the front of the library's roof (see figure). Find the height of the pole.

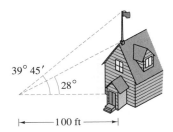

39° 45′

28°

100 ft

FIGURE FOR 24

25. *Navigation*   An airplane flying at 550 miles per hour has a bearing of N 52° E. After flying for 1.5 hours, how far north and how far east has the plane traveled from its point of departure?

26. *Navigation*   A ship leaves port at noon and has a bearing of S 27° W. If the ship is sailing at 20 knots, how many nautical miles south and how many nautical miles west has the ship traveled by 6:00 P.M.?

27. *Navigation*   A ship is 45 miles east and 30 miles south of port. If the captain wants to sail directly to port, what bearing should be taken?

28. *Navigation*   A plane is 120 miles north and 85 miles east of an airport. If the pilot wants to fly directly to the airport, what bearing should be taken?

29. *Surveying* A surveyor wishes to find the distance across a swamp (see figure). The bearing from A to B is N 32° W. The surveyor walks 50 yards from A, and at point C the bearing to B is N 68° W. (a) Find the bearing from A to C. (b) Find the distance from A to B.

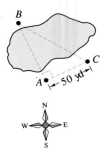

FIGURE FOR 29

30. *Location of a Fire* Two fire towers are 20 miles apart, tower A being due west of tower B. A fire is spotted from the towers, and the bearings from A and B are N 76° E and N 56° W (see figure). Find the distance d of the fire from the line segment AB. [*Hint:* Use the fact that d = 20/(cot 14° + cot 34°).]

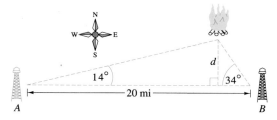

FIGURE FOR 30

31. *Distance Between Ships* An observer in a lighthouse 300 feet above sea level spots two ships directly offshore. The angles of depression to the ships are 4° and 6.5° (see figure). How far apart are the ships?

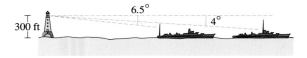

FIGURE FOR 31

32. *Distance Between Towns* A passenger in an airplane flying at 30,000 feet sees two towns directly to the left of the airplane. The angles of depression to the towns are 28° and 55° (see figure). How far apart are the towns?

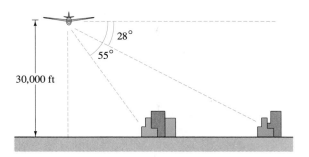

FIGURE FOR 32

33. *Altitude of a Plane* A plane is observed approaching your home, and you assume it is traveling at 550 miles per hour. If the angle of elevation of the plane is 16° at one time, and 1 minute later the angle is 57°, approximate the altitude of the plane.

34. *Height of a Mountain* In traveling across flat land, you notice a mountain directly in front of you. The angle of elevation (to the peak) is 3.5°. After you drive 13 miles closer to the mountain, the angle of elevation is 9°. Approximate the height of the mountain.

35. *Length* A regular pentagon is inscribed in a circle of radius 25 inches. Find the length of the sides of the pentagon.

36. *Length* A regular hexagon is inscribed in a circle of radius 25 inches. Find the length of the sides of the hexagon.

37. *Wrench Size* Use the figure to find the distance y across the flat sides of the hexagonal nut as a function of r.

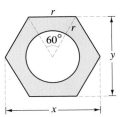

FIGURE FOR 37

38. *Bolt Circle*   The figure shows a circular sheet 25 cm in diameter containing 12 equally spaced bolt holes. Determine the straight-line distance between the centers of adjacent bolt holes.

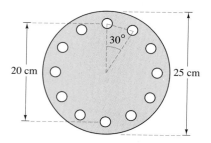

FIGURE FOR 38

Trusses   In Exercises 39 and 40, find the lengths of all the pieces of the truss.

39.

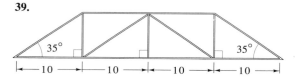

40.

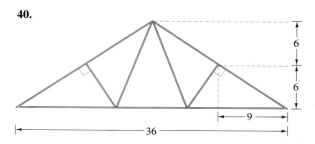

Harmonic Motion   In Exercises 41–44, for the simple harmonic motion described by the given trigonometric function, find (a) the maximum displacement, (b) the frequency, and (c) the least possible value of t for which $d = 0$.

41. $d = 4 \cos 8\pi t$ **42.** $d = \frac{1}{2} \cos 20\pi t$

43. $d = \frac{1}{16} \sin 120\pi t$ **44.** $d = \frac{1}{64} \sin 792\pi t$

45. *Tuning Fork*   A point on the end of a tuning fork moves in simple harmonic motion described by $d = a \sin \omega t$. Find ω given that the tuning fork for middle C has a frequency of 264 vibrations per second.

46. *Wave Motion*   A buoy oscillates in simple harmonic motion as waves go past. At a given time it is noted that the buoy moves a total of 3.5 feet from its low point to its high point, and that it returns to its high point every 10 seconds. Write an equation that describes the motion of the buoy if, at $t = 0$, it is at its high point.

CHAPTER 2 · REVIEW EXERCISES

In Exercises 1–4, sketch the angle in standard position, and list one positive and one negative coterminal angle.

1. $\dfrac{11\pi}{4}$

2. $\dfrac{2\pi}{9}$

3. $-110°$

4. $-405°$

In Exercises 5–8, convert the angle measurement to decimal form. (Round your result to two decimal places.)

5. $135°\ 16'\ 45''$

6. $-234°\ 50''$

7. $5°\ 22'\ 53''$

8. $280°\ 8'\ 50''$

In Exercises 9–12, convert the angle measurement to $D°\ M'\ S''$ form.

9. $135.27°$

10. $25.1°$

11. $-85.15°$

12. $-327.85°$

In Exercises 13–16, convert the angle measurement from radians to degrees. (Round your result to two decimal places.)

13. $\dfrac{5\pi}{7}$

14. $-\dfrac{3\pi}{5}$

15. -3.5

16. 1.75

In Exercises 17–20, convert the angle measurement from degrees to radians. (Round your result to four decimal places.)

17. $480°$

18. $-16.5°$

19. $-33°\ 45'$

20. $84°\ 15'$

In Exercises 21–24, find the reference angle for the given angle.

21. $252°$

22. $640°$

23. $-\dfrac{6\pi}{5}$

24. $\dfrac{17\pi}{3}$

In Exercises 25–30, find the six trigonometric functions of the angle θ (in standard position) whose terminal side passes through the point.

25. $(12, 16)$

26. $(x, 4x)$

27. $(-7, 2)$

28. $(4, -8)$

29. $(-4, -6)$

30. $\left(\frac{2}{3}, \frac{5}{2}\right)$

In Exercises 31–34, find the remaining five trigonometric functions of θ satisfying the given conditions. (*Hint:* Sketch a right triangle.)

31. $\sec\theta = \frac{6}{5}$, $\tan\theta < 0$

32. $\tan\theta = -\frac{12}{5}$, $\sin\theta > 0$

33. $\sin\theta = \frac{3}{8}$, $\cos\theta < 0$

34. $\cos\theta = -\frac{2}{5}$, $\sin\theta > 0$

In Exercises 35–40, evaluate the trigonometric function without using a calculator.

35. $\tan\dfrac{\pi}{3}$

36. $\sec\dfrac{\pi}{4}$

37. $\sin\dfrac{5\pi}{3}$

38. $\cot\left(\dfrac{5\pi}{6}\right)$

39. $\cos 495°$

40. $\csc 270°$

In Exercises 41–44, use a calculator to evaluate the trigonometric function. (Round your result to two decimal places.)

41. $\tan 33°$

42. $\csc 105°$

43. $\sec\dfrac{12\pi}{5}$

44. $\sin\left(-\dfrac{\pi}{9}\right)$

In Exercises 45–48, find two values of θ in degrees $(0° \le \theta < 360°)$ and in radians $(0 \le \theta < 2\pi)$ without using a calculator.

45. $\cos\theta = -\dfrac{\sqrt{2}}{2}$

46. $\sec\theta$ is undefined

47. $\csc\theta = -2$

48. $\tan\theta = \dfrac{\sqrt{3}}{3}$

In Exercises 49–52, find two values of θ in degrees $(0° \le \theta < 360°)$ and in radians $(0 \le \theta < 2\pi)$ by using a calculator.

49. $\sin\theta = 0.8387$

50. $\cot\theta = -1.5399$

51. $\sec\theta = -1.0353$

52. $\csc\theta = 11.4737$

In Exercises 53–56, write an algebraic expression for the given expression.

53. $\sec[\arcsin(x - 1)]$ **✱54.** $\tan\left(\arccos\dfrac{x}{2}\right)$

55. $\sin\left(\arccos\dfrac{x^2}{4 - x^2}\right)$ **✱56.** $\csc(\arcsin 10x)$

In Exercises 57–70, sketch a graph of the function through two full periods. (Use a graphing utility to verify your result.)

57. $y = 3 \cos 2\pi x$ **✱58.** $y = -2 \sin \pi x$

59. $f(x) = 5 \sin \dfrac{2x}{5}$ **✱60.** $f(x) = 8 \cos\left(-\dfrac{x}{4}\right)$

61. $f(x) = -\dfrac{1}{4} \cos \dfrac{\pi x}{4}$ **✱62.** $f(x) = -\tan \dfrac{\pi x}{4}$

63. $g(t) = \frac{5}{2} \sin(t - \pi)$ **✱64.** $g(t) = 3 \cos(t + \pi)$

65. $h(t) = \tan\left(t - \dfrac{\pi}{4}\right)$ **✱66.** $h(t) = \sec\left(t - \dfrac{\pi}{4}\right)$

67. $f(t) = \csc\left(3t - \dfrac{\pi}{2}\right)$ **68.** $f(t) = 3 \csc\left(2t + \dfrac{\pi}{4}\right)$

69. $f(\theta) = \cot \dfrac{\pi\theta}{8}$

70. $E(t) = 110 \cos\left(120\pi t - \dfrac{\pi}{3}\right)$

In Exercises 71–80, use a graphing utility to graph the function.

71. $f(x) = \dfrac{x}{4} - \sin x$ **72.** $g(x) = 3\left(\sin \dfrac{\pi x}{3} + 1\right)$

73. $y = \dfrac{x}{3} + \cos \pi x$ **74.** $y = 4 - \dfrac{x}{4} + \cos \pi x$

75. $h(\theta) = \theta \sin \pi\theta$ **76.** $f(t) = 2.5e^{-t/4} \sin 2\pi t$

77. $y = \arcsin \dfrac{x}{2}$ **78.** $y = 2 \arccos x$

79. $f(x) = \dfrac{\pi}{2} + \arctan x$ **80.** $f(x) = \arccos(x - \pi)$

In Exercises 81–84, use a graphing utility to graph the function. Use the graph to determine if the function is periodic. If the function is periodic, find any relative maximum or minimum points through one period.

81. $f(x) = e^{\sin x}$ **82.** $g(x) = \sin e^x$

83. $g(x) = 2 \sin x \cos^2 x$ **84.** $h(x) = 4 \sin^2 x \cos^2 x$

85. *Altitude of a Triangle*   Find the altitude of the triangle in the figure.

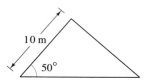

FIGURE FOR 85

86. *Angle of Elevation*   The height of a radio transmission tower is 225 feet, and it casts a shadow of length 105 feet (see figure). Find the angle of elevation of the Sun.

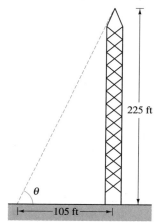

FIGURE FOR 86

87. *Shuttle Height*   An observer 2.5 miles from the launch pad of a space shuttle measures the angle of elevation to the base of the vehicle to be 28° soon after lift-off (see figure). How high is the shuttle at that instant? (Assume the shuttle is still moving vertically.)

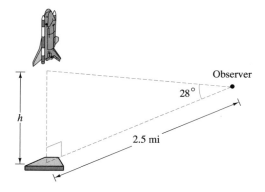

FIGURE FOR 87

88. *Distance*   From city A to city B, a plane flies 650 miles at a bearing of N 48° E. From city B to city C, the plane flies 810 miles at a bearing of S 65° E. Find the distance from A to C and the bearing from A to C.

89. *Railroad Grade*   A train travels 2.5 miles on a straight track with a grade of 1° 10′ (see figure). What is the vertical rise of the train in that distance?

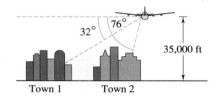

FIGURE FOR 89

90. *Distance Between Towns*   A passenger in an airplane flying at 35,000 feet sees two towns directly to the left of the airplane. The angles of depression to the towns are 32° and 76° (see figure). How far apart are the towns?

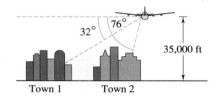

FIGURE FOR 90

91. Using calculus, it can be shown that the sine and cosine functions can be approximated by the polynomials

$$\sin x \approx x - \frac{x^3}{3!} + \frac{x^5}{5!} - \frac{x^7}{7!}$$

and

$$\cos x \approx 1 - \frac{x^2}{2!} + \frac{x^4}{4!} - \frac{x^6}{6!},$$

where x is in radians.

(a) Use a graphing utility to graph the sine function and its polynomial approximation in the same viewing rectangle.

(b) Use a graphing utility to graph the cosine function and its polynomial approximation in the same viewing rectangle.

(c) Study the patterns in the polynomial approximations of the sine and cosine functions and guess the next term in each. Then repeat parts (a) and (b). Do you think your guesses were correct? How did the accuracy of the approximations change when additional terms were added?

CHAPTERS 1 AND 2 · CUMULATIVE TEST

Take this test as you would take a test in class. After you are done, check your work against the answers in the back of the book.

In Exercises 1 and 2, solve the equation.

1. $1 - \frac{1}{2}(x - 3) = \frac{3}{4}$ **2.** $3x^2 + x - 10 = 0$

In Exercises 3–6, sketch the graph of the equation without the aid of a graphing utility.

3. $x - 3y + 12 = 0$ **4.** $y = x^2 - 4x + 1$

5. $y = \sqrt{4 - x}$ **6.** $y = 3 - |x - 2|$

7. A line passes through the point $(4, 9)$ with slope $m = -\frac{2}{3}$. Find the coordinates of three additional points on the line. (There are many correct answers.)

8. Find an equation of the line passing through the points $(-\frac{1}{2}, 1)$ and $(3, 8)$.

9. Evaluate (if possible) the function $f(x) = x/(x - 2)$ at the specified values of the independent variable.
   (a) $f(6)$ (b) $f(2)$ (c) $f(2t)$ (d) $f(s + 2)$

10. Express the area A of an equilateral triangle as a function of the length s of its sides.

11. Express the angle $4\pi/9$ in degree measure and sketch the angle in standard position.

12. Express the angle $-120°$ in radian measure as a multiple of π and sketch the angle in standard position.

13. The terminal side of an angle θ in standard position passes through the point $(12, 5)$. Evaluate the six trigonometric functions of the angle.

14. If $\cos t = -\frac{2}{3}$ where $\pi/2 < t < \pi$, find $\sin t$ and $\tan t$.

15. Use a calculator to approximate two values of $\theta(0° \le \theta < 360°)$ such that $\sec \theta = 2.125$. Round your answer to two decimal places.

16. Sketch a graph of each of the functions through two periods.
   (a) $y = -3 \sin 2x$ (b) $f(x) = 2 \cos\left(x - \frac{\pi}{2}\right)$
   (c) $g(x) = \tan \frac{\pi x}{2}$ (d) $h(t) = \sec t$

17. Write a sentence describing the relationship between the graphs of the functions $f(x) = \sin x$ and g.
   (a) $g(x) = 10 + \sin x$ (b) $g(x) = \sin \frac{\pi x}{2}$
   (c) $g(x) = \sin\left(x + \frac{\pi}{4}\right)$ (d) $g(x) = -\sin x$

18. Consider the function $f(x) = \sin 3x - 2 \cos x$.
   (a) Use a graphing utility to graph the function.
   (b) Approximate (accurate to one decimal place) the zero of the function in the interval $[0, 3]$.
   (c) Approximate (accurate to one decimal place) the maximum value of the function in the interval $[0, 3]$.

19. Evaluate the expression without the aid of a calculator.
   (a) $\arcsin \frac{1}{2}$ (b) $\arctan \sqrt{3}$

20. Write an expression that is equivalent to $\sin(\arccos 2x)$.

21. From a point on the ground 600 feet from the foot of a cliff, the angle of elevation to the top of the cliff is $32° \ 30'$. How high is the cliff?

ANALYTIC TRIGONOMETRY

3.1 APPLICATIONS OF FUNDAMENTAL IDENTITIES

Introduction / Some Uses of the Fundamental Identities

Introduction

In Chapter 2 you studied the basic definitions, properties, graphs, and applications of the individual trigonometric functions. In this chapter, you will learn how to use the fundamental identities to perform the following.

1. Evaluate trigonometric functions.
2. Simplify trigonometric expressions.
3. Develop additional trigonometric identities.
4. Solve trigonometric equations.

You will also make use of many algebraic skills, such as finding special products, factoring, performing operations with fractional expressions, rationalizing denominators, and solving equations.

As you study this chapter, remember that the best problem solvers are those who can solve problems by a variety of means. For instance, in this chapter many of the problems can be solved graphically, analytically, and numerically (using a table). Even after you have solved a problem one way, we encourage you to check your solution by solving the problem another way.

For convenience, we have summarized the fundamental trigonometric identities on the next page. Because these identities are used so frequently, it would be a good idea to commit them to memory.

FUNDAMENTAL TRIGONOMETRIC IDENTITIES

Reciprocal Identities

$$\sin u = \frac{1}{\csc u} \qquad \sec u = \frac{1}{\cos u} \qquad \tan u = \frac{1}{\cot u}$$

$$\csc u = \frac{1}{\sin u} \qquad \cos u = \frac{1}{\sec u} \qquad \cot u = \frac{1}{\tan u}$$

Quotient Identities

$$\tan u = \frac{\sin u}{\cos u} \qquad \cot u = \frac{\cos u}{\sin u}$$

Pythagorean Identities

$$\sin^2 u + \cos^2 u = 1 \qquad 1 + \tan^2 u = \sec^2 u \qquad 1 + \cot^2 u = \csc^2 u$$

Cofunction Identities

$$\sin\left(\frac{\pi}{2} - u\right) = \cos u \quad \sec\left(\frac{\pi}{2} - u\right) = \csc u \quad \tan\left(\frac{\pi}{2} - u\right) = \cot u$$

$$\cos\left(\frac{\pi}{2} - u\right) = \sin u \quad \csc\left(\frac{\pi}{2} - u\right) = \sec u \quad \cot\left(\frac{\pi}{2} - u\right) = \tan u$$

Even-Odd Identities

$$\sin(-u) = -\sin u \qquad \sec(-u) = \sec u \qquad \tan(-u) = -\tan u$$

$$\csc(-u) = -\csc u \qquad \cos(-u) = \cos u \qquad \cot(-u) = -\cot u$$

REMARK   The Pythagorean identities are sometimes used in radical forms such as

$$\sin u = \pm\sqrt{1 - \cos^2 u} \quad \text{or}$$
$$\tan u = \pm\sqrt{\sec^2 u - 1},$$

where the sign depends on the choice of u.

Some Uses of the Fundamental Identities

One common use of trigonometric identities is to use given values of one or more trigonometric functions to find the values of the other trigonometric functions.

EXAMPLE 1   Using Identities to Evaluate a Function

Use the conditions $\sec u = -\frac{3}{2}$ and $\tan u > 0$ to find the values of all six trigonometric functions.

SOLUTION

Using a reciprocal identity, you can write

$$\cos u = \frac{1}{\sec u} = \frac{1}{-\dfrac{3}{2}} = -\frac{2}{3}$$

By a Pythagorean identity, you obtain

$$\sin^2 u = 1 - \cos^2 u = 1 - \left(-\frac{2}{3}\right)^2 = 1 - \frac{4}{9} = \frac{5}{9}.$$

Because sec $u < 0$ and tan $u > 0$, it follows that u lies in Quadrant III. Moreover, because sin u is negative when u is in Quadrant III, choose the negative root to obtain sin $u = -\sqrt{5}/3$. Now, knowing the values of the sine and cosine, you can find the values of all six trigonometric functions.

$$\sin u = -\frac{\sqrt{5}}{3} \qquad\qquad \csc u = \frac{1}{\sin u} = -\frac{3}{\sqrt{5}}$$

$$\cos u = -\frac{2}{3} \qquad\qquad \sec u = -\frac{3}{2}$$

$$\tan u = \frac{\sin u}{\cos u} = \frac{-\dfrac{\sqrt{5}}{3}}{-\dfrac{2}{3}} = \frac{\sqrt{5}}{2} \qquad \cot u = \frac{1}{\tan u} = \frac{2}{\sqrt{5}}$$

Compare this approach with the triangle approach in Example 2, Section 2.3.

The next four examples use algebra and the fundamental identities to simplify, factor, and/or combine trigonometric *expressions* such as

$$\cot x - \cos x \sin x, \quad \frac{1 - \sin x}{\cos^2 x}, \quad \text{and} \quad \frac{\tan x}{\sec x - 1}.$$

EXAMPLE 2   Simplifying a Trigonometric Expression

Simplify the expression $\sin x \cos^2 x - \sin x$.

SOLUTION

First factor out a common monomial factor and then use a fundamental identity.

$$\begin{aligned}
\sin x \cos^2 x - \sin x &= \sin x(\cos^2 x - 1) &&\textit{Monomial factor}\\
&= -\sin x(1 - \cos^2 x)\\
&= -\sin x(\sin^2 x) &&\textit{Pythagorean Identity}\\
&= -\sin^3 x &&\textit{Multiply}
\end{aligned}$$

You can graphically confirm this result with a graphing utility. Graph $y = \sin x \cos^2 x - \sin x$ and $y = -\sin^3 x$ on the same display screen, and notice that the two graphs coincide, as shown in Figure 3.1.

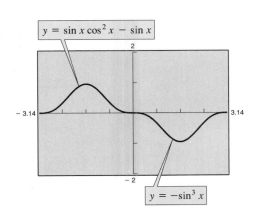

$y = \sin x \cos^2 x - \sin x$

$y = -\sin^3 x$

FIGURE 3.1

EXAMPLE 3   Factoring Trigonometric Expressions

Factor the following.

A. $\sec^2 \theta - 1$ **B.** $4 \tan^2 \theta + \tan \theta - 3$

SOLUTION

A. Using the difference-of-two-squares pattern, you can factor the expression as

$$\sec^2 \theta - 1 = (\sec \theta - 1)(\sec \theta + 1).$$

B. This expression has the polynomial form, $ax^2 + bx + c$, where $x = \tan \theta$, and it factors as

$$4 \tan^2 \theta + \tan \theta - 3 = (4 \tan \theta - 3)(\tan \theta + 1).$$

On occasion, factoring or simplifying can best be done by first rewriting the expression in terms of just *one* trigonometric function or in terms of *sine and cosine alone.*

EXAMPLE 4   Factoring a Trigonometric Expression

Factor the expression $\csc^2 x - \cot x - 3$.

SOLUTION

As given, this expression cannot be factored. If, however, you use the identity $\csc^2 x = 1 + \cot^2 x$ to rewrite the expression in terms of the cotangent alone, you can factor to obtain

$$\csc^2 x - \cot x - 3 = (1 + \cot^2 x) - \cot x - 3 \qquad \textit{Pythagorean Identity}$$
$$= \cot^2 x - \cot x - 2 \qquad \textit{Combine terms}$$
$$= (\cot x - 2)(\cot x + 1). \qquad \textit{Factor}$$

EXAMPLE 5   Combining Fractional Expressions

Perform the indicated addition and simplify the result.

$$\frac{\sin \theta}{1 + \cos \theta} + \frac{\cos \theta}{\sin \theta}$$

SOLUTION

$$\frac{\sin\theta}{1+\cos\theta} + \frac{\cos\theta}{\sin\theta} = \frac{(\sin\theta)(\sin\theta) + (\cos\theta)(1+\cos\theta)}{(1+\cos\theta)(\sin\theta)}$$

$$= \frac{\sin^2\theta + \cos^2\theta + \cos\theta}{(1+\cos\theta)\sin\theta} \qquad \textit{Multiply}$$

$$= \frac{1+\cos\theta}{(1+\cos\theta)\sin\theta} \qquad \textit{Pythagorean Identity}$$

$$= \frac{1}{\sin\theta} \qquad \textit{Cancel common factor}$$

$$= \csc\theta \qquad \textit{Reciprocal Identity}$$

EXAMPLE 6   **Verifying Trigonometric Identities Graphically**

Use a graphing utility to determine which of the following is an identity.

A. $\cos 3x \overset{?}{=} 4\cos^3 x - 3\cos x$ **B.** $\cos 3x \overset{?}{=} \sin\left(3x - \frac{\pi}{2}\right)$

SOLUTION

A. Using a graphing utility, you can see that the graphs of $y = \cos 3x$ and $y = 4\cos^3 x - 3\cos x$ appear to coincide, as shown in Figure 3.2(a). Therefore, this appears to be an identity.

B. From the graphs shown in Figure 3.2(b), you can see that this is not an identity.

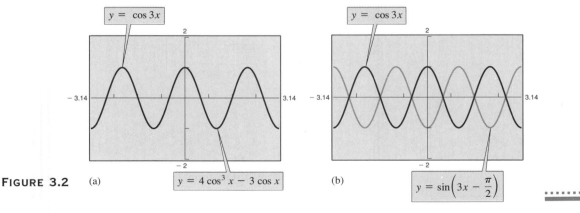

FIGURE 3.2   (a)   (b)

The last two examples of this section involve techniques for rewriting expressions into forms that are useful in calculus.

EXAMPLE 7   Rewriting a Trigonometric Expression

Rewrite

$$\frac{1}{1 + \sin x}$$

so that it is *not* in fractional form.

SOLUTION

$$\frac{1}{1 + \sin x} = \frac{1}{1 + \sin x} \cdot \frac{1 - \sin x}{1 - \sin x}$$ *Multiply numerator and denominator by* $(1 - \sin x)$

$$= \frac{1 - \sin x}{1 - \sin^2 x}$$ *Multiply*

$$= \frac{1 - \sin x}{\cos^2 x}$$ *Pythagorean Identity*

$$= \frac{1}{\cos^2 x} - \frac{\sin x}{\cos^2 x}$$ *Separate fractions*

$$= \frac{1}{\cos^2 x} - \frac{\sin x}{\cos x} \cdot \frac{1}{\cos x}$$

$$= \sec^2 x - \tan x \sec x$$ *Identities*

EXAMPLE 8   Trigonometric Substitution

Use the substitution $x = 2 \tan \theta$ to express $\sqrt{4 + x^2}$ as a trigonometric function of θ where $0 < \theta < \pi/2$.

SOLUTION

Letting $x = 2 \tan \theta$ produces the following.

$$\sqrt{4 + x^2} = \sqrt{4 + (2 \tan \theta)^2}$$

$$= \sqrt{4(1 + \tan^2 \theta)}$$

$$= \sqrt{4 \sec^2 \theta}$$ *Pythagorean Identity*

$$= 2 \sec \theta$$ $\sec \theta > 0$ *for* $0 < \theta < \dfrac{\pi}{2}$

Figure 3.3 shows a right triangle illustration of the trigonometric substitution in Example 8. For $0 < \theta < \pi/2$, you obtain opp $= x$, adj $= 2$, and hyp $= \sqrt{4 + x^2}$. Thus, you can write

$$\sec \theta = \frac{\sqrt{4 + x^2}}{2} \quad \text{or} \quad \sqrt{4 + x^2} = 2 \sec \theta.$$

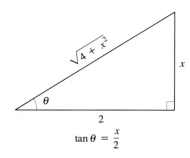

$$\tan \theta = \frac{x}{2}$$

FIGURE 3.3

DISCUSSION PROBLEM	Most people find the Pythagorean Identity involving sine and cosine to be fairly easy to remember: $\sin^2 u + \cos^2 u = 1$. The one involving tangent and secant, however, tends to give some people trouble. They can't remember if the identity is

REMEMBERING TRIGONOMETRIC IDENTITIES

$$1 + \tan^2 u \overset{?}{=} \sec^2 u$$

or

$$1 + \sec^2 u \overset{?}{=} \tan^2 u.$$

Which of these two is the correct Pythagorean Identity involving tangent and secant? Write a short paragraph describing how a person can remember (or derive) this identity.

WARM-UP

The following warm-up exercises involve skills that were covered in earlier sections. You will use these skills in the exercise set for this section.

In Exercises 1 and 2, use a right triangle to evaluate the other five trigonometric functions of the acute angle θ.

1. $\tan \theta = \frac{3}{2}$ **2.** $\sec \theta = 3$

In Exercises 3 and 4, determine the exact value of the six trigonometric functions of θ. Assume the given point is on the terminal side of an angle θ in standard position.

3. $(7, -3)$ **4.** $(-10, 5)$

In Exercises 5–8, simplify the expression.

5. $\sqrt{1 - \left(\dfrac{\sqrt{3}}{2}\right)^2}$ **6.** $\sqrt{\left(\dfrac{3}{4}\right)^2 + 1}$

7. $\sqrt{1 + \left(\dfrac{3}{8}\right)^2}$ **8.** $\sqrt{1 - \left(\dfrac{\sqrt{5}}{3}\right)^2}$

In Exercises 9 and 10, perform the indicated operations and simplify the result.

9. $\dfrac{4}{1 + x} + \dfrac{x}{4}$ **10.** $\dfrac{3}{1 - x} - \dfrac{5}{1 + x}$

SECTION 3.1 · EXERCISES

In Exercises 1–10, use the fundamental identities to evaluate (if possible) the other trigonometric functions.

1. $\sin x = \dfrac{1}{2}, \quad \cos x = \dfrac{\sqrt{3}}{2}$

2. $\tan x = \dfrac{\sqrt{3}}{3}, \quad \cos x = -\dfrac{\sqrt{3}}{2}$

3. $\sec \theta = \sqrt{2}, \quad \sin \theta = -\dfrac{\sqrt{2}}{2}$

4. $\csc \theta = \dfrac{5}{3}, \quad \tan \theta = \dfrac{3}{4}$

5. $\sin(-x) = -\dfrac{2}{3}, \quad \tan x = -\dfrac{2\sqrt{5}}{5}$

6. $\cos\left(\dfrac{\pi}{2} - x\right) = \dfrac{3}{5}, \quad \cos x = \dfrac{4}{5}$

7. $\tan \theta = 2, \quad \sin \theta < 0$

8. $\sec \theta = -3, \quad \tan \theta < 0$

9. $\sin \theta = -1, \quad \cot \theta = 0$

10. $\tan \theta$ is undefined, $\quad \sin \theta > 0$

In Exercises 11–16, match the trigonometric expression with one of the following.

(a) -1 (b) $\cos x$ (c) $\cot x$

(d) 1 (e) $-\tan x$ (f) $\sin x$

11. $\sec x \cos x$

12. $\dfrac{\sin(-x)}{\cos(-x)}$

13. $\tan^2 x - \sec^2 x$

14. $\dfrac{1 - \cos^2 x}{\sin x}$

15. $\cot x \sin x$

16. $\dfrac{\sin\left(\dfrac{\pi}{2} - x\right)}{\cos\left(\dfrac{\pi}{2} - x\right)}$

In Exercises 17–22, match the trigonometric expression with one of the following.

(a) $\csc x$ (b) $\tan x$ (c) $\sin^2 x$

(d) $\sin x \tan x$ (e) $\sec^2 x$ (f) $\sec^2 x + \tan^2 x$

17. $\sin x \sec x$

18. $\cos^2 x(\sec^2 x - 1)$

19. $\dfrac{\sec^2 x - 1}{\sin^2 x}$

20. $\cot x \sec x$

21. $\sec^4 x - \tan^4 x$

22. $\dfrac{\cos^2\left(\dfrac{\pi}{2} - x\right)}{\cos x}$

In Exercises 23–36, use fundamental identities to simplify the expression. Use a graphing utility to verify your result. (See Example 2.)

23. $\tan \phi \csc \phi$

24. $\sin \phi(\csc \phi - \sin \phi)$

25. $\cos \beta \tan \beta$

26. $\sec \alpha \dfrac{\sin \alpha}{\tan \alpha}$

27. $\dfrac{\cot x}{\csc x}$

28. $\dfrac{\csc \theta}{\sec \theta}$

29. $\sec^2 x(1 - \sin^2 x)$

30. $\dfrac{1}{\tan^2 x + 1}$

31. $\dfrac{\sin(-x)}{\cos x}$

32. $\dfrac{\tan^2 \theta}{\sec^2 \theta}$

33. $\cos\left(\dfrac{\pi}{2} - x\right)\sec x$

34. $\cot\left(\dfrac{\pi}{2} - x\right)\cos x$

35. $\dfrac{\cos^2 y}{1 - \sin y}$

36. $\cos t(1 + \tan^2 t)$

In Exercises 37–44, factor the expression and use fundamental identities to simplify. Use a graphing utility to verify your result.

37. $\tan^2 x - \tan^2 x \sin^2 x$

38. $\sec^2 x \tan^2 x + \sec^2 x$

39. $\sin^2 x \sec^2 x - \sin^2 x$

40. $\dfrac{\sec^2 x - 1}{\sec x - 1}$

41. $\tan^4 x + 2 \tan^2 x + 1$

42. $1 - 2 \cos^2 x + \cos^4 x$

43. $\sin^4 x - \cos^4 x$

44. $\csc^3 x - \csc^2 x - \csc x + 1$

In Exercises 45–48, perform the multiplication and use fundamental identities to simplify.

45. $(\sin x + \cos x)^2$

46. $(\cot x + \csc x)(\cot x - \csc x)$

47. $(\sec x + 1)(\sec x - 1)$

48. $(3 - 3 \sin x)(3 + 3 \sin x)$

In Exercises 49–52, perform the addition or subtraction and use fundamental identities to simplify.

49. $\dfrac{1}{1 + \cos x} + \dfrac{1}{1 - \cos x}$

50. $\dfrac{1}{\sec x + 1} - \dfrac{1}{\sec x - 1}$

51. $\dfrac{\cos x}{1 + \sin x} + \dfrac{1 + \sin x}{\cos x}$

52. $\tan x - \dfrac{\sec^2 x}{\tan x}$

In Exercises 53–56, rewrite the expression so that it is *not* in fractional form.

53. $\dfrac{\sin^2 y}{1 - \cos y}$ **54.** $\dfrac{5}{\tan x + \sec x}$

55. $\dfrac{3}{\sec x - \tan x}$ **56.** $\dfrac{\tan^2 x}{\csc x + 1}$

In Exercises 57–60, use a graphing utility to determine whether or not the equation is an identity. (See Example 6.)

57. $\csc x = \cot \dfrac{x}{2} - \cot x$

58. $\sec^2 x \csc x = \sec^2 x + \csc x$

59. $\sin 2x = 2 \sin x$

60. $\sin x \tan x = \csc x - \cos x$

61. Use a graphing utility to determine the values of x in the interval $[0, 2\pi]$ for which (a) $\cos x = \sqrt{1 - \sin^2 x}$, and (b) $\cos x = -\sqrt{1 - \sin^2 x}$.

62. Use a graphing utility to determine which of the six trigonometric functions is equal to the expression

$\sin x \tan x + \cos x.$

In Exercises 63–72, use the specified trigonometric substitution to write the algebraic expression as a trigonometric function of θ, where $0 < \theta < \pi/2$.

63. $\sqrt{25 - x^2}$, $x = 5 \sin \theta$

64. $\sqrt{16 - 4x^2}$, $x = 2 \sin \theta$

65. $\sqrt{x^2 - 9}$, $x = 3 \sec \theta$

66. $\sqrt{x^2 - 4}$, $x = 2 \sec \theta$

67. $\sqrt{x^2 + 25}$, $x = 5 \tan \theta$

68. $\sqrt{x^2 + 100}$, $x = 10 \tan \theta$

69. $\sqrt{1 - (x - 1)^2}$, $x - 1 = \sin \theta$

70. $\sqrt{1 - e^{2x}}$, $e^x = \sin \theta$

71. $\sqrt{(9 + x^2)^3}$, $x = 3 \tan \theta$

72. $\sqrt{(x^2 - 16)^3}$, $x = 4 \sec \theta$

In Exercises 73 and 74, determine the values of θ, $0 \le \theta < 2\pi$, for which the equation is true.

73. $\sec \theta = \sqrt{1 + \tan^2 \theta}$

74. $\sin \theta = -\sqrt{1 - \cos^2 \theta}$

In Exercises 75 and 76, rewrite the expression as a single logarithm and simplify the result.

75. $\ln |\cos \theta| - \ln |\sin \theta|$

76. $\ln |\cot t| + \ln(1 + \tan^2 t)$

In Exercises 77–80, determine whether or not the equation is an identity, and give a reason for your answer.

77. $\dfrac{\sin k\theta}{\cos k\theta} = \tan \theta$, k is constant **78.** $\dfrac{1}{5 \cos \theta} = 5 \sec \theta$

79. $\sin \theta \csc \theta = 1$ **80.** $\sin \theta \csc \phi = 1$

In Exercises 81–84, use a calculator to demonstrate the identity for the given values of θ.

81. $\csc^2 \theta - \cot^2 \theta = 1$
  (a) $\theta = 132°$ (b) $\theta = \dfrac{2\pi}{7}$

82. $\tan^2 \theta + 1 = \sec^2 \theta$
  (a) $\theta = 346°$ (b) $\theta = 3.1$

83. $\cos\left(\dfrac{\pi}{2} - \theta\right) = \sin \theta$
  (a) $\theta = 80°$ (b) $\theta = 0.8$

84. $\sin(-\theta) = -\sin \theta$
  (a) $\theta = 250°$ (b) $\theta = \dfrac{1}{2}$

85. Express each of the other trigonometric functions of θ in terms of $\sin \theta$.

86. Express each of the other trigonometric functions of θ in terms of $\cos \theta$.

3.2   VERIFYING TRIGONOMETRIC IDENTITIES

Introduction   /   Verifying Trigonometric Identities

Introduction

In the previous section, you learned how to rewrite trigonometric expressions in equivalent forms. In this section, you will learn to prove or verify trigonometric identities. (In the next section, you will learn to solve trigonometric equations.) The key to verifying identities and solving equations is the ability to use the fundamental identities and the rules of algebra to rewrite trigonometric expressions.

Before going on, let's review some distinctions among expressions, equations, and identities. An *expression* has no equal sign. It is merely a combination of terms. When simplifying expressions, you use an equal sign only to indicate the equivalence of the original expression and the new form. An *equation* is a statement containing an equal sign that is true for a specific set of values. In this sense, it is really a *conditional* equation. For example, the equation

$$\sin x = -1$$

is true only for $x = (3\pi/2) \pm 2n\pi$. Hence, it is a conditional equation. On the other hand, an equation that is true for all real values in the domain of the variable is an *identity*. For example, the familiar equation

$$\sin^2 x = 1 - \cos^2 x$$

is true for all real numbers x. Hence, it is an identity.

Although there are similarities, proving that a trigonometric equation is an identity is quite different from solving an equation. There is no well-defined set of rules to follow in verifying trigonometric identities, and the process is best learned by practice. However, the following guidelines should be helpful.

GUIDELINES FOR VERIFYING TRIGONOMETRIC IDENTITIES

1. Work with one side of the equation at a time. It is often better to work with the more complicated side first.
2. Look for opportunities to factor an expression, add fractions, square a binomial, or create a monomial denominator.
3. Look for opportunities to use the fundamental identities. Note which functions are in the final expression you want. Sines and cosines pair up well, as do secants and tangents, and cosecants and cotangents.
4. If the preceding guidelines do not help, try converting all terms to sines and cosines.
5. Do not just sit and stare at the problem. Try something! Even paths that lead to dead ends can give you insights.

Verifying Trigonometric Identities

EXAMPLE 1   Verifying a Trigonometric Identity

Verify the identity

$$\frac{\sec^2 \theta - 1}{\sec^2 \theta} = \sin^2 \theta.$$

SOLUTION

Technology Note _____

A graphing utility can be used to confirm trigonometric identities graphically. For instance, in Example 1 you can graph

$$y = \frac{\sec^2 \theta - 1}{\sec^2 \theta} \text{ and } y = \sin^2 \theta$$

on the same viewing rectangle and observe that the graphs are the same. You can also use this technique to confirm intermediate steps in your verification process.

Start with the left side, because it is more complicated.

$$\frac{\sec^2 \theta - 1}{\sec^2 \theta} = \frac{(\tan^2 \theta + 1) - 1}{\sec^2 \theta} \qquad \textit{Pythagorean Identity}$$

$$= \frac{\tan^2 \theta}{\sec^2 \theta} \qquad \textit{Simplify}$$

$$= \tan^2 \theta (\cos^2 \theta) \qquad \textit{Reciprocal Identity}$$

$$= \frac{\sin^2 \theta}{\cos^2 \theta} (\cos^2 \theta) \qquad \textit{Tangent Identity}$$

$$= \sin^2 \theta \qquad \textit{Reduce}$$

ALTERNATIVE SOLUTION

Sometimes it is helpful to separate a fraction into two parts.

$$\frac{\sec^2 \theta - 1}{\sec^2 \theta} = \frac{\sec^2 \theta}{\sec^2 \theta} - \frac{1}{\sec^2 \theta} \qquad \textit{Separate fractions}$$

$$= 1 - \cos^2 \theta \qquad \textit{Reciprocal Identity}$$

$$= \sin^2 \theta \qquad \textit{Pythagorean Identity}$$

As you can see in Example 1, there can be more than one way to verify an identity. Your method may differ from that used by your instructor or fellow students. Here is a good chance to be creative and establish your own style, but try to be as efficient as possible.

EXAMPLE 2   Combining Fractions Before Using Identities

Verify the identity

$$\frac{1}{1 - \sin \alpha} + \frac{1}{1 + \sin \alpha} = 2 \sec^2 \alpha.$$

SOLUTION

$$\frac{1}{1 - \sin \alpha} + \frac{1}{1 + \sin \alpha} = \frac{1 + \sin \alpha + 1 - \sin \alpha}{(1 - \sin \alpha)(1 + \sin \alpha)} \qquad \textit{Add fractions}$$

$$= \frac{2}{1 - \sin^2 \alpha} \qquad \textit{Simplify}$$

$$= \frac{2}{\cos^2 \alpha} \qquad \textit{Pythagorean Identity}$$

$$= 2 \sec^2 \alpha \qquad \textit{Reciprocal Identity}$$

Technology Note _____

One way to verify the result of Example 3 is to graph

$y_1 = ((\tan x)^2 + 1)((\cos x)^2 - 1)$

$y_2 = -(\tan x)^2$

on the "trig" viewing rectangle, $-2\pi \le x \le 2\pi$, $-3 \le y \le 3$. Now graph $y_3 = y_1 - y_2 + 1$. What do you observe? Why is this better than simply graphing $y_3 = y_1 - y_2$?

EXAMPLE 3   **Verifying a Trigonometric Identity**

Verify the identity $(\tan^2 x + 1)(\cos^2 x - 1) = -\tan^2 x$.

SOLUTION

By applying identities before multiplying, you obtain the following.

$$(\tan^2 x + 1)(\cos^2 x - 1) = (\sec^2 x)(-\sin^2 x) \qquad \textit{Pythagorean identities}$$

$$= -\frac{\sin^2 x}{\cos^2 x} \qquad \textit{Reciprocal Identity}$$

$$= -\left(\frac{\sin x}{\cos x}\right)^2 \qquad \textit{Rule of exponents}$$

$$= -\tan^2 x \qquad \textit{Tangent Identity}$$

EXAMPLE 4   **Converting to Sines and Cosines**

Verify the identity $\tan x + \cot x = \sec x \csc x$.

SOLUTION

In this case, there appear to be no fractions to add, no products to find, and no opportunity to use one of the Pythagorean identities. Notice, however, what happens when you convert the left side to sines and cosines.

$$\tan x + \cot x = \frac{\sin x}{\cos x} + \frac{\cos x}{\sin x} \qquad \textit{Identities}$$

$$= \frac{\sin^2 x + \cos^2 x}{\cos x \sin x} \qquad \textit{Add fractions}$$

$$= \frac{1}{\cos x \sin x} \qquad \sin^2 x + \cos^2 x = 1$$

$$= \frac{1}{\cos x} \cdot \frac{1}{\sin x} \qquad \textit{Product of fractions}$$

$$= \sec x \csc x \qquad \textit{Reciprocal identities}$$

EXAMPLE 5   Verifying a Trigonometric Identity

Verify the identity $\sec y + \tan y = \dfrac{\cos y}{1 - \sin y}$.

SOLUTION

Work with the *right* side. Note that you can create a monomial denominator by multiplying the numerator and denominator by $(1 + \sin y)$.

$$\frac{\cos y}{1 - \sin y} = \frac{\cos y}{1 - \sin y}\left(\frac{1 + \sin y}{1 + \sin y}\right) \qquad \textit{Multiply numerator and}$$
$$\textit{denominator by } (1 + \sin y)$$

$$= \frac{\cos y + \cos y \sin y}{1 - \sin^2 y}$$

$$= \frac{\cos y + \cos y \sin y}{\cos^2 y} \qquad \textit{Pythagorean Identity}$$

$$= \frac{\cos y}{\cos^2 y} + \frac{\cos y \sin y}{\cos^2 y} \qquad \textit{Separate fractions}$$

$$= \frac{1}{\cos y} + \frac{\sin y}{\cos y} \qquad \textit{Reduce}$$

$$= \sec y + \tan y \qquad \textit{Identities}$$

So far in this section, you have been verifying trigonometric identities by working with one side of the equation and converting to the form given on the other side. On occasion, it is practical to work with each side *separately,* to obtain one common form equivalent to both sides.

EXAMPLE 6   Working with Each Side Separately

Verify the identity

$$\frac{\cot^2 \theta}{1 + \csc \theta} = \frac{1 - \sin \theta}{\sin \theta}.$$

SOLUTION

Working with the left side, you have

$$\frac{\cot^2 \theta}{1 + \csc \theta} = \frac{\csc^2 \theta - 1}{1 + \csc \theta} \qquad \cot^2 \theta = \csc^2 \theta - 1$$

$$= \frac{(\csc \theta - 1)(\csc \theta + 1)}{1 + \csc \theta} \qquad \textit{Factor}$$

$$= \csc \theta - 1. \qquad \textit{Reduce}$$

Now, simplifying the right side, you have

$$\frac{1 - \sin \theta}{\sin \theta} = \frac{1}{\sin \theta} - \frac{\sin \theta}{\sin \theta} = \csc \theta - 1.$$

The identity is verified, because both sides are equal to $\csc \theta - 1$.

In Example 7, powers of trigonometric functions are rewritten as more complicated sums or products of trigonometric functions. This is a common procedure in calculus.

EXAMPLE 7 Two Examples from Calculus

Verify the identities.

A. $\tan^4 x = \tan^2 x \sec^2 x - \tan^2 x$
B. $\sin^3 x \cos^4 x = (\cos^4 x - \cos^6 x)\sin x$

SOLUTION

Note the use of the Pythagorean identities in the verifications.

A. $\tan^4 x = (\tan^2 x)(\tan^2 x)$ *Separate factors*
       $= \tan^2 x (\sec^2 x - 1)$ *Pythagorean Identity*
       $= \tan^2 x \sec^2 x - \tan^2 x$ *Multiply*
B. $\sin^3 x \cos^4 x = \sin^2 x \cos^4 x \sin x$ *Separate factors*
       $= (1 - \cos^2 x)\cos^4 x \sin x$ *Pythagorean Identity*
       $= (\cos^4 x - \cos^6 x)\sin x$ *Multiply*

DISCUSSION PROBLEM
··
YOU BE THE INSTRUCTOR

Suppose you are tutoring a student in trigonometry. After working several home-work problems, your student becomes discouraged because of an inability to get answers that agree with those in the back of the textbook. Which of the following answers has your student actually gotten right? Which are wrong? Why?

Student's Answers	*Text's Answers*
(a) $\dfrac{1}{2 \csc x}$	$\dfrac{1}{2} \sin x$
(b) $(1 - \cos x)(1 + \cos x)$	$\sin^2 x$
(c) $\dfrac{\sin^2 x}{1 - \cos x}$	$1 + \cos x$
(d) $\cot^2 x + 2$	$1 + \csc^2 x$
(e) $(\sec x - \tan x)(\sec x + \tan x)$	1

WARM-UP

The following warm-up exercises involve skills that were covered in earlier sections. You will use these skills in the exercise set for this section.

In Exercises 1–6, factor each expression and, if possible, simplify the results.

1. (a) $x^2 - x^2 y^2$
(b) $\sin^2 x - \sin^2 x \cos^2 x$

2. (a) $x^2 + x^2 y^2$
(b) $\cos^2 x + \cos^2 x \tan^2 x$

3. (a) $x^4 - 1$
(b) $\tan^4 x - 1$

4. (a) $z^3 + 1$
(b) $\tan^3 x + 1$

5. (a) $x^3 - x^2 + x - 1$
(b) $\cot^3 x - \cot^2 x + \cot x - 1$

6. (a) $x^4 - 2x^2 + 1$
(b) $\sin^4 x - 2 \sin^2 x + 1$

In Exercises 7–10, perform the additions or subtractions and, if possible, simplify the results.

7. (a) $\dfrac{y^2}{x} - x$

(b) $\dfrac{\csc^2 x}{\cot x} - \cot x$

8. (a) $1 - \dfrac{1}{x^2}$

(b) $1 - \dfrac{1}{\sec^2 x}$

9. (a) $\dfrac{y}{1+z} + \dfrac{1+z}{y}$

(b) $\dfrac{\sin x}{1 + \cos x} + \dfrac{1 + \cos x}{\sin x}$

10. (a) $\dfrac{y}{z} - \dfrac{z}{1+y}$

(b) $\dfrac{\tan x}{\sec x} - \dfrac{\sec x}{1 + \tan x}$

SECTION 3.2 · EXERCISES

In Exercises 1–50, verify the identity, and confirm it with a graphing utility.

1. $\sin t \csc t = 1$

2. $\tan y \cot y = 1$

3. $(1 + \sin \alpha)(1 - \sin \alpha) = \cos^2 \alpha$

4. $\cot^2 y (\sec^2 y - 1) = 1$

5. $\cos^2 \beta - \sin^2 \beta = 1 - 2 \sin^2 \beta$

6. $\cos^2 \beta - \sin^2 \beta = 2 \cos^2 \beta - 1$

7. $\tan^2 \theta + 4 = \sec^2 \theta + 3$

8. $2 - \sec^2 z = 1 - \tan^2 z$

9. $\sin^2 \alpha - \sin^4 \alpha = \cos^2 \alpha - \cos^4 \alpha$

10. $\cos x + \sin x \tan x = \sec x$

11. $\dfrac{\sec^2 x}{\tan x} = \sec x \csc x$

12. $\dfrac{\cot^3 t}{\csc t} = \cos t (\csc^2 t - 1)$

13. $\dfrac{\cot^2 t}{\csc t} = \csc t - \sin t$

14. $\dfrac{1}{\sin x} - \sin x = \dfrac{\cos^2 x}{\sin x}$

15. $\sin^{1/2} x \cos x - \sin^{5/2} x \cos x = \cos^3 x \sqrt{\sin x}$

16. $\sec^6 x (\sec x \tan x) - \sec^4 x (\sec x \tan x) = \sec^5 x \tan^3 x$

17. $\dfrac{1}{\sec x \tan x} = \csc x - \sin x$

18. $\dfrac{\sec \theta - 1}{1 - \cos \theta} = \sec \theta$

19. $\csc x - \sin x = \cos x \cot x$

20. $\dfrac{\sec x + \tan x}{\sec x - \tan x} = (\sec x + \tan x)^2$

21. $\dfrac{1}{\tan x} + \dfrac{1}{\cot x} = \tan x + \cot x$

22. $\dfrac{1}{\sin x} - \dfrac{1}{\csc x} = \csc x - \sin x$

23. $\dfrac{\cos \theta \cot \theta}{1 - \sin \theta} - 1 = \csc \theta$

24. $\cos x - \dfrac{\cos x}{1 - \tan x} = \dfrac{\sin x \cos x}{\sin x - \cos x}$

25. $2 \sec^2 x - 2 \sec^2 x \sin^2 x - \sin^2 x - \cos^2 x = 1$

26. $\csc x(\csc x - \sin x) + \dfrac{\sin x - \cos x}{\sin x} + \cot x = \csc^2 x$

27. $2 + \cos^2 x - 3 \cos^4 x = \sin^2 x(2 + 3 \cos^2 x)$

28. $4 \tan^4 x + \tan^2 x - 3 = \sec^2 x(4 \tan^2 x - 3)$

29. $\sec^4 \theta - \tan^4 \theta = 1 + 2 \tan^2 \theta$

30. $\csc^4 \theta - \cot^4 \theta = 2 \csc^2 \theta - 1$

31. $\dfrac{\sin \beta}{1 - \cos \beta} = \dfrac{1 + \cos \beta}{\sin \beta}$

32. $\dfrac{\cot \alpha}{\csc \alpha - 1} = \dfrac{\csc \alpha + 1}{\cot \alpha}$

33. $\cos\left(\dfrac{\pi}{2} - x\right)\csc x = 1$

34. $\dfrac{\cos\left(\dfrac{\pi}{2} - x\right)}{\sin\left(\dfrac{\pi}{2} - x\right)}$

35. $\dfrac{\csc(-x)}{\sec(-x)} = -\cot x$

36. $(1 + \sin y)[1 + \sin(-y)] = \cos^2 y$

37. $\dfrac{\cos(-\theta)}{1 + \sin(-\theta)} = \sec \theta + \tan \theta$

38. $\dfrac{1 + \sec(-\theta)}{\sin(-\theta) + \tan(-\theta)} = -\csc \theta$

39. $\sin^2 x + \sin^2\left(\dfrac{\pi}{2} - x\right) = 1$

40. $\sec^2 y - \cot^2\left(\dfrac{\pi}{2} - y\right) = 1$

41. $\sqrt{\dfrac{1 + \sin \theta}{1 - \sin \theta}} = \dfrac{1 + \sin \theta}{|\cos \theta|}$

42. $\sqrt{\dfrac{1 - \cos \theta}{1 + \cos \theta}} = \dfrac{1 - \cos \theta}{|\sin \theta|}$

43. $\dfrac{\sin x \cos y + \cos x \sin y}{\cos x \cos y - \sin x \sin y} = \dfrac{\tan x + \tan y}{1 - \tan x \tan y}$

44. $\dfrac{\tan x + \tan y}{1 - \tan x \tan y} = \dfrac{\cot x + \cot y}{\cot x \cot y - 1}$

45. $\dfrac{\tan x + \cot y}{\tan x \cot y} = \tan y + \cot x$

46. $\dfrac{\cos x - \cos y}{\sin x + \sin y} + \dfrac{\sin x - \sin y}{\cos x + \cos y} = 0$

47. $\ln|\tan \theta| = \ln|\sin \theta| - \ln|\cos \theta|$

48. $\ln|\sec \theta| = -\ln|\cos \theta|$

49. $-\ln(1 + \cos \theta) = \ln(1 - \cos \theta) - 2 \ln|\sin \theta|$

50. $-\ln|\sec \theta + \tan \theta| = \ln|\sec \theta - \tan \theta|$

In Exercises 51–54, explain why the equation is *not* an identity, and find one value of the variable for which the equation is not true.

51. $\sin \theta = \sqrt{1 - \cos^2 \theta}$

52. $\tan \theta = \sqrt{\sec^2 \theta - 1}$

53. $\sqrt{\tan^2 x} = \tan x$

54. $\sqrt{\sin^2 x + \cos^2 x} = \sin x + \cos x$

55. *Friction* The force acting on a stationary object weighing W units on an inclined plane positioned at an angle of θ with the horizontal is modeled by

$$\mu W \cos \theta = W \sin \theta,$$

where μ is the coefficient of friction (see figure). Solve the equation for μ and simplify the result.

FIGURE FOR 55

56. *Rate of Change* The rate of change of the function $f(x) = \sin x + \csc x$ with respect to change in the variable x is given by the expression $\cos x - \csc x \cot x$. Show that the expression for the rate of change can also be given by $-\cos x \cot^2 x$.

3.3   SOLVING TRIGONOMETRIC EQUATIONS

Introduction / Equations of Quadratic Type / Functions Involving Multiple
Angles / Using Inverse Functions and a Calculator

Introduction

In this section, you will switch from *verifying* trigonometric identities to
solving trigonometric equations. To see the difference, consider the two
equations

$$\sin^2 x + \cos^2 x = 1$$

and

$$\sin x = 1.$$

The first equation is an identity because it is true for *all* real values of x. The
second equation, however, is true only for *some* values of x. When you find
these values, you are solving the equation.

EXAMPLE 1   Solving a Trigonometric Equation

Solve $2 \sin x - 1 = 0$.

SOLUTION

$$\begin{aligned}
2 \sin x - 1 &= 0 & &\text{\textit{Original equation}} \\
2 \sin x &= 1 & &\text{\textit{Add 1 to both sides}} \\
\sin x &= \frac{1}{2} & &\text{\textit{Divide both sides by 2}}
\end{aligned}$$

To solve for x, note that the equation $\sin x = \frac{1}{2}$ has solutions $x = \pi/6$ and
$x = 5\pi/6$ in the interval $[0, 2\pi)$. Moreover, because $\sin x$ has a period of 2π,
there are infinitely many other solutions, which can be written as

$$x = \frac{\pi}{6} + 2n\pi \quad \text{and} \quad x = \frac{5\pi}{6} + 2n\pi, \quad \text{\textit{General solution}}$$

where n is an integer, as shown in Figure 3.4. This is called the **general form**
of the solution.

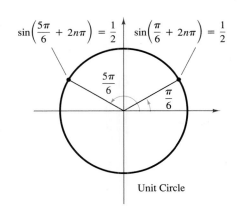

$$\sin\left(\frac{5\pi}{6} + 2n\pi\right) = \frac{1}{2} \qquad \sin\left(\frac{\pi}{6} + 2n\pi\right) = \frac{1}{2}$$

Unit Circle

FIGURE 3.4

Another way to see that the equation $\sin x = \frac{1}{2}$ has infinitely many solutions is indicated in Figure 3.5, which shows the graphs of $y = \sin x$ and $y = \frac{1}{2}$. For $0 \le x < 2\pi$, the solutions are $x = \pi/6$ and $x = 5\pi/6$. Any angles that are coterminal with $\pi/6$ or $5\pi/6$ will also be solutions of the equation.

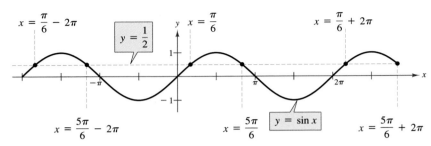

FIGURE 3.5

EXAMPLE 2   Collecting Like Terms

Solve $\sin x + \sqrt{2} = -\sin x$.

SOLUTION

$$\sin x + \sqrt{2} = -\sin x \qquad \textit{Original equation}$$

$$\sin x + \sin x = -\sqrt{2} \qquad \textit{Add sin x to both sides}$$
$$\textit{and subtract } \sqrt{2} \textit{ from both sides}$$

$$2 \sin x = -\sqrt{2} \qquad \textit{Collect like terms}$$

$$\sin x = -\frac{\sqrt{2}}{2} \qquad \textit{Divide both sides by 2}$$

Because $\sin x$ has a period of 2π, you can begin by finding all solutions in the interval $[0, 2\pi)$. These are $x = 5\pi/4$ and $x = 7\pi/4$. Next, add $2n\pi$ to each of these solutions to obtain the general form

$$x = \frac{5\pi}{4} + 2n\pi \quad \text{and} \quad x = \frac{7\pi}{4} + 2n\pi, \qquad \textit{General solution}$$

where n is an integer. The graph of $y = 2 \sin x + \sqrt{2}$, shown in Figure 3.6, confirms this result.

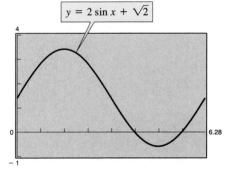

FIGURE 3.6

EXAMPLE 3   Extracting Square Roots

Solve $3 \tan^2 x - 1 = 0$.

SOLUTION

$$3 \tan^2 x - 1 = 0 \qquad \text{\textit{Original equation}}$$

$$3 \tan^2 x = 1 \qquad \text{\textit{Add 1 to both sides}}$$

$$\tan^2 x = \frac{1}{3} \qquad \text{\textit{Divide both sides by 3}}$$

$$\tan x = \pm \frac{1}{\sqrt{3}} \qquad \text{\textit{Extract square roots}}$$

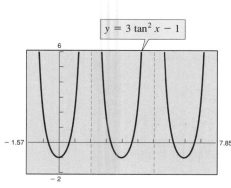

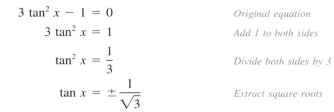

FIGURE 3.7

Because $\tan x$ has a period of π, you can begin by finding all solutions in the interval $[0, \pi)$. These are $x = \pi/6 \approx 0.52$ and $x = 5\pi/6 \approx 2.62$. Next, add $n\pi$ to each to obtain the general form

$$x = \frac{\pi}{6} + n\pi \quad \text{and} \quad x = \frac{5\pi}{6} + n\pi, \qquad \text{\textit{General solution}}$$

where n is an integer. The graph of $y = 3 \tan^2 x - 1$, shown in Figure 3.7, confirms these results.

When two or more functions occur in the same equation, collect all terms on one side and try to separate the functions by factoring.

EXAMPLE 4   Factoring

Solve $\cot x \cos^2 x = 2 \cot x$.

SOLUTION

$$\cot x \cos^2 x = 2 \cot x \qquad \text{\textit{Original equation}}$$

$$\cot x \cos^2 x - 2 \cot x = 0 \qquad \text{\textit{Subtract 2 cot x from both sides}}$$

$$\cot x (\cos^2 x - 2) = 0 \qquad \text{\textit{Factor left side}}$$

By setting each of these factors equal to zero, you obtain the following.

$$\cot x = 0 \qquad \text{and} \qquad \cos^2 x - 2 = 0$$
$$x = \frac{\pi}{2} \qquad \qquad \qquad \cos^2 x = 2$$
$$\cos x = \pm \sqrt{2}$$

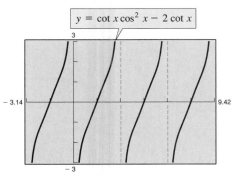

FIGURE 3.8

No solution is obtained from $\cos x = \pm \sqrt{2}$, because $\pm \sqrt{2}$ are outside the range of the cosine function. Therefore, the general form of the solution is obtained by adding multiples of π to $x = \pi/2$, to obtain

$$x = \frac{\pi}{2} + n\pi, \qquad \text{\textit{General solution}}$$

where n is an integer. The graph of $y = \cot x \cos^2 x - 2 \cot x$, shown in Figure 3.8, confirms these results.

Equations of Quadratic Type

Many trigonometric equations are of quadratic type. Here are two examples.

Quadratic in sin x	*Quadratic in sec x*
$2 \sin^2 x - \sin x - 1 = 0$	$\sec^2 x - 3 \sec x - 2 = 0$
$2(\sin x)^2 - (\sin x) - 1 = 0$	$(\sec x)^2 - 3(\sec x) - 2 = 0$

To solve equations of this type, factor the quadratic or, if this is not possible, use the Quadratic Formula.

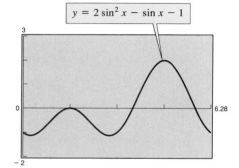

$y = 2 \sin^2 x - \sin x - 1$

FIGURE 3.9

EXAMPLE 5   Factoring an Equation of Quadratic Type

Solve $2 \sin^2 x - \sin x - 1 = 0$.

SOLUTION

The graph of $y = 2 \sin^2 x - \sin x - 1$, shown in Figure 3.9, indicates that there are three solutions in the interval $[0, 2\pi)$. Treating the equation as a quadratic in sin x and factoring, you can obtain the following.

$$2 \sin^2 x - \sin x - 1 = 0 \qquad \textit{Original equation}$$
$$(2 \sin x + 1)(\sin x - 1) = 0 \qquad \textit{Factor}$$

Setting each factor equal to zero produces the following solutions.

$$2 \sin x + 1 = 0 \qquad \text{and} \qquad \sin x - 1 = 0$$

$$\sin x = -\frac{1}{2} \qquad\qquad\qquad \sin x = 1$$

$$x = \frac{7\pi}{6}, \frac{11\pi}{6} \qquad\qquad\qquad x = \frac{\pi}{2}$$

The general solution is

$$x = \frac{7\pi}{6} + 2n\pi, \qquad x = \frac{11\pi}{6} + 2n\pi, \qquad x = \frac{\pi}{2} + 2n\pi,$$

where n is an integer.

EXAMPLE 6   Writing in Terms of a Single Trigonometric Function

Solve $2 \sin^2 x + 3 \cos x - 3 = 0$.

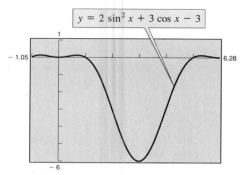

$y = 2 \sin^2 x + 3 \cos x - 3$

FIGURE 3.10

SOLUTION

The graph of $y = 2 \sin^2 x + 3 \cos x - 3$, shown in Figure 3.10, indicates that there are three solutions in the interval $[0, 2\pi)$. Proceeding algebraically, you can write the following.

$2 \sin^2 x + 3 \cos x - 3 = 0$	*Original equation*
$2(1 - \cos^2 x) + 3 \cos x - 3 = 0$	*Pythagorean Identity*
$2 \cos^2 x - 3 \cos x + 1 = 0$	*Multiply both sides by -1*
$(2 \cos x - 1)(\cos x - 1) = 0$	*Factor*

By setting each factor equal to zero, you can find the solutions in the interval $[0, 2\pi)$ to be $x = 0$, $x = \pi/3$, and $x = 5\pi/3$. The general solution is therefore

$$x = 2n\pi, \qquad x = \frac{\pi}{3} + 2n\pi, \qquad x = \frac{5\pi}{3} + 2n\pi, \qquad \textit{General solution}$$

where n is an integer.

Squaring both sides of an equation can introduce extraneous solutions, as indicated in Example 7.

EXAMPLE 7   **Squaring and Converting to Quadratic Type**

$\cos x + 1 = \sin x$	*Original equation*
$\cos^2 x + 2 \cos x + 1 = \sin^2 x$	*Square both sides*
$\cos^2 x + 2 \cos x + 1 = 1 - \cos^2 x$	*Identity*
$2 \cos^2 x + 2 \cos x = 0$	*Collect terms*
$2 \cos x(\cos x + 1) = 0$	*Factor*

Setting each factor equal to zero produces the following.

$$2 \cos x = 0 \qquad \text{and} \qquad \cos x + 1 = 0$$
$$\cos x = 0 \qquad\qquad\qquad \cos x = -1$$
$$x = \frac{\pi}{2}, \frac{3\pi}{2} \qquad\qquad\qquad x = \pi$$

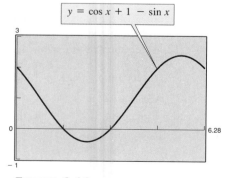

$y = \cos x + 1 - \sin x$

FIGURE 3.11

Because you squared both sides of the original equation, you must check for extraneous solutions. Of the three possible solutions, $x = 3\pi/2$ turns out to be extraneous. (Check this.) Thus, in the interval $[0, 2\pi)$, the only two solutions are $x = \pi/2$ and $x = \pi$. You can confirm this from the graph of $y = \cos x + 1 - \sin x$, shown in Figure 3.11. The general solution is $x = \pi/2 + 2n\pi$ and $x = \pi + 2n\pi$, where n is an integer.

Functions Involving Multiple Angles

EXAMPLE 8   Functions of Multiple Angles

Solve $2 \cos 3t - 1 = 0$.

SOLUTION

The graph of $y = 2 \cos 3t - 1$, shown in Figure 3.12, indicates that there are six solutions in the interval $[0, 2\pi)$. Proceeding algebraically, you can write the following.

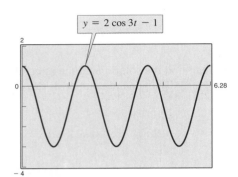

$y = 2 \cos 3t - 1$

FIGURE 3.12

$2 \cos 3t - 1 = 0$	*Original equation*
$2 \cos 3t = 1$	*Add 1 to both sides*
$\cos 3t = \dfrac{1}{2}$	*Divide both sides by 2*

In the interval $[0, 2\pi)$, you know that $3t = \pi/3$ and $3t = 5\pi/3$, so that

$$3t = \frac{\pi}{3} + 2n\pi \quad \text{and} \quad 3t = \frac{5\pi}{3} + 2n\pi$$

$$t = \frac{\pi}{9} + \frac{2n\pi}{3} \quad \text{and} \quad t = \frac{5\pi}{9} + \frac{2n\pi}{3}, \quad \text{\textit{General solution}}$$

where n is an integer. The six solutions in the interval $[0, 2\pi)$ are as follows.

$$\frac{\pi}{9}, \quad \frac{7\pi}{9}, \quad \frac{13\pi}{9}, \quad \frac{5\pi}{9}, \quad \frac{11\pi}{9}, \quad \frac{17\pi}{9}$$

EXAMPLE 9   Functions of Multiple Angles

Solve $3 \tan(x/2) + 3 = 0$.

SOLUTION

The graph of $y = 3 \tan(x/2) + 3$, shown in Figure 3.13, indicates that there is only one solution in the interval $[0, 2\pi)$. Using the zoom and trace features, you can determine that the solution is $x \approx 4.712$. Algebraically, you can solve the equation as follows.

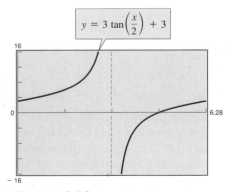

$y = 3 \tan\left(\dfrac{x}{2}\right) + 3$

FIGURE 3.13

$3 \tan \dfrac{x}{2} + 3 = 0$	*Original equation*
$3 \tan \dfrac{x}{2} = -3$	*Subtract 3 from both sides*
$\tan \dfrac{x}{2} = -1$	*Divide both sides by 3*

In the interval $[0, \pi)$, you know that $x/2 = 3\pi/4$, so that

$$\frac{x}{2} = \frac{3\pi}{4} + n\pi$$

$$x = \frac{3\pi}{2} + 2n\pi,$$ *General solution*

where n is an integer.

Using Inverse Functions and a Calculator

EXAMPLE 10   Using Inverse Functions

Solve $\sec^2 x - 2 \tan x = 4$.

SOLUTION

$$\sec^2 x - 2 \tan x = 4$$ *Original equation*

$$1 + \tan^2 x - 2 \tan x - 4 = 0$$ *Pythagorean Identity*

$$\tan^2 x - 2 \tan x - 3 = 0$$ *Combine like terms*

$$(\tan x - 3)(\tan x + 1) = 0$$ *Factor*

Setting each factor equal to zero produces two solutions in the interval $(-\pi/2, \pi/2)$.

$$\tan x = 3, \qquad \tan x = -1$$

$$x = \arctan 3 \qquad x = -\frac{\pi}{4}$$

Adding multiples of π (the period of the tangent) produces the general solution

$$x = \arctan 3 + n\pi \quad \text{and} \quad x = -\frac{\pi}{4} + n\pi,$$ *General solution*

where n is an integer. Note that the graph of $y = \tan^2 x - 2 \tan x - 3$, shown in Figure 3.14, has the two x-intercepts

$$x = \arctan 3 \approx 1.2490 \quad \text{and} \quad x = -\frac{\pi}{4} \approx -0.7854$$

in the interval $(-\pi/2, \pi/2)$.

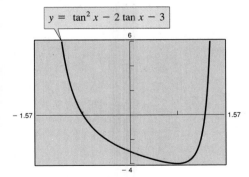

$y = \tan^2 x - 2 \tan x - 3$

FIGURE 3.14

When a calculator is used for arcsin x, arccos x, and arctan x, the displayed solution may need to be adjusted to obtain solutions in the desired interval.

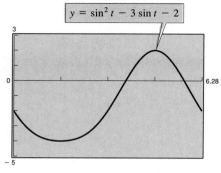

$y = \sin^2 t - 3 \sin t - 2$

FIGURE 3.15

EXAMPLE 11   Using the Quadratic Formula

Solve $\sin^2 t - 3 \sin t - 2 = 0$ in the interval $[0, 2\pi)$.

SOLUTION

The graph of $y = \sin^2 t - 3 \sin t - 2$, shown in Figure 3.15, indicates that there are two solutions in the interval $[0, 2\pi)$. To find these solutions algebraically, you can use the Quadratic Formula as follows.

$$\sin^2 t - 3 \sin t - 2 = 0 \qquad \text{\textit{Original equation}}$$

$$\sin t = \frac{-(-3) \pm \sqrt{(-3)^2 - 4(1)(-2)}}{2(1)} \qquad \text{\textit{Quadratic Formula}}$$

$$= \frac{3 \pm \sqrt{17}}{2} \qquad \text{\textit{Simplify}}$$

$$\approx 3.561553 \quad \text{or} \quad -0.5615528$$

Because the range of the sine function is $[-1, 1]$, the equation $\sin t = 3.561553$ has no solution. To solve the equation $\sin t = -0.5615528$, use a calculator and the inverse sine function to obtain

$$t \approx \arcsin(-0.5615528) \approx -0.5962612.$$

Note that this solution is not in the interval $[0, 2\pi)$. In this interval, the two solutions are $t \approx \pi + 0.60 \approx 3.74$ and $t \approx 2\pi - 0.60 \approx 5.68$.

For the next example, there is no analytic way to solve the equation: we rely solely on a graphing utility to approximate the solutions.

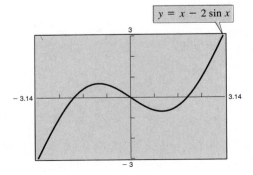

$y = x - 2 \sin x$

FIGURE 3.16

EXAMPLE 12   Approximating Solutions

Approximate the solutions of $x = 2 \sin x$.

SOLUTION

The graph of $y = x - 2 \sin x$ is shown in Figure 3.16. From the graph, you can see that one solution is $x = 0$. Using the zoom and trace features of a graphing utility, you can approximate the positive solution to be $x \approx 1.8955$. Finally, by symmetry, you can approximate the negative solution to be $x \approx -1.8955$.

FIGURE 3.17

EXAMPLE 13   Surface Area of a Honeycomb

The surface area of a honeycomb is given by the equation

$$S = 6hs + \frac{3}{2}s^2\left(\frac{\sqrt{3} - \cos\theta}{\sin\theta}\right), \qquad 0 \le \theta \le 90°,$$

where $h = 2.4$ inches, $s = 0.75$ inches, and θ is the angle indicated in Figure 3.17.

A. What value of θ gives a surface area of 12 square inches?

B. What value of θ gives the minimum surface area?

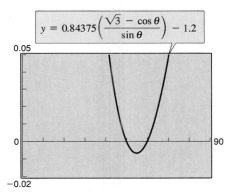

FIGURE 3.18

SOLUTION

A. Let $h = 2.4$, $s = 0.75$, and $S = 12$.

$$10.8 + 0.84375\left(\frac{\sqrt{3} - \cos\theta}{\sin\theta}\right) = 12$$

$$0.84375\left(\frac{\sqrt{3} - \cos\theta}{\sin\theta}\right) - 1.2 = 0$$

The graph of

$$y = 0.84375\left(\frac{\sqrt{3} - \cos\theta}{\sin\theta}\right) - 1.2$$

is shown in Figure 3.18. Using the zoom and trace features, you can determine that $\theta \approx 59.9°$ and $\approx 49.9°$.

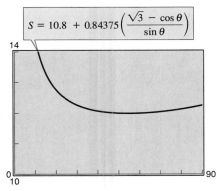

FIGURE 3.19

B. Graph the function

$$S = 10.8 + 0.84375\left(\frac{\sqrt{3} - \cos\theta}{\sin\theta}\right),$$

as shown in Figure 3.19. You can zoom in on the minimum point on the graph, which occurs at $\theta \approx 54.7°$. By using calculus, it can be shown that

$$\theta = \arccos\left(\frac{1}{\sqrt{3}}\right) \approx 54.7356$$

is the exact minimum value.

DISCUSSION PROBLEM

EQUATIONS WITH NO SOLUTIONS

One of the following three equations has solutions and the other two don't. Which two equations do not have solutions?

(a) $\sin^2 x - 5 \sin x + 6 = 0$
(b) $\sin^2 x - 4 \sin x + 6 = 0$
(c) $\sin^2 x - 5 \sin x - 6 = 0$

Can you find conditions involving the constants b and c that will guarantee that the equation $\sin^2 x + b \sin x + c = 0$ has at least one solution?

WARM-UP

The following warm-up exercises involve skills that were covered in earlier sections. You will use these skills in the exercise set for this section.

In Exercises 1–6, find the values of θ in the interval $0 \le \theta < 2\pi$ that satisfy the equation.

1. $\cos \theta = -\dfrac{1}{2}$ **2.** $\sin \theta = \dfrac{\sqrt{3}}{2}$ **3.** $\cos \theta = \dfrac{\sqrt{2}}{2}$

4. $\sin \theta = -\dfrac{\sqrt{2}}{2}$ **5.** $\tan \theta = \sqrt{3}$ **6.** $\tan \theta = -1$

In Exercises 7–10, solve for x.

7. $\dfrac{x}{3} + \dfrac{x}{5} = 1$ **8.** $2x(x + 3) - 5(x + 3) = 0$

9. $2x^2 - 4x - 5 = 0$ **10.** $\dfrac{1}{x} = \dfrac{x}{2x + 3}$

SECTION 3.3 · EXERCISES

In Exercises 1–6, verify that the given values of x are solutions of the equation.

1. $2 \cos x - 1 = 0$
   (a) $x = \dfrac{\pi}{3}$ (b) $x = \dfrac{5\pi}{3}$

2. $\csc x - 2 = 0$
   (a) $x = \dfrac{\pi}{6}$ (b) $x = \dfrac{5\pi}{6}$

3. $3 \tan^2 2x - 1 = 0$
   (a) $x = \dfrac{\pi}{12}$ (b) $x = \dfrac{5\pi}{12}$

4. $2 \cos^2 4x - 1 = 0$
   (a) $x = \dfrac{\pi}{16}$ (b) $x = \dfrac{3\pi}{16}$

5. $2 \sin^2 x - \sin x - 1 = 0$
   (a) $x = \dfrac{\pi}{2}$ (b) $x = \dfrac{7\pi}{6}$

6. $\sec^4 x - 4 \sec^2 x = 0$
   (a) $x = \dfrac{2\pi}{3}$ (b) $x = \dfrac{5\pi}{3}$

In Exercises 7–20, solve the equation. Verify your result with a graphing utility.

7. $2 \cos x + 1 = 0$ **8.** $2 \sin x - 1 = 0$

9. $\sqrt{3} \csc x - 2 = 0$ **10.** $\tan x + 1 = 0$

11. $2 \sin^2 x = 1$ **12.** $\tan^2 x = 3$

13. $3 \sec^2 x - 4 = 0$ **14.** $\csc^2 x - 2 = 0$

15. $\tan x(\tan x - 1) = 0$ **16.** $\cos x(2 \cos x + 1) = 0$

17. $\sin x (\sin x + 1) = 0$

18. $4 \sin^2 x - 3 = 0$

19. $\sin^2 x = 3 \cos^2 x$

20. $(3 \tan^2 x - 1)(\tan^2 x - 3) = 0$

In Exercises 21–32, find all solutions of the equation in the interval $[0, 2\pi)$. Verify your results with a graphing utility.

21. $\sec x \csc x - 2 \csc x = 0$

22. $\sec^2 x - \sec x - 2 = 0$

23. $2 \sin^2 x + 3 \sin x + 1 = 0$

24. $3 \tan^3 x - \tan x = 0$

25. $2 \sec^2 x + \tan^2 x - 3 = 0$

26. $2 \sin^2 x = 2 + \cos x$

27. $2 \sin x + \csc x = 0$

28. $\csc x + \cot x = 1$

29. $\sin 2x = -\dfrac{\sqrt{3}}{2}$

30. $\tan 3x = 1$

31. $\cos \dfrac{x}{2} = \dfrac{\sqrt{2}}{2}$

32. $\sec 4x = 2$

In Exercises 33–36, solve the algebraic and trigonometric equations. Restrict the solutions of the trigonometric equations to the interval $[0, 2\pi)$.

33. $6y^2 - 13y + 6 = 0$
$6 \cos^2 x - 13 \cos x + 6 = 0$

34. $y^2 + y - 20 = 0$
$\sin^2 x + \sin x - 20 = 0$

35. $y^2 - 8y + 13 = 0$
$\tan^2 x - 8 \tan x + 13 = 0$

36. $2y^2 + 6y - 1 = 0$
$2 \cos^2 x + 6 \cos x - 1 = 0$

In Exercises 37–50, use a graphing utility to approximate all solutions of the equation in the interval $[0, 2\pi)$.

37. $2 \cos x - \sin x = 0$

38. $4 \sin^3 x - 2 \sin x - 1 = 0$

39. $\dfrac{1 + \sin x}{\cos x} + \dfrac{\cos x}{1 + \sin x} = 4$

40. $\dfrac{\cos x \cot x}{1 - \sin x} = 3$

41. $2 \sin x - x = 0$

42. $x \cos x - 1 = 0$

43. $\sec^2 x + 0.5 \tan x - 1 = 0$

44. $\csc^2 x + 0.5 \cot x - 5 = 0$

45. $2 \tan^2 x + 7 \tan x - 15 = 0$

46. $12 \cos^2 x + 5 \cos x - 3 = 0$

47. $12 \sin^2 x - 13 \sin x + 3 = 0$

48. $3 \tan^2 x + 4 \tan x - 4 = 0$

49. $\sin^2 x + 2 \sin x - 1 = 0$

50. $4 \cos^2 x - 4 \cos x - 1 = 0$

Extrema of a Function In Exercises 51 and 52, (a) use a graphing utility to graph the function f and approximate the maximum and minimum points on the graph in the interval $[0, 2\pi]$. (b) Solve the given trigonometric equation and verify that its solutions are the x-coordinates of the maximum and minimum points of f.

Function	*Trigonometric Equation*
51. $f(x) = \sin x + \cos x$	$\cos x - \sin x = 0$
52. $f(x) = 2 \sin x + \cos 2x$	$2 \cos x - 4 \sin x \cos x = 0$

53. *Picard Iteration* Use the following procedure to solve the equation $\cos x = x$. Begin by calculating $\cos 0.5$. Then take the cosine of your result. Repeat this procedure over and over until you observe that the result does not change. Describe another way to solve the same equation.

54. *How Many Solutions?* Use a graphing utility to graph $y = \sin(1/x)$.

(a) How many solutions does the equation

$$\sin \frac{1}{x} = 0$$

have in the interval $[-1, 1]$? Describe one or more viewing rectangles that can be used to help support your conclusion.

(b) Does the equation $\sin(1/x) = 0$ have a greatest solution? If so, approximate the solution. If not, explain.

55. *Maximum Area* The area of a rectangle inscribed in one arch of the graph of $y = \cos x$, as shown in the figure, is given by

$$A = 2x \cos x, \qquad -\frac{\pi}{2} < x < \frac{\pi}{2}.$$

Use a graphing utility to graph the area function, and approximate the area of the largest inscribed rectangle.

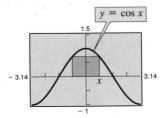

FIGURE FOR 55

56. *Sales*   The monthly sales (in thousands of units) of a seasonal product are approximated by

$$S = 74.50 + 43.75 \sin \frac{\pi t}{6},$$

where t is the time in months, with $t = 1$ corresponding to January (see figure). Determine the months when sales exceed 100,000 units.

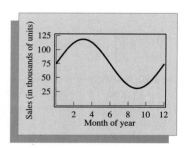

57. *Harmonic Motion*   A weight is oscillating on the end of a spring. The position of the weight relative to the point of equilibrium is given by

$y = \frac{1}{4}(\cos 8t - 3 \sin 8t)$,

where y is the displacement in feet and t is the time in seconds (see figure). Find the times when the weight is at the point of equilibrium ($y = 0$) for $0 \le t \le 1$.

58. *Damped Harmonic Motion*   The displacement from equilibrium of a weight oscillating on the end of a spring is given by

$y = 1.56e^{-0.22t} \cos 4.9t$,

where y is the displacement in feet and t is the time in seconds (see figure). Use a graphing utility to graph the displacement function for $0 \le t \le 10$. Find the time beyond which the displacement does not exceed 1 inch from equilibrium.

59. *Projectile Motion*   A batted baseball leaves the bat at an angle of θ with the horizontal and an initial velocity of $v_0 = 100$ feet per second. The ball is caught by an outfielder 300 feet from home plate (see figure). Find θ if the range r of a projectile is given by

$r = \frac{1}{32}v_0^2 \sin 2\theta$.

60. *Projectile Motion*   A marksman intends to hit a target at a distance of 1000 yards with a gun that has a muzzle velocity of 1200 feet per second (see figure). Neglecting air resistance, determine the minimum elevation of the gun if the range is given by

$r = \frac{1}{32}v_0^2 \sin 2\theta$.

A number c is a **fixed point** of a function f if $f(c) = c$. For example, 2 is a fixed point of $f(x) = x^3 - x - 4$. Finding a fixed point of f is equivalent to finding a solution to the equation $f(x) = x$, or $f(x) - x = 0$.

In Exercises 61–64, find the smallest positive fixed point of the following functions.

61. $f(x) = \tan \dfrac{\pi x}{4}$

62. $f(x) = e^{-x}$

63. $f(x) = \dfrac{4x - x^2 + 2}{4}$

64. $f(x) = \dfrac{5x + 2 - x^3}{5}$

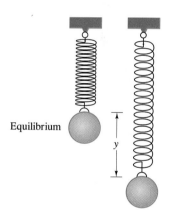

FIGURE FOR 57 AND 58

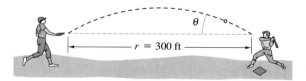

FIGURE FOR 59

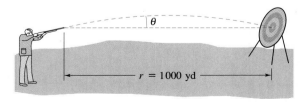

FIGURE FOR 60

3.4   SUM AND DIFFERENCE FORMULAS
....
Introduction   /   Using Sum and Difference Formulas

Introduction

In this and the following section, you will study several trigonometric identities that are important in scientific applications. We begin with six sum and difference formulas that express trigonometric functions of $(u \pm v)$ as functions of u and v alone.

SUM AND DIFFERENCE FORMULAS

$$\sin(u + v) = \sin u \cos v + \cos u \sin v$$

$$\cos(u + v) = \cos u \cos v - \sin u \sin v$$

$$\tan(u + v) = \frac{\tan u + \tan v}{1 - \tan u \tan v}$$

$$\sin(u - v) = \sin u \cos v - \cos u \sin v$$

$$\cos(u - v) = \cos u \cos v + \sin u \sin v$$

$$\tan(u - v) = \frac{\tan u - \tan v}{1 + \tan u \tan v}$$

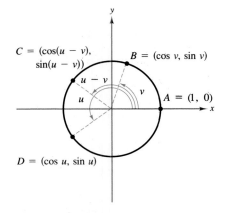

FIGURE 3.20

PROOF

We prove only the formulas for $\cos(u \pm v)$. In Figure 3.20, let A be the point $(1, 0)$ and then use u and v to locate the points $B = (\cos v, \sin v)$, $C = (\cos(u - v), \sin(u - v))$, and $D = (\cos u, \sin u)$ on the unit circle. For convenience, assume that $0 < v < u < 2\pi$. From Figure 3.21, note that arcs AC and BD have the same length. Hence, *line segments AC and BD are also equal in length*. Let the length of $AC = d_1$, and the length of $BD = d_2$. Then $d_1 = d_2$ and $d_1{}^2 = d_2{}^2$. By the Distance Formula, you have

$$d_1{}^2 = [1 - \cos(u - v)]^2 + [0 - \sin(u - v)]^2$$
$$= 1 - 2\cos(u - v) + \cos^2(u - v) + \sin^2(u - v)$$
$$= 2 - 2\cos(u - v)$$

and

$$d_2{}^2 = (\cos u - \cos v)^2 + (\sin u - \sin v)^2$$
$$= \cos^2 u - 2\cos u \cos v + \cos^2 v + \sin^2 u - 2\sin u \sin v + \sin^2 v$$
$$= (\sin^2 u + \cos^2 u) + (\sin^2 v + \cos^2 v) - 2\cos u \cos v - 2\sin u \sin v$$
$$= 2 - 2\cos u \cos v - 2\sin u \sin v.$$

Equating $d_1{}^2$ and $d_2{}^2$ yields

$$2 - 2\cos(u - v) = 2 - 2\cos u \cos v - 2\sin u \sin v$$
$$-2\cos(u - v) = -2(\cos u \cos v + \sin u \sin v)$$
$$\cos(u - v) = \cos u \cos v + \sin u \sin v.$$

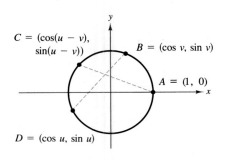

FIGURE 3.21

REMARK   Note that $\sin(u + v) \neq$ $\sin u + \sin v$. Similar statements can be made for $\cos(u + v)$ and $\tan(u + v)$.

The formula for $\cos(u + v)$ can be established by considering $u + v = u - (-v)$ and using the formula just derived to obtain

$$\cos(u + v) = \cos[u - (-v)]$$
$$= \cos u \cos(-v) + \sin u \sin(-v) = \cos u \cos v - \sin u \sin v.$$

D I S C O V E R Y

(a) Use a graphing utility to graph $y = \cos(x + 2)$ and $y = \cos x + \cos 2$ in the same viewing rectangle. What can you conclude from the graphs? Is it true that $\cos(x + 2) = \cos x + \cos 2$?

(b) Use a graphing utility to graph $y = \sin(x + 4)$ and $y = \sin x + \sin 4$ in the same viewing rectangle. What can you conclude from the graphs? Is it true that $\sin(x + 4) = \sin x + \sin 4$?

Using Sum and Difference Formulas

EXAMPLE 1   **Evaluating a Trigonometric Function**

Find the exact value of $\cos 75°$.

SOLUTION

To find the *exact* value of $\cos 75°$, use the fact that $75° = 30° + 45°$. Consequently, the formula for $\cos(u + v)$ yields

$$\cos 75° = \cos(30° + 45°)$$
$$= \cos 30° \cos 45° - \sin 30° \sin 45°$$
$$= \frac{\sqrt{3}}{2}\left(\frac{\sqrt{2}}{2}\right) - \frac{1}{2}\left(\frac{\sqrt{2}}{2}\right) = \frac{\sqrt{6} - \sqrt{2}}{4}.$$

Check this with a calculator. You will find that $\cos 75° \approx 0.259 \approx (\sqrt{6} - \sqrt{2})/4$.

EXAMPLE 2   **Evaluating a Trigonometric Function**

Find the exact value of

$$\cos \frac{\pi}{12}.$$

SOLUTION

Using the fact that

$$\frac{\pi}{12} = \frac{\pi}{3} - \frac{\pi}{4},$$

together with the formula for $\cos(u - v)$, you obtain

$$\cos \frac{\pi}{12} = \cos\left(\frac{\pi}{3} - \frac{\pi}{4}\right)$$
$$= \cos \frac{\pi}{3} \cos \frac{\pi}{4} + \sin \frac{\pi}{3} \sin \frac{\pi}{4}$$
$$= \frac{1}{2}\left(\frac{\sqrt{2}}{2}\right) + \frac{\sqrt{3}}{2}\left(\frac{\sqrt{2}}{2}\right) = \frac{\sqrt{2} + \sqrt{6}}{4}.$$

EXAMPLE 3 Evaluating a Trigonometric Expression

Find the exact value of sin 42° cos 12° − cos 42° sin 12°.

SOLUTION

Recognizing that this expression fits the formula for $\sin(u - v)$, you can write

$$\sin 42° \cos 12° - \cos 42° \sin 12° = \sin(42° - 12°) = \sin 30° = \frac{1}{2}.$$

EXAMPLE 4 Evaluating a Trigonometric Expression

Find the exact value of

$$\frac{\tan 80° + \tan 55°}{1 - \tan 80° \tan 55°}.$$

SOLUTION

From the formula for $\tan(u + v)$, you have

$$\frac{\tan 80° + \tan 55°}{1 - \tan 80° \tan 55°} = \tan(80° + 55°) = \tan 135° = -\tan 45° = -1.$$

EXAMPLE 5 An Application of a Difference Formula

Find $\cos(u - v)$ given that

$$\cos u = -\frac{15}{17}, \quad \pi < u < \frac{3\pi}{2}, \quad \text{and} \quad \sin v = \frac{4}{5}, \quad 0 < v < \frac{\pi}{2}.$$

SOLUTION

Using the given values for cos u and sin v, you can sketch angles u and v, as shown in Figure 3.22. This implies that

$$\cos v = \frac{3}{5} \quad \text{and} \quad \sin u = -\frac{8}{17}.$$

Therefore,

$$\cos(u - v) = \cos u \cos v + \sin u \sin v$$

$$= \left(-\frac{15}{17}\right)\left(\frac{3}{5}\right) + \left(-\frac{8}{17}\right)\left(\frac{4}{5}\right) = -\frac{77}{85}.$$

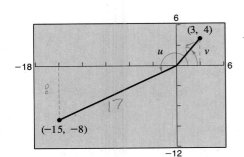

FIGURE 3.22

EXAMPLE 6   Proving a Cofunction Identity

Use the formula for $\cos(u - v)$ to prove the cofunction identity

$$\cos\left(\frac{\pi}{2} - x\right) = \sin x.$$

SOLUTION

Using the difference formula $\cos(u - v) = \cos u \cos v + \sin u \sin v$ produces

$$\cos\left(\frac{\pi}{2} - x\right) = \cos\frac{\pi}{2}\cos x + \sin\frac{\pi}{2}\sin x$$

$$= (0)\cos x + (1)\sin x$$

$$= \sin x.$$

You should verify this identity by graphing

$$y = \cos\left(\frac{\pi}{2} - x\right)$$

and $y = \sin x$ on the same screen.

Sum and difference formulas can be used to derive **reduction formulas** involving expressions such as

$$\sin\left(\theta + \frac{n\pi}{2}\right) \quad \text{and} \quad \cos\left(\theta + \frac{n\pi}{2}\right),$$

where n is an integer.

EXAMPLE 7   Deriving Reduction Formulas

Simplify

$$\cos\left(\theta - \frac{3\pi}{2}\right).$$

SOLUTION

Using the formula $\cos(u - v) = \cos u \cos v + \sin u \sin v$ produces

$$\cos\left(\theta - \frac{3\pi}{2}\right) = \cos\theta\cos\frac{3\pi}{2} + \sin\theta\sin\frac{3\pi}{2}$$

$$= (\cos\theta)(0) + (\sin\theta)(-1)$$

$$= -\sin\theta.$$

EXAMPLE 8   Solving a Trigonometric Equation

Solve

$$\sin\left(x + \frac{\pi}{4}\right) + \sin\left(x - \frac{\pi}{4}\right) = -1$$

in the interval $[0, 2\pi)$.

SOLUTION

Using sum and difference formulas, you can rewrite the given equation as

$$\sin x \cos \frac{\pi}{4} + \cos x \sin \frac{\pi}{4} + \sin x \cos \frac{\pi}{4} - \cos x \sin \frac{\pi}{4} = -1$$

$$2 \sin x \cos \frac{\pi}{4} = -1$$

$$2(\sin x)\left(\frac{\sqrt{2}}{2}\right) = -1$$

$$\sin x = -\frac{1}{\sqrt{2}}$$

$$\sin x = -\frac{\sqrt{2}}{2}.$$

Therefore, the only solutions in the interval $[0, 2\pi)$ are $x = 5\pi/4$ and $x = 7\pi/4$. The graph of

$$y = 1 + \sin\left(x + \frac{\pi}{4}\right) + \sin\left(x - \frac{\pi}{4}\right),$$

shown in Figure 3.23, confirms these two solutions.

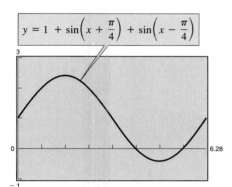

$$y = 1 + \sin\left(x + \frac{\pi}{4}\right) + \sin\left(x - \frac{\pi}{4}\right)$$

FIGURE 3.23

Example 9 shows how a sum formula can be used to rewrite a trigonometric expression in a form that is useful in calculus.

EXAMPLE 9   An Application from Calculus

Verify that

$$\frac{\sin(x + h) - \sin x}{h} = (\cos x)\left(\frac{\sin h}{h}\right) - (\sin x)\left(\frac{1 - \cos h}{h}\right),$$

where $h \neq 0$.

SOLUTION

$$\frac{\sin(x + h) - \sin x}{h} = \frac{\sin x \cos h + \cos x \sin h - \sin x}{h}$$

$$= \frac{\cos x \sin h - \sin x(1 - \cos h)}{h}$$

$$= (\cos x)\left(\frac{\sin h}{h}\right) - (\sin x)\left(\frac{1 - \cos h}{h}\right)$$

**DISCUSSION
PROBLEM**
....
**VERIFYING A
SUM FORMULA**

At the beginning of this section, we listed a proof of the formula for $\cos(u - v)$. Show how you can use this formula together with the identity

$$\sin x = \cos\left(\frac{\pi}{2} - x\right)$$

to prove the formula for $\sin(u + v)$. *Hint:* Start by writing

$$\sin(u + v) = \cos\left[\frac{\pi}{2} - (u + v)\right].$$

WARM-UP
....

The following warm-up exercises involve skills that were covered in earlier sections. You will use these skills in the exercise set for this section.

In Exercises 1–4, use the given information to find $\sin \theta$.

1. $\tan \theta = \frac{1}{3}$, θ in Quadrant I

2. $\cot \theta = \frac{3}{5}$, θ in Quadrant III

3. $\cos \theta = \frac{3}{4}$, θ in Quadrant IV

4. $\sec \theta = -3$, θ in Quadrant II

In Exercises 5 and 6, find all solutions in the interval $[0, 2\pi)$.

5. $\sin x = \frac{\sqrt{2}}{2}$

6. $\cos x = 0$

In Exercises 7–10, simplify the expression.

7. $\tan x \sec^2 x - \tan x$

8. $\dfrac{\cos x \csc x}{\tan x}$

9. $\dfrac{\cos x}{1 - \sin x} - \tan x$

10. $\dfrac{\cos^4 x - \sin^4 x}{\cos^2 x}$

SECTION 3.4 · EXERCISES

In Exercises 1–10, use the sum and difference identities to find the exact values of the sine, cosine, and tangent of the given angle.

1. $75° = 30° + 45°$

2. $15° = 45° - 30°$

3. $105° = 60° + 45°$

4. $165° = 135° + 30°$

5. $195° = 225° - 30°$

6. $255° = 300° - 45°$

7. $\dfrac{11\pi}{12} = \dfrac{3\pi}{4} + \dfrac{\pi}{6}$

8. $\dfrac{7\pi}{12} = \dfrac{\pi}{3} + \dfrac{\pi}{4}$

9. $\dfrac{17\pi}{12} = \dfrac{9\pi}{4} - \dfrac{5\pi}{6}$

10. $-\dfrac{\pi}{12} = \dfrac{\pi}{6} - \dfrac{\pi}{4}$

In Exercises 11–20, use the sum and difference identities to write the expression as the sine, cosine, or tangent of an angle.

11. $\cos 25° \cos 15° - \sin 25° \sin 15°$

12. $\sin 140° \cos 50° + \cos 140° \sin 50°$

13. $\sin 230° \cos 30° - \cos 230° \sin 30°$

14. $\cos 20° \cos 30° + \sin 20° \sin 30°$

15. $\dfrac{\tan 325° - \tan 86°}{1 + \tan 325° \tan 86°}$

16. $\dfrac{\tan 140° - \tan 60°}{1 + \tan 140° \tan 60°}$

17. $\sin 3 \cos 1.2 - \cos 3 \sin 1.2$

18. $\cos \dfrac{\pi}{7} \cos \dfrac{\pi}{5} - \sin \dfrac{\pi}{7} \sin \dfrac{\pi}{5}$

19. $\dfrac{\tan 2x + \tan x}{1 - \tan 2x \tan x}$ $TAN\ 3X$

20. $\cos 3x \cos 2y + \sin 3x \sin 2y$ $cos(3x-2y)$

In Exercises 21–24, find the exact value of the trigonometric function given that

$\sin u = \dfrac{5}{13}, \quad 0 < u < \dfrac{\pi}{2}$ and

$\cos v = -\dfrac{3}{5}, \quad \dfrac{\pi}{2} < v < \pi.$

21. $\sin(u + v)$

22. $\cos(v - u)$

23. $\cos(u + v)$

24. $\sin(u - v)$

In Exercises 25–28, find the exact value of the trigonometric function given that

$\sin u = \dfrac{7}{25}, \quad \dfrac{\pi}{2} < u < \pi$ and

$\cos v = \dfrac{4}{5}, \quad \dfrac{3\pi}{2} < v < 2\pi.$

25. $\cos(u + v)$

26. $\sin(u + v)$

27. $\sin(v - u)$

28. $\cos(u - v)$

In Exercises 29–42, verify the identity.

29. $\sin\left(\dfrac{\pi}{2} + x\right) = \cos x$

30. $\sin(3\pi - x) = \sin x$

31. $\sin\left(\dfrac{\pi}{6} + x\right) = \dfrac{1}{2}(\cos x + \sqrt{3} \sin x)$

32. $\cos\left(\dfrac{5\pi}{4} - x\right) = -\dfrac{\sqrt{2}}{2}(\cos x + \sin x)$

33. $\cos(\pi - \theta) + \sin\left(\dfrac{\pi}{2} + \theta\right) = 0$

34. $\tan\left(\dfrac{\pi}{4} - \theta\right) = \dfrac{1 - \tan \theta}{1 + \tan \theta}$

35. $\cos(x + y) \cos(x - y) = \cos^2 x - \sin^2 y$

36. $\sin(x + y) \sin(x - y) = \sin^2 x - \sin^2 y$

37. $\sin(x + y) + \sin(x - y) = 2 \sin x \cos y$

38. $\cos(x + y) + \cos(x - y) = 2 \cos x \cos y$

39. $\cos(n\pi + \theta) = (-1)^n \cos \theta, \quad n$ is an integer

40. $\sin(n\pi + \theta) = (-1)^n \sin \theta, \quad n$ is an integer

41. $a \sin B\theta + b \cos B\theta = \sqrt{a^2 + b^2} \sin(B\theta + C)$, where $C = \arctan b/a, a > 0$

42. $a \sin B\theta + b \cos B\theta = \sqrt{a^2 + b^2} \cos(B\theta - C)$, where $C = \arctan a/b, b > 0$

In Exercises 43–46, use a graphing utility to verify the identity. (Graph both members of the equation and show that the graphs coincide.)

43. $\cos\left(\dfrac{3\pi}{2} - x\right) = -\sin x$

44. $\cos(\pi + x) = -\cos x$

45. $\sin\left(\dfrac{3\pi}{2} + \theta\right) + \sin(\pi - \theta) = \sin\theta - \cos\theta$

46. $\tan(\pi + \theta) = \tan\theta$

In Exercises 47–50, use the formulas given in Exercises 41 and 42 to write the trigonometric expression in the following forms. Use a graphing utility to verify your results.

(a) $\sqrt{a^2 + b^2}\,\sin(B\theta + C)$

(b) $\sqrt{a^2 + b^2}\,\cos(B\theta - C)$

47. $\sin\theta + \cos\theta$ **48.** $3\sin 2\theta + 4\cos 2\theta$
49. $12\sin 3\theta + 5\cos 3\theta$ **50.** $\sin 2\theta - \cos 2\theta$

In Exercises 51 and 52, use the formulas given in Exercises 41 and 42 to write the trigonometric expression in the form $a\sin B\theta + b\cos B\theta$. Use a graphing utility to verify your result.

51. $2\sin\left(\theta + \dfrac{\pi}{4}\right)$ **52.** $5\cos\left(\theta + \dfrac{3\pi}{4}\right)$

In Exercises 53 and 54, write the trigonometric expression as an algebraic expression.

53. $\sin(\arcsin x + \arccos x)$ **54.** $\sin(\arctan 2x - \arccos x)$

In Exercises 55–58, find all solutions of the equation in the interval $[0, 2\pi)$.

55. $\sin\left(x + \dfrac{\pi}{3}\right) + \sin\left(x - \dfrac{\pi}{3}\right) = 1$

56. $\sin\left(x + \dfrac{\pi}{6}\right) - \sin\left(x - \dfrac{\pi}{6}\right) = \dfrac{1}{2}$

57. $\cos\left(x + \dfrac{\pi}{4}\right) + \cos\left(x - \dfrac{\pi}{4}\right) = 1$

58. $\tan(x + \pi) - \cos\left(x + \dfrac{\pi}{2}\right) = 0$

In Exercises 59 and 60, use a graphing utility to approximate all solutions of the equation in the interval $[0, 2\pi)$.

59. $\cos\left(x + \dfrac{\pi}{4}\right) - \cos\left(x - \dfrac{\pi}{4}\right) = 1$

60. $\tan(x + \pi) + 2\sin(x + \pi) = 0$

61. *Standing Waves*   The equation of a standing wave is obtained by adding the displacements of two waves traveling in opposite directions (see figure). Assume that each of the waves has amplitude A, period T, and wavelength λ. If the models for these waves are

$$y_1 = A\cos\left[2\pi\left(\dfrac{t}{T} - \dfrac{x}{\lambda}\right)\right] \quad \text{and} \quad y_2 = A\cos\left[2\pi\left(\dfrac{t}{T} + \dfrac{x}{\lambda}\right)\right],$$

show that

$$y_1 + y_2 = 2A\cos\dfrac{2\pi t}{T}\cos\dfrac{2\pi x}{\lambda}.$$

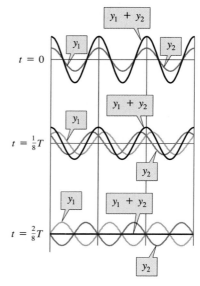

FIGURE FOR 61

62. *Harmonic Motion*   A weight is attached to a spring suspended vertically from a ceiling. When a driving force is applied to the system, the weight moves vertically from its equilibrium position, and this motion is described by the model

$$y = \tfrac{1}{3}\sin 2t + \tfrac{1}{4}\cos 2t,$$

where y is the distance from equilibrium measured in feet and t is the time in seconds.

(a) Write the model in the form

$$y = \sqrt{a^2 + b^2}\,\sin(Bt + C).$$

(See Exercise 41.)

(b) Use a graphing utility to graph the model.
(c) Find the amplitude of the oscillations of the weight.
(d) Find the frequency of the oscillations of the weight.

63. Verify the following identity used in calculus.

$$\frac{\cos(x + h) - \cos x}{h} = \cos x\left(\frac{\cos h - 1}{h}\right) - \sin x\left(\frac{\sin h}{h}\right)$$

64. Use the sum formulas for the sine and cosine to derive the formula

$$\tan(u + v) = \frac{\tan u + \tan v}{1 - \tan u \tan v}.$$

3.5   MULTIPLE-ANGLE AND PRODUCT-TO-SUM FORMULAS

Multiple-Angle Formulas / Power-Reducing Formulas / Half-Angle Formulas / Product-to-Sum Formulas

Multiple-Angle Formulas

In this section, you will study four other categories of trigonometric identities. The first involves functions of multiple angles such as $\sin ku$ and $\cos ku$. The second involves squares of trigonometric functions such as $\sin^2 u$. The third involves functions of half-angles such as $\sin u/2$, and the fourth involves products of trigonometric functions such as $\sin u \cos v$.

> **DOUBLE-ANGLE FORMULAS**
>
> $$\sin 2u = 2 \sin u \cos u$$
> $$\cos 2u = \cos^2 u - \sin^2 u = 2 \cos^2 u - 1 = 1 - 2 \sin^2 u$$
> $$\tan 2u = \frac{2 \tan u}{1 - \tan^2 u}$$

PROOF

To prove the first formula, let $v = u$ in the formula for $\sin(u + v)$, and obtain

$$\begin{aligned} \sin 2u &= \sin(u + u) \\ &= \sin u \cos u + \cos u \sin u \\ &= 2 \sin u \cos u. \end{aligned}$$

The other double-angle formulas can be proved in a similar way.

DISCOVERY

(a) Use a graphing utility to graph $y = \cos(2x)$ and $y = 2 \cos x$ in the same viewing rectangle. What can you conclude from the graphs? Is it true that $\cos(2x) = 2 \cos x$?

(b) Use a graphing utility to graph $y = \sin(4x)$ and $y = 4 \sin x$ in the same viewing rectangle. What can you conclude from the graphs? Is it true that $\sin(4x) = 4 \sin x$?

REMARK   Note that $\sin 2u \neq 2 \sin u$. Similar statements can be made for $\cos 2u$ and $\tan 2u$.

y = 2 cos x + sin 2x

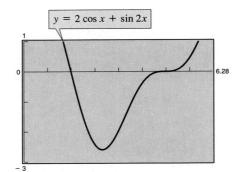

FIGURE 3.24

EXAMPLE 1   Solving a Trigonometric Equation

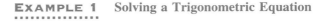

EXAMPLE 1   Solving a Trigonometric Equation

Solve $2 \cos x + \sin 2x = 0$.

SOLUTION

The graph of $y = 2 \cos x + \sin 2x$, shown in Figure 3.24, suggests that there are two solutions in the interval $[0, 2\pi)$. To find these solutions analytically, begin by rewriting the equation so that it involves functions of x (rather than $2x$). Then factor and solve as usual.

$2 \cos x + \sin 2x = 0$	*Original equation*
$2 \cos x + 2 \sin x \cos x = 0$	*Double-angle formula*
$2 \cos x (1 + \sin x) = 0$	*Factor*
$\cos x = 0, \qquad 1 + \sin x = 0$	*Set factors to zero*
$x = \dfrac{\pi}{2}, \dfrac{3\pi}{2} \qquad x = \dfrac{3\pi}{2}$	*Solutions in* $[0, 2\pi)$

Therefore, the general solution is

$$x = \frac{\pi}{2} + 2n\pi \quad \text{and} \quad x = \frac{3\pi}{2} + 2n\pi,$$

where n is an integer.

EXAMPLE 2   Locating Relative Minimums and Maximums

Find the relative minimums and relative maximums of $y = 4 \cos^2 x - 2$ in the interval $[0, \pi]$.

SOLUTION

Using a double-angle identity, you can rewrite the given function as

$$y = 4 \cos^2 x - 2 = 2(2 \cos^2 x - 1) = 2 \cos 2x.$$

Using the techniques discussed in Section 2.5, you can recognize that the graph of this function has an amplitude of 2 and a period of π, as shown in Figure 3.25. The key points in the interval $[0, \pi]$ are as follows.

Maximum	Intercept	Minimum	Intercept	Maximum
$(0, 2)$	$\left(\dfrac{\pi}{4}, 0\right)$	$\left(\dfrac{\pi}{2}, -2\right)$	$\left(\dfrac{3\pi}{4}, 0\right)$	$(\pi, 2)$

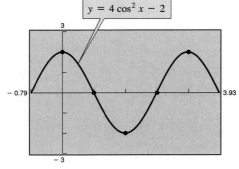

FIGURE 3.25

EXAMPLE 3   Evaluating Functions Involving Double Angles

Use the fact that

$$\cos \theta = \frac{5}{13}, \qquad \frac{3\pi}{2} < \theta < 2\pi$$

to find $\sin 2\theta$, $\cos 2\theta$, and $\tan 2\theta$.

SOLUTION

In Figure 3.26, you can see that $\sin \theta = y/r = -12/13$. Consequently, you can write the following

$$\sin 2\theta = 2 \sin \theta \cos \theta = 2\left(\frac{-12}{13}\right)\left(\frac{5}{13}\right) = -\frac{120}{169}$$

$$\cos 2\theta = 2 \cos^2 \theta - 1 = 2\left(\frac{25}{169}\right) - 1 = -\frac{119}{169}$$

$$\tan 2\theta = \frac{\sin 2\theta}{\cos 2\theta} = \frac{120}{119}$$

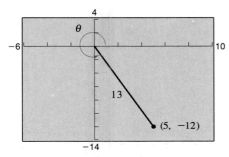

FIGURE 3.26

The double-angle formulas are not restricted to angles 2θ and θ. Other *double* combinations such as 4θ and 2θ are also valid. Here are two examples.

$$\sin 4\theta = 2 \sin 2\theta \cos 2\theta \quad \text{and} \quad \cos 6\theta = \cos^2 3\theta - \sin^2 3\theta$$

EXAMPLE 4   Deriving a Triple-Angle Formula

Express $\sin 3x$ in terms of $\sin x$.

SOLUTION

Consider that $3x = 2x + x$ produces

$$\begin{aligned}
\sin 3x &= \sin(2x + x) \\
&= \sin 2x \cos x + \cos 2x \sin x \\
&= 2 \sin x \cos x \cos x + (1 - 2 \sin^2 x) \sin x \\
&= 2 \sin x \cos^2 x + \sin x - 2 \sin^3 x \\
&= 2 \sin x (1 - \sin^2 x) + \sin x - 2 \sin^3 x \\
&= 2 \sin x - 2 \sin^3 x + \sin x - 2 \sin^3 x \\
&= 3 \sin x - 4 \sin^3 x.
\end{aligned}$$

Try graphing $y = \sin 3x$ and $y = 3 \sin x - 4 \sin^3 x$ in the same viewing rectangle. Their graphs should coincide.

Power-Reducing Formulas

The double-angle formulas can be used to obtain the following **power-reducing formulas.**

POWER-REDUCING FORMULAS

$$\sin^2 u = \frac{1 - \cos 2u}{2} \qquad \cos^2 u = \frac{1 + \cos 2u}{2}$$

$$\tan^2 u = \frac{1 - \cos 2u}{1 + \cos 2u}$$

PROOF

The first two formulas can be verified by solving for $\sin^2 u$ and $\cos^2 u$, respectively, in the double-angle formulas

$$\cos 2u = 1 - 2 \sin^2 u \quad \text{and} \quad \cos 2u = 2 \cos^2 u - 1.$$

The third formula can be verified using the fact that

$$\tan^2 u = \frac{\sin^2 u}{\cos^2 u}.$$

Example 5 shows a typical power reduction that is used in calculus.

EXAMPLE 5   Reducing the Power of a Trigonometric Function

Rewrite $\sin^4 x$ as a sum involving first powers of the cosines of multiple angles.

SOLUTION

Note the repeated use of power-reducing formulas in the following procedure.

$$\sin^4 x = (\sin^2 x)^2 = \left(\frac{1 - \cos 2x}{2} \right)^2$$

$$= \frac{1}{4}(1 - 2 \cos 2x + \cos^2 2x)$$

$$= \frac{1}{4}\left(1 - 2 \cos 2x + \frac{1 + \cos 4x}{2} \right)$$

$$= \frac{1}{4} - \frac{1}{2} \cos 2x + \frac{1}{8} + \frac{1}{8} \cos 4x$$

$$= \frac{3}{8} - \frac{1}{2} \cos 2x + \frac{1}{8} \cos 4x$$

$$= \frac{1}{8}(3 - 4 \cos 2x + \cos 4x)$$

Half-Angle Formulas

You can derive some useful alternative forms of the power-reducing formulas by replacing u with $u/2$. The results are **half-angle formulas.**

HALF-ANGLE FORMULAS

$$\sin \frac{u}{2} = \pm \sqrt{\frac{1 - \cos u}{2}}$$

$$\cos \frac{u}{2} = \pm \sqrt{\frac{1 + \cos u}{2}}$$

$$\tan \frac{u}{2} = \frac{1 - \cos u}{\sin u} = \frac{\sin u}{1 + \cos u}$$

The signs of $\sin u/2$ and $\cos u/2$ depend on the quadrant in which $u/2$ lies.

EXAMPLE 6   Using a Half-Angle Formula

Find the exact value of $\sin 105°$.

SOLUTION

Begin by noting that $105°$ is half of $210°$. Then, using the half-angle formula for $\sin(u/2)$ and the fact that $105°$ lies in Quadrant II, you have

$$\sin 105° = \sqrt{\frac{1 - \cos 210°}{2}}$$

$$= \sqrt{\frac{1 - (-\cos 30°)}{2}}$$

$$= \sqrt{\frac{1 + (\sqrt{3}/2)}{2}}$$

$$= \frac{\sqrt{2 + \sqrt{3}}}{2}.$$

$$\sqrt{1 + \frac{\sqrt{3}}{2}} \qquad \sqrt{\frac{1}{2} + \frac{\sqrt{3}}{4}} \quad \sqrt{\frac{2}{4} + \frac{\sqrt{3}}{4}}$$

$$\sqrt{\frac{2 + \sqrt{3}}{4}}$$

$$\frac{\sqrt{2 + \sqrt{3}}}{2}$$

Choose the positive square root because $\sin \theta$ is positive in Quadrant II. Use your calculator to evaluate $\sin 105°$ and $\sqrt{2 + \sqrt{3}}/2$ to see that both values are approximately 0.9659258.

EXAMPLE 7   Solving a Trigonometric Equation

Solve

$$2 - \sin^2 x = 2 \cos^2 \frac{x}{2}.$$

SOLUTION

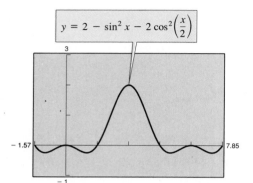

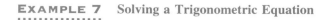

$$y = 2 - \sin^2 x - 2 \cos^2 \left(\frac{x}{2}\right)$$

FIGURE 3.27

The graph of $y = 2 - \sin^2 x - 2 \cos^2(x/2)$, shown in Figure 3.27, suggests that there are three solutions in the interval $[0, 2\pi)$. To find these analytically, solve the given equation as follows.

$$2 - \sin^2 x = 2 \cos^2 \frac{x}{2} \qquad \textit{Original equation}$$

$$2 - \sin^2 x = 2\left(\frac{1 + \cos x}{2}\right) \qquad \textit{Half-angle formula}$$

$$2 - \sin^2 x = 1 + \cos x \qquad \textit{Simplify}$$

$$2 - (1 - \cos^2 x) = 1 + \cos x \qquad \textit{Pythagorean Identity}$$

$$\cos^2 x - \cos x = 0 \qquad \textit{Simplify}$$

$$\cos x(\cos x - 1) = 0 \qquad \textit{Factor}$$

By setting the factors $\cos x$ and $(\cos x - 1)$ equal to zero, you can find that the solutions in the interval $[0, 2\pi)$ are $x = \pi/2 \approx 1.57$, $x = 3\pi/2 \approx 4.71$, and $x = 0$. Therefore, the general solution is

$$x = 2n\pi, \quad x = \frac{\pi}{2} + 2n\pi, \quad \text{and} \quad x = \frac{3\pi}{2} + 2n\pi,$$

where n is an integer.

Product-to-Sum Formulas

Each of the following **product-to-sum formulas** is easily verified using the sum and difference formulas discussed in the preceding section.

PRODUCT-TO-SUM FORMULAS

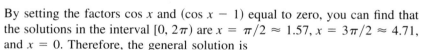

$$\sin u \sin v = \frac{1}{2}[\cos(u - v) - \cos(u + v)]$$

$$\cos u \cos v = \frac{1}{2}[\cos(u - v) + \cos(u + v)]$$

$$\sin u \cos v = \frac{1}{2}[\sin(u + v) + \sin(u - v)]$$

$$\cos u \sin v = \frac{1}{2}[\sin(u + v) - \sin(u - v)]$$

EXAMPLE 8 Writing Products as Sums

Rewrite $\cos 5x \sin 4x$ as a sum or difference.

SOLUTION

$$\cos 5x \sin 4x = \frac{1}{2}[\sin(5x + 4x) - \sin(5x - 4x)] = \frac{1}{2}\sin 9x - \frac{1}{2}\sin x$$

Occasionally, it is useful to reverse the procedure and write a sum of trigonometric functions as a product. This can be accomplished with the following **sum-to-product formulas.**

SUM-TO-PRODUCT FORMULAS

$$\sin x + \sin y = 2 \sin\left(\frac{x + y}{2}\right) \cos\left(\frac{x - y}{2}\right)$$

$$\sin x - \sin y = 2 \cos\left(\frac{x + y}{2}\right) \sin\left(\frac{x - y}{2}\right)$$

$$\cos x + \cos y = 2 \cos\left(\frac{x + y}{2}\right) \cos\left(\frac{x - y}{2}\right)$$

$$\cos x - \cos y = -2 \sin\left(\frac{x + y}{2}\right) \sin\left(\frac{x - y}{2}\right)$$

EXAMPLE 9 Using a Sum-to-Product Formula

Find the exact value of $\cos 195° + \cos 105°$.

SOLUTION

Using the appropriate sum-to-product formula produces

$$\cos 195° + \cos 105° = 2 \cos\left(\frac{195° + 105°}{2}\right) \cos\left(\frac{195° - 105°}{2}\right)$$

$$= 2 \cos 150° \cos 45°$$

$$= 2\left(-\frac{\sqrt{3}}{2}\right)\left(\frac{\sqrt{2}}{2}\right)$$

$$= -\frac{\sqrt{6}}{2}.$$

EXAMPLE 10   Solving a Trigonometric Equation

Solve $\sin 5x + \sin 3x = 0$.

SOLUTION

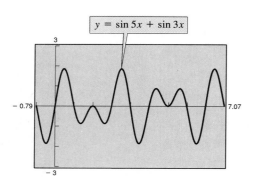

FIGURE 3.28

The graph of $y = \sin 5x + \sin 3x$, shown in Figure 3.28, suggests that there are eight solutions in the interval $[0, 2\pi)$. To find these analytically, proceed as follows.

$$\sin 5x + \sin 3x = 0 \qquad \textit{Original equation}$$

$$2 \sin\left(\frac{5x + 3x}{2}\right) \cos\left(\frac{5x - 3x}{2}\right) = 0 \qquad \textit{Sum-to-product formula}$$

$$2 \sin 4x \cos x = 0 \qquad \textit{Simplify}$$

By setting the factor $\sin 4x$ equal to zero, you find that the solutions in the interval $[0, 2\pi)$ are

$$x = 0, \quad \frac{\pi}{4}, \quad \frac{\pi}{2}, \quad \frac{3\pi}{4}, \quad \pi, \quad \frac{5\pi}{4}, \quad \frac{3\pi}{2}, \quad \frac{7\pi}{4}.$$

Moreover, the equation $\cos x = 0$ yields no additional solutions, and you can conclude the solutions are of the form

$$x = \frac{n\pi}{4},$$

where n is an integer.

EXAMPLE 11   Verifying a Trigonometric Identity

Verify the identity

$$\frac{\sin t + \sin 3t}{\cos t + \cos 3t} = \tan 2t.$$

SOLUTION

Using appropriate sum-to-product formulas, you obtain

$$\frac{\sin t + \sin 3t}{\cos t + \cos 3t} = \frac{2 \sin 2t \cos(-t)}{2 \cos 2t \cos(-t)} = \frac{\sin 2t}{\cos 2t} = \tan 2t.$$

DISCUSSION PROBLEM

DERIVING A TRIPLE-ANGLE FORMULA

In Example 4, you saw how to derive a formula for sin 3x. Show how you can derive a similar formula for cos 3x. That is, find a formula that expresses cos 3x in terms of cos x, and then verify your formula graphically.

WARM-UP

The following warm-up exercises involve skills that were covered in earlier sections. You will use these skills in the exercise set for this section.

In Exercises 1 and 2, factor the trigonometric expression.

1. $2 \sin x + \sin x \cos x$

2. $\cos^2 x - \cos x - 2$

In Exercises 3–6, find all solutions of the equation in the interval $[0, 2\pi)$.

3. $\sin 2x = 0$

4. $\cos 2x = 0$

5. $\cos \dfrac{x}{2} = 0$

6. $\sin \dfrac{x}{2} = 0$

In Exercises 7–10, simplify the expression.

7. $\dfrac{1 - \cos \dfrac{\pi}{4}}{2}$

8. $\dfrac{1 + \cos \dfrac{\pi}{3}}{2}$

9. $\dfrac{2 \sin 3x \cos x}{2 \cos 3x \cos x}$

10. $(1 - 2 \sin^2 x) \cos x - 2 \sin^2 x \cos x$

SECTION 3.5 · EXERCISES

In Exercises 1–10, use a graphing utility to graph the function and approximate its zeros in the interval $[0, 2\pi)$. If possible, find the exact values of the zeros algebraically.

1. $f(x) = \sin 2x - \sin x$

2. $f(x) = \sin 2x + \cos x$

3. $g(x) = 4 \sin x \cos x - 1$

4. $g(x) = \sin 2x \sin x - \cos x$

5. $h(x) = \cos 2x - \cos x$

6. $h(x) = \cos 2x + \sin x$

7. $y = \tan 2x - \cot x$

8. $y = \tan 2x - 2 \cos x$

9. $h(t) = \sin 4t + 2 \sin 2t$

10. $f(s) = (\sin 2s + \cos 2s)^2 - 1$

In Exercises 11–14, use a double-angle identity to rewrite the function. Use a graphing utility to graph both versions of the function and note that the graphs coincide. Identify the relative minimums and relative maximums over the interval $[0, 2\pi)$.

11. $f(x) = 6 \sin x \cos x$

12. $g(x) = 4 \sin x \cos x + 2$

13. $g(x) = 4 - 8 \sin^2 x$

14. $f(x) = (\cos x + \sin x)(\cos x - \sin x)$

In Exercises 15–20, find the exact values of sin 2u, cos 2u, and tan 2u by using the double-angle formulas.

15. $\sin u = \dfrac{3}{5}, \quad 0 < u < \dfrac{\pi}{2}$

16. $\cos u = -\dfrac{2}{3}, \quad \dfrac{\pi}{2} < u < \pi$

17. $\tan u = \dfrac{1}{2}, \quad \pi < u < \dfrac{3\pi}{2}$

18. $\cot u = -4, \quad \dfrac{3\pi}{2} < u < 2\pi$

19. $\sec u = -\dfrac{5}{2}, \quad \dfrac{\pi}{2} < u < \pi$

20. $\csc u = 3, \quad \dfrac{\pi}{2} < u < \pi$

In Exercises 21–26, use the power-reducing formulas to write the expression in terms of the first power of the cosine.

21. $\cos^4 x$ **22.** $\sin^4 x$

23. $\sin^2 x \cos^2 x$ **24.** $\cos^6 x$

25. $\sin^2 x \cos^4 x$ **26.** $\sin^4 x \cos^2 x$

In Exercises 27–32, use the half-angle formulas to determine the exact values of the sine, cosine, and tangent of the given angle.

27. $105°$ **28.** $165°$

29. $112° \, 30'$ **30.** $67° \, 30'$

31. $\dfrac{\pi}{8}$ **32.** $\dfrac{\pi}{12}$

In Exercises 33–38, find the exact values of $\sin(u/2)$, $\cos(u/2)$, and $\tan(u/2)$ by using the half-angle formulas.

33. $\sin u = \dfrac{5}{13}, \quad \dfrac{\pi}{2} < u < \pi$

34. $\cos u = \dfrac{3}{5}, \quad 0 < u < \dfrac{\pi}{2}$

35. $\tan u = -\dfrac{5}{8}, \quad \dfrac{3\pi}{2} < u < 2\pi$

36. $\cot u = 3, \quad \pi < u < \dfrac{3\pi}{2}$

37. $\csc u = -\dfrac{5}{3}, \quad \pi < u < \dfrac{3\pi}{2}$

38. $\sec u = -\dfrac{7}{2}, \quad \dfrac{\pi}{2} < u < \pi$

In Exercises 39–42, use the half-angle formulas to simplify the expression.

39. $\sqrt{\dfrac{1 - \cos 6x}{2}}$ **40.** $\sqrt{\dfrac{1 + \cos 4x}{2}}$

41. $-\sqrt{\dfrac{1 - \cos 8x}{1 + \cos 8x}}$ **42.** $-\sqrt{\dfrac{1 - \cos(x - 1)}{2}}$

In Exercises 43–46, use a graphing utility to graph the function and approximate its zeros in the interval $[0, 2\pi)$. If possible, find the exact values of the zeros algebraically.

43. $f(x) = \sin \dfrac{x}{2} + \cos x$

44. $h(x) = \sin \dfrac{x}{2} + \cos x - 1$

45. $h(x) = \cos \dfrac{x}{2} - \sin x$ **46.** $g(x) = \tan \dfrac{x}{2} - \sin x$

In Exercises 47–56, use the product-to-sum formulas to write the product as a sum.

47. $6 \sin \dfrac{\pi}{4} \cos \dfrac{\pi}{4}$ **48.** $4 \sin \dfrac{\pi}{3} \cos \dfrac{5\pi}{6}$

49. $\sin 5\theta \cos 3\theta$ **50.** $3 \sin 2\alpha \sin 3\alpha$

51. $5 \cos(-5\beta) \cos 3\beta$ **52.** $\cos 2\theta \cos 4\theta$

53. $\sin(x + y) \sin(x - y)$ **54.** $\sin(x + y) \cos(x - y)$

55. $\sin(\theta + \pi) \cos(\theta - \pi)$ **56.** $10 \cos 75° \cos 15°$

In Exercises 57–66, use the sum-to-product formulas to write the sum or difference as a product.

57. $\sin 60° + \sin 30°$ **58.** $\cos 120° + \cos 30°$

59. $\cos \dfrac{3\pi}{4} - \cos \dfrac{\pi}{4}$ **60.** $\sin 5\theta - \sin 3\theta$

61. $\cos 6x + \cos 2x$ **62.** $\sin x + \sin 5x$

63. $\sin(\alpha + \beta) - \sin(\alpha - \beta)$

64. $\cos\left(\theta + \dfrac{\pi}{2}\right) - \cos\left(\theta - \dfrac{\pi}{2}\right)$

65. $\cos(\phi + 2\pi) + \cos \phi$

66. $\sin\left(x + \dfrac{\pi}{2}\right) + \sin\left(x - \dfrac{\pi}{2}\right)$

In Exercises 67–70, use a graphing utility to graph the function and approximate its zeros in the interval $[0, 2\pi)$. If possible, find the exact values of the zeros algebraically.

67. $g(x) = \sin 6x + \sin 2x$

68. $h(x) = \cos 2x - \cos 6x$

69. $f(x) = \dfrac{\cos 2x}{\sin 3x - \sin x} - 1$

70. $f(x) = \sin^2 3x - \sin^2 x$

In Exercises 71–88, verify the identity. Use a graphing utility to confirm the identity.

71. $\csc 2\theta = \dfrac{\csc \theta}{2 \cos \theta}$

72. $\sec 2\theta = \dfrac{\sec^2 \theta}{2 - \sec^2 \theta}$

73. $\cos^2 2\alpha - \sin^2 2\alpha = \cos 4\alpha$

74. $\cos^4 x - \sin^4 x = \cos 2x$

75. $(\sin x + \cos x)^2 = 1 + \sin 2x$

76. $\sin \dfrac{\alpha}{3} \cos \dfrac{\alpha}{3} = \dfrac{1}{2} \sin \dfrac{2\alpha}{3}$

77. $\cos 3\beta = \cos^3 \beta - 3 \sin^2 \beta \cos \beta$

78. $\sin 4\beta = 4 \sin \beta \cos \beta (1 - \sin^2 2\beta)$

79. $1 + \cos 10y = 2 \cos^2 5y$

80. $\dfrac{\cos 3\beta}{\cos \beta} = 1 - 4 \sin^2 \beta$

81. $\sec \dfrac{u}{2} = \pm \sqrt{\dfrac{2 \tan u}{\tan u + \sin u}}$

82. $\tan \dfrac{u}{2} = \csc u - \cot u$

83. $\dfrac{\cos 4x + \cos 2x}{\sin 4x + \sin 2x} = \cot 3x$

84. $\dfrac{\cos 3x - \cos x}{\sin 3x - \sin x} = -\tan 2x$

85. $\dfrac{\cos 4x - \cos 2x}{2 \sin 3x} = -\sin x$

86. $\dfrac{\sin x \pm \sin y}{\cos x + \cos y} = \tan \dfrac{x \pm y}{2}$

87. $\dfrac{\cos t + \cos 3t}{\sin 3t - \sin t} = \cot t$

88. $\sin\left(\dfrac{\pi}{6} + x\right) + \sin\left(\dfrac{\pi}{6} - x\right) = \cos x$

In Exercises 89 and 90, sketch the graph of the function by using the power-reducing formulas.

89. $f(x) = \sin^2 x$

90. $f(x) = \cos^2 x$

In Exercises 91 and 92, write the trigonometric expression as an algebraic expression.

91. $\sin(2 \arcsin x)$

92. $\cos(2 \arccos x)$

In Exercises 93 and 94, prove the product-to-sum formula.

93. $\cos u \sin v = \frac{1}{2}[\sin(u + v) - \sin(u - v)]$

94. $\cos u \cos v = \frac{1}{2}[\cos(u - v) + \cos(u + v)]$

95. *Area* The lengths of the two equal sides of an isosceles triangle are 10 feet (see figure). The angle between the two sides is θ.

(a) Express the area of the triangle as a function of $\theta/2$.

(b) Express the area of the triangle as a function of θ and determine the value of θ so the area is maximum.

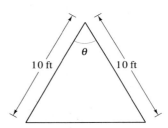

FIGURE FOR 95

96. *Projectile Motion* The range of a projectile fired at an angle θ with the horizontal and with an initial velocity of v_0 feet per second is given by

$$r = \frac{1}{32} v_0^2 \sin 2\theta,$$

where r is measured in feet. Determine the expression for the range in terms of θ.

CHAPTER 3 · REVIEW EXERCISES

In Exercises 1–10, simplify the trigonometric expression.

1. $\dfrac{1}{\cot^2 x + 1}$

2. $\dfrac{\sin 2\alpha}{\cos^2 \alpha - \sin^2 \alpha}$

3. $\dfrac{\sin^2 \alpha - \cos^2 \alpha}{\sin^2 \alpha - \sin \alpha \cos \alpha}$

4. $\dfrac{\sin^3 \beta + \cos^3 \beta}{\sin \beta + \cos \beta}$

5. $\cos^2 \beta + \cos^2 \beta \tan^2 \beta$

6. $\dfrac{\sin \theta}{1 + \cos \theta} + \dfrac{1 + \cos \theta}{\sin \theta}$

7. $\tan^2 \theta(\csc^2 \theta - 1)$

8. $\dfrac{2 \tan(x + 1)}{1 - \tan^2(x + 1)}$

9. $1 - 4 \sin^2 x \cos^2 x$

10. $\sqrt{\dfrac{1 - \cos^2 x}{1 + \cos x}}$

In Exercises 11–30, verify the identity.

11. $\tan x(1 - \sin^2 x) = \frac{1}{2} \sin 2x$

12. $\cos x(\tan^2 x + 1) = \sec x$

13. $\sec^2 x \cot x - \cot x = \tan x$

14. $\sin^3 \theta + \sin \theta \cos^2 \theta = \sin \theta$

15. $\sin^5 x \cos^2 x = (\cos^2 x - 2 \cos^4 x + \cos^6 x) \sin x$

16. $\cos^3 x \sin^2 x = (\sin^2 x - \sin^4 x) \cos x$

17. $\sin 3\theta \sin \theta = \frac{1}{2}(\cos 2\theta - \cos 4\theta)$

18. $\sin 3x \cos 2x = \frac{1}{2}(\sin 5x + \sin x)$

19. $\sqrt{\dfrac{1 - \sin \theta}{1 + \sin \theta}} = \dfrac{1 - \sin \theta}{|\cos \theta|}$

20. $\sqrt{1 - \cos x} = \dfrac{|\sin x|}{\sqrt{1 + \cos x}}$

21. $\cos 3x = 4 \cos^3 x - 3 \cos x$

22. $\cos\left(x + \dfrac{\pi}{2}\right) = -\sin x$

23. $\cot\left(\dfrac{\pi}{2} - x\right) = \tan x$

24. $\sin(\pi - x) = \sin x$

25. $\dfrac{\sec x - 1}{\tan x} = \tan \dfrac{x}{2}$

26. $\dfrac{2 \cos 3x}{\sin 4x - \sin 2x} = \csc x$

27. $\dfrac{\cos 3x - \cos x}{\sin 3x - \sin x} = -\tan 2x$

28. $1 - \cos 2x = 2 \sin^2 x$

29. $2 \sin y \cos y \sec 2y = \tan 2y$

30. $\dfrac{\sin(\alpha + \beta)}{\cos \alpha \cos \beta} = \tan \alpha + \tan \beta$

In Exercises 31–36, use a graphing utility to verify the identity. (Graph each member of the equation, and note that the graphs coincide.)

31. $\sin\left(x - \dfrac{3\pi}{2}\right) = \cos x$

32. $\sin 4x = 8 \cos^3 x \sin x - 4 \cos x \sin x$

33. $\tan^2 x = \dfrac{1 - \cos 2x}{1 + \cos 2x}$

34. $\cos^2 5x - \cos^2 x = -\sin 4x \sin 6x$

35. $1 + \cos 2x + \cos 4x + \cos 6x = 4 \cos x \cos 2x \cos 3x$

36. $\sin 2x + \sin 4x - \sin 6x = 4 \sin x \sin 2x \sin 3x$

In Exercises 37–40, find the exact value of the trigonometric function by using the sum, difference, or half-angle formulas.

37. $\sin \dfrac{5\pi}{12} = \sin\left(\dfrac{2\pi}{3} - \dfrac{\pi}{4}\right)$

38. $\cos 285° = \cos(225° + 60°)$

39. $\cos(157° \, 30') = \cos \dfrac{315°}{2}$

40. $\sin \dfrac{3\pi}{8} = \sin\left[\dfrac{1}{2}\left(\dfrac{3\pi}{4}\right)\right]$

In Exercises 41–46, find the exact value of the trigonometric function given that $\sin u = \frac{3}{4}$, $\cos v = -\frac{5}{13}$, and u and v are in Quadrant II.

41. $\sin(u + v)$

42. $\tan(u + v)$

43. $\cos(u - v)$

44. $\sin 2v$

45. $\cos \dfrac{u}{2}$

46. $\tan 2v$

In Exercises 47–50, determine if the statement is true or false. If it is false, make the necessary correction.

47. If $\dfrac{\pi}{2} < \theta < \pi$, then $\cos \dfrac{\theta}{2} < 0$.

48. $\sin(x + y) = \sin x + \sin y$

49. $4 \sin(-x) \cos(-x) = -2 \sin 2x$

50. $4 \sin 45° \cos 15° = 1 + \sqrt{3}$

In Exercises 51–60, use a graphing utility to obtain a graph of the function and approximate its zeros in the interval $[0, 2\pi)$. If possible, find the exact values of the zeros algebraically.

51. $f(x) = \sin x - \tan x$

52. $f(x) = \csc x - 2 \cot x$

53. $g(x) = \sin 2x + \sqrt{2} \sin x$

54. $g(x) = \cos 4x - 7 \cos 2x - 8$

55. $h(x) = \cos^2 x + \sin x - 1$

56. $h(x) = \sin 4x - \sin 2x$

57. $y = \dfrac{1 + \sin x}{\cos x} + \dfrac{\cos x}{1 + \sin x} - 4$

58. $y = \cos x - \cos \dfrac{x}{2}$

59. $g(t) = \tan^3 t - \tan^2 t + 3 \tan t - 3$

60. $h(s) = \sin s + \sin 3s + \sin 5s$

In Exercises 61 and 62, write the trigonometric expression as a product.

61. $\cos 3\theta + \cos 2\theta$

62. $\sin\left(x + \dfrac{\pi}{4}\right) - \sin\left(x - \dfrac{\pi}{4}\right)$

In Exercises 63 and 64, write the trigonometric expression as a sum or difference.

63. $\sin 3\alpha \sin 2\alpha$

64. $\cos \dfrac{x}{2} \cos \dfrac{x}{4}$

In Exercises 65 and 66, write the trigonometric expression as an algebraic expression.

65. $\cos(2 \arccos 2x)$

66. $\sin(2 \arctan x)$

67. *Rate of Change*   The rate of change of the function $f(x) = 2\sqrt{\sin x}$ with respect to change in the variable x is given by the expression $\cos x / \sqrt{\sin x}$. Show that the expression for the rate of change can also be given by $\cot x \sqrt{\sin x}$.

68. *Projectile Motion*   A baseball leaves the hand of the first baseman at an angle of θ with the horizontal and an initial velocity of $v_0 = 80$ feet per second. The ball is caught by the second baseman 100 feet away. Find θ if the range r of a projectile is given by

$$r = \tfrac{1}{32} v_0^2 \sin 2\theta.$$

69. *Harmonic Motion*   A weight is attached to a spring suspended vertically from a ceiling. When a driving force is applied to the system, the weight moves vertically from its equilibrium position. This motion is described by the model.

$$y = 1.5 \sin 8t - 0.5 \cos 8t,$$

where y is the distance from equilibrium measured in feet and t is the time in seconds.

(a) Write the model in the form

$$y = \sqrt{a^2 + b^2} \, \sin(Bt + C).$$

(b) Use a graphing utility to obtain a graph of the model.

(c) Find the amplitude of the oscillations of the weight.

(d) Find the frequency of the oscillations of the weight.

70. *Volume*   A trough for feeding cattle is 16 feet long, and its cross sections are isosceles triangles with the two equal sides being 18 inches in length (see figure). The angle between the two equal sides is θ.

(a) Express the volume of the trough as a function of $\theta/2$.

(b) Express the volume of the trough as a function of θ and determine the value of θ so the volume is maximum.

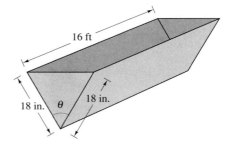

FIGURE FOR 70

ADDITIONAL TOPICS IN TRIGONOMETRY

4.1 LAW OF SINES

Law of Sines / The Ambiguous Case (SSA) / Application / The Area of an Oblique Triangle

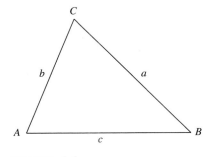

FIGURE 4.1

Law of Sines

In Chapter 2 you studied techniques for solving right triangles. In this section and the next, you will solve **oblique triangles**—triangles that have no right angles. As standard notation, we label the angles of a triangle as A, B, and C, and their opposite sides as a, b, and c, as shown in Figure 4.1.

To solve an oblique triangle, you need to know the measure of at least one side and any two other parts of the triangle—either two sides, two angles, or one angle and one side. This breaks down into four cases.

1. Two angles and any side (AAS or ASA).
2. Two sides and an angle opposite one of them (SSA).
3. Three sides (SSS).
4. Two sides and their included angle (SAS).

The first two cases can be solved using the **Law of Sines,** whereas the last two cases require the **Law of Cosines** (to be discussed in Section 4.2).

253

REMARK   The Law of Sines can also be written in the reciprocal form

$$\frac{\sin A}{a} = \frac{\sin B}{b} = \frac{\sin C}{c}.$$

LAW OF SINES

If ABC is a triangle with sides a, b, and c, then

$$\frac{a}{\sin A} = \frac{b}{\sin B} = \frac{c}{\sin C}.$$

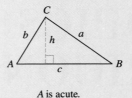

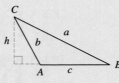

A is acute.   A is obtuse.

Oblique Triangles

When using a calculator with the Law of Sines, remember to store all intermediate calculations. By not rounding until the final result, you minimize the round-off error.

EXAMPLE 1   Given Two Angles and One Side—AAS

Given a triangle with $C = 102.3°$, $B = 28.7°$, and $b = 27.4$ feet, as shown in Figure 4.2, find the remaining angle and sides.

SOLUTION

The third angle of the triangle is

$$A = 180° - B - C = 180° - 28.7° - 102.3° = 49.0°.$$

By the Law of Sines, you have

$$\frac{a}{\sin 49°} = \frac{b}{\sin 28.7°} = \frac{c}{\sin 102.3°}.$$

Because $b = 27.4$, you obtain

$$a = \frac{27.4}{\sin 28.7°}(\sin 49°) \approx 43.06 \text{ feet}$$

and

$$c = \frac{27.4}{\sin 28.7°}(\sin 102.3°) \approx 55.75 \text{ feet}.$$

Note that the ratio $27.4/\sin 28.7°$ occurs in both solutions, and you can save time by storing this result for repeated use.

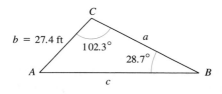

FIGURE 4.2

When solving triangles, a careful sketch is useful as a quick test for the feasibility of an answer. Remember that the longest side lies opposite the largest angle, and the shortest side lies opposite the smallest angle of a triangle.

EXAMPLE 2   Given Two Angles and One Side—ASA

A pole tilts *toward* the sun at an 8° angle from vertical, and it casts a 22-foot shadow. The angle of elevation from the tip of the shadow to the top of the pole is 43°. How tall is the pole?

SOLUTION

In Figure 4.3, note that $A = 43°$, and $B = 90° + 8° = 98°$. Thus, the third angle is

$$C = 180° - A - B = 180° - 43° - 98° = 39°.$$

By the Law of Sines, you have

$$\frac{a}{\sin 43°} = \frac{c}{\sin 39°}.$$

Because $c = 22$ feet, the length of the pole is

$$a = \frac{22}{\sin 39°}(\sin 43°) \approx 23.84 \text{ feet.}$$

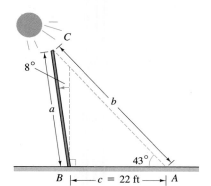

FIGURE 4.3

REMARK   For practice, try reworking Example 2 for a pole that tilts *away* from the sun under the same conditions.

The Ambiguous Case (SSA)

In Examples 1 and 2, you saw that two angles (whose sum is less than 180°) and one side determine a unique triangle. However, if two sides and one opposite angle are given, three possible situations can occur: (1) no such triangle exists, (2) one such triangle exists, or (3) two distinct triangles may satisfy the conditions. The possibilities in this *ambiguous* case (SSA) are summarized in the following table.

The Ambiguous Case (SSA) (given: *a*, *b*, and *A*)

	A Is Acute				A Is Obtuse	
Sketch ($h = b \sin A$)						
Necessary Condition	$a < h$	$a = h$	$a > b$	$h < a < b$	$a \leq b$	$a > b$
Triangles Possible	None	One	One	Two	None	One

To determine which of the possibilities holds for a given pair of sides and opposite angle, it helps to make a sketch.

EXAMPLE 3   Single-Solution Case—SSA

Given a triangle with $a = 22$ inches, $b = 12$ inches, and $A = 42°$, find the remaining side and angles.

SOLUTION

Because A is acute and $a > b$, you know that B is also acute and there is only one triangle that satisfies the given conditions, as shown in Figure 4.4. Thus, by the Law of Sines, you have

$$\frac{22}{\sin 42°} = \frac{12}{\sin B},$$

which implies that

$$\sin B = 12\left(\frac{\sin 42°}{22}\right) \approx 0.3649803$$

$$B \approx 21.41°.$$

Now you can determine that $C \approx 180° - 42° - 21.41° = 116.59°$, and the remaining side is given by

$$\frac{c}{\sin 116.59°} = \frac{22}{\sin 42°}$$

$$c = \sin 116.59°\left(\frac{22}{\sin 42°}\right) \approx 29.40 \text{ inches.}$$

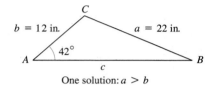

$b = 12$ in. C $a = 22$ in.

A $42°$ c B

One solution: $a > b$

FIGURE 4.4

EXAMPLE 4   No-Solution Case—SSA

Show that there is no triangle that satisfies either of the following conditions.

A. $a = 15,$ $b = 25,$ $A = 85°$
B. $a = 15.2,$ $b = 20,$ $A = 110°$

SOLUTION

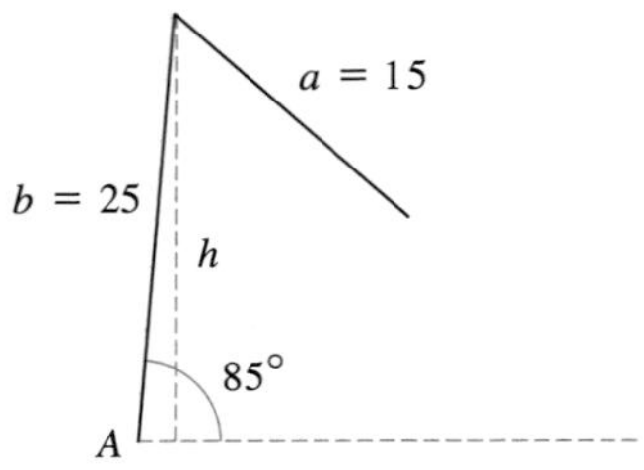

A. Begin by making the sketch shown in Figure 4.5. From this figure it appears that no triangle is formed. You can verify this using the Law of Sines.

$a = 15$

$b = 25$

h

$85°$

A

No solution: $a < h$

FIGURE 4.5

$$\frac{a}{\sin A} = \frac{b}{\sin B}$$

$$\frac{15}{\sin 85°} = \frac{25}{\sin B}$$

$$\sin B = 25\left(\frac{\sin 85°}{15}\right) \approx 1.660 > 1$$

This contradicts the fact that $|\sin B| \leq 1$. Hence, no triangle can be formed having sides $a = 15$ and $b = 25$ and an angle of $A = 85°$.

B. Because A is obtuse and $a = 15.2$ is less than $b = 20$, you can conclude that there is *no solution,* as shown in Figure 4.6. Try using the Law of Sines to verify this.

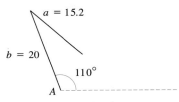

No solution: $a < b$ and $A > 90°$

FIGURE 4.6

EXAMPLE 5   Two-Solution Case—SSA

Find two triangles for which $a = 12$ meters, $b = 31$ meters, and $A = 20.5°$.

SOLUTION

To begin, note that

$$h = b \sin A = 31(\sin 20.5°) \approx 10.86 \text{ meters.}$$

Hence, $h < a < b$, and you can conclude that there are two possible triangles. By the Law of Sines, you obtain

$$\frac{a}{\sin A} = \frac{b}{\sin B},$$

which implies that

$$\sin B = b\left(\frac{\sin A}{a}\right) = 31\left(\frac{\sin 20.5°}{12}\right) \approx 0.9047 \text{ meters.}$$

There are two angles $B_1 \approx 64.8°$ and $B_2 \approx 115.2°$ between $0°$ and $180°$ whose sine is 0.9047. For $B_1 \approx 64.8°$, you obtain

$$C \approx 180° - 20.5° - 64.8° = 94.7°$$

$$c = \frac{a}{\sin A}(\sin C) = \frac{12}{\sin 20.5°}(\sin 94.7°) \approx 34.15.$$

For $B_2 \approx 115.2°$, you obtain

$$C \approx 180° - 20.5° - 115.2° = 44.3°$$

$$c = \frac{a}{\sin A}(\sin C) = \frac{12}{\sin 20.5°}(\sin 44.3°) \approx 23.93 \text{ meters.}$$

The resulting triangles are shown in Figures 4.7 and 4.8.

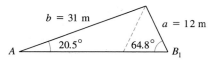

FIGURE 4.7

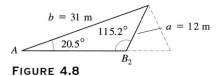

FIGURE 4.8

Application

EXAMPLE 6   **An Application of the Law of Sines**

The course for a boat race starts at point A and proceeds in the direction S 52° W to point B, then in the direction S 40° E to point C, and finally back to A, as shown in Figure 4.9. The point C lies 8 kilometers directly south of point A. Approximate the total distance of the race course.

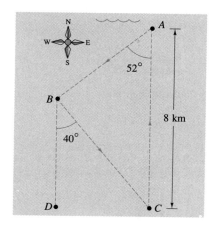

FIGURE 4.9

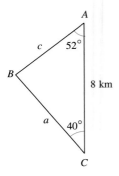

FIGURE 4.10

SOLUTION

Because lines *BD* and *AC* are parallel, it follows that $\angle BCA = \angle DBC$. Consequently, triangle *ABC* has the measures shown in Figure 4.10. For angle *B*, you have

$$B = 180° - 52° - 40° = 88°.$$

Using the Law of Sines

$$\frac{a}{\sin 52°} = \frac{b}{\sin 88°} = \frac{c}{\sin 40°},$$

you can let $b = 8$ and obtain the following.

$$a = \frac{8}{\sin 88°}(\sin 52°) \approx 6.308$$

$$c = \frac{8}{\sin 88°}(\sin 40°) \approx 5.145$$

Finally, the total length of the course is approximately

Length $\approx 8 + 6.308 + 5.145 = 19.453$ kilometers.

The Area of an Oblique Triangle

The procedure used to prove the Law of Sines leads to a simple formula for the area of an oblique triangle. Referring to Figure 4.11, note that each triangle has a height of

$$h = b \sin A.$$

Consequently, the area of each triangle is given by

$$\text{Area} = \frac{1}{2}(\text{base})(\text{height}) = \frac{1}{2}(c)(b \sin A) = \frac{1}{2}bc \sin A.$$

By similar arguments, you can develop the formula

$$\text{Area} = \frac{1}{2}ab \sin C = \frac{1}{2}ac \sin B.$$

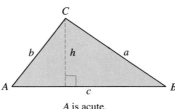

A is acute.

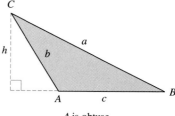

A is obtuse.

Oblique Triangles

FIGURE 4.11

AREA OF AN OBLIQUE TRIANGLE

The area of any triangle is given by one-half the product of the lengths of two sides times the sine of their included angle. That is,

$$\text{Area} = \frac{1}{2}bc \sin A = \frac{1}{2}ab \sin C = \frac{1}{2}ac \sin B.$$

EXAMPLE 7   Finding the Area of an Oblique Triangle

Find the area of a triangular lot having two sides of lengths 90 meters and 52 meters and an included angle of 102°.

SOLUTION

Consider $a = 90$ m, $b = 52$ m, and angle $C = 102°$, as shown in Figure 4.12. Then the area of the triangle is

$$\text{Area} = \frac{1}{2}ab \sin C = \frac{1}{2}(90)(52)(\sin 102°) \approx 2289 \text{ square meters.}$$

$b = 52$ m

$102°$

C $a = 90$ m

FIGURE 4.12

DISCUSSION PROBLEM

SOLVING RIGHT TRIANGLES

In this section, you have been using the Law of Sines to solve *oblique* triangles. Can the Law of Sines also be used to solve a right triangle? If so, write a short paragraph explaining how to use the Law of Sines to solve the following two triangles. Is there an easier way to solve these triangles?

(a) (AAS)

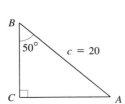

(b) (ASA)

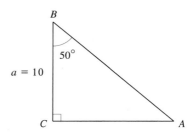

WARM-UP

The following warm-up exercises involve skills that were covered in earlier sections. You will use these skills in the exercise set for this section.

In Exercises 1–6, solve the indicated *right* triangle. (In the right triangle, a and b are the lengths of the sides and c is the length of the hypotenuse.)

1. $a = 3$, $c = 6$ **2.** $a = 5$, $b = 5$ **3.** $b = 15$, $c = 17$

4. $A = 42°$, $a = 7.5$ **5.** $B = 10°$, $b = 4$ **6.** $B = 72° 15'$, $c = 150$

In Exercises 7 and 8, find the altitude of the triangle.

7.

8.
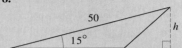

In Exercises 9 and 10, solve the equation for x.

9. $\dfrac{2}{\sin 30°} = \dfrac{9}{x}$ **10.** $\dfrac{100}{\sin 72°} = \dfrac{x}{\sin 60°}$

SECTION 4.1 · EXERCISES

In Exercises 1–16, use the given information to find the remaining sides and angles of the triangle labeled as shown in Figure 4.1.

1.
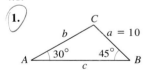

2.
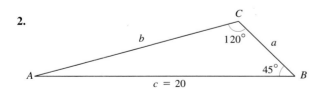

3.
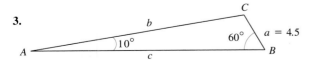

4.
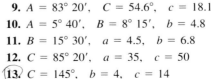

5. $A = 36°$, $a = 8$, $b = 5$

6. $A = 60°$, $a = 9$, $c = 10$

7. $A = 150°$, $C = 20°$, $a = 200$

8. $A = 24.3°$, $C = 54.6°$, $c = 2.68$

9. $A = 83° 20'$, $C = 54.6°$, $c = 18.1$

10. $A = 5° 40'$, $B = 8° 15'$, $b = 4.8$

11. $B = 15° 30'$, $a = 4.5$, $b = 6.8$

12. $C = 85° 20'$, $a = 35$, $c = 50$

13. $C = 145°$, $b = 4$, $c = 14$

14. $A = 100°$, $a = 125$, $c = 10$

15. $A = 110° 15'$, $a = 48$, $b = 16$

16. $B = 2° 45'$, $b = 6.2$, $c = 5.8$

In Exercises 17–22, use the given information to find (if possible) the remaining sides and angles of the triangle. If two solutions exist, find both.

17. $A = 58°$, $a = 4.5$, $b = 12.8$
18. $A = 58°$, $a = 11.4$, $b = 12.8$
19. $A = 58°$, $a = 4.5$, $b = 5$
20. $A = 58°$, $a = 42.4$, $b = 50$
21. $A = 110°$, $a = 125$, $b = 200$
22. $A = 110°$, $a = 125$, $b = 100$

In Exercises 23 and 24, find values for b such that the triangle has (a) one solution, (b) two solutions, and (c) no solution.

23. $A = 36°$, $a = 5$ **24.** $A = 60°$, $a = 10$

In Exercises 25–30, find the area of the triangle having the indicated sides and angle.

25. $C = 120°$, $a = 4$, $b = 6$
26. $B = 72° \ 30'$, $a = 105$, $c = 64$
27. $A = 43° \ 45'$, $b = 57$, $c = 85$
28. $A = 5° \ 15'$, $b = 4.5$, $c = 22$
29. $B = 130°$, $a = 62$, $c = 20$
30. $C = 84° \ 30'$, $a = 16$, $b = 20$

31. *Streetlight Design*   Find the length d of the brace required to support the streetlight in the figure.

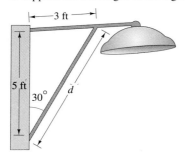

FIGURE FOR 31

32. *Height*   Because of the prevailing winds, a tree grew so that it was leaning 6° from the vertical. At a point 100 feet away from the tree, the angle of elevation to the top of the tree is 22° 50′ (see figure). Find the height h of the tree.

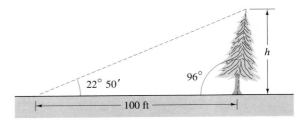

FIGURE FOR 32

33. *Bridge Design*   A bridge is to be built across a small lake from B to C (see figure). The bearing from B to C is S 41° W. From a point A, 100 yards from B, the bearings to B and C are S 74° E and S 28° E, respectively. Find the distance from B to C.

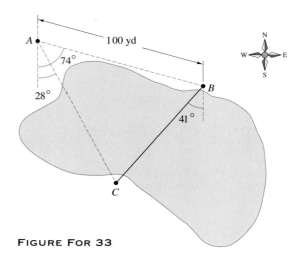

FIGURE FOR 33

34. *Railroad Track Design*   The circular arc of a railroad curve has a chord of length 3000 feet, and a central angle of 40° (see figure). Find (a) the radius r of the circular arc and (b) the length s of the circular arc.

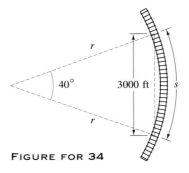

FIGURE FOR 34

35. *Altitude*   The angles of elevation to an airplane from two points *A* and *B* on level ground are 51° and 68°, respectively. The points *A* and *B* are 6 miles apart, and the airplane is between these positions in the same vertical plane. Find the altitude of the airplane.

36. *Altitude*   The angles of elevation to an airplane from two points *A* and *B* on level ground are 51° and 68°, respectively. The points *A* and *B* are 2.5 miles apart, and the airplane is east of both points in the same vertical plane. Find the altitude of the plane.

37. *Locating a Fire*   Two fire towers *A* and *B* are 18.5 miles apart. The bearing from *A* to *B* is N 65° E. A fire is spotted by the ranger in each tower, and its bearings from *A* and *B* are N 28° E and N 16.5° W, respectively (see figure). Find the distance of the fire from each tower.

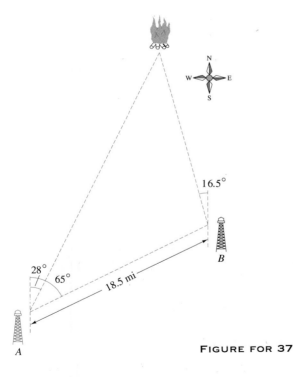

FIGURE FOR 37

38. *Distance*   A boat is sailing due east parallel to the shoreline at a speed of 10 miles per hour. At a given time, the bearing to the lighthouse is S 72° E, and 15 minutes later the bearing is S 66° E (see figure). Find the distance from the boat to the shoreline if the lighthouse is at the shoreline.

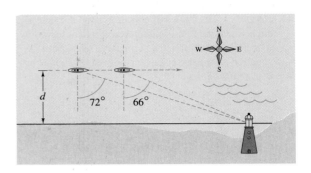

FIGURE FOR 38

39. *Distance*   A family is traveling due west on a road that passes a famous landmark. At a given time, the bearing to the landmark is N 62° W, and after the family travels 5 miles farther the bearing is N 38° W. What is the closest the family will come to the landmark while on the road?

40. *Engine Design*   The connecting rod in a certain engine is 6 inches long, and the radius of the crankshaft is $1\frac{1}{2}$ inches (see figure). The spark plug fires at 5° before top dead center. How far is the piston from the top of its stroke at this time?

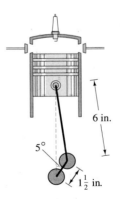

FIGURE FOR 33

41. *Verification of Testimony*   The following information about a triangular parcel of land is given at a zoning board meeting: "One side is 450 feet long, and another is 120 feet long. The angle opposite the shorter side is 30°." Could this information be correct?

42. *Distance*   The angles of elevation to an airplane, θ and ϕ, are being continuously monitored at two observation points A and B, which are 2 miles apart (see figure). Write an equation giving the distance d between the plane and point B in terms of θ and ϕ.

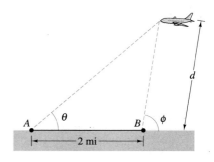

FIGURE FOR 42

4.2   LAW OF COSINES

Law of Cosines   /   Heron's Formula

Law of Cosines

Two cases remain in the list of conditions needed to solve an oblique triangle—SSS and SAS. The Law of Sines does not work in either of these cases. To see why, consider the three ratios given in the Law of Sines.

$$\frac{a}{\sin A} = \frac{b}{\sin B} = \frac{c}{\sin C}$$

To use the Law of Sines, you must know at least one side and its opposite angle. If you are given three sides (SSS), or two sides and their included angle (SAS), none of the above ratios would be complete. In such cases you must rely on the **Law of Cosines.**

> **LAW OF COSINES**
>
> If ABC is a triangle with sides a, b, and c, then the following equations are valid.
>
> *Standard Form*
>
> $$a^2 = b^2 + c^2 - 2bc \cos A$$
>
> $$b^2 = a^2 + c^2 - 2ac \cos B$$
>
> $$c^2 = a^2 + b^2 - 2ab \cos C$$
>
> *Alternative Form*
>
> $$\cos A = \frac{b^2 + c^2 - a^2}{2bc}$$
>
> $$\cos B = \frac{a^2 + c^2 - b^2}{2ac}$$
>
> $$\cos C = \frac{a^2 + b^2 - c^2}{2ab}$$

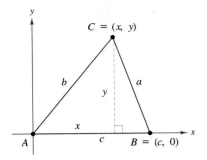

FIGURE 4.13

PROOF

We prove only the first equation for a triangle that has three acute angles, as shown in Figure 4.13. In the figure, note that vertex B has coordinates $(c, 0)$. Furthermore, C has coordinates (x, y), where $x = b \cos A$ and $y = b \sin A$. Because a is the distance from C to B, you have

$$a = \sqrt{(x - c)^2 + (y - 0)^2}$$
$$a^2 = (b \cos A - c)^2 + (b \sin A)^2$$
$$= b^2 \cos^2 A - 2bc \cos A + c^2 + b^2 \sin^2 A$$
$$= b^2(\sin^2 A + \cos^2 A) + c^2 - 2bc \cos A.$$

Using the identity $\sin^2 A + \cos^2 A = 1$ produces

$$a^2 = b^2 + c^2 - 2bc \cos A.$$

Similar arguments can be used to establish the other two equations.

Note that if $A = 90°$ in Figure 4.13, then $\cos A = 0$, and the first form of the Law of Cosines becomes the Pythagorean Theorem.

$$a^2 = b^2 + c^2$$

Thus, the Pythagorean Theorem is actually just a special case of the more general Law of Cosines.

EXAMPLE 1   Given Three Sides of a Triangle—SSS

Find the three angles of the triangle whose sides have lengths $a = 8$ feet, $b = 19$ feet, and $c = 14$ feet.

SOLUTION

It is a good idea first to find the angle opposite the longest side—side b in this case (see Figure 4.14). Using the Law of Cosines produces

$$\cos B = \frac{a^2 + c^2 - b^2}{2ac} = \frac{8^2 + 14^2 - 19^2}{2(8)(14)} \approx -0.45089.$$

Because $\cos B$ is negative, you know B is an *obtuse* angle given by $B \approx 116.80°$. At this point you could use the Law of Cosines to find $\cos A$ and $\cos C$. However, knowing that $B \approx 116.80°$, it is simpler to use the Law of Sines to obtain the following.

$$\frac{b}{\sin B} = \frac{a}{\sin A}$$

$$\sin A = a\left(\frac{\sin B}{b}\right) \approx 8\left(\frac{\sin 116.80°}{19}\right) \approx 0.37582$$

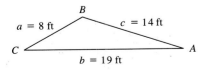

FIGURE 4.14

Because B is obtuse, you know that A must be acute, because a triangle can have, at most, one obtuse angle. Thus, $A \approx 22.08°$ and

$$C \approx 180° - 22.08° - 116.80° = 41.12°.$$

Do you see why it was wise to find the largest angle *first* in Example 1? Knowing the cosine of an angle, you can determine whether the angle is acute or obtuse. That is,

$\cos \theta > 0$ for $0° < \theta \not< 90°$ *Acute*

$\cos \theta < 0$ for $90° < \theta < 180°.$ *Obtuse*

So, in Example 1, once you found that B was obtuse, you could determine that A and C were both acute. If the largest angle is acute, then the remaining two angles will be acute also.

EXAMPLE 2   Given Two Sides and the Included Angle—SAS

The pitcher's mound on a softball field is 46 feet from home plate, and the distance between the bases is 60 feet, as shown in Figure 4.15. How far is the pitcher's mound from first base? (Note that the pitcher's mound is *not* halfway between home plate and second base.)

SOLUTION

From triangle HPF, you can see that $H = 45°$ (line HP bisects the right angle at H), $f = 46$, and $p = 60$. Using the Law of Cosines for this SAS case produces

$$\begin{aligned} h^2 &= f^2 + p^2 - 2fp \cos H \\ &= 46^2 + 60^2 - 2(46)(60) \cos 45° \\ &\approx 1812.8. \end{aligned}$$

Therefore, the approximate distance from the pitcher's mound to first base is

$$h \approx \sqrt{1812.8} \approx 42.58 \text{ feet.}$$

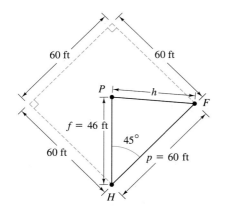

FIGURE 4.15

EXAMPLE 3   Given Two Sides and the Included Angle—SAS

A ship travels 60 miles due east, then adjusts its course 15° northward, as shown in Figure 4.16. After traveling 80 miles in that direction, how far is the ship from its point of departure?

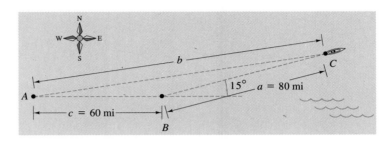

FIGURE 4.16

SOLUTION

You have $c = 60$, $B = 180° - 15° = 165°$, and $a = 80$. Consequently, by the Law of Cosines, you obtain

$$b^2 = a^2 + c^2 - 2ac \cos B$$
$$= 80^2 + 60^2 - 2(80)(60) \cos 165° \approx 19{,}273.$$

Therefore, the distance b is

$$b \approx \sqrt{19{,}273} \approx 138.8 \text{ miles.}$$

Heron's Formula

The following formula for the area of a triangle is credited to the Greek mathematician Heron (c. 100 B.C.).

HERON'S AREA FORMULA

Given any triangle with sides of lengths a, b, and c, the area of the triangle is

$$\text{Area} = \sqrt{s(s - a)(s - b)(s - c)},$$

where $s = (a + b + c)/2$ is one-half the perimeter of the triangle.

EXAMPLE 4   Using Heron's Area Formula

Find the area of the triangular region having sides of lengths $a = 47$ yards, $b = 58$ yards, and $c = 78.6$ yards.

SOLUTION

Because

$$s = \frac{1}{2}(a + b + c) = \frac{183.6}{2} = 91.8,$$

Heron's Formula yields

$$\text{Area} = \sqrt{s(s - a)(s - b)(s - c)}$$
$$= \sqrt{91.8(44.8)(33.8)(13.2)}$$
$$\approx 1354.58 \text{ square yards.}$$

DISCUSSION PROBLEM

THE AREA OF A TRIANGLE

We have now discussed three different formulas for the area of a triangle.

Standard formula: Area $= \frac{1}{2}bh$

Oblique triangle: Area $= \frac{1}{2}bc \sin A = \frac{1}{2}ab \sin C = \frac{1}{2}ac \sin B$

Heron's formula: Area $= \sqrt{s(s - a)(s - b)(s - c)}$, $s = \frac{1}{2}(a + b + c)$

Use the most appropriate formula to find the area of each of the following triangles. Show your work and give your reason for choosing each formula.

(a)

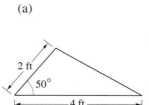

(b)

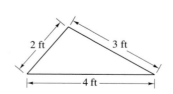

(c)

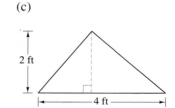

WARM-UP

The following warm-up exercises involve skills that were covered in earlier sections. You will use these skills in the exercise set for this section.

In Exercises 1 and 2, simplify the expression.

1. $\sqrt{(7 - 3)^2 + [1 - (-5)]^2}$

2. $\sqrt{[-2 - (-5)]^2 + (12 - 6)^2}$

In Exercises 3 and 4, find the distance between the two points.

3. $(4, -2), (8, 10)$

4. $(1, 3), (7, 12)$

In Exercises 5 and 6, find the area of the triangle.

5.

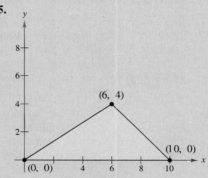

6.

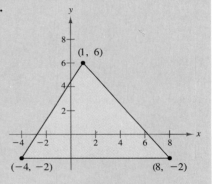

In Exercises 7–10, find (if possible) the remaining sides and angles of the triangle.

7. $A = 10°, \quad C = 100°, \quad b = 25$

8. $A = 20°, \quad C = 90°, \quad c = 100$

9. $B = 30°, \quad b = 6.5, \quad c = 15$

10. $A = 30°, \quad b = 6.5, \quad a = 10$

SECTION 4.2 · EXERCISES

In Exercises 1–14, use the Law of Cosines to solve the triangle.

1.

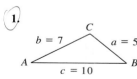

2.

3.

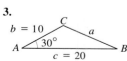

4.

5. $a = 9, \quad b = 12, \quad c = 15$

6. $a = 55, \quad b = 25, \quad c = 72$

7. $a = 75.4, \quad b = 52, \quad c = 52$

8. $a = 1.42, \quad b = 0.75, \quad c = 1.25$

9. $A = 120°, \quad b = 3, \quad c = 10$

10. $A = 55°, \quad b = 3, \quad c = 10$

11. $B = 8° \, 45', \quad a = 25, \quad c = 15$

12. $B = 75° \, 20', \quad a = 6.2, \quad c = 9.5$

13. $C = 125° \, 40', \quad a = 32, \quad b = 32$

14. $C = 15°, \quad a = 6.25, \quad b = 2.15$

In Exercises 15–20, complete the table by solving the parallelogram shown in the figure. (The lengths of the diagonals are c and d.)

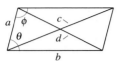

FIGURE FOR 15–20

	a	b	c	d	θ	ϕ
15.	4	6			30°	
16.	25	35				120°
17.	10	14	20			
18.	40	60		80		
19.	10		18	12		
20.		25	50	35		

In Exercises 21–26, use Heron's Formula to find the area of the triangle.

21. $a = 5, \quad b = 7, \quad c = 10$
22. $a = 2.5, \quad b = 10.2, \quad c = 9$
23. $a = 12, \quad b = 15, \quad c = 9$
24. $a = 75.4, \quad b = 52, \quad c = 52$
25. $a = 20, \quad b = 20, \quad c = 10$
26. $a = 4.25, \quad b = 1.55, \quad c = 3.00$

27. *Area* The lengths of the sides of a triangular parcel of land are approximately 400 feet, 500 feet, and 700 feet. Approximate the area of the parcel.

28. *Area* The lengths of two adjacent sides of a parallelogram are 4 yards and 6 yards. Find the area of the parallelogram if the angle between the two sides is 30°.

29. *Navigation* A boat race is run along a triangular course marked by buoys A, B, and C. The race starts with the boats headed west. The other two sides of the course lie to the north of the first side, and their lengths are 3500 feet and 6500 feet (see figure). Find the bearings for the last two legs of the race.

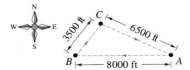

FIGURE FOR 29

30. *Navigation* A plane flies 675 miles from A to B with a bearing of N 75° E. Then it flies 540 miles from B to C with a bearing of N 32° E (see figure). Find the straight-line distance and bearing from C to A.

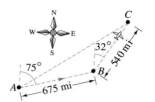

FIGURE FOR 30

31. *Distance* Two ships leave a port at 9 A.M. One travels at a bearing of N 53° W at 12 miles per hour and the other at a bearing of S 67° W at 16 miles per hour. Approximately how far apart are they at noon that day?

32. *Distance* A 100-foot vertical tower is to be erected on the side of a hill that makes an 8° angle with the horizontal (see figure). Find the length of each of the two guy wires that will be anchored 75 feet uphill and downhill from the base of the tower.

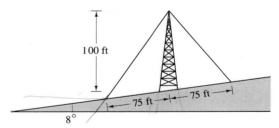

FIGURE FOR 32

33. *Surveying* To approximate the length of a marsh, a surveyor walks 950 feet from point A to point B, then turns 80° and walks 800 feet to point C (see figure). Approximate the length $\overline{AC}$ of the marsh.

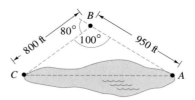

FIGURE FOR 33

34. *Surveying*   A triangular parcel of land has 375 feet of frontage, and the other boundaries have lengths of 250 feet and 300 feet. What angles does the frontage make with the two other boundaries?

35. *Streetlight Design*   Determine the angle θ in the design of the streetlight as shown in the figure.

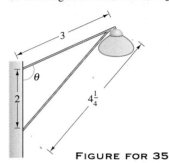

FIGURE FOR 35

36. *Aircraft Tracking*   In order to determine the distance between two aircraft, a tracking station continuously determines the distance to each aircraft and the angle α between them. Determine the distance a between the planes when $\alpha = 42°$, $b = 35$ miles, and $c = 20$ miles.

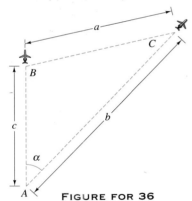

FIGURE FOR 36

37. *Engineering*   If Q is the midpoint of the line segment $\overline{PR}$, find the lengths of the line segments $\overline{PQ}$, $\overline{QS}$, and $\overline{RS}$ on the truss rafter shown in the figure.

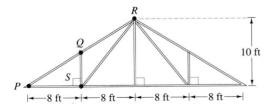

FIGURE FOR 37

38. *Paper Manufacturing*   In a certain process with continuous paper, the paper passes across three rollers of radii 3 inches, 4 inches, and 6 inches (see figure). The centers of the 3-inch and 6-inch rollers are d inches apart, and the length of the arc in contact with the paper on the 4-inch roller is s inches. Complete the following table.

d (inches)	9	10	12	13	14	15	16
θ (degrees)							
s (inches)							

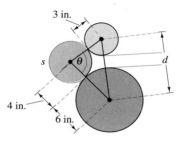

FIGURE FOR 38

39. *Navigation*   On a certain map, Orlando is 7 inches due south of Niagara Falls, Denver is 10.75 inches from Orlando, and Denver is 9.25 inches from Niagara Falls (see figure).
(a) Find the bearing of Denver from Orlando.
(b) Find the bearing of Denver from Niagara Falls.

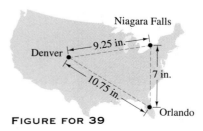

FIGURE FOR 39

40. *Navigation*   On a certain map, Minneapolis is 6.5 inches due west of Albany, Phoenix is 8.5 inches from Minneapolis, and Phoenix is 14.5 inches from Albany (see figure).
(a) Find the bearing of Minneapolis from Phoenix.
(b) Find the bearing of Albany from Phoenix.

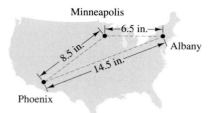

FIGURE FOR 40

41. *Baseball*   In a (square) baseball diamond with 90-foot sides, the pitcher's mound is 60 feet from home plate.
(a) How far is it from the pitcher's mound to third base?
(b) When a runner is halfway from second to third, how far is the runner from the pitcher's mound?

42. *Baseball*   The baseball player in center field is playing approximately 330 feet from the television camera that is behind home plate. A batter hits a fly ball that goes to the wall 420 feet from the camera (see figure). Approximate the number of feet that the center fielder had to run to make the catch if the camera turned 9° in following the play.

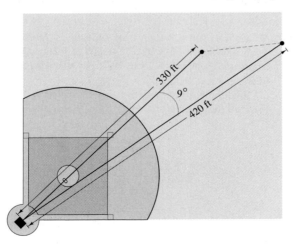

FIGURE FOR 42

43. *Awning Design*   A retractable awning lowers at an angle of 50° from the top of a patio door that is 7 feet high (see figure). Find the length *x* of the awning if no direct sunlight is to enter the door when the angle of elevation of the sun is greater than 65°.

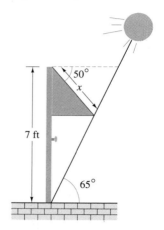

FIGURE FOR 43

44. *Circumscribed and Inscribed Circles*   Let *R* and *r* be the radii of the circumscribed and inscribed circles of a triangle *ABC*, respectively, and let $s = (a + b + c)/2$ (see figure). Prove the following.

(a) $2R = \dfrac{a}{\sin A} = \dfrac{b}{\sin B} = \dfrac{c}{\sin C}$

(b) $r = \sqrt{\dfrac{(s - a)(s - b)(s - c)}{s}}$

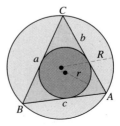

R = radius of the large circle

r = radius of the small circle

FIGURE FOR 44

Circumscribed and Inscribed Circles   In Exercises 45 and 46, use the results of Exercise 44.

45. Given the triangle with $a = 25$, $b = 55$, and $c = 72$, find the areas of (a) the triangle, (b) the circumscribed circle, and (c) the inscribed circle.

46. Find the length of the largest circular track that can be built on a triangular piece of property whose sides measure 200 feet, 250 feet, and 325 feet.

47. Use the Law of Cosines to prove that

$$\frac{1}{2}bc(1 + \cos A) = \frac{a + b + c}{2} \cdot \frac{-a + b + c}{2}.$$

48. Use the Law of Cosines to prove that

$$\frac{1}{2}bc(1 - \cos A) = \frac{a - b + c}{2} \cdot \frac{a + b - c}{2}.$$

4.3   VECTORS IN THE PLANE

Vectors in the Plane / Component Form of a Vector / Vector Operations / Unit Vectors / Direction Angles / Applications of Vectors

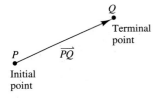

FIGURE 4.17

Vectors in the Plane

Many quantities in geometry and physics, such as area, time, and temperature, can be represented by a single real number. Other quantities, such as force and velocity, involve both *magnitude* and *direction* and cannot be completely characterized by a single real number. To represent such a quantity, we use a **directed line segment,** as shown in Figure 4.17. The directed line segment $\overrightarrow{PQ}$ has **initial point** P and **terminal point** Q, and we denote its **length** by $\|PQ\|$.

Two directed line segments that have the same length (or magnitude) and direction are called **equivalent.** For example, the directed line segments in Figure 4.18 are all equivalent. The set of all directed line segments that are equivalent to a given directed line segment $\overrightarrow{PQ}$ is a **vector v in the plane,** and we write $\mathbf{v} = \overrightarrow{PQ}$. Vectors are denoted by lowercase, boldface letters such as $\mathbf{u}$, $\mathbf{v}$, and $\mathbf{w}$.

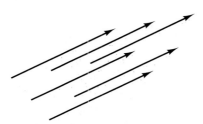

Equivalent Directed
Line Segments

FIGURE 4.18

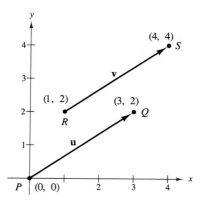

FIGURE 4.19

EXAMPLE 1 Vector Representation by Directed Line Segments

The vector **u** is represented by the directed line segment from $P = (0, 0)$ to $Q = (3, 2)$, and the vector **v** is represented by the directed line segment from $R = (1, 2)$ to $S = (4, 4)$, as shown in Figure 4.19. Show that $\mathbf{u} = \mathbf{v}$.

SOLUTION

From the distance formula, you can see that $\overrightarrow{PQ}$ and $\overrightarrow{RS}$ have the *same length.*

$$\|\overrightarrow{PQ}\| = \sqrt{(3 - 0)^2 + (2 - 0)^2} = \sqrt{13}$$
$$\|\overrightarrow{RS}\| = \sqrt{(4 - 1)^2 + (4 - 2)^2} = \sqrt{13}$$

Moreover, both line segments have the *same direction* because they are both directed toward the upper right on lines having a slope of $\frac{2}{3}$. Thus, $\overrightarrow{PQ}$ and $\overrightarrow{RS}$ have the same length and direction, and you can conclude that $\mathbf{u} = \mathbf{v}$.

Component Form of a Vector

The directed line segment whose initial point is the origin is often the most convenient representative of a set of equivalent directed line segments. This representative of the vector **v** is in **standard position.**

A vector whose initial point is at the origin $(0, 0)$ can be uniquely represented by the coordinates of its terminal point (v_1, v_2). We call this the **component form of a vector v** and write

$$\mathbf{v} = \langle v_1, v_2 \rangle.$$

The coordinates v_1 and v_2 are the **components** of **v**. If both the initial point and the terminal point lie at the origin, then **v** is the **zero vector** and is denoted by $\mathbf{0} = \langle 0, 0 \rangle$. To convert directed line segments to component form, use the following procedure.

COMPONENT FORM OF A VECTOR

The component form of the vector with initial point $P = (p_1, p_2)$ and terminal point $Q = (q_1, q_2)$ is

$$\overrightarrow{PQ} = \langle q_1 - p_1, q_2 - p_2 \rangle = \langle v_1, v_2 \rangle = \mathbf{v}.$$

The **length** (or magnitude) of **v** is given by

$$\|\mathbf{v}\| = \sqrt{(q_1 - p_1)^2 + (q_2 - p_2)^2} = \sqrt{v_1{}^2 + v_2{}^2}.$$

If $\|\mathbf{v}\| = 1$, then **v** is a **unit vector.** Moreover, $\|\mathbf{v}\| = 0$ if and only if **v** is the **zero vector 0.**

Two vectors $\mathbf{u} = \langle u_1, u_2 \rangle$ and $\mathbf{v} = \langle v_1, v_2 \rangle$ are **equal** if and only if $u_1 = v_1$ and $u_2 = v_2$. For instance, in Example 1, the vector $\mathbf{u}$ from $P = (0, 0)$ to $Q = (3, 2)$ is

$$\mathbf{u} = \overrightarrow{PQ} = \langle 3 - 0, 2 - 0 \rangle = \langle 3, 2 \rangle,$$

and the vector $\mathbf{v}$ from $R = (1, 2)$ to $S = (4, 4)$ is

$$\mathbf{v} = \overrightarrow{RS} = \langle 4 - 1, 4 - 2 \rangle = \langle 3, 2 \rangle,$$

which shows that $\mathbf{u}$ and $\mathbf{v}$ are equal.

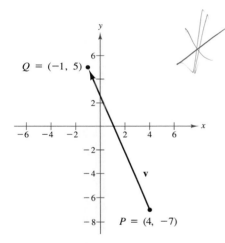

Component form of $\mathbf{v}$:
$\mathbf{v} = \langle -5, 12 \rangle$

FIGURE 4.20

EXAMPLE 2   Finding the Component Form and Length of a Vector

Find the component form and length of the vector $\mathbf{v}$ that has initial point $(4, -7)$ and terminal point $(-1, 5)$.

SOLUTION

Let $P = (4, -7) = (p_1, p_2)$ and $Q = (-1, 5) = (q_1, q_2)$. The components of $\mathbf{v} = \langle v_1, v_2 \rangle$ are given by

$$v_1 = q_1 - p_1 = -1 - 4 = -5$$
$$v_2 = q_2 - p_2 = 5 - (-7) = 12.$$

Thus, $\mathbf{v} = \langle -5, 12 \rangle$, and the length of $\mathbf{v}$ is

$$\| \mathbf{v} \| = \sqrt{(-5)^2 + 12^2} = \sqrt{169} = 13,$$

as shown in Figure 4.20.

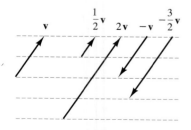

Scalar Multiplication of $\mathbf{v}$

FIGURE 4.21

Vector Operations

The two basic vector operations are **scalar multiplication** and **vector addition**. (In this text, we use the term **scalar** to mean a real number.) Geometrically, the product of a vector $\mathbf{v}$ and a scalar k is the vector that is $|k|$ times as long as $\mathbf{v}$. If k is positive, then $k\mathbf{v}$ has the same direction as $\mathbf{v}$, and if k is negative, then $k\mathbf{v}$ has the opposite direction of $\mathbf{v}$, as shown in Figure 4.21.

To add two vectors geometrically, position them (without changing length or direction) so that the initial point of one coincides with the terminal point of the other. The sum **u** + **v** is formed by joining the initial point of the second vector **v** with the terminal point of the first vector **u**, as shown in Figure 4.22. Because the vector **u** + **v** is the diagonal of a parallelogram having **u** and **v** as its adjacent sides, we call this the **parallelogram law** for vector addition.

Vector addition and scalar multiplication can also be defined using components of vectors.

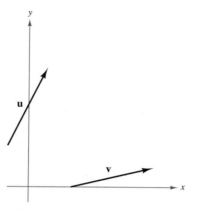

To find **u** + **v**,

FIGURE 4.22

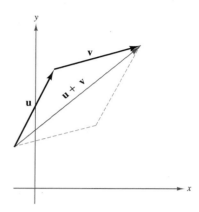

move the initial point of **v** to the terminal point of **u**.

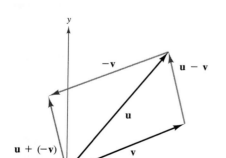

Vector Subtraction

FIGURE 4.23

DEFINITION OF VECTOR ADDITION AND SCALAR MULTIPLICATION

Let $\mathbf{u} = \langle u_1, u_2 \rangle$ and $\mathbf{v} = \langle v_1, v_2 \rangle$ be vectors and let k be a scalar (a real number). Then the **sum** of **u** and **v** is the vector

$$\mathbf{u} + \mathbf{v} = \langle u_1 + v_1, u_2 + v_2 \rangle, \qquad \textit{Sum}$$

and the **scalar multiple** of k times **u** is the vector

$$k\mathbf{u} = k\langle u_1, u_2 \rangle = \langle ku_1, ku_2 \rangle. \qquad \textit{Scalar multiple}$$

The **negative** of $\mathbf{v} = \langle v_1, v_2 \rangle$ is

$$-\mathbf{v} = (-1)\mathbf{v} = \langle -v_1, -v_2 \rangle, \qquad \textit{Negative}$$

and the **difference** of **u** and **v** is

$$\mathbf{u} - \mathbf{v} = \mathbf{u} + (-\mathbf{v}) = \langle u_1 - v_1, u_2 - v_2 \rangle. \qquad \textit{Difference}$$

To represent $\mathbf{u} - \mathbf{v}$ graphically, we use directed line segments with the *same* initial points. The difference $\mathbf{u} - \mathbf{v}$ is the vector from the terminal point of **v** to the terminal point of **u**, as shown in Figure 4.23.

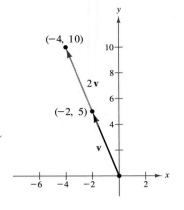

(a)

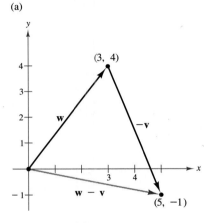

(b)

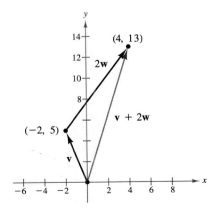

(c)

FIGURE 4.24

EXAMPLE 3 **Vector Operations**

Let $\mathbf{v} = \langle -2, 5 \rangle$ and $\mathbf{w} = \langle 3, 4 \rangle$, and find the following vectors.

A. $2\mathbf{v}$
B. $\mathbf{w} - \mathbf{v}$
C. $\mathbf{v} + 2\mathbf{w}$

SOLUTION

Sketches of the vectors are shown in Figure 4.24.

A. Because $\mathbf{v} = \langle -2, 5 \rangle$, you have

$$2\mathbf{v} = \langle 2(-2), 2(5) \rangle = \langle -4, 10 \rangle.$$

B. The difference of $\mathbf{w}$ and $\mathbf{v}$ is given by

$$\mathbf{w} - \mathbf{v} = \langle 3 - (-2), 4 - 5 \rangle = \langle 5, -1 \rangle.$$

C. Because $2\mathbf{w} = \langle 6, 8 \rangle$, it follows that

$$\begin{aligned} \mathbf{v} + 2\mathbf{w} &= \langle -2, 5 \rangle + \langle 6, 8 \rangle \\ &= \langle -2 + 6, 5 + 8 \rangle \\ &= \langle 4, 13 \rangle. \end{aligned}$$

Vector addition and scalar multiplication share many of the properties of ordinary arithmetic.

PROPERTIES OF VECTOR ADDITION AND SCALAR MULTIPLICATION

Let $\mathbf{u}$, $\mathbf{v}$, and $\mathbf{w}$ be vectors, and let c and d be scalars. Then the following properties are true.

1. $\mathbf{u} + \mathbf{v} = \mathbf{v} + \mathbf{u}$
2. $(\mathbf{u} + \mathbf{v}) + \mathbf{w} = \mathbf{u} + (\mathbf{v} + \mathbf{w})$
3. $\mathbf{u} + \mathbf{0} = \mathbf{u}$
4. $\mathbf{u} + (-\mathbf{u}) = \mathbf{0}$
5. $c(d\mathbf{u}) = (cd)\mathbf{u}$
6. $(c + d)\mathbf{u} = c\mathbf{u} + d\mathbf{u}$
7. $c(\mathbf{u} + \mathbf{v}) = c\mathbf{u} + c\mathbf{v}$
8. $1(\mathbf{u}) = \mathbf{u}$, $0(\mathbf{u}) = \mathbf{0}$
9. $\|c\mathbf{v}\| = |c| \, \|\mathbf{v}\|$

REMARK   Property 9 can be stated as follows: The length of the vector $c\,\mathbf{v}$ is the absolute value of c times the length of $\mathbf{v}$.

Unit Vectors

In many applications of vectors, it is useful to find a unit vector that has the same direction as a given nonzero vector $\mathbf{v}$. To do this, divide $\mathbf{v}$ by its length to obtain

$$\mathbf{u} = \frac{\mathbf{v}}{\|\mathbf{v}\|} = \left(\frac{1}{\|\mathbf{v}\|}\right)\mathbf{v}. \qquad \textit{Unit vector}$$

Note that $\mathbf{u}$ is a scalar multiple of $\mathbf{v}$. The vector $\mathbf{u}$ has length 1 and the same direction as $\mathbf{v}$. We call $\mathbf{u}$ a **unit vector in the direction of v.**

EXAMPLE 4   **Finding a Unit Vector**

Find a unit vector in the direction of $\mathbf{v} = \langle -2, 5 \rangle$ and verify that the result has length 1.

SOLUTION

The unit vector in the direction of $\mathbf{v}$ is

$$\frac{\mathbf{v}}{\|\mathbf{v}\|} = \frac{\langle -2, 5 \rangle}{\sqrt{(-2)^2 + (5)^2}} = \frac{1}{\sqrt{29}}\langle -2, 5 \rangle = \left\langle \frac{-2}{\sqrt{29}}, \frac{5}{\sqrt{29}} \right\rangle.$$

This vector has length 1 because

$$\sqrt{\left(\frac{-2}{\sqrt{29}}\right)^2 + \left(\frac{5}{\sqrt{29}}\right)^2} = \sqrt{\frac{4}{29} + \frac{25}{29}} = \sqrt{\frac{29}{29}} = 1.$$

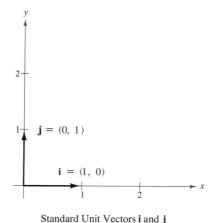

$\mathbf{j} = \langle 0, 1 \rangle$

$\mathbf{i} = \langle 1, 0 \rangle$

Standard Unit Vectors $\mathbf{i}$ and $\mathbf{j}$

FIGURE 4.25

The unit vectors $\langle 1, 0 \rangle$ and $\langle 0, 1 \rangle$ are the **standard unit vectors** and are denoted by

$$\mathbf{i} = \langle 1, 0 \rangle \quad \text{and} \quad \mathbf{j} = \langle 0, 1 \rangle,$$

as shown in Figure 4.25. $\big($Note that the lowercase letter $\mathbf{i}$ is written in boldface to distinguish it from the imaginary number $i = \sqrt{-1}$.$\big)$ These vectors can be used to represent any vector $\mathbf{v} = \langle v_1, v_2 \rangle$ as follows.

$$\mathbf{v} = \langle v_1, v_2 \rangle = v_1\langle 1, 0 \rangle + v_2\langle 0, 1 \rangle = v_1\mathbf{i} + v_2\mathbf{j}$$

The scalars v_1 and v_2 are the **horizontal and vertical components of v,** respectively. The vector sum $v_1\mathbf{i} + v_2\mathbf{j}$ is a **linear combination** of the vectors $\mathbf{i}$ and $\mathbf{j}$. Any vector in the plane can be expressed as a linear combination of the standard unit vectors $\mathbf{i}$ and $\mathbf{j}$.

EXAMPLE 5   **Representing a Vector as a Linear Combination of Unit Vectors**

Let $\mathbf{u}$ be the vector with initial point $(2, -5)$ and terminal point $(-1, 3)$. Write $\mathbf{u}$ as a linear combination of the standard unit vectors $\mathbf{i}$ and $\mathbf{j}$.

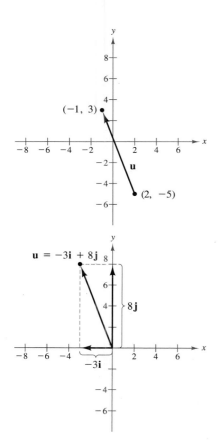

FIGURE 4.26

SOLUTION

$$\begin{aligned}
\mathbf{u} &= \langle -1 - 2,\ 3 - (-5) \rangle \\
&= \langle -3,\ 8 \rangle \\
&= -3\mathbf{i} + 8\mathbf{j}
\end{aligned}$$

This result is shown graphically in Figure 4.26.

EXAMPLE 6   Vector Operations

Let $\mathbf{u} = -3\mathbf{i} + 8\mathbf{j}$ and $\mathbf{v} = 2\mathbf{i} - \mathbf{j}$. Find $2\mathbf{u} - 3\mathbf{v}$.

SOLUTION

$$\begin{aligned}
2\mathbf{u} - 3\mathbf{v} &= 2(-3\mathbf{i} + 8\mathbf{j}) - 3(2\mathbf{i} - \mathbf{j}) \\
&= -6\mathbf{i} + 16\mathbf{j} - 6\mathbf{i} + 3\mathbf{j} \\
&= -12\mathbf{i} + 19\mathbf{j}
\end{aligned}$$

Direction Angles

If $\mathbf{u}$ is a *unit vector* such that θ is the angle (measured counterclockwise) from the positive x-axis to $\mathbf{u}$, then the terminal point of $\mathbf{u}$ lies on the unit circle and you have

$$\mathbf{u} = \langle \cos \theta,\ \sin \theta \rangle = (\cos \theta)\mathbf{i} + (\sin \theta)\mathbf{j},$$

as shown in Figure 4.27. We call θ the **direction angle** of the vector $\mathbf{u}$.

Suppose that $\mathbf{u}$ is a unit vector with direction angle θ. If $\mathbf{v}$ is any vector that makes an angle θ with the positive x-axis, then it has the same direction as $\mathbf{u}$ and you can write

$$\mathbf{v} = \|\mathbf{v}\| \langle \cos \theta,\ \sin \theta \rangle = \|\mathbf{v}\| (\cos \theta)\mathbf{i} + \|\mathbf{v}\| (\sin \theta)\mathbf{j}.$$

For instance, the vector $\mathbf{v}$ of length 3 making an angle of 30° with the positive x-axis is given by

$$\mathbf{v} = 3(\cos 30°)\mathbf{i} + 3(\sin 30°)\mathbf{j} = \frac{3\sqrt{3}}{2}\mathbf{i} + \frac{3}{2}\mathbf{j},$$

where $\|\mathbf{v}\| = 3$.

Because $\mathbf{v} = a\mathbf{i} + b\mathbf{j} = \|\mathbf{v}\| (\cos \theta)\mathbf{i} + \|\mathbf{v}\| (\sin \theta)\mathbf{j}$, it follows that the direction angle θ for $\mathbf{v}$ is determined from

$$\tan \theta = \frac{\sin \theta}{\cos \theta} = \frac{\|\mathbf{v}\| \sin \theta}{\|\mathbf{v}\| \cos \theta} = \frac{b}{a}.$$

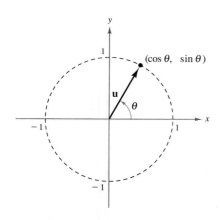

FIGURE 4.27

EXAMPLE 7   Finding Direction Angles of Vectors

Find the direction angles of the vectors.

A. u = 3**i** + 3**j**
B. v = 3**i** − 4**j**

SOLUTION

A. The direction angle is given by

$$\tan \theta = \frac{b}{a} = \frac{3}{3} = 1.$$

Therefore, $\theta = 45°$, as shown in Figure 4.28.

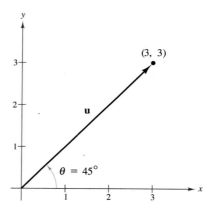

FIGURE 4.28

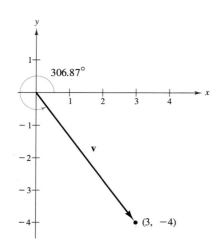

FIGURE 4.29

B. The direction angle is given by

$$\tan \theta = \frac{b}{a} = \frac{-4}{3}.$$

Moreover, because **v** = 3**i** − 4**j** lies in Quadrant IV, θ lies in Quadrant IV and its reference angle is

$$\theta' = \left| \arctan\left(-\frac{4}{3}\right) \right| \approx |-53.13°| = 53.13°.$$

Therefore, $\theta \approx 360° - 53.13° = 306.87°$, as shown in Figure 4.29.

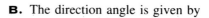

Applications of Vectors

Many applications of vectors involve the use of triangles and trigonometry in their solutions.

EXAMPLE 8   **Finding Component Form, Given Magnitude and Direction**

Find the component form of the vector that represents the velocity of an airplane descending at a speed of 100 miles per hour at an angle 30° below horizontal, as shown in Figure 4.30.

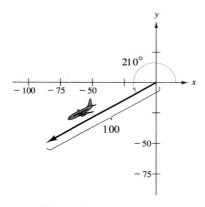

FIGURE 4.30

SOLUTION

The velocity vector **v** has a magnitude of 100 and a direction angle of $\theta = 210°$. Hence, the component form of **v** is

$$\mathbf{v} = \|\mathbf{v}\| (\cos \theta)\,\mathbf{i} + \|\mathbf{v}\| (\sin \theta)\,\mathbf{j}$$
$$= 100(\cos 210°)\,\mathbf{i} + 100(\sin 210°)\,\mathbf{j}$$
$$= 100\left(\frac{-\sqrt{3}}{2}\right)\mathbf{i} + 100\left(\frac{-1}{2}\right)\mathbf{j}$$
$$= -50\sqrt{3}\,\mathbf{i} - 50\mathbf{j}$$
$$= \langle -50\sqrt{3}, -50 \rangle.$$

You should check to see that $\|\mathbf{v}\| = 100$.

For an object to be in *equilibrium,* it must be at rest and the sum of all force vectors acting on the object must be the zero vector.

EXAMPLE 9 An Equilibrium Problem

A person is standing on a tightrope, as shown in Figure 4.31. The total weight of the person and the balancing pole is 200 pounds. Find the force (or tension) on each end of the tightrope.

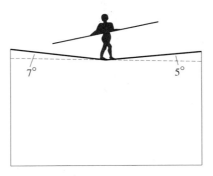

FIGURE 4.31

SOLUTION

Begin by sketching a diagram in which all force vectors are in standard position (with initial points at the origin), as shown in Figure 4.32. In the diagram,

$$\mathbf{w} = -200\mathbf{j} \qquad \text{\textit{Weight of person}}$$

represents the force of the person. (Note that this force is acting straight down.) The forces corresponding to the left and right ends of the rope are as follows.

$$\mathbf{u} = -\cos 7° \,\|\mathbf{u}\| \, \mathbf{i} + \sin 7° \,\|\mathbf{u}\| \, \mathbf{j} \qquad \text{\textit{Left end of rope}}$$
$$\mathbf{v} = \cos 5° \,\|\mathbf{v}\| \, \mathbf{i} + \sin 5° \,\|\mathbf{v}\| \, \mathbf{j} \qquad \text{\textit{Right end of rope}}$$

Because all three forces must sum to zero, you can write the following.

$$\begin{aligned}
\mathbf{0} &= \mathbf{u} + \mathbf{v} + \mathbf{w} \\
&= -\cos 7° \,\|\mathbf{u}\| \, \mathbf{i} + \sin 7° \,\|\mathbf{u}\| \, \mathbf{j} + \cos 5° \,\|\mathbf{v}\| \, \mathbf{i} + \sin 5° \,\|\mathbf{v}\| \, \mathbf{j} - 200\mathbf{j} \\
&= (-\cos 7° \,\|\mathbf{u}\| + \cos 5° \,\|\mathbf{v}\|) \, \mathbf{i} + (\sin 7° \,\|\mathbf{u}\| + \sin 5° \,\|\mathbf{v}\| - 200) \, \mathbf{j} \\
&= 0\mathbf{i} + 0\mathbf{j}
\end{aligned}$$

Because two vectors are equal only if their corresponding components are equal, you can write the following system of equations.

$$-\cos 7° \,\|\mathbf{u}\| + \cos 5° \,\|\mathbf{v}\| = 0$$
$$\sin 7° \,\|\mathbf{u}\| + \sin 5° \,\|\mathbf{v}\| = 200$$

The solution of this system is

$$\|\mathbf{u}\| \approx 958.3 \text{ pounds} \quad \text{and} \quad \|\mathbf{v}\| \approx 954.8 \text{ pounds.}$$

These two quantities represent the forces acting on the left and right ends of the rope. Are you surprised at the magnitudes of these forces?

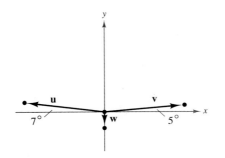

FIGURE 4.32

DISCUSSION PROBLEM

SOLVING AN EQUILIBRIUM PROBLEM

In Example 9, suppose the tightrope was deflected with different angles, as shown below. Would the tension at each end of the rope be greater or less than the tensions found in Example 9? Explain your reasoning.

(a)

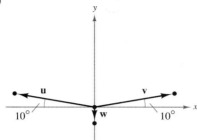

(b)

WARM-UP

The following warm-up exercises involve skills that were covered in earlier sections. You will use these skills in the exercise set for this section.

In Exercises 1 and 2, find the distance between the points.

1. $(-2, 6), (5, -15)$ **2.** $(0, 0), (-3, -7)$

In Exercises 3 and 4, find an equation of the line through the two points.

3. $(3, 1), (-2, 4)$ **4.** $(-2, -3), (4, 5)$

In Exercises 5 and 6, find an angle θ $(0 \le \theta \le 360°)$ whose vertex is at the origin and whose terminal side passes through the given point.

5. $(-2, 5)$ **6.** $(4, -3)$

In Exercises 7–10, find the sine and cosine of the angle θ.

7. $\theta = 30°$ **8.** $\theta = 120°$ **9.** $\theta = 300°$ **10.** $\theta = 210°$

SECTION 4.3 · EXERCISES

In Exercises 1–6, use the figure to sketch a graph of the specified vector.

1. $-\mathbf{u}$ **2.** $3\mathbf{v}$

3. $\mathbf{u} + \mathbf{v}$ **4.** $\mathbf{u} + 2\mathbf{v}$

5. $\mathbf{u} - \mathbf{v}$ **6.** $\mathbf{v} - \frac{1}{2}\mathbf{u}$

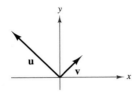

FIGURE FOR 1–6

In Exercises 7–16, find the component form and the magnitude of the vector $\mathbf{v}$.

7.
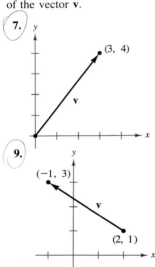
$(3, 4)$

8.
$(-4, -2)$

9.
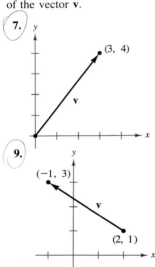
$(-1, 3)$
$(2, 1)$

10.
$(3, 4)$
$(-1, -2)$

11.
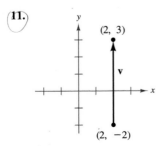
$(2, 3)$
$(2, -2)$

12.
$(-4, -2)$ $(3, -2)$

13. Initial point: $(-1, 5)$
Terminal point: $(15, 2)$

14. Initial point: $(1, 11)$
Terminal point: $(9, 3)$

15. Initial point: $(-3, -5)$
Terminal point: $(5, -1)$

16. Initial point: $(-3, 11)$
Terminal point: $(9, 40)$

In Exercises 17–26, find (a) $\mathbf{u} + \mathbf{v}$, (b) $\mathbf{u} - \mathbf{v}$, and (c) $2\mathbf{u} - 3\mathbf{v}$.

17. $\mathbf{u} = \langle 1, 2 \rangle$, $\mathbf{v} = \langle 3, 1 \rangle$

18. $\mathbf{u} = \langle 2, 3 \rangle$, $\mathbf{v} = \langle 4, 0 \rangle$

19. $\mathbf{u} = \langle -2, 3 \rangle$, $\mathbf{v} = \langle -2, 1 \rangle$

20. $\mathbf{u} = \langle 0, 1 \rangle$, $\mathbf{v} = \langle 0, -1 \rangle$

21. $\mathbf{u} = \langle 4, -2 \rangle$, $\mathbf{v} = \langle 0, 0 \rangle$

22. $\mathbf{u} = \langle 0, 0 \rangle$, $\mathbf{v} = \langle 2, 1 \rangle$

23. $\mathbf{u} = \mathbf{i} + \mathbf{j}$, $\mathbf{v} = 2\mathbf{i} - 3\mathbf{j}$

24. $\mathbf{u} = 2\mathbf{i} - \mathbf{j}$, $\mathbf{v} = -\mathbf{i} + \mathbf{j}$

25. $\mathbf{u} = 2\mathbf{i}$, $\mathbf{v} = \mathbf{j}$

26. $\mathbf{u} = 3\mathbf{j}$, $\mathbf{v} = 2\mathbf{i}$

In Exercises 27–30, find the magnitude and direction angle of the vector $\mathbf{v}$.

27. $\mathbf{v} = 5\langle \cos 30°, \sin 30° \rangle$

28. $\mathbf{v} = 8\langle \cos 135°, \sin 135° \rangle$

29. $\mathbf{v} = 6\mathbf{i} - 6\mathbf{j}$

30. $\mathbf{v} = -2\mathbf{i} + 5\mathbf{j}$

In Exercises 31–38, sketch $\mathbf{v}$ and find its component form. (Assume θ is measured counterclockwise from the x-axis to the vector.)

31. $\|\mathbf{v}\| = 3$, $\theta = 0°$

32. $\|\mathbf{v}\| = 1$, $\theta = 45°$

33. $\|\mathbf{v}\| = 1$, $\theta = 150°$

34. $\|\mathbf{v}\| = \frac{5}{2}$, $\theta = 45°$

35. $\|\mathbf{v}\| = 3\sqrt{2}$, $\theta = 150°$

36. $\|\mathbf{v}\| = 8$, $\theta = 90°$

37. $\|\mathbf{v}\| = 2$, $\mathbf{v}$ in the direction $\mathbf{i} + 3\mathbf{j}$

38. $\|\mathbf{v}\| = 3$, $\mathbf{v}$ in the direction $3\mathbf{i} + 4\mathbf{j}$

In Exercises 39–44, find the component form of $\mathbf{v}$ and sketch the specified vector operations geometrically, where

$$\mathbf{u} = 2\mathbf{i} - \mathbf{j} \quad \text{and} \quad \mathbf{w} = \mathbf{i} + 2\mathbf{j}.$$

39. $\mathbf{v} = \frac{3}{2}\mathbf{u}$ **40.** $\mathbf{v} = \mathbf{u} + \mathbf{w}$

41. $\mathbf{v} = \mathbf{u} + 2\mathbf{w}$ **42.** $\mathbf{v} = -\mathbf{u} + \mathbf{w}$

43. $\mathbf{v} = \frac{1}{2}(3\mathbf{u} + \mathbf{w})$ **44.** $\mathbf{v} = \mathbf{u} - 2\mathbf{w}$

In Exercises 45–48, find the component form of the sum of the vectors **u** and **v** with direction angles θ_u and θ_v, respectively.

45. $\|\mathbf{u}\| = 5$, $\quad \theta_\mathbf{u} = 0°$
$\quad\|\mathbf{v}\| = 5$, $\quad \theta_\mathbf{v} = 90°$

46. $\|\mathbf{u}\| = 2$, $\quad \theta_\mathbf{u} = 30°$
$\quad\|\mathbf{v}\| = 2$, $\quad \theta_\mathbf{v} = 90°$

47. $\|\mathbf{u}\| = 20$, $\quad \theta_\mathbf{u} = 45°$
$\quad\|\mathbf{v}\| = 50$, $\quad \theta_\mathbf{v} = 180°$

48. $\|\mathbf{u}\| = 35$, $\quad \theta_\mathbf{u} = 25°$
$\quad\|\mathbf{v}\| = 50$, $\quad \theta_\mathbf{v} = 120°$

In Exercises 49–52, find a unit vector in the direction of the given vector.

49. $\mathbf{v} = 4\mathbf{i} - 3\mathbf{j}$

50. $\mathbf{v} = \mathbf{i} + \mathbf{j}$

51. $\mathbf{v} = 2\mathbf{j}$

52. $\mathbf{v} = \mathbf{i} - 2\mathbf{j}$

PAGE 264

In Exercises 53–56, use the Law of Cosines to find the angle α between the given vectors. (Assume $0° \leq \alpha \leq 180°$.)

53. $\mathbf{v} = \mathbf{i} + \mathbf{j}$, $\quad \mathbf{w} = 2(\mathbf{i} - \mathbf{j})$

54. $\mathbf{v} = 3\mathbf{i} + \mathbf{j}$, $\quad \mathbf{w} = 2\mathbf{i} - \mathbf{j}$

55. $\mathbf{v} = \mathbf{i} + \mathbf{j}$, $\quad \mathbf{w} = 3\mathbf{i} - \mathbf{j}$

56. $\mathbf{v} = \mathbf{i} + 2\mathbf{j}$, $\quad \mathbf{w} = 2\mathbf{i} - \mathbf{j}$

In Exercises 57 and 58, find the angle between the forces, given the magnitude of their resultant (vector sum). (*Hint:* Write one force as a vector in the direction of the positive x-axis and the other as a vector at an angle θ with the positive x-axis.)

57. Force one: 45 pounds
Force two: 60 pounds
Resultant force: 90 pounds

58. Force one: 3000 pounds
Force two: 1000 pounds
Resultant force: 3750 pounds

59. *Resultant Force* Forces with magnitudes of 35 pounds and 50 pounds act on a hook (see figure). The angle between the two forces is 30°. Find the direction and magnitude of the resultant (vector sum) of these two forces.

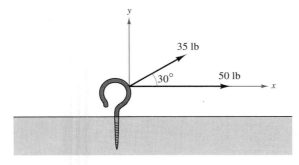

FIGURE FOR 59

60. *Resultant Force* Forces with magnitudes of 500 pounds and 200 pounds act on a machine part at angles of 30° and −45°, respectively, with the x-axis (see figure). Find the direction and magnitude of the resultant (vector sum) of these forces.

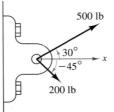

FIGURE FOR 60

61. *Resultant Force* Three forces with magnitudes of 75 pounds, 100 pounds, and 125 pounds act on an object at angles of 30°, 45°, and 120°, respectively, with the positive x-axis. Find the direction and magnitude of the resultant of these forces.

62. *Resultant Force* Three forces with magnitudes of 70 pounds, 40 pounds, and 60 pounds act on an object at angles of −30°, 45°, and 135°, respectively, with the positive x-axis. Find the direction and magnitude of the resultant of these forces.

63. *Horizontal and Vertical Components of Velocity* A ball is thrown with an initial velocity of 80 feet per second, at an angle of 50° with the horizontal (see figure). Find the vertical and horizontal components of the velocity.

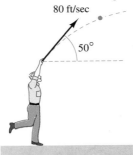

FIGURE FOR 63

64. *Horizontal and Vertical Components of Velocity* A gun with a muzzle velocity of 1200 feet per second is fired at an angle of 6° with the horizontal. Find the vertical and horizontal components of the velocity.

Cable Tension In Exercises 65 and 66, use the figure to determine the tension in each cable supporting the given load.

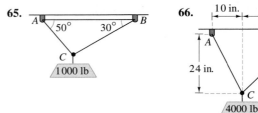

67. *Barge Towing*   A loaded barge is being towed by two tugboats, and the magnitude of the resultant is 6000 pounds directed along the axis of the barge (see figure). Find the tension in the tow lines if they each make a 20° angle with the axis of the barge.

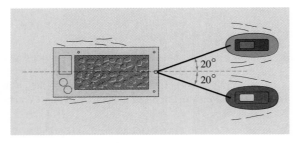

FIGURE FOR 67

68. *Shared Load*   To carry a 100-pound cylindrical weight, two people lift on the ends of short ropes that are tied to an eyelet on the top center of the cylinder. Find the tension in the ropes if they each make a 30° angle with the vertical (see figure).

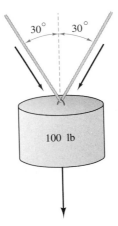

FIGURE FOR 68

69. *Navigation*   An airplane is flying in the direction S 32° E, with an air speed of 540 miles per hour. Because of the wind, the plane's ground speed and direction are 500 miles per hour and S 40° E, respectively (see figure). Find the direction and speed of the wind.

70. *Navigation*   An airplane's velocity with respect to the air is 580 miles per hour, and it is headed N 58° W. The wind, at the altitude of the plane, is from the southwest and has velocity of 60 miles per hour (see figure). What is the true direction of the plane, and what is its speed with respect to the ground?

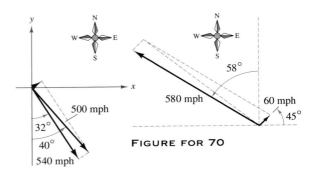

FIGURE FOR 69

FIGURE FOR 70

71. *Work*   A heavy implement is pulled 10 feet across a floor, using a force of 85 pounds. Find the work done if the direction of the force is 60° above the horizontal (see figure). (Use the formula for work, $W = FD$, where F is the component of the force in the direction of motion and D is the distance.)

72. *Tether Ball*   A tether ball weighing 1 pound is pulled outward from the pole by a horizontal force **u** until the rope makes a 30° angle with the pole (see figure). Determine the resulting tension in the rope and the magnitude of **u**.

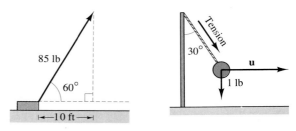

FIGURE FOR 71   **FIGURE FOR 72**

73. *Technology*   Enter the following program in a graphing utility. (This program is for the TI-81; other versions are listed in the appendix.) Use $-6 \le x \le 6$ and $-2 \le y \le 6$, and enter $A = 5$, $B = 2$, $C = -4$, and $D = 3$. Explain what the program does.

```
:Disp "ENTER (A, B)"      :Line (0, 0, A, B)
:Disp "ENTER A"           :Line (0, 0, C, D)
:Input A                  :A+C→E
:Disp "ENTER B"           :B+D→F
:Input B                  :Line (0, 0, E, F)
:Disp "ENTER (C, D)"      :Line (A, B, E, F)
:Disp "ENTER C"           :Line (C, D, E, F)
:Input C                  :Pause
:Disp "ENTER D"           :ClrDraw
:Input D                  :End
```

4.4   THE DOT PRODUCT OF TWO VECTORS

The Dot Product of Two Vectors   /   Angle Between Two Vectors   /   Finding Vector Components   /   Work

The Dot Product of Two Vectors

So far you have studied two vector operations—vector addition and multiplication by a scalar—each of which yields another vector. In this section you will study a third vector operation, the **dot product.** This product yields a scalar, rather than a vector.

DEFINITION OF DOT PRODUCT

The **dot product** of $\mathbf{u} = \langle u_1, u_2 \rangle$ and $\mathbf{v} = \langle v_1, v_2 \rangle$ is

$$\mathbf{u} \cdot \mathbf{v} = u_1 v_1 + u_2 v_2.$$

PROPERTIES OF THE DOT PRODUCT

Let $\mathbf{u}$, $\mathbf{v}$, and $\mathbf{w}$ be vectors in the plane and let c be a scalar.

1. $\mathbf{u} \cdot \mathbf{v} = \mathbf{v} \cdot \mathbf{u}$
2. $\mathbf{0} \cdot \mathbf{v} = 0$
3. $\mathbf{u} \cdot (\mathbf{v} + \mathbf{w}) = \mathbf{u} \cdot \mathbf{v} + \mathbf{u} \cdot \mathbf{w}$
4. $\mathbf{v} \cdot \mathbf{v} = \| \mathbf{v} \|^2$
5. $c(\mathbf{u} \cdot \mathbf{v}) = c\mathbf{u} \cdot \mathbf{v} = \mathbf{u} \cdot c\mathbf{v}$

PROOF

To prove the first property, let $\mathbf{u} = \langle u_1, u_2 \rangle$ and $\mathbf{v} = \langle v_1, v_2 \rangle$. Then

$$\mathbf{u} \cdot \mathbf{v} = u_1 v_1 + u_2 v_2 = v_1 u_1 + v_2 u_2 = \mathbf{v} \cdot \mathbf{u}.$$

For the fourth property, let $\mathbf{v} = \langle v_1, v_2 \rangle$. Then

$$\begin{aligned}
\mathbf{v} \cdot \mathbf{v} &= v_1{}^2 + v_2{}^2 \\
&= \left(\sqrt{v_1{}^2 + v_2{}^2} \right)^2 \\
&= \| \mathbf{v} \|^2.
\end{aligned}$$

Proofs of the other properties are left to you.

D I S C O V E R Y

The following program is written for a Texas Instruments TI-81 graphing calculator. (Program steps for other programmable calculators are given in the appendix.) The program sketches two vectors $\mathbf{u} = \langle a, b \rangle$ and $\mathbf{v} = \langle c, d \rangle$ in standard position. Then, the program finds the angle between vectors. For instance, for the vectors $\mathbf{u} = \langle 5, 3 \rangle$ and $\mathbf{v} = \langle -6, 4 \rangle$ the calculator display will be as follows. (Before running the program, set the viewing rectangle as shown and set the mode to degrees.) Explain how the program works.

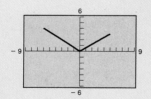

TI-81 Program

```
:Disp "ENTER (A, B)"
:Disp "ENTER A"
:Input A
:Disp "ENTER B"
:Input B
:Disp "ENTER (C, D)"
:Disp "ENTER C"
:Input C
:Disp "ENTER D"
:Input D
:Line (0, 0, A, B)
:Line (0, 0, C, D)
:Pause
:AC+BD→E
:√ (A²+B²) →U
:√ (C²+D²) →V
:cos⁻¹ (E/(UV))→T
:Disp T
:ClrDraw
:End
```

EXAMPLE 1 Finding Dot Products

Find the following dot products.

A. $\langle 4, 5 \rangle \cdot \langle 2, 3 \rangle$ **B.** $\langle 2, -1 \rangle \cdot \langle 1, 2 \rangle$ **C.** $(-5\mathbf{i} - 2\mathbf{j}) \cdot (3\mathbf{i} + 2\mathbf{j})$

SOLUTION

A. $\langle 4, 5 \rangle \cdot \langle 2, 3 \rangle = 4(2) + 5(3) = 8 + 15 = 23$
B. $\langle 2, -1 \rangle \cdot \langle 1, 2 \rangle = 2(1) + (-1)(2) = 2 - 2 = 0$
C. $(-5\mathbf{i} - 2\mathbf{j}) \cdot (3\mathbf{i} + 2\mathbf{j}) = (-5)(3) + (-2)(2)$
$$= -15 - 4$$
$$= -19$$

REMARK   In Example 1, be sure you see that the dot product of two vectors is a scalar (a real number), not a vector. Moreover, notice that the dot product can be positive, zero, or negative.

EXAMPLE 2 Using Properties of Dot Products

Let $\mathbf{u} = \langle -1, 3 \rangle$, $\mathbf{v} = \langle 2, -4 \rangle$, and $\mathbf{w} = \langle 1, -2 \rangle$. Find the following products.

A. $(\mathbf{u} \cdot \mathbf{v}) \mathbf{w}$ **B.** $\mathbf{u} \cdot (2\mathbf{v})$

SOLUTION

Begin by finding the dot product of $\mathbf{u}$ and $\mathbf{v}$.

$$\mathbf{u} \cdot \mathbf{v} = \langle -1, 3 \rangle \cdot \langle 2, -4 \rangle = (-1)(2) + 3(-4) = -14$$

A. $(\mathbf{u} \cdot \mathbf{v}) \mathbf{w} = -14\langle 1, -2 \rangle = \langle -14, 28 \rangle$
B. $\mathbf{u} \cdot (2\mathbf{v}) = 2(\mathbf{u} \cdot \mathbf{v}) = 2(-14) = -28$

Notice that the first product is a vector, whereas the second is a scalar. Can you see why?

EXAMPLE 3 Dot Product and Length

The dot product of $\mathbf{u}$ with itself is 5. What is the length of $\mathbf{u}$?

SOLUTION

Because $\|\mathbf{u}\|^2 = \mathbf{u} \cdot \mathbf{u} = 5$, it follows that

$$\|\mathbf{u}\| = \sqrt{\mathbf{u} \cdot \mathbf{u}} = \sqrt{5}.$$

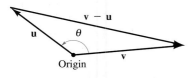

Angle Between Two Vectors

FIGURE 4.33

Angle Between Two Vectors

The **angle between two nonzero vectors** is the angle θ, $0 \le \theta \le \pi$, between their respective standard position vectors, as shown in Figure 4.33. This angle can be found using the dot product. (Note that we do not define the angle between the zero vector and another vector.)

ANGLE BETWEEN TWO VECTORS

If θ is the angle between two nonzero vectors $\mathbf{u}$ and $\mathbf{v}$, then

$$\cos \theta = \frac{\mathbf{u} \cdot \mathbf{v}}{\|\mathbf{u}\| \, \|\mathbf{v}\|}.$$

PROOF

Consider the triangle determined by vectors $\mathbf{u}$, $\mathbf{v}$, and $\mathbf{v} - \mathbf{u}$, as shown in Figure 4.33. By the Law of Cosines, you can write

$$\|\mathbf{v} - \mathbf{u}\|^2 = \|\mathbf{u}\|^2 + \|\mathbf{v}\|^2 - 2\|\mathbf{u}\| \, \|\mathbf{v}\| \cos \theta$$

$$(\mathbf{v} - \mathbf{u}) \cdot (\mathbf{v} - \mathbf{u}) = \|\mathbf{u}\|^2 + \|\mathbf{v}\|^2 - 2\|\mathbf{u}\| \, \|\mathbf{v}\| \cos \theta$$

$$(\mathbf{v} - \mathbf{u}) \cdot \mathbf{v} - (\mathbf{v} - \mathbf{u}) \cdot \mathbf{u} = \|\mathbf{u}\|^2 + \|\mathbf{v}\|^2 - 2\|\mathbf{u}\| \, \|\mathbf{v}\| \cos \theta$$

$$\mathbf{v} \cdot \mathbf{v} - \mathbf{u} \cdot \mathbf{v} - \mathbf{v} \cdot \mathbf{u} + \mathbf{u} \cdot \mathbf{u} = \|\mathbf{u}\|^2 + \|\mathbf{v}\|^2 - 2\|\mathbf{u}\| \, \|\mathbf{v}\| \cos \theta$$

$$\|\mathbf{v}\|^2 - 2\mathbf{u} \cdot \mathbf{v} + \|\mathbf{u}\|^2 = \|\mathbf{u}\|^2 + \|\mathbf{v}\|^2 - 2\|\mathbf{u}\| \, \|\mathbf{v}\| \cos \theta$$

$$-2\mathbf{u} \cdot \mathbf{v} = -2\|\mathbf{u}\| \, \|\mathbf{v}\| \cos \theta$$

$$\cos \theta = \frac{\mathbf{u} \cdot \mathbf{v}}{\|\mathbf{u}\| \, \|\mathbf{v}\|}.$$

EXAMPLE 4   Finding the Angle Between Two Vectors

Find the angle between $\mathbf{u} = \langle 4, 3 \rangle$ and $\mathbf{v} = \langle 3, 5 \rangle$.

SOLUTION

$$\cos \theta = \frac{\mathbf{u} \cdot \mathbf{v}}{\|\mathbf{u}\| \, \|\mathbf{v}\|} = \frac{\langle 4, 3 \rangle \cdot \langle 3, 5 \rangle}{\|\langle 4, 3 \rangle\| \, \|\langle 3, 5 \rangle\|} = \frac{27}{5\sqrt{34}}$$

$$\frac{(4 \cdot 3) + (3 \cdot 5)}{\sqrt{4^2 + 3^2} \, \bullet \, \sqrt{3^2 + 5^2}}$$

This implies that the angle between the two vectors is

$$\theta = \arccos \frac{27}{5\sqrt{34}} \approx 22.2°,$$

as shown in Figure 4.34.

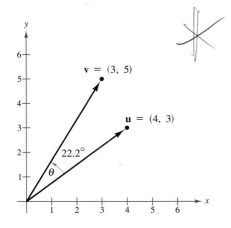

FIGURE 4.34

Rewriting the expression for the angle between two vectors in the form

$$\mathbf{u} \cdot \mathbf{v} = \|\mathbf{u}\| \|\mathbf{v}\| \cos \theta \qquad \textit{Alternative form of dot product}$$

produces an alternative way to calculate the dot product. From this form, you can see that because $\|\mathbf{u}\|$ and $\|\mathbf{v}\|$ are always positive, $\mathbf{u} \cdot \mathbf{v}$ and $\cos \theta$ will always have the same sign. Figure 4.35 shows the five possible orientations of two vectors.

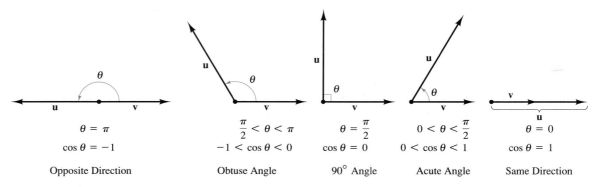

| Opposite Direction | Obtuse Angle | 90° Angle | Acute Angle | Same Direction |

FIGURE 4.35

DEFINITION OF ORTHOGONAL VECTORS

The vectors $\mathbf{u}$ and $\mathbf{v}$ are **orthogonal** if $\mathbf{u} \cdot \mathbf{v} = 0$.

The terms "orthogonal" and "perpendicular" mean essentially the same thing—meeting at right angles. By definition, however, the zero vector is orthogonal to every vector $\mathbf{u}$, because $\mathbf{0} \cdot \mathbf{u} = 0$.

EXAMPLE 5   Determining Orthogonal Vectors

Are the vectors $\mathbf{u} = \langle 2, -3 \rangle$ and $\mathbf{v} = \langle 6, 4 \rangle$ orthogonal?

SOLUTION

Begin by finding the dot product of the two vectors.

$$\begin{aligned} \mathbf{u} \cdot \mathbf{v} &= \langle 2, -3 \rangle \cdot \langle 6, 4 \rangle \\ &= 2(6) + (-3)(4) \\ &= 0 \end{aligned}$$

Because the dot product is 0, the two vectors are orthogonal, as shown in Figure 4.36.

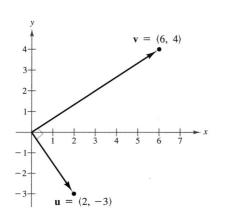

FIGURE 4.36

Finding Vector Components

You have already seen applications in which two vectors are added to produce a resultant vector. Many applications in physics and engineering pose the reverse problem—decomposing a given vector into the sum of two **vector components.**

Consider a boat on an inclined ramp, as shown in Figure 4.37. The force **F** due to gravity pulls the boat *down* the ramp and *against* the ramp. These two orthogonal forces, **w**₁ and **w**₂, are vector components of **F**. That is,

$$\mathbf{F} = \mathbf{w}_1 + \mathbf{w}_2.$$ *Vector components of* **F**

The negative of component **w**₁ represents the force needed to keep the boat from rolling down the ramp, whereas **w**₂ represents the force that the tires must withstand against the ramp. A procedure for finding **w**₁ and **w**₂ is shown below.

FIGURE 4.37

DEFINITION OF VECTOR COMPONENTS

Let **u** and **v** be nonzero vectors such that

$$\mathbf{u} = \mathbf{w}_1 + \mathbf{w}_2,$$

where **w**₁ and **w**₂ are orthogonal and **w**₁ is parallel to **v**, as shown in Figure 4.38. The vectors **w**₁ and **w**₂ are called **vector components** of **u**. The vector **w**₁ is the **projection** of **u** onto **v** and is denoted by

$$\mathbf{w}_1 = \text{proj}_{\mathbf{v}}\, \mathbf{u}.$$

The vector **w**₂ is given by $\mathbf{w}_2 = \mathbf{u} - \mathbf{w}_1$.

From this definition, you can see that it is easy to find the component **w**₂ once you have found the projection of **u** onto **v**. To find the projection, you can use the dot product, as shown below.

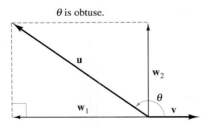

θ is acute.

θ is obtuse.

$\mathbf{w}_1 = \text{proj}_{\mathbf{v}}\, \mathbf{u} = \text{projection of } \mathbf{u} \text{ onto } \mathbf{v}.$
$\mathbf{w}_2 = \text{vector component of}$
$\qquad \mathbf{u} \text{ orthogonal to } \mathbf{v}.$

FIGURE 4.38

PROJECTION OF U ONTO V

Let **u** and **v** be nonzero vectors. The projection of **u** onto **v** is

$$\text{proj}_{\mathbf{v}}\, \mathbf{u} = \left(\frac{\mathbf{u} \cdot \mathbf{v}}{\|\mathbf{v}\|^2} \right) \mathbf{v}.$$

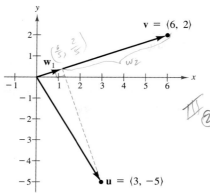

FIGURE 4.39

EXAMPLE 6   Decomposing a Vector into Components

Find the projection of $u = \langle 3, -5 \rangle$ onto $v = \langle 6, 2 \rangle$. Then write u as the sum of two orthogonal vectors, one of which is $\text{proj}_v\, u$.

SOLUTION

The projection of u onto v is

$$w_1 = \text{proj}_v\, u = \left(\frac{u \cdot v}{\|v\|^2}\right) v = \left(\frac{8}{40}\right)\langle 6, 2 \rangle = \left\langle \frac{6}{5}, \frac{2}{5}\right\rangle,$$

as shown in Figure 4.39. The other component, w_2, is

$$w_2 = u - w_1 = \langle 3, -5 \rangle - \left\langle \frac{6}{5}, \frac{2}{5}\right\rangle = \left\langle \frac{9}{5}, -\frac{27}{5}\right\rangle.$$

Thus, $u = w_1 + w_2 = \left\langle \frac{6}{5}, \frac{2}{5}\right\rangle + \left\langle \frac{9}{5}, -\frac{27}{5}\right\rangle = \langle 3, -5 \rangle.$

EXAMPLE 7   Finding a Force

A 600-pound boat sits on a ramp inclined at 30°, as shown in Figure 4.40. What force is required to keep the boat from rolling down the ramp?

SOLUTION

Because the force due to gravity is vertical and downward, you can represent the gravitational force by the vector

$$F = -600j. \qquad \textit{Force due to gravity}$$

To find the force required to keep the boat from rolling down the ramp, project F onto a unit vector v in the direction of the ramp, as follows.

$$v = \cos 30°\, i + \sin 30°\, j = \frac{\sqrt{3}}{2}i + \frac{1}{2}j \quad \textit{Unit vector along ramp}$$

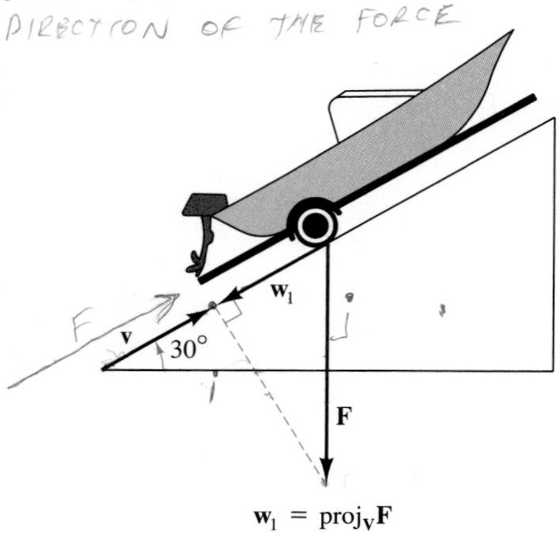

Therefore, the projection of F onto v is given by

$$w_1 = \text{proj}_v\, F = \left(\frac{F \cdot v}{\|v\|^2}\right) v$$

$$= (F \cdot v)\, v = (-600)\left(\frac{1}{2}\right) v$$

$$= -300\left(\frac{\sqrt{3}}{2}i + \frac{1}{2}j\right).$$

The magnitude of this force is 300, and therefore a force of 300 pounds is required to keep the boat from rolling down the ramp.

$w_1 = \text{proj}_v F$

FIGURE 4.40

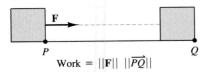

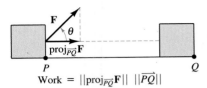

$$\text{Work} = \|\mathbf{F}\|\,\|\overrightarrow{PQ}\|$$

(a) Force acts along the line of motion.

$$\text{Work} = \|\text{proj}_{\overrightarrow{PQ}}\mathbf{F}\|\,\|\overrightarrow{PQ}\|$$

(b) Force acts at angle θ with the line of motion.

FIGURE 4.41

Work

The work W done by a constant force $\mathbf{F}$ acting along the line of motion of an object is given by

$$W = (\text{magnitude of force})(\text{distance}) = \|\mathbf{F}\|\,\|\overrightarrow{PQ}\|,$$

as shown in Figure 4.41(a). If the constant force $\mathbf{F}$ is not directed along the line of motion, then you can see from Figure 4.41(b) that the work W done by the force is

$$W = \|\text{proj}_{\overrightarrow{PQ}}\mathbf{F}\|\,\|\overrightarrow{PQ}\| = (\cos\theta)\|\mathbf{F}\|\,\|\overrightarrow{PQ}\| = \mathbf{F}\cdot\overrightarrow{PQ}.$$

This notion of work is summarized in the following definition.

DEFINITION OF WORK

The **work** W done by a constant force $\mathbf{F}$ as its point of application moves along the vector $\overrightarrow{PQ}$ is given by either of the following.

1. $W = \|\text{proj}_{\overrightarrow{PQ}}\mathbf{F}\|\,\|\overrightarrow{PQ}\|$ *Projection form*
2. $W = \mathbf{F}\cdot\overrightarrow{PQ}$ *Dot product form*

EXAMPLE 8   Finding Work

To close a sliding door, a person pulls on a rope with a constant force of 50 pounds at a constant angle of 60°, as shown in Figure 4.42. Find the work done in moving the door 12 feet to its closed position.

SOLUTION

Using a projection, you can calculate the work as follows.

$$\begin{aligned}
W &= \|\text{proj}_{\overrightarrow{PQ}}\mathbf{F}\|\,\|\overrightarrow{PQ}\| \\
&= (\cos 60°)\|\mathbf{F}\|\,\|\overrightarrow{PQ}\| \\
&= \frac{1}{2}(50)(12) \\
&= 300 \text{ ft-lb}
\end{aligned}$$

Thus, the work done is 300 foot-pounds.

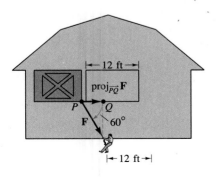

FIGURE 4.42

DISCUSSION PROBLEM

THE SIGN OF THE DOT PRODUCT

In this section, you were given the alternative form of the dot product of two vectors.

$$\mathbf{u} \cdot \mathbf{v} = \|\mathbf{u}\| \, \|\mathbf{v}\| \cos \theta$$

Use this form to determine the sign of the dot product of **u** and **v** for the given angle. Explain your reasoning.

(a) 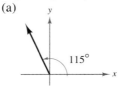   (b)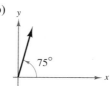

WARP-UP

The following warm-up exercises involve skills that were covered in earlier sections. You will use these skills in the exercise set for this section.

In Exercises 1–4, find (a) $\mathbf{u} + 2\mathbf{v}$, and (b) $\|\mathbf{u}\|$.

1. $\mathbf{u} = \langle 6, -3 \rangle$ **2.** $\mathbf{u} = \left\langle \frac{3}{8}, \frac{4}{5} \right\rangle$
$\mathbf{v} = \langle -10, -1 \rangle$ $\mathbf{v} = \left\langle \frac{5}{2}, -\frac{1}{10} \right\rangle$

3. $\mathbf{u} = 4\mathbf{i} - 16\mathbf{j}$ **4.** $\mathbf{u} = 0.5\mathbf{i} + 1.4\mathbf{j}$
$\mathbf{v} = -5\mathbf{i} + 10\mathbf{j}$ $\mathbf{v} = 4.1\mathbf{i} - 1.8\mathbf{j}$

In Exercises 5–8, find the values of θ in the interval $0 \le \theta < 2\pi$ that satisfy the equation. If the exact value is not known, round the result to two decimal places.

5. $\cos \theta = -\frac{1}{2}$ **6.** $\cos \theta = 0$

7. $\cos \theta = 0.5403$ **8.** $\cos \theta = -0.9689$

In Exercises 9 and 10, find a unit vector (a) in the direction of **u** and (b) in the direction opposite that of **u**.

9. $\mathbf{u} = \langle 120, -50 \rangle$ **10.** $\mathbf{u} = \left\langle \frac{4}{5}, \frac{1}{3} \right\rangle$

SECTION 4.4 · EXERCISES

In Exercises 1–6, find the dot product of **u** and **v**.

1. $\mathbf{u} = \langle 6, 2 \rangle$ **2.** $\mathbf{u} = \langle 4, 10 \rangle$ **5.** $\mathbf{u} = 4\mathbf{i} - 2\mathbf{j}$ **6.** $\mathbf{u} = 2\mathbf{i} + 5\mathbf{j}$
$\mathbf{v} = \langle 2, -3 \rangle$ $\mathbf{v} = \langle -3, 1 \rangle$ $\mathbf{v} = \mathbf{i} + 3\mathbf{j}$ $\mathbf{v} = 9\mathbf{i} - 3\mathbf{j}$

3. $\mathbf{u} = \langle 3, -3 \rangle$ **4.** $\mathbf{u} = \mathbf{j}$
$\mathbf{v} = \langle 0, 5 \rangle$ $\mathbf{v} = \mathbf{j}$

In Exercises 7–10, use the vectors $\mathbf{u} = \langle 2, 2 \rangle$ and $\mathbf{v} = \langle -3, 4 \rangle$ to find the indicated quantity. State whether the result is a vector or a scalar.

7. $\mathbf{u} \cdot \mathbf{u}$ **8.** $\|\mathbf{u}\|^2$

9. $(\mathbf{u} \cdot \mathbf{v}) \mathbf{v}$ **10.** $\mathbf{u} \cdot (2\mathbf{v})$

In Exercises 11 and 12, find $\mathbf{u} \cdot \mathbf{v}$, where θ is the angle between $\mathbf{u}$ and $\mathbf{v}$.

11. $\|\mathbf{u}\| = 4$, $\|\mathbf{v}\| = 10$, $\theta = \dfrac{2\pi}{3}$

12. $\|\mathbf{u}\| = 100$, $\|\mathbf{v}\| = 250$, $\theta = \dfrac{\pi}{6}$

13. *Revenue* The vector $\mathbf{u} = \langle 1245, 2600 \rangle$ gives the number of units of two products produced by a company. The vector $\mathbf{v} = \langle 12.20, 8.50 \rangle$ gives the price (in dollars) of each unit, respectively. Find the dot product $\mathbf{u} \cdot \mathbf{v}$, and explain what information it gives.

14. Repeat Exercise 13 after increasing the prices by 5%.

In Exercises 15–22, find the angle θ between the given vectors.

15. $\mathbf{u} = \langle 1, 0 \rangle$
$\mathbf{v} = \langle 0, -2 \rangle$

16. $\mathbf{u} = \langle 4, 4 \rangle$
$\mathbf{v} = \langle 2, 0 \rangle$

17. $\mathbf{u} = \langle 3, 4 \rangle$
$\mathbf{v} = \langle 5, 3 \rangle$

18. $\mathbf{u} = \langle -2, 5 \rangle$
$\mathbf{v} = \langle 1, 4 \rangle$

19. $\mathbf{u} = 2\mathbf{i} + 3\mathbf{j}$
$\mathbf{v} = -2\mathbf{i} + 2\mathbf{j}$

20. $\mathbf{u} = \mathbf{i} - 6\mathbf{j}$
$\mathbf{v} = 2\mathbf{i} - 12\mathbf{j}$

21. $\mathbf{u} = \cos\left(\dfrac{\pi}{3}\right) \mathbf{i} + \sin\left(\dfrac{\pi}{3}\right) \mathbf{j}$
$\mathbf{v} = \cos\left(\dfrac{3\pi}{4}\right) \mathbf{i} + \sin\left(\dfrac{3\pi}{4}\right) \mathbf{j}$

22. $\mathbf{u} = \cos\left(\dfrac{\pi}{4}\right) \mathbf{i} + \sin\left(\dfrac{\pi}{4}\right) \mathbf{j}$
$\mathbf{v} = \cos\left(\dfrac{\pi}{2}\right) \mathbf{i} + \sin\left(\dfrac{\pi}{2}\right) \mathbf{j}$

In Exercises 23–26, use a graphing utility to sketch the vectors and find the degree measure of the angle between the vectors.

23. $\mathbf{u} = 3\mathbf{i} + 4\mathbf{j}$
$\mathbf{v} = -7\mathbf{i} + 5\mathbf{j}$

24. $\mathbf{u} = -6\mathbf{i} - 3\mathbf{j}$
$\mathbf{v} = -8\mathbf{i} + 4\mathbf{j}$

25. $\mathbf{u} = 5\mathbf{i} + 5\mathbf{j}$
$\mathbf{v} = -6\mathbf{i} + 6\mathbf{j}$

26. $\mathbf{u} = 2\mathbf{i} - 3\mathbf{j}$
$\mathbf{v} = 4\mathbf{i} + 3\mathbf{j}$

In Exercises 27–32, determine whether $\mathbf{u}$ and $\mathbf{v}$ are orthogonal, parallel, or neither.

27. $\mathbf{u} = \langle 4, 6 \rangle$
$\mathbf{v} = \langle 3, -2 \rangle$

28. $\mathbf{u} = \langle 4, 14 \rangle$
$\mathbf{v} = \langle 6, 21 \rangle$

29. $\mathbf{u} = \langle -12, 30 \rangle$
$\mathbf{v} = \langle \frac{1}{2}, -\frac{5}{4} \rangle$

30. $\mathbf{u} = \langle 15, 45 \rangle$
$\mathbf{v} = \langle -5, 12 \rangle$

31. $\mathbf{u} = -\frac{1}{4}(3\mathbf{i} - \mathbf{j})$
$\mathbf{v} = 5\mathbf{i} + 6\mathbf{j}$

32. $\mathbf{u} = \cos 20° \, \mathbf{i} + \sin 20° \, \mathbf{j}$
$\mathbf{v} = \sin 70° \, \mathbf{i} + \cos 70° \, \mathbf{j}$

In Exercises 33–38, find the projection of $\mathbf{u}$ onto $\mathbf{v}$. Then find the vector component of $\mathbf{u}$ orthogonal to $\mathbf{v}$.

33. $\mathbf{u} = \langle 3, 4 \rangle$
$\mathbf{v} = \langle 8, 2 \rangle$

34. $\mathbf{u} = \langle 4, -2 \rangle$
$\mathbf{v} = \langle 10, 2 \rangle$

35. $\mathbf{u} = \langle 4, 2 \rangle$
$\mathbf{v} = \langle 1, -2 \rangle$

36. $\mathbf{u} = \langle 0, 3 \rangle$
$\mathbf{v} = \langle 2, 15 \rangle$

37. $\mathbf{u} = \langle 2, 1 \rangle$
$\mathbf{v} = \langle 0, 8 \rangle$

38. $\mathbf{u} = \langle -5, -1 \rangle$
$\mathbf{v} = \langle -1, 1 \rangle$

39. *Braking Load* A truck with a gross weight of 26,000 pounds is parked on a 10° slope (see figure). Assume the only force to overcome is that due to gravity.
(a) Find the force required to keep the truck from rolling down the hill.
(b) Find the force perpendicular to the hill.

40. *Work* An object is pulled 20 feet across a floor using a force of 45 pounds. Find the work done if the direction of the force is 30° above the horizontal (see figure).

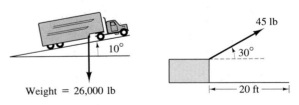

Weight = 26,000 lb

FIGURE FOR 39   **FIGURE FOR 40**

41. *Work* A tractor pulls a log 2500 feet and the tension in the cable connecting the tractor and log is approximately 3600 pounds. Approximate the work done if the direction of the force is 35° above the horizontal.

42. Prove Property 2 of the dot product: $\mathbf{0} \cdot \mathbf{v} = 0$.

43. Prove Property 3 of the dot product: $\mathbf{u} \cdot (\mathbf{v} + \mathbf{w}) = \mathbf{u} \cdot \mathbf{v} + \mathbf{u} \cdot \mathbf{w}$.

44. Prove Property 5 of the dot product: $c(\mathbf{u} \cdot \mathbf{v}) = c\mathbf{u} \cdot \mathbf{v} = \mathbf{u} \cdot c\mathbf{v}$.

4.5 THE THREE-DIMENSIONAL COORDINATE SYSTEM

The Three-Dimensional Coordinate System / The Distance and Midpoint
Formulas / The Equation of a Sphere

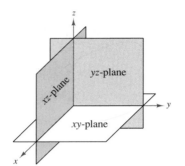

FIGURE 4.43

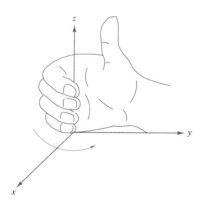

FIGURE 4.44

The Three-Dimensional Coordinate System

Recall that the Cartesian plane is determined by two perpendicular number lines called the x-axis and the y-axis. These axes together with their point of intersection (the origin) provide a two-dimensional coordinate system for identifying points in a plane. To identify a point in space, you must introduce a third dimension to the model. The geometry of this three-dimensional model is called **solid analytic geometry.**

You can construct a **three-dimensional coordinate system** by passing a z-axis perpendicular to both the x- and y-axes at the origin. Figure 4.43 shows the positive portion of each coordinate axis. Taken as pairs, the axes determine three **coordinate planes:** the **xy-plane,** the **xz-plane,** and the **yz-plane.** These three coordinate planes separate the three-dimensional coordinate system into eight **octants.** The first octant is the one for which all three coordinates are positive. In this three-dimensional system, a point P in space is determined by an ordered triple (x, y, z), where x, y, and z are as follows.

x = directed distance from yz-plane to P
y = directed distance from xz-plane to P
z = directed distance from xy-plane to P

A three-dimensional coordinate system can have either a **left-handed** or a **right-handed** orientation. In this text, we work exclusively with right-handed systems, as shown in Figure 4.44.

EXAMPLE 1 Plotting Points in Space

Plot the following points in space.

A. $(2, -3, 3)$ **B.** $(-2, 6, 2)$ **C.** $(1, 4, 0)$ **D.** $(2, 2, -3)$

SOLUTION

To plot the point $(2, -3, 3)$, notice that $x = 2$, $y = -3$, and $z = 3$. To help visualize the point, locate the point $(2, -3)$ in the xy-plane (denoted by a cross in Figure 4.45). The point $(2, -3, 3)$ lies 3 units above the cross. The other three points are also shown in Figure 4.45.

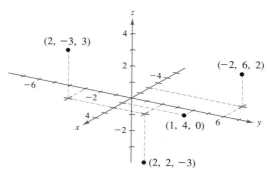

FIGURE 4.45

The Distance and Midpoint Formulas

Many of the formulas established for the two-dimensional coordinate system can be extended to three dimensions. For example, to find the distance between two points in space, you can use the Pythagorean Theorem twice, as shown in Figure 4.46. By doing this, you will obtain the formula for the distance between two points in space.

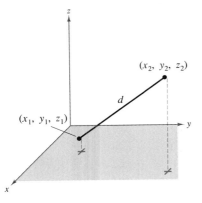

d = distance between two points

FIGURE 4.46

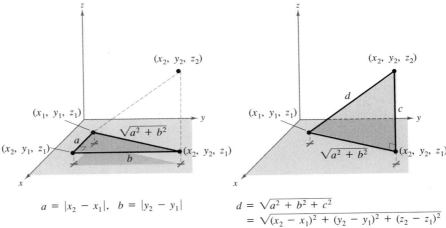

$a = |x_2 - x_1|, \quad b = |y_2 - y_1|$

$d = \sqrt{a^2 + b^2 + c^2}$
$\quad = \sqrt{(x_2 - x_1)^2 + (y_2 - y_1)^2 + (z_2 - z_1)^2}$

DISTANCE FORMULA IN SPACE

The distance between the points (x_1, y_1, z_1) and (x_2, y_2, z_2) is

$$d = \sqrt{(x_2 - x_1)^2 + (y_2 - y_1)^2 + (z_2 - z_1)^2}.$$

EXAMPLE 2 **Finding the Distance Between Two Points**

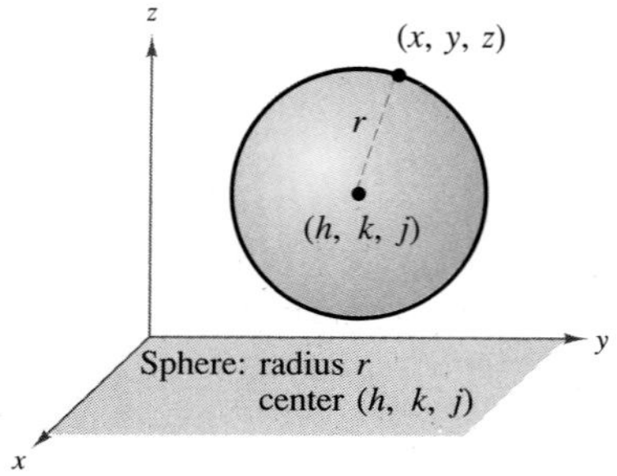

Find the distance between $(1, 0, 2)$ and $(2, 4, -3)$.

SOLUTION

$$
\begin{aligned}
d &= \sqrt{(x_2 - x_1)^2 + (y_2 - y_1)^2 + (z_2 - z_1)^2} \\
&= \sqrt{(2 - 1)^2 + (4 - 0)^2 + (-3 - 2)^2} \quad \textit{Substitute} \\
&= \sqrt{1 + 16 + 25} \quad \textit{Simplify} \\
&= \sqrt{42} \quad \textit{Simplify}
\end{aligned}
$$

Notice the similarity between the Distance Formulas in the plane and in space, and the similarity between the standard equation of a circle and a sphere. The Midpoint Formulas in the plane and in space are also similar.

MIDPOINT FORMULA IN SPACE

The midpoint of the line segment joining the points (x_1, y_1, z_1) and (x_2, y_2, z_2) is

$$\left(\frac{x_1 + x_2}{2}, \frac{y_1 + y_2}{2}, \frac{z_1 + z_2}{2}\right).$$

EXAMPLE 3 **Using the Midpoint Formula**

Find the midpoint of the line segment joining $(5, -2, 3)$ and $(0, 4, 4)$.

SOLUTION

Using the Midpoint Formula, the midpoint is

$$\left(\frac{5 + 0}{2}, \frac{-2 + 4}{2}, \frac{3 + 4}{2}\right) = \left(\frac{5}{2}, 1, \frac{7}{2}\right),$$

as shown in Figure 4.47.

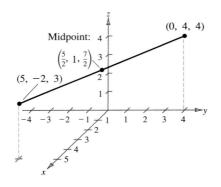

Midpoint: $\left(\frac{5}{2}, 1, \frac{7}{2}\right)$

$(0, 4, 4)$

$(5, -2, 3)$

FIGURE 4.47

The Equation of a Sphere

A **sphere** with center at (h, k, j) and radius r is defined to be the set of all points (x, y, z) such that the distance between (x, y, z) and (h, k, j) is r, as shown in Figure 4.48. Using the Distance Formula, this condition can be written as

$$\sqrt{(x - h)^2 + (y - k)^2 + (z - j)^2} = r.$$

By squaring both sides of this equation, you obtain the standard equation of a sphere.

(x, y, z)

r

(h, k, j)

Sphere: radius r
center (h, k, j)

FIGURE 4.48

STANDARD EQUATION OF A SPHERE

The **standard equation of a sphere** whose center is (h, k, j) and whose radius is r is

$$(x - h)^2 + (y - k)^2 + (z - j)^2 = r^2.$$

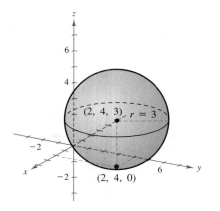

FIGURE 4.49

EXAMPLE 4   Finding the Equation of a Sphere

Find the standard equation for the sphere whose center is $(2, 4, 3)$ and whose radius is 3. Does this sphere intersect the xy-plane?

SOLUTION

$$(x - h)^2 + (y - k)^2 + (z - j)^2 = r^2 \qquad \textit{Standard equation}$$
$$(x - 2)^2 + (y - 4)^2 + (z - 3)^2 = 3^2 \qquad \textit{Substitute}$$

From the graph shown in Figure 4.49, you can see that the center of the sphere lies 3 units above the xy-plane. Because the sphere has a radius of 3, you can conclude that it does intersect the xy-plane—at the point $(2, 4, 0)$.

EXAMPLE 5   Finding the Equation of a Sphere

Find the equation of the sphere that has the points $(3, -2, 6)$ and $(-1, 4, 2)$ as endpoints of a diameter.

SOLUTION

By the Midpoint Rule, the center of the sphere is

$$\left(\frac{3 - 1}{2}, \frac{-2 + 4}{2}, \frac{6 + 2}{2} \right) = (1, 1, 4).$$

By the Distance Formula, the radius is

$$r = \sqrt{(3 - 1)^2 + (-2 - 1)^2 + (6 - 4)^2} = \sqrt{17}.$$

Therefore, the standard equation of the sphere is

$$(x - 1)^2 + (y - 1)^2 + (z - 4)^2 = 17.$$

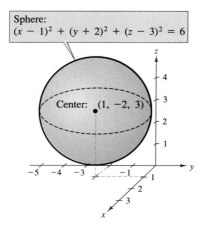

Sphere:
$(x - 1)^2 + (y + 2)^2 + (z - 3)^2 = 6$

Center: $(1, -2, 3)$

FIGURE 4.50

EXAMPLE 6   **Finding the Center and Radius of a Sphere**

Find the center and radius of the sphere whose equation is

$$x^2 + y^2 + z^2 - 2x + 4y - 6z + 8 = 0.$$

SOLUTION

You can obtain the standard equation of this sphere by completing the square, as follows.

$$x^2 + y^2 + z^2 - 2x + 4y - 6z + 8 = 0$$
$$(x^2 - 2x +) + (y^2 + 4y +) + (z^2 - 6z +) = -8$$
$$(x^2 - 2x + 1) + (y^2 + 4y + 4) + (z^2 - 6z + 9) = -8 + 1 + 4 + 9$$
$$(x - 1)^2 + (y + 2)^2 + (z - 3)^2 = 6$$

Therefore, the center of the sphere is $(1, -2, 3)$, and its radius is $\sqrt{6}$, as shown in Figure 4.50.

Note in Example 6 that the points satisfying the equation of the sphere are "surface points," not "interior points." In general, the collection of points satisfying an equation involving x, y, and z is called a **surface in space.**

Finding the intersection of a surface with one of the three coordinate planes (or with a plane parallel to one of the three coordinate planes) helps visualize the surface. Such an intersection is called a **trace** of the surface. For example, the xy-trace of a surface consists of all points that are common to both the surface *and* the xy-plane. Similarly, the xz-trace of a surface consists of all points that are common to both the surface and the xz-plane.

EXAMPLE 7   **Finding a Trace of a Surface**

Sketch the xy-trace of the sphere whose equation is

$$(x - 3)^2 + (y - 2)^2 + (z + 4)^2 = 5^2.$$

SOLUTION

To find the xy-trace of this surface, use the fact that every point in the xy-plane has a z-coordinate of zero. This means that if you substitute $z = 0$ into the given equation, the resulting equation will represent the intersection of the surface with the xy-plane.

$$(x - 3)^2 + (y - 2)^2 + (0 + 4)^2 = 25$$
$$(x - 3)^2 + (y - 2)^2 + 16 = 25$$
$$(x - 3)^2 + (y - 2)^2 = 9$$
$$(x - 3)^2 + (y - 2)^2 = 3^2$$

From this form, you can see that the xy-trace is a circle of radius 3, as shown in Figure 4.51.

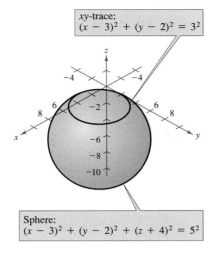

xy-trace:
$(x - 3)^2 + (y - 2)^2 = 3^2$

Sphere:
$(x - 3)^2 + (y - 2)^2 + (z + 4)^2 = 5^2$

FIGURE 4.51

DISCUSSION PROBLEM

....

COMPARING TWO AND THREE DIMENSIONS

In this section, you saw similarities between formulas in two-dimensional coordinate geometry and three-dimensional coordinate geometry. In two-dimensional coordinate geometry, the graph of the equation

$$ax + by + c = 0$$

is a line. In three-dimensional coordinate geometry, what is the graph of equation

$$ax + by + cz = 0?$$

Is it a line? Explain your reasoning.

WARM-UP

....

The following warm-up exercises involve skills that were covered in earlier sections. You will use these skills in the exercise set for this section.

In Exercises 1–6, find the distance between A and B and the midpoint of the line segment joining A and B.

1. $A(0, 0)$, $B(5, 12)$ 2. $A(-4, 1)$, $B(3, 8)$
3. $A(1, 6)$, $B(6, -1)$ 4. $A(0.5, 0.8)$, $B(-1.2, -2.4)$
5. $A\left(\frac{1}{2}, 1\right)$, $B\left(\frac{5}{2}, \frac{1}{2}\right)$ 6. $A\left(\frac{4}{3}, \frac{3}{2}\right)$, $B\left(\frac{2}{3}, 1\right)$

In Exercises 7–10, find the standard equation of the circle satisfying the given conditions.

7. Center: $(4, -5)$; Radius: 4   8. Center: $(-1, 3)$; Radius: 1
9. Endpoints of a diameter: $(1, 4)$, $(5, 2)$ 10. Endpoints of a diameter: $(-1, 0)$, $(0, 6)$

SECTION 4.5 · EXERCISES

In Exercises 1–4, plot the points on the same three-dimensional coordinate system.

1. (a) $(1, 2, 4)$
   (b) $(1, -2, 1)$
2. (a) $(3, -1, 4)$
   (b) $(2, 4, -3)$
3. (a) $(5, -3, 4)$
   (b) $(5, -3, -4)$
4. (a) $(2, 0, 5)$
   (b) $(0, 3, -5)$

In Exercises 5–8, find the lengths of the triangle with the indicated vertices, and determine whether the triangle is a right triangle, an isosceles triangle, or neither.

5. $(0, 0, 0)$
   $(3, 3, 2)$
   $(3, -6, 2)$
6. $(3, 1, 2)$
   $(5, -1, 1)$
   $(1, 3, 1)$
7. $(2, -1, 0)$
   $(6, 0, 3)$
   $(0, 2, 3)$
8. $(6, 0, 0)$
   $(0, 3, 0)$
   $(0, 0, -2)$

In Exercises 9–12, find the coordinates of the midpoint of the line segment joining the given points.

9. $(3, -6, 10)$. $(-3, 2, 2)$
10. $(2, -2, -8)$, $(5, 6, 18)$
11. $(4, -2, 5)$, $(-4, 2, 8)$
12. $(-3, 5, 7)$, $(-6, 4, 10)$

In Exercises 13–18, find the standard equation of the sphere.

13. Center: $(0, 4, 3)$; Radius: 4
14. Center: $(1, -2, 3)$; Radius: 5
15. Center: $(-3, 7, 5)$; Diameter: 10
16. Center: $(0, 5, -9)$; Diameter: 6
17. Endpoints of a diameter: $(3, 0, 0)$, $(0, 0, 6)$
18. Endpoints of a diameter: $(2, -2, 2)$, $(-1, 4, 6)$

In Exercises 19–24, find the center and radius of the sphere.

19. $x^2 + y^2 + z^2 - 4x + 2y - 6z + 10 = 0$
20. $x^2 + y^2 + z^2 - 6x + 4y + 9 = 0$
21. $x^2 + y^2 + z^2 + 4x - 8z + 19 = 0$
22. $x^2 + y^2 + z^2 - 8y - 6z + 13 = 0$
23. $9x^2 + 9y^2 + 9z^2 - 18x - 6y - 72z + 73 = 0$
24. $2x^2 + 2y^2 + 2z^2 - 2x - 6y - 4z + 5 = 0$

In Exercises 25 and 26, sketch the graph of the equation and sketch the specified traces.

25. $x^2 + y^2 + z^2 = 16$
    (a) yz-trace   (b) xy-trace
26. $x^2 + y^2 + (z - 3)^2 = 9$
    (a) yz-trace   (b) xz-trace

In Exercises 27–30, use a graphing utility to graph the sphere. (*Hint:* Solve for z and graph the two resulting expressions in x and y.)

27. $x^2 + y^2 + z^2 - 16 = 0$
28. $x^2 + y^2 + z^2 - 4y - 4 = 0$
29. $(x - 3)^2 + (y - 4)^2 + (z - 5)^2 = 4$
30. $x^2 + y^2 + z^2 + 6y - 8z + 21 = 0$

4.6   VECTORS IN SPACE

Vectors in Space / Parallel Vectors / Application

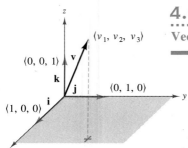

Standard Unit Vectors in Space

FIGURE 4.52

Vectors in Space

Physical forces and velocities are not confined to the plane, so it is natural to extend the concept of vectors from two-dimensional space to three-dimensional space. In space, vectors are denoted by ordered triples

$$\mathbf{v} = \langle v_1, v_2, v_3 \rangle. \qquad \textit{Component form}$$

The **zero vector** is denoted by $\mathbf{0} = \langle 0, 0, 0 \rangle$. Using the unit vectors $\mathbf{i} = \langle 1, 0, 0 \rangle$, $\mathbf{j} = \langle 0, 1, 0 \rangle$, and $\mathbf{k} = \langle 0, 0, 1 \rangle$ in the direction of the positive z-axis, the **standard unit vector notation** for $\mathbf{v}$ is

$$\mathbf{v} = v_1 \mathbf{i} + v_2 \mathbf{j} + v_3 \mathbf{k}, \qquad \textit{Unit vector form}$$

as shown in Figure 4.52. If $\mathbf{v}$ is represented by the directed line segment from $P(p_1, p_2, p_3)$ to $Q(q_1, q_2, q_3)$, as shown in Figure 4.53, the component form of $\mathbf{v}$ is given by subtracting the coordinates of the initial point from the coordinates of the terminal point, as follows.

$$\mathbf{v} = \langle v_1, v_2, v_3 \rangle = \langle q_1 - p_1, q_2 - p_2, q_3 - p_3 \rangle$$

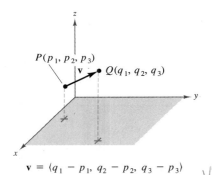

$$\mathbf{v} = \langle q_1 - p_1, q_2 - p_2, q_3 - p_3 \rangle$$

FIGURE 4.53

VECTORS IN SPACE

1. *Equality of vectors:* Two vectors are equal if and only if their corresponding components are equal.
2. *Length of a vector:* The length of $\mathbf{u} = \langle u_1, u_2, u_3 \rangle$ is
$$\|\mathbf{u}\| = \sqrt{u_1^2 + u_2^2 + u_3^2}.$$
3. *Unit vector:* A unit vector $\mathbf{v}$ in the direction of $\mathbf{u}$ is given by
$$\mathbf{v} = \frac{\mathbf{u}}{\|\mathbf{u}\|}, \qquad \mathbf{u} \neq \mathbf{0}.$$
4. *Vector addition:* The sum of the vectors $\mathbf{u} = \langle u_1, u_2, u_3 \rangle$ and $\mathbf{v} = \langle v_1, v_2, v_3 \rangle$ is
$$\mathbf{v} + \mathbf{u} = \langle v_1 + u_1, v_2 + u_2, v_3 + u_3 \rangle.$$
5. *Scalar multiplication:* The scalar multiple of the real number c and the vector $\mathbf{u} = \langle u_1, u_2, u_3 \rangle$ is
$$c\mathbf{u} = \langle cu_1, cu_2, cu_3 \rangle.$$
6. *Dot product:* The dot product of the vectors $\mathbf{u} = \langle u_1, u_2, u_3 \rangle$ and $\mathbf{v} = \langle v_1, v_2, v_3 \rangle$ is
$$\mathbf{u} \cdot \mathbf{v} = u_1 v_1 + u_2 v_2 + u_3 v_3.$$

REMARK Note how similar these definitions are to the corresponding definitions for vectors in the plane. The properties of vector operations discussed in Sections 4.3 and 4.4 are also valid for vectors in space.

EXAMPLE 1   Finding the Component Form of a Vector

Find the component form and length of the vector **v** having initial point (3, 4, 2) and terminal point (3, 6, 4). Then find a unit vector in the direction of **v**.

SOLUTION

The component form of **v** is

$$\mathbf{v} = \langle 3 - 3, 6 - 4, 4 - 2 \rangle = \langle 0, 2, 2 \rangle,$$

which implies that its length is

$$\|\mathbf{v}\| = \sqrt{0^2 + 2^2 + 2^2} = \sqrt{8} = 2\sqrt{2}.$$

 The unit vector in the direction of **v** is

$$\mathbf{u} = \frac{\mathbf{v}}{\|\mathbf{v}\|} = \frac{1}{2\sqrt{2}} \langle 0, 2, 2 \rangle = \left\langle 0, \frac{1}{\sqrt{2}}, \frac{1}{\sqrt{2}} \right\rangle.$$

$$U = \frac{1}{\|V\|}(V)$$

Technology Note

Some graphing utilities have the capability to perform vector operations, such as the dot product. For example, on the TI-85, you can use the VECTR menu to verify the dot product in Example 2 as follows.

dot ([0, 3, −2], [4, −2, 3])

EXAMPLE 2   Finding the Dot Product of Two Vectors

Find the following dot product.

$$\langle 0, 3, -2 \rangle \cdot \langle 4, -2, 3 \rangle$$

SOLUTION

$$\begin{aligned} \langle 0, 3, -2 \rangle \cdot \langle 4, -2, 3 \rangle &= 0(4) + 3(-2) + (-2)(3) \\ &= 0 - 6 - 6 \\ &= -12 \end{aligned}$$

 EXAMPLE 3   Standard Unit Vector Notation

Write the vector **v** = 2**j** − 6**k** in component form.

SOLUTION

Because **i** is missing, its component is 0 and

$$\mathbf{v} = 2\mathbf{j} - 6\mathbf{k} = \langle 0, 2, -6 \rangle.$$

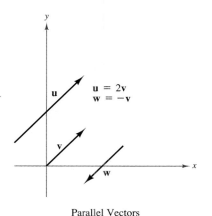

Parallel Vectors

FIGURE 4.54

Parallel Vectors

Recall from the definition of scalar multiplication that positive scalar multiples of a nonzero vector **v** have the same direction as **v**, whereas negative multiples have the direction opposite that of **v**. In general, two nonzero vectors **u** and **v** are **parallel** if there is some scalar c such that **u** $= c$**v**. For example, in Figure 4.54, the vectors **u**, **v**, and **w** are parallel because **u** $= 2$**v** and **w** $= -$**v**.

EXAMPLE 4   **Parallel Vectors**

Vector **w** has initial point $(1, -2, 0)$ and terminal point $(3, 2, 1)$. Which of the following vectors is parallel to **w**?

A. **u** $= \langle 4, 8, 2 \rangle$ **B.** **v** $= \langle 4, 8, 4 \rangle$

SOLUTION

Begin by writing **w** in component form.

$$\mathbf{w} = \langle 3 - 1, 2 + 2, 1 - 0 \rangle = \langle 2, 4, 1 \rangle$$

A. Because

$$\begin{aligned}
\mathbf{u} &= \langle 4, 8, 2 \rangle \\
&= 2\langle 2, 4, 1 \rangle \\
&= 2\mathbf{w},
\end{aligned}$$

you can conclude that **u** *is* parallel to **w**.

B. In this case, you need to find a scalar c such that

$$\langle 4, 8, 4 \rangle = c\langle 2, 4, 1 \rangle.$$

However, equating corresponding components produces $c = 2$ for the first two components and $c = 4$ for the third. Hence, the equation has no solution, and the vectors are *not* parallel.   ▬▬▬▬▬▬

You can use vectors to determine whether three points are collinear (lie on the same line). The points P, Q, and R are collinear if and only if the vectors $\overrightarrow{PQ}$ and $\overrightarrow{PR}$ are parallel.

EXAMPLE 5   **Using Vectors to Determine Collinear Points**

Determine whether the following points lie on the same line.

$$P(2, -1, 4), \quad Q(5, 4, 6), \quad \text{and} \quad R(-4, -11, 0)$$

SOLUTION

The component forms of $\overrightarrow{PQ}$ and $\overrightarrow{PR}$ are

$$\overrightarrow{PQ} = \langle 5 - 2, 4 + 1, 6 - 4 \rangle = \langle 3, 5, 2 \rangle$$

and

$$\overrightarrow{PR} = \langle -4 - 2, -11 + 1, 0 - 4 \rangle = \langle -6, -10, -4 \rangle.$$

Because $\overrightarrow{PR} = -2\overrightarrow{PQ}$, you can conclude that $\overrightarrow{PQ}$ and $\overrightarrow{PR}$ are parallel. Therefore, the points P, Q, and R lie on the same line.

EXAMPLE 6   **Finding the Terminal Point of a Vector**

The initial point of the vector $\mathbf{v} = 4\mathbf{i} + 2\mathbf{j} - \mathbf{k}$ is $P(3, -1, 6)$. What is the terminal point of this vector?

SOLUTION

Using the component form of the vector whose initial point is P and whose terminal point is Q, you can write

$$\overrightarrow{PQ} = \langle q_1 - p_1, q_2 - p_2, q_3 - p_3 \rangle$$
$$= \langle q_1 - 3, q_2 + 1, q_3 - 6 \rangle$$
$$= \langle 4, 2, -1 \rangle.$$

This implies that

$$q_1 - 3 = 4, \quad q_2 + 1 = 2, \quad \text{and} \quad q_3 - 6 = -1.$$

The solutions of these three equations are $q_1 = 7$, $q_2 = 1$, and $q_3 = 5$. Thus, the terminal point is $Q(7, 1, 5)$.

Application

In Section 4.3, you saw how to use vectors to solve an equilibrium problem in a plane. The next example shows how to use vectors to solve an equilibrium problem in space.

EXAMPLE 7   **Solving an Equilibrium Problem**

A weight of 480 pounds is supported by three ropes. As shown in Figure 4.55, the weight is located at $S(0, 2, -1)$. The ropes are tied to the points $P(2, 0, 0)$, $Q(0, 4, 0)$, and $R(-2, 0, 0)$. Find the force (or tension) on each rope.

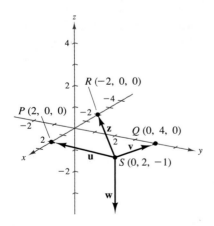

FIGURE 4.55

SOLUTION

The (downward) force of the weight is represented by the vector $\mathbf{w} = \langle 0, 0, -480 \rangle$. The force vectors corresponding to the ropes are as follows.

$$\mathbf{u} = \|\mathbf{u}\| \frac{\overrightarrow{SP}}{\|\overrightarrow{SP}\|} = \|\mathbf{u}\| \frac{\langle 2 - 0, 0 - 2, 0 + 1 \rangle}{3} = \|\mathbf{u}\| \left\langle \frac{2}{3}, -\frac{2}{3}, \frac{1}{3} \right\rangle$$

$$\mathbf{v} = \|\mathbf{v}\| \frac{\overrightarrow{SQ}}{\|\overrightarrow{SQ}\|} = \|\mathbf{v}\| \frac{\langle 0 - 0, 4 - 2, 0 + 1 \rangle}{\sqrt{5}} = \|\mathbf{v}\| \left\langle 0, \frac{2}{\sqrt{5}}, \frac{1}{\sqrt{5}} \right\rangle$$

$$\mathbf{z} = \|\mathbf{z}\| \frac{\overrightarrow{SR}}{\|\overrightarrow{SR}\|} = \|\mathbf{z}\| \frac{\langle -2 - 0, 0 - 2, 0 + 1 \rangle}{3} = \|\mathbf{z}\| \left\langle -\frac{2}{3}, -\frac{2}{3}, \frac{1}{3} \right\rangle$$

For the system to be in equilibrium, it must be true that

$$\mathbf{u} + \mathbf{v} + \mathbf{z} + \mathbf{w} = \mathbf{0} \quad \text{or} \quad \mathbf{u} + \mathbf{v} + \mathbf{z} = -\mathbf{w}.$$

This yields the following system of linear equations.

$$\frac{2}{3}\|\mathbf{u}\| \qquad\qquad -\frac{2}{3}\|\mathbf{z}\| = 0$$

$$-\frac{2}{3}\|\mathbf{u}\| + \frac{2}{\sqrt{5}}\|\mathbf{v}\| - \frac{2}{3}\|\mathbf{z}\| = 0$$

$$\frac{1}{3}\|\mathbf{u}\| + \frac{1}{\sqrt{5}}\|\mathbf{v}\| + \frac{1}{3}\|\mathbf{z}\| = 480$$

The solution of this system is

$$\|\mathbf{u}\| \approx 360.0, \quad \|\mathbf{v}\| = 536.7, \quad \text{and} \quad \|\mathbf{z}\| \approx 360.0.$$

Thus, the rope attached at point P has about 360 pounds of tension, the rope attached at point Q has about 536.7 pounds of tension, and the rope attached at point R has about 360 pounds of tension.

DISCUSSION PROBLEM

·······················

PROPERTIES OF VECTORS IN SPACE

The properties of vectors in the plane that you studied earlier in the text also apply to vectors in space. Choose one of the following properties. Then show how the properties of real numbers can be used to prove the property for vectors in space.

1. $\mathbf{u} + \mathbf{v} = \mathbf{v} + \mathbf{u}$

2. $(\mathbf{u} + \mathbf{v}) + \mathbf{w} = \mathbf{u} + (\mathbf{v} + \mathbf{w})$

3. $\mathbf{u} + \mathbf{0} = \mathbf{u}$

4. $\mathbf{u} + (-\mathbf{u}) = \mathbf{0}$

5. $c(d\mathbf{u}) = (cd)\mathbf{u}$

6. $(c + d)\mathbf{u} = c\mathbf{u} + d\mathbf{u}$

7. $c(\mathbf{u} + \mathbf{v}) = c\mathbf{u} + c\mathbf{v}$

8. $1(\mathbf{u}) = \mathbf{u}$

9. $0(\mathbf{u}) = \mathbf{0}$

10. $\|c\mathbf{u}\| = |c|\,\|\mathbf{u}\|$

WARM-UP

The following warm-up exercises involve skills that were covered in earlier sections. You will use these skills in the exercise set for this section.

In Exercises 1–4, find the component forms of the vectors $\vec{AB}, \vec{AC}$, and $\vec{AB} + \vec{AC}$. Sketch the three vectors.

1. $A(0, 0)$
   $B(1, -2)$
   $C(2, 4)$

2. $A(0, 0)$
   $B(-4, -2)$
   $C(-2, 3)$

3. $A(6, 4)$
   $B(0, -1)$
   $C(-2, 4)$

4. $A(-3, -5)$
   $B(4, -1)$
   $C(1, 3)$

In Exercises 5 and 6, find a unit vector in the direction of **u**.

5. $\mathbf{u} = \langle 5, -12 \rangle$

6. $\mathbf{u} = \langle 3 \cos 38°, 3 \sin 38° \rangle$

In Exercises 7–10, find (a) $\mathbf{u} \cdot \mathbf{v}$ and (b) the angle between **u** and **v**.

7. $\mathbf{u} = \langle 4, 0 \rangle$
   $\mathbf{v} = \langle 2, -2 \rangle$

8. $\mathbf{u} = \langle 3, 4 \rangle$
   $\mathbf{v} = \langle -3, 4 \rangle$

9. $\mathbf{u} = \langle 3, 4 \rangle$
   $\mathbf{v} = \langle 4, -3 \rangle$

10. $\mathbf{u} = \langle -2, 8 \rangle$
   $\mathbf{v} = \langle 3, -12 \rangle$

SECTION 4.6 · EXERCISES

In Exercises 1–4, (a) find the component form of the vector **v** and (b) sketch the vector with its initial point at the origin.

1.

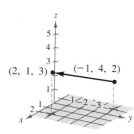

2.

3.

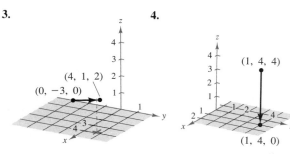

4.

In Exercises 5 and 6, write the component form of **v**. Then write the component forms of (a) a vector parallel to **v** and (b) a vector in the opposite direction of **v**. (There are many correct answers.)

5. Initial point: $(-1, -2, 1)$
   Terminal point: $(3, 2, 5)$

6. Initial point: $(-4, 5, 5)$
   Terminal point: $(4, 0, 0)$

In Exercises 7–12, find the vector **z**, given $\mathbf{u} = \langle -1, 3, 2 \rangle$, $\mathbf{v} = \langle 1, -2, -2 \rangle$, and $\mathbf{w} = \langle 5, 0, -5 \rangle$.

7. $\mathbf{z} = \mathbf{u} - 2\mathbf{v}$

8. $\mathbf{z} = 3\mathbf{u} - \mathbf{v} + \mathbf{w}$

9. $\mathbf{z} = 2\mathbf{u} + 8\mathbf{v} - \mathbf{w}$

10. $\mathbf{z} = -7\mathbf{u} + \mathbf{v} - \frac{1}{5}\mathbf{w}$

11. $2\mathbf{z} - 4\mathbf{u} = \mathbf{w}$

12. $\mathbf{u} + \mathbf{v} - 2\mathbf{w} + \mathbf{z} = 0$

In Exercises 13 and 14, find the length of **v**. P 204

13. $\mathbf{v} = \langle 4, 1, 4 \rangle$

14. $\mathbf{v} = 4\mathbf{i} - 3\mathbf{j} - 7\mathbf{k}$

In Exercises 15 and 16, find a unit vector in the direction of **u**.

15. $\mathbf{u} = 8\mathbf{i} + 3\mathbf{j} - \mathbf{k}$

16. $\mathbf{u} = -3\mathbf{i} + 5\mathbf{j} + 10\mathbf{k}$ P 204

In Exercises 17–20, find the dot product of **u** and **v**.

17. $\mathbf{u} = \langle 3, -3, 5 \rangle$
   $\mathbf{v} = \langle 0, 5, 3 \rangle$

18. $\mathbf{u} = \langle 4, 4, -1 \rangle$
   $\mathbf{v} = \langle 2, -5, -8 \rangle$

19. $\mathbf{u} = 4\mathbf{i} - 2\mathbf{j} + \mathbf{k}$
   $\mathbf{v} = \mathbf{i} + 3\mathbf{j} - \mathbf{k}$

20. $\mathbf{u} = 2\mathbf{i} + 5\mathbf{j} - 3\mathbf{k}$
   $\mathbf{v} = 9\mathbf{i} - 3\mathbf{j} + \mathbf{k}$

In Exercises 21–24, find the angle θ between the given vectors.

21. $\mathbf{u} = \langle 0, 2, 2 \rangle$
   $\mathbf{v} = \langle 3, 0, -4 \rangle$

22. $\mathbf{u} = \langle 4, 1, 2 \rangle$
   $\mathbf{v} = \langle 2, -4, 1 \rangle$

23. $\mathbf{u} = 10\mathbf{i} + 40\mathbf{j}$
   $\mathbf{v} = -3\mathbf{j} + 8\mathbf{k}$

24. $\mathbf{u} = \mathbf{i} - 6\mathbf{j} + 2\mathbf{k}$
   $\mathbf{v} = 2\mathbf{i} - 4\mathbf{j} - 3\mathbf{k}$

In Exercises 25–28, determine whether **u** and **v** are orthogonal, parallel, or neither.

25. $\mathbf{u} = \langle -12, 6, 15 \rangle$
   $\mathbf{v} = \langle 8, -4, -10 \rangle$

26. $\mathbf{u} = \langle 6, -3, 3 \rangle$
   $\mathbf{v} = \langle 1, 5, 3 \rangle$

27. $\mathbf{u} = \frac{3}{4}\mathbf{i} - \frac{1}{2}\mathbf{j} + 2\mathbf{k}$
   $\mathbf{v} = 4\mathbf{i} + 10\mathbf{j} + \mathbf{k}$

28. $\mathbf{u} = -7\mathbf{i} - 14\mathbf{j} + 21\mathbf{k}$
   $\mathbf{v} = \mathbf{i} + 2\mathbf{j} - \mathbf{k}$

In Exercises 29 and 30, find the projection $\mathbf{w}_1$ of **u** onto **v**. Then write **u** as the sum of $\mathbf{w}_1$ and $\mathbf{w}_2$, where $\mathbf{w}_2$ is orthogonal to $\mathbf{w}_1$.

29. $\mathbf{u} = \langle 2, 1, 0 \rangle$
   $\mathbf{v} = \langle 0, 8, 3 \rangle$

30. $\mathbf{u} = \langle -5, -1, 2 \rangle$
   $\mathbf{v} = \langle -1, 1, 1 \rangle$

In Exercises 31–34, use vectors to determine whether the given points lie in a straight line.

31. $(1, 3, 2)$
   $(-1, 2, 5)$
   $(3, 4, -1)$

32. $(0, 4, 4)$
   $(-1, 5, 6)$
   $(-2, 6, 7)$

33. $(5, 4, 1)$
   $(7, 3, -1)$
   $(4, 5, 3)$

34. $(-2, 7, 4)$
   $(-4, 8, 1)$
   $(0, 6, 7)$

In Exercises 35 and 36, the vector **v** and its initial point are given. Find the terminal point.

35. $\mathbf{v} = \langle 2, -4, 7 \rangle$
   Initial point: $(1, 5, 0)$

36. $\mathbf{v} = \langle 4, \frac{3}{2}, -\frac{1}{4} \rangle$
   Initial point: $\left(-2, 1, -\frac{3}{2}\right)$

In Exercises 37 and 38, write the component form of **v**.

37. $\|\mathbf{v}\| = 4$, **v** lies in the *yz*-plane and makes an angle of 45° with the positive *y*-axis.

38. $\|\mathbf{v}\| = 10$, **v** lies in the *xz*-plane and makes an angle of 60° with the positive *z*-axis.

39. The initial and terminal points of the vector **v** are (x_1, y_1, z_1) and (x, y, z), respectively. Describe the set of all points (x, y, z) such that $\|\mathbf{v}\| = 9$.

40. *Light Installation*   The lights in an auditorium are 30-pound disks with a radius of 24 inches. Each disk is supported by three equally spaced 60-inch wires from the ceiling (see figure). Find the tension in each wire.

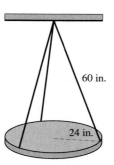

60 in.

24 in.

FIGURE FOR 40

Work   In Exercises 41 and 42, find the work done in moving a particle from P to Q if the force is given by **F**.

41. $P(0, 0, 0)$, $Q(10, 5, 4)$, $\mathbf{F} = \langle 3, 2, 7 \rangle$

42. $P(2, 4, 0)$, $Q(-4, 4, 10)$, $\mathbf{F} = -6\mathbf{i} + 2\mathbf{j} + 6\mathbf{k}$

4.7   THE CROSS PRODUCT OF TWO VECTORS

The Cross Product / Geometric Properties of the Cross Product / The
Triple Scalar Product

The Cross Product

Many applications in physics, engineering, and geometry involve finding a
vector in space that is orthogonal to two given vectors. In this section you will
study a product that will yield such a vector. It is called the **cross product,** and
it is most conveniently defined and calculated using the standard unit vector
form.

DEFINITION OF CROSS PRODUCT OF TWO VECTORS IN SPACE

Let $\mathbf{u} = u_1 \mathbf{i} + u_2 \mathbf{j} + u_3 \mathbf{k}$ and $\mathbf{v} = v_1 \mathbf{i} + v_2 \mathbf{j} + v_3 \mathbf{k}$ be vectors in
space. The **cross product** of $\mathbf{u}$ and $\mathbf{v}$ is the vector

$$\mathbf{u} \times \mathbf{v} = (u_2 v_3 - u_3 v_2)\, \mathbf{i} - (u_1 v_3 - u_3 v_1)\, \mathbf{j} + (u_1 v_2 - u_2 v_1)\, \mathbf{k}.$$

REMARK   Be sure you see that this definition applies only to three-dimensional vectors. The cross product is not defined for two-dimensional vectors.   ▪▪▪▪▪▪

A convenient way to calculate $\mathbf{u} \times \mathbf{v}$ is to use the following *determinant
form* with cofactor expansion. (This 3×3 determinant form is used simply to
help remember the formula for the cross product—it is technically not a
determinant, because the entries of the corresponding matrix are not all real
numbers.)

$$\mathbf{u} \times \mathbf{v} = \begin{vmatrix} \mathbf{i} & \mathbf{j} & \mathbf{k} \\ u_1 & u_2 & u_3 \\ v_1 & v_2 & v_3 \end{vmatrix} \quad \begin{array}{l} \leftarrow \text{Put } \mathbf{u} \text{ in Row 2.} \\ \leftarrow \text{Put } \mathbf{v} \text{ in Row 3.} \end{array}$$

$$= \begin{vmatrix} u_2 & u_3 \\ v_2 & v_3 \end{vmatrix} \mathbf{i} - \begin{vmatrix} u_1 & u_3 \\ v_1 & v_3 \end{vmatrix} \mathbf{j} + \begin{vmatrix} u_1 & u_2 \\ v_1 & v_2 \end{vmatrix} \mathbf{k}$$

$$= (u_2 v_3 - u_3 v_2)\, \mathbf{i} - (u_1 v_3 - u_3 v_1)\, \mathbf{j} + (u_1 v_2 - u_2 v_1)\, \mathbf{k}$$

Note the minus sign in front of the **j**-component.

Technology Note

Some graphing utilities have the capability to perform vector operations, such as the cross product. For example, on the TI-85, you can use the VECTR menu to verify the cross product in Example 1 as follows.

cross ([1, 2, 1], [3, 1, 2])

EXAMPLE 1   Finding Cross Products

Given $\mathbf{u} = \mathbf{i} + 2\mathbf{j} + \mathbf{k}$ and $\mathbf{v} = 3\mathbf{i} + \mathbf{j} + 2\mathbf{k}$, find the following.

A. $\mathbf{u} \times \mathbf{v}$ **B.** $\mathbf{v} \times \mathbf{u}$ **C.** $\mathbf{v} \times \mathbf{v}$

SOLUTION

A. $\mathbf{u} \times \mathbf{v} = \begin{vmatrix} \mathbf{i} & \mathbf{j} & \mathbf{k} \\ 1 & 2 & 1 \\ 3 & 1 & 2 \end{vmatrix} = \begin{vmatrix} 2 & 1 \\ 1 & 2 \end{vmatrix} \mathbf{i} - \begin{vmatrix} 1 & 1 \\ 3 & 2 \end{vmatrix} \mathbf{j} + \begin{vmatrix} 1 & 2 \\ 3 & 1 \end{vmatrix} \mathbf{k}$

$$= (4 - 1)\,\mathbf{i} - (2 - 3)\,\mathbf{j} + (1 - 6)\,\mathbf{k}$$
$$= 3\mathbf{i} + \mathbf{j} - 5\mathbf{k}$$

B. $\mathbf{v} \times \mathbf{u} = \begin{vmatrix} \mathbf{i} & \mathbf{j} & \mathbf{k} \\ 3 & 1 & 2 \\ 1 & 2 & 1 \end{vmatrix} = \begin{vmatrix} 1 & 2 \\ 2 & 1 \end{vmatrix} \mathbf{i} - \begin{vmatrix} 3 & 2 \\ 1 & 1 \end{vmatrix} \mathbf{j} + \begin{vmatrix} 3 & 1 \\ 1 & 2 \end{vmatrix} \mathbf{k}$

$$= (1 - 4)\,\mathbf{i} - (3 - 2)\,\mathbf{j} + (6 - 1)\,\mathbf{k}$$
$$= -3\mathbf{i} - \mathbf{j} + 5\mathbf{k}$$

Note that this result is the negative of that in part **a**.

C. $\mathbf{v} \times \mathbf{v} = \begin{vmatrix} \mathbf{i} & \mathbf{j} & \mathbf{k} \\ 3 & 1 & 2 \\ 3 & 1 & 2 \end{vmatrix} = \mathbf{0}$

The results obtained in Example 1 suggest some interesting *algebraic* properties of the cross product. For instance

$$\mathbf{u} \times \mathbf{v} = -(\mathbf{v} \times \mathbf{u}) \quad \text{and} \quad \mathbf{v} \times \mathbf{v} = \mathbf{0}.$$

These properties, and several others, are summarized in the following list.

ALGEBRAIC PROPERTIES OF THE CROSS PRODUCT

Let $\mathbf{u}$, $\mathbf{v}$, and $\mathbf{w}$ be vectors in space and let c be a scalar.

1. $\mathbf{u} \times \mathbf{v} = -(\mathbf{v} \times \mathbf{u})$
2. $\mathbf{u} \times (\mathbf{v} + \mathbf{w}) = (\mathbf{u} \times \mathbf{v}) + (\mathbf{u} \times \mathbf{w})$
3. $c(\mathbf{u} \times \mathbf{v}) = (c\,\mathbf{u}) \times \mathbf{v} = \mathbf{u} \times (c\,\mathbf{v})$
4. $\mathbf{u} \times \mathbf{0} = \mathbf{0} \times \mathbf{u} = \mathbf{0}$
5. $\mathbf{u} \times \mathbf{u} = \mathbf{0}$
6. $\mathbf{u} \cdot (\mathbf{v} \times \mathbf{w}) = (\mathbf{u} \times \mathbf{v}) \cdot \mathbf{w}$

Geometric Properties of the Cross Product

The first property listed on the previous page indicates that the cross product is *not commutative*. In particular, this property indicates that the vectors $\mathbf{u} \times \mathbf{v}$ and $\mathbf{v} \times \mathbf{u}$ have equal lengths but opposite directions. The following list gives some other *geometric* properties of the cross product of two vectors.

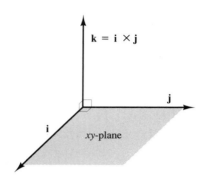

$\mathbf{k} = \mathbf{i} \times \mathbf{j}$

$\mathbf{j}$

$\mathbf{i}$ *xy*-plane

Right-handed Systems

FIGURE 4.56

GEOMETRIC PROPERTIES OF THE CROSS PRODUCT

Let $\mathbf{u}$ and $\mathbf{v}$ be nonzero vectors in space, and let θ be the angle between $\mathbf{u}$ and $\mathbf{v}$.

1. $\mathbf{u} \times \mathbf{v}$ is orthogonal to both $\mathbf{u}$ and $\mathbf{v}$.
2. $\|\mathbf{u} \times \mathbf{v}\| = \|\mathbf{u}\| \|\mathbf{v}\| \sin \theta$.
3. $\mathbf{u} \times \mathbf{v} = \mathbf{0}$ if and only if $\mathbf{u}$ and $\mathbf{v}$ are scalar multiples.
4. $\|\mathbf{u} \times \mathbf{v}\| =$ area of parallelogram having $\mathbf{u}$ and $\mathbf{v}$ as sides.

Both $\mathbf{u} \times \mathbf{v}$ and $\mathbf{v} \times \mathbf{u}$ are perpendicular to the plane determined by $\mathbf{u}$ and $\mathbf{v}$. One way to remember the orientation of the vectors $\mathbf{u}$, $\mathbf{v}$, and $\mathbf{u} \times \mathbf{v}$ is to compare them with the unit vectors $\mathbf{i}$, $\mathbf{j}$, and $\mathbf{k} = \mathbf{i} \times \mathbf{j}$, as shown in Figure 4.56. The three vectors $\mathbf{u}$, $\mathbf{v}$, and $\mathbf{u} \times \mathbf{v}$ form a *right-handed system*.

EXAMPLE 2   Using the Cross Product

Find a unit vector that is orthogonal to both

$$\mathbf{u} = 3\mathbf{i} - 4\mathbf{j} + \mathbf{k} \quad \text{and} \quad \mathbf{v} = -3\mathbf{i} + 6\mathbf{j}.$$

SOLUTION

The cross product $\mathbf{u} \times \mathbf{v}$, as shown in Figure 4.57, is orthogonal to both $\mathbf{u}$ and $\mathbf{v}$.

$$\mathbf{u} \times \mathbf{v} = \begin{vmatrix} \mathbf{i} & \mathbf{j} & \mathbf{k} \\ 3 & -4 & 1 \\ -3 & 6 & 0 \end{vmatrix} = -6\mathbf{i} - 3\mathbf{j} + 6\mathbf{k}$$

Because

$$\|\mathbf{u} \times \mathbf{v}\| = \sqrt{(-6)^2 + (-3)^2 + 6^2} = \sqrt{81} = 9,$$

a unit vector orthogonal to both $\mathbf{u}$ and $\mathbf{v}$ is

$$\frac{\mathbf{u} \times \mathbf{v}}{\|\mathbf{u} \times \mathbf{v}\|} = -\frac{2}{3}\mathbf{i} - \frac{1}{3}\mathbf{j} + \frac{2}{3}\mathbf{k}.$$

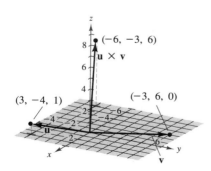

$(-6, -3, 6)$

$\mathbf{u} \times \mathbf{v}$

$(3, -4, 1)$ $(-3, 6, 0)$

$\mathbf{u}$

$\mathbf{v}$

FIGURE 4.57

In Example 2, note that you could have used the cross product $\mathbf{v} \times \mathbf{u}$ to form a unit vector that is orthogonal to both $\mathbf{u}$ and $\mathbf{v}$. With that choice, you would have obtained the negative of the unit vector found in the example.

The fourth geometric property of cross products states that $\|\mathbf{u} \times \mathbf{v}\|$ is the area of the parallelogram that has $\mathbf{u}$ and $\mathbf{v}$ as adjacent sides. A simple example of this is given by the unit square with adjacent sides of $\mathbf{i}$ and $\mathbf{j}$. Because

$$\mathbf{i} \times \mathbf{j} = \mathbf{k}$$

and $\|\mathbf{k}\| = 1$, it follows that the square has an area of 1. This geometric property of the cross product is illustrated further in the next example.

EXAMPLE 3   Geometric Application of the Cross Product

Show that the quadrilateral with vertices at the following points is a parallelogram. Then find the area of the parallelogram.

$$A = (5, 2, 0) \qquad B = (2, 6, 1)$$
$$C = (2, 4, 7) \qquad D = (5, 0, 6)$$

SOLUTION

From Figure 4.58 you can see that the sides of the quadrilateral correspond to the following four vectors.

$$\overrightarrow{AB} = -3\mathbf{i} + 4\mathbf{j} + \mathbf{k}$$
$$\overrightarrow{CD} = 3\mathbf{i} - 4\mathbf{j} - \mathbf{k} = -\overrightarrow{AB}$$
$$\overrightarrow{AD} = 0\mathbf{i} - 2\mathbf{j} + 6\mathbf{k}$$
$$\overrightarrow{CB} = 0\mathbf{i} + 2\mathbf{j} - 6\mathbf{k} = -\overrightarrow{AD}$$

Because $\overrightarrow{AB}$ is parallel to $\overrightarrow{CD}$ and $\overrightarrow{AD}$ is parallel to $\overrightarrow{CB}$, it follows that the quadrilateral is a parallelogram with $\overrightarrow{AB}$ and $\overrightarrow{AD}$ as adjacent sides. Moreover, because

$$\overrightarrow{AB} \times \overrightarrow{AD} = \begin{vmatrix} \mathbf{i} & \mathbf{j} & \mathbf{k} \\ -3 & 4 & 1 \\ 0 & -2 & 6 \end{vmatrix} = 26\mathbf{i} + 18\mathbf{j} + 6\mathbf{k},$$

the area of the parallelogram is

$$\|\overrightarrow{AB} \times \overrightarrow{AD}\| = \sqrt{1036} \approx 32.19.$$

Is the parallelogram a rectangle? You can tell whether it is by finding the angle between the vectors $\overrightarrow{AB}$ and $\overrightarrow{AD}$.

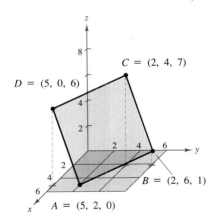

FIGURE 4.58

The Triple Scalar Product

For vectors **u**, **v**, and **w** in space, the dot product of **u** and **v** $\times$ **w** is called the **triple scalar product** of **u**, **v**, and **w**.

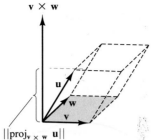

v $\times$ w

u

w

v

$\|\text{proj}_{\mathbf{v} \times \mathbf{w}} \mathbf{u}\|$

Area of base = $\|\mathbf{v} \times \mathbf{w}\|$
Volume of
parallelepiped = $|\mathbf{u} \cdot (\mathbf{v} \times \mathbf{w})|$

FIGURE 4.59

THE TRIPLE SCALAR PRODUCT

The **triple scalar product** of **u**, **v**, and **w** is given by

$$\mathbf{u} \cdot (\mathbf{v} \times \mathbf{w}) = \begin{vmatrix} u_1 & u_2 & u_3 \\ v_1 & v_2 & v_3 \\ w_1 & w_2 & w_3 \end{vmatrix}.$$

If the vectors **u**, **v**, and **w** do not lie in the same plane, then the triple scalar product **u** $\cdot$ (**v** $\times$ **w**) can be used to determine the volume of the parallelepiped with **u**, **v**, and **w** as adjacent edges, as shown in Figure 4.59.

GEOMETRIC PROPERTY OF TRIPLE SCALAR PRODUCT

The volume V of a parallelepiped with vectors **u**, **v**, and **w** as adjacent edges is given by

$$V = |\mathbf{u} \cdot (\mathbf{v} \times \mathbf{w})|.$$

EXAMPLE 4   Volume by the Triple Scalar Product

Find the volume of the parallelepiped having $\mathbf{u} = 3\mathbf{i} - 5\mathbf{j} + \mathbf{k}$, $\mathbf{v} = 2\mathbf{j} - 2\mathbf{k}$, and $\mathbf{w} = 3\mathbf{i} + \mathbf{j} + \mathbf{k}$ as adjacent edges, as shown in Figure 4.60.

SOLUTION

The volume of the parallelepiped is

$$V = |\mathbf{u} \cdot (\mathbf{v} \times \mathbf{w})|$$

$$= \begin{vmatrix} 3 & -5 & 1 \\ 0 & 2 & -2 \\ 3 & 1 & 1 \end{vmatrix}$$

$$= 3 \begin{vmatrix} 2 & -2 \\ 1 & 1 \end{vmatrix} - (-5) \begin{vmatrix} 0 & -2 \\ 3 & 1 \end{vmatrix} + (1) \begin{vmatrix} 0 & 2 \\ 3 & 1 \end{vmatrix}$$

$$= 3(4) + 5(6) + 1(-6)$$

$$= 36.$$

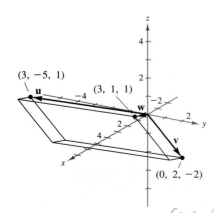

$(3, -5, 1)$
u
$(3, 1, 1)$
w
v
$(0, 2, -2)$

FIGURE 4.60

DISCUSSION PROBLEM

COPLANAR VECTORS

Consider three vectors **u**, **v**, and **w** with the same initial point. It can be shown that they are coplanar if and only if

$$\mathbf{u} \cdot (\mathbf{v} \times \mathbf{w}) = \begin{vmatrix} u_1 & u_2 & u_3 \\ v_1 & v_2 & v_3 \\ w_1 & w_2 & w_3 \end{vmatrix} = 0.$$

Explain why this test for coplanar vectors is valid.

WARM-UP

The following warm-up exercises involve skills that were covered in earlier sections. You will use these skills in the exercise set for this section.

In Exercises 1–4, find (a) $\mathbf{u} \cdot \mathbf{v}$, (b) $\|\mathbf{u}\|^2$, and (c) $(\mathbf{u} \cdot \mathbf{v})\,\mathbf{v}$.

1. $\mathbf{u} = \langle 0, 8 \rangle$
$\mathbf{v} = \langle 6, 8 \rangle$

2. $\mathbf{u} = \langle 25, 15 \rangle$
$\mathbf{v} = \langle -3, 5 \rangle$

3. $\mathbf{u} = \langle 12, -6, 16 \rangle$
$\mathbf{v} = \langle 9, -\frac{9}{2}, 12 \rangle$

4. $\mathbf{u} = 4\mathbf{i} + 3\mathbf{k}$
$\mathbf{v} = -2\mathbf{i} + 6\mathbf{j}$

In Exercises 5–7, find any values of the constant k so the vectors **u** and **v** are orthogonal.

5. $\mathbf{u} = \langle 15, k \rangle$
$\mathbf{v} = \langle 6, 12 \rangle$

6. $\mathbf{u} = \langle k, 9 \rangle$
$\mathbf{v} = \langle -4, 8 \rangle$

7. $\mathbf{u} = \langle -2, k, 7 \rangle$
$\mathbf{v} = \langle 8, k, 1 \rangle$

In Exercises 8–10, evaluate the determinant.

8. $\begin{vmatrix} 10 & 4 \\ -6 & 2 \end{vmatrix}$

9. $\begin{vmatrix} 4 & -7 \\ -2 & 1 \end{vmatrix}$

10. $\begin{vmatrix} 8 & 0 & 2 \\ -3 & -2 & 1 \\ 0 & 4 & 1 \end{vmatrix}$

SECTION 4.7 · EXERCISES

In Exercises 1–4, find the cross product of the given unit vectors and sketch your result.

1. $\mathbf{i} \times \mathbf{j}$

2. $\mathbf{j} \times \mathbf{k}$

3. $\mathbf{i} \times \mathbf{k}$

4. $\mathbf{k} \times \mathbf{i}$

In Exercises 5–16, find $\mathbf{u} \times \mathbf{v}$ and show that it is orthogonal to both **u** and **v**.

5. $\mathbf{u} = \langle 1, -4, 0 \rangle$
$\mathbf{v} = \langle 2, 6, 0 \rangle$

6. $\mathbf{u} = \langle -3, 2, 3 \rangle$
$\mathbf{v} = \langle 0, 1, 0 \rangle$

7. $\mathbf{u} = \langle 7, -5, 2 \rangle$
$\mathbf{v} = \langle -1, 4, -1 \rangle$

8. $\mathbf{u} = \langle -5, 5, 11 \rangle$
$\mathbf{v} = \langle 2, 2, 3 \rangle$

9. $\mathbf{u} = \langle 2, 4, 3 \rangle$
$\mathbf{v} = \langle 0, -2, 1 \rangle$

10. $\mathbf{u} = \langle 4, -2, 6 \rangle$
$\mathbf{v} = \langle -1, 5, 7 \rangle$

11. $\mathbf{u} = 6\mathbf{i} + 2\mathbf{j} + \mathbf{k}$
$\mathbf{v} = \mathbf{i} + 3\mathbf{j} - 2\mathbf{k}$

12. $\mathbf{u} = 6\mathbf{k}$
$\mathbf{v} = -\mathbf{i} + 3\mathbf{j} + \mathbf{k}$

13. $\mathbf{u} = \mathbf{i} + \frac{3}{2}\mathbf{j} - \frac{5}{2}\mathbf{k}$
$\mathbf{v} = \frac{1}{2}\mathbf{i} - \frac{3}{4}\mathbf{j} + \frac{1}{4}\mathbf{k}$

14. $\mathbf{u} = \frac{2}{3}\mathbf{i}$
$\mathbf{v} = \frac{1}{3}\mathbf{i} + 3\mathbf{k}$

15. $\mathbf{u} = 6\mathbf{i} - 5\mathbf{j} + \mathbf{k}$
$\mathbf{v} = \frac{1}{3}\mathbf{i} - \frac{1}{3}\mathbf{j} + \frac{2}{3}\mathbf{k}$

16. $\mathbf{u} = -\mathbf{i} + \mathbf{k}$
$\mathbf{v} = \mathbf{j} - 2\mathbf{k}$

In Exercises 17–22, find a unit vector orthogonal to **u** and **v**.

17. **u** = 3**i** + **j**
   v = **j** + **k**

18. **u** = **i** + 2**j**
   v = **i** − 3**j**

19. **u** = −2**i** + **j** + 3**k**
   v = **i** + 4**j** + 6**k**

20. **u** = 7**i** − 14**j** + 5**k**
   v = 14**i** + 28**j** − 15**k**

21. **u** = **i** + **j** − **k**
   v = **i** + **j** + **k**

22. **u** = **i** − 2**j** + 2**k**
   v = 2**i** − **j** − 2**k**

In Exercises 23–28, find the area of the parallelogram that has the vectors as adjacent sides.

23. **u** = **k**
   v = **i** + **k**

24. **u** = **i** + 2**j** + 2**k**
   v = **i** + **k**

25. **u** = 3**i** + 4**j** + 6**k**
   v = 2**i** − **j** + 5**k**

26. **u** = ⟨−2, 3, 2⟩
   v = ⟨1, 2, 4⟩

27. **u** = ⟨2, 2, −3⟩
   v = ⟨0, 2, 3⟩

28. **u** = ⟨4, −3, 2⟩
   v = ⟨5, 0, 1⟩

In Exercises 29 and 30, verify that the points are the vertices of a parallelogram and find its area.

29. $A(2, -1, 4)$, $B(3, 1, 2)$, $C(0, 5, 6)$, $D(-1, 3, 8)$

30. $A(3, 5, 0)$, $B(-1, 8, 5)$, $C(1, 3, 11)$, $D(5, 0, 6)$

In Exercises 31–34, find the area of the triangle with the given vertices. (The area of the triangle having **u** and **v** as adjacent sides is $\frac{1}{2}\|\mathbf{u} \times \mathbf{v}\|$.)

31. $(0, 0, 0)$, $(4, -2, 6)$, $(-4, 0, 3)$

32. $(1, -4, 3)$, $(2, 0, 2)$, $(-2, 2, 0)$

33. $(2, 3, -5)$, $(-2, -2, 0)$, $(3, 0, 6)$

34. $(2, 4, 0)$, $(-2, -4, 0)$, $(0, 0, 4)$

In Exercises 35 and 36, find **u** · (**v** × **w**).

35. **u** = ⟨2, 3, 3⟩, **v̇** = ⟨4, 4, 0⟩, **w** = ⟨0, 0, 4⟩

36. **u** = ⟨20, 10, 10⟩, **v** = ⟨1, 4, 4⟩, **w** = ⟨0, 2, 2⟩

In Exercises 37 and 38, use the triple scalar product to find the volume of the parallelepiped having adjacent edges **u**, **v**, and **w**.

37. **u** = 2**i**, **v** = 2**j**, **w** = 2**i** + 2**j** + **k**

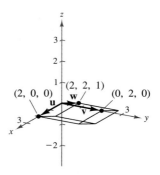

FIGURE FOR 37

38. **u** = ⟨1, 1, 3⟩, **v** = ⟨0, 3, 3⟩,
   w = ⟨3, 0, 3⟩

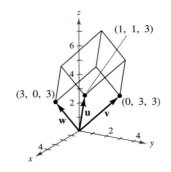

FIGURE FOR 38

In Exercises 39 and 40, find the volume of the parallelepiped with the given vertices.

39. $(0, 0, 0)$, $(4, 0, 0)$, $(4, -2, 3)$, $(0, -2, 3)$, $(4, 5, 3)$, $(0, 5, 3)$, $(0, 3, 6)$, $(4, 3, 6)$

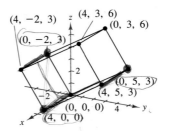

FIGURE FOR 39

40. $(3, 0, 0), (4, 1, 2), (3, -1, 4), (2, -2, 2), (-1, 5, 4), (0, 6, 6),$
$(-1, 4, 8), (-2, 3, 6)$

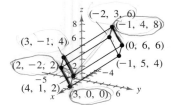

FIGURE FOR 40

In Exercises 41 and 42, prove the property of the cross product where $\mathbf{u} = \langle u_1, u_2, u_3 \rangle$ and $\mathbf{v} = \langle v_1, v_2, v_3 \rangle$.

41. $\mathbf{u} \times \mathbf{u} = \mathbf{0}$

42. $\mathbf{u} \times \mathbf{v}$ is orthogonal to both $\mathbf{u}$ and $\mathbf{v}$.

43. Consider the vectors $\mathbf{u} = \langle \cos \beta, \sin \beta, 0 \rangle$ and $\mathbf{v} = \langle \cos \alpha, \sin \alpha, 0 \rangle$ where $\alpha > \beta$. Find the cross product of the vectors and use the result to prove the identity
$$\sin(\alpha - \beta) = \sin \alpha \cos \beta - \cos \alpha \sin \beta.$$

CHAPTER 4 · REVIEW EXERCISES

In Exercises 1–16, solve the triangle (if possible). If two solutions exist, list both.

1. $a = 5, b = 8, c = 10$

2. $a = 6, b = 9, C = 45°$

3. $A = 12°, B = 58°, a = 5$

4. $B = 110°, C = 30°, c = 10.5$

5. $B = 110°, a = 4, c = 4$

6. $a = 80, b = 60, c = 100$

7. $A = 75°\ a = 2.5, b = 16.5$

8. $A = 130°, a = 50, b = 30$

9. $B = 115°, a = 7, b = 14.5$

10. $C = 50°, a = 25, c = 22$

11. $A = 15°, a = 5, b = 10$

12. $B = 150°, a = 64, b = 10$

13. $B = 150°, a = 10, c = 20$

14. $a = 2.5, b = 15.0, c = 4.5$

15. $B = 25°, a = 6.2, b = 4$

16. $B = 90°, a = 5, c = 12$

In Exercises 17–20, find the area of the triangle.

17. $a = 4, b = 5, c = 7$ **18.** $a = 15, b = 8, c = 10$

19. $A = 27°, b = 5, c = 8$ **20.** $B = 80°, a = 4, c = 8$

21. *Height of a Tree*   Find the height of a tree that stands on a hillside of slope $32°$ (from the horizontal) if from a point 75 feet down the hill from the tree the angle of elevation to the top of the tree is $48°$ (see figure)

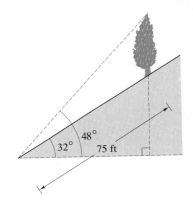

FIGURE FOR 21

22. *Surveying*   To approximate the length of a marsh, a surveyor walks 450 meters from point A to point B. Then the surveyor turns $65°$ and walks 325 meters to point C. Approximate the length AC of the marsh (see figure).

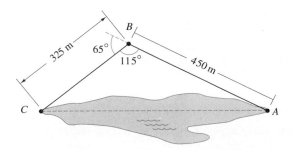

FIGURE FOR 22

23. *Height* From a certain distance, the angle of elevation to the top of a building is 17°. At a point 50 meters closer to the building, the angle of elevation is 31°. Approximate the height of the building.

24. *River Width* Determine the width of a river that flows due east, if a tree on the opposite bank has a bearing of N 22° 30′ E and, after walking 400 feet downstream, a surveyor finds the tree has a bearing of N 15° W.

25. *Navigation* Two planes leave an airport at approximately the same time. One is flying at 425 miles per hour at a bearing of N 5° W, and the other is flying at 530 miles per hour at a bearing of N 67° E (see figure). How far apart are the planes after flying for 2 hours?

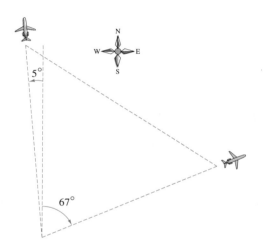

FIGURE FOR 25

26. *Geometry* The lengths of the diagonals of a parallelogram are 10 feet and 16 feet. Find the lengths of the sides of the parallelogram if the diagonals intersect at an angle of 28°.

In Exercises 27 and 28, find the component forms of the vectors $\overrightarrow{AB}$, $\overrightarrow{AC}$, and $\overrightarrow{AB} + \overrightarrow{AC}$. Sketch the three vectors.

27. $A(2, 1)$, $B(5, 0)$, $C(-2, 0)$

28. $A(4, 6)$, $B(0, 5)$, $C(-2, 0)$

In Exercises 29–32, find the component form of the vector **v** satisfying the given conditions.

29. Initial point: $(0, 10)$
Terminal point: $(7, 3)$

30. Initial point: $(1, 5)$
Terminal point; $(15, 9)$

31. $\|\mathbf{v}\| = 8$, $\theta = 120°$

32. $\|\mathbf{v}\| = \frac{1}{2}$, $\theta = 225°$

In Exercises 33 and 34, sketch **v** and find its component form. (Assume θ is measured counterclockwise from the x-axis to the vector.)

33. $\|\mathbf{v}\| = 3$, $\theta = 135°$ **34.** $\|\mathbf{v}\| = 5$, $\theta = 300°$

In Exercises 35–38 find the component form of the specified vector and sketch its graph, given that $\mathbf{u} = 6\mathbf{i} - 5\mathbf{j}$ and $\mathbf{v} = 10\mathbf{i} + 3\mathbf{j}$.

35. $\dfrac{1}{\|\mathbf{u}\|}\mathbf{u}$ **36.** $3\mathbf{v}$

37. $4\mathbf{u} - 5\mathbf{v}$ **38.** $\frac{1}{2}\mathbf{v}$

In Exercises 39 and 40, (a) $\mathbf{u} + \mathbf{v}$, (b) $\mathbf{u} - \mathbf{v}$, and (c) $2\mathbf{u} - 3\mathbf{v}$.

39. $\mathbf{u} = \frac{7}{2}\mathbf{i} - \mathbf{j}$, $\mathbf{v} = -\mathbf{i} + 2\mathbf{j}$ **40.** $\mathbf{u} = -3\mathbf{i} + 5\mathbf{j}$, $\mathbf{v} = 10\mathbf{i}$

In Exercises 41 and 42, find the length and direction angle of the vector **v**.

41. $\mathbf{v} = 2\mathbf{i} - 2\sqrt{3}\,\mathbf{j}$ **42.** $\mathbf{v} = -3\mathbf{i} - 3\mathbf{j}$

43. *Resultant Force* Find the direction and magnitude of the resultant of the three forces shown in the figure.

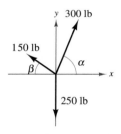

$\tan \beta = \frac{3}{4}$ $\tan \alpha = \frac{12}{5}$

FIGURE FOR 43

44. *Resultant Force* Forces with magnitudes of 85 pounds and 50 pounds act on a single point. Find the magnitude of the resultant if the angle between the forces is 15°.

45. *Rope Tension* A 100-pound weight is supported by two ropes, as shown in the figure. Find the tension in each rope.

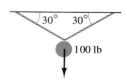

FIGURE FOR 45

46. *Cable Tension* In a manufacturing process, an electric hoist lifts 200-pound ingots (see figure). Find the tension in the supporting cables.

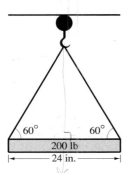

60°   60°
200 lb
24 in.

FIGURE FOR 46

47. *Navigation* An airplane has an air speed of 450 miles per hour at a bearing of N 30° E. If the wind velocity is 20 miles per hour from the west, find the ground speed and the direction of the plane.

48. *Braking Force* A 500-pound motorcycle is headed up a hill inclined at 12°. What force is required to keep the motorcycle from rolling back down the hill when stopped at a red light?

In Exercises 49 and 50, find the standard form of the equation of the sphere that has the given points as the endpoints of a diameter.

49. $(0, 0, 4)$, $(4, 6, 0)$ **50.** $(3, 4, 2)$, $(5, 8, 1)$

In Exercises 51 and 52, find a unit vector in the direction of $\overrightarrow{PQ}$.

51. $P(7, -4, 3)$ **52.** $P(0, 3, -1)$
   $Q(3, 2, 10)$ $Q(5, -8, 6)$

In Exercises 53 and 54, determine if the vectors are orthogonal, parallel, or neither.

53. $\langle 39, -12, 21 \rangle$ **54.** $\langle 8, 5, -8 \rangle$
   $\langle -26, 8, -14 \rangle$ $\langle -2, 4, \frac{1}{2} \rangle$

In Exercises 55–58, find the angle θ between the vectors **u** and **v**.

55. $\mathbf{u} = 32\left[\cos\left(\dfrac{7\pi}{4}\right) \mathbf{i} + \sin\left(\dfrac{7\pi}{4}\right) \mathbf{j} \right]$

   $\mathbf{v} = 14\left[\cos\left(\dfrac{5\pi}{6}\right) \mathbf{i} + \sin\left(\dfrac{5\pi}{6}\right) \mathbf{j} \right]$

56. $\mathbf{u} = \langle -6, -3, 18 \rangle$
   $\mathbf{v} = \langle 4, 2, -12 \rangle$

57. $\mathbf{v} = \langle 2\sqrt{2}, 1, 2, \rangle$ **58.** $\mathbf{v} = \langle 4, 5, 2 \rangle$

In Exercises 59–62, find $\text{proj}_\mathbf{v}\, \mathbf{u}$.

59. $\mathbf{u} = \langle -4, 3 \rangle$, $\mathbf{v} = \langle -8, -2 \rangle$
60. $\mathbf{u} = \langle 5, 6 \rangle$, $\mathbf{v} = \langle 10, 0 \rangle$
61. $\mathbf{u} = \langle 2, 7, 4 \rangle$, $\mathbf{v} = \langle 1, -1, 0 \rangle$
62. $\mathbf{u} = \langle -3, 5, 1 \rangle$, $\mathbf{v} = \langle -5, 2, 6 \rangle$

In Exercises 63 and 64, find $\mathbf{u} \times \mathbf{v}$.

63. $\mathbf{u} = \langle -2, 8, 2 \rangle$, $\mathbf{v} = \langle 1, 1, -1 \rangle$
64. $\mathbf{u} = \langle 20, 15, 5 \rangle$, $\mathbf{v} = \langle 5, -3, 0 \rangle$

In Exercises 65–69, let $\mathbf{u} = \langle 1, -2, -1 \rangle$, $\mathbf{v} = \langle 2, 4, 0 \rangle$, and $\mathbf{w} = \langle 3, 4, 5 \rangle$.

65. Show that $\mathbf{u} \cdot \mathbf{u} = \|\mathbf{u}\|^2$.
66. Determine a unit vector perpendicular to the vectors **v** and **w**.
67. Show that $\mathbf{u} \times \mathbf{v} = -(\mathbf{v} \times \mathbf{u})$.
68. Show that $\mathbf{u} \times (\mathbf{v} + \mathbf{w}) = (\mathbf{u} \times \mathbf{v}) + (\mathbf{u} \times \mathbf{w})$.
69. *Volume* Find the volume of the solid whose edges are **u**, **v**, and **w**.
70. *Volume* Use the triple scalar product to find the volume of the parallelepiped with vertices $(3, 2, -1)$, $(2, 5, 1)$, $(3, 2, 8)$, $(4, -1, 6)$, $(-5, 4, 2)$, $(-6, 7, 4)$, $(-5, 4, 11)$, and $(-4, 1, 9)$.
71. *Machine Design* A component used to channel grain into the end of an elevator has the shape and dimensions shown in the figure. In fabricating the part, it is necessary to know the angle θ between the adjacent sides. Find the angle.

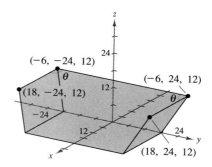

FIGURE FOR 71

COMPLEX NUMBERS

5.1 COMPLEX NUMBERS

The Imaginary Unit i / Operations with Complex Numbers / Complex Conjugates and Division / Applications

The Imaginary Unit i

Some quadratic equations have no real solutions. For instance, the quadratic equation $x^2 + 1 = 0$ has no real solution because there is no real number x that can be squared to produce -1. To overcome this deficiency, mathematicians created an expanded system of numbers using the **imaginary unit i,** defined as

$$i = \sqrt{-1},$$

where $i^2 = -1$. Adding real numbers to real multiples of this imaginary unit produces the set of **complex numbers.** Each complex number can be written in the **standard form,** $a + bi$.

DEFINITION OF A COMPLEX NUMBER

For real numbers a and b, the number

$a + bi$

is a **complex number.** If $a = 0$ and $b \neq 0$, then the complex number bi is an **imaginary number.**

The set of real numbers is a subset of the set of complex numbers because every real number a can be written as a complex number using $b = 0$. That is, for every real number a, you can write $a = a + 0i$.

Two complex numbers $a + bi$ and $c + di$, written in standard form, are **equal** to each other

$$a + bi = c + di \qquad \text{\textit{Equality of two complex numbers}}$$

if and only if $a = c$ and $b = d$.

Operations with Complex Numbers

To add (or subtract) two complex numbers, you add (or subtract) the real and imaginary parts of the numbers separately.

ADDITION AND SUBTRACTION OF COMPLEX NUMBERS

If $a + bi$ and $c + di$ are two complex numbers written in standard form, then their sum and difference are defined as follows.

$$\textit{Sum:} \qquad (a + bi) + (c + di) = (a + c) + (b + d)i$$
$$\textit{Difference:} \ (a + bi) - (c + di) = (a - c) + (b - d)i$$

The **additive identity** in the complex number system is zero (the same as in the real number system). Furthermore, the **additive inverse** of the complex number $a + bi$ is

$$-(a + bi) = -a - bi. \qquad \text{\textit{Additive inverse}}$$

Thus, $(a + bi) + (-a - bi) = 0 + 0i = 0$.

EXAMPLE 1   Adding and Subtracting Complex Numbers

Write the sums and differences in standard form.

A. $(3 - i) + (2 + 3i)$
B. $2i + (-4 - 2i)$
C. $3 - (-2 + 3i) + (-5 + i)$

SOLUTION

A.
$$\begin{aligned}
(3 - i) + (2 + 3i) &= 3 - i + 2 + 3i & \text{\textit{Remove parentheses}} \\
&= 3 + 2 - i + 3i & \text{\textit{Group real and imaginary terms}} \\
&= (3 + 2) + (-1 + 3)i \\
&= 5 + 2i & \text{\textit{Standard form}}
\end{aligned}$$

B.
$$\begin{aligned}
2i + (-4 - 2i) &= 2i - 4 - 2i & \text{\textit{Remove parentheses}} \\
&= -4 + 2i - 2i & \text{\textit{Group real and imaginary terms}} \\
&= -4 & \text{\textit{Standard form}}
\end{aligned}$$

C.
$$\begin{aligned}
3 - (-2 + 3i) + (-5 + i) &= 3 + 2 - 3i - 5 + i \\
&= 3 + 2 - 5 - 3i + i \\
&= 0 - 2i \\
&= -2i
\end{aligned}$$

Technology Note

Some graphing utilities can perform operations with complex numbers. A complex number $a + bi$ is usually expressed as an ordered pair (a, b). For example, on the TI-85 or the HP 48G, the complex number $2 + 3i$ would be entered as $(2, 3)$. You can verify the result of Example 1 (A) by evaluating the expression

$(3, -1) + (2, 3).$

Your result should be the ordered pair $(5, 2)$.

REMARK   Note in Example 1 (B) that the sum of two complex numbers can be a real number.

REMARK   When performing operations with complex numbers, it is sometimes necessary to evaluate powers of the imaginary unit i. The first several powers of i are as follows.

$i^1 = i$

$i^2 = -1$

$i^3 = i(i^2) = i(-1) = -i$

$i^4 = (i^2)(i^2) = (-1)(-1) = 1$

$i^5 = i(i^4) = i(1) = i$

$i^6 = (i^2)(i^4) = (-1)(1) = -1$

$i^7 = (i^3)(i^4) = (-i)(1) = -i$

$i^8 = (i^4)(i^4) = (1)(1) = 1$

Note how the pattern of values i, -1, $-i$, and 1 repeats for powers greater than 4.

Many of the properties of real numbers are valid for complex numbers as well. Here are some.

Associative Property of Addition and Multiplication
Commutative Property of Addition and Multiplication
Distributive Property of Multiplication over Addition

Notice how these properties are used when two complex numbers are multiplied.

$$
\begin{aligned}
(a + bi)(c + di) &= a(c + di) + bi(c + di) &&\textit{Distributive}\\
&= ac + (ad)i + (bc)i + (bd)i^2 &&\textit{Distributive}\\
&= ac + (ad)i + (bc)i + (bd)(-1) &&i^2 = -1\\
&= ac - bd + (ad)i + (bc)i &&\textit{Commutative}\\
&= (ac - bd) + (ad + bc)i &&\textit{Distributive}
\end{aligned}
$$

Rather than trying to memorize this multiplication rule, just remember how the distributive property is used to multiply two complex numbers. The procedure is similar to multiplying two polynomials and combining like terms (as in the FOIL method).

EXAMPLE 2   Multiplying Complex Numbers

Find the products.

A. $(i)(-3i)$ **B.** $(2 - i)(4 + 3i)$ **C.** $(3 + 2i)(3 - 2i)$

SOLUTION

A. $(i)(-3i) = -3i^2 = -3(-1) = 3$

B.
$$
\begin{aligned}
(2 - i)(4 + 3i) &= 8 + 6i - 4i - 3i^2 &&\textit{Binomial product}\\
&= 8 + 6i - 4i - 3(-1) &&i^2 = -1\\
&= 8 + 3 + 6i - 4i &&\textit{Collect terms}\\
&= 11 + 2i &&\textit{Standard form}
\end{aligned}
$$

C.
$$
\begin{aligned}
(3 + 2i)(3 - 2i) &= 9 - 6i + 6i - 4i^2\\
&= 9 - 4(-1)\\
&= 9 + 4\\
&= 13
\end{aligned}
$$

Complex Conjugates and Division

Notice in Example 2 (C) that the product of two complex numbers can be a real number. This occurs with pairs of complex numbers of the form $a + bi$ and $a - bi$, called **complex conjugates.** In general, the product of two complex conjugates can be written as follows.

$$(a + bi)(a - bi) = a^2 - abi + abi - b^2i^2$$
$$= a^2 - b^2(-1)$$
$$= a^2 + b^2$$

Complex conjugates can be used to divide one complex number by another. That is, to find the quotient

$$\frac{a + bi}{c + di}, \qquad c \text{ and } d \text{ not both zero,}$$

multiply the numerator and denominator by the conjugate of the denominator to obtain

$$\frac{a + bi}{c + di} = \frac{a + bi}{c + di}\left(\frac{c - di}{c - di}\right)$$
$$= \frac{(ac + bd) + (bc - ad)i}{c^2 + d^2}.$$

This procedure is demonstrated in Examples 3 and 4.

EXAMPLE 3   Dividing Complex Numbers

Write the complex number $\dfrac{1}{1 + i}$ in standard form.

SOLUTION

$$\frac{1}{1 + i} = \frac{1}{1 + i}\left(\frac{1 - i}{1 - i}\right)$$
$$= \frac{1 - i}{1^2 - i^2}$$
$$= \frac{1 - i}{1 - (-1)}$$
$$= \frac{1 - i}{2}$$
$$= \frac{1}{2} - \frac{1}{2}i$$

Technology Note

If your graphing utility can perform operations with complex numbers, try verifying Example 3. On a TI-85 you could evaluate the expression

$(1, 1)^{-1}$

whereas on the HP 48G you could enter the ordered pair $(1, 1)$ onto the stack and then press the $1/x$ key.

In both cases, the answer should be $(0.5, -0.5)$.

EXAMPLE 4　Dividing Complex Numbers
·················

Write the complex number $\dfrac{2 + 3i}{4 - 2i}$ in standard form.

SOLUTION

$$
\begin{aligned}
\frac{2 + 3i}{4 - 2i} &= \frac{2 + 3i}{4 - 2i}\left(\frac{4 + 2i}{4 + 2i}\right) \\
&= \frac{8 + 4i + 12i + 6i^2}{16 - 4i^2} \\
&= \frac{8 - 6 + 16i}{16 + 4} \\
&= \frac{1}{20}(2 + 16i) \\
&= \frac{1}{10} + \frac{4}{5}i
\end{aligned}
$$

Using the Quadratic Formula to solve a quadratic equation can produce a result such as $\sqrt{-3}$, which is not a real number. By factoring out $i = \sqrt{-1}$, you can write this number in standard form.

$$\sqrt{-3} = \sqrt{3(-1)} = \sqrt{3}\sqrt{-1} = \sqrt{3}\,i$$

The number $\sqrt{3}\,i$ is the principal square root of -3.

PRINCIPAL SQUARE ROOT OF A NEGATIVE NUMBER

If a is a positive number, then the **principal square root** of the negative number $-a$ is defined as

$$\sqrt{-a} = \sqrt{a}\,i.$$

Technology Note

If your graphing utility permits complex arithmetic, you can verify that $\sqrt{-5}\sqrt{-5} \neq \sqrt{(-5)(-5)}$. On the TI-85, you would have $\sqrt{(-5, 0)}\sqrt{(-5, 0)} = (-5, 0)$ and $\sqrt{(-5, 0)(-5, 0)} = (5, 0)$.

In this definition you are using the rule $\sqrt{ab} = \sqrt{a}\sqrt{b}$, for $a > 0$ and $b < 0$. This rule is not valid if *both* a and b are negative. For example,

$$\sqrt{-5}\sqrt{-5} = \left(\sqrt{5}\,i\right)\left(\sqrt{5}\,i\right) = 5i^2 = -5$$

whereas

$$\sqrt{(-5)(-5)} = \sqrt{25} = 5.$$

Consequently, $\sqrt{(-5)(-5)} \neq \sqrt{-5}\sqrt{-5}$.

REMARK　When working with square roots of negative numbers, be sure to convert to standard form *before* multiplying.

EXAMPLE 5   **Writing Complex Numbers in Standard Form**

A. $\sqrt{-3}\sqrt{-12} = \sqrt{3}\,i\,\sqrt{12}\,i = \sqrt{36}\,i^2 = 6(-1) = -6$

B. $\sqrt{-48} - \sqrt{-27} = \sqrt{48}\,i - \sqrt{27}\,i = 4\sqrt{3}\,i - 3\sqrt{3}\,i = \sqrt{3}\,i$

C. $(-1 + \sqrt{-3})^2 = (-1 + \sqrt{3}\,i)^2$

$$= (-1)^2 - 2\sqrt{3}\,i + (\sqrt{3})^2(i^2)$$
$$= 1 - 2\sqrt{3}\,i + 3(-1)$$
$$= -2 - 2\sqrt{3}\,i$$

Applications

Just as every real number corresponds to a point on the real number line, every complex number corresponds to a point in the *complex plane*, as shown in Figure 5.1. In this figure, note that the vertical axis is called the *imaginary axis* and the horizontal axis is called the *real axis*.

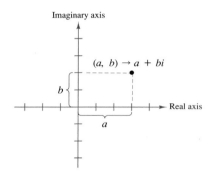

FIGURE 5.1

EXAMPLE 6   **Plotting Complex Numbers**

Plot the following complex numbers in the complex plane.

A. $2 + 3i$ **B.** $-1 + 2i$ **C.** 4

SOLUTION

A. To plot the complex number $2 + 3i$, move (from the origin) two units to the right on the real axis and then three units up, as shown in Figure 5.2. In other words, plotting the complex number $2 + 3i$ in the complex plane is comparable to plotting the point $(2, 3)$ in the Cartesian plane.

B. The complex number $-1 + 2i$ corresponds to the point $(-1, 2)$, as shown in Figure 5.2.

C. The complex number 4 corresponds to the point $(4, 0)$, as shown in Figure 5.2.

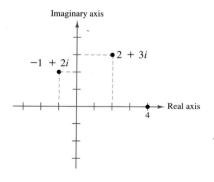

FIGURE 5.2

In the hands of a person who understands "fractal geometry," the complex plane can become an easel on which stunning pictures, called **fractals,** can be drawn. The most famous such picture is called the **Mandelbrot Set,** named after the Polish-born mathematician Benoit Mandelbrot. To draw the Mandelbrot Set, consider the sequence of numbers

$$c, \; c^2 + c, \; (c^2 + c)^2 + c, \; [(c^2 + c)^2 + c]^2 + c, \ldots .$$

The behavior of this sequence depends on the value of the complex number c. For some values of c this sequence is **bounded,** and for other values it is **unbounded.** If the sequence is bounded, then the complex number c is in the Mandelbrot Set, and if the sequence is unbounded, then the complex number c is not in the Mandelbrot Set.

EXAMPLE 7   Members of the Mandelbrot Set

A. The complex number -2 is in the Mandelbrot Set because for $c = -2$, the corresponding Mandelbrot sequence is

$$-2, 2, 2, 2, 2, 2, \ldots ,$$

which is bounded.

B. The complex number i is also in the Mandelbrot Set because for $c = i$, the corresponding Mandelbrot sequence is

$$i, \; -1 + i, \; -i, \; -1 + i, \; -i, \; -1 + i, \ldots ,$$

which is bounded.

C. The complex number $1 + i$ is *not* in the Mandelbrot Set because for $c = 1 + i$, the corresponding Mandelbrot sequence is

$$1 + i, \; 1 + 3i, \; -7 + 7i, \; 1 - 97i, \; -9407 - 193i,$$
$$88454401 + 3631103i, \ldots ,$$

which is unbounded.

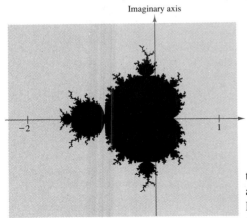

Imaginary axis

Real axis

-2

1

FIGURE 5.3

Within this definition, a picture of the Mandelbrot Set would have only two colors—one color for points that are in the set (the sequence is bounded) and one color for points that are outside the set (the sequence is not bounded). Figure 5.3 shows a black and blue picture of the Mandelbrot Set. The points that are colored black are in the Mandelbrot Set, and the points that are colored blue are not.

To add more interest to the picture, computer scientists discovered that the points that are not in the Mandelbrot Set can be assigned a variety of colors, depending on how "quickly" their sequences diverge. Figure 5.4 shows three different appendages of the Mandelbrot Set. (The black portions of the picture represent points that are in the Mandelbrot Set.)

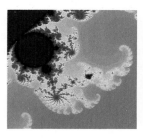  

FIGURE 5.4

DISCUSSION PROBLEM

....

CUBE ROOTS OF UNITY

In the real number system, the equation $x^3 = 1$ has 1 as its only solution. However, in the complex number system, this equation has three solutions.

$$1, \quad \frac{-1 + \sqrt{3}\,i}{2}, \quad \text{and} \quad \frac{-1 - \sqrt{3}\,i}{2}$$

Try cubing each of these numbers to show that each number has the property that $x^3 = 1$.

WARM-UP

....

The following warm-up exercises involve skills that were covered in earlier sections. You will use these skills in the exercise set for this section.

In Exercises 1–8, simplify the expression.

1. $\sqrt{12}$ **2.** $\sqrt{500}$

3. $\sqrt{20} - \sqrt{5}$ **4.** $\sqrt{27} - \sqrt{243}$

5. $\sqrt{24}\sqrt{6}$ **6.** $2\sqrt{18}\sqrt{32}$

7. $\dfrac{1}{\sqrt{3}}$ **8.** $\dfrac{2}{\sqrt{2}}$

In Exercises 9 and 10, solve the quadratic equation.

9. $x^2 + x - 1 = 0$ **10.** $x^2 + 2x - 1 = 0$

SECTION 5.1 · EXERCISES

1. Write the first 16 positive integer powers of i (that is, i, i^2, $i^3, \ldots, i^{16}$), and express each as i, $-i$, 1, or -1.

2. Express each of the following powers of i as i, $-i$, 1, or -1.
   (a) i^{40} (b) i^{25}
   (c) i^{50} (d) i^{67}

In Exercises 3–6, find real numbers a and b so that the equation is true.

3. $a + bi = -10 + 6i$ 4. $a + bi = 13 + 4i$
5. $(a - 1) + (b + 3)i = 5 + 8i$
6. $(a + 6) + 2bi = 6 - 5i$

In Exercises 7–18, write the complex number in standard form.

7. $4 + \sqrt{-9}$ 8. $3 + \sqrt{-16}$
9. $2 - \sqrt{-27}$ 10. $1 + \sqrt{-8}$
11. $\sqrt{-75}$ 12. 45
13. $-6i + i^2$ 14. $-4i^2 + 2i$
15. $-5i^5$ 16. $(-i)^3$
17. 8   18. $\left(\sqrt{-4}\right)^2 - 5$

In Exercises 19–26, perform the indicated addition or subtraction and write the result in standard form.

19. $(5 + i) + (6 - 2i)$ 20. $(13 - 2i) + (-5 + 6i)$
21. $(8 - i) - (4 - i)$ 22. $(3 + 2i) - (6 + 13i)$
23. $\left(-2 + \sqrt{-8}\right) + \left(5 - \sqrt{-50}\right)$
24. $\left(8 + \sqrt{-18}\right) - \left(4 + 3\sqrt{2}\,i\right)$
25. $-\left(\frac{3}{2} + \frac{5}{2}i\right) + \left(\frac{5}{3} + \frac{11}{3}i\right)$
26. $(1.6 + 3.2i) + (-5.8 + 4.3i)$

In Exercises 27–30, write the conjugate of the complex number, and find the product of the number and its conjugate.

27. $5 + 3i$ 28. $9 - 12i$
29. $-2 - \sqrt{5}\,i$ 30. $20i$

In Exercises 31–54, perform the specified operation and write the result in standard form.

31. $\sqrt{-6} \cdot \sqrt{-2}$ 32. $\sqrt{-5} \cdot \sqrt{-10}$
33. $\left(\sqrt{-10}\right)^2$ 34. $\left(\sqrt{-75}\right)^2$
35. $(1 + i)(3 - 2i)$ 36. $(6 - 2i)(2 - 3i)$
37. $6i(5 - 2i)$ 38. $-8i(9 + 4i)$

39. $\left(\sqrt{14} + \sqrt{10}\,i\right)\left(\sqrt{14} - \sqrt{10}\,i\right)$
40. $\left(3 + \sqrt{-5}\right)\left(7 - \sqrt{-10}\right)$
41. $(4 + 5i)^2$ 42. $(2 - 3i)^2$
43. $(2 + 3i)^2 + (2 - 3i)^2$ 44. $(1 - 2i)^2 - (1 + 2i)^2$

45. $\dfrac{4}{4 - 5i}$ 46. $\dfrac{3}{1 - i}$
47. $\dfrac{2 + i}{2 - i}$ 48. $\dfrac{8 - 7i}{1 - 2i}$
49. $\dfrac{6 - 7i}{i}$ 50. $\dfrac{8 + 20i}{2i}$
51. $\dfrac{5}{(1 + i)^3}$ 52. $\dfrac{1}{i^3}$
53. $\dfrac{(21 - 7i)(4 + 3i)}{2 - 5i}$ 54. $\dfrac{(2 - 3i)(5i)}{2 + 3i}$

In Exercises 55–60, plot the given complex number in the complex plane.

55. $-2 + i$ 56. i
57. 3   58. $-2 - 3i$
59. $1 - 2i$ 60. $-2i$

In Exercises 61–66, find the first six terms in the following sequence.

$c, \quad c^2 + c, \quad (c^2 + c)^2 + c, \quad [(c^2 + c)^2 + c]^2 + c, \quad \ldots$

From these terms, do you think the given complex number is in the Mandelbrot Set?

61. $c = 0$ 62. $c = 2$
63. $c = \frac{1}{2}i$ 64. $c = -i$
65. $c = 1$ 66. $c = -1$

67. Cube the complex numbers 2, $-1 + \sqrt{3}\,i$, and $-1 - \sqrt{3}\,i$. What do you notice?

68. Raise the complex numbers 2, -2, $2i$, and $-2i$ to the fourth power. What do you notice?

69. Prove that the sum of a complex number $a + bi$ and its conjugate is a real number.

70. Prove that the difference of a complex number $a + bi$ and its conjugate is an imaginary number.

71. Prove that the product of a complex number $a + bi$ and its conjugate is a real number.

72. Prove that the conjugate of the product of two complex numbers $a_1 + b_1 i$ and $a_2 + b_2 i$ is the product of their conjugates.

5.2   COMPLEX SOLUTIONS OF EQUATIONS

The Number of Solutions of a Polynomial Equation   /   Finding Solutions of
Polynomial Equations   /   Finding Zeros of Polynomial Functions

The Number of Solutions of a Polynomial Equation

The Fundamental Theorem of Algebra tells us that a polynomial equation of
degree n has precisely n solutions in the complex number system. These
solutions can be real or complex and may be repeated.

EXAMPLE 1   Zeros of Polynomial Functions

A. The first-degree equation

$$x - 2 = 0$$

has exactly *one* solution: $x = 2$.

B. Counting multiplicity, the second-degree equation

$$x^2 - 6x + 9 = 0$$
$$(x - 3)(x - 3) = 0$$

has exactly *two* solutions: $x = 3$ and $x = 3$.

C. The third-degree equation

$$x^3 + 4x = 0$$
$$x(x - 2i)(x + 2i) = 0$$

has exactly *three* solutions: $x = 0$, $x = -2i$, and $x = 2i$.

D. The fourth-degree equation

$$x^4 - 1 = 0$$
$$(x - 1)(x + 1)(x - i)(x + i) = 0$$

has exactly *four* solutions: $x = -1$, $x = 1$, $x = -i$, and $x = i$.

In Section 1.2 we pointed out that the discriminant $(b^2 - 4ac)$ can be used
to classify the two solutions given by the Quadratic Formula. In particular, if
the discriminant is negative, the quadratic equation $ax^2 + bx + c$ has two
complex solutions given by

$$x = \frac{-b \pm \sqrt{b^2 - 4ac}}{2a}.$$

EXAMPLE 2   Using the Discriminant

Use the discriminant to determine the number of real solutions of each of the
following quadratic equations.

A. $4x^2 - 20x + 25 = 0$ **B.** $13x^2 + 7x + 1 = 0$
C. $5x^2 = 8x$

SOLUTION

A. For the standard quadratic form $4x^2 - 20x + 25 = 0$, you have $a = 4$, $b = -20$, and $c = 25$. Therefore, the discriminant is

$$b^2 - 4ac = 400 - 4(4)(25) = 400 - 400 = 0.$$

Because the discriminant is zero, the equation has one repeated solution.

B. In this case, $a = 13$, $b = 7$, and $c = 1$, and so

$$b^2 - 4ac = 49 - 4(13)(1) = 49 - 52 = -3.$$

Because the discriminant is negative, there are no real solutions.

C. In standard form the equation is $5x^2 - 8x = 0$, and so $a = 5$, $b = -8$, and $c = 0$. Thus, the discriminant is

$$b^2 - 4ac = 64 - 4(5)(0) = 64.$$

Because the discriminant is positive, there are two real solutions.

Finding Solutions of Polynomial Equations

Finding the solution of a first- or second-degree polynomial equation is straightforward. The solution of the first-degree equation $ax + b = 0$ is simply $x = -b/a$, $a \neq 0$. To find the solutions of the second-degree equation $ax^2 + bx + c = 0$, try factoring or applying the Quadratic Formula (it will work for *any* quadratic equation).

EXAMPLE 3   Solving a Quadratic Equation

Use the Quadratic Formula to solve the following equation. If the discriminant is negative, list the complex solutions in $a + bi$ form.

$$6x^2 - 2x + 5 = 0$$

SOLUTION

Here you have $a = 6$, $b = -2$, and $c = 5$, and thus you will find the solutions to be

$$x = \frac{-b \pm \sqrt{b^2 - 4ac}}{2a} = \frac{-(-2) \pm \sqrt{4 - 4(6)(5)}}{2(6)}$$

$$= \frac{2 \pm \sqrt{-116}}{12}$$

$$= \frac{2 \pm 2\sqrt{-29}}{12}.$$

In $a + bi$ form, these complex solutions are

$$x = \frac{1}{6} + \frac{\sqrt{29}}{6}i \quad \text{and} \quad x = \frac{1}{6} - \frac{\sqrt{29}}{6}i.$$

In Example 3, note that the two complex solutions are **conjugates.** That is, they are of the form

$$a + bi \quad \text{and} \quad a - bi.$$

This is not a coincidence. In fact, the following rule tells us that if a polynomial equation (with real coefficients) has one complex solution, $a + bi$, then it must also have the conjugate $a - bi$ as a solution.

REMARK Be sure you see that complex solutions occur in conjugate pairs only for polynomial equations that have *real coefficients*. For instance, this rule applies to the equation $x^2 + 1 = 0$, but not to the equation $x - i = 0$.

COMPLEX SOLUTIONS OCCUR IN CONJUGATE PAIRS

If $a + bi$, $b \neq 0$ is a solution of a polynomial equation with real coefficients, then the conjugate $a - bi$ is also a solution of the equation.

EXAMPLE 4   Solving a Polynomial Equation

Find all solutions of $x^4 - x^2 - 20 = 0$.

SOLUTION

$$x^4 - x^2 - 20 = 0$$
$$(x^2 - 5)(x^2 + 4) = 0$$
$$(x + \sqrt{5})(x - \sqrt{5})(x^2 + 4) = 0$$
$$(x + \sqrt{5})(x - \sqrt{5})(x + 2i)(x - 2i) = 0$$

By setting each of these factors equal to 0, you find the solutions to be

$$x = -\sqrt{5}, \quad x = \sqrt{5}, \quad x = -2i, \quad \text{and} \quad x = 2i.$$

Find Zeros of Polynomial Functions

The problem of finding the zeros of a polynomial function is essentially the same problem as finding the solutions of a polynomial equation. For instance, the zeros of the polynomial function $f(x) = 3x^2 - 4x + 5$ are simply the solutions of the polynomial equation

$$3x^2 - 4x + 5 = 0.$$

EXAMPLE 5   Finding the Zeros of a Polynomial Function

Find all zeros of

$$f(x) = x^4 - 3x^3 + 6x^2 + 2x - 60$$

given that $1 + 3i$ is a zero of f.

SOLUTION

Because complex zeros occur in conjugate pairs, you know that $1 - 3i$ is also a zero of f. This means that both

$$x - (1 + 3i) \quad \text{and} \quad x - (1 - 3i)$$

are factors of $f(x)$. Multiplying these two factors produces

$$[x - (1 + 3i)][x - (1 - 3i)] = [(x - 1) - 3i][(x - 1) + 3i]$$
$$= (x - 1)^2 - 9i^2$$
$$= x^2 - 2x + 10.$$

Now, using long division, you can divide $x^2 - 2x + 10$ into $f(x)$, as follows.

$$
\begin{array}{r}
x^2 - x - 6 \\
x^2 - 2x + 10{\overline{\smash{\big)}\,x^4 - 3x^3 + 6x^2 + 2x - 60}} \\
\underline{x^4 - 2x^3 + 10x^2} \\
-x^3 - 4x^2 + 2x \\
\underline{-x^3 + 2x^2 - 10x} \\
-6x^2 + 12x - 60 \\
\underline{-6x^2 + 12x - 60} \\
0
\end{array}
$$

Therefore,

$$f(x) = (x^2 - 2x + 10)(x^2 - x - 6)$$
$$= (x^2 - 2x + 10)(x - 3)(x + 2)$$

and you can conclude that the zeros of f are

$$x = 1 + 3i, \quad x = 1 - 3i, \quad x = 3, \quad \text{and} \quad x = -2.$$

EXAMPLE 6   Finding a Polynomial with Given Zeros

Find a fourth-degree polynomial function, with real coefficients, that has -1, -1, and $3i$ as zeros.

Technology Note

Some graphing utilities have a built-in capability for finding all the zeros of a polynomial function. For instance, you can verify the result of Example 5 on the TI-85 using the POLY menu. Enter order=4 and then supply the appropriate coefficients. Press the SOLVE key and after a few seconds you will see the four zeros expressed as ordered pairs.

SOLUTION

Because $3i$ is a zero, you know that $-3i$ is also a zero. Thus, $f(x)$ can be written as

$$f(x) = a(x + 1)(x + 1)(x - 3i)(x + 3i).$$

For simplicity, let $a = 1$ and obtain

$$f(x) = (x^2 + 2x + 1)(x^2 + 9)$$
$$= x^4 + 2x^3 + 10x^2 + 18x + 9.$$

EXAMPLE 7   **Finding a Polynomial with Given Zeros**

Find a cubic polynomial function f, with real coefficients, that has 2 and $1 - i$ as zeros, and such that $f(1) = 3$.

SOLUTION

Because $1 - i$ is a zero of f, so is $1 + i$. Therefore,

$$f(x) = a(x - 2)[x - (1 - i)][x - (1 + i)]$$
$$= a(x - 2)[(x - 1) + i][(x - 1) - i]$$
$$= a(x - 2)[(x - 1)^2 - i^2]$$
$$= a(x - 2)(x^2 - 2x + 2)$$
$$= a(x^3 - 4x^2 + 6x - 4).$$

To find the value of a, use the fact that $f(1) = 3$ and obtain $f(1) = a(1 - 4 + 6 - 4) = 3$. Thus, $a = -3$ and you conclude that

$$f(x) = -3(x^3 - 4x^2 + 6x - 4)$$
$$= -3x^3 + 12x^2 - 18x + 12.$$

DISCUSSION PROBLEM

SOLUTIONS AND ZEROS

Write a paragraph explaining the relationship among the solutions of a polynomial equation

$$a_n x_n + a_{n-1} x^{n-1} + \cdots + a_2 x^2 + a_1 x + a_0 = 0,$$

the zeros of the corresponding polynomial function

$$f(x) = a_n x_n + a_{n-1} x^{n-1} + \cdots + a_2 x^2 + a_1 x + a_0,$$

and the x-intercepts of the graph of f.

WARM-UP

The following warm-up exercises involve skills that were covered in earlier sections. You will use these skills in the exercise set for this section.

In Exercises 1–4, write each complex number in standard form and give its complex conjugate.

1. $4 - \sqrt{-29}$

2. $-5 - \sqrt{-144}$

3. $-1 + \sqrt{-32}$

4. $6 + \sqrt{-\frac{1}{4}}$

In Exercises 5–10, perform the indicated operations and write the answers in standard form.

5. $(-3 + 6i) - (10 - 3i)$

6. $(12 - 4i) + 20i$

7. $(4 - 2i)(3 + 7i)$

8. $(2 - 5i)(2 + 5i)$

9. $\dfrac{1 + i}{1 - i}$

10. $(3 + 2i)^3$

SECTION 5.2 · EXERCISES

In Exercises 1–4, determine the number of solutions of the equation in the complex number system.

1. $x^3 - 4x + 5 = 0$

2. $2x^6 + 3x^3 - 10 = 0$

3. $25 - x^4 = 0$

4. $12 - x + 3x^2 - 3x^5 = 0$

In Exercises 5–12, use the discriminant to determine the number of real solutions of the quadratic equation.

5. $4x^2 - 4x + 1 = 0$

6. $2x^2 - x - 1 = 0$

7. $3x^2 + 4x + 1 = 0$

8. $x^2 + 2x + 4 = 0$

9. $2x^2 - 5x + 5 = 0$

10. $3x^2 - 6x + 3 = 0$

11. $\frac{1}{5}x^2 + \frac{6}{5}x - 8 = 0$

12. $\frac{1}{3}x^2 - 5x + 25 = 0$

In Exercises 13–24, solve the equation. List any complex solutions in the form $a + bi$.

13. $x^2 - 5 = 0$

14. $3x^2 - 1 = 0$

15. $(x + 5)^2 - 6 = 0$

16. $16 - (x - 1)^2 = 0$

17. $x^2 - 8x + 16 = 0$

18. $4x^2 + 4x + 1 = 0$

19. $x^2 + 2x + 5 = 0$

20. $54 + 16x - x^2 = 0$

21. $4x^2 - 4x + 5 = 0$

22. $4x^2 - 4x + 21 = 0$

23. $230 + 20x - 0.5x^2 = 0$

24. $6 - (x - 1)^2 = 0$

In Exercises 25–30, find all the zeros of the function and write the polynomial as a product of linear factors.

25. $f(z) = z^2 - 2z + 2$

26. $f(x) = x^2 - x + 56$

27. $h(x) = x^2 - 4x + 1$

28. $g(x) = x^2 + 10x + 23$

29. $f(x) = x^4 - 81$

30. $f(y) = y^4 - 625$

In Exercises 31–56, use the zero of the function as an aid in finding all its zeros. Write the polynomial as a product of linear factors. Check the real zeros by graphing the function with a graphing utility.

Function	Zero
31. $g(x) = x^3 - 6x^2 + 13x - 10$	2
32. $f(x) = x^3 - 2x^2 - 11x + 52$	-4
33. $f(t) = t^3 - 3t^2 - 15t + 125$	-5
34. $f(x) = x^3 + 11x^2 + 39x + 29$	-1
35. $f(x) = x^3 + 24x^2 + 214x + 740$	$-7 + 5i$
36. $f(s) = 2s^3 - 5s^2 + 12s - 5$	$1 + 2i$
37. $f(x) = 16x^3 - 20x^2 - 4x + 15$	$1 - \frac{1}{2}i$
38. $f(x) = 9x^3 - 15x^2 + 11x - 5$	$\frac{1}{3} - \frac{2}{3}i$
39. $h(x) = x^3 - x + 6$	$1 - \sqrt{2}\,i$
40. $h(x) = x^3 - 9x^2 + 27x + 35$	$-2 + \sqrt{3}\,i$
41. $f(x) = 5x^3 - 9x^2 + 28x + 6$	$-\frac{1}{5}$
42. $g(x) = 3x^3 - 4x^2 + 8x + 8$	$-\frac{2}{3}$

Function	*Zero*
43. $g(x) = x^4 - 4x^3 + 8x^2 - 16x + 16$	$2i$
44. $h(x) = x^4 + 6x^3 + 10x^2 + 6x + 9$	i
45. $f(x) = x^4 + 10x^2 + 9$	$-3i$
46. $f(x) = x^4 + 29x^2 + 100$	$-5i$
47. $f(x) = 2x^3 + 3x^2 + 50x + 75$	$r = 5i$
48. $f(x) = x^3 + x^2 + 9x + 9$	$r = 3i$
49. $f(x) = 2x^4 - x^3 + 7x^2 - 4x - 4$	$r = 2i$
50. $g(x) = x^3 - 7x^2 - x + 87$	$r = 5 + 2i$
51. $g(x) = 4x^3 + 23x^2 + 34x - 10$	$r = -3 + i$
52. $h(x) = 3x^3 - 4x^2 + 8x + 8$	$r = 1 - \sqrt{3}i$
53. $f(x) = x^4 + 3x^3 - 5x^2 - 21x + 22$	$r = -3 + \sqrt{2}i$
54. $f(x) = x^3 + 4x^2 + 14x + 20$	$r = -1 - 3i$
55. $h(x) = 8x^3 - 14x^2 + 18x - 9$	$r = \left(1 - \sqrt{5}i\right)/2$
56. $f(x) = 25x^3 - 55x^2 - 54x - 18$	$r = \left(-2 + \sqrt{2}i\right)/5$

In Exercises 57–60, write the polynomial in completely factored form.

57. $f(x) = x^4 + 6x^2 - 27$

58. $f(x) = x^4 - 2x^3 - 3x^2 + 12x - 18$
(*Hint*: One factor is $x^2 - 6$.)

59. $f(x) = x^4 - 4x^3 + 5x^2 - 2x - 6$
(*Hint*: One factor is $x^2 - 2x - 2$.)

60. $f(x) = x^4 - 3x^3 - x^2 - 12x - 20$
(*Hint*: One factor is $x^2 + 4$.)

In Exercises 61–70, find a polynomial with integer coefficients that has the given zeros.

61. $1, 5i, -5i$

62. $4, 3i, -3i$

63. $2, 4 + i, 4 - i$

64. $6, -5 + 2i, -5 - 2i$

65. $i, -i, 6i, -6i$

66. $2, 2, 2, 4i, -4i$

67. $-5, -5, 1 + \sqrt{3}i$

68. $\frac{2}{3}, -1, 3 + \sqrt{2}i$

69. $\frac{3}{4}, -2, -\frac{1}{2} + i$

70. $0, 0, 4, 1 + i$

71. *Maximum Height* A baseball is thrown upward from ground level with an initial velocity of 48 feet per second, and its height h in feet is given by

$$h = -16t^2 + 48t, \qquad 0 \le t \le 3,$$

where t is the time in seconds. Suppose you are told the ball reaches a height of 64 feet. Explain why this is not possible.

72. *Profit* The demand equation for a certain product is given by $p = 140 - 0.0001x$, where p is the unit price (in dollars) of the product and x is the number of units produced and sold. The cost equation for the product is $C = 80x + 150,000$, where C is the total cost (in dollars) and x is the number of units produced. The total profit obtained by producing and selling x units is given by

$$P = R - C = xp - C.$$

Suppose you are working in the marketing department and are asked to determine a price p that would yield a profit of 9 million dollars. Explain why this is not possible.

73. Find a quadratic function f (with integer coefficients) that has $\pm \sqrt{b}i$ as zeros. Assume that b is a positive integer.

5.3 TRIGONOMETRIC FORM OF A COMPLEX NUMBER

The Complex Plane / Trigonometric Form of a Complex Number / Multiplication and Division of Complex Numbers

The Complex Plane

In this section you will study the trigonometric form of a complex number. The convenience of this form will not be fully apparent until DeMoivre's Theorem is introduced in Section 5.4.

Just as real numbers can be represented by points on the real number line, you can represent a complex number

$$z = a + bi$$

as the point (a, b) in the **complex plane.** The horizontal axis is the **real axis** and the vertical axis is the **imaginary axis,** as shown in Figure 5.5.

The **absolute value** of the complex number $a + bi$ is defined to be the distance between the origin $(0, 0)$ and the point (a, b), as shown in Figure 5.6.

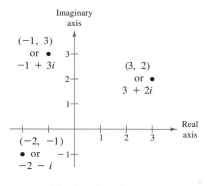

FIGURE 5.5

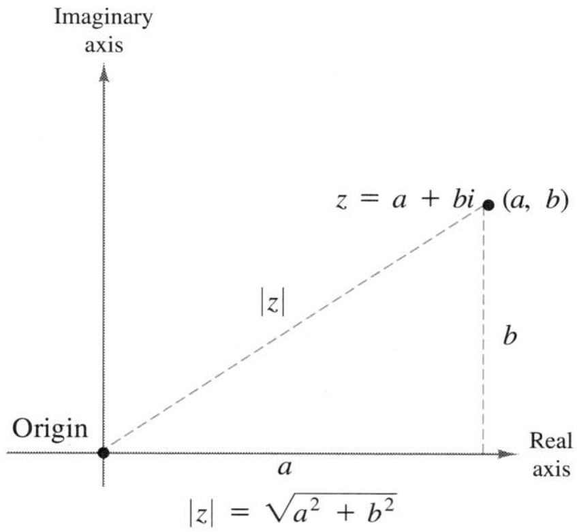

The Complex Plane

$|z| = \sqrt{a^2 + b^2}$

FIGURE 5.6

REMARK   Note that if the complex number $a + bi$ happens to be a real number (that is, if $b = 0$), then this definition agrees with that given for the absolute value of a real number.

$$|a + 0i| = \sqrt{a^2 + 0^2} = \sqrt{a^2} = |a|$$

DEFINITION OF THE ABSOLUTE VALUE OF A COMPLEX NUMBER

The **absolute value** of the complex number $z = a + bi$ is given by

$$|a + bi| = \sqrt{a^2 + b^2}.$$

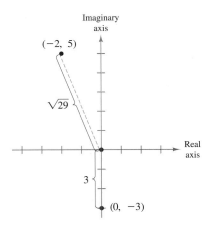

FIGURE 5.7

EXAMPLE 1   **Finding the Absolute Value of a Complex Number**

Plot the points corresponding to the following complex numbers and find the absolute value of each.

A. $z = -3i$ **B.** $z = -2 + 5i$

SOLUTION

The points are shown in Figure 5.7.

A. The complex number $z = 0 + (-3)i$ has an absolute value of

$$|z| = \sqrt{0^2 + (-3)^2} = \sqrt{9} = 3.$$

B. The complex number $z = -2 + 5i$ has an absolute value of

$$|z| = \sqrt{(-2)^2 + 5^2} = \sqrt{29}.$$

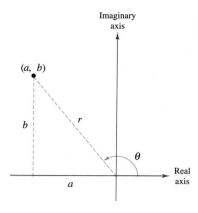

Complex Number: $a + bi$

FIGURE 5.8

REMARK The trigonometric form of a complex number is also called the **polar form.** Because there are infinitely many choices for θ, the trigonometric form of a complex number is not unique. Normally, we use θ values in the interval $0 \le \theta < 2\pi$, though on occasion we may use $\theta < 0$.

Trigonometric Form of a Complex Number

In Section 5.1, you learned how to add, subtract, multiply, and divide complex numbers. To work effectively with *powers* and *roots* of complex numbers, it is helpful to write complex numbers in **trigonometric form.** In Figure 5.8, consider the nonzero complex number $a + bi$. By letting θ be the angle from the positive x-axis (measured counterclockwise) to the line segment connecting the origin and the point (a, b), you can write

$$a = r \cos \theta \quad \text{and} \quad b = r \sin \theta,$$

where $r = \sqrt{a^2 + b^2}$. Consequently, you have

$$a + bi = (r \cos \theta) + (r \sin \theta)i,$$

from which you obtain the following **trigonometric form of a complex number.**

TRIGONOMETRIC FORM OF A COMPLEX NUMBER

Let $z = a + bi$ be a complex number. The **trigonometric form** of z is

$$z = r(\cos \theta + i \sin \theta),$$

where $a = r \cos \theta$, $b = r \sin \theta$, $r = \sqrt{a^2 + b^2}$, and $\tan \theta = b/a$. The number r is the **modulus** of z, and θ is an **argument** of z.

EXAMPLE 2   **Writing Complex Numbers in Trigonometric Form**

Write the complex numbers in trigonometric form.

A. $z = -2 - 2\sqrt{3}i$ **B.** $z = 6 + 2i$

Use radian measure for part (A) and degree measure for part (B).

SOLUTION

The graphs of the complex numbers are shown in Figure 5.9.

A. The absolute value of z is

$$r = |-2 - 2\sqrt{3}i| = \sqrt{(-2)^2 + (-2\sqrt{3})^2} = \sqrt{16} = 4,$$

and the angle θ is given by

$$\tan \theta = \frac{b}{a} = \frac{-2\sqrt{3}}{-2} = \sqrt{3}.$$

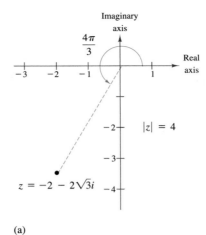

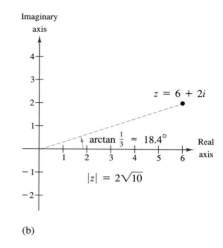

(a)   (b)

FIGURE 5.9

Because $\tan \pi/3 = \sqrt{3}$ and $z = -2 - 2\sqrt{3}i$ lies in Quadrant III, choose θ to be $\theta = \pi + \pi/3 = 4\pi/3$. Thus, the trigonometric form is

$$z = r(\cos \theta + i \sin \theta) = 4\left(\cos \frac{4\pi}{3} + i \sin \frac{4\pi}{3}\right).$$

B. Here you have $r = |6 + 2i| = 2\sqrt{10}$, with θ given by

$$\tan \theta = \frac{2}{6} = \frac{1}{3} \qquad \qquad \theta \text{ in Quadrant I}$$

$$\theta = \arctan \frac{1}{3} \approx 18.4°.$$

Therefore, the trigonometric form of z is

$$z = r(\cos \theta + i \sin \theta)$$
$$= 2\sqrt{10}\left[\cos\left(\arctan \frac{1}{3}\right) + i \sin\left(\arctan \frac{1}{3}\right)\right]$$
$$\approx 2\sqrt{10}(\cos 18.4° + i \sin 18.4°).$$

EXAMPLE 3   **Writing a Complex Number in Standard Form**

Write the complex number in standard form $a + bi$.

$$z = \sqrt{8}\left[\cos\left(-\frac{\pi}{3}\right) + i \sin\left(-\frac{\pi}{3}\right)\right]$$

SOLUTION

Because $\cos(-\pi/3) = 1/2$ and $\sin(-\pi/3) = -\sqrt{3}/2$, you can write

$$z = \sqrt{8}\left[\cos\left(-\frac{\pi}{3}\right) + i \sin\left(-\frac{\pi}{3}\right)\right]$$

$$= \sqrt{8}\left(\frac{1}{2} - \frac{\sqrt{3}}{2}i\right)$$

$$= 2\sqrt{2}\left(\frac{1}{2} - \frac{\sqrt{3}}{2}i\right)$$

$$= \sqrt{2} - \sqrt{6}\,i.$$

Multiplication and Division of Complex Numbers

The trigonometric form adapts nicely to multiplication and division of complex numbers. Suppose you are given two complex numbers

$$z_1 = r_1(\cos\theta_1 + i\sin\theta_1) \quad \text{and} \quad z_2 = r_2(\cos\theta_2 + i\sin\theta_2).$$

The product of z_1 and z_2 is

$$z_1 z_2 = r_1 r_2(\cos\theta_1 + i\sin\theta_1)(\cos\theta_2 + i\sin\theta_2)$$

$$= r_1 r_2[(\cos\theta_1 \cos\theta_2 - \sin\theta_1 \sin\theta_2)$$

$$+ i(\sin\theta_1 \cos\theta_2 + \cos\theta_1 \sin\theta_2)].$$

Using the sum and difference formulas for cosine and sine, you can rewrite this equation as

$$z_1 z_2 = r_1 r_2[\cos(\theta_1 + \theta_2) + i\sin(\theta_1 + \theta_2)].$$

This establishes the first part of the following rule.

PRODUCT AND QUOTIENT OF TWO COMPLEX NUMBERS

Let $z_1 = r_1(\cos\theta_1 + i\sin\theta_1)$ and $z_2 = r_2(\cos\theta_2 + i\sin\theta_2)$ be complex numbers.

$$z_1 z_2 = r_1 r_2[\cos(\theta_1 + \theta_2) + i\sin(\theta_1 + \theta_2)] \qquad \textit{Product}$$

$$\frac{z_1}{z_2} = \frac{r_1}{r_2}[\cos(\theta_1 - \theta_2) + i\sin(\theta_1 - \theta_2)], \quad z_2 \neq 0 \qquad \textit{Quotient}$$

Note that this rule says that to multiply two complex numbers, you multiply moduli and add arguments, whereas to divide two complex numbers you divide moduli and subtract arguments.

EXAMPLE 4   Multiplying Complex Numbers in Trigonometric Form

Find the product of the following complex numbers.

$$z_1 = 2\left(\cos \frac{2\pi}{3} + i \sin \frac{2\pi}{3}\right) \qquad z_2 = 8\left(\cos \frac{11\pi}{6} + i \sin \frac{11\pi}{6}\right)$$

Express your answer in standard form.

SOLUTION

$$z_1 z_2 = 2\left(\cos \frac{2\pi}{3} + i \sin \frac{2\pi}{3}\right) \cdot 8\left(\cos \frac{11\pi}{6} + i \sin \frac{11\pi}{6}\right)$$

$$= 16\left[\cos\left(\frac{2\pi}{3} + \frac{11\pi}{6}\right) + i \sin\left(\frac{2\pi}{3} + \frac{11\pi}{6}\right)\right]$$

$$= 16\left(\cos \frac{5\pi}{2} + i \sin \frac{5\pi}{2}\right)$$

$$= 16\left(\cos \frac{\pi}{2} + i \sin \frac{\pi}{2}\right)$$

$$= 16[0 + i(1)] = 16i$$

Check this result by first converting to the standard forms $z_1 = -1 + \sqrt{3}\,i$ and $z_2 = 4\sqrt{3} - 4i$ and then multiplying algebraically, as in Section 5.1.

EXAMPLE 5   Dividing Complex Numbers in Trigonometric Form

Find z_1/z_2 for the two complex numbers

$$z_1 = 24(\cos 300° + i \sin 300°) \qquad z_2 = 8(\cos 75° + i \sin 75°).$$

Express your answer in standard form.

SOLUTION

$$\frac{z_1}{z_2} = \frac{24(\cos 300° + i \sin 300°)}{8(\cos 75° + i \sin 75°)}$$

$$= \frac{24}{8}[\cos(300° - 75°) + i \sin(300° - 75°)]$$

$$= 3(\cos 225° + i \sin 225°)$$

$$= 3\left[\left(-\frac{\sqrt{2}}{2}\right) + i\left(-\frac{\sqrt{2}}{2}\right)\right]$$

$$= -\frac{3\sqrt{2}}{2} - \frac{3\sqrt{2}}{2}i$$

DISCUSSION PROBLEM

ADDITION OF TWO COMPLEX NUMBERS

In Section 4.3, we gave a graphical interpretation of addition of two vectors in the plane, called the parallelogram law for vector addition. A similar graphical interpretation can be given for addition of two complex numbers. Write a paragraph describing this interpretation, and illustrate your result with the complex numbers $z_1 = 1 + 2i$ and $z_2 = 3 + i$.

WARM-UP

The following warm-up exercises involve skills that were covered in earlier sections. You will use these skills in the exercise set for this section.

In Exercises 1–4, write the complex number in standard form.

1. $-5 - \sqrt{-100}$ **2.** $7 + \sqrt{-54}$

3. $-4i + i^2$ **4.** $3i^3$

In Exercises 5–10, perform the indicated operations and write your result in standard form.

5. $(3 - 10i) - (-3 + 4i)$ **6.** $\left(2 + \sqrt{-50}\right) + \left(4 - \sqrt{2}i\right)$

7. $(4 - 2i)(-6 + i)$ **8.** $(3 - 2i)(3 + 2i)$

9. $\dfrac{1 + 4i}{1 - i}$ **10.** $\dfrac{3 - 5i}{2i}$

SECTION 5.3 · EXERCISES

In Exercises 1–6, represent the complex number graphically and find its absolute value.

1. $-5i$
2. -5
3. $-4 + 4i$
4. $5 - 12i$
5. $6 - 7i$
6. $-8 + 3i$

In Exercises 7–10, express the complex number in trigonometric form.

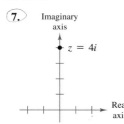

7.

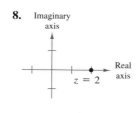

8.

9.

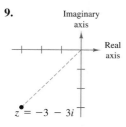

10.

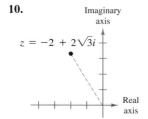

In Exercises 11–26, represent the complex number graphically, and find the trigonometric form of the number.

11. $3 - 3i$
12. $-2 - 2i$
13. $\sqrt{3} + i$
14. $-1 + \sqrt{3}i$
15. $-2(1 + \sqrt{3}i)$
16. $\frac{5}{2}(\sqrt{3} - i)$
17. $6i$
18. 4
19. $-7 + 4i$
20. $3 - i$
21. 7
22. $-2i$
23. $1 + 6i$
24. $2\sqrt{2} - i$
25. $-3 - i$
26. $1 + 3i$

In Exercises 27–36, represent the complex number graphically, and find the standard form of the number.

27. $2(\cos 150° + i \sin 150°)$
28. $5(\cos 135° + i \sin 135°)$
29. $\frac{3}{2}(\cos 300° + i \sin 300°)$
30. $\frac{3}{4}(\cos 315° + i \sin 315°)$
31. $3.75\left(\cos \dfrac{3\pi}{4} + i \sin \dfrac{3\pi}{4}\right)$
32. $8\left(\cos \dfrac{\pi}{12} + i \sin \dfrac{\pi}{12}\right)$
33. $4\left(\cos \dfrac{3\pi}{2} + i \sin \dfrac{3\pi}{2}\right)$
34. $7(\cos 0° + i \sin 0°)$
35. $3[\cos(18° \, 45') + i \sin(18° \, 45')]$
36. $6[\cos(230° \, 30') + i \sin(230° \, 30')]$

In Exercises 37–48, perform the indicated operation and leave the result in trigonometric form.

37. $\left[3\left(\cos \dfrac{\pi}{3} + i \sin \dfrac{\pi}{3}\right)\right]\left[4\left(\cos \dfrac{\pi}{6} + i \sin \dfrac{\pi}{6}\right)\right]$
38. $\left[\dfrac{3}{2}\left(\cos \dfrac{\pi}{2} + i \sin \dfrac{\pi}{2}\right)\right]\left[6\left(\cos \dfrac{\pi}{4} + i \sin \dfrac{\pi}{4}\right)\right]$
39. $[\frac{5}{3}(\cos 140° + i \sin 140°)][\frac{2}{3}(\cos 60° + i \sin 60°)]$
40. $[0.5(\cos 100° + i \sin 100°)][0.8(\cos 300° + i \sin 300°)]$
41. $[0.45(\cos 310° + i \sin 310°)][0.60(\cos 200° + i \sin 200°)]$
42. $(\cos 5° + i \sin 5°)(\cos 20° + i \sin 20°)$
43. $\dfrac{2(\cos 120° + i \sin 120°)}{4(\cos 40° + i \sin 40°)}$
44. $\dfrac{\cos 40° + i \sin 40°}{\cos 10° + i \sin 10°}$
45. $\dfrac{\cos(5\pi/3) + i \sin (5\pi/3)}{\cos \pi + i \sin \pi}$
46. $\dfrac{5(\cos 4.3 + i \sin 4.3)}{4(\cos 2.1 + i \sin 2.1)}$
47. $\dfrac{12(\cos 52° + i \sin 52°)}{3(\cos 110° + i \sin 110°)}$
48. $\dfrac{9(\cos 20° + i \sin 20°)}{5(\cos 75° + i \sin 75°)}$

In Exercises 49–54, (a) give the trigonometric form of the complex number, (b) perform the indicated operation using the trigonometric form, and (c) perform the indicated operation using the standard form and check your result with the answer to part (b).

49. $(2 + 2i)(1 - i)$

50. $\left(\sqrt{3} + i\right)(1 + i)$

51. $-2i(1 + i)$

52. $\dfrac{3 + 4i}{1 - \sqrt{3}i}$

53. $\dfrac{5}{2 + 3i}$

54. $\dfrac{4i}{-4 + 2i}$

55. Given two complex numbers $z_1 = r_1(\cos \theta_1 + i \sin \theta_1)$ and $z_2 = r_2(\cos \theta_2 + i \sin \theta_2)$, $z_2 \neq 0$, prove that

$$\frac{z_1}{z_2} = \frac{r_1}{r_2}[\cos(\theta_1 - \theta_2) + i \sin(\theta_1 - \theta_2)].$$

56. Show that the complex conjugate of $z = r(\cos \theta + i \sin \theta)$ is $\bar{z} = r[\cos(-\theta) + i \sin(-\theta)]$.

57. Use the trigonometric form of z and $\bar{z}$ in Exercise 56 to find (a) $z\bar{z}$ and (b) $z/\bar{z}$, $z \neq 0$.

58. Show that the negative of $z = r(\cos \theta + i \sin \theta)$ is $-z = r[\cos(\theta + \pi) + i \sin(\theta + \pi)]$.

In Exercises 59 and 60, sketch the graph of all complex numbers z satisfying the given condition.

59. $|z| = 2$

60. $\theta = \pi/6$

5.4 DeMoivre's Theorem and Nth Roots

Powers of Complex Numbers / Roots of Complex Numbers

Powers of Complex Numbers

In this section you will study procedures for finding powers and roots of complex numbers. To begin, consider the complex number (in trigonometric form) $z = r(\cos \theta + i \sin \theta)$. Repeated use of the multiplication rule from the previous section yields

$$z = r(\cos \theta + i \sin \theta)$$
$$z^2 = r(\cos \theta + i \sin \theta)r(\cos \theta + i \sin \theta)$$
$$\quad = r^2(\cos 2\theta + i \sin 2\theta)$$
$$z^3 = z^2(z) = r^2(\cos 2\theta + i \sin 2\theta)r(\cos \theta + i \sin \theta)$$
$$\quad = r^3(\cos 3\theta + i \sin 3\theta).$$

Similarly,

$$z^4 = r^4(\cos 4\theta + i \sin 4\theta)$$
$$z^5 = r^5(\cos 5\theta + i \sin 5\theta)$$
$$\vdots$$

This pattern leads to the following important theorem, which is named after the French mathematician Abraham DeMoivre (1667–1754).

> ### DeMOIVRE'S THEOREM
>
> If $z = r(\cos\ \theta + i\ \sin\ \theta)$ is a complex number and n is a positive integer, then
>
> $$z^n = [r(\cos\ \theta + i\ \sin\ \theta)]^n = r^n(\cos\ n\theta + i\ \sin\ n\theta).$$

EXAMPLE 1 **Finding Powers of a Complex Number**

Use DeMoivre's Theorem to find $(-1 + \sqrt{3}i)^{12}$.

SOLUTION

First convert to trigonometric form.

$$-1 + \sqrt{3}i = 2\left(\cos\ \frac{2\pi}{3} + i\ \sin\ \frac{2\pi}{3}\right)$$

Then, by DeMoivre's Theorem, you have

$$(-1 + \sqrt{3}i)^{12} = \left[2\left(\cos\ \frac{2\pi}{3} + i\ \sin\ \frac{2\pi}{3}\right)\right]^{12}$$

$$= 2^{12}\left[\cos\left(12\cdot\frac{2\pi}{3}\right) + i\ \sin\left(12\cdot\frac{2\pi}{3}\right)\right]$$

$$= 4096(\cos\ 8\pi + i\ \sin\ 8\pi)$$

$$= 4096(1 + 0)$$

$$= 4096.$$

Are you surprised to see a real number as the answer?

Roots of Complex Numbers

Recall that a consequence of the Fundamental Theorem of Algebra is that a polynomial equation of degree n has n solutions in the complex number system. Hence, an equation such as $x^6 = 1$ has six solutions, which can be found by factoring and using the Quadratic Formula.

$$x^6 - 1 = (x^3 - 1)(x^3 + 1)$$
$$= (x - 1)(x^2 + x + 1)(x + 1)(x^2 - x + 1) = 0$$

Consequently, the solutions are

$$x = \pm1, \quad x = \frac{-1 \pm \sqrt{3}i}{2}, \quad \text{and} \quad x = \frac{1 \pm \sqrt{3}i}{2}.$$

Each of these numbers is a *sixth root of* 1.

DEFINITION OF *N*TH ROOT OF A COMPLEX NUMBER

The complex number $u = a + bi$ is an ***n*th root** of the complex number z if

$$z = u^n = (a + bi)^n.$$

To find a formula for an nth root of a complex number, let u be an nth root of z, where

$$u = s(\cos \beta + i \sin \beta) \quad \text{and} \quad z = r(\cos \theta + i \sin \theta).$$

By DeMoivre's Theorem and the fact that $u^n = z$, you have

$$s^n(\cos n\beta + i \sin n\beta) = r(\cos \theta + i \sin \theta).$$

Now, taking the absolute value of both sides of this equation, it follows that $s^n = r$. Substituting back into the previous equation and dividing by r produces

$$\cos n\beta + i \sin n\beta = \cos \theta + i \sin \theta.$$

Thus, it follows that

$$\cos n\beta = \cos \theta \quad \text{and} \quad \sin n\beta = \sin \theta.$$

Because both sine and cosine have a period of 2π, these last two equations have solutions if and only if the angles differ by a multiple of 2π. Consequently, there must exist an integer k such that

$$n\beta = \theta + 2\pi k$$
$$\beta = \frac{\theta + 2\pi k}{n}.$$

By substituting this value for β into the trigonometric form of u, you obtain the result stated in the following theorem.

REMARK Note that when k exceeds $n - 1$ the roots begin to repeat. For instance, if $k = n$, the angle

$$\frac{\theta + 2\pi n}{n} = \frac{\theta}{n} + 2\pi$$

is coterminal with θ/n, which is also obtained when $k = 0$.

*N*TH ROOTS OF A COMPLEX NUMBER

For a positive integer n, the complex number $z = r(\cos \theta + i \sin \theta)$ has exactly n distinct nth roots given by

$$\sqrt[n]{r}\left(\cos \frac{\theta + 2\pi k}{n} + i \sin \frac{\theta + 2\pi k}{n} \right),$$

where $k = 0, 1, 2, \ldots, n - 1$.

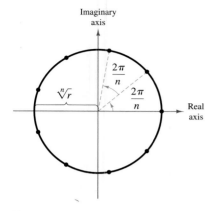

FIGURE 5.10

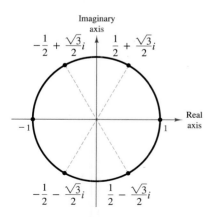

FIGURE 5.11

This formula for the nth roots of a complex number z has a nice geometric interpretation, as shown in Figure 5.10. Note that because the nth roots of z all have the same magnitude $\sqrt[n]{r}$, they all lie on a circle of radius $\sqrt[n]{r}$ with center at the origin. Furthermore, the n roots are equally spaced along the circle, because successive nth roots have arguments that differ by $2\pi/n$.

EXAMPLE 2   **Finding nth Roots of a Real Number**

Find all the sixth roots of 1.

SOLUTION

First, write 1 in the trigonometric form $1 = 1(\cos 0 + i \sin 0)$. Then, by the nth root formula, with $n = 6$ and $r = 1$, the roots have the form

$$\sqrt[6]{1}\left(\cos \frac{0 + 2\pi k}{6} + i \sin \frac{0 + 2\pi k}{6}\right),$$

or simply $\cos(\pi k/3) + i \sin(\pi k/3)$. Thus, for $k = 0, 1, 2, 3, 4, 5$, the sixth roots are as follows.

$$\cos 0 + i \sin 0 = 1$$

$$\cos \frac{\pi}{3} + i \sin \frac{\pi}{3} = \frac{1}{2} + \frac{\sqrt{3}}{2}i$$

$$\cos \frac{2\pi}{3} + i \sin \frac{2\pi}{3} = -\frac{1}{2} + \frac{\sqrt{3}}{2}i$$

$$\cos \pi + i \sin \pi = -1$$

$$\cos \frac{4\pi}{3} + i \sin \frac{4\pi}{3} = -\frac{1}{2} - \frac{\sqrt{3}}{2}i$$

$$\cos \frac{5\pi}{3} + i \sin \frac{5\pi}{3} = \frac{1}{2} - \frac{\sqrt{3}}{2}i$$

See Figure 5.11.

In Figure 5.11, notice that the roots obtained in Example 2 all have a magnitude of 1 and are equally spaced around this unit circle. Also, notice that the complex roots occur in conjugate pairs, as previously discussed in Section 5.2. We refer to the special case of the n distinct nth roots of 1 as the **nth roots of unity.**

EXAMPLE 3 Finding the nth Roots of a Complex Number

Find the three cube roots of $z = -2 + 2i$.

SOLUTION

Because z lies in the second quadrant, the trigonometric form of z is

$$z = -2 + 2i = \sqrt{8}(\cos 135° + i \sin 135°).$$

By our formula for nth roots, the cube roots have the form

$$\sqrt[6]{8}\left(\cos \frac{135° + 360°k}{3} + i \sin \frac{135° + 360°k}{3}\right).$$

Finally, for $k = 0, 1, 2$, you obtain the roots

$$\sqrt{2}(\cos 45° + i \sin 45°) = 1 + i$$
$$\sqrt{2}(\cos 165° + i \sin 165°) \approx -1.3660 + 0.3660i$$
$$\sqrt{2}(\cos 285° + i \sin 285°) \approx 0.3660 - 1.3660i.$$

REMARK Note that the roots do *not* occur in conjugate pairs. Do you see why?

EXAMPLE 4 Finding the Roots of a Polynomial Equation

Solve $x^4 + 16 = 0$.

SOLUTION

The given equation can be written as

$$x^4 = -16 \quad \text{or} \quad x^4 = 16(\cos \pi + i \sin \pi),$$

which means that you can solve the equation by finding the four fourth roots of -16. Each of these roots has the form

$$\sqrt[4]{16}\left(\cos \frac{\pi + 2\pi k}{4} + i \sin \frac{\pi + 2\pi k}{4}\right).$$

Finally, using $k = 0, 1, 2, 3$, you obtain the roots

$$2\left(\cos \frac{\pi}{4} + i \sin \frac{\pi}{4}\right) = 2\left(\frac{\sqrt{2}}{2} + \frac{\sqrt{2}}{2}i\right) = \sqrt{2} + \sqrt{2}i$$

$$2\left(\cos \frac{3\pi}{4} + i \sin \frac{3\pi}{4}\right) = 2\left(-\frac{\sqrt{2}}{2} + \frac{\sqrt{2}}{2}i\right) = -\sqrt{2} + \sqrt{2}i$$

$$2\left(\cos \frac{5\pi}{4} + i \sin \frac{5\pi}{4}\right) = 2\left(-\frac{\sqrt{2}}{2} - \frac{\sqrt{2}}{2}i\right) = -\sqrt{2} - \sqrt{2}i$$

$$2\left(\cos \frac{7\pi}{4} + i \sin \frac{7\pi}{4}\right) = 2\left(\frac{\sqrt{2}}{2} - \frac{\sqrt{2}}{2}i\right) = \sqrt{2} - \sqrt{2}i.$$

DISCUSSION PROBLEM

....

A FAMOUS MATHEMATICAL FORMULA

In this section you studied DeMoivre's Theorem, which gives a formula for raising a complex number to a positive integer power. Another famous formula that involves complex numbers and powers is called Euler's Formula, after the German mathematician Leonhard Euler (1707–1783). This formula states that

$$e^{a+bi} = e^a(\cos b + i \sin b).$$

Although the interpretation of this formula is beyond the scope of this text, we decided to include it because it gives rise to one of the most wonderful equations in mathematics.

$$e^{\pi i} + 1 = 0$$

This elegant equation relates the five most famous numbers in mathematics, 0, 1, π, e, and i, in a single equation. Show how Euler's Formula can be used to derive this equation.

WARM-UP

....

The following warm-up exercises involve skills that were covered in earlier sections. You will use these skills in the exercise set for this section.

In Exercises 1 and 2, simplify the expression.

1. $\sqrt[3]{54}$ **2.** $\sqrt[4]{16 + 48}$

In Exercises 3–6, write the complex number in trigonometric form.

3. $-5 + 5i$ **4.** $-3i$ **5.** -12 **6.** 12

In Exercises 7–10, perform the indicated operation. Leave the result in trigonometric form.

7. $\left(\cos \dfrac{\pi}{4} + i \sin \dfrac{\pi}{4}\right)\left(\cos \dfrac{\pi}{2} + i \sin \dfrac{\pi}{2}\right)$

8. $\left(\cos \dfrac{\pi}{12} + i \sin \dfrac{\pi}{12}\right)\left(\cos \dfrac{5\pi}{6} + i \sin \dfrac{5\pi}{6}\right)$

9. $\dfrac{6[\cos(2\pi/3) + i \sin(2\pi/3)]}{3[\cos(\pi/6) + i \sin(\pi/6)]}$

10. $\dfrac{2(\cos 55° + i \sin 55°)}{3(\cos 10° + i \sin 10°)}$

SECTION 5.4 · EXERCISES

In Exercises 1–12, use DeMoivre's Theorem to find the indicated power of the complex number. Express the result in standard form.

1. $(1 + i)^5$ **2.** $(2 + 2i)^6$

3. $(-1 + i)^{10}$ **4.** $(1 - i)^{12}$

5. $2(\sqrt{3} + i)^7$ **6.** $4(1 - \sqrt{3}i)^3$

7. $[5(\cos 20° + i \sin 20°)]^3$

8. $[3(\cos 150° + i \sin 150°)]^4$

9. $\left(\cos \dfrac{5\pi}{4} + i \sin \dfrac{5\pi}{4}\right)^{10}$

10. $\left[2\left(\cos \dfrac{\pi}{2} + i \sin \dfrac{\pi}{2}\right)\right]^8$

11. $[5(\cos 3.2 + i \sin 3.2)]^4$

12. $(\cos 0 + i \sin 0)^{20}$

In Exercises 13–24, (a) use DeMoivre's Theorem to find the indicated roots of the complex number, (b) represent each of the roots graphically, and (c) express each of the roots in standard form.

13. Square roots of: $9(\cos 120° + i \sin 120°)$

14. Square roots of: $16(\cos 60° + i \sin 60°)$

15. Fourth roots of: $16\left(\cos \dfrac{4\pi}{3} + i \sin \dfrac{4\pi}{3}\right)$

16. Fifth roots of: $32\left(\cos \dfrac{5\pi}{6} + i \sin \dfrac{5\pi}{6}\right)$

17. Square roots of: $-25i$

18. Fourth roots of: $625i$

19. Cube roots of: $-\dfrac{125}{2}\left(1 + \sqrt{3}i\right)$

20. Cube roots of: $-8\sqrt{2}(1 - i)$

21. Cube roots of: 8

22. Fourth roots of: i

23. Fifth roots of: 1

24. Cube roots of: 1000

In Exercises 25–32, find all the solutions of the equation and represent the solutions on a unit circle.

25. $x^4 - i = 0$ **26.** $x^3 + 1 = 0$

27. $x^5 + 243 = 0$ **28.** $x^4 - 81 = 0$ ← TEST QUESTION

29. $x^3 + 64i = 0$ **30.** $x^6 - 64i = 0$

31. $x^3 - (1 - i) = 0$ **32.** $x^4 + (1 + i) = 0$

CHAPTER 5 · REVIEW EXERCISES

In Exercises 1–12, perform the indicated operations and write the result in standard form.

1. $(7 + 5i) + (-4 + 2i)$ **2.** $-(6 - 2i) + (-8 + 3i)$

3. $\left(\dfrac{\sqrt{2}}{2} - \dfrac{\sqrt{2}}{2}i\right) - \left(\dfrac{\sqrt{2}}{2} + \dfrac{\sqrt{2}}{2}i\right)$

4. $(13 - 8i) - 5i$

5. $5i(13 - 8i)$ **6.** $(1 + 6i)(5 - 2i)$

7. $(10 - 8i)(2 - 3i)$ **8.** $i(6 + i)(3 - 2i)$

9. $\dfrac{6 + i}{i}$ **10.** $\dfrac{3 + 2i}{5 + i}$

11. $\dfrac{4}{(-3i)}$ **12.** $\dfrac{1}{(2 + i)^4}$

In Exercises 13–20, use the discriminant to determine the number of real solutions of the equation.

13. $6x^2 + x - 2 = 0$ **14.** $20x^2 - 21x + 4 = 0$

15. $9x^2 - 12x + 4 = 0$ **16.** $3x^2 + 5x + 3 = 0$

17. $0.13x^2 - 0.45x + 0.65 = 0$

18. $4x^2 + \frac{4}{3}x + \frac{1}{9} = 0$

19. $15 + 2x - x^2 = 0$ **20.** $x(x + 12) = -46$

In Exercises 21–26, find all the zeros of the function.

21. $g(x) = x^2 - 2x$ **22.** $f(x) = 6x - x^2$

23. $f(x) = x^2 + 8x + 10$ **24.** $h(x) = 3 + 4x - x^2$

25. $r(x) = 2x^2 + 2x + 3$ **26.** $s(x) = 2x^2 + 5x + 4$

In Exercises 27–34, use the given zero of the function as an aid in finding all its zeros. Write the polynomial as a product of linear factors. Check the real zeros by graphing the function with a graphing utility.

Function	Zero
27. $f(x) = 4x^3 - 11x^2 + 10x - 3$	1
28. $f(x) = 10x^3 + 21x^2 - x - 6$	-2
29. $f(x) = x^3 + 3x^2 - 5x + 25$	-5
30. $g(x) = x^3 - 5x^2 - 9x + 45$	-3
31. $h(x) = x^3 - 18x^2 + 106x - 200$	$7 + i$
32. $f(x) = 5x^3 - 4x^2 + 20x - 16$	$2i$
33. $f(x) = x^4 + 5x^3 + 2x^2 - 50x - 84$	$-3 + \sqrt{5}i$
34. $g(x) = x^4 - 8x^3 + 24x^2 - 36x + 27$	$1 + \sqrt{2}i$

In Exercises 35–38, find a polynomial with integer coefficients that has the given zeros.

35. $-1, -1, \frac{1}{3}, -\frac{1}{2}$

36. $5, 1 - \sqrt{2}, 1 + \sqrt{2}$

37. $\frac{2}{3}, 4, \sqrt{3}i, -\sqrt{3}i$

38. $2, -3, 1 - 2i, 1 + 2i$

In Exercises 39–42, find the trigonometric form of the complex number.

39. $5 - 5i$

40. $-3\sqrt{3} + 3i$

41. $5 + 12i$

42. -7

In Exercises 43–46, write the complex number in standard form.

43. $100(\cos 240° + i \sin 240°)$

44. $24(\cos 330° + i \sin 330°)$

45. $13(\cos 0 + i \sin 0)$

46. $8\left(\cos \dfrac{5\pi}{6} + i \sin \dfrac{5\pi}{6}\right)$

In Exercises 47–50, (a) express the two complex numbers in trigonometric form, and (b) use the trigonometric form to find $z_1 z_2$ and z_1/z_2.

47. $z_1 = -6, \quad z_2 = 5i$

48. $z_1 = 2\sqrt{3} - 2i, \quad z_2 = -10i$

49. $z_1 = -3(1 + i), \quad z_2 = 2(\sqrt{3} + i)$

50. $z_1 = 5i, \quad z_2 = 2(1 - i)$

In Exercises 51–54, use DeMoivre's Theorem to find the indicated power of the complex number. Express the result in standard form.

51. $\left[5\left(\cos \dfrac{\pi}{12} + i \sin \dfrac{\pi}{12}\right)\right]^4$

52. $\left[2\left(\cos \dfrac{4\pi}{15} + i \sin \dfrac{4\pi}{15}\right)\right]^5$

53. $(2 + 3i)^6$

54. $(1 - i)^8$

In Exercises 55–58, use DeMoivre's Theorem to find the roots of the complex number.

55. Sixth roots of: $-729i$ **56.** Fourth roots of: 256

57. Cube roots of: -1 **58.** Fourth roots of: $-1 + i$

In Exercises 59–62, find all solutions to the equation and represent the solutions graphically.

59. $x^4 + 81 = 0$ **60.** $x^5 - 32 = 0$

61. $(x^3 - 1)(x^2 + 1) = 0$ **62.** $x^3 + 8i = 0$

CHAPTERS 3–5 · CUMULATIVE TEST

Take this test as you would take a test in class. After you are done, check your work against the answers in the back of the book.

1. Add and simplify: $\dfrac{1 + \cos \beta}{\sin \beta} + \dfrac{\sin \beta}{1 + \cos \beta}$

2. Use the trigonometric substitution $x = 5 \sec \theta$ to write the expression $\sqrt{x^2 - 25}$ as a trigonometric expression of θ where $0 < \theta < \pi/2$. Simplify the expression.

3. Find all solutions to the equation $2 \cos^2 x - \cos x = 0$ in the interval $[0, 2\pi)$.

4. Given $\sin x = \frac{2}{3}$, find the exact value of $\sin(2x)$, where x is a first-quadrant angle.

5. Evaluate $\cos 105° = \cos(135° - 30°)$ *without* the aid of a calculator.

6. Find the remaining sides and angles of each triangle.

(a)

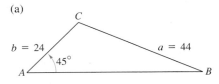

(b)

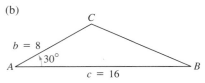

7. In a (square) baseball diamond with 90-foot sides, the pitcher's mound is 60 feet from home plate. How far is a runner from the pitcher's mound when the runner is halfway from third to home?

8. Find the component form of **u** if $\|\mathbf{u}\| = 3$ and $\theta = 30°$. (Assume the angle is measured counterclockwise from the x-axis to the vector.)

9. A projectile is fired with an initial velocity of 120 feet per second at an angle of 15° with the horizontal. Find the vertical and horizontal components of the velocity accurate to two decimal places.

10. Find $\mathbf{u} \cdot \mathbf{v}$ and $\mathbf{u} \times \mathbf{v}$ if $\mathbf{u} = \langle -3, 4, 1 \rangle$ and $\mathbf{v} = \langle 5, 0, 2 \rangle$.

11. Find the angle θ between the vectors $\mathbf{u} = \langle 2, 2, 0 \rangle$ and $\mathbf{v} = \langle -2, 0, 4 \rangle$.

12. Find an equation of the sphere with $(0, 4, 0)$ and $(0, 0, 3)$ as endpoints of a diameter.

13. Perform the specified operation and write the result in standard form.

    (a) $\sqrt{-6} \cdot \sqrt{-9}$ (b) $(4 - 2i) - (3.2 - 5i)$

    (c) $(3 - 2i)(5 + 4i)$ (d) $\dfrac{5}{4 - i}$

14. Find all the zeros of the function $f(x) = 2x^3 - 11x^2 + 38x - 39$.

15. Write the complex number $z = 2(1 - i)$ in trigonometric form and find z^3.

16. Represent the complex number $4(\cos 210° + i \sin 210°)$ graphically and find the standard form of the number.

17. Find the cube roots of 27.

CHAPTER 6

EXPONENTIAL AND LOGARITHMIC FUNCTIONS

6.1 EXPONENTIAL FUNCTIONS AND THEIR GRAPHS

Exponential Functions / Graphs of Exponential Functions / The Natural Base e / Compound Interest / Other Applications

Exponential Functions

In this chapter you will study two types of nonalgebraic functions—*exponential* functions and *logarithmic* functions. These functions are **transcendental functions.**

Exponential functions are used in describing economic and physical phenomena such as compound interest, population growth, memory retention, and decay of radioactive material. Exponential functions involve a *constant base* and a *variable exponent* such as $f(x) = 2^x$ or $g(x) = 3^{-x}$.

DEFINITION OF EXPONENTIAL FUNCTION

The **exponential function** f **with base** a is denoted by

$$f(x) = a^x,$$

where $a > 0$, $a \neq 1$, and x is any real number.

REMARK The base $a = 1$ is not used because it yields $f(x) = 1^x = 1$, which is a constant function, not an exponential function.

You already know how to evaluate a^x for integer and rational values of x. For example, $8^3 = 512$ and $8^{2/3} = 4$. However, to evaluate a^x for any real number x, you need to interpret forms with *irrational* exponents, such as $a^{\sqrt{2}}$

353

and 8^{π}. A technical definition of such forms is beyond the scope of this text. For now, it is sufficient to think of $a^{\sqrt{2}}$ as that value that has the successively closer approximations

$$a^{1.4}, a^{1.41}, a^{1.414}, a^{1.4142}, a^{1.41421}, a^{1.414214}, \ldots$$

The properties of exponents can be extended to cover exponential functions.

EXAMPLE 1   Using a Calculator to Evaluate Exponential Expressions

Number	Display	Rounded to 3 Decimal Places
A. $(1.085)^3$	1.277289125	1.277
B. $12^{5/7}$	5.899887726	5.900
C. $2^{-\pi}$	0.1133147323	0.113

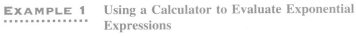

Graphs of Exponential Functions

EXAMPLE 2   Graphs of $y = a^x$

On the same coordinate plane, graph the following functions.

A. $f(x) = 2^x$ **B.** $g(x) = 4^x$

SOLUTION

Table 6.1 lists some values for each function, and Figure 6.1 shows their graphs. Note that both graphs are decreasing. Moreover, the graph of $g(x) = 4^x$ is increasing more rapidly than the graph of $f(x) = 2^x$.

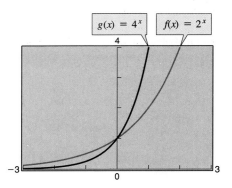

FIGURE 6.1

TABLE 6.1

x	-2	-1	0	1	2	3
A. $f(x) = 2^x$	$\frac{1}{4}$	$\frac{1}{2}$	1	2	4	8
B. $g(x) = 4^x$	$\frac{1}{16}$	$\frac{1}{4}$	1	4	16	64

EXAMPLE 3   Graphs of $y = a^{-x}$

On the same coordinate plane, graph the following functions.

A. $F(x) = 2^{-x} = \left(\frac{1}{2}\right)^x$ **B.** $G(x) = 4^{-x} = \left(\frac{1}{4}\right)^x$

REMARK   Examples 2 and 3 illustrate that the function $f(x) = a^x$ *increases* if $a > 1$ and *decreases* if $0 < a < 1$.

SOLUTION

Table 6.2 lists some values for each function, and Figure 6.2 shows their graphs. Note that both graphs are decreasing. Moreover, the graph of $G(x) = 4^{-x}$ is decreasing more rapidly than the graph of $F(x) = 2^{-x}$.

TABLE 6.2

x	-3	-2	-1	0	1	2
A. $F(x) = 2^{-x}$	8	4	2	1	$\frac{1}{2}$	$\frac{1}{4}$
B. $G(x) = 4^{-x}$	64	16	4	1	$\frac{1}{4}$	$\frac{1}{16}$

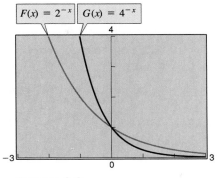

FIGURE 6.2

DISCOVERY

Use a graphing utility to graph $y = a^x$ with $a = 2, 3$, and 5 on the same viewing rectangle. (Use a viewing rectangle in which $-2 \le x \le 1$ and $0 \le y \le 2$.) How do the graphs compare with each other? Which graph is on the top in the interval $(-\infty, 0)$? Which is on the bottom? Which graph is on the top in the interval $(0, \infty)$? Which is on the bottom?

Repeat this experiment with the graphs of $y = a^{-x}$ with $a = 2, 3$, and 5.

The graphs in Figures 6.1 and 6.2 are typical of the exponential functions a^x and a^{-x}. They have $(0, 1)$ as their y-intercept, they have the x-axis as a horizontal asymptote, and they are continuous. The basic characteristics of these exponential functions are summarized in Figure 6.3. Try using a graphing utility to verify the characteristics listed in this summary.

Graph of $y = a^x$
- Domain: $(-\infty, \infty)$
- Range: $(0, \infty)$
- Intercept: $(0, 1)$
- Increasing
- x-axis is a horizontal asymptote ($a^x \to 0$ as $x \to -\infty$)
- Continuous

Graph of $y = a^{-x}$
- Domain: $(-\infty, \infty)$
- Range: $(0, \infty)$
- Intercept: $(0, 1)$
- Decreasing
- x-axis is a horizontal asymptote ($a^{-x} \to 0$ as $x \to \infty$)
- Continuous
- Reflection of graph of $y = a^x$ about y-axis

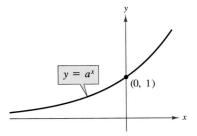

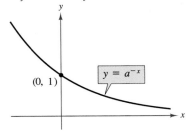

Characteristics of the Exponential Functions a^x and $a^{-x}(a > 1)$

FIGURE 6.3

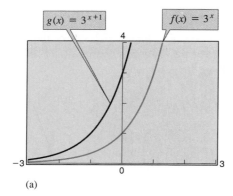

(a)

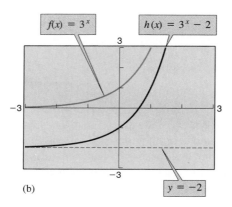

(b)

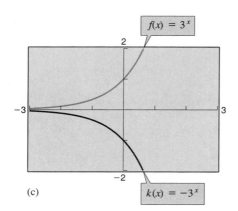

(c)

FIGURE 6.4

EXAMPLE 4   **Graphing Exponential Functions**

Compare the graph of each of the following with the graph of $f(x) = 3^x$. Identify the domain and range of each function.

A. $g(x) = 3^{x+1}$
B. $h(x) = 3^x - 2$
C. $k(x) = -3^x$

SOLUTION

A. Because $g(x) = 3^{x+1} = f(x + 1)$, the graph of g can be obtained by shifting the graph of f one unit to the left, as shown in Figure 6.4(a). The domain is $(-\infty, \infty)$ and the range is $(0, \infty)$.
B. Because $h(x) = 3^x - 2 = f(x) - 2$, the graph of h can be obtained by shifting the graph of f down two units, as shown in Figure 6.4(b). The domain is $(-\infty, \infty)$ and the range is $(-2, \infty)$.
C. Because $k(x) = -3^x = -f(x)$, the graph of k can be obtained by reflecting the graph of f in the x-axis, as shown in Figure 6.4(c). The domain is $(-\infty, \infty)$ and the range is $(-\infty, 0)$.

The Natural Base e

For many applications, the convenient choice for a base is the irrational number

$$e \approx 2.71828 \ldots,$$

called the **natural base**. The function $f(x) = e^x$ is the **natural exponential function**. Its graph is shown in Figure 6.5. Be sure you see that for the exponential function $f(x) = e^x$, e is the constant $2.71828 \ldots$, whereas x is the variable.

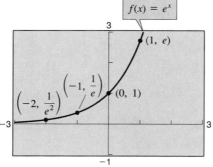

FIGURE 6.5

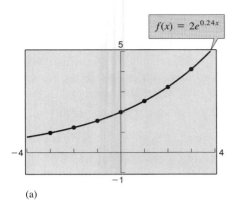

$f(x) = 2e^{0.24x}$

(a)

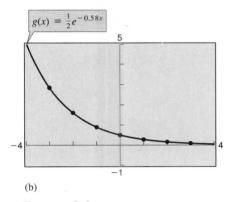

$g(x) = \frac{1}{2}e^{-0.58x}$

(b)

FIGURE 6.6

EXAMPLE 5   **Evaluating the Natural Exponential Function**

Number	Display	Rounded to 3 Decimal Places
A. e^2	7.389056099	7.389
B. e^{-1}	0.3678794412	0.368
C. $e^{0.48}$	1.616074402	1.616

EXAMPLE 6   **Graphing a Natural Exponential Function**

Graph the following natural exponential functions.

A. $f(x) = 2e^{0.24x}$ **B.** $g(x) = \frac{1}{2}e^{-0.58x}$

SOLUTION

In Figure 6.6, note that each graph has the x-axis as a horizontal asymptote.

EXAMPLE 7   **Approximation of the Number e**

Evaluate the expression

$$\left(1 + \frac{1}{x}\right)^x$$

for several large values of x to see that the values approach $e \approx 2.71828$ as x increases without bound.

SOLUTION

Using a calculator, you can complete Table 6.3. From this table, it seems reasonable to conclude that

$$\left(1 + \frac{1}{x}\right)^x \to e \quad \text{as} \quad x \to \infty.$$

You can further confirm this conclusion by using a graphing utility to graph $f(x) = [1 + (1/x)]^x$ and $y = e$ on the same display, as shown in Figure 6.7. Note that as x increases, the graph of f gets closer and closer to the line given by $y = e$.

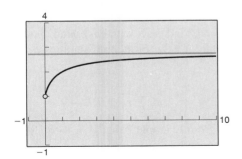

FIGURE 6.7

TABLE 6.3

x	10	100	1000	10,000	100,000	1,000,000
$\left(1 + \frac{1}{x}\right)^x$	2.59374	2.70481	2.71692	2.71815	2.71827	2.71828

D I S C O V E R Y

Use a graphing utility to graph
$y = (1 + x)^{1/x}$.

What is the domain of this
function? Describe the behavior
of the graph near $x = 0$. Is there
a y-intercept? How does the
behavior of the graph near $x = 0$
relate to the result of Example 7?

Compound Interest

One of the most familiar examples of exponential growth is that of an investment earning *continuously compounded interest*. Suppose a principal P is invested at an annual percentage rate r, compounded once a year. If the interest is added to the principal at the end of the year, then the balance is

$$P_1 = P + Pr = P(1 + r).$$

This pattern of multiplying the previous principal by $1 + r$ is then repeated each successive year, as shown in Table 6.4.

TABLE 6.4

Time in years	Balance after each compounding
0	$P = P$
1	$P_1 = P(1 + r)$
2	$P_2 = P_1(1 + r) = P(1 + r)(1 + r) = P(1 + r)^2$
3	$P_3 = P_2(1 + r) = P(1 + r)^2(1 + r) = P(1 + r)^3$
.	.
.	.
.	.
n	$P_n = P(1 + r)^n$

To accommodate more frequent (quarterly, monthly, or daily) compounding of interest, let n be the number of compoundings per year and let t be the number of years. Then the rate per compounding is r/n, and the account balance after t years is

$$A = P\left(1 + \frac{r}{n}\right)^{nt}. \qquad \text{\textit{Amount with n compoundings per year}}$$

If you let the number of compoundings, n, increase without bound, you approach **continuous compounding.** In the formula for n compoundings per year, let $m = n/r$. This produces

$$A = P\left(1 + \frac{r}{n}\right)^{nt} = P\left(1 + \frac{1}{m}\right)^{mrt} = P\left[\left(1 + \frac{1}{m}\right)^m\right]^{rt}.$$

As m increases without bound, you know from Example 7 that $[1 + (1/m)]^m$ approaches e. Hence, for continuous compounding, it follows that

$$P\left[\left(1 + \frac{1}{m}\right)^m\right]^{rt} \rightarrow P(e)^{rt},$$

and you can write $A = Pe^{rt}$. This result is part of the reason that e is the "natural" choice for a base of an exponential function.

FORMULAS FOR COMPOUND INTEREST

After t years, the balance A in an account with principal P and annual percentage rate r (expressed as a decimal) is given by the following formulas.

1. For n compoundings per year: $A = P\left(1 + \dfrac{r}{n}\right)^{nt}$

2. For continuous compounding: $A = Pe^{rt}$

EXAMPLE 8   Finding the Balance for Compound Interest

A sum of $9000 is invested at an annual percentage rate of 8.5%, compounded annually. Find the balance in the account after 3 years.

SOLUTION

In this case, $P = 9000$, $r = 8.5\% = 0.085$, $n = 1$, and $t = 3$. Using the formula

$$A = P\left(1 + \frac{r}{n}\right)^{nt},$$

you have

$$A = 9000(1 + 0.085)^3 = 9000(1.085)^3 \approx \$11,495.60.$$

The graph of the balance in the account after t years is shown in Figure 6.8.

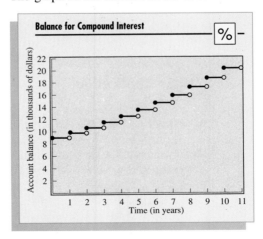

FIGURE 6.8

EXAMPLE 9 Compounding n Times and Compounding Continuously

A total of $12,000 is invested at an annual percentage rate of 9%. Find the balance after 5 years if it is compounded

A. quarterly.
B. continuously.

SOLUTION

A. For quarterly compoundings, you have $n = 4$. Thus, in 5 years at 9%, the balance is

$$A = P\left(1 + \frac{r}{n}\right)^{nt}$$

$$= 12,000\left(1 + \frac{0.09}{4}\right)^{4(5)}$$

$$= \$18,726.11.$$

B. Compounding continuously, the balance is

$$A = Pe^{rt} = 12,000e^{0.09(5)} = \$18,819.75.$$

Note that continuous compounding yields

$$\$18,819.75 - \$18,726.11 = \$93.64$$

more than quarterly compounding.

Other Applications

EXAMPLE 10 An Application Involving Radioactive Decay

Let y represent the mass of a particular radioactive element whose half-life is 25 years. After t years, the mass (in grams) is given by

$$y = 10\left(\frac{1}{2}\right)^{t/25}.$$

A. What is the initial mass (when $t = 0$)?
B. How much of the initial mass is present after 80 years?

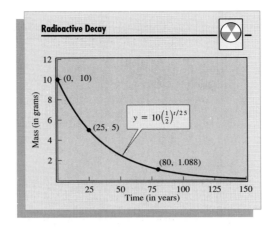

FIGURE 6.9

*Technology Note*_____

When will the mass be four grams in Example 10? You can easily determine this by graphing

$y_1 = 10(.5)^{t/25}$

$y_2 = 4$

on the same viewing rectangle. You should be able to verify that the point of intersection is approximately $t = 33$.

SOLUTION

A. When $t = 0$, the mass is

$$y = 10\left(\frac{1}{2}\right)^0 = 10(1) = 10 \text{ grams.}$$

B. When $t = 80$, the mass is

$$y = 10\left(\frac{1}{2}\right)^{80/25} = 10(0.5)^{3.2} \approx 1.088 \text{ grams.}$$

The graph of this function is shown in Figure 6.9.

EXAMPLE 11   Population Growth

The approximate number of fruit flies in an experimental population after t hours is given by

$$Q(t) = 20e^{0.03t}, \qquad t \geq 0.$$

A. Find the initial number of fruit flies in the population.
B. How large is the population of fruit flies after 72 hours?
C. Sketch the graph of Q.

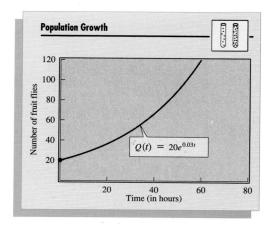

FIGURE 6.10

A. To find the initial population, evaluate $Q(t)$ at $t = 0$.

$$Q(0) = 20e^{0.03(0)} = 20e^0 = 20(1) = 20 \text{ flies}$$

B. After 72 hours, the population size is

$$Q(72) = 20e^{(0.03)(72)} = 20e^{2.16} \approx 173 \text{ flies.}$$

C. The graph of Q is shown in Figure 6.10.

DISCUSSION PROBLEM

EXPONENTIAL GROWTH

Consider the following sequences of numbers.

Sequence 1:   2, 4, 6, 8, 10, 12, . . . , $2n$

Sequence 2:   2, 4, 8, 16, 32, 64, . . . , 2^n

The first sequence, $f(n) = 2n$, is an example of **linear growth.** The second sequence, $f(n) = 2^n$, is an example of **exponential growth.** Which of the following sequences represents linear growth and which represents exponential growth? Can you find a linear function and an exponential function that represent the two sequences?

(a) 3, 6, 9, 12, 15, . . .
(b) 3, 9, 27, 81, 243, . . .

Later you will see that sequences that represent linear growth are *arithmetic sequences* and sequences that represent exponential growth are *geometric sequences.*

WARM-UP

The following warm-up exercises involve skills that were covered in earlier sections. You will use these skills in the exercise set for this section.

In Exercises 1–10, use the properties of exponents to simplify the expression.

1. $5^{2x}(5^{-x})$

2. $3^{-x}(3^{3x})$

3. $\dfrac{4^{5x}}{4^{2x}}$

4. $\dfrac{10^{2x}}{10^x}$

5. $(4^x)^2$

6. $(4^{2x})^5$

7. $\left(\dfrac{2^x}{3^x}\right)^{-1}$

8. $(4^{6x})^{1/2}$

9. $(2^{3x})^{-1/3}$

10. $(16^x)^{1/4}$

SECTION 6.1 · EXERCISES

In Exercises 1–10, use a calculator to evaluate the expression. Round your result to three decimal places.

1. $(3.4)^{5.6}$

2. $5000(2^{-1.5})$

3. $1000(1.06)^{-5}$

4. $(1.005)^{400}$

5. $5^{-\pi}$

6. $\sqrt[3]{4395}$

7. $100^{\sqrt{2}}$

8. $e^{1/2}$

9. $e^{-3/4}$

10. $e^{3.2}$

In Exercises 11–18, match the exponential function with its graph, and describe the given viewing rectangle. [The graphs are labeled (a), (b), (c), (d), (e), (f), (g), and (h).]

11. $f(x) = 3^x$

12. $f(x) = -3^x$

13. $f(x) = 3^{-x}$

14. $f(x) = -3^{-x}$

15. $f(x) = 3^x - 4$

16. $f(x) = 3^x + 1$

17. $f(x) = -3^{x-2}$

18. $f(x) = 3^{x-2}$

(c)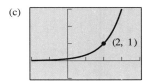

(d)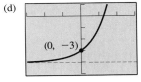

(e)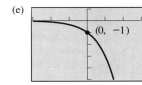

(f)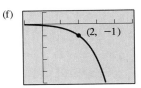

(g)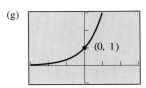

(h)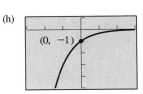

19. Use a graphing utility to graph $y = 3^x$ and $y = 4^x$ on the same viewing rectangle. Use the graphs to solve the inequality

$$4^x < 3^x.$$

20. Use a graphing utility to graph $y = \left(\tfrac{1}{2}\right)^x$ and $y = \left(\tfrac{1}{4}\right)^x$ on the same viewing rectangle. Use the graphs to solve the inequality

$$\left(\tfrac{1}{4}\right)^x < \left(\tfrac{1}{2}\right)^x.$$

(a)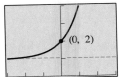

(b)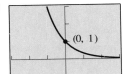

In Exercises 21–30, sketch the graph of the exponential function *by hand*. Then use a graphing utility to confirm your graph.

21. $g(x) = 5^x$

22. $f(x) = \left(\frac{3}{2}\right)^x$

23. $f(x) = \left(\frac{1}{5}\right)^x = 5^{-x}$

24. $h(x) = \left(\frac{3}{2}\right)^{-x}$

25. $h(x) = 5^{x-2}$

26. $g(x) = \left(\frac{3}{2}\right)^{x+2}$

27. $g(x) = 5^{-x} - 3$

28. $f(x) = \left(\frac{3}{2}\right)^{-x} + 2$

29. $y = 3^{x-2} + 1$

30. $y = 4^{x+1} - 2$

In Exercises 31–36, use a graphing utility to graph the exponential function.

31. $y = 1.08^{-5x}$

32. $y = 1.08^{5x}$

33. $s(t) = 2e^{0.12t}$

34. $s(t) = 3e^{-0.2t}$

35. $g(x) = 1 + e^{-x}$

36. $h(x) = e^{x-2}$

37. Graph the following functions in the same viewing rectangle of a graphing utility.
   (a) $f(x) = 3^x$
   (b) $g(x) = f(x - 2) = 3^{x-2}$
   (c) $h(x) = -\frac{1}{2}f(x) = -\frac{1}{2}3^x$
   (d) $q(x) = f(-x) + 3 = 3^{-x} + 3$

38. Use a graphing utility to graph each of the following functions. Use the graphs to determine any asymptotes of the functions.
   (a) $f(x) = \dfrac{8}{1 + e^{-0.5x}}$ (b) $g(x) = \dfrac{8}{1 + e^{-0.5/x}}$

39. Use a graphing utility to graph each of the following functions. Use the graphs to determine where each function is increasing and decreasing, and approximate any relative maximum or minimum values of each function.
   (a) $f(x) = x^2 e^{-x}$ (b) $g(x) = x2^{3-x}$

40. Use a graphing utility to demonstrate that
$$\left(1 + \frac{0.5}{x}\right)^x \rightarrow e^{0.5}$$
as x increases without bound.

In Exercises 41–44, complete the table to determine the balance A for P dollars invested at rate r for t years and compounded n times per year.

n	1	2	3	12	365	Continuous compounding
A						

41. $P = \$2500$, $r = 12\%$, $t = 10$ years

42. $P = \$1000$, $r = 10\%$, $t = 10$ years

43. $P = \$2500$, $r = 12\%$, $t = 20$ years

44. $P = \$1000$, $r = 10\%$, $t = 40$ years

In Exercises 45–48, complete the table to determine the amount of money P that should be invested at rate r to produce a final balance of \$100,000 in t years.

t	1	10	20	30	40	50
P						

45. $r = 9\%$, compounded continuously

46. $r = 12\%$, compounded continuously

47. $r = 10\%$, compounded monthly

48. $r = 7\%$, compounded daily

49. *Trust Fund* On the day of your grandchild's birth, you deposited \$25,000 in a trust fund that pays 8.75% interest, compounded continuously. Determine the balance in this account on your grandchild's 25th birthday.

50. *Trust Fund* Suppose you deposit \$5000 in a trust fund that pays 7.5% interest, compounded continuously. In the trust fund, you specify that the balance will be given to the college from which you graduated after the money has earned interest for 50 years. How much will your college receive after 50 years?

51. *Demand Function* The demand equation for a certain product is
$$p = 500 - 0.5e^{0.004x}.$$
Find the price p for a demand of (a) $x = 1000$ units and (b) $x = 1500$ units. Describe the graph of p. What is the domain? What is the range?

52. *Demand Function* The demand equation for a certain product is
$$p = 5000\left(1 - \frac{4}{4 + e^{-0.002x}}\right).$$
Find the price p for a demand of (a) $x = 100$ units and (b) $x = 500$ units. Describe the graph of p. What is the domain? What is the range?

53. *Bacteria Growth* A certain type of bacteria increases according to the model
$$P(t) = 100e^{0.2197t},$$
where t is the time in hours. Find (a) $P(0)$, (b) $P(5)$, and (c) $P(10)$.

54. *Population Growth* The population of a town increases according to the model
$$P(t) = 2500e^{0.0293t},$$
where t is the time in years, with $t = 0$ corresponding to 1990. Use the model to approximate the population in (a) 1995, (b) 2000, and (c) 2010.

55. *Radioactive Decay*   Let Q represent the mass of radium (^{226}Ra) whose half-life is 1620 years. The quantity of radium present after t years is given by

$$Q = 25 \left(\tfrac{1}{2}\right)^{t/1620}.$$

(a) Determine the initial quantity (when $t = 0$).
(b) Determine the quantity present after 1000 years.
(c) Use a graphing utility to graph this function over the interval $t = 0$ to $t = 5000$.

56. *Radioactive Decay*   Let Q represent the mass of carbon-14 (^{14}C) whose half-life is 5730 years. The quantity of carbon-14 present after t years is given by

$$Q = 10\left(\tfrac{1}{2}\right)^{t/5730}.$$

(a) Determine the initial quantity (when $t = 0$).
(b) Determine the quantity present after 2000 years.
(c) Sketch the graph of this function over the interval $t = 0$ to $t = 10,000$.

57. *Forest Defoliation*   To estimate the amount of defoliation caused by the gypsy moth during a given year, a forester counts the number of egg masses on $\frac{1}{40}$ of an acre the preceding fall. The percentage of defoliation y is approximated by

$$y = \frac{300}{3 + 17e^{-1.57x}},$$

where x is the number of egg masses in thousands.

(a) Use a graphing utility to graph the function.
(b) Estimate the percentage of defoliation if 2000 egg masses are counted.
(c) Estimate the number of egg masses that existed if you observe that approximately $\frac{2}{3}$ of a forest is defoliated. (*Source:* Department of Environmental Resources)

58. *Inflation*   If the annual rate of inflation averages 5% over the next 10 years, then the approximate cost C of goods or services during any year in that decade will be given by

$$C(t) = P(1.05)^t,$$

where t is the time in years and P is the present cost. If the price of an oil change for your car is presently $19.95, estimate the price 10 years from now.

59. *Depreciation*   After t years, the value of a car that cost $20,000 is given by

$$V(t) = 20,000 \left(\tfrac{3}{4}\right)^t.$$

Graph the function and determine the value of the car 2 years after it was purchased.

60. Given the exponential function $f(x) = a^x$, show that
(a) $f(u + v) = f(u) \cdot f(v)$ (b) $f(2x) = [f(x)]^2$.

6.2   LOGARITHMIC FUNCTIONS AND THEIR GRAPHS

Logarithmic Functions   /   Graphs of Logarithmic Functions   /
The Natural Logarithmic Function   /   Application

Logarithmic Functions

In Section 1.6, you learned that if a function has the property that no horizontal line intersects its graph more than once, then the function must have an inverse. By looking back at the graphs of the exponential functions introduced in Section 6.1, you will see that every function of the form $f(x) = a^x$ passes the "horizontal line test," and therefore must have an inverse. This inverse function is the **logarithmic function with base a.**

REMARK The equations $y = \log_a x$ and $a^y = x$ are equivalent. The first equation is in logarithmic form and the second is in exponential form.

> ## DEFINITION OF LOGARITHMIC FUNCTION
>
> For $x > 0$ and $0 < a \neq 1$,
>
> $y = \log_a x$ if and only if $a^y = x$.
>
> The function $f(x) = \log_a x$ is the **logarithmic function with base a.**

When evaluating logarithms, remember that *a logarithm is an exponent.* This means that $\log_a x$ is the exponent to which a must be raised to obtain x. For instance, $\log_2 8 = 3$ because 2 must be raised to the third power to obtain 8.

EXAMPLE 1 Evaluating Logarithms

A. $\log_2 32 = 5$ because $2^5 = 32$.

B. $\log_3 27 = 3$ because $3^3 = 27$.

C. $\log_4 2 = \dfrac{1}{2}$ because $4^{1/2} = \sqrt{4} = 2$.

D. $\log_{10} \dfrac{1}{100} = -2$ because $10^{-2} = \dfrac{1}{10^2} = \dfrac{1}{100}$.

E. $\log_3 1 = 0$ because $3^0 = 1$.

F. $\log_2 2 = 1$ because $2^1 = 2$.

The logarithmic function with base 10 is the **common logarithmic function.** On most calculators, this function is denoted by *log*. You can tell whether this key denotes base 10 by evaluating log 10. The display should be 1.

Technology Note _____

Some graphing utilities do not give you an error message when you evaluate $\log_{10}(-2)$. For example, on the TI-85 and the HP 48G, you obtain the ordered pair

(.301029995664, 1.36437635384),

which represents the complex number

$0.301029995664 + 1.36437635384i$.

However, in this text we will still consider the domain of the logarithm function to be the set of positive real numbers.

EXAMPLE 2 Evaluating Logarithms on a Calculator

Number	Display	Rounded to 3 Decimal Places
A. $\log_{10} 54$	1.73239376	1.732
B. $2 \log_{10} 2.5$	0.7958800173	0.796
C. $\log_{10}(-2)$	ERROR	

Note that most calculators display an error message when you try to evaluate $\log_{10}(-2)$. The reason for this is that the domain of every logarithmic function is the set of *positive real numbers.*

PROPERTIES OF LOGARITHMS

1. $\log_a 1 = 0$ because 0 is the power to which a must be raised to obtain 1.
2. $\log_a a = 1$ because 1 is the power to which a must be raised to obtain a.
3. $\log_a a^x = x$ because x is the power to which a must be raised to obtain a^x.

Graphs of Logarithmic Functions

To sketch the graph of $y = \log_a x$, you can use the fact that the graphs of inverse functions are reflections of each other in the line $y = x$.

EXAMPLE 3   Graphs of Exponential and Logarithmic Functions

On the same coordinate plane, sketch graphs of the following functions by hand.

A. $f(x) = 2^x$
B. $g(x) = \log_2 x$

SOLUTION

A. For $f(x) = 2^x$, construct a table of values.

x	-2	-1	0	1	2	3
$f(x) = 2^x$	$\frac{1}{4}$	$\frac{1}{2}$	1	2	4	8

By plotting these points and connecting them with a smooth curve, you obtain the graph shown in Figure 6.11.

B. Because $g(x) = \log_2 x$ is the inverse of $f(x) = 2^x$, the graph of g is obtained by reflecting the graph of f in the line $y = x$, as shown in Figure 6.11. This reflection is formed by interchanging the x- and y-coordinates of the points in the above table of values.

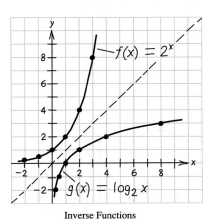

Inverse Functions

FIGURE 6.11

EXAMPLE 4 Sketching the Graph of a Logarithmic Function

Sketch the graph of $f(x) = \log_{10} x$ by hand. Then verify your result with a graphing utility.

SOLUTION

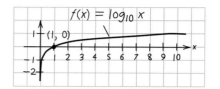

FIGURE 6.12

Begin by making a table of values. Note that some of the values can be obtained without a calculator, whereas others require a calculator. Plot the corresponding points and sketch the graph in Figure 6.12.

	Without a calculator				With a calculator		
x	$\frac{1}{100}$	$\frac{1}{10}$	1	10	2	5	8
$\log_{10} x$	-2	-1	0	1	0.301	0.699	0.903

Compare this graph with that obtained with a graphing utility, and note that the domain and range are $(0, \infty)$ and $(-\infty, \infty)$, respectively.

DISCOVERY

Use a graphing utility to graph $y = \log_{10} x$ and $y = 8$ on the same viewing rectangle. Do the graphs intersect? If so, find a viewing rectangle that shows the point of intersection. What is the point of intersection?

The graph in Figure 6.12 is typical of functions of the form $f(x) = \log_a x$, where $a > 1$. They have one x-intercept, $(1, 0)$, and one vertical asymptote, $x = 0$. Notice how slowly the graph rises for $x > 1$. In Figure 6.12 you would need to move out to $x = 1000$ before the graph would rise to $y = 3$. The basic characteristics of logarithmic graphs are summarized in Figure 6.13.

In Example 5, the graph of $\log_a x$ is used to sketch the graphs of functions of the form $y = b \pm \log_a(x + c)$. The function $f(x) = \log_a(bx + c)$ has a domain that consists of all x such that $bx + c > 0$. The vertical asymptote occurs when $bx + c = 0$, and the x-intercept occurs when $bx + c = 1$.

REMARK In Figure 6.13, note that the vertical asymptote occurs at $x = 0$, where $\log_a x$ is *undefined*. As x gets close to 0, the value of $\log_a$ approaches negative infinity.

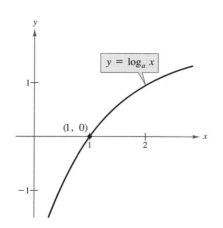

Graph of $y = \log_a x,\ a > 1$
- Domain: $(0, \infty)$
- Range: $(-\infty, \infty)$
- Intercept: $(1, 0)$
- Increasing
- y-axis is a vertical asymptote
  ($\log_a x \to -\infty$ as $x \to 0^+$)
- Continuous
- Reflection of graph of $y = a^x$
  about the line $y = x$

FIGURE 6.13

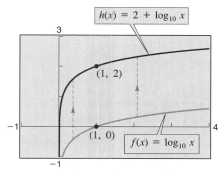

(a) Right shift of 1 unit
Vertical asymptote is $x = 1$.

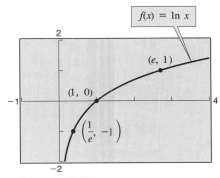

(b) Upward shift of 2 units
Vertical asymptote remains $x = 0$.

FIGURE 6.14

EXAMPLE 5   Sketching the Graphs of Logarithmic Functions

Compare the graphs of the following functions with the graph of $f(x) = \log_{10} x$.

A. $g(x) = \log_{10}(x - 1)$ **B.** $h(x) = 2 + \log_{10} x$

SOLUTION

The graph of each of these functions is similar to the graph of $f(x) = \log_{10} x$, as shown in Figure 6.14.

A. Because $g(x) = \log_{10}(x - 1) = f(x - 1)$, the graph of g can be obtained by shifting the graph of f one unit to the right.

B. Because $h(x) = 2 + \log_{10} x = 2 + f(x)$, the graph of h can be obtained by shifting the graph of f two units up.

The Natural Logarithmic Function

As with exponential functions, the most widely used base for logarithmic functions is the number e. The logarithmic function with base e is called the **natural logarithmic function.** It is denoted by the symbol $\ln x$, read as "el en of x."

THE NATURAL LOGARITHMIC FUNCTION

The function defined by

$$f(x) = \log_e x = \ln x, \qquad x > 0$$

is called the **natural logarithmic function.**

The three properties of logarithms listed earlier in this section are also valid for natural logarithms.

PROPERTIES OF NATURAL LOGARITHMS

1. $\ln 1 = 0$ because 0 is the power to which e must be raised to obtain 1.
2. $\ln e = 1$ because 1 is the power to which e must be raised to obtain e.
3. $\ln e^x = x$ because x is the power to which e must be raised to obtain e^x.

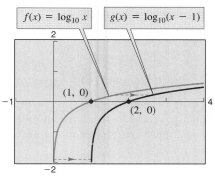

FIGURE 6.15

The graph of the natural logarithmic function is shown in Figure 6.15.

EXAMPLE 6   Evaluating the Natural Logarithmic Function
. . . .

A. $\ln \dfrac{1}{e} = \ln e^{-1} = -1$ *Property 3*

B. $\ln e^2 = 2$ *Property 3*

On most calculators, the natural logarithm is denoted by *ln*. Try using your calculator to evaluate the natural logarithmic expressions shown in the next example.

EXAMPLE 7   Evaluating the Natural Logarithmic Function
. . . .

Number	Display	Rounded to 3 Decimal Places
A. $\ln 2$	0.6931471806	0.693
B. $\ln 0.3$	-1.203972804	-1.204
C. $\ln e$	1	1.000
D. $\ln(-1)$	ERROR	

Be sure you see that $\ln(-1)$ gives an error. This occurs because the domain of $\ln x$ is the set of positive real numbers. (See Figure 6.15.) Hence, $\ln(-1)$ is undefined.

EXAMPLE 8   Finding the Domains of Logarithmic Functions
. . . .

Find the domains of the following functions.

A. $f(x) = \ln(x - 2)$
B. $g(x) = \ln(2 - x)$
C. $h(x) = \ln x^2$

Then use a graphing utility to graph each function.

SOLUTION

A. Because $\ln(x - 2)$ is defined only if $x - 2 > 0$, it follows that the domain of f is $(2, \infty)$. The graph of f is shown in Figure 6.16(a).
B. Because $\ln(2 - x)$ is defined only if $2 - x > 0$, it follows that the domain of g is $(-\infty, 2)$. The graph of g is shown in Figure 6.16(b).
C. Because $\ln x^2$ is defined only if $x^2 > 0$, it follows that the domain of h is all real numbers except $x = 0$. Note that $\ln x^2$ means $\ln(x^2)$. The graph of h is shown in Figure 6.16(c).

REMARK   Note how the graphing utility confirms that the domains are correct.

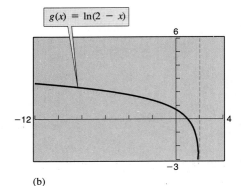

(a)

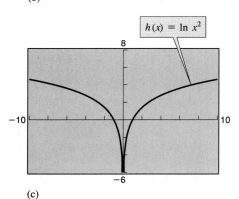

(b)

(c)

FIGURE 6.16

Applications

EXAMPLE 9   Human Memory Model

Students participating in a psychological experiment attended several lectures on a subject. Every month for a year after that, the students were tested to see how much of the material they remembered. The average scores for the group were given by the *human memory model*

$$f(t) = 75 - 6 \ln(t + 1), \qquad 0 \le t \le 12,$$

where *t* is the time in months.

A. What was the average score on the original ($t = 0$) exam?
B. What was the average score at the end of $t = 2$ months?
C. What was the average score at the end of $t = 6$ months?
D. Choose an appropriate viewing rectangle and sketch the graph of *f*.

SOLUTION

A. The original average score was

$$f(0) = 75 - 6 \ln(0 + 1) = 75 - 6(0) = 75.$$

B. After 2 months, the average score was

$$f(2) = 75 - 6 \ln 3 \approx 75 - 6(1.0986) \approx 68.4.$$

C. After 6 months, the average score was

$$f(6) = 75 - 6 \ln 7 \approx 75 - 6(1.9459) \approx 63.3.$$

D. The graph of *f* is shown in Figure 6.17.

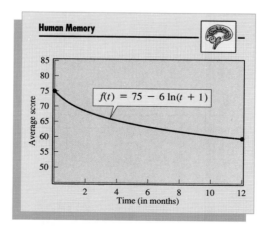

FIGURE 6.17

DISCUSSION PROBLEM

THE GRAPH OF A LOGARITHMIC FUNCTION

Because the range of the logarithmic function $f(x) = \log_a x$ is $(-\infty, \infty)$, you can make the value of $\log_a x$ as large as you want. Can you find values of x that satisfy the following equations?

(a) $\log_{10} x = 10$ (b) $\log_{10} x = 1000$ (c) $\log_{10} x = 10,000,000$

WARM-UP

The following warm-up exercises involve skills that were covered in earlier sections. You will use these skills in the exercise set for this section.

In Exercises 1–4, solve for x.

1. $2^x = 8$ **2.** $4^x = 1$
3. $10^x = 0.1$ **4.** $e^x = e$

In Exercises 5 and 6, evaluate the expression. (Round your result to three decimal places.)

5. e^2 **6.** e^{-1}

In Exercises 7–10, describe how the graph of g is related to the graph of f.

7. $g(x) = f(x + 2)$ **8.** $g(x) = -f(x)$
9. $g(x) = -1 + f(x)$ **10.** $g(x) = f(-x)$

SECTION 6.2 · EXERCISES

In Exercises 1–12, evaluate the expression without using a calculator.

1. $\log_2 16$ **2.** $\log_2 \frac{1}{8}$
3. $\log_{16} 4$ **4.** $\log_{27} 9$
5. $\log_7 1$ **6.** $\log_{10} 1000$
7. $\log_{10} 0.01$ **8.** $\log_{10} 10$
9. $\ln e^3$ **10.** $\ln 1$
11. $\log_a a^2$ **12.** $\log_a \frac{1}{a}$

In Exercises 13–20, use the definition of a logarithm to rewrite the exponential equation as a logarithmic equation. For instance, the logarithmic form of $2^3 = 8$ is $\log_2 8 = 3$.

13. $5^3 = 125$ **14.** $8^2 = 64$
15. $81^{1/4} = 3$ **16.** $9^{3/2} = 27$
17. $6^{-2} = \frac{1}{36}$ **18.** $10^{-3} = 0.001$
19. $e^3 = 20.0855 \ldots$ **20.** $e^0 = 1$

In Exercises 21–26, use a calculator to evaluate the logarithm. (Round your result to three decimal places.)

21. $\log_{10} 345$

22. $\log_{10}(\frac{4}{5})$

23. $\log_{10} 0.48$

24. $\log_{10} 12.5$

25. $\ln 18.42$

26. $\ln\left(\sqrt{5} - 2\right)$

In Exercises 27–30, graph f and g on the same coordinate plane to demonstrate that one is the inverse of the other.

27. $f(x) = 3^x$, $g(x) = \log_3 x$

28. $f(x) = 5^x$, $g(x) = \log_5 x$

29. $f(x) = e^x$, $g(x) = \ln x$

30. $f(x) = 10^x$, $g(x) = \log_{10} x$

In Exercises 31–36, use the graph of $y = \ln x$ to match the function with its graph, and describe the given viewing rectangle. [The graphs are labeled, (a), (b), (c), (d), (e), and (f).]

31. $f(x) = \ln x + 2$

32. $f(x) = -\ln x$

33. $f(x) = -\ln(x + 2)$

34. $f(x) = \ln(x - 1)$

35. $f(x) = \ln(1 - x)$

36. $f(x) = -\ln(-x)$

(a)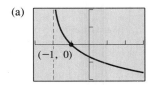

(b)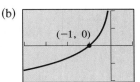

(c)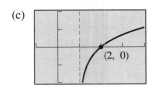

(d)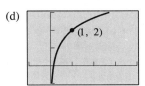

(e)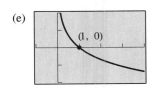

(f)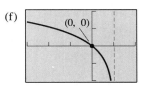

In Exercises 37–42, find the domain, vertical asymptote, and x-intercept of the logarithmic function, and sketch its graph.

37. $f(x) = \log_4 x$

38. $f(x) = -\log_6(x + 2)$

39. $y = -\log_3(x + 2)$

40. $y = \log_{10}\left(\frac{x}{5}\right)$

41. $f(x) = \ln(x - 2)$

42. $g(x) = \ln(-x)$

In Exercises 43 and 44, use a graphing utility to graph the function. Use the graph to determine the intervals in which the function is increasing and decreasing and approximate any relative maximum or minimum values of the function.

43. $f(x) = \frac{x}{2} - \ln\frac{x}{4}$

44. $g(x) = \frac{12 \ln x}{x}$

45. Use a graphing utility to graph f and g on the same screen. Then determine which is increasing at the greater rate for "large" values of x. What can you conclude about the rate of growth of the natural logarithmic function?

(a) $f(x) = \ln x$, $g(x) = \sqrt{x}$

(b) $f(x) = \ln x$, $g(x) = \sqrt[4]{x}$

46. The table of values was obtained by evaluating a function. Determine which of the statements may be true and which must be false.

x	1	2	8
y	0	1	3

(a) y is an exponential function of x.

(b) y is a logarithmic function of x.

(c) x is an exponential function of y.

(d) y is linear function of x.

47. *Human Memory Model* Students in a mathematics class were given an exam and then tested monthly with an equivalent exam. The average score for the class was given by the human memory model

$$f(t) = 80 - 17 \log_{10}(t + 1), \qquad 0 \le t \le 12,$$

where t is the time in months.

(a) What was the average score on the original exam $(t = 0)$?

(b) What was the average score after 4 months?

(c) What was the average score after 10 months?

48. *Population Growth* The population of a town will double in

$$t = \frac{10 \ln 2}{\ln 67 - \ln 50} \text{ years.}$$

Find t.

49. *World Population Growth* The time in years required for the world population to double if it is increasing at a continuous rate of r is given by

$$t = \frac{\ln 2}{r}.$$

Complete the table.

r	0.005	0.010	0.015	0.020	0.025	0.030
t						

50. *Investment Time* A principal P, invested at $9\frac{1}{2}\%$ and compounded continuously, increases to an amount K times the original principal after t years, where t is given by

$$t = \frac{\ln K}{0.095}.$$

(a) Complete the table.

K	1	2	4	6	8	10	12
t							

(b) Use a graphing utility to graph this function.

Ventilation Rates In Exercises 51 and 52, use the model

$$y = 80.4 - 11 \ln x, \qquad 100 \le x \le 1500,$$

which approximates the minimum required ventilation rate in terms of the air space per child in a public school classroom. In the model, x is the air space per child in cubic feet and y is the ventilation rate in cubic feet per minute.

51. Use a graphing utility to graph the function and approximate the required ventilation rate if there is 300 cubic feet of air space per child.

52. A classroom is designed for 30 students. The air-conditioning system in the room has the capacity of moving 450 cubic feet of air per minute.
(a) Determine the ventilation rate per child assuming the room is filled to capacity.
(b) Use the graph from Exercise 51 to estimate the air space required per child.
(c) Determine the minimum number of square feet of floor space required for the room if the ceiling height is 30 feet.

Monthly Payment In Exercises 53–56, use the model

$$t = \frac{5.315}{-6.7968 + \ln x}, \qquad 1000 < x,$$

which approximates the length of a home mortgage of $120,000 at 10% in terms of the monthly payment. In the model, t is the length of the mortgage in years and x is the monthly payment in dollars (see figure).

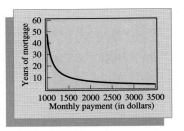

FIGURE FOR 53–56

53. Use the model to approximate the length of a home mortgage (for $120,000 at 10%) that has a monthly payment of $1167.41.

54. Use the model to approximate the length of a home mortgage (for $120,000 at 10%) that has a monthly payment of $1068.45.

55. Approximate the total amount paid over the term of a mortgage with a monthly payment of $1167.41.

56. Approximate the total amount paid over the term of a mortgage with a monthly payment of $1068.45.

57. *Work* The work (in foot-pounds) done in compressing an initial volume of 9 cubic feet at a pressure of 15 pounds per square inch to a volume of 3 cubic feet is

$$W = 19,440(\ln 9 - \ln 3).$$

Find W.

58. *Sound Intensity*   The relationship between the number of decibels β and the intensity of a sound I in watts per square meter is given by

$$\beta = 10 \log_{10}\left(\frac{I}{10^{-16}}\right).$$

Determine the number of decibels of a sound with an intensity of 10^{-4} watts per square meter.

59. (a) Use a calculator to complete the table for the function

$$f(x) = \frac{\ln x}{x}.$$

x	1	5	10	10^2	10^4	10^6
$f(x)$						

(b) Use the table in part (a) to determine what $f(x)$ approaches as x increases without bound.

(c) Use a graphing utility to confirm the result of part (b).

60. Answer the following for the function $f(x) = \log_{10} x$. Do not use a calculator.

(a) What is the domain of f?

(b) Find f^{-1}.

(c) If x is a real number between 1000 and 10,000, determine the interval in which $f(x)$ will be found.

(d) Determine the interval in which x will be found if $f(x)$ is negative.

(e) If $f(x)$ is increased by one unit, then x must have been increased by what factor?

(f) If $f(x_1) = 3n$ and $f(x_2) = n$, find the ratio of x_1 to x_2.

61. Use a graphing utility to determine whether $y = 2 \ln x$ and $y = \ln x^2$ are identical. Explain your reasoning.

6.3   PROPERTIES OF LOGARITHMS

Change of Base   /   Properties of Logarithms   /
Rewriting Logarithmic Expressions

Change of Base

Most calculators have only two types of "log keys," one for common logarithms (base 10) and one for natural logarithms (base e). Although common logs and natural logs are the most frequently used, you may occasionally need to evaluate logarithms to other bases. To do this, the following *change of base formula* is useful. (This formula is derived in Example 9 in Section 6.4.)

REMARK   One way to look at the change of base formula is that logarithms to base a are simply *constant multiples* of logarithms to base b. The constant multiplier is $1/(\log_b a)$.

CHANGE OF BASE FORMULA

Let a, b, and x be positive real numbers such that $a \neq 1$ and $b \neq 1$. Then $\log_a x$ is given by

$$\log_a x = \frac{\log_b x}{\log_b a}.$$

*Technology Note*_____

Most graphing utilities have two log keys: *ln* is the natural logarithmic function and *log* is the logarithm to base 10. You can graph logarithms to other bases, $y = \log_a(x)$, by using the change of base formula.

$$y = \log_a(x) = \frac{\ln x}{\ln a} = \frac{\log x}{\log a}$$

EXAMPLE 1 Changing Bases

Use *common logarithms* to evaluate the following.

A. $\log_4 30$ **B.** $\log_2 14$

SOLUTION

A. Using the change of base formula with $a = 4$, $b = 10$, and $x = 30$, convert to common logarithms and obtain

$$\log_4 30 = \frac{\log_{10} 30}{\log_{10} 4} \approx \frac{1.47712}{0.60206} \approx 2.4534.$$

B. Using the change of base formula with $a = 2$, $b = 10$, and $x = 14$, convert to common logarithms and obtain

$$\log_2 14 = \frac{\log_{10} 14}{\log_{10} 2} \approx \frac{1.14613}{0.30103} \approx 3.8074.$$

EXAMPLE 2 Changing Bases

Use *natural logarithms* to evaluate the following.

A. $\log_4 30$ **B.** $\log_2 14$

SOLUTION

A. Using the change of base formula with $a = 4$, $b = e$, and $x = 30$, convert to natural logarithms and obtain

$$\log_4 30 = \frac{\ln 30}{\ln 4} \approx \frac{3.40120}{1.38629} \approx 2.4534.$$

B. Using the change of base formula with $a = 2$, $b = e$, and $x = 14$, convert to natural logarithms and obtain

$$\log_2 14 = \frac{\ln 14}{\ln 2} \approx \frac{2.63906}{0.693147} \approx 3.8074.$$

Note that the results agree with those obtained in Example 1, using common logarithms.

D I S C O V E R Y

Use a graphing utility to graph $y = \ln x$ and $y = \ln x/\ln a = \log_a x$ with $a = 2$, 3, and 5 on the same viewing rectangle. (Use a viewing rectangle in which $0 \le x \le 10$ and $-4 \le y \le 4$.) On the interval $(0, 1)$, which graph is on top? Which is on the bottom? On the interval $(1, \infty)$, which graph is on top? Which is on the bottom?

Properties of Logarithms

You know from the previous section that the logarithmic function with base a is the *inverse* of the exponential function with base a. Thus, it makes sense that the properties of exponents should have corresponding properties involving logarithms. For instance, the exponential property $a^0 = 1$ corresponds to the logarithmic property $\log_a 1 = 0$.

In this section you will learn how to use the logarithmic properties that correspond to the following three exponential properties.

1. $a^n a^m = a^{n+m}$
2. $\dfrac{a^n}{a^m} = a^{n-m}$
3. $(a^n)^m = a^{nm}$

PROPERTIES OF LOGARITHMS

Let a be a positive number such that $a \neq 1$, and let n be a real number. If u and v are positive real numbers, then the following properties are true.

Base a Logarithm

1. $\log_a(uv) = \log_a u + \log_a v$
2. $\log_a \dfrac{u}{v} = \log_a u - \log_a v$
3. $\log_a u^n = n \log_a u$

Natural Logarithm

1. $\ln(uv) = \ln u + \ln v$
2. $\ln \dfrac{u}{v} = \ln u - \ln v$
3. $\ln u^n = n \ln u$

PROOF

We give a proof of Property 1 and leave the other two proofs for you. To prove Property 1, let

$$x = \log_a u \quad \text{and} \quad y = \log_a v.$$

The corresponding exponential forms of these two equations are

$$a^x = u \quad \text{and} \quad a^y = v.$$

Multiplying u and v produces $uv = a^x a^y = a^{x+y}$. The corresponding logarithmic form of $uv = a^{x+y}$ is

$$\log_a(uv) = x + y.$$

Hence, $\log_a(uv) = \log_a u + \log_a v$.

REMARK There is no general property that can be used to rewrite $\log_a(u \pm v)$. Specifically,

$$\log_a(x + y) \quad \text{DOES NOT EQUAL} \quad \log_a x + \log_a y.$$

EXAMPLE 3   Using Properties of Logarithms

Given $\ln 2 \approx 0.693$, $\ln 3 \approx 1.099$, and $\ln 7 \approx 1.946$, use the properties of logarithms to approximate the following. Then use a calculator to verify the result.

A. $\ln 6$

B. $\ln \dfrac{7}{27}$

SOLUTION

A. $\ln 6 = \ln(2 \cdot 3)$

$\qquad\quad = \ln 2 + \ln 3$ *Property 1*

$\qquad\quad \approx 0.693 + 1.099$

$\qquad\quad = 1.792$

B. $\ln \dfrac{7}{27} = \ln 7 - \ln 27$ *Property 2*

$\qquad\quad\ = \ln 7 - \ln 3^3$

$\qquad\quad\ = \ln 7 - 3 \ln 3$ *Property 3*

$\qquad\quad\ \approx 1.946 - 3(1.099)$

$\qquad\quad\ = -1.351$

EXAMPLE 4   Using Properties of Logarithms

Use the properties of logarithms to verify that

$$-\ln \frac{1}{2} = \ln 2.$$

SOLUTION

$$-\ln \frac{1}{2} = -\ln(2^{-1}) = -(-1)\ln 2 = \ln 2$$

Try verifying this result on your calculator.

Rewriting Logarithmic Expressions

The properties of logarithms are useful for rewriting logarithmic expressions in forms that simplify the operations of algebra. This is true because they convert complicated products, quotients, and exponential forms into simpler sums, differences, and products, respectively.

EXAMPLE 5   Rewriting the Logarithm of a Product

Use the properties of logarithms to rewrite

$$\log_{10} 5x^3 y$$

as the sum of logarithms.

SOLUTION

$$\begin{aligned} \log_{10} 5x^3 y &= \log_{10} 5 + \log_{10} x^3 y \qquad \text{\textit{Property 1}} \\ &= \log_{10} 5 + \log_{10} x^3 + \log_{10} y \qquad \text{\textit{Property 1}} \\ &= \log_{10} 5 + 3 \log_{10} x + \log_{10} y \qquad \text{\textit{Property 3}} \end{aligned}$$

EXAMPLE 6   Rewriting the Logarithm of a Quotient

Use the properties of logarithms to rewrite

$$\ln \frac{\sqrt{3x - 5}}{7}$$

as the sum and/or difference of logarithms.

SOLUTION

$$\ln \frac{\sqrt{3x - 5}}{7} = \ln(3x - 5)^{1/2} - \ln 7 \qquad \text{\textit{Property 2}}$$

$$= \frac{1}{2} \ln(3x - 5) - \ln 7 \qquad \text{\textit{Property 3}}$$

Examples 5 and 6 use the properties of logarithms to *expand* logarithmic expressions. Examples 7 and 8 reverse the procedure by using properties of logarithms to *condense* logarithmic expressions.

Technology Note _____

When you rewrite a logarithmic expression, be careful to check that the domains are the same. For example, the domain of $\ln x^4$ is all $x \neq 0$, whereas the domain of $4 \ln x$ is $x > 0$. Verify this on your graphing utility by graphing the two functions $y_1 = \ln x^4$ and $y_2 = 4 \ln x$.

EXAMPLE 7   Condensing a Logarithmic Expression

Rewrite the following expression as the logarithm of a single quantity.

$$\frac{1}{2} \log_{10} x - 3 \log_{10}(x + 1)$$

SOLUTION

$$\frac{1}{2} \log_{10} x - 3 \log_{10}(x + 1) = \log_{10} x^{1/2} - \log_{10}(x + 1)^3$$

$$= \log_{10} \frac{\sqrt{x}}{(x + 1)^3}$$

EXAMPLE 8   Condensing a Logarithmic Expression

Rewrite the following expression as the logarithm of a single quantity.

$$2 \ln(x + 2) - \ln x$$

SOLUTION

$$2 \ln(x + 2) - \ln x = \ln(x + 2)^2 - \ln x$$
$$= \ln \frac{(x + 2)^2}{x}$$

When expanding or condensing logarithmic expressions, you should compare the domain of the original expression with the domain of the expanded or condensed expression. For instance, the domain of $\ln x^2$ is all nonzero real numbers, whereas the domain of $2 \ln x$ is all positive real numbers.

DISCUSSION PROBLEM

DEMONSTRATING PROPERTIES OF LOGARITHMS

Use a calculator to demonstrate that

$$\frac{\ln x}{\ln y} \neq \ln \frac{x}{y} = \ln x - \ln y$$

by completing the table.

x	y	$\dfrac{\ln x}{\ln y}$	$\ln \dfrac{x}{y}$	$\ln x - \ln y$
1	2			
3	4			
10	5			
4	0.5			

WARM-UP

The following warm-up exercises involve skills that were covered in earlier sections. You will use these skills in the exercise set for this section.

In Exercises 1–4, evaluate the expression without using a calculator.

1. $\log_7 49$ **2.** $\log_2(\frac{1}{32})$ **3.** $\ln \dfrac{1}{e^2}$ **4.** $\log_{10} 0.001$

In Exercises 5–8, simplify the expression.

5. $e^2 e^3$ **6.** $\dfrac{e^2}{e^3}$ **7.** $(e^2)^3$ **8.** $(e^2)^0$

In Exercises 9 and 10, rewrite the expression in exponential form.

9. $\dfrac{1}{x^2}$ **10.** $\sqrt{x}$

SECTION 6.3 · EXERCISES

In Exercises 1–4, use the change of base formula to write the logarithm as a quotient of common logarithms. For instance, $\log_2 3 = (\log_{10} 3)/(\log_{10} 2)$.

1. $\log_3 5$ **2.** $\log_4 10$ **3.** $\log_2 x$ **4.** $\ln 5$

In Exercises 5–8, use the change of base formula to write the logarithm as a quotient of natural logarithms. For instance, $\log_2 3 = (\ln 3)/(\ln 2)$.

5. $\log_3 5$ **6.** $\log_4 10$ **7.** $\log_2 x$ **8.** $\log_{10} 5$

In Exercises 9–16, evaluate the logarithm using the change of base formula. Do the problem twice, once with common logarithms and once with natural logarithms. Round your result to three decimal places.

9. $\log_3 7$ **10.** $\log_7 4$
11. $\log_{1/2} 4$ **12.** $\log_4 0.55$
13. $\log_9 0.4$ **14.** $\log_{20} 125$
15. $\log_{15} 1250$ **16.** $\log_{1/3} 0.015$

In Exercises 17–36, use the properties of logarithms to write the expression as a sum, difference, and/or constant multiple of logarithms.

17. $\log_{10} 5x$ **18.** $\log_{10} 10z$
19. $\log_{10} \dfrac{5}{x}$ **20.** $\log_{10} \dfrac{y}{2}$
21. $\log_8 x^4$ **22.** $\log_6 z^{-3}$
23. $\ln \sqrt{z}$ **24.** $\ln \sqrt[3]{t}$
25. $\ln xyz$ **26.** $\ln \dfrac{xy}{z}$
27. $\ln \sqrt{a-1}$ **28.** $\ln\left(\dfrac{x^2-1}{x^3}\right)$
29. $\ln z(z-1)^2$ **30.** $\ln \sqrt{\dfrac{x^2}{y^3}}$
31. $\ln \sqrt[3]{\dfrac{x}{y}}$ **32.** $\ln \dfrac{x}{\sqrt{x^2+1}}$
33. $\ln \dfrac{x^4 \sqrt{y}}{z^5}$ **34.** $\ln \sqrt{x^2(x+2)}$
35. $\log_b \dfrac{x^2}{y^2 z^3}$ **36.** $\log_b \dfrac{\sqrt{xy^4}}{z^4}$

In Exercises 37–56, write the expression as the logarithm of a single quantity.

37. $\ln x + \ln 2$

38. $\ln y + \ln z$

39. $\log_4 z - \log_4 y$

40. $\log_5 8 - \log_5 t$

41. $2 \log_2(x + 4)$

42. $-4 \log_6 2x$

43. $\frac{1}{3} \log_3 5x$

44. $\frac{3}{2} \log_7(z - 2)$

45. $\ln x - 3 \ln(x + 1)$

46. $2 \ln 8 + 5 \ln z$

47. $\ln(x - 2) - \ln(x + 2)$

48. $3 \ln x + 2 \ln y - 4 \ln z$

49. $\ln x - 2[\ln(x + 2) + \ln(x - 2)]$

50. $4[\ln z + \ln(z + 5)] - 2 \ln(z - 5)$

51. $\frac{1}{3}[2 \ln(x + 3) + \ln x - \ln(x^2 - 1)]$

52. $2[\ln x - \ln(x + 1) - \ln(x - 1)]$

53. $\frac{1}{3}[\ln y + 2 \ln(y + 4)] - \ln(y - 1)$

54. $\frac{1}{2}[\ln(x + 1) + 2 \ln(x - 1)] + 3 \ln x$

55. $2 \ln 3 - \frac{1}{2} \ln(x^2 + 1)$

56. $\frac{3}{2} \ln 5t^6 - \frac{3}{4} \ln t^4$

In Exercises 57–66, approximate the logarithm using the properties of logarithms, given $\log_b 2 \approx 0.3562$, $\log_b 3 \approx 0.5646$, and $\log_b 5 \approx 0.8271$.

57. $\log_b 6$

58. $\log_b(\frac{5}{3})$

59. $\log_b 40$ $(40 = 2^3 \cdot 5)$

60. $\log_b 18$

61. $\log_b \dfrac{\sqrt{2}}{2}$

62. $\log_b \sqrt[3]{75}$

63. $\log_b \sqrt{5b}$

64. $\log_b(3b^2)$

65. $\log_b \dfrac{(4.5)^3}{\sqrt{3}}$

66. $\log_b 1$

In Exercises 67–72, find the exact value of the logarithm.

67. $\log_3 9$

68. $\log_6 \sqrt[3]{6}$

69. $\log_4 16^{1.2}$

70. $\log_5(\frac{1}{125})$

71. $\ln e^{4.5}$

72. $\ln \sqrt[4]{e^3}$

In Exercises 73–80, use the properties of logarithms to simplify the logarithmic expression.

73. $\log_4 8$

74. $\log_5(\frac{1}{15})$

75. $\log_7 \sqrt{70}$

76. $\log_2(4^2 \cdot 3^4)$

77. $\log_5(\frac{1}{250})$

78. $\log_{10}(\frac{9}{300})$

79. $\ln(5e^6)$

80. $\ln \dfrac{6}{e^2}$

81. *Sound Intensity*   The relationship between the number of decibels β and the intensity of a sound I in watts per square meter is given by

$$\beta = 10 \log_{10}\left(\frac{I}{10^{-16}}\right).$$

Use properties of logarithms to write the formula in simpler form, and determine the number of decibels of a sound with an intensity of 10^{-10} watts per square meter.

82. Approximate the natural logarithms of as many integers as possible between 1 and 20 given that $\ln 2 \approx 0.6931$, $\ln 3 \approx 1.0986$, and $\ln 5 \approx 1.6094$.

83. Use a graphing utility to graph

$$f(x) = \ln \frac{x}{2}, \qquad g(x) = \frac{\ln x}{\ln 2}, \qquad h(x) = \ln x - \ln 2$$

in the same viewing rectangle. Which two functions have identical graphs?

84. Prove that $\log_b \dfrac{u}{v} = \log_b u - \log_b v$.

85. Prove that $\log_b u^n = n \log_b u$.

6.4 SOLVING EXPONENTIAL AND LOGARITHMIC EQUATIONS

Introduction / Solving Exponential Equations / Solving Logarithmic Equations / Approximating Solutions / Application

Introduction

So far in this chapter, you have studied the definitions, graphs, and properties of exponential and logarithmic functions. In this section, you will study procedures for *solving equations* involving these exponential and logarithmic functions. As a simple example, consider the exponential equation $2^x = 32$. You can solve this equation by rewriting it as $2^x = 2^5$, which implies that $x = 5$. Although this method works in some cases, it does not work for an equation as simple as $e^x = 7$. To solve for x in this case, you can take the natural logarithm of both sides to obtain

$e^x = 7$	*Original equation*
$\ln e^x = \ln 7$	*Take ln of both sides*
$x = \ln 7.$	*Solution*

Technology Note

When solving exponential and logarithmic equations you need to be aware that you can introduce "extraneous solutions" during the solution process. Show how your graphing utility can help you eliminate the extraneous solution in the following example.

$2 \ln x = \ln 4$
$\ln x^2 = \ln 4$
$x^2 = 4$
$x = \pm 2$
$x = 2$ ($x = -2$ is extraneous)

GUIDELINES FOR SOLVING EXPONENTIAL AND LOGARITHMIC EQUATIONS

1. *To solve an exponential equation,* first isolate the exponential expression, then take the logarithm of both sides and solve for the variable.
2. *To solve a logarithmic equation,* rewrite the equation in exponential form and solve for the variable.

Note that these two guidelines are based on the **inverse properties** of exponential and logarithmic functions.

Base a	*Base e*
1. $\log_a a^x = x$	$\ln e^x = x$
2. $a^{\log_a x} = x$	$e^{\ln x} = x$

Solving Exponential Equations

EXAMPLE 1 Solving an Exponential Equation

$e^x = 72$	*Original equation*
$\ln e^x = \ln 72$	*Take ln of both sides*
$x = \ln 72$	*Inverse property of logs and exponents*
$x \approx 4.277$	

The solution is $x = \ln 72$. Check this solution in the original equation.

EXAMPLE 2 Solving an Exponential Equation

Solve $e^x + 5 = 60$.

SOLUTION

$e^x + 5 = 60$	*Original equation*
$e^x = 55$	*Subtract 5 from both sides*
$\ln e^x = \ln 55$	*Take ln of both sides*
$x = \ln 55$	*Inverse property of logs and exponents*
$x \approx 4.007$	

The solution is $x = \ln 55$. Check this solution in the original equation. The graph of $y = e^x - 55$ is shown in Figure 6.18. Use the zoom feature of your graphing utility to find the x-intercept and thus confirm the solution.

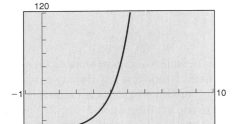

FIGURE 6.18

EXAMPLE 3 Solving an Exponential Equation

$4e^{2x} = 5$	*Original equation*
$e^{2x} = \dfrac{5}{4}$	*Divide both sides by 4*
$\ln e^{2x} = \ln \dfrac{5}{4}$	*Take ln of both sides*
$2x = \ln \dfrac{5}{4}$	*Inverse property of logs and exponents*
$x = \dfrac{1}{2} \ln \dfrac{5}{4}$	*Divide both sides by 2*
$x \approx 0.112$	

The solution is $x = \frac{1}{2} \ln \frac{5}{4}$. Check this solution in the original equation. The graph of $y = 4e^{2x} - 5$, shown in Figure 6.19, helps to confirm this solution.

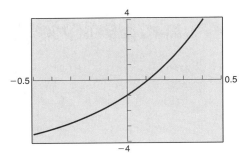

FIGURE 6.19

When an equation involves two or more exponential expressions, you can still use a procedure similar to that demonstrated in the first three examples. However, the algebra is a bit more complicated and a graphical approach is often easier. Study the next example carefully.

EXAMPLE 4   Solving an Exponential Equation

Solve for x in the equation $e^{2x} - 3e^x + 2 = 0$.

SOLUTION

The graph of the function $f(x) = e^{2x} - 3e^x + 2$ in Figure 6.20 indicates that there are two solutions: one near $x = 0$ and one near $x = 0.7$. You can verify that $x = 0$ is indeed a solution by noting that $f(0) = e^{2(0)} - 3e^0 + 2 = 1 - 3 + 2 = 0$. Similarly, you can use the zoom and trace features to determine that $x \approx 0.693$. You can also solve this problem algebraically.

$$e^{2x} - 3e^x + 2 = 0 \qquad \textit{Original equation}$$
$$(e^x)^2 - 3e^x + 2 = 0 \qquad \textit{Quadratic form}$$
$$(e^x - 2)(e^x - 1) = 0 \qquad \textit{Factor}$$
$$e^x - 2 = 0 \qquad e^x - 1 = 0 \qquad \textit{Set factors to zero}$$
$$e^x = 2 \qquad\qquad e^x = 1$$
$$x = \ln 2 \qquad\qquad x = 0 \qquad \textit{Solutions}$$

The equation has two solutions: $x = \ln 2 \approx 0.693$ and $x = 0$, which confirms the graphical analysis.

Examples 1 through 4 all deal with exponential equations in which the base is e. The same approach can be used to solve exponential equations involving other bases, as shown in Example 5.

EXAMPLE 5   A Base Other Than e

Solve $2^x = 10$.

SOLUTION

$$2^x = 10 \qquad \textit{Original equation}$$
$$\ln 2^x = \ln 10 \qquad \textit{Take log of both sides}$$
$$x \ln 2 = \ln 10 \qquad \textit{Property of logarithms}$$
$$x = \frac{\ln 10}{\ln 2} \qquad \textit{Divide both sides by ln 2}$$

The equation has one solution: $x = \ln 10/\ln 2 \approx 3.32$. Check this solution in the original equation. The graph of $y = 2^x - 10$, shown in Figure 6.21, helps to confirm this solution. (*Note:* Using the change of base formula, the solution could be written as $x = \log_2 10$.)

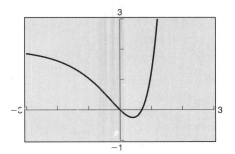

FIGURE 6.20

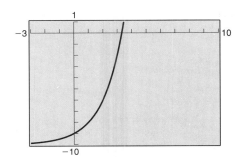

FIGURE 6.21

Solving Logarithmic Equations

To solve a logarithmic equation such as $\ln x = 3$, you can write the equation in exponential form as follows.

$$\ln x = 3 \qquad \textit{Logarithmic form}$$

$$e^{\ln x} = e^3 \qquad \textit{Exponentiate both sides}$$

$$x = e^3 \qquad \textit{Exponential form}$$

This procedure is called *exponentiating* both sides of an equation. It is applied after isolating the logarithmic expression.

EXAMPLE 6   Solving a Logarithmic Equation

Solve $5 + 2 \ln x = 4$.

SOLUTION

$$5 + 2 \ln x = 4 \qquad \textit{Original equation}$$

$$2 \ln x = -1 \qquad \textit{Subtract 5 from both sides}$$

$$\ln x = -\frac{1}{2} \qquad \textit{Divide both sides by 2}$$

$$e^{\ln x} = e^{-1/2} \qquad \textit{Exponentiate both sides}$$

$$x = e^{-1/2} \qquad \textit{Inverse property of exponents and logs}$$

$$x \approx 0.607$$

The equation has one solution: $x = e^{-1/2}$. Check this solution in the original equation. The graph of $y = 1 + 2 \ln x$, shown in Figure 6.22, helps to confirm this solution.

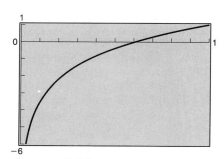

FIGURE 6.22

EXAMPLE 7   Solving a Logarithmic Equation

Solve $2 \ln 3x = 4$.

SOLUTION

$$2 \ln 3x = 4 \qquad \textit{Original equation}$$

$$\ln 3x = 2 \qquad \textit{Divide both sides by 2}$$

$$e^{\ln 3x} = e^2 \qquad \textit{Exponentiate both sides}$$

$$3x = e^2 \qquad \textit{Inverse property of exponents and logs}$$

$$x = \frac{1}{3}e^2 \qquad \textit{Divide both sides by 3}$$

$$x \approx 2.463$$

The equation has one solution: $x = \frac{1}{3}e^2$. Check this solution in the original equation. The graph of $y = -4 + 2 \ln 3x$, shown in Figure 6.23, helps to confirm this solution.

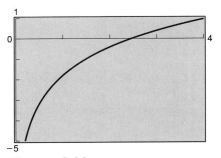

FIGURE 6.23

Complicated equations involving logarithmic expressions can be solved using a graphing utility, as demonstrated in Example 8.

*Technology Note*_____

When solving logarithmic equations algebraically, extraneous solutions are often produced. Using a graphing utility can help you determine which solutions are extraneous.

EXAMPLE 8   Solving a Logarithmic Equation

Solve for x in the equation $\ln(x - 2) + \ln(2x - 3) = 2 \ln x$.

SOLUTION

The graph of the function $f(x) = \ln(x - 2) + \ln(2x - 3) - 2 \ln x$, shown in Figure 6.24, indicates that the only zero is $x = 6$. This can be verified by substituting $x = 6$ into the original equation. You could also solve this equation algebraically. Notice that in this case the technique produces an extraneous solution.

$\ln(x - 2) + \ln(2x - 3) = 2 \ln x$	*Original equation*
$\ln(x - 2)(2x - 3) = \ln x^2$	*Properties of logarithms*
$\ln(2x^2 - 7x + 6) = \ln x^2$	
$e^{\ln(2x^2 - 7x + 6)} = e^{\ln x^2}$	*Exponentiate both sides*
$2x^2 - 7x + 6 = x^2$	*Inverse property of exponents and logs*
$x^2 - 7x + 6 = 0$	*Quadratic form*
$(x - 6)(x - 1) = 0$	*Factor*
$x - 6 = 0 \;\longrightarrow\; x = 6$	*Set 1st factor equal to 0*
$x - 1 = 0 \;\longrightarrow\; x = 1$	*Set 2nd factor equal to 0*

Finally, by checking these two "solutions" in the original equation, you can conclude that $x = 1$ is not valid. Can you see why? Thus, the only solution is $x = 6$.

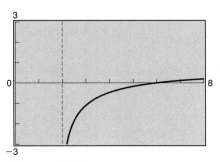

FIGURE 6.24

EXAMPLE 9 The Change of Base Formula

Prove the change of base formula given in Section 6.3.

$$\log_a x = \frac{\log_b x}{\log_b a}$$

SOLUTION

Begin by letting $y = \log_a x$ and writing the equivalent exponential form $a^y = x$. Now, taking the logarithm *with base b* of both sides produces the following.

$$\log_b a^y = \log_b x$$
$$y \log_b a = \log_b x$$
$$y = \frac{\log_b x}{\log_b a}$$
$$\log_a x = \frac{\log_b x}{\log_b a}$$

When solving exponential or logarithmic equations, the following properties are useful.

1. $x = y$ if and only if $\log_a x = \log_a y$.
2. $x = y$ if and only if $a^x = a^y$, $a > 0$, $a \neq 1$.

Can you see where these properties were used in the examples in this section?

Approximating Solutions

Equations that involve combinations of algebraic functions, exponential functions, and/or logarithmic functions can be very difficult to solve by algebraic procedures. Here again you can take advantage of a graphing utility.

EXAMPLE 10 Approximating the Solution of an Equation

Approximate the solutions of $\ln x = x^2 - 2$.

SOLUTION

To begin, use a graphing utility to graph

$$y = -x^2 + 2 + \ln x,$$

as shown in Figure 6.25. From this graph, you can see that the equation has two solutions. Next, using the zoom and trace features, you can approximate the two solutions to be $x \approx 0.138$ and $x \approx 1.564$.

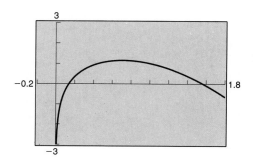

FIGURE 6.25

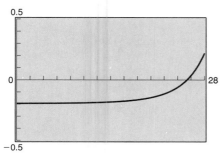

FIGURE 6.26

Applications

EXAMPLE 11   Waste Processed for Energy Recovery

From 1960 to 1986, the amount of municipal waste processed for energy recovery in the United States can be approximated by the equation

$$y = 0.00643e^{0.00533t^2},$$

where y is the amount of waste (in pounds per person) that was processed for energy recovery and t represents the year, with $t = 0$ corresponding to 1960. According to this model, during which year did the amount of waste reach 0.2 pound? (*Source:* Franklin Associates *Characterization of Municipal Solid Waste in U.S.*)

SOLUTION

To solve for t in the equation $0.00643e^{0.00533t^2} = 0.2$, use a graphing utility to graph

$$y = 0.00643e^{0.00533t^2} - 0.2$$

on the domain $0 \leq t \leq 28$. From the graph (Figure 6.26), you can see that there is a zero between $t = 25$ and $t = 26$. Using the zoom and trace features, you can determine that $t \approx 25.4$ years. Because $t = 0$ represents 1960, it follows that the amount of waste would have reached 0.2 pound per person in 1985.

DISCUSSION PROBLEM

VERIFYING INVERSE RELATIONSHIPS

Use a graphing utility to verify the following inverse relationships between logarithmic and exponential functions.

Base 10	*Base e*
1. $\log_{10} 10^x = x$	$\ln e^x = x$
2. $10^{\log_{10} x} = x$	$e^{\ln x} = x$

You can do this by sketching the graph of each relationship. For instance, to verify that $\ln e^x = x$, try sketching the graphs of $y = \ln e^x$ and $y = x$. The two graphs should be identical. Be sure to determine the domain of each function.

WARM-UP

The following warm-up exercises involve skills that were covered in earlier sections. You will use these skills in the exercise set for this section.

In Exercises 1–6, solve for x.

1. $x \ln 2 = \ln 3$ **2.** $(x - 1)\ln 4 = 2$ **3.** $2xe^2 = e^3$
4. $4xe^{-1} = 8$ **5.** $x^2 - 4x + 5 = 0$ **6.** $2x^2 - 3x + 1 = 0$

In Exercises 7–10, simplify the expression.

7. $\log_{10} 100^x$ **8.** $\log_4 64^x$ **9.** $\ln e^{2x}$ **10.** $\ln e^{-x^2}$

SECTION 6.4 · EXERCISES

In Exercises 1–10, solve for x.

1. $4^x = 16$ **2.** $3^x = 243$
3. $7^x = \frac{1}{49}$ **4.** $8^x = 4$
5. $(\frac{3}{4})^x = \frac{27}{64}$ **6.** $3^{x-1} = 27$
7. $\log_4 x = 3$ **8.** $\log_5 5x = 2$
9. $\log_{10} x = -1$ **10.** $\ln(2x - 1) = 0$

In Exercises 11–16, apply the inverse properties of $\ln x$ and e^x to simplify the expression.

11. $\ln e^{x^2}$ **12.** $\ln e^{2x-1}$
13. $e^{\ln(5x+2)}$ **14.** $-1 + \ln e^{2x}$
15. $e^{\ln x^2}$ **16.** $-8 + e^{\ln x^3}$

In Exercises 17–36, solve the exponential equation algebraically. (Round your result to three decimal places.)

17. $e^x = 10$ **18.** $4e^x = 91$
19. $7 - 2e^x = 5$ **20.** $-14 + 3e^x = 11$
21. $e^{3x} = 12$ **22.** $e^{2x} = 50$
23. $500e^{-x} = 300$ **24.** $1000e^{-4x} = 75$
25. $e^{2x} - 4e^x - 5 = 0$ **26.** $e^{2x} - 5e^x + 6 = 0$
27. $20(100 - e^{x/2}) = 500$ **28.** $\dfrac{400}{1 + e^{-x}} = 200$
29. $10^x = 42$ **30.** $10^x = 570$
31. $3^{2x} = 80$ **32.** $6^{5x} = 3000$
33. $5^{-t/2} = 0.20$ **34.** $4^{-3t} = 0.10$
35. $\left(1 + \dfrac{0.10}{12}\right)^{12t} = 2$ **36.** $2^{3-x} = 565$

In Exercises 37–44, use a graphing utility to solve the exponential equation. (Round your result to three decimal places.)

37. $3e^{3x/2} = 962$ **38.** $6e^{1-x} = 25$
39. $e^{0.09t} = 3$ **40.** $e^{0.125t} = 8$
41. $8(10^{3x}) = 12$ **42.** $3(5^{x-1}) = 21$
43. $\left(1 + \dfrac{0.065}{365}\right)^{365t} = 4$ **44.** $\dfrac{3000}{2 + e^{2x}} = 2$

In Exercises 45–60, solve the logarithmic equation algebraically. (Round your result to three decimal places.)

45. $\ln x = -3$ **46.** $\ln x = 2$
47. $\ln 2x = 2.4$ **48.** $3 \ln 5x = 10$
49. $\ln \sqrt{x + 2} = 1$ **50.** $\ln(x + 1)^2 = 2$
51. $\ln x + \ln(x - 2) = 1$ **52.** $\ln x + \ln(x + 3) = 1$
53. $\log_{10}(z - 3) = 2$ **54.** $\log_{10} x^2 = 6$
55. $\log_{10}(x + 4) - \log_{10} x = \log_{10}(x + 2)$
56. $\log_4 x - \log_4(x - 1) = \frac{1}{2}$
57. $\log_3 x + \log_3(x^2 - 8) = \log_3 8x$
58. $\log_2 x + \log_2(x + 2) = \log_2(x + 6)$
59. $\ln(x + 5) = \ln(x - 1) - \ln(x + 1)$
60. $\ln(x + 1) - \ln(x - 2) = \ln x^2$

In Exercises 61–64, use a graphing utility to solve the logarithmic equation. (Round your result to three decimal places.)

61. $2 \ln x = 7$ **62.** $\ln 4x = 1$
63. $\ln x + \ln(x^2 + 1) = 8$
64. $\log_{10} 8x - \log_{10}\left(1 + \sqrt{x}\right) = 2$

Compound Interest   In Exercises 65 and 66, find the time required for a $1000 investment to double at interest rate r, compounded continuously.

65. $r = 0.085$ **66.** $r = 0.12$

Compound Interest   In Exercises 67 and 68, find the time required for a $1000 investment to triple at interest rate r, compounded continuously.

67. $r = 0.085$ **68.** $r = 0.12$

69. *Demand Function*   The demand equation for a product is

$$p = 500 - 0.5(e^{0.004x}).$$

Find the demand x for a price of (a) $p = \$350$ and (b) $p = \$300$.

70. *Demand Function*   The demand equation for a product is

$$p = 5000\left(1 - \frac{4}{4 + e^{-0.002x}}\right).$$

Find the demand x for a price of (a) $p = \$600$ and (b) $p = \$400$.

71. *Forest Yield*   The yield V (in millions of cubic feet per acre) for a forest at age t years is given by

$$V = 6.7e^{-48.1/t}.$$

(a) Use a graphing utility to graph the function.

(b) Determine the horizontal asymptote of the function. Interpret its meaning in the context of the problem.

(c) Find the time necessary to have a yield of 1.3 million cubic feet.

72. *Trees per Acre*   The number of trees per acre N of a certain species is approximated by the model

$$N = 68 \cdot 10^{-0.04x}, \qquad 5 \le x \le 40,$$

where x is the average diameter of the trees 3 feet above the ground. Use the model to approximate the average diameter of the trees in a test plot when $N = 21$.

73. *Average Heights*   The percentage of American males between the ages of 18 and 24 who are no more than x inches tall is given by

$$m(x) = \frac{100}{1 + e^{-0.6114(x-69.71)}}, \qquad 60 \le x \le 80,$$

where m is the percentage and x is the height in inches. (*Source:* U.S. National Center for Health Statistics) The function giving the percentages f for females for the same ages is given by

$$f(x) = \frac{100}{1 + e^{-0.66607(x-64.51)}}, \qquad 55 \le x \le 75.$$

(a) Use a graphing utility to graph each function in the same viewing rectangle.

(b) Determine the horizontal asymptotes of the functions.

(c) What is the median height of each sex?

74. *Human Memory Model*   In a group project in learning theory, a mathematical model for the proportion P of correct responses after n trials was found to be

$$P = \frac{0.83}{1 + e^{-0.2n}}.$$

(a) Use a graphing utility to graph the function.

(b) Determine the horizontal asymptotes of the function. Interpret the meaning of the upper asymptote in the context of this problem.

(c) After how many trials will 60% of the responses be correct?

6.5 APPLICATIONS OF EXPONENTIAL AND LOGARITHMIC FUNCTIONS

Compound Interest / Growth and Decay /
Logistics Growth Models / Logarithmic Models

Compound Interest

In this section, you will study four basic types of applications: (1) Compound Interest, (2) Growth and Decay, (3) Logistics Models, and (4) Intensity Models. The problems presented in this section involve the full range of solution techniques studied in this chapter.

EXAMPLE 1   Doubling Time for an Investment

An investment is made in a trust fund at an annual percentage rate of 9.5%, compounded quarterly. How long will it take for the investment to double in value?

SOLUTION

For quarterly compounding, use the formula

$$A = P\left(1 + \frac{r}{4}\right)^{4t}.$$

Using $r = 0.095$, the time required for the investment to double is given by solving for t in the equation $2P = A$.

$$2P = P\left(1 + \frac{0.095}{4}\right)^{4t} \qquad \textit{2P = A}$$

$$2 = (1.02375)^{4t} \qquad \textit{Divide both sides by P}$$

$$\ln 2 = \ln(1.02375)^{4t} \qquad \textit{Take ln of both sides}$$

$$\ln 2 = 4t \ln(1.02375)$$

$$t = \frac{\ln 2}{4 \ln(1.02375)} \approx 7.4$$

Therefore, it will take approximately 7.4 years for the investment to double in value with quarterly compounding.

Try reworking Example 1 using continuous compounding. To do this you will need to solve the equation

$$2P = Pe^{0.095t}.$$

The solution is $t \approx 7.3$ years, which makes sense because the principal should double more quickly with continuous compounding than with quarterly compounding.

DISCOVERY

A person deposits $1000 in an account that pays 9.5% per year, compounded quarterly. The balance after t years is

$$A = 1000\left(1 + \frac{0.095}{4}\right)^{4t}.$$

Another person deposits $1000 in an account that pays 9.5% per year, compounded continuously. The balance after t years is $A = 1000e^{0.095t}$. Use a graphing utility to graph both equations on the same viewing rectangle. (Use a viewing rectangle in which $0 \le x \le 8$ and $1000 \le y \le 2200$.) What do you conclude? Which of the two accounts will reach a balance of $2000 first? Justify your answer by zooming in near $x = 7$.

EXAMPLE 2   Finding an Annual Percentage Rate

An investment of $10,000 is compounded continuously. What annual percentage rate will produce a balance of $25,000 in 10 years?

SOLUTION

Use the formula $A = Pe^{rt}$ with $P = 10,000$, $A = 25,000$, and $t = 10$, and solve the equation for r.

$$10,000e^{10r} = 25,000$$
$$e^{10r} = 2.5$$
$$10r = \ln 2.5$$
$$r = \frac{1}{10}\ln 2.5 \approx 0.0916$$

Thus, the annual percentage rate must be approximately 9.16%. You can verify this result using the zoom and trace features of a graphing utility. By graphing $y = 10,000e^{10x}$ and $y = 25,000$ in the same viewing rectangle with an x scale of 0.1, you can confirm that $r \approx 0.0916$, as shown in Figure 6.27.

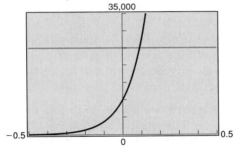

FIGURE 6.27

EXAMPLE 3   The Effective Yield for an Investment

A deposit is compounded continuously at an annual percentage rate of 7.5%. Find the **effective yield.** That is, find the simple interest rate that would yield the same balance at the end of 1 year.

SOLUTION

Using the formula $A = Pe^{rt}$ with $r = 0.075$ and $t = 1$, the balance at the end of 1 year is

$$A = Pe^{0.075(1)}$$
$$\approx P(1.0779)$$
$$= P(1 + 0.0779). \qquad\qquad A = P(1 + r)$$

Because the formula for simple interest after 1 year is $A = P(1 + r)$, it follows that the effective yield is approximately 7.79%.

Growth and Decay

The balance in an account earning *continuously compounded* interest is one example of a quantity that increases over time according to the **exponential growth model**

$$Q(t) = Ce^{kt}.$$

In general, $Q(t)$ is the size of the "population" at any time t, C is the original population (when $t = 0$), and k is a constant determined by the rate of growth. If $k > 0$, the population *grows* (increases) over time, and if $k < 0$ it *decays* (decreases) over time. Example 11 in Section 6.1 is an example of population growth. It may help to remember this growth model as *(Then)* = *(Now)* (e^{kt}).

EXAMPLE 4   Exponential Decay

Radioactive iodine is a by-product of some types of nuclear reactors. Its **half-life** is 60 days. That is, after 60 days, a given amount of radioactive iodine will have decayed to half the original amount. Suppose a contained nuclear accident occurs and gives off an initial amount C of radioactive iodine.

A. Write an equation for the amount of radioactive iodine present at any time t following the accident.
B. How long will it take for the radioactive iodine to decay to a level of 20% of the original amount?

SOLUTION

A. We first need to find the rate k in the exponential model $Q(t) = Ce^{kt}$. Knowing that half the original amount remains after $t = 60$ days, you obtain

$$Q(60) = Ce^{k(60)} = \frac{1}{2}C$$

$$e^{60k} = \frac{1}{2}$$

$$60k = -\ln 2$$

$$k = \frac{-\ln 2}{60} \approx -0.0116.$$

Thus, the exponential model is

$$Q(t) = Ce^{-0.0116t}.$$

Technology Note

You can use your graphing utility to
solve part (b) by graphing

$y_1 = e^{-0.0116t}$

$y_2 = 0.2$

on the viewing rectangle
$0 \le x \le 180, 0 \le y \le 1$ and finding
the point of intersection.

B. The time required to decay to 20% of the original amount is given by

$$Q(t) = Ce^{-0.0116t} = (0.2)C$$
$$e^{-0.0116t} = 0.2$$
$$-0.0116t = \ln 0.2$$
$$t = \frac{\ln 0.2}{-0.0116} \approx 139 \text{ days.}$$

In living organic material, the ratio of radioactive carbon isotopes (carbon-14) to the number of nonradioactive carbon isotopes (carbon-12) is about 1 to 10^{12}. When organic material dies, its carbon-12 content remains fixed, whereas its radioactive carbon-14 begins to decay with a half-life of about 5700 years. To estimate the age of dead organic material, scientists use the following formula, which denotes the ratio of carbon-14 to carbon-12 present at any time t (in years).

$$R = \frac{1}{10^{12}} e^{-t/8223}$$

The graph of R is shown in Figure 6.28. Note that R decreases as the time t increases.

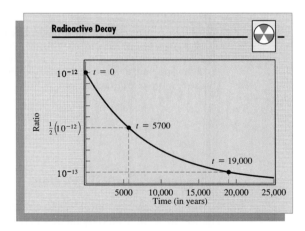

FIGURE 6.28

EXAMPLE 5   Carbon Dating

Suppose the carbon-14/carbon-12 ratio of a newly discovered fossil is

$$R = \frac{1}{10^{13}}.$$

Estimate the age of the fossil.

SOLUTION

In the carbon dating model, substitute the given value of R to obtain the following.

$$\frac{1}{10^{12}} e^{-t/8223} = R \qquad \text{\textit{Original model}}$$

$$\frac{e^{-t/8223}}{10^{12}} = \frac{1}{10^{13}} \qquad \text{\textit{Let R equal }} 1/10^{13}$$

$$e^{-t/8223} = \frac{1}{10} \qquad \text{\textit{Multiply both sides by }} 10^{12}$$

$$\ln e^{-t/8223} = \ln \frac{1}{10} \qquad \text{\textit{Take log of both sides}}$$

$$-\frac{t}{8223} \approx -2.3026 \qquad \text{\textit{Inverse property of logs and exponents}}$$

$$t \approx 18{,}934 \qquad \text{\textit{Multiply both sides by 8223}}$$

Thus, to the nearest thousand years, you can estimate the age of the fossil to be 19,000 years. You could obtain this solution graphically by graphing the equation $y = e^{-t/8223} - \frac{1}{10}$, as indicated in Figure 6.29. ·········

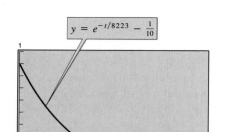

$y = e^{-t/8223} - \frac{1}{10}$

(18,934, 0)

25,000

FIGURE 6.29

REMARK   The carbon dating model in Example 5 assumes that the carbon-14/carbon-12 ratio was one part in 10,000,000,000,000. Suppose an error in measurement occurred and the actual ratio was only one part in 8,000,000,000,000. The fossil age corresponding to the actual ratio would then be approximately 17,000 years. Try checking this result. ·····

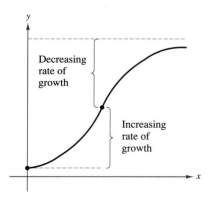

Decreasing
rate of
growth

Increasing
rate of
growth

Logistics Curve

FIGURE 6.30

Logistics Growth Models

Some populations initially have rapid growth, followed by a declining rate of growth, as indicated by the graph in Figure 6.30. One model for describing this type of growth pattern is the **logistics curve** given by the function

$$y = \frac{a}{1 + be^{-(t-c)/d}},$$

where y is the population size and t is the time. An example would be a bacteria culture allowed to grow initially under ideal conditions, followed by less favorable conditions that inhibit growth, such as overcrowding. A logistics growth curve is also called a **sigmoidal curve.**

EXAMPLE 6   Spread of a Virus

On a college campus of 5000 students, one student returned from vacation with a contagious flu virus. The spread of the virus through the student body is given by

$$y = \frac{5000}{1 + 4999e^{-0.8t}},$$

where y is the total number infected after t days. The college will cancel classes when 40% or more of the students are ill.

A. How many are infected after 5 days?
B. After how many days will the college cancel classes?

SOLUTION

A. After 5 days, the number of students infected is

$$y = \frac{5000}{1 + 4999e^{-0.8(5)}} = \frac{5000}{1 + 4999e^{-4}} \approx 54.$$

B. In this case, the number of students infected is $(0.40)(5000) = 2000$. Therefore, you can solve for t in the following equation.

$$2000 = \frac{5000}{1 + 4999e^{-0.8t}}$$

Using a graphing utility, graph $y = 2000$ and $y = 5000/(1 + 4999e^{-0.8t})$ and find the point of intersection. In Figure 6.31, you can see that the point of intersection occurs near $t \approx 10.1$. Hence, after 10 days, at least 40% of the students will be infected, and the college will cancel classes.

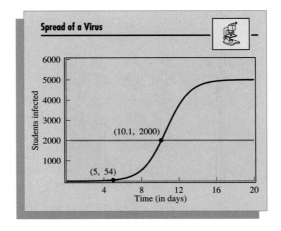

FIGURE 6.31

Logarithmic Models

Sound and shock waves can be measured by the **intensity model**

$$S = K \log_{10} \frac{I}{I_0},$$

where I is the intensity of the stimulus wave, I_0 is the **threshold intensity** (the smallest value of I that can be detected by the listening device), and K determines the units in which S is measured. Sound heard by the human ear is measured in decibels. One **decibel** is considered to be the smallest detectable difference in the loudness of two sounds.

EXAMPLE 7 Magnitude of Earthquakes

On the Richter Scale, the magnitude R of an earthquake of intensity I is given by

$$R = \log_{10} \frac{I}{I_0},$$

where $I_0 = 1$ is the minimum intensity used for comparison. Find the intensity per unit of area for the following earthquakes. (Intensity is a measure of the wave energy of an earthquake.)

A. San Francisco in 1906, $R = 8.6$
B. Mexico City in 1978, $R = 7.85$
C. San Francisco Bay Area in 1989, $R = 7.1$

SOLUTION

A. Because $I_0 = 1$ and $R = 8.6$, you have

$$8.6 = \log_{10} I$$
$$I = 10^{8.6} \approx 398,107,171.$$

B. For Mexico City, you have $7.85 = \log_{10} I$, and

$$I = 10^{7.85} \approx 70,795,000.$$

C. For $R = 7.1$, you have $7.1 = \log_{10} I$, and

$$I = 10^{7.1} \approx 12,589,254$$

Note that an increase of 1.5 units on the Richter Scale (from 7.1 to 8.6) represents an intensity change by a factor of

$$\frac{398,107,171}{12,589,254} \approx 31.6.$$

In other words, the "great San Francisco earthquake" in 1906 had a magnitude that was about 32 times greater than the one in 1989.

DISCUSSION PROBLEM

COMPARING POPULATION MODELS

The population (in millions) of the United States from 1800 to 1990 is given in the accompanying table.

t	0	1	2	3	4	5	6	7	8	9
Year	1800	1810	1820	1830	1840	1850	1860	1870	1880	1890
Population	5.31	7.23	9.64	12.87	17.07	23.19	31.44	39.82	50.16	62.95

t	10	11	12	13	14	15	16	17	18	19
Year	1900	1910	1920	1930	1940	1950	1960	1970	1980	1990
Population	75.99	91.97	105.71	122.78	131.67	151.33	179.32	203.30	226.55	250.00

Using a statistical procedure called *least squares regression analysis,* we found the best quadratic and exponential models for this data. Which of the following two equations is a better model for the population of the United States between 1800 and 1990? Write a short paragraph describing the method you used to reach your conclusion.

Quadratic Model	*Exponential Model*
$P = 0.662t^2 + 0.211t + 6.165$	$P = 7.7899e^{0.2013t}$

WARM-UP

The following warm-up exercises involve skills that were covered in earlier sections. You will use these skills in the exercise set for this section.

In Exercises 1–6, sketch the graph of the equation.

1. $y = 2^{0.25x}$ **2.** $y = 2^{-0.25x}$

3. $y = 4 \log_2 x$ **4.** $y = \ln(x - 3)$

5. $y = e^{-x^2/5}$ **6.** $y = \dfrac{2}{1 + e^{-x}}$

In Exercises 7–10, solve the given equation for x. (Round your result to three decimal places.)

7. $3e^{2x} = 7$ **8.** $2e^{-0.2x} = 0.002$

9. $4 \ln 5x = 14$ **10.** $6 \ln 2x = 12$

SECTION 6.5 · EXERCISES

Compound Interest In Exercises 1–10, complete the table for a savings account in which interest is compounded continuously.

	Initial Investment	Annual % Rate	Effective Yield	Time to Double	Amount After 10 Years
1.	$1000	12%			
2.	$20,000	$10\frac{1}{2}$%			
3.	$750			$7\frac{3}{4}$ yr	
4.	$10,000			5 yr	
5.	$500				$1292.85
6.	$2000		4.5%		
7.		11%			$19,205.00
8.		8%			$20,000.00
9.	$5000		8.33%		
10.	$250		12.19%		

Compound Interest In Exercises 11 and 12, determine the principal P that must be invested at rate r, compounded monthly, so that $500,000 will be available for retirement in t years.

11. $r = 7\frac{1}{2}\%$, $t = 20$ **12.** $r = 12\%$, $t = 40$

Compound Interest In Exercises 13 and 14, determine the time necessary for $1000 to double if it is invested at interest rate r compounded (a) annually, (b) monthly, (c) daily, and (d) continuously.

13. $r = 11\%$ **14.** $r = 10\frac{1}{2}\%$

15. *Compound Interest* Complete the table for the time t necessary for P dollars to triple if interest is compounded continuously at rate r.

r	2%	4%	6%	8%	10%	12%
t						

16. *Compound Interest* Complete the table for the time t necessary for P dollars to triple if interest is compounded annually at rate r.

r	2%	4%	6%	8%	10%	12%
t						

17. *Comparing Investments* If $1 is invested in an account over a 10-year period, the amount in the account is given by

$$A = 1 + 0.075t \quad \text{or} \quad A = e^{0.07t}$$

depending on whether it is simple interest at $7\frac{1}{2}$% or continuous compound interest at 7%. Use a graphing utility to graph each function in the same viewing rectangle, and determine which grows at the higher rate.

18. *Comparing Investments* If $1 is invested in an account over a 10-year period, the amount in the account is given by

$$A = 1 + 0.06t \quad \text{or} \quad A = \left(1 + \frac{0.055}{365}\right)^{365t}$$

depending on whether it is simple interest at 6% or compound interest at $5\frac{1}{2}$% compounded daily. Use a graphing utility to graph each function in the same viewing rectangle, and determine which grows at the higher rate.

Radioactive Decay In Exercises 19–24, complete the table for the given radioactive isotope.

	Isotope	Half-life (Years)	Initial Quantity	Amount After 1000 Years	Amount After 10,000 Years
19.	^{226}Ra	1620	10 g		
20.	^{226}Ra	1620		1.5 g	
21.	^{14}C	5730			2 g
22.	^{14}C	5730	3 g		
23.	^{230}Pu	24,360		2.1 g	
24.	^{230}Pu	24,360			0.4 g

In Exercises 25–28, find the constant k such that the exponential function $y = Ce^{kt}$ passes through the given points on the graph.

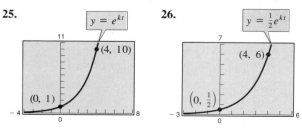

25. $y = e^{kt}$, (4, 10), (0, 1)   **26.** $y = \frac{1}{2}e^{kt}$, (4, 6), $\left(0, \frac{1}{2}\right)$

27.

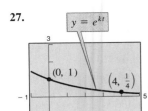

28.

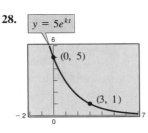

29. *Population*   The population P of a city is given by

$$P = 105{,}300e^{0.015t},$$

where t is the time in years, with $t = 0$ corresponding to 1990. What was the population in 1990? According to this model, in what year will the city have a population of 150,000?

30. *Population*   The population P of a city is given by

$$P = 240{,}360e^{0.012t},$$

where t is the time in years, with $t = 0$ corresponding to 1990. What was the population in 1990? According to this model, in what year will the city have a population of 250,000?

31. *Population*   The population P of a city is given by

$$P = 2500e^{kt},$$

where t is the time in years, with $t = 0$ corresponding to the year 1990. In 1945, the population was 1350. Find the value of k and use this result to predict the population in the year 2010.

32. *Population*   The population P of a city is given by

$$P = 140{,}500e^{kt},$$

where t is the time in years, with $t = 0$ corresponding to the year 1990. In 1960, the population was 100,250. Find the value of k and use this result to predict the population in the year 2000.

33. *Population*   The population of Dhaka, Bangladesh was 4.22 million in 1990 and its projected population for the year 2000 is 6.49 million. Find the exponential growth model $y = Ce^{kt}$ for the population growth of Dhaka by letting $t = 0$ correspond to 1990. Use the model to predict the population of the city in 2010. (*Source:* United Nations)

34. *Population*   The population of Houston, Texas was 2.30 million in 1990 and its projected population for the year 2000 is 2.65 million. Find the exponential growth model $y = Ce^{kt}$ for the population growth of Houston by letting $t = 0$ correspond to 1990. Use the model to predict the population of the city in 2010. (*Source:* U. S. Bureau of the Census)

35. *Bacteria Growth*   The number of bacteria N in a culture is given by the model

$$N = 100e^{kt},$$

where t is the time in hours, with $t = 0$ corresponding to the time when $N = 100$. If $N = 300$ when $t = 5$, estimate the time required for the population to double in size.

36. *Bacteria Growth*   The number of bacteria N in a culture is given by the model

$$N = 250e^{kt},$$

where t is the time in hours, with $t = 0$ corresponding to the time when $N = 250$. If $N = 280$ when $t = 10$, estimate the time required for the population to double in size.

37. *Radioactive Decay*   The half-life of radioactive radium (^{226}Ra) is 1620 years. What percentage of a present amount of radioactive radium will remain after 100 years?

38. *Radioactive Decay* ^{14}C dating assumes that the carbon dioxide on earth today has the same radioactive content as it did centuries ago. If this is true, then the amount of ^{14}C absorbed by a tree that grew several centuries ago should be the same as the amount of ^{14}C absorbed by a tree growing today. A piece of ancient charcoal contains only 15% as much of the radioactive carbon as a piece of modern charcoal. How long ago was the tree burned to make the ancient charcoal if the half-life of ^{14}C is 5730 years?

39. *Depreciation*   A certain car that cost $22,000 new has a depreciated value of $16,500 after 1 year. Find the value of the car when it is 3 years old by using the exponential model $y = Ce^{kt}$.

40. *Depreciation*   A computer that cost $4600 new has a depreciated value of $3000 after 2 years. Find the value of the computer after 3 years by using the exponential model $y = Ce^{kt}$.

41. *Sales*   The sales S (in thousands of units) of a new product after it has been on the market t years are given by

$$S(t) = 100(1 - e^{kt}).$$

(a) Find S as a function of t if 15,000 units have been sold after 1 year.

(b) How many units will be sold after 5 years?

42. *Learning Curve*   The management at a factory has found that the maximum number of units a worker can produce in a day is 30. The learning curve for the number of units N produced per day after a new employee has worked t days is given by $N = 30(1 - e^{kt})$. After 20 days on the job, a particular worker produced 19 units.

(a) Find the learning curve for this worker (that is, find the value of k).

(b) How many days should pass before this worker is producing 25 units per day?

43. *Women's Heights* The distribution of the heights of American women between the ages of 25 and 34 can be approximated by the function

$$p = 0.166e^{-(x-64.5)^2/11.5},$$

where x is the height in inches. Use a graphing utility to graph this function. What is the average height of women in this age bracket? (*Source:* U.S. National Center for Health Statistics)

44. *Men's Heights* The distribution of the heights of American men between the ages of 25 and 34 can be approximated by the function

$$p = 0.193e^{-(x-70)^2/14.32},$$

where x is the height in inches. Use a graphing utility to graph this function. What is the average height of men in this age bracket? (*Source:* U.S. National Center for Health Statistics)

45. *Stocking a Lake with Fish* A certain lake was stocked with 500 fish and the fish population increased according to the logistics curve

$$p(t) = \frac{10,000}{1 + 19e^{-t/5}},$$

where t is the time in months.
(a) Use a graphing utility to graph the function. Determine the larger of the two horizontal asymptotes and interpret its meaning in the context of the problem.
(b) Estimate the fish population after 5 months.
(c) After how many months will the fish population be 2000?

46. *Endangered Species* A conservation organization releases 100 animals of an endangered species into a game preserve. The organization believes that the preserve has a carrying capacity of 1000 animals and that the growth of the herd will be modeled by the logistics curve

$$p(t) = \frac{1000}{1 + 9e^{-0.1656t}},$$

where t is the time in months.
(a) Use a graphing utility to graph the function. Determine the larger of the two horizontal asymptotes and interpret its meaning in the context of the problem.
(b) Estimate the population after 5 months.
(c) After how many months will the population be 500?

47. *Sales and Advertising* The sales S (in thousands of units) of a product after x hundred dollars is spent on advertising is given by

$$S = 10(1 - e^{kx}).$$

(a) Find S as a function of x if 2500 units are sold when $500 is spent on advertising.
(b) Estimate the number of units that will be sold if advertising expenditures are raised to $700.

48. *Sales and Advertising* After discontinuing all advertising for a certain product in 1988, the manufacturer noted that sales began to drop according to the model

$$S = \frac{500,000}{1 + 0.6e^{kt}},$$

where S represents the number of units sold and t represents the year, with $t = 0$ corresponding to 1988.
(a) Find k if the company sold 300,000 units in 1990.
(b) According to this model, what will sales be in 1993?

Earthquake Magnitudes In Exercises 49 and 50, use the Richter Scale (see Example 7) for measuring the magnitude of earthquakes.

49. Find the magnitude R of an earthquake of intensity I (let $I_0 = 1$).
(a) $I = 80,500,000$
(b) $I = 48,275,000$

50. Find the intensity I of an earthquake measuring R on the Richter Scale (let $I_0 = 1$).
(a) Mexico City in 1985, $R = 8.1$
(b) Los Angeles in 1971, $R = 6.7$

Intensity of Sound In Exercises 51–54, use the following information to determine the level of sound (in decibels) for the given sound intensity. The level of sound β, in decibels, with an intensity of I is given by

$$\beta(I) = 10 \log_{10} \frac{I}{I_0},$$

where I_0 is an intensity of 10^{-16} watts per square centimeter, corresponding roughly to the faintest sound that can be heard by the human ear.

51. (a) $I = 10^{-14}$ watts per square centimeter (faint whisper)
(b) $I = 10^{-9}$ watts per square centimeter (busy street corner)
(c) $I = 10^{-6.5}$ watts per square centimeter (air hammer)
(d) $I = 10^{-4}$ watts per square centimeter (threshold of pain)

52. (a) $I = 10^{-13}$ watts per square centimeter (whisper)
    (b) $I = 10^{-7.5}$ watts per square centimeter (DC-8 4 miles from takeoff)
    (c) $I = 10^{-7}$ watts per square centimeter (diesel truck at 25 feet)
    (d) $I = 10^{-4.5}$ watts per square centimeter (auto horn at 3 feet)

53. *Noise Level* Due to the installation of noise suppression materials, the noise level in an auditorium was reduced from 93 to 80 decibels. Find the percentage decrease in the intensity level of the noise because of the installation of these materials.

54. *Noise Level* Due to the installation of a muffler, the noise level in an engine was reduced from 88 to 72 decibels. Find the percentage decrease in the intensity level of the noise because of the installation of the muffler.

Acidity In Exercises 55–60, use the acidity model given by

$$pH = -\log_{10}[H^+],$$

where acidity (pH) is a measure of the hydrogen ion concentration $[H^+]$ (measured in moles of hydrogen per liter) of a solution.

55. Find the pH if $[H^+] = 2.3 \times 10^{-5}$.
56. Find the pH if $[H^+] = 11.3 \times 10^{-6}$.
57. Compute $[H^+]$ for a solution in which pH = 5.8.
58. Compute $[H^+]$ for a solution in which pH = 3.2.
59. A certain fruit has a pH of 2.5 and an antacid tablet has a pH of 9.5. The hydrogen ion concentration of the fruit is how many times the concentration of the tablet?
60. If the pH of a solution is decreased by one unit, the hydrogen ion concentration is increased by what factor?

61. *Home Mortgage* An $80,000 home mortgage for 35 years at $9\frac{1}{2}\%$ has a monthly payment of $657.28. Part of the monthly payment goes for the interest charge on the unpaid balance and the remainder of the payment is used to reduce the principal. The amount that goes for interest is given by

$$u = M - \left(M - \frac{Pr}{12}\right)\left(1 + \frac{r}{12}\right)^{12t},$$

and the amount that goes toward reduction of the principal is given by

$$v = \left(M - \frac{Pr}{12}\right)\left(1 + \frac{r}{12}\right)^{12t}.$$

In these formulas, P is the size of the mortgage, r is the interest rate, M is the monthly payment, and t is the time in years.

(a) Use a graphing utility to graph each function in the same viewing rectangle. (The viewing rectangle should show all 35 years of mortgage payments.)
(b) In the early years of the mortgage, the larger part of the monthly payment goes for what purpose? Approximate the time when the monthly payment is evenly divided between interest and principal reduction.

62. *Home Mortgage* The total interest paid on a home mortgage of P dollars, at interest rate r for t years, is given by

$$u = P\left[\frac{rt}{1 - \left(\dfrac{1}{1 + r/12}\right)^{12t}} - 1\right].$$

Consider an $80,000 home mortgage at $9\frac{1}{2}\%$.
(a) Use a graphing utility to graph the total interest function.
(b) Approximate the length of the mortgage when the total interest paid is the same as the size of the mortgage. Is it possible that some people are paying twice as much in interest charges as the size of the mortgage?

63. *Estimating the Time of Death* At 8:30 A.M., a coroner was called to the home of a person who had died during the night. In order to estimate the time of death, the coroner took the person's temperature twice. At 9:00 A. M., the temperature was 85.7°, and at 9:30 A. M., the temperature was 82.8°. From these two temperatures, the coroner was able to determine that

$$t = -2.5 \ln \frac{T - 70}{98.6 - 70},$$

where t is the time in hours that has elapsed since the person died and T is the temperature (in degrees Fahrenheit) of the person's body at 9:00 A. M. (The person had a normal body temperature of 98.6° at death, and the room temperature was a constant 70°. This formula is derived from a general cooling principle called Newton's Law of Cooling.) Use this formula to estimate the time of death of the person.

64. *Population Growth* In Exercises 33 and 34, you can see that the populations of Dhaka and Houston are growing at different rates. What constant in the equation $y = Ce^{kt}$ is affected by these different growth rates? Discuss the relationship between the different growth rates and the magnitude of the constant.

6.6 NONLINEAR MODELS

Classifying Scatter Plots / Fitting Nonlinear Models to Data / Applications

Classifying Scatter Plots

In a typical real-life situation, data are collected and written as a set of ordered pairs. The graph of such a set is called a **scatter plot.** A scatter plot can be used to give you an idea of which type of model can be used to best fit a set of data.

EXAMPLE 1 Classifying Scatter Plots

Sketch a scatter plot for each set of data. Then decide whether the data could be best modeled by an exponential model, $y = ae^{bx}$, or a logarithmic model, $y = a + b \ln x$.

A. (0.9, 1.9), (1.3, 2.4), (1.3, 2.2), (1.4, 2.4), (1.6, 2.7), (1.8, 3.0), (2.1, 3.4), (2.1, 3.3), (2.5, 4.2), (2.9, 5.1), (3.2, 6.0), (3.3, 6.2), (3.6, 7.3), (4.0, 8.8), (4.2, 9.8), (4.3, 10.2)

B. (0.9, 3.2), (1.3, 4.0), (1.3, 3.8), (1.4, 4.2), (1.6, 4.5), (1.8, 4.8), (2.1, 5.1), (2.1, 5.0), (2.5, 5.5), (2.9, 5.8), (3.2, 6.0), (3.3, 6.2), (3.6, 6.2), (4.0, 6.6), (4.2, 6.7), (4.3, 6.8)

SOLUTION

Begin by entering the data into a graphing utility. You should obtain the scatter plots shown in Figure 6.32.

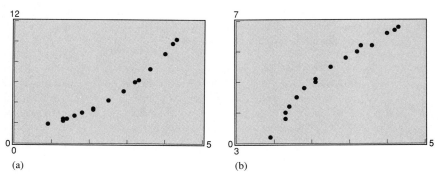

(a)   (b)

FIGURE 6.32

From the scatter plots, it appears that the data in part (A) can be modeled by an exponential function and the data in part (B) can be modeled by a logarithmic function.

Fitting Nonlinear Models to Data

Once you have used a scatter plot to determine the type of model to be fit to a set of data, there are several ways that you can actually find the model. Each method is best used by applying built-in least squares regression formulas with a computer or calculator, rather than with hand calculations.

The classic method is to transform one or both of the coordinates of the data points, so that the resulting points can be fit with a linear model. For instance, you can fit an *exponential* model to a set of points of the form (x, y) by fitting a *linear* model to points of the form $(x, \ln y)$.

EXAMPLE 2 Fitting an Exponential Model to Data

Find an exponential model for the points given in Example 1(A).

(0.9, 1.9), (1.3, 2.4), (1.3, 2.2), (1.4, 2.4), (1.6, 2.7), (1.8, 3.0),
(2.1, 3.4), (2.1, 3.3), (2.5, 4.2), (2.9, 5.1), (3.2, 6.0), (3.3, 6.2),
(3.6, 7.3), (4.0, 8.8), (4.2, 9.8), (4.3, 10.2)

SOLUTION

Begin by transforming the points by taking the natural logarithm of each y-coordinate, as shown in Table 6.5.

TABLE 6.5

x	0.9	1.3	1.3	1.4	1.6	1.8	2.1	2.1
y	1.9	2.4	2.2	2.4	2.7	3.0	3.4	3.3
$\ln y$	0.642	0.875	0.788	0.875	0.993	1.099	1.224	1.194

x	2.5	2.9	3.2	3.3	3.6	4.0	4.2	4.3
y	4.2	5.1	6.0	6.2	7.3	8.8	9.8	10.2
$\ln y$	1.435	1.629	1.792	1.825	1.988	2.175	2.282	2.322

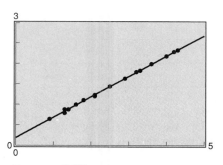

FIGURE 6.33

Next, use a graphing utility to sketch a scatter plot of the points $(x, \ln y)$, as shown in Figure 6.33. Notice that the points appear to fit a linear model. By applying the least squares regression formulas to the transformed points, you obtain

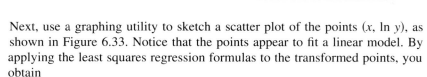

$$\ln y = 0.1847 + 0.4984x \qquad \text{\textit{Least squares regression "line"}}$$
$$y = e^{0.1847 + 0.4984x} \qquad \text{\textit{Exponentiate both sides}}$$
$$y = 1.203e^{0.4984x} \qquad \text{\textit{Exponential model}}$$

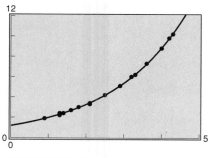

FIGURE 6.34

Figure 6.34 shows the graph of this model with the original data points.

You can fit a *logarithmic* model to a set of points of the form (x, y) by fitting a *linear* model to points of the form $(\ln x, y)$, as illustrated in Example 3.

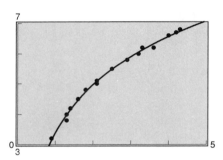

FIGURE 6.35

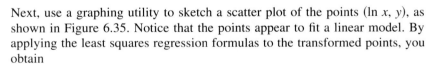

FIGURE 6.36

EXAMPLE 3 Fitting a Logarithmic Model to Data

Find a logarithmic model for the points given in Example 1(B).

(0.9, 3.2), (1.3, 4.0), (1.3, 3.8), (1.4, 4.2), (1.6, 4.5), (1.8, 4.8),
(2.1, 5.1), (2.1, 5.0), (2.5, 5.5), (2.9, 5.8), (3.2, 6.0), (3.3, 6.2),
(3.6, 6.2), (4.0, 6.6), (4.2, 6.7), (4.3, 6.8)

SOLUTION

Begin by transforming the points by taking the natural logarithm of each x-coordinate, as shown in Table 6.6.

TABLE 6.6

x	0.9	1.3	1.3	1.4	1.6	1.8	2.1	2.1
$\ln x$	-0.105	0.262	0.262	0.336	0.470	0.588	0.742	0.742
y	3.2	4.0	3.8	4.2	4.5	4.8	5.1	5.0

x	2.5	2.9	3.2	3.3	3.6	4.0	4.2	4.3
$\ln x$	0.916	1.065	1.163	1.194	1.281	1.386	1.435	1.459
y	5.5	5.8	6.0	6.2	6.2	6.6	6.7	6.8

Next, use a graphing utility to sketch a scatter plot of the points $(\ln x, y)$, as shown in Figure 6.35. Notice that the points appear to fit a linear model. By applying the least squares regression formulas to the transformed points, you obtain

$$y = 3.377 + 2.301 \ln x.$$ *Logarithmic model*

Figure 6.36 shows the graph of this model with the original data points.

Technology Note

Most graphing utilities have built-in programs that will fit nonlinear models to data. Try using such a program to duplicate the results of Examples 2 and 3. When we did that, we obtained the model

$y = 1.203(1.6459)^x$

or $y = 1.203e^{0.4984x}$

as a model for Example 2 and

$y = 3.377 + 2.301 \ln x$

as a model for Example 3. (Both models have a correlation coefficient of over 0.99, which indicates that the fit is very good.)

Applications

EXAMPLE 4 Finding an Exponential Model

The total amount A (in billions of dollars) spent on health care in the United States from 1970 through 1989 is shown in Table 6.7. Find a model for the data. Then use the model to predict the amount spent in 1998. In the table, t represents the year, with $t = 0$ corresponding to 1970. (*Source*: U.S. Health Care Financing Administration)

TABLE 6.7

t	0	1	2	3	4	5	6	7	8	9
A	74.4	82.3	92.3	102.5	116.1	132.9	152.2	172.0	193.4	216.6

t	10	11	12	13	14	15	16	17	18	19
A	249.1	288.6	323.8	356.1	387.0	420.1	452.3	492.5	544.0	604.1

SOLUTION

Begin by entering the data into a computer or graphing calculator. Then, use the computer to sketch a scatter plot of the data, as shown in Figure 6.37. From the scatter plot, it appears that an exponential model is a good fit. After running the exponential regression program, you should obtain

$$A = 76.26(1.12)^t \quad \text{or} \quad A = 76.25e^{0.1133t}.$$

(The correlation coefficient is $r = 0.997$, which implies that the model is a good fit to the data.) From the model, you can see that the amount spent on health care from 1970 through 1989 had an average annual increase of 12%. From this model, you can predict the 1998 amount to be

$$A = 76.26(1.12)^{28} \approx 1821.4 \text{ billion dollars,}$$

which is over three times the amount spent in 1989.

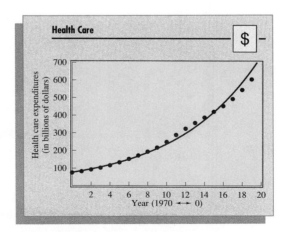

FIGURE 6.37

There are many other types of nonlinear models. One type is called a power model,

$$y = ax^b. \qquad \textit{Power model}$$

To fit data to this type of model, you can use a built-in program on a computer or calculator. Or you can transform the points (x, y) to the form $(\ln x, \ln y)$ and fit a linear model to the transformed points. This technique is illustrated in Example 5.

EXAMPLE 5   Using Logarithms to Find a Power Model

Table 6.8 gives the mean distance x and the period y of the six planets that are closest to the Sun. In the table the mean distance is given in terms of astronomical units (where the earth's mean distance is defined to be 1.0), and the period is given in years. Find a model for these data.

TABLE 6.8

Planet	Mercury	Venus	Earth	Mars	Jupiter	Saturn
Period, y	0.241	0.615	1.0	1.881	11.861	29.457
Distance, x	0.387	0.723	1.0	1.523	5.203	9.541

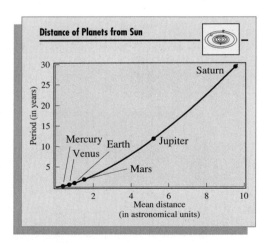

FIGURE 6.38

SOLUTION

Figure 6.38 shows a scatter plot for the points in the table. From the plot, it is not clear what type of model would best fit the data. (An exponential model has a correlation coefficient of only 0.895. With scientific data, one would

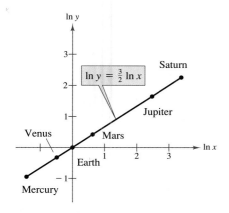

FIGURE 6.39

expect a better correlation.) To help classify the type of scatter plot, you can take the logarithm of each coordinate so that you obtain points of the form $(\ln x, \ln y)$, as shown in Table 6.9.

TABLE 6.9

Planet	Mercury	Venus	Earth	Mars	Jupiter	Saturn
ln y	−1.423	−0.486	0	0.632	2.473	3.383
ln x	−0.949	−0.324	0.0	0.421	1.649	2.256

Figure 6.39 shows a scatter plot for the transformed points. Note that the points fit a linear model. The least squares regression line for the transformed points is

$$\ln y = \frac{3}{2} \ln x \quad \text{or} \quad y = x^{3/2}.$$

Sine and cosine waves have many real-life applications, including electric circuits, musical tones, radio waves, tides, sunrises and sunsets, and weather patterns. The next example shows how to fit a cosine function to the pattern of water depth from high tide to low tide.

EXAMPLE 6   Finding a Trigonometric Model

Throughout the day, the depth of water at the end of a boat dock varies with the tides. Table 6.10 shows the depth (in meters) of water at various times during the morning hours.

TABLE 6.10

t (time)	Midnight	2 A.M.	4 A.M.	6 A.M.	8 A.M.	10 A.M.	Noon
y (depth)	2.55	3.80	4.40	3.80	2.55	1.80	2.27

A. Create a trigonometric equation that models the depth of the water at time t hours after midnight.

B. Find the depth of the water at 9 A.M. and at 3 P.M.

C. If a boat needs at least 3 meters of water to moor at the dock, during what time periods after noon can it safely dock?

SOLUTION

Because high tide occurs earlier than low tide, consider using a *cosine* model such as $y = a \cos bt$. There is a horizontal shift because high tide occurs after midnight ($t = 0$), and there is a vertical shift because the water depth is always greater than zero. Therefore, the model has the form

$$y = a \cos b(x - c) + d$$

Horizontal   Vertical
shift   shift

A. First you can make a scatter plot, as shown in Figure 6.40, to help find the amplitude and period.

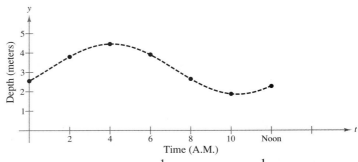

FIGURE 6.40

Amplitude: $a = \dfrac{1}{2}(\text{high} - \text{low}) = \dfrac{1}{2}(4.4 - 1.8) = 1.3$

Period: $p = 2(\text{time of low tide} - \text{time of high tide})$

$$= 2(10 - 4) = 12$$

$$p = \frac{2\pi}{b} = 12 \quad \rightarrow \quad b \approx 0.524$$

Horizontal shift: $c = 4$, because high tide occurs 4 hours after midnight.

Vertical shift: $d = \dfrac{1}{2}(\text{high tide} + \text{low tide}) = \dfrac{1}{2}(4.4 + 1.8) = 3.1$

Therefore, an equation that models the depth y at t hours after midnight is

$$y = a \cos b(t - c) + d = 1.3 \cos 0.524(t - 4) + 3.1.$$

B. The depth of the water at 9 A.M. ($t = 9$) is

$$y = 1.3 \cos 0.524(9 - 4) + 3.1 \approx 1.97 \text{ m}$$

At 3 P.M. ($t = 15$) the depth is

$$y = 1.3 \cos 0.524(15 - 4) + 3.1 \approx 4.23 \text{ m}.$$

C. You can use a graphing utility to obtain the graph shown in Figure 6.41. Then graph the line $y = 3$ on the same axes and find that it intersects the curve at times after noon of

$$t \approx 12.9 \text{ (1:12 P.M.)} \quad \text{and} \quad t \approx 19.1 \text{ (7:48 P.M.).}$$

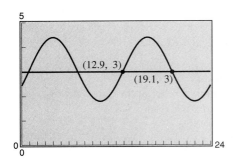

FIGURE 6.41

DISCUSSION PROBLEM

RESEARCH PROJECT

Use your school's library or some other reference source to find data that can be modeled with a *nonlinear* model. Find the model that you think best fits the data and explain how you could use the model to answer questions about the data.

WARM-UP

The following warm-up exercises involve skills that were covered in earlier sections. You will use these skills in the exercise set for this section.

In Exercises 1–6, sketch a graph of the equation.

1. $y = 10e^{-0.2x}$ **2.** $y = 2e^{0.2x}$

3. $y = 3 + 0.4 \ln x$ **4.** $y = 15 - 2 \ln x$

5. $y = 600(1.08)^t$ **6.** $y = 1000(0.8)^t$

In Exercises 7–10, find an equation of the regression line through the points.

7. $(-2, 4)$, $(-1, 3)$, $(0, 3)$, $(1, 2)$, $(2, 1)$

8. $(-6, 0)$, $(-3, 4)$, $(3, 6)$, $(6, 9)$

9. $(0, 10)$, $(1, 14)$, $(2, 17)$, $(3, 22)$

10. $(0, 8)$, $(2, 7)$, $(3, 5)$, $(5, 4)$, $(8, 0)$

SECTION 6.6 · EXERCISES

In Exercises 1–6, sketch a scatter plot for the data set. Decide whether the data could be best modeled by a linear model, $y = mx + b$, an exponential model, $y = ae^{bx}$, or a logarithmic model, $y = a + b \ln x$.

1. $(1, 2.0)$, $(1.5, 3.5)$, $(2, 4.0)$, $(4, 5.8)$, $(6, 7.0)$, $(8, 7.8)$

2. $(1, 5.8)$, $(1.5, 6.0)$, $(2, 6.5)$, $(4, 7.6)$, $(6, 8.9)$, $(8, 10.0)$

3. $(1, 4.4)$, $(1.5, 4.7)$, $(2, 5.5)$, $(4, 9.9)$, $(6, 18.1)$, $(8, 33.0)$

4. $(1, 11.0)$, $(1.5, 9.6)$, $(2, 8.2)$, $(4, 4.5)$, $(6, 2.5)$, $(8, 1.4)$

5. $(1, 7.5)$, $(1.5, 7.0)$, $(2, 6.8)$, $(4, 5.0)$, $(6, 3.5)$, $(8, 2.0)$

6. $(1, 5.0)$, $(1.5, 6.0)$, $(2, 6.4)$, $(4, 7.8)$, $(6, 8.6)$, $(8, 9.0)$

In Exercises 7–10, use a graphing utility to find an exponential model, $y = ae^{bx}$, through the points. Sketch a scatter plot and the exponential model.

7. $(0, 4)$, $(1, 5)$, $(2, 6)$, $(3, 8)$, $(4, 12)$

8. $(0, 6)$, $(2, 8.9)$, $(4, 20.0)$, $(6, 34.3)$, $(8, 61.1)$, $(10, 120.5)$

9. $(0, 10.0)$, $(1, 6.1)$, $(2, 4.2)$, $(3, 3.8)$, $(4, 3.6)$

10. $(-3, 120.2)$, $(0, 80.5)$, $(3, 64.8)$, $(6, 58.2)$, $(10, 55.0)$

In Exercises 11–14, use a graphing utility to find a logarithmic model, $y = a + b \ln x$, through the points. Sketch a scatter plot and the logarithmic model.

11. (1, 2), (2, 3), (3, 3.5), (4, 4), (5, 4.1), (6, 4.2), (7, 4.5)
12. (1, 8.5), (2, 11.4), (4, 12.8), (6, 13.6), (8, 14.2), (10, 14.6)
13. (1, 10), (2, 6), (3, 6), (4, 5), (5, 3), (6, 2)
14. (3, 14.6), (6, 11.0), (9, 9.0), (12, 7.6), (15, 6.5)

In Exercises 15–18, use a graphing utility to find a power model, $y = ax^b$, through the points. Sketch a scatter plot and the power model.

15. (1, 2.0), (2, 3.4), (5, 6.7), (6, 7.3), (10, 12.0)
16. (0.5, 1.0), (2, 12.5), (4, 33.2), (6, 65.7), (8, 98.5), (10, 150.0)
17. (1, 10.0), (2, 4.0), (3, 0.7), (4, 0.1)
18. (2, 450), (4, 385), (6, 345), (8, 332), (10, 312), (12, 300)

In Exercises 19 and 20, use a computer or calculator to (a) fit an exponential function $y = ab^x$ to the data, and (b) plot the data and the exponential function.

19. *Breaking Strength*   The breaking strength y (in tons) of steel cable of diameter d (in inches) is given in the accompanying table.

d	0.50	0.75	1.00	1.25	1.50	1.75
y	9.85	21.8	38.3	59.2	84.4	114.0

20. *World Population*   The world population y (in billions) for the 10 years from 1982 through 1991 is given in the accompanying table where $x = 2$ corresponds to 1982.

x	2	3	4	5	6
y	4.60	4.68	4.77	4.85	4.94

x	7	8	9	10	11
y	5.02	5.11	5.20	5.33	5.42

In Exercises 21 and 22, use a computer or calculator to (a) fit a logarithmic function $y = a + b \ln t$ to the data, and (b) plot the data and the logarithmic function.

21. *Water Pollution*   The amount y (in billions of dollars) spent on water pollution abatement in constant 1982 dollars in the United States from 1981 through 1989 is given in the following table. In the table, $t = 1$ represents the year 1981. (*Source:* U.S. Bureau of Economic Analysis)

t	1	2	3	4	5
y	22.0	21.2	21.5	23.3	24.7

t	6	7	8	9
y	26.4	28.0	27.2	28.8

22. *Rail Travel Receipts*   The receipts y (in millions of dollars) of the railroad industry in the United States from 1981 through 1990 is given in the following table. In the table, $t = 1$ represents the year 1981. (*Source:* U.S. Travel Data Center)

t	1	2	3	4	5
y	429	436	481	524	563

t	6	7	8	9	10
y	592	639	738	842	888

23. *Lumber*   The accompanying table gives the total domestic production x, and the domestic consumption y, of lumber in the United States from 1983 through 1990. The measurements are in millions of board feet. (*Source:* Current Industrial Reports) Find a power model for the relationship between production and consumption.

x	34.6	37.1	36.4	42.0
y	48.7	52.7	53.5	57.4

x	44.9	44.6	43.6	43.9
y	61.8	58.8	58.8	54.5

24. *Oceanography*   Rework the tide problem in Example 6, assuming that the first high tide occurs at 3:00 A.M. with a depth of 4.0 meters and the first low tide occurs at 9:36 A.M. with a depth of 1.4 meters.

25. *Oceanography*   The north coast of the Atlantic Ocean has some very high tide locations. At a dock in the Bay of Fundy, the depth of water at low tide was 3 feet. Six hours and 24 minutes later, at high tide, the depth was 60 feet. Find a trigonometric equation that models the depth of the water at time t after high tide.

26. *Temperature*   The figure shows the variation in average daily temperature T in Louisville, Kentucky. Find a trigonometric equation that models this temperature variation.

27. *Astronomy*   The scatter plot shows the time of sunsets at Erie, Pennsylvania on either the 21st or 22nd day of each month of a year. Find a trigonometric equation that models the data in the scatter plot.

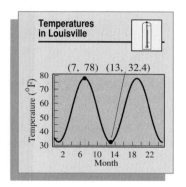

FIGURE FOR 26

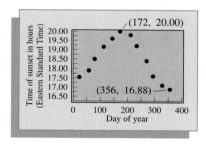

FIGURE FOR 27

CHAPTER 6 · REVIEW EXERCISES

In Exercises 1–6, match the function with its graph and describe the given viewing rectangle. [The graphs are labeled (a), (b), (c), (d), (e), and (f).]

1. $f(x) = 2^x$

2. $f(x) = 2^{-x}$

3. $f(x) = -2^x$

4. $f(x) = 2^x + 1$

5. $f(x) = \log_2 x$

6. $f(x) = \log_2(x - 1)$

(a)
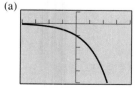

(b)
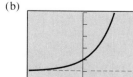

(c)
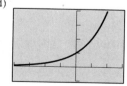

(d)
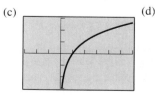

(e)
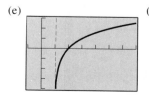

(f)
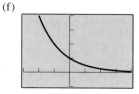

In Exercises 7–14, sketch the graph of the function.

7. $f(x) = 6^x$

8. $f(x) = 0.3^x$

9. $g(x) = 6^{-x}$

10. $g(x) = 0.3^{-x}$

11. $h(x) = e^{-x/2}$

12. $h(x) = 2 - e^{-x/2}$

13. $f(x) = e^{x+2}$

14. $s(t) = 4e^{-2/t}, \ t > 0$

In Exercises 15 and 16, complete the table to determine the balance A for P dollars invested at rate r for t years and compounded n times per year.

n	1	2	4	12	365	Continuous compounding
A						

15. $P = \$3500, \ r = 10.5\%, \ t = 10$ years

16. $P = \$2000, \ r = 12\%, \ t = 30$ years

In Exercises 17 and 18, complete the following table to determine the amount of money P that should be invested at rate r to produce a final balance of \$200,000 in t years.

t	1	10	20	30	40	50
P						

17. $r = 8\%$, compounded continuously

18. $r = 10\%$, compounded monthly

19. *Trust Fund*   On the day your child was born you deposited \$50,000 in a trust fund that pays 8.75% interest, compounded continuously. Determine the balance in the account at the time of your child's 35th birthday.

20. *Depreciation*   After t years, the value of a car that cost \$14,000 is given by

$$V(t) = 14,000(\tfrac{3}{4})^t.$$

Use a graphing utility to graph the function and determine the value of the car 2 years after it was purchased.

21. *Drug Decomposition*   A solution of a certain drug contained 500 units per milliliter when prepared. It was analyzed after 40 days and found to contain 300 units per milliliter. Assuming that the rate of decomposition is proportional to the amount present, the equation giving the amount A after t days is

$$A = 500e^{-0.013t}.$$

Use this model to find A when $t = 60$. Solve this problem algebraically and graphically. Which method do you prefer?

22. *Waiting Times*   The average time between incoming calls at a switchboard is 3 minutes. The probability of waiting less than t minutes until the next incoming call is approximated by the model

$$P(t) = 1 - e^{-t/3}.$$

If a call has just come in, find the probability that the next call will be within

(a) $\frac{1}{2}$ minute.   (b) 2 minutes.   (c) 5 minutes.

Solve the problem algebraically and graphically. Which method do you prefer?

23. *Fuel Efficiency*   A certain automobile gets 28 miles per gallon of gasoline for speeds up to 50 miles per hour. Over 50 miles per hour, the number of miles per gallon drops at the rate of 12% for each 10 miles per hour. If s is the speed and y is the number of miles per gallon, then

$$y = 28e^{0.6-0.012s}, \quad s \geq 50.$$

Use this function to complete the table. Write a short paragraph describing the results.

Speed	50	55	60	65	70
Miles per gallon					

24. *Inflation*   If the annual rate of inflation averages 4.5% over the next 10 years, then the approximate cost C of goods or services during any year in that decade will be given by

$$C(t) = P(1.05)^t,$$

where t is the time in years and P is the present cost. If the price of a tire for your car is presently \$69.95, estimate the price 10 years from now.

In Exercises 25–30, sketch the graph of the function.

25. $g(x) = \log_3 x$ **26.** $g(x) = \log_5 x$

27. $f(x) = \ln x + 3$ **28.** $f(x) = \ln(x - 3)$

29. $h(x) = \ln(e^{x-1})$ **30.** $f(x) = \frac{1}{4} \ln x$

In Exercises 31 and 32, use the definition of a logarithm to write the given equation in logarithmic form.

31. $4^3 = 64$ **32.** $25^{3/2} = 125$

In Exercises 33–40, evaluate the expression.

33. $\log_{10} 1000$

34. $\log_9 3$

35. $\log_3 \frac{1}{9}$

36. $\log_4 \frac{1}{16}$

37. $\ln e^7$

38. $\log_a \dfrac{1}{a}$

39. $\ln 1$

40. $\ln e^{-3}$

In Exercises 41–44, evaluate the logarithm using the change of base formula. Do each problem twice, once with common logarithms and once with natural logarithms. (Round your result to three decimal places.)

41. $\log_4 9$

42. $\log_{1/2} 5$

43. $\log_{12} 200$

44. $\log_3 0.28$

In Exercises 45–50, use the properties of logarithms to write the expression as a sum, difference, and/or multiple of logarithms.

45. $\log_5 5x^2$

46. $\log_7 \dfrac{\sqrt{x}}{4}$

47. $\log_{10} \dfrac{5\sqrt{y}}{x^2}$

48. $\ln \left| \dfrac{x - 1}{x + 1} \right|$

49. $\ln[(x^2 + 1)(x - 1)]$

50. $\ln \sqrt[5]{\dfrac{4x^2 - 1}{4x^2 + 1}}$

In Exercises 51–56, write the expression as the logarithm of a single quantity.

51. $\log_2 5 + \log_2 x$

52. $\log_6 y - 2 \log_6 z$

53. $\frac{1}{2} \ln |2x - 1| - 2 \ln |x + 1|$

54. $5 \ln |x - 2| - \ln |x + 2| - 3 \ln |x|$

55. $\ln 3 + \frac{1}{3} \ln(4 - x^2) - \ln x$

56. $3[\ln x - 2 \ln(x^2 + 1)] + 2 \ln 5$

In Exercises 57–60, determine whether the statement or equation is true or false. Explain how to use a graphing utility to support your conclusion.

57. The domain of $f(x) = \ln x$ is the set of all real numbers.

58. $\log_b b^{2x} = 2x$

59. $\ln(x + 2) = \ln x + \ln 2$

60. $e^{x-1} = \dfrac{e^x}{e}$

In Exercises 61–64, approximate the logarithm using the properties of logarithms given $\log_b 2 \approx 0.3562$, $\log_b 3 \approx 0.5646$, and $\log_b 5 \approx 0.8271$.

61. $\log_b 25$

62. $\log_b(\frac{25}{9})$

63. $\log_b \sqrt{3}$

64. $\log_b 30$

65. *Snow Removal* The number of miles s of roads cleared of snow is approximated by the model

$$s = 25 - \dfrac{13 \ln \dfrac{h}{12}}{\ln 3}, \quad 2 \le h \le 15,$$

where h is the depth of the snow in inches. Use this model to find s when $h = 10$ inches.

66. *Climb Rate* The time t, in minutes, required for a small plane to climb to an altitude of h feet is given by

$$t = 50 \log_{10} \frac{18,000}{18,000 - h},$$

where 18,000 feet is the plane's absolute ceiling. Find the time for the plane to climb to an altitude of 4000 feet.

In Exercises 67–72, solve the exponential equation. (Round your result to three decimal places.)

67. $e^x = 12$

68. $e^{3x} = 25$

69. $3e^{-5x} = 132$

70. $14e^{3x+2} = 560$

71. $e^{2x} - 7e^x + 10 = 0$

72. $e^{2x} - 6x + 8 = 0$

In Exercises 73–76, solve the logarithmic equation. (Round your result to three decimal places.)

73. $\ln 3x = 8.2$

74. $2 \ln 4x = 15$

75. $\ln x - \ln 3 = 2$

76. $\ln \sqrt{x + 1} = 2$

In Exercises 77–80, find the exponential function $y = Ce^{kt}$ that passes through the two points.

77. $(0, 2), (4, 3)$

78. $(0, \frac{1}{2}), (5, 5)$

79. $(0, 4), (5, \frac{1}{2})$

80. $(0, 2), (5, 1)$

81. *Demand Function* The demand equation for a certain product is given by

$$p = 500 - 0.5e^{0.004x}.$$

Find the demand x for a price of (a) $p = \$450$ and (b) $p = \$400$.

82. *Typing Speed* In a typing class, the average number of words per minute typed after t weeks of lessons was found to be

$$N = \frac{157}{1 + 5.4e^{-0.12t}}.$$

Find the time necessary to type (a) 50 words per minute and (b) 75 words per minute.

83. *Compound Interest* A deposit of $750 is made in a savings account for which the interest is compounded continuously. The balance will double in $7\frac{3}{4}$ years.
(a) What is the annual percentage rate for this account?
(b) Find the balance in the account after 10 years.
(c) Find the effective yield.

84. *Compound Interest* A deposit of $10,000 is made in a savings account for which the interest is compounded continuously. The balance will double in 5 years.
(a) What is the annual percentage rate for this account?
(b) Find the balance after 1 year.
(c) Find the effective yield.

85. *Sound Intensity* The relationship between the number of decibels β and the intensity of a sound I in watts per square centimeter is given by

$$\beta = 10 \log_{10}\left(\frac{I}{10^{-16}}\right).$$

Determine the intensity of a sound in watts per square centimeter if the decibel level is 125.

86. *Earthquake Magnitudes* On the Richter Scale, the magnitude R of an earthquake of intensity I is given by

$$R = \log_{10} \frac{I}{I_0},$$

where $I_0 = 1$ is the minimum intensity used for comparison. Find the intensity per unit of area for the following values of R.
(a) $R = 8.4$ (b) $R = 6.85$ (c) $R = 9.1$

87. Use a graphing utility to find an exponential model $y = ae^{bx}$ through the points $(0, 250), (4, 135), (6, 92), (10, 67)$. Sketch a scatter plot and the exponential model.

88. The data in the accompanying table give the yield y (in milligrams) of a chemical reaction after t minutes.

t	1	2	3	4
y	1.5	7.4	10.2	13.4

t	5	6	7	8
y	15.8	16.3	18.2	18.3

(a) Use a graphing utility to find the least squares regression line for the data.
(b) Use a graphing utility to fit the logarithmic equation $y = a + b \ln x$ to the data.
(c) Create a scatter plot of the data and sketch the graphs of the equations found in parts (a) and (b). Which is a better model of the data?

SOME TOPICS IN ANALYTIC GEOMETRY

7.1 LINES
......
Inclination of a Line / The Angle Between Two Lines / The Distance Between a Point and a Line

Inclination of a Line

In Section 1.5 we saw that the graph of the linear equation $y = mx + b$ is a nonvertical line with slope m and y-intercept at $(0, b)$. There, we described the slope of a line as the rate of change in y with respect to x. In this section, we look at the slope of a line in terms of the angle of inclination of the line.

Every nonhorizontal line must intersect the x-axis. The angle formed by such an intersection determines the **inclination** of the line, as specified in the following definition.

DEFINITION OF INCLINATION

The **inclination** of a nonhorizontal line is the positive angle θ (less than 180°) measured counterclockwise from the x-axis to the line (see figure).

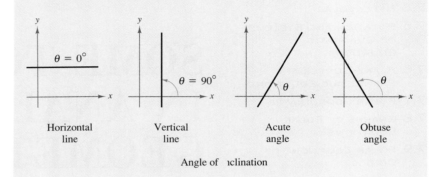

| Horizontal line | Vertical line | Acute angle | Obtuse angle |

Angle of inclination

The inclination of a line is related to its slope in the following manner.

INCLINATION AND SLOPE

If a nonvertical line has inclination θ and slope m, then

$$m = \tan \theta.$$

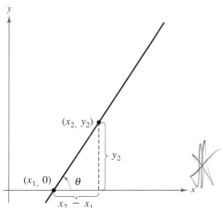

FIGURE 7.1

PROOF

If $m = 0$, then the line is horizontal and $\theta = 0$. Thus, the result is true for horizontal lines because $m = 0 = \tan 0$.

If the line has a positive slope, then it will intersect the x-axis. We label this point $(x_1, 0)$, as shown in Figure 7.1. If (x_2, y_2) is a second point on the line, then the slope is given by

$$m = \frac{y_2 - 0}{x_2 - x_1} = \frac{y_2}{x_2 - x_1} = \tan \theta.$$

We leave the case in which the line has a negative slope for you to prove.

EXAMPLE 1   Finding the Inclination of a Line

Find the inclination of the line given by $2x + 3y = 6$.

SOLUTION

The slope of this line is $m = -\frac{2}{3}$. Thus, its inclination is determined from the equation

$$\tan \theta = -\frac{2}{3}.$$

From Figure 7.2, we see that $90° < \theta < 180°$. This means that

$$\theta = 180° + \arctan\left(-\frac{2}{3}\right)$$
$$\approx 180° - 33.69°$$
$$\approx 146.31°.$$

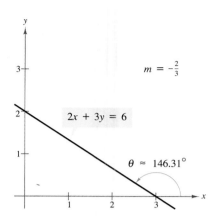

$m = -\frac{2}{3}$

$2x + 3y = 6$

$\theta \approx 146.31°$

FIGURE 7.2

The Angle Between Two Lines

Two distinct lines in a plane either are parallel or intersect. If they intersect, then their intersection forms two pairs of opposing angles, as shown in Figure 7.3. The measure of the smaller of these is called the **angle between the two lines.** As shown in Figure 7.3, we can use the inclinations of the two lines to find the angle between the two lines. Specifically, if two lines have inclinations θ_1 and θ_2, then the angle between the two lines is

$$\theta = \theta_2 - \theta_1,$$

where $\theta_1 < \theta_2$.

We can use the formula for the tangent of the difference of two angles

$$\tan \theta = \tan(\theta_2 - \theta_1) = \frac{\tan \theta_2 - \tan \theta_1}{1 + \tan \theta_1 \tan \theta_2}$$

to obtain the following convenient formula for the angle between two lines.

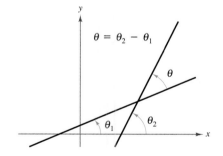

$\theta = \theta_2 - \theta_1$

θ

θ_1 θ_2

FIGURE 7.3

ANGLE BETWEEN TWO LINES

If two nonperpendicular lines have slopes m_1 and m_2, then the angle between the two lines is given by

$$\tan \theta = \left| \frac{m_2 - m_1}{1 + m_1 m_2} \right|.$$

EXAMPLE 2 Finding the Angle Between Two Lines

Find the angle between the following two lines.

$$\text{Line 1:} \quad 2x - y - 4 = 0$$
$$\text{Line 2:} \quad 3x + 4y - 12 = 0$$

SOLUTION

The two lines have slopes of

$$m_1 = 2 \quad \text{and} \quad m_2 = -\frac{3}{4},$$

respectively. Thus, the angle between the two lines is given by

$$\tan \theta = \left| \frac{m_2 - m_1}{1 + m_1 m_2} \right|$$

$$= \left| \frac{(-3/4) - 2}{1 + (-3/4)(2)} \right|$$

$$= \left| \frac{-11/4}{-2/4} \right|$$

$$= \frac{11}{2}.$$

Finally, you conclude that the angle is

$$\theta = \arctan \frac{11}{2} \approx 79.70°,$$

as shown in Figure 7.4.

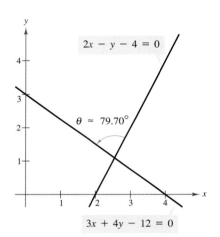

FIGURE 7.4

The Distance Between a Point and a Line

Finding the distance between a line and a point not on the line is an application of perpendicular lines. We define this distance to be the length of the perpendicular line segment joining the point to the given line, as shown in Figure 7.5.

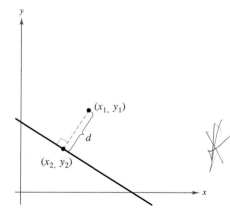

FIGURE 7.5

DISTANCE BETWEEN A POINT AND A LINE

The distance between the point (x_1, y_1) and the line given by $Ax + By + C = 0$ is

$$d = \frac{|Ax_1 + By_1 + C|}{\sqrt{A^2 + B^2}}.$$

PROOF

For simplicity's sake, we assume that the given line is neither horizontal nor vertical. By writing the equation $Ax + By + C = 0$ in slope-intercept form

$$y = -\frac{A}{B}x - \frac{C}{B},$$

we see that the line has a slope of

$$m = -\frac{A}{B}.$$

Thus, the slope of the line passing through (x_1, y_1) and perpendicular to the given line is B/A, and its equation is

$$y - y_1 = \frac{B}{A}(x - x_1).$$

These two lines intersect at the point (x_2, y_2), where

$$x_2 = \frac{B(Bx_1 - Ay_1) - AC}{A^2 + B^2} \quad \text{and} \quad y_2 = \frac{A(-Bx_1 + Ay_1) - BC}{A^2 + B^2}.$$

Finally, the distance between (x_1, y_1) and (x_2, y_2) is

$$d = \sqrt{(x_2 - x_1)^2 + (y_2 - y_1)^2}$$

$$= \sqrt{\left(\frac{B^2x_1 - ABy_1 - AC}{A^2 + B^2} - x_1\right)^2 + \left(\frac{-ABx_1 + A^2y_1 - BC}{A^2 + B^2} - y_1\right)^2}$$

$$= \sqrt{\frac{A^2(Ax_1 + By_1 + C)^2 + B^2(Ax_1 + By_1 + C)^2}{(A^2 + B^2)^2}}$$

$$= \frac{|Ax_1 + By_1 + C|}{\sqrt{A^2 + B^2}}.$$

EXAMPLE 3   Finding the Distance Between a Point and a Line

Find the distance between the point $(4, 1)$ and the line $y = 2x + 1$.

SOLUTION

The general form of the given equation is

$$-2x + y - 1 = 0.$$

Hence, the distance between the point and the line is

$$d = \frac{|-2(4) + 1(1) - 1|}{\sqrt{(-2)^2 + 1^2}} = \frac{8}{\sqrt{5}} \approx 3.58.$$

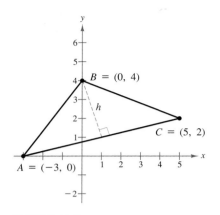

FIGURE 7.6

EXAMPLE 4   An Application of Two Distance Formulas

Figure 7.6 shows a triangle with vertices $A = (-3, 0)$, $B = (0, 4)$, and $C = (5, 2)$.

A. Find the altitude from vertex B to side AC.

B. Find the area of the triangle.

SOLUTION

A. To find the altitude, you can use the formula for the distance between line AC and the point $(0, 4)$. The equation of line AC is obtained as follows.

$$\text{Slope:} \qquad m = \frac{2 - 0}{5 + 3} = \frac{1}{4}$$

$$\text{Equation:} \qquad y - 0 = \frac{1}{4}(x + 3)$$

$$x - 4y + 3 = 0$$

Therefore, the distance between this line and the point $(0, 4)$ is

$$\text{Altitude} = h = \frac{|1(0) - 4(4) + 3|}{\sqrt{1^2 + (-4)^2}} = \frac{13}{\sqrt{17}}.$$

B. Using the formula for the distance between two points, you find the length of the base AC to be

$$b = \sqrt{(5 + 3)^2 + (2 - 0)^2} = \sqrt{68} = 2\sqrt{17}.$$

Finally, the area of the triangle in Figure 7.6 is

$$A = \frac{1}{2}bh = \frac{1}{2}(2\sqrt{17})\left(\frac{13}{\sqrt{17}}\right) = 13.$$

DISCUSSION PROBLEM

INCLINATION AND THE ANGLE BETWEEN TWO LINES

Write a paragraph explaining why the inclination of a line can be an angle that is larger than 90°, but the angle between two lines cannot be larger than 90°. Explain whether the following statement is true or false: "The inclination of a line is the angle between the line and the x-axis."

WARM-UP

The following warm-up exercises involve skills that were covered in earlier sections. You will use these skills in the exercise set for this section.

In Exercises 1 and 2, find the distance between the points.

1. $(-2, 0)$, $(3, 8)$ **2.** $(0, 4)$, $(4, -2)$

In Exercises 3–6, find the slope of the line passing through the pair of points.

3. $(-5, 1)$, $(4, 10)$ **4.** $(0, 2)$, $(7, 8)$

5. $(2, 12)$, $(10, 0)$ **6.** $(-4, 4)$, $(6, -1)$

In Exercises 7–10, find an equation of the line passing through the point with the specified slope.

	Point	Slope
7.	$(0, 3)$	$m = \frac{2}{3}$
8.	$(2, -5)$	$m = \frac{7}{2}$
9.	$(3, 20)$	$m = -4$
10.	$(-6, 4)$	$m = -\frac{3}{4}$

SECTION 7.1 · EXERCISES

In Exercises 1–8, find the slope of the line with the given inclination θ.

1.

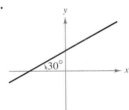

2.

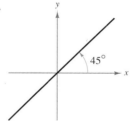

3.

4.

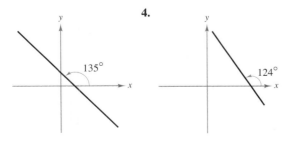

5. $\theta = 38.2°$ **6.** $\theta = 75.4°$

7. $\theta = 110°$ **8.** $\theta = 145.5°$

In Exercises 9–12, find the inclination θ of the line with the given slope.

ARCTAN 2 = θ

9. $m = -1$ **10.** $m = 2$

11. $m = \frac{3}{4}$ **12.** $m = -\frac{5}{2}$

In Exercises 13–16, find the inclination θ of the line passing through the pair of points.

13. $(6, 1)$, $(10, 8)$ **14.** $(-1, -4)$, $(7, 12)$

15. $(-2, 20)$, $(10, 0)$ **16.** $(0, 100)$, $(50, 0)$

In Exercises 17–20, find the inclination θ of the given line.

17. $5x - y + 3 = 0$ **18.** $4x + 5y - 9 = 0$

19. $5x + 3y = 0$ **20.** $x - y - 10 = 0$

In Exercises 21–30, find the angle θ between the two lines.

21. $2x + y - 4 = 0$
$x - y - 2 = 0$

22. $x + 3y - 2 = 0$
$x - 2y + 3 = 0$

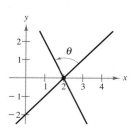

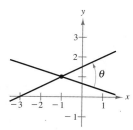

23. $x - y = 0$
$3x - 2y + 1 = 0$

24. $2x - y - 2 = 0$
$4x + 3y - 24 = 0$

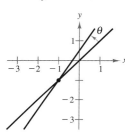

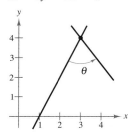

25. $x - 2y - 5 = 0$
$6x + 2y - 7 = 0$

26. $5x + 3y - 18 = 0$
$2x - 6y + 1 = 0$

27. $x + 2y - 4 = 0$
$x - 2y - 1 = 0$

28. $3x - 5y - 2 = 0$
$2x + 5y - 13 = 0$

29. $0.05x - 0.03y - 0.21 = 0$
$0.07x + 0.02y - 0.16 = 0$

30. $0.02x - 0.05y + 0.19 = 0$
$0.03x + 0.04y - 0.52 = 0$

In Exercises 31–38, find the distance between the point and the line.

Point	Line
31. $(0, 0)$	$4x + 3y - 10 = 0$
32. $(0, 0)$	$2x - y - 4 = 0$
33. $(2, 3)$	$4x + 3y - 10 = 0$
34. $(-2, 1)$	$x - y - 2 = 0$
35. $(6, 2)$	$x + 1 = 0$
36. $(10, 8)$	$y - 4 = 0$
37. $(0, 8)$	$6x - y = 0$
38. $(4, 2)$	$x - y - 20 = 0$

In Exercises 39 and 40, find the distance between the parallel lines. (*Hint:* Find the coordinates of any point on the one line and then find the distance between that point and the second line.)

39. $x + y - 1 = 0$
$x + y - 5 = 0$

40. $3x - 4y - 1 = 0$
$3x - 4y - 10 = 0$

41. *Grade of a Road*   A straight road rises with an inclination of 6.5° from the horizontal. Find the slope of the road and the change in elevation after driving 2 miles.

42. *Grade of a Road*   A straight road rises with an inclination of 4.5° from the horizontal. Find the slope of the road and the change in elevation after driving 3 miles.

43. *Mountain Climbing*   A group of mountain climbers are located in a mountain pass between two peaks (see figure). The angles of elevation to the two peaks are 48° and 63°. A range finder shows that the distances to the respective peaks are 3250 feet and 6700 feet.
(a) Find the angle between the two lines of sight to the peaks.
(b) Approximate the amount of vertical climb that is necessary to reach the summit of each peak.

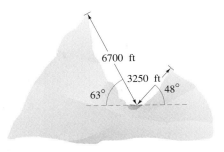

FIGURE FOR 43

44. *Conveyor Design*   A moving conveyor is built to rise 1 meter for each 3 meters of horizontal change.
(a) Find the inclination of the conveyor.
(b) Suppose the conveyor runs between two floors in a factory. Find the length of the conveyor if the vertical distance between floors is 10 feet.

45. *Pitch of a Roof* There is a rise of 2 feet for every horizontal change of 3 feet on a roof. Find the inclination of the roof.

46. *Pitch of a Roof* There is a rise of 3 feet for every horizontal change of 5 feet on a roof. Find the inclination of the roof.

Angle Measurement In Exercises 51–54, find the slope of each side of the triangle and use the slopes to find the magnitudes of the interior angles.

51.

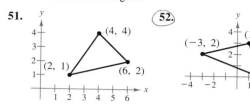

52.

Area In Exercises 47–50, (a) find the altitude from vertex *B* of a triangle to the side *AC*, and (b) find the area of the triangle.

47. $A = (0, 0)$, $B = (1, 5)$, $C = (3, 1)$

48. $A = (0, 0)$, $B = (4, 5)$, $C = (5, -2)$

49. $A = \left(-\frac{1}{2}, \frac{1}{2}\right)$, $B = (2, 3)$, $C = \left(\frac{5}{2}, 0\right)$

50. $A = (-4, -5)$, $B = (3, 10)$, $C = (6, 10)$

53.   **54.**

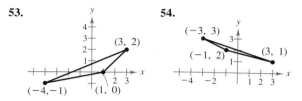

7.2   INTRODUCTION TO CONICS: PARABOLAS

Conics / Parabolas / Application

Conics

Conic sections were discovered during the classical Greek period (600 to 300 B.C.). This early Greek study was concerned largely with the geometrical properties of conics. It was not until the early 17th century that the broad applicability of conics became apparent, and they then played a prominent role in the early development of calculus.

A **conic section** (or simply **conic**) is the intersection of a plane and a double-napped cone. Notice in Figure 7.7 that in the formation of the four basic conics, the intersecting plane does not pass through the vertex of the cone. When the plane does pass through the vertex, the resulting figure is a **degenerate conic,** as shown in Figure 7.8.

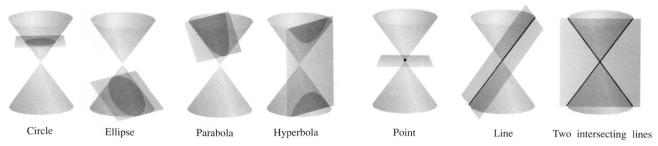

| Circle | Ellipse | Parabola | Hyperbola | | Point | Line | Two intersecting lines |

FIGURE 7.7   **FIGURE 7.8**   Degenerate Conics

Parabolas

There are several ways to begin a study of conics. You could define conics as the intersections of planes and cones, as the Greeks did, or you could define them algebraically in terms of the general second-degree equation

$$Ax^2 + Bxy + Cy^2 + Dx + Ey + F = 0.$$

However, we will use a third approach, in which each of the conics is defined as a collection of points satisfying a certain geometric property. For example, in Section 1.3, you saw how the definition of a circle as *the collection of all points (x, y) that are equidistant from a fixed point (h, k)* led easily to the standard equation of a circle,

$$(x - h)^2 + (y - k)^2 = r^2.$$

In this and the following two sections, we give similar definitions for the other three types of conics.

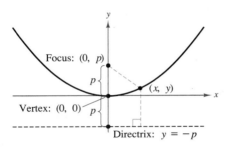

FIGURE 7.9

DEFINITION OF A PARABOLA

A **parabola** is the set of all points (x, y) that are equidistant from a fixed line (**directrix**) and a fixed point (**focus**) not on the line.

The midpoint between the focus and the directrix is the **vertex,** and the line passing through the focus and the vertex is the **axis** of the parabola. Note in Figure 7.9 that a parabola is symmetric with respect to its axis.

REMARK   Be sure you understand that the term *parabola* is a technical term used in mathematics and does not simply refer to any U-shaped curve.

STANDARD EQUATION OF A PARABOLA

The **standard form** of the equation of a parabola with vertex at (h, k) is as follows.

$$(x - h)^2 = 4p(y - k), \qquad p \neq 0 \quad \text{\textit{Vertical axis, directrix: } } y = k - p$$
$$(y - k)^2 = 4p(x - h), \qquad p \neq 0 \quad \text{\textit{Horizontal axis, directrix: } } x = h - p$$

The focus lies on the axis p units (*directed distance*) from the vertex. If the vertex is at the origin $(0, 0)$, the equation takes one of the following forms.

$$x^2 = 4py \qquad\qquad \text{\textit{Vertical axis}}$$
$$y^2 = 4px \qquad\qquad \text{\textit{Horizontal axis}}$$

PROOF

We prove only the case for which the directrix is parallel to the x-axis and the focus lies above the vertex, as shown in Figure 7.10(a). If (x, y) is any point on the parabola, then by definition it is equidistant from the focus $(h, k + p)$ and the directrix $y = k - p$, and we have

$$\sqrt{(x - h)^2 + [y - (k + p)]^2} = y - (k - p)$$
$$(x - h)^2 + [y - (k + p)]^2 = [y - (k - p)]^2$$
$$(x - h)^2 + y^2 - 2y(k + p) + (k + p)^2 = y^2 - 2y(k - p) + (k - p)^2$$
$$(x - h)^2 - 2py + 2pk = 2py - 2pk$$
$$(x - h)^2 = 4p(y - k).$$

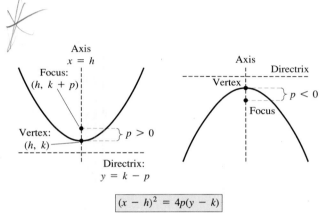

Axis
$x = h$
Focus:
$(h, k + p)$

Vertex:
(h, k)
$\Big\} p > 0$
Directrix:
$y = k - p$

$(x - h)^2 = 4p(y - k)$

(a) Vertical axis: $p > 0$

Axis

Directrix
Vertex

$\Big\} p < 0$

Focus

(b) Vertical axis: $p < 0$

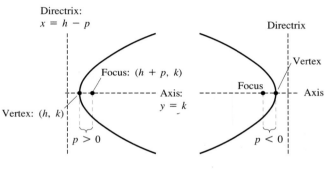

Directrix:
$x = h - p$

Focus: $(h + p, k)$

Axis:
$y = k$

Vertex: (h, k)

$p > 0$

Directrix

Vertex

Focus

Axis

$p < 0$

$(y - k)^2 = 4p(x - h)$

(c) Horizontal axis: $p > 0$

(d) Horizontal axis: $p < 0$

Parabolic Orientations

FIGURE 7.10

REMARK   By expanding the standard equation in Example 1, you obtain the more common quadratic form $y = \frac{1}{12}(x^2 - 4x + 16)$.

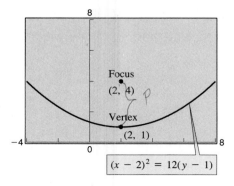

Focus
$(2, 4)$

Vertex
$(2, 1)$

$(x - 2)^2 = 12(y - 1)$

FIGURE 7.11

EXAMPLE 1   Finding the Standard Equation of a Parabola

Find the standard form of the equation of the parabola with vertex $(2, 1)$ and focus $(2, 4)$.

SOLUTION

Because the axis of the parabola is vertical, consider the equation

$$(x - h)^2 = 4p(y - k),$$

where $h = 2$, $k = 1$, and $p = 4 - 1 = 3$. Thus, the standard form is

$$(x - 2)^2 = 12(y - 1).$$

The graph of this parabola is shown in Figure 7.11.

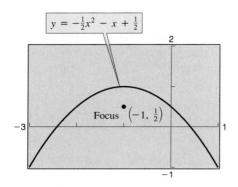

FIGURE 7.12

EXAMPLE 2 **Finding the Focus of a Parabola**

Find the focus of the parabola given by

$$y = -\frac{1}{2}x^2 - x + \frac{1}{2}.$$

SOLUTION

To find the focus, convert to standard form by completing the square.

$y = -\dfrac{1}{2}x^2 - x + \dfrac{1}{2}$	*Original equation*
$-2y = x^2 + 2x - 1$	*Multiply by* -2
$1 - 2y = x^2 + 2x$	*Group terms*
$2 - 2y = x^2 + 2x + 1$	*Add 1 to both sides*
$-2(y - 1) = (x + 1)^2$	*Standard form*

Comparing this equation with $(x - h)^2 = 4p(y - k)$, you can conclude that $h = -1$, $k = 1$, and $p = -\frac{1}{2}$. Because p is negative, the parabola opens downward, as shown in Figure 7.12. Therefore, the focus of the parabola is

$$(h, k + p) = \left(-1, \frac{1}{2}\right). \qquad Focus$$

EXAMPLE 3 **Vertex at the Origin**

Find the standard equation of the parabola with vertex at the origin and focus at (2, 0).

SOLUTION

The axis of the parabola is horizontal, passing through the origin and (2, 0), as shown in Figure 7.13. Thus, the standard form is

$$y^2 = 4px,$$

where $h = k = 0$ and $p = 2$. Therefore, the equation is

$$y^2 = 8x.$$

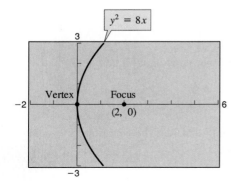

FIGURE 7.13

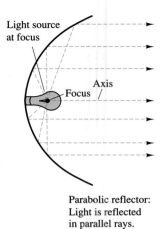

Parabolic reflector:
Light is reflected
in parallel rays.

FIGURE 7.14

Application

A line segment that passes through the focus of a parabola and has endpoints on the parabola is called a **focal chord.** The specific focal chord perpendicular to the axis of the parabola is called the **latus rectum.**

Parabolas occur in a wide variety of applications. For instance, a parabolic reflector can be formed by revolving a parabola about its axis. The resulting surface has the property that all incoming rays parallel to the axis are reflected through the focus of the parabola; this is the principle behind the construction of the parabolic mirrors used in reflecting telescopes. Conversely, the light rays emanating from the focus of a parabolic reflector used in a flashlight are all reflected parallel to one another, as shown in Figure 7.14.

A line is **tangent** to a parabola at a point on the parabola if the line intersects, but does not cross, the parabola at that point. Tangent lines to parabolas have special properties related to the use of parabolas in constructing reflective surfaces.

REFLECTIVE PROPERTY OF A PARABOLA

The tangent line to a parabola at a point P makes equal angles with the following two lines (see figure).

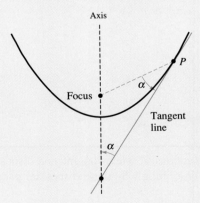

1. The line passing through P and the focus.
2. The axis of the parabola.

Technology Note

In Example 4, use a graphing utility to graph the parabola and the tangent line on the same screen. Then zoom-in to the point $(1, 1)$. You should be able to zoom-in close enough so that the parabola and the tangent line are indistinguishable. Why?

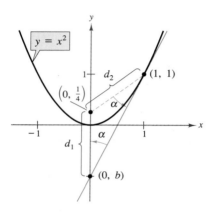

FIGURE 7.15

EXAMPLE 4 Finding the Tangent Line at a Point on a Parabola

Find the equation of the tangent line to the parabola given by $y = x^2$ at the point $(1, 1)$.

SOLUTION

For this parabola, $p = \frac{1}{4}$ and the focus is $(0, \frac{1}{4})$, as shown in Figure 7.15. You can find the y-intercept $(0, b)$ of the tangent line by equating the lengths of the two sides of the isosceles triangle

$$d_1 = \frac{1}{4} - b$$

and

$$d_2 = \sqrt{(1 - 0)^2 + (1 - \tfrac{1}{4})^2} = \frac{5}{4},$$

as shown in Figure 7.15. Setting $d_1 = d_2$ produces

$$\frac{1}{4} - b = \frac{5}{4}$$

$$b = -1.$$

Thus, the slope of the tangent line is

$$m = \frac{1 - (-1)}{1 - 0} = 2,$$

and its slope-intercept equation is

$$y = 2x - 1.$$

DISCUSSION PROBLEM

TELEVISION ANTENNA DISHES

Cross sections of television antenna dishes are parabolic in shape. Write a paragraph describing why these dishes are parabolic.

WARM-UP

The following warm-up exercises involve skills that were covered in earlier sections. You will use these skills in the exercise set for this section.

In Exercises 1–4, expand and simplify the expression.

1. $(x - 5)^2 - 20$ **2.** $(x + 3)^2 - 1$ **3.** $10 - (x + 4)^2$ **4.** $4 - (x - 2)^2$

In Exercises 5–8, complete the square for the quadratic expression.

5. $x^2 + 6x + 8$ **6.** $x^2 - 10x + 21$ **7.** $-x^2 + 2x + 1$ **8.** $-2x^2 + 4x - 2$

In Exercises 9 and 10, find an equation of the line passing through the given point with the specified slope.

9. $m = -\frac{2}{3}$, $(1, 6)$ **10.** $m = \frac{3}{4}$, $(3, -2)$

SECTION 7.2 · EXERCISES

In Exercises 1–6, match the equation with its graph, and describe the given viewing rectangle. [The graphs are labeled (a), (b), (c), (d), (e), and (f).]

1. $y^2 = 4x$ ℮   **2.** $x^2 = -2y$ F

3. $x^2 = 8y$ ℀   **4.** $y^2 = -12x$ ℭ

5. $(y - 1)^2 = 4(x - 2)$ D   **6.** $(x + 3)^2 = -2(y - 2)$ B

(a)   (b)

(c)   (d)

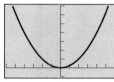
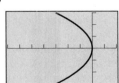

(e)   (f)

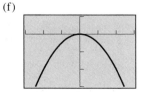

In Exercises 7–22, find the vertex, focus, and directrix of the parabola, and sketch its graph.

7. $y = 4x^2$ **8.** $y = 2x^2$

9. $y^2 = -6x$ **10.** $y^2 = 3x$

11. $x^2 + 8y = 0$ **12.** $x + y^2 = 0$

13. $(x - 1)^2 + 8(y + 2) = 0$

14. $(x + 3) + (y - 2)^2 = 0$

15. $\left(y + \frac{1}{2}\right)^2 = 2(x - 5)$

16. $\left(x + \frac{1}{2}\right)^2 = 4(y - 3)$

17. $y = \frac{1}{4}(x^2 - 2x + 5)$

18. $4x - y^2 - 2y - 33 = 0$

19. $y^2 - 4y - 4x = 0$

20. $y^2 + 6y + 8x + 25 = 0$

21. $y^2 + 4y + 8x - 12 = 0$

22. $x^2 + 4x + 4y - 4 = 0$

In Exercises 23–26, find the vertex, focus, and directrix of the parabola, and use a graphing utility to graph the parabola.

23. $y = -\frac{1}{6}(x^2 + 4x - 2)$

24. $x^2 - 2x + 8y + 9 = 0$

25. $y^2 + x + y = 0$

26. $y^2 - 4x - 4 = 0$

In Exercises 27–36, find an equation of the specified parabola.

27. Vertex: (0, 0)
Focus: $(0, -\frac{3}{2})$

28. Vertex: (0, 0)
Focus: $(-2, 0)$

29. Vertex: (0, 0)
Directrix: $x = 3$

30. Vertex: (3, 2)
Focus: (1, 2)

31. Vertex: (0, 4)
Directrix: $y = 2$

32. Vertex: (0, 0)
Directrix: $y = 4$

33. Axis: Parallel to y-axis
Passes through the points: (0, 3), (3, 4), (4, 11)

34. Axis: Parallel to x-axis
Passes through the points: (4, -2), (0, 0), (3, -3)

35.

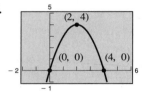

36.

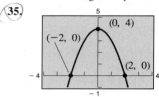

Norman /

37. *Satellite Antenna* The receiver in a parabolic television dish antenna is 3 feet from the vertex and is located at the focus (see figure). Find an equation of a cross section of the reflector. (Assume that the dish is directed upward and the vertex is at the origin.)

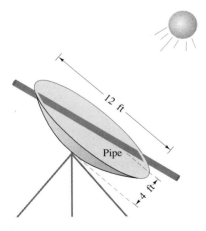

FIGURE FOR 37

38. *Solar Collector* A solar collector for heating water is constructed with a sheet of stainless steel that is formed into the shape of a parabola (see figure). The water flows through a pipe that passes through the focus of the parabola. At what distance is the pipe from the vertex?

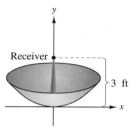

FIGURE FOR 38

39. *Suspension Bridge* Each cable of a suspension bridge is suspended (in the shape of a parabola) between two towers that are 400 feet apart and 50 feet above the roadway (see figure). The cables touch the roadway midway between the towers.
(a) Find an equation for the parabolic shape of each cable.
(b) Find the length of the vertical supporting cable when $x = 100$.

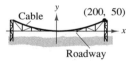

FIGURE FOR 39

40. *Beam Deflection* A simply supported beam that is 100 feet long has a load concentrated at its center (see figure). The deflection of the beam at its center is 3 inches. If the shape of the deflected beam is parabolic, find the equation of the parabola. (Assume the vertex is at the origin.)

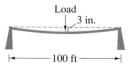

FIGURE FOR 40

41. *Escape Velocity*   A satellite in a 100-mile-high circular orbit around the earth has a velocity of approximately 17,500 miles per hour. If this velocity is multiplied by $\sqrt{2}$, the satellite will have the minimum velocity necessary to escape the earth's gravity and it will follow a parabolic path with the center of the earth as the focus (see figure).
(a) Find the escape velocity of the satellite.
(b) Find the equation of its path (assume that the radius of the earth is 4000 miles).

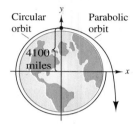

FIGURE FOR 41

42. *Highway Design*   Highway engineers design a parabolic curve for an entrance ramp from a straight street to an interstate highway (see figure). Find an equation of the parabola.

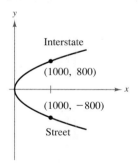

FIGURE FOR 42

Projectile Motion   In Exercises 43–45, consider the path of a projectile projected horizontally with a velocity of v feet per second at a height of s feet, where the model for the path is given by

$$y = -\frac{16}{v^2}x^2 + s.$$

In this model, air resistance is disregarded and y is the height (in feet) of the projectile t seconds after its release.

43. A ball is thrown horizontally from the top of a 75-foot tower with a velocity of 32 feet per second.
(a) Find the equation of the parabolic path.
(b) How far does the ball travel horizontally before striking the ground?

44. A ball is thrown horizontally from the top of a 100-foot tower with a velocity of 32 feet per second.
(a) Find the equation of the parabolic path.
(b) How far does the ball travel horizontally before striking the ground?

45. A bomber flying due east at 550 miles per hour at an altitude of 42,000 feet releases a bomb. Determine how far the bomb travels horizontally before striking the ground.

46. Find the equation of the tangent line to the parabola $y = ax^2$ at $x = x_0$. Prove that the x-intercept of this tangent line is $(x_0/2, 0)$.

In Exercises 47–50, find an equation of the tangent line to the parabola at the given point and find the x-intercept of the line.

47. $y = \frac{1}{2}x^2$, $(4, 8)$ **48.** $y = \frac{1}{2}x^2$, $\left(-3, \frac{9}{2}\right)$

49. $y = -2x^2$, $(-1, -2)$ **50.** $y = -2x^2$, $(3, -18)$

7.3   ELLIPSES

Ellipses   /   Application   /   Eccentricity

Ellipses

The second type of conic is called an **ellipse,** and is defined as follows.

DEFINITION OF AN ELLIPSE

An **ellipse** is the set of all points (x, y) the sum of whose distances from two distinct fixed points **(foci)** is constant.

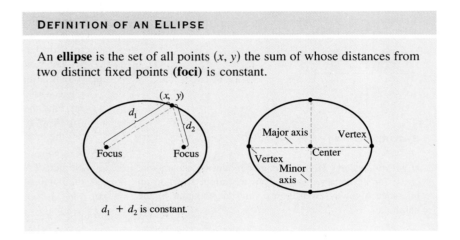

$d_1 + d_2$ is constant.

The line through the foci intersects the ellipse at two points **(vertices).** The chord joining the vertices is the **major axis,** and its midpoint is the **center** of the ellipse. The chord perpendicular to the major axis at the center is the **minor axis** of the ellipse.

You can visualize the definition of an ellipse by imagining two thumbtacks placed at the foci, as shown in Figure 7.16. If the ends of a fixed length of string are fastened to the thumbtacks and the string is drawn taut with a pencil, the path traced by the pencil will be an ellipse.

FIGURE 7.16

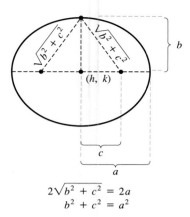

$$2\sqrt{b^2 + c^2} = 2a$$
$$b^2 + c^2 = a^2$$

FIGURE 7.17

To derive the standard form of the equation of an ellipse, consider the ellipse in Figure 7.17 with the following points.

Center: (h, k)

Vertices: $(h \pm a, k)$

Foci: $(h \pm c, k)$

The sum of the distances from any point on the ellipse to the two foci is constant. At a vertex, this constant sum is

$$(a + c) + (a - c) = 2a, \qquad \textit{Length of major axis}$$

or simply the length of the major axis. Now, if you let (x, y) be *any* point on the ellipse, the sum of the distances between (x, y) and the two foci must also be $2a$. That is,

$$\sqrt{[x - (h - c)]^2 + (y - k)^2} + \sqrt{[x - (h + c)]^2 + (y - k)^2} = 2a.$$

Finally, in Figure 7.17, you can see that $b^2 = a^2 - c^2$, which implies that the equation of the ellipse is

$$b^2(x - h)^2 + a^2(y - k)^2 = a^2b^2$$
$$\frac{(x - h)^2}{a^2} + \frac{(y - k)^2}{b^2} = 1.$$

Had you chosen a vertical major axis, you would have obtained a similar equation. Both results are summarized as follows.

STANDARD EQUATION OF AN ELLIPSE

The standard form of the equation of an ellipse, with center (h, k) and major and minor axes of lengths $2a$ and $2b$, where $0 < b < a$, is

$$\frac{(x - h)^2}{a^2} + \frac{(y - k)^2}{b^2} = 1 \qquad \textit{Major axis is horizontal}$$

$$\frac{(x - h)^2}{b^2} + \frac{(y - k)^2}{a^2} = 1. \qquad \textit{Major axis is vertical}$$

The foci lie on the major axis, c units from the center, with $c^2 = a^2 - b^2$. If the center is at the origin $(0, 0)$, the equation takes one of the following forms.

$$\frac{x^2}{a^2} + \frac{y^2}{b^2} = 1 \qquad \textit{Major axis is horizontal}$$

$$\frac{x^2}{b^2} + \frac{y^2}{a^2} = 1 \qquad \textit{Major axis is vertical}$$

Figure 7.18 shows both the vertical and horizontal orientations for an ellipse.

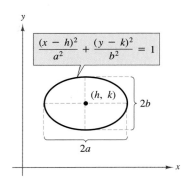
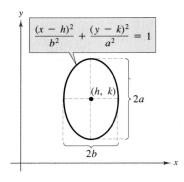

FIGURE 7.18

EXAMPLE 1   **Finding the Standard Equation of an Ellipse**

Find the standard form of the equation of the ellipse having foci at (0, 1) and (4, 1) and a major axis of length 6, as shown in Figure 7.19.

SOLUTION

Because the foci occur at (0, 1) and (4, 1), the center of the ellipse is (2, 1). This implies that the distance from the center to one of the foci is $c = 2$, and since $2a = 6$ you know that $a = 3$. Now, using $c^2 = a^2 - b^2$, you have

$$b = \sqrt{a^2 - c^2} = \sqrt{9 - 4} = \sqrt{5}.$$

Because the major axis is horizontal, the standard equation is

$$\frac{(x - 2)^2}{9\,a} + \frac{(y - 1)^2}{5\,b} = 1.$$

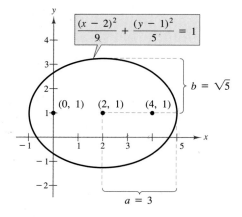

FIGURE 7.19

Technology Note

Most graphing utilities have a "parametric mode." Try using the parametric mode to graph the ellipse given by $x = 2 + 3 \cos t$ and $y = 1 + \sqrt{4} \sin t$. How does the result compare with the graph given in Figure 7.19? (*Hint:* Set your graphing utility to radian mode before graphing the ellipse.)

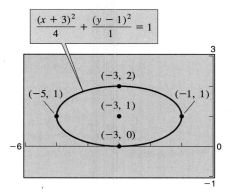

FIGURE 7.20

EXAMPLE 2 Writing an Equation in Standard Form

Sketch the graph of the ellipse whose equation is

$$x^2 + 4y^2 + 6x - 8y + 9 = 0.$$

SOLUTION

Begin by writing the given equation is standard form.

$$
\begin{aligned}
x^2 + 4y^2 + 6x - 8y + 9 &= 0 &&\text{\textit{Original equation}} \\
(x^2 + 6x + \quad) + (4y^2 - 8y + \quad) &= -9 &&\text{\textit{Group terms}} \\
(x^2 + 6x + \quad) + 4(y^2 - 2y + \quad) &= -9 &&\text{\textit{Factor 4 out of y-term}} \\
(x^2 + 6x + 9) + 4(y^2 - 2y + 1) &= -9 + 9 + 4 &&\text{\textit{Add 9 and 4 to both sides}} \\
(x + 3)^2 + 4(y - 1)^2 &= 4 &&\text{\textit{Completed square form}} \\
\frac{(x + 3)^2}{4} + \frac{(y - 1)^2}{1} &= 1 &&\text{\textit{Standard form}}
\end{aligned}
$$

From the standard form, you can see that the center occurs at $(h, k) = (-3, 1)$. Because the denominator of the x-term is $a^2 = 2^2$, locate the endpoints of the major axis two units to the right and left of the center. Similarly, because the denominator of the y-term is $b^2 = 1^2$, locate the endpoints of the minor axis one unit up and down from the center. The graph of this ellipse is shown in Figure 7.20.

EXAMPLE 3 Analyzing an Ellipse

Find the center, vertices, and foci of the ellipse given by

$$4x^2 + y^2 - 8x + 4y - 8 = 0.$$

SOLUTION

By completing the square, you can write the given equation in standard form.

$$
\begin{aligned}
4x^2 + y^2 - 8x + 4y - 8 &= 0 \\
4(x^2 - 2x + 1) + (y^2 + 4y + 4) &= 8 + 4 + 4 \\
4(x - 1)^2 + (y + 2)^2 &= 16 \\
\frac{(x - 1)^2}{4} + \frac{(y + 2)^2}{16} &= 1
\end{aligned}
$$

Thus, the major axis is vertical, where $h = 1$, $k = -2$, $a = 4$, $b = 2$, and $c = \sqrt{16 - 4} = 2\sqrt{3}$. Therefore, you have the following.

Center: $(1, -2)$ Vertices: $(1, -6)$ Foci: $(1, -2 - 2\sqrt{3})$
$(1, 2)$ $(1, -2 + 2\sqrt{3})$

The graph of the ellipse is shown in Figure 7.21.

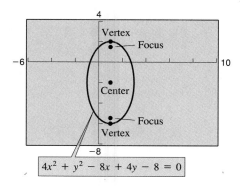

FIGURE 7.21

Technology Note _____

You can also use a graphing utility to graph an ellipse by graphing the upper and lower portions on the same viewing rectangle. For example, to graph the ellipse in Example 3, first solve for y

$$\frac{(x-1)^2}{4} + \frac{(y+2)^2}{16} = 1$$

$$(y+2)^2 = 16\left(1 - \frac{(x-1)^2}{4}\right)$$

$$y + 2 = \pm 4\sqrt{1 - \frac{(x-1)^2}{4}}$$

$$y = -2 \pm 4\sqrt{1 - \frac{(x-1)^2}{4}}$$

If you then graph the two equations

$$y_1 = -2 + 4\sqrt{1 - \frac{(x-1)^2}{4}}$$

$$y_2 = -2 - 4\sqrt{1 - \frac{(x-1)^2}{4}}$$

you will obtain the ellipse in Figure 7.21.

REMARK If the constant term in the equation in Example 3 had been $F \geq 8$, you would have obtained one of the following degenerate cases.

1. Single point: $\dfrac{(x-1)^2}{4} + \dfrac{(y+2)^2}{16} = 0$ $F = 8$

2. No solution points: $\dfrac{(x-1)^2}{4} + \dfrac{(y+2)^2}{16} < 0$ $F > 8$

Application

Ellipses have many practical and aesthetic uses. For instance, machine gears, supporting arches, and acoustical designs often involve elliptical shapes. The orbits of satellites and planets are also ellipses. Example 4 investigates the elliptical orbit of the moon about the earth.

EXAMPLE 4   **An Application Involving an Elliptical Orbit**

The moon travels about the earth in an elliptical orbit with the earth at one focus, as shown in Figure 7.22. The major and minor axes of the orbit have lengths of 768,806 kilometers and 767,746 kilometers, respectively. Find the greatest and least distances (the apogee and perigee) from the earth's center to the moon's center.

SOLUTION

Because $2a = 768,806$ and $2b = 767,746$, you have $a = 384,403$, $b = 383,873$, and

$$c = \sqrt{a^2 - b^2} \approx 20,179.$$

Therefore, the greatest distance between the center of the earth and the center of the moon is

$$a + c \approx 404,582 \text{ km,}$$

and the least distance is

$$a - c \approx 364,224 \text{ km.}$$

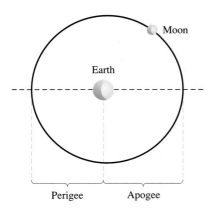

FIGURE 7.22

Eccentricity

One of the reasons it was difficult for early astronomers to detect that the orbits of the planets are ellipses is that the foci of the planetary orbits are relatively close to their centers, thus making the orbits nearly circular. To measure the ovalness of an ellipse, we use the concept of **eccentricity.**

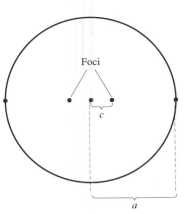

Foci

c

a

(a) $\dfrac{c}{a}$ is small.

> ### DEFINITION OF ECCENTRICITY
>
> The **eccentricity** e of an ellipse is given by the ratio
>
> $$e = \frac{c}{a}.$$

To see how this ratio is used to describe the shape of an ellipse, note that since the foci of an ellipse are located along the major axis between the vertices and the center, it follows that

$$0 < c < a.$$

For an ellipse that is nearly circular, the foci are close to the center and the ratio c/a is small, as shown in Figure 7.23(a). On the other hand, for an elongated ellipse, the foci are close to the vertices, and the ratio c/a is close to 1, as shown in Figure 7.23(b).

REMARK Note that $0 < e < 1$ for every ellipse.

The orbit of the moon has an eccentricity of $e = 0.0549$, and the eccentricities of the nine planetary orbits are as follows.

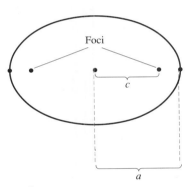

Foci

c

a

(b) $\dfrac{c}{a}$ is close to 1.

FIGURE 7.23

Mercury:	$e = 0.2056$	Saturn:	$e = 0.0543$
Venus:	$e = 0.0068$	Uranus:	$e = 0.0460$
Earth:	$e = 0.0167$	Neptune:	$e = 0.0082$
Mars:	$e = 0.0934$	Pluto:	$e = 0.2481$
Jupiter:	$e = 0.0484$		

DISCUSSION PROBLEM

....

IS A CIRCLE AN ELLIPSE?

If $a = b = r$, then the equation

$$\frac{x^2}{a^2} + \frac{y^2}{b^2} = 1$$

can be rewritten as

$$x^2 + y^2 = r^2.$$

Does this imply that a circle is an ellipse? Why or why not?

WARM-UP

The following warm-up exercises involve skills that were covered in earlier sections. You will use these skills in the exercise set for this section.

In Exercises 1–4, sketch a graph of the equation.

1. $x^2 = 9y$ **2.** $y^2 = 9x$ **3.** $y^2 = -9x$ **4.** $x^2 = -9y$

In Exercises 5–8, find the unknown in the equation $c^2 = a^2 - b^2$. (Assume a, b, and c are positive.)

5. $a = 13, b = 5$ **6.** $a = \sqrt{10}, c = 3$ **7.** $b = 6, c = 8$ **8.** $a = 7, b = 5$

In Exercises 9 and 10, simplify the compound fraction.

9. $\dfrac{x^2}{\frac{1}{4}} + \dfrac{y^2}{\frac{1}{3}}$ **10.** $\dfrac{(x-1)^2}{\frac{4}{9}} + \dfrac{(y+2)^2}{\frac{1}{9}}$

SECTION 7.3 · EXERCISES

In Exercises 1–6, match the equation with its graph, and describe the given viewing rectangle. [The graphs are labeled (a), (b), (c), (d), (e), and (f).]

1. $\dfrac{x^2}{1} + \dfrac{y^2}{9} = 1$ E

2. $\dfrac{x^2}{9} + \dfrac{y^2}{1} = 1$ a

3. $\dfrac{x^2}{9} + \dfrac{y^2}{4} = 1$ C

4. $\dfrac{y^2}{9} + \dfrac{x^2}{9} = 1$ b

5. $\dfrac{(x-2)^2}{16} + \dfrac{(y+1)^2}{4} = 1$ F
4   2

6. $\dfrac{(x+2)^2}{4} + \dfrac{(y+2)^2}{25} = 1$ D

(e) 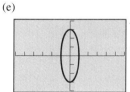   (f)

(a) 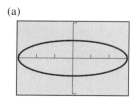   (b)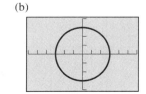

(c) 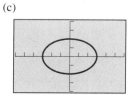   (d)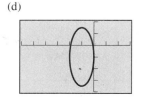

In Exercises 7–22, find the center, foci, vertices, and eccentricity of the ellipse and sketch its graph.

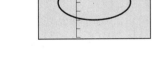

7. $\dfrac{x^2}{25} + \dfrac{y^2}{16} = 1$ **8.** $\dfrac{x^2}{144} + \dfrac{y^2}{169} = 1$

9. $\dfrac{x^2}{16} + \dfrac{y^2}{25} = 1$ **10.** $\dfrac{x^2}{169} + \dfrac{y^2}{144} = 1$

11. $\dfrac{x^2}{9} + \dfrac{y^2}{5} = 1$ **12.** $\dfrac{x^2}{28} + \dfrac{y^2}{64} = 1$

13. $x^2 + 4y^2 = 4$ **14.** $5x^2 + 3y^2 = 15$

15. $3x^2 + 2y^2 = 6$ **16.** $5x^2 + 7y^2 = 70$

17. $\dfrac{(x-1)^2}{9} + \dfrac{(y-5)^2}{25} = 1$

18. $(x+2)^2 + \dfrac{(y+4)^2}{\frac{1}{4}} = 1$

19. $9x^2 + 4y^2 + 36x - 24y + 36 = 0$
20. $9x^2 + 4y^2 - 36x + 8y + 31 = 0$
21. $16x^2 + 25y^2 - 32x + 50y + 31 = 0$
22. $9x^2 + 25y^2 - 36x - 50y + 61 = 0$

In Exercises 23–28, find the center, foci, and vertices of the ellipse. Use a graphing utility to graph the ellipse. (Explain how you used the utility to obtain the graph.)

23. $4x^2 + y^2 = 1$
24. $16x^2 + 25y^2 = 1$
25. $12x^2 + 20y^2 - 12x + 40y - 37 = 0$
26. $36x^2 + 9y^2 + 48x - 36y + 43 = 0$
27. $x^2 + 2y^2 - 3x + 4y + 0.25 = 0$
28. $2x^2 + y^2 + 4.8x - 6.4y + 3.12 = 0$

In Exercises 29–36, find an equation of the specified ellipse.

29. Vertices: $(\pm 6, 0)$
    Foci: $(\pm 5, 0)$
30. Vertices: $(0, \pm 8)$
    Foci: $(0, \pm 4)$
31. Vertices: $(0, \pm 2)$
    Minor axis of length 2
32. Vertices: $(0, 2)$, $(4, 2)$
    Minor axis of length 2
33. Foci: $(0, 0)$, $(0, 8)$
    Major axis of length 16
34. Vertices: $(0, \pm 5)$
    Solution point: $(4, 2)$
35. Center: $(3, 2)$, $a = 3c$
    Foci: $(1, 2)$, $(5, 2)$
36. Vertices: $(\pm 5, 0)$
    Eccentricity: $\frac{3}{5}$

37. *Fireplace Arch*   A fireplace arch is to be constructed in the shape of a semi-ellipse. The opening is to have a height of 2 feet at the center and a width of 5 feet along the base (see figure). The contractor draws the outline of the ellipse by the method shown in Figure 7.16. Where should the tacks be placed and what should be the length of the piece of string?

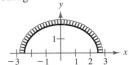

FIGURE FOR 37

38. *Mountain Tunnel*   A semi-elliptical arch over a tunnel for a road through a mountain has a major axis of 100 feet, and its height at the center is 30 feet (see figure). Determine the height of the arch 5 feet from the edge of the tunnel.

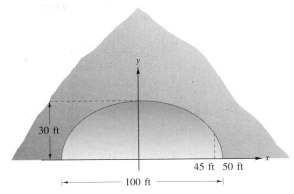

FIGURE FOR 38

39. Sketch a graph of the ellipse that consists of all points (x, y) such that the sum of the distances between (x, y) and two fixed points is 16 units and the foci are located at the centers of the two sets of concentric circles in the figure.

40. A line segment through a focus with endpoints on the ellipse and perpendicular to the major axis is called a **latus rectum** of the ellipse. Therefore, an ellipse has two latus recta. Knowing the length of the latus recta is helpful in sketching an ellipse because it yields other points on the curve (see figure). Show that the length of each latus rectum is $2b^2/a$.

FIGURE FOR 39

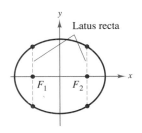

FIGURE FOR 40

In Exercises 41–44, sketch the graph of the ellipse, making use of the latus recta (see Exercise 40).

41. $\dfrac{x^2}{4} + \dfrac{y^2}{1} = 1$ **42.** $\dfrac{x^2}{9} + \dfrac{y^2}{16} = 1$

43. $9x^2 + 4y^2 = 36$ **44.** $5x^2 + 3y^2 = 15$

45. *Orbit of the Earth*   The earth moves in an elliptical orbit with the sun at one of the foci (see figure). The length of half of the major axis is 92.957×10^6 miles and the eccentricity is 0.017. Find the smallest distance (*perihelion*) and the greatest distance (*aphelion*) of the earth from the sun.

46. *Orbit of Pluto*   The planet Pluto moves in an elliptical orbit with the sun at one of the foci (see figure). The length of half of the major axis is 3.666×10^9 miles and the eccentricity is 0.248. Find the smallest distance and the greatest distance of Pluto from the sun.

47. *Orbit of Saturn*   The planet Saturn moves in an elliptical orbit with the sun at one of the foci (see figure). The smallest distance and the greatest distance of the planet from the sun are 1.3495×10^9 kilometers and 1.5045×10^9 kilometers, respectively. Find the eccentricity of the orbit.

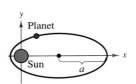

FIGURE FOR 45–47

48. *Satellite Orbit*   If the apogee and the perigee (see Example 4) of an elliptical orbit of an earth satellite are given by A and P, respectively, show that the eccentricity of the orbit is given by

$$e = \frac{A - P}{A + P}.$$

49. *Sputnik I*   The first artificial satellite to orbit the earth was Sputnik I (launched by the Soviet Union in 1957). Its highest point above the earth's surface was 583 miles, and its lowest point was 132 miles. Assume that the center of the earth is the focus of the elliptical orbit and the radius of the earth is 4000 miles. Find the eccentricity of the orbit.

50. *Explorer 18*   On November 26, 1963, the United States launched Explorer 18. Its low and high points over the surface of the earth were 119 miles and 122,000 miles, respectively. Find the eccentricity of its elliptical orbit.

51. Show that the equation of an ellipse can be written as

$$\frac{(x - h)^2}{a^2} + \frac{(y - k)^2}{a^2(1 - e^2)} = 1.$$

Note that as e approaches zero the ellipse approaches a circle of radius a.

7.4 HYPERBOLAS

Hyperbolas / Asymptotes of a Hyperbola / Applications

Hyperbolas

The definition of a hyperbola parallels that of an ellipse. The difference is that for an ellipse the *sum* of the distances between the foci and a point on the ellipse is fixed, while for a hyperbola the *difference* of these distances is fixed.

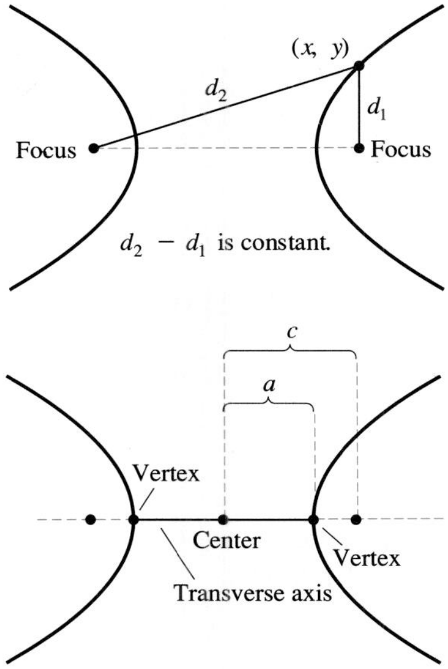

$d_2 - d_1$ is constant.

FIGURE 7.24

DEFINITION OF A HYPERBOLA

A **hyperbola** is the set of all points (x, y) the difference of whose distances from two distinct fixed points (foci) is constant (see Figure 7.24).

Every hyperbola has two disconnected parts **(branches).** The line through the two foci intersects a hyperbola at two points **(vertices).** The line segment connecting the vertices is the **transverse axis,** and the midpoint of the transverse axis is the **center** of the hyperbola.

The development of the standard form of the equation of a hyperbola is similar to that of an ellipse, and we list the following result without proof.

STANDARD EQUATION OF A HYPERBOLA

The standard form of the equation of a hyperbola with center at (h, k) is

$$\frac{(x - h)^2}{a^2} - \frac{(y - k)^2}{b^2} = 1 \qquad \textit{Transverse axis is horizontal}$$

$$\frac{(y - k)^2}{a^2} - \frac{(x - h)^2}{b^2} = 1. \qquad \textit{Transverse axis is vertical}$$

The vertices are a units from the center, and the foci are c units from the center. Moreover, $b^2 = c^2 - a^2$. If the center of the hyperbola is at the origin $(0, 0)$, the equation takes one of the following forms.

$$\frac{x^2}{a^2} - \frac{y^2}{b^2} = 1 \qquad \textit{Transverse axis is horizontal}$$

$$\frac{y^2}{a^2} - \frac{x^2}{b^2} = 1 \qquad \textit{Transverse axis is vertical}$$

Figure 7.25 shows both the horizontal and vertical orientations of a hyperbola.

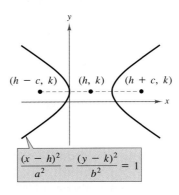

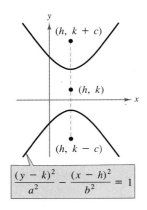

FIGURE 7.25   Standard Equations of Hyperbolas

EXAMPLE 1   **Finding the Standard Equation of a Hyperbola**

Find the standard form of the equation of the hyperbola with foci at $(-1, 2)$ and $(5, 2)$ and vertices at $(0, 2)$ and $(4, 2)$.

SOLUTION

By the Midpoint Formula, the center of the hyperbola occurs at the point $(2, 2)$. Furthermore, $c = 3$ and $a = 2$, and it follows that

$$b^2 = 3^2 - 2^2 = 9 - 4 = 5.$$

Thus, the equation of the hyperbola is

$$\frac{(x-2)^2}{4} - \frac{(y-2)^2}{5} = 1.$$

Figure 7.26 shows the graph of the hyperbola.

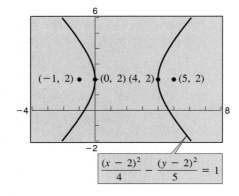

FIGURE 7.26

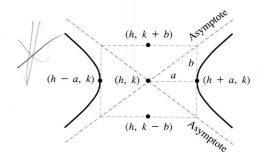

FIGURE 7.27

Asymptotes of a Hyperbola

An important aid in sketching the graph of a hyperbola is the determination of its **asymptotes,** as shown in Figure 7.27. Each hyperbola has two asymptotes that intersect at the center of the hyperbola. The asymptotes pass through the vertices of a rectangle of dimensions $2a$ by $2b$, with its center at (h, k). The line segment of length $2b$ joining $(h, k + b)$ and $(h, k - b)$ is referred to as the **conjugate axis** of the hyperbola. The following result identifies the equations for the asymptotes.

ASYMPTOTES OF A HYPERBOLA

For a *horizontal* transverse axis, the equations of the asymptotes are

$$y = k + \frac{b}{a}(x - h) \quad \text{and} \quad y = k - \frac{b}{a}(x - h).$$

For a *vertical* transverse axis, the equations of the asymptotes are

$$y = k + \frac{a}{b}(x - h) \quad \text{and} \quad y = k - \frac{a}{b}(x - h).$$

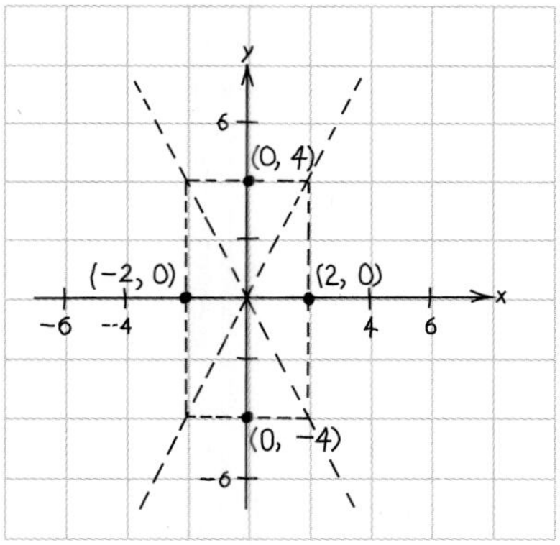

FIGURE 7.28

EXAMPLE 2   Using Asymptotes to Sketch a Hyperbola

Sketch the hyperbola whose equation is $4x^2 - y^2 = 16$.

SOLUTION

$$4x^2 - y^2 = 16 \qquad \text{\textit{Original equation}}$$

$$\frac{4x^2}{16} - \frac{y^2}{16} = \frac{16}{16}$$

$$\frac{x^2}{2^2} - \frac{y^2}{4^2} = 1 \qquad \text{\textit{Standard form}}$$

From this, it follows that the transverse axis is horizontal and the vertices occur at $(-2, 0)$ and $(2, 0)$. Moreover, the ends of the conjugate axis occur at $(0, -4)$ and $(0, 4)$, and you can sketch the rectangle shown in Figure 7.28. Finally, by drawing the asymptotes through the corners of this rectangle, you can complete the sketch, as shown in Figure 7.29.

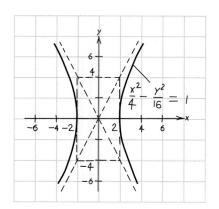

FIGURE 7.29

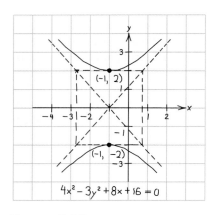

FIGURE 7.30

REMARK If the constant term F in the equation in Example 3 had been $F = -4$ instead of 16, you would have obtained the following degenerate case (two intersecting lines).

$$\frac{y^2}{4} - \frac{(x + 1)^2}{3} = 0$$

EXAMPLE 3   Finding the Asymptotes of a Hyperbola

Sketch the hyperbola given by $4x^2 - 3y^2 + 8x + 16 = 0$ and find the equations of its asymptotes.

SOLUTION

$$4x^2 - 3y^2 + 8x + 16 = 0 \qquad \textit{Original equation}$$
$$4(x^2 + 2x) - 3y^2 = -16$$
$$-4(x^2 + 2x + 1) + 3y^2 = 16 - 4$$
$$-4(x + 1)^2 + 3y^2 = 12$$
$$\frac{y^2}{4} - \frac{(x + 1)^2}{3} = 1 \qquad \textit{Standard form}$$

From this equation it follows that the hyperbola is centered at $(-1, 0)$ and has vertices at $(-1, 2)$ and $(-1, -2)$, and that the ends of the conjugate axis occur at $\left(-1 - \sqrt{3}, 0\right)$ and $\left(-1 + \sqrt{3}, 0\right)$. To sketch the graph of the hyperbola, draw a rectangle through the vertices and the ends of the conjugate axis. The asymptotes are the lines passing through the corners of the rectangle, as shown in Figure 7.30. Finally, using $a = 2$ and $b = \sqrt{3}$, it follows that the equations of the asymptotes are

$$y = \frac{2}{\sqrt{3}}(x + 1) \quad \text{and} \quad y = -\frac{2}{\sqrt{3}}(x + 1).$$

EXAMPLE 4   Using Asymptotes to Find the Standard Equation

Find the standard form of the equation of the hyperbola having vertices at $(3, -5)$ and $(3, 1)$ and asymptotes $y = 2x - 8$ and $y = -2x + 4$, as shown in Figure 7.31.

SOLUTION

By the Midpoint Formula, the center of the hyperbola is at $(3, -2)$. Furthermore, the hyperbola has a vertical transverse axis with $a = 3$. From the given equation, you can determine the slopes of the asymptotes to be

$$m_1 = 2 = \frac{a}{b} \quad \text{and} \quad m_2 = -2 = -\frac{a}{b},$$

and since $a = 3$, you can conclude that $b = \frac{3}{2}$. Thus, the standard equation is

$$\frac{(y + 2)^2}{9} - \frac{(x - 3)^2}{\frac{9}{4}} = 1.$$

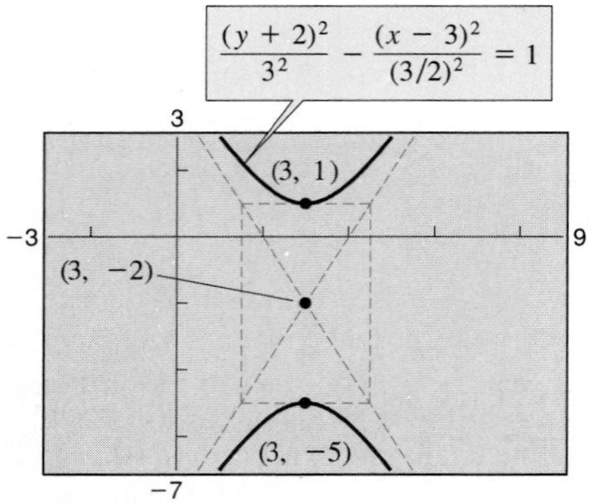

FIGURE 7.31

As with ellipses, the **eccentricity** of a hyperbola is $e = c/a$, and because $c > a$ it follows that $e > 1$. If the eccentricity is large, the branches of the hyperbola are nearly flat. If the eccentricity is close to 1, the branches of the hyperbola are more pointed, as shown in Figure 7.32.

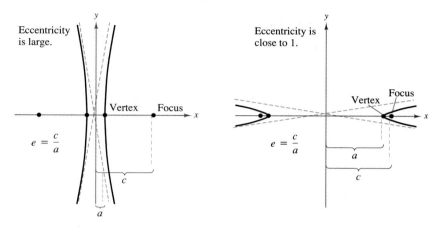

FIGURE 7.32

Applications

The following application was developed during World War II. It shows how the properties of hyperbolas can be used in radar and other detection systems.

EXAMPLE 5   An Application Involving Hyperbolas

Two microphones, 1 mile apart, record an explosion. Microphone *A* received the sound 2 seconds before microphone *B*. Where did the explosion occur?

SOLUTION

Assuming sound travels at 1100 feet per second, you know that the explosion took place 2200 feet farther from *B* than from *A*, as shown in Figure 7.33. The locus of all points that are 2200 feet closer to *A* than to *B* is one branch of the hyperbola $(x^2/a^2) - (y^2/b^2) = 1$, where

$$c = \frac{5280}{2} = 2640 \quad \text{and} \quad a = \frac{2200}{2} = 1100.$$

Thus, $b^2 = c^2 - a^2 = 5{,}759{,}600$ and you can conclude that the explosion occurred somewhere on the right branch of the hyperbola given by

$$\frac{x^2}{1{,}210{,}000} - \frac{y^2}{5{,}759{,}600} = 1.$$

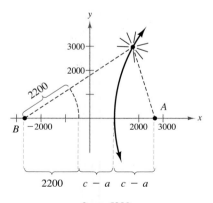

$$2c = 5280$$
$$2200 + 2(c - a) = 5280$$

FIGURE 7.33

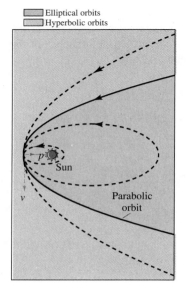

Elliptical orbits
Hyperbolic orbits

FIGURE 7.34

In Example 5, you were able to determine only the hyperbola on which the explosion occurred, but not the exact location of the explosion. If, however, you had received the sound from a third position C, then two other hyperbolas would have been determined. The exact location of the explosion would have been the point where these three hyperbolas intersected.

Another interesting application of conic sections involves the orbits of comets in our solar system. Of the 610 comets identified prior to 1970, 245 have elliptical orbits, 295 have parabolic orbits, and 70 have hyperbolic orbits. The center of the sun is a focus of each of these orbits, and each orbit has a vertex at the point where the comet is closest to the sun, as shown in Figure 7.34. Undoubtedly there have been many comets with parabolic or hyperbolic orbits that were not identified. We only get to see such comets *once*. Comets with elliptical orbits, such as Halley's comet, are the only ones that remain in our solar system.

If p is the distance between the vertex and the focus in meters, and v is the velocity of the comet at the vertex in meters per second, then the type of orbit is determined as follows.

1. Ellipse: $v < \sqrt{2GM/p}$
2. Parabola: $v = \sqrt{2GM/p}$
3. Hyperbola: $v > \sqrt{2GM/p}$

In each of these equations, $M \approx 1.991 \times 10^{30}$ kilograms (the mass of the sun) and $G \approx 6.67 \times 10^{-11}$ cubic meters per gram-second squared.

We conclude this section with a procedure for classifying a conic using the coefficients in the general form of its equation.

CLASSIFYING A CONIC FROM ITS GENERAL EQUATION

The graph of $Ax^2 + Cy^2 + Dx + Ey + F = 0$ is one of the following (except in degenerate cases).

1. Circle: $A = C$
2. Parabola: $AC = 0$ *$A = 0$ or $C = 0$, but not both.*
3. Ellipse: $AC > 0$ *A and C have like signs.*
4. Hyperbola: $AC < 0$ *A and C have unlike signs.*

EXAMPLE 6   Classifying Conics from General Equations
....

A. For the general equation

$$4x^2 - 9x + y - 5 = 0,$$

you have $AC = 4(0) = 0$. Thus, the graph is a parabola.

B. For the general equation

$$4x^2 - y^2 + 8x - 6y + 4 = 0,$$

you have $AC = 4(-1) < 0$. Thus, the graph is a hyperbola.

C. For the general equation

$$2x^2 + 4y^2 - 4x + 12y = 0,$$

you have $AC = (2)(4) > 0$. Thus, the graph is an ellipse.

DISCUSSION PROBLEM

HYPERBOLAS IN APPLICATIONS

At the beginning of Section 7.2, we mentioned that each type of conic section can be formed by the intersection of a plane and a double-napped cone. Three examples of how such intersections can occur in physical situations are shown below.

Identify the cone and hyperbola (or portion of a hyperbola) in each of the three situations. Can you think of other examples of physical situations in which hyperbolas are formed?

WARM-UP

The following warm-up exercises involve skills that were covered in earlier sections. You will use these skills in the exercise set for this section.

In Exercises 1 and 2, find the distance between the two points.

1. $(4, 1), (10, 6)$ **2.** $(-1, 5), (3, -2)$

In Exercises 3–6, sketch the graphs of the lines on the same set of coordinate axes.

3. $y = \pm\frac{1}{2}x$ **4.** $y = 3 \pm \frac{1}{2}x$

5. $y = 3 \pm \frac{1}{2}(x - 4)$ **6.** $y = \pm\frac{1}{2}(x - 4)$

In Exercises 7–10, identify the graph of the equation.

7. $x^2 + 4y = 4$ **8.** $x^2 + 4y^2 = 4$

9. $4x^2 + 4y^2 = 4$ **10.** $x + 4y^2 = 4$

SECTION 7.4 · EXERCISES

In Exercises 1–6, match the equation with its graph, and describe the given viewing rectangle. [The graphs are labeled (a), (b), (c), (d), (e), and (f).]

1. $\dfrac{x^2}{9} - \dfrac{y^2}{4} = 1$ e

2. $\dfrac{y^2}{9} - \dfrac{x^2}{4} = 1$

3. $\dfrac{y^2}{1} - \dfrac{x^2}{16} = 1$

4. $\dfrac{y^2}{16} - \dfrac{x^2}{1} = 1$

5. $\dfrac{(x - 2)^2}{9} - \dfrac{y^2}{4} = 1$ D

6. $\dfrac{(x + 1)^2}{16} - \dfrac{(y - 3)^2}{9} = 1$ B

(a)

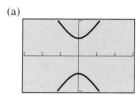

(b)

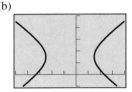

(c)

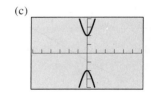

(d)

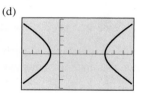

(e)

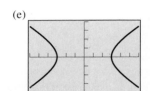

(f)

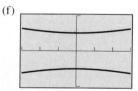

In Exercises 7–22, find the center, foci, and vertices of the hyperbola and sketch its graph using asymptotes as an aid.

7. $x^2 - y^2 = 1$

8. $\dfrac{x^2}{9} - \dfrac{y^2}{16} = 1$

9. $y^2 - \dfrac{x^2}{4} = 1$

10. $\dfrac{y^2}{9} - \dfrac{x^2}{1} = 1$

11. $\dfrac{y^2}{25} - \dfrac{x^2}{144} = 1$

12. $\dfrac{x^2}{36} - \dfrac{y^2}{4} = 1$

13. $5y^2 = 4x^2 + 20$

14. $7x^2 - 3y^2 = 21$

15. $\dfrac{(x - 1)^2}{4} - \dfrac{(y + 2)^2}{1} = 1$

16. $\dfrac{(x + 1)^2}{144} - \dfrac{(y - 4)^2}{25} = 1$

17. $(y + 6)^2 - (x - 2)^2 = 1$

18. $\dfrac{(y - 1)^2}{\frac{1}{4}} - \dfrac{(x + 3)^2}{\frac{1}{9}} = 1$

19. $9x^2 - y^2 - 36x - 6y + 18 = 0$

20. $x^2 - 9y^2 + 36y - 72 = 0$

21. $x^2 - 9y^2 + 2x - 54y - 80 = 0$

22. $16y^2 - x^2 + 2x + 64y + 63 = 0$

In Exercises 23–28, find the center, foci, and vertices of the hyperbola and sketch a graph of the hyperbola and its asymptotes with the aid of a graphing utility.

23. $2x^2 - 3y^2 = 6$

24. $3y^2 = 5x^2 + 15$

25. $9y^2 - x^2 + 2x + 54y + 62 = 0$

26. $9x^2 - y^2 + 54x + 10y + 55 = 0$

27. $3x^2 - 2y^2 - 6x - 12y - 27 = 0$

28. $3y^2 - x^2 + 6x - 12y = 0$

In Exercises 29–38, find an equation of the specified hyperbola.

29. Vertices: $(0, \pm 2)$
Foci: $(0, \pm 4)$

30. Vertices: $(0, \pm 3)$
Asymptotes: $y = \pm 3x$

31. Vertices: $(\pm 1, 0)$
Asymptotes: $y = \pm 3x$

32. Vertices: $(0, \pm 3)$
Solution point: $(-2, 5)$

33. Vertices: $(2, 0), (6, 0)$
Foci: $(0, 0), (8, 0)$

34. Vertices: $(4, 1), (4, 9)$
Foci: $(4, 0), (4, 10)$

35. Vertices: $(2, \pm 3)$
Solution point: $(0, 5)$

36. Vertices: $(\pm 2, 1)$
Solution point: $(4, 3)$

37. Vertices: $(0, 2), (6, 2)$
Asymptotes: $y = \frac{2}{3}x$
$y = 4 - \frac{2}{3}x$

38. Vertices: $(3, 0), (3, 4)$
Asymptotes: $y = \frac{2}{3}x$
$y = 4 - \frac{2}{3}x$

39. *Sound Location* Three listening stations located at $(4400, 0)$, $(4400, 1100)$, and $(-4400, 0)$ monitor an explosion. If the latter two stations detect the explosion 1 second and 5 seconds after the first, respectively, determine the coordinates of the explosion. (Assume that the coordinate system is measured in feet and that sound travels at 1100 feet per second.)

40. *LORAN*   Long-distance radio navigation for aircraft and ships uses synchronized pulses transmitted by widely separated transmitting stations. These pulses travel at the speed of light (186,000 miles per second). The difference in the times of arrival of these pulses at an aircraft or ship is constant on a hyperbola having the transmitting stations as foci. Assume that two stations, 300 miles apart, are positioned on the rectangular coordinate system at points with coordinates $(-150, 0)$ and $(150, 0)$ and that a ship is traveling on a path with coordinates $(x, 75)$ (see figure). Find the x-coordinate of the position of the ship if the time difference between the pulses from the transmitting stations is 1000 microseconds (0.001 second).

FIGURE FOR 40

41. *Hyperbolic Mirror*   A hyperbolic mirror (used in some telescopes) has the property that a light ray directed at the focus will be reflected to the other focus (see figure). The focus of a hyperbolic mirror has coordinates $(12, 0)$. Find the vertex of the mirror if its mount has coordinates $(12, 12)$.

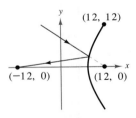

FIGURE FOR 41

42. Sketch a graph of the hyperbola that consists of all points (x, y) such that the difference of the distances between (x, y) and two fixed points is 10 units. The foci are located at the centers of the two sets of concentric circles in the figure.

FIGURE FOR 42

In Exercises 43–50, classify the graph of each equation as a circle, a parabola, an ellipse, or a hyperbola.

43. $x^2 + y^2 - 6x + 4y + 9 = 0$

44. $x^2 + 4y^2 - 6x + 16y + 21 = 0$

45. $4x^2 - y^2 - 4x - 3 = 0$

46. $y^2 - 4y - 4x = 0$

47. $4x^2 + 3y^2 + 8x - 24y + 51 = 0$

48. $4y^2 - 2x^2 - 4y - 8x - 15 = 0$

49. $25x^2 - 10x - 200y - 119 = 0$

50. $4x^2 + 4y^2 - 16y + 15 = 0$

7.5 ROTATION AND SYSTEMS OF QUADRATIC EQUATIONS

Rotation / Invariants Under Rotation / Systems of Quadratic Equations

Rotation

In Section 7.4 you saw that the equation of a conic with its axis parallel to one of the coordinate axes has a standard form that can be written in the general form

$$Ax^2 + Cy^2 + Dx + Ey + F = 0.$$ *Horizontal or vertical axis*

In this section you will study the equations of conics whose axes are rotated so that they are not parallel to either the x-axis or the y-axis. The general equation for such conics contains an *xy-term*.

$$Ax^2 + Bxy + Cy^2 + Dx + Ey + F = 0$$ *Equation in xy-plane*

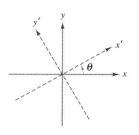

Rotated: x'-axis
y'-axis

FIGURE 7.35

To eliminate this *xy*-term, you can use a procedure called **rotation of axes.** The objective is to rotate the x- and y-axes until they are parallel to the axes of the conic. The rotated axes are denoted as the x'-axis and the y'-axis, as shown in Figure 7.35. The equation of the conic in the new $x'y'$-plane will have the form

$$A'(x')^2 + C'(y')^2 + D'x' + E'y' + F' = 0.$$ *Equation in x'y'-plane*

Because this equation has no *xy*-term, you can obtain a standard form by completing the square.

The following result identifies how much to rotate the axes to eliminate the *xy*-term and also the equations for determining the new coefficients A', C', D', E', and F'.

ROTATION OF AXES TO ELIMINATE AN XY-TERM

The general second-degree equation $Ax^2 + Bxy + Cy^2 + Dx + Ey + F = 0$ can be rewritten as

$$A'(x')^2 + C'(y')^2 + D'x' + E'y' + F' = 0$$

by rotating the coordinate axes through an angle θ, where

$$\cot 2\theta = \frac{A - C}{B}.$$

The coefficients of the new equation are obtained by making the substitutions

$$x = x' \cos \theta - y' \sin \theta \quad \text{and} \quad y = x' \sin \theta + y' \cos \theta.$$

PROOF

You need to discover how the coordinates in the xy-system are related to the coordinates in the $x'y'$-system. To do this, choose a point $P = (x, y)$ in the original system and attempt to find its coordinates (x', y') in the rotated system. In either system, the distance r between the point P and the origin is the same; thus, the equations for x, y, x', and y' are those given in Figure 7.36. Using the formulas for the sine and cosine of the difference of two angles, you have the following.

$$x' = r\cos(\alpha - \theta)$$
$$= r(\cos\alpha\cos\theta + \sin\alpha\sin\theta)$$
$$= r\cos\alpha\cos\theta + r\sin\alpha\sin\theta$$
$$= x\cos\theta + y\sin\theta$$
$$y' = r\sin(\alpha - \theta)$$
$$= r(\sin\alpha\cos\theta - \cos\alpha\sin\theta)$$
$$= r\sin\alpha\cos\theta - r\cos\alpha\sin\theta$$
$$= y\cos\theta - x\sin\theta$$

Solving this system for x and y yields

$$x = x'\cos\theta - y'\sin\theta$$
$$y = x'\sin\theta + y'\cos\theta.$$

Finally, by substituting these values for x and y into the original equation and collecting terms, you obtain

$$A' = A\cos^2\theta + B\cos\theta\sin\theta + C\sin^2\theta$$
$$C' = A\sin^2\theta - B\cos\theta\sin\theta + C\cos^2\theta$$
$$D' = D\cos\theta + E\sin\theta$$
$$E' = -D\sin\theta + E\cos\theta$$
$$F' = F.$$

Now, in order to eliminate the $x'y'$-term, you must select θ so that $B' = 0$, as follows.

$$B' = 2(C - A)\sin\theta\cos\theta + B(\cos^2\theta - \sin^2\theta)$$
$$= (C - A)\sin 2\theta + B\cos 2\theta$$
$$= B(\sin 2\theta)\left(\frac{C - A}{B} + \cot 2\theta\right) = 0, \quad \sin 2\theta \neq 0$$

If $B = 0$, no rotation is necessary, because the xy-term is not present in the original equation. If $B \neq 0$, the only way to make $B' = 0$ is to let

$$\cot 2\theta = \frac{A - C}{B}, \quad B \neq 0.$$

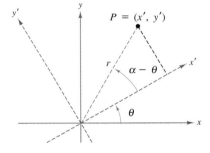

Rotated: $x' = r\cos(\alpha - \theta)$
$\qquad\quad y' = r\sin(\alpha - \theta)$

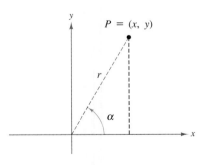

Original: $x = r\cos\alpha$
$\qquad\quad\ y = r\sin\alpha$

FIGURE 7.36

EXAMPLE 1   Rotation of Axes for a Hyperbola
....

Write the equation $xy - 1 = 0$ in standard form.

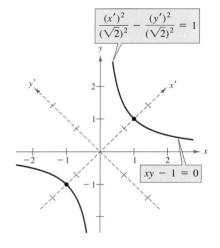

$$\frac{(x')^2}{(\sqrt{2})^2} - \frac{(y')^2}{(\sqrt{2})^2} = 1$$

$xy - 1 = 0$

Vertices:
In $x'y'$-system: $(\sqrt{2}, 0)$, $(-\sqrt{2}, 0)$
In xy-system: $(1, 1)$, $(-1, -1)$

FIGURE 7.37

SOLUTION

Because $A = 0$, $B = 1$, and $C = 0$, you have

$$\cot 2\theta = \frac{A - C}{B} = 0 \rightarrow 2\theta = \frac{\pi}{2} \rightarrow \theta = \frac{\pi}{4},$$

which implies that

$$x = x' \cos \frac{\pi}{4} - y' \sin \frac{\pi}{4}$$

$$= x'\left(\frac{\sqrt{2}}{2}\right) - y'\left(\frac{\sqrt{2}}{2}\right) = \frac{x' - y'}{\sqrt{2}}$$

and

$$y = x' \sin \frac{\pi}{4} + y' \cos \frac{\pi}{4}$$

$$= x'\left(\frac{\sqrt{2}}{2}\right) + y'\left(\frac{\sqrt{2}}{2}\right) = \frac{x' + y'}{\sqrt{2}}.$$

The equation in the $x'y'$-system is obtained by substituting these expressions into the equation $xy - 1 = 0$.

$$\left(\frac{x' - y'}{\sqrt{2}}\right)\left(\frac{x' + y'}{\sqrt{2}}\right) - 1 = 0$$

$$\frac{(x')^2 - (y')^2}{2} - 1 = 0$$

$$\frac{(x')^2}{(\sqrt{2})^2} - \frac{(y')^2}{(\sqrt{2})^2} = 1 \qquad \textit{Standard form}$$

This is the equation of a hyperbola centered at the origin with vertices at $\left(\pm\sqrt{2}, 0\right)$ in the $x'y'$-system, as shown in Figure 7.37. To find the coordinates of the vertices in the xy-system, substitute the coordinates $\left(\pm\sqrt{2}, 0\right)$ into the equations

$$x = \frac{x' - y'}{\sqrt{2}} \quad \text{and} \quad y = \frac{x' + y'}{\sqrt{2}}.$$

This substitution yields the vertices $(1, 1)$ and $(-1, -1)$ in the xy-system. Note also that the asymptotes of the hyperbola have equations $y' = \pm x'$, which correspond to the original x- and y-axes.
■■■■■■■■■

REMARK   Remember that the substitutions

$$x = x' \cos \theta - y' \sin \theta$$

and

$$y = x' \sin \theta + y' \cos \theta$$

were developed to eliminate the $x'y'$-term in the rotated system. You can use this as a check on your work. In other words, if your final equation contains an $x'y'$-term, you know that you have made a mistake.

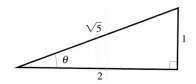

FIGURE 7.38

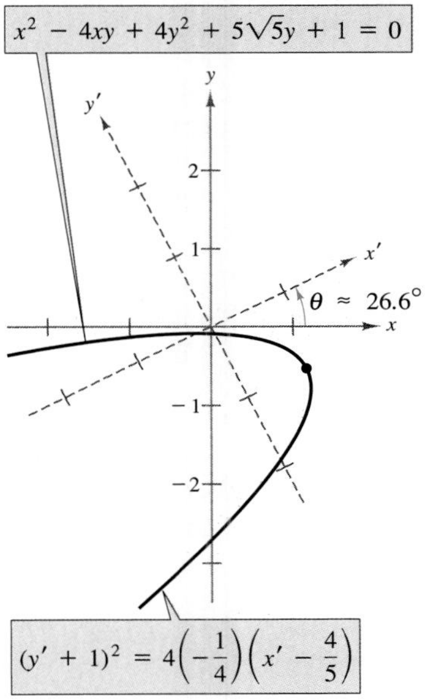

$x^2 - 4xy + 4y^2 + 5\sqrt{5}y + 1 = 0$

$$(y' + 1)^2 = 4\left(-\frac{1}{4}\right)\left(x' - \frac{4}{5}\right)$$

Vertex:

In $x'y'$-system: $\left(\dfrac{4}{5}, -1\right)$

In xy-system: $\left(\dfrac{13}{5\sqrt{5}}, -\dfrac{6}{5\sqrt{5}}\right)$

FIGURE 7.39

EXAMPLE 2   Rotation of Axes for a Parabola

Sketch the graph of $x^2 - 4xy + 4y^2 + 5\sqrt{5}\,y + 1 = 0$.

SOLUTION

Because $A = 1$, $B = -4$, and $C = 4$, you have

$$\cot 2\theta = \frac{A - C}{B} = \frac{1 - 4}{-4} = \frac{3}{4}.$$

Using the identity $\cot 2\theta = \dfrac{3}{4} = \dfrac{\cot^2 \theta - 1}{2 \cot \theta}$

$$\cot 2\theta = \frac{3}{4} = \frac{\cot^2 \theta - 1}{2 \cot \theta}$$

from which you obtain the equation

$$4 \cot^2 \theta - 4 = 6 \cot \theta$$
$$4 \cot^2 \theta - 6 \cot \theta - 4 = 0$$
$$(2 \cot \theta - 4)(2 \cot \theta + 1) = 0.$$

Considering $0 < \theta < \pi/2$, you have $2 \cot \theta = 4$. Thus,

$$\cot \theta = 2 \rightarrow \theta \approx 26.6°.$$

From the triangle in Figure 7.38 you obtain $\sin \theta = 1/\sqrt{5}$ and $\cos \theta = 2/\sqrt{5}$. Consequently, you should use the substitutions

$$x = x' \cos \theta - y' \sin \theta = x'\left(\frac{2}{\sqrt{5}}\right) - y'\left(\frac{1}{\sqrt{5}}\right) = \frac{2x' - y'}{\sqrt{5}}$$

$$y = x' \sin \theta + y' \cos \theta = x'\left(\frac{1}{\sqrt{5}}\right) + y'\left(\frac{2}{\sqrt{5}}\right) = \frac{x' + 2y'}{\sqrt{5}}.$$

Substituting these expressions into the original equation, you have

$$x^2 - 4xy + 4y^2 + 5\sqrt{5}\,y + 1 = 0$$

$$\left(\frac{2x' - y'}{\sqrt{5}}\right)^2 - 4\left(\frac{2x' - y'}{\sqrt{5}}\right)\left(\frac{x' + 2y'}{\sqrt{5}}\right) + 4\left(\frac{x' + 2y'}{\sqrt{5}}\right)^2 + 5\sqrt{5}\left(\frac{x' + 2y'}{\sqrt{5}}\right) + 1 = 0,$$

which simplifies as follows.

$$5(y')^2 + 5x' + 10y' + 1 = 0$$

$$5(y' + 1)^2 = -5x' + 4 \qquad \textit{Complete the square}$$

$$(y' + 1)^2 = (-1)\left(x' - \frac{4}{5}\right) \qquad \textit{Standard form}$$

The graph of this equation is a parabola with its vertex at $\left(\frac{4}{5}, -1\right)$. Its axis is parallel to the x'-axis in the $x'y'$-system, as shown in Figure 7.39.

Invariants Under Rotation

In the rotation of axes theorem listed at the beginning of this section, note that the constant term $F' = F$ is the same in both equations, and is said to be **invariant under rotation.** The next theorem lists some other rotation invariants.

ROTATION INVARIANTS

The rotation of coordinate axes through an angle θ that transforms the equation $AX^2 + Bxy + Cy^2 + Dx + Ey + F = 0$ into the form

$$A'(x')^2 + C'(y')^2 + D'x' + E'y' + F' = 0$$

has the following rotation invariants.

1. $F = F'$
2. $A + C = A' + C'$
3. $B^2 - 4AC = (B')^2 - 4A'C'$

You can use these results to classify the graph of a second-degree equation *with* an xy-term in much the same way as was done for second-degree equations *without* an xy-term. Note that, because $B' = 0$, the invariant $B^2 - 4AC$ reduces to

$$B^2 - 4AC = -4A'C'. \qquad \textit{Discriminant}$$

This quantity is the **discriminant** of the equation

$$Ax^2 + Bxy + Cy^2 + Dx + Ey + F = 0.$$

Now, from the classification procedure given in Section 7.4, you know that the sign of $A'C'$ determines the type of graph for the equation.

$$A'(x')^2 + C'(y')^2 + D'x' + E'y' + F' = 0.$$

Consequently, the sign of $B^2 - 4AC$ will determine the type of graph for the original equation, as shown in the following classification.

CLASSIFICATION OF CONICS BY THE DISCRIMINANT

The graph of the equation $Ax^2 + Bxy + Cy^2 + Dx + Ey + F = 0$ is, except in degenerate cases, determined by its discriminant, as follows.

1. Ellipse or circle: $B^2 - 4AC < 0$
2. Parabola: $B^2 - 4AC = 0$
3. Hyperbola: $B^2 - 4AC > 0$

EXAMPLE 3   Using the Discriminant

Classify the graph of each of the following equations.

A. $4xy - 9 = 0$
B. $2x^2 - 3xy + 2y^2 - 2x = 0$
C. $x^2 - 6xy + 9y^2 - 2y + 1 = 0$
D. $3x^2 + 8xy + 4y^2 - 7 = 0$

SOLUTION

A. Because $B^2 - 4AC = 16 - 0 > 0$, the graph is a hyperbola.
B. Because $B^2 - 4AC = 9 - 16 < 0$, the graph is a circle or an ellipse.
C. Because $B^2 - 4AC = 36 - 36 = 0$, the graph is a parabola.
D. Because $B^2 - 4AC = 64 - 48 > 0$, the graph is a hyperbola.

Systems of Quadratic Equations

To find the points of intersection of two conics, you can use elimination or substitution, as demonstrated in Examples 4 and 5.

EXAMPLE 4   Solving a Quadratic System by Elimination

Solve the system of quadratic equations.

$$x^2 + y^2 - 16x + 39 = 0 \qquad \textit{Equation 1}$$
$$x^2 - y^2 - 9 = 0 \qquad \textit{Equation 2}$$

SOLUTION

You can eliminate the y^2-term by adding the two equations. The resulting equation can then be solved for x.

$$2x^2 - 16x + 30 = 0$$
$$2(x - 3)(x - 5) = 0$$

There are two real solutions: $x = 3$ and $x = 5$. The corresponding y-values are $y = 0$ and $y = \pm 4$. Thus the graphs of the equations have three points of intersection: $(3, 0)$, $(5, 4)$, and $(5, -4)$, as shown in Figure 7.40.

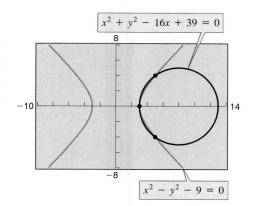

$x^2 + y^2 - 16x + 39 = 0$

$x^2 - y^2 - 9 = 0$

FIGURE 7.40

EXAMPLE 5 Solving a Quadratic System by Substitution

Find the points of intersection of the graphs of the system.

$$x^2 + 4y^2 - 4x - 8y + 4 = 0 \qquad \textit{Equation 1}$$
$$x^2 + 4y - 4 = 0 \qquad \textit{Equation 2}$$

SOLUTION

Because the second equation has no y^2-term, solve that equation for y to obtain $y = 1 - \frac{1}{4}x^2$. Next, substitute this expression for y in the first equation and solve for x.

$$x^2 + 4y^2 - 4x - 8y + 4 = 0$$
$$x^2 + 4(1 - \tfrac{1}{4}x^2)^2 - 4x - 8(1 - \tfrac{1}{4}x^2) + 4 = 0$$
$$x^2 + 4 - 2x^2 + \tfrac{1}{4}x^4 - 4x - 8 + 2x^2 + 4 = 0$$
$$\tfrac{1}{4}x^4 + x^2 - 4x = 0$$
$$x^4 + 4x^2 - 16x = 0$$
$$x(x - 2)(x^2 + 2x + 8) = 0$$

The last equation has only two real solutions, $x = 0$ and $x = 2$. The corresponding values of y are $y = 1$ and $y = 0$. Thus the original system of equations has two solutions: $(0, 1)$ and $(2, 0)$, as shown in Figure 7.41.

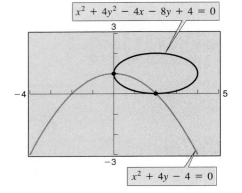

$x^2 + 4y^2 - 4x - 8y + 4 = 0$

$x^2 + 4y - 4 = 0$

FIGURE 7.41

DISCUSSION PROBLEM

· ·

CLASSIFYING A GRAPH AS A HYPERBOLA

Is the graph of the rational function

$$y = \frac{1}{x}$$

a hyperbola? Write a short paragraph that justifies your answer.

WARM-UP

The following warm-up exercises involve skills that were covered in earlier sections. You will use these skills in the exercise set for this section.

In Exercises 1–6, match the equation with its graph. [The graphs are labeled (a), (b), (c), (d), (e), and (f).]

1. $\dfrac{x^2}{1} + \dfrac{y^2}{4} = 1$ **2.** $x^2 = 4y$

3. $\dfrac{x^2}{1} - \dfrac{y^2}{4} = 1$ **4.** $(x - 1)^2 + y^2 = 4$

5. $y^2 = -4x$ **6.** $x^2 - 5y^2 = -5$

(a)

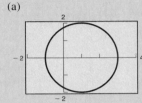

(b)

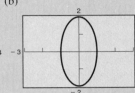

(c)

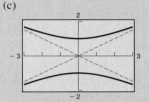

(d)

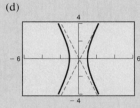

(e)

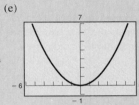

(f)

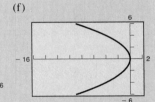

In Exercises 7 and 8, evaluate the trigonometric functions and rewrite the expression.

7. $x \cos \dfrac{\pi}{3} - y \sin \dfrac{\pi}{3}$

8. $x \sin\left(-\dfrac{\pi}{6}\right) + y \cos\left(-\dfrac{\pi}{6}\right)$

In Exercises 9 and 10, expand the expression.

9. $\left(\dfrac{2x - 3y}{\sqrt{13}}\right)^2$

10. $\left(\dfrac{x - \sqrt{2}y}{\sqrt{3}}\right)^2$

SECTION 7.5 · EXERCISES

In Exercises 1–12, rotate the axes to eliminate the xy-term. Sketch the graph of the resulting equation, showing both sets of axes.

1. $xy + 1 = 0$
2. $xy - 4 = 0$
3. $x^2 - 10xy + y^2 + 1 = 0$
4. $xy + x - 2y + 3 = 0$
5. $xy - 2y - 4x = 0$
6. $13x^2 + 6\sqrt{3}\,xy + 7y^2 - 16 = 0$
7. $5x^2 - 2xy + 5y^2 - 12 = 0$
8. $2x^2 - 3xy - 2y^2 + 10 = 0$
9. $3x^2 - 2\sqrt{3}\,xy + y^2 + 2x + 2\sqrt{3}\,y = 0$
10. $16x^2 - 24xy + 9y^2 - 60x - 80y + 100 = 0$
11. $9x^2 + 24xy + 16y^2 + 90x - 130y = 0$
12. $9x^2 + 24xy + 16y^2 + 80x - 60y = 0$

In Exercises 13–18, use a graphing utility to graph the conic. Determine the angle θ through which the axes are rotated. Explain how you used the utility to obtain the graph.

13. $x^2 + xy + y^2 = 10$
14. $x^2 - 4xy + 2y^2 = 6$
15. $17x^2 + 32xy - 7y^2 = 75$
16. $40x^2 + 36xy + 25y^2 = 52$
17. $32x^2 + 50xy + 7y^2 = 52$
18. $4x^2 - 12xy + 9y^2 + \left(4\sqrt{13} - 12\right)x$
$- \left(6\sqrt{13} + 8\right)y = 91$

In Exercises 19–26, use the discriminant to determine whether the graph of the equation is a parabola, an ellipse, or a hyperbola.

19. $16x^2 - 24xy + 9y^2 - 30x - 40y = 0$
20. $x^2 - 4xy - 2y^2 - 6 = 0$
21. $13x^2 - 8xy + 7y^2 - 45 = 0$
22. $2x^2 + 4xy + 5y^2 + 3x - 4y - 20 = 0$
23. $x^2 - 6xy - 5y^2 + 4x - 22 = 0$
24. $36x^2 - 60xy + 25y^2 + 9y = 0$
25. $x^2 + 4xy + 4y^2 - 5x - y - 3 = 0$
26. $x^2 + xy + 4y^2 + x + y = 0$

In Exercises 27–34, use a graphing utility to graph the equations and find any points of intersection of the graphs by the method of elimination.

27. $-x^2 + y^2 + 4x - 6y + 4 = 0$
$x^2 + y^2 - 4x - 6y + 12 = 0$
28. $-x^2 - y^2 - 8x + 20y - 7 = 0$
$x^2 + 9y^2 + 8x + 4y + 7 = 0$
29. $-4x^2 - y^2 - 32x + 24y - 64 = 0$
$4x^2 + y^2 + 56x - 24y + 304 = 0$
30. $x^2 - 4y^2 - 20x - 64y - 172 = 0$
$16x^2 + 4y^2 - 320x + 64y + 1600 = 0$
31. $x^2 - y^2 - 12x + 12y - 36 = 0$
$x^2 + y^2 - 12x - 12y + 36 = 0$
32. $x^2 + 4y^2 - 2x - 8y + 1 = 0$
$-x^2 + 2x - 4y - 1 = 0$
33. $-16x^2 - y^2 + 24y - 80 = 0$
$16x^2 + 25y^2 - 400 = 0$
34. $16x^2 - y^2 + 16y - 128 = 0$
$y^2 - 48x - 16y - 32 = 0$

In Exercises 35–40, use a graphing utility to graph the equations and find any points of intersection of the graphs by the method of substitution.

35. $x^2 + y^2 - 25 = 0$
$9x - 4y^2 = 0$
36. $4x^2 + 9y^2 - 36y = 0$
$x^2 + 9y - 27 = 0$
37. $x^2 + 2y^2 - 4x + 6y - 5 = 0$
$x + y + 5 = 0$
38. $x^2 + 2y^2 - 4x + 6y - 5 = 0$
$x^2 - 4x - y + 4 = 0$
39. $xy + x - 2y + 3 = 0$
$x^2 + 4y^2 - 9 = 0$
40. $5x^2 - 2xy + 5y^2 - 12 = 0$
$x + y - 1 = 0$

41. Show that the equation $x^2 + y^2 = r^2$ is invariant under rotation of axes.

7.6 PLANE CURVES AND PARAMETRIC EQUATIONS
Plane Curves / Sketching a Plane Curve / Eliminating the Parameter /
Finding Parametric Equations for a Graph

Plane Curves

Up to this point, you have been representing a graph by a single equation involving the *two* variables x and y. In this section, you will study situations in which it is useful to introduce a *third* variable to represent a curve in the plane.

To see the usefulness of this procedure, consider the path followed by an object that is propelled into the air at an angle of 45°. If the initial velocity of the object is 48 feet per second, it can be shown that it follows the parabolic path given by

$$y = -\frac{x^2}{72} + x,$$ *Rectangular equation*

as shown in Figure 7.42. However, this equation does not tell the whole story.

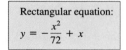

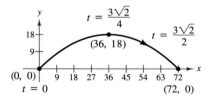

Curvilinear Motion:
two variables for position
one variable for time

FIGURE 7.42

Although it does tell us *where* the object has been, it doesn't tell us *when* the object was at a given point (x, y) on the path. To determine this time, we introduce a third variable t, which we call a **parameter.** It is possible to write both x and y as functions of t to obtain the **parametric equations**

$$x = 24\sqrt{2}\,t \quad \text{and} \quad y = -16t^2 + 24\sqrt{2}\,t. \quad \textit{Parametric equations}$$

From this set of equations we can determine that at time $t = 0$, the object is at the point $(0, 0)$. Similarly, at time $t = 1$, the object is at the point $(24\sqrt{2}, 24\sqrt{2} - 16)$, and so on.

For this particular motion problem, x and y are continuous functions of t, and we call the resulting path a **plane curve.** (Recall that a *continuous function* is one whose graph can be traced without lifting the pencil from the paper.)

DEFINITION OF A PLANE CURVE

If f and g are continuous functions of t on an interval I, then the set of ordered pairs $(f(t), g(t))$ is a **plane curve** C. The equations $x = f(t)$ and $y = g(t)$ are **parametric equations** for C, and t is the **parameter.**

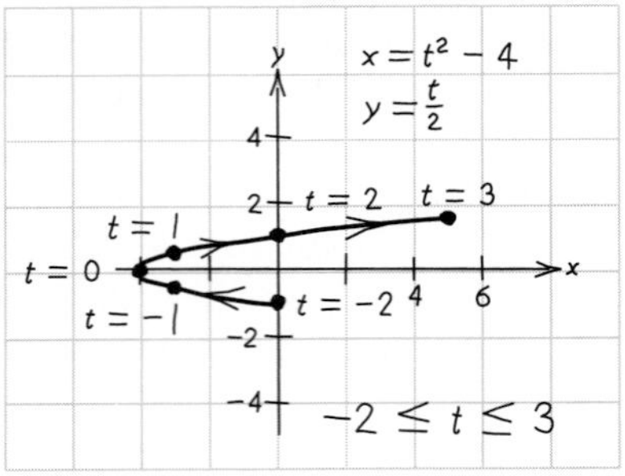

Sketching a Plane Curve

One way to sketch a curve represented by a pair of parametric equations is to plot points in the xy plane. Each set of coordinates (x, y) is determined from a value chosen for the parameter t. By plotting the resulting points in the order of *increasing* values of t, you trace the curve in a specific direction. This is called the **orientation** of the curve.

EXAMPLE 1   Sketching a Curve

Sketch the curve described by the parametric equations

$$x = t^2 - 4 \quad \text{and} \quad y = \frac{t}{2}, \qquad -2 \le t \le 3.$$

SOLUTION

Using values of t in the given interval, the parametric equations yield the points (x, y) shown in the table.

t	-2	-1	0	1	2	3
x	0	-3	-4	-3	0	5
y	-1	$-\frac{1}{2}$	0	$\frac{1}{2}$	1	$\frac{3}{2}$

By plotting these points in the order of increasing t, we obtain the curve shown in Figure 7.43. Note that the arrows on the curve indicate its orientation as t increases from -2 to 3.

The graph shown in Figure 7.43 does not define y as a function of x. This points out one benefit of parametric equations—they can be used to represent graphs that are more general than graphs of functions.

FIGURE 7.43

DISCOVERY

Set your graphing utility to parametric mode and enter the two functions

$X_{1T} = T^2 - 4$

$Y_{1T} = T/2$

Use the range settings $-2 \le T \le 3$, Tstep $= .1$, $-6 \le x \le 10$, and $-4 \le y \le 4$. You should obtain the same curve as in Figure 7.43. Trace along the curve and observe how the t values determine the x and y coordinates of the point. What is the orientation of the curve? What happens if you change Tstep to .01?

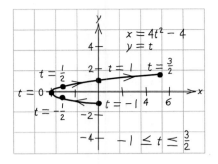

FIGURE 7.44

Two different sets of parametric equations can have the same graph. For example, the set of parametric equations

$$x = 4t^2 - 4 \quad \text{and} \quad y = t, \quad -1 \le t \le \frac{3}{2}$$

has the same graph as the set given in Example 1. However, by comparing the values of t in Figures 7.43 and 7.44, you can see that this second graph is traced out more *rapidly* (considering t as time) than the first graph. Thus, in applications, different parametric representations can be used to represent various *speeds* at which objects travel along a given path.

Another way to display a curve represented by a pair of parametric equations is to use a graphing utility. When you do this, be sure to set the graphing utility to parametric mode.

EXAMPLE 2   **Using a Graphing Utility in Parametric Mode**

Use a graphing utility to graph the curves represented by the parametric equations. For which curve is y a function of x?

A. $x = \cos^3 t$ **B.** $x = t$
  $y = \sin^3 t$ $y = t^3$

SOLUTION

Begin by setting the graphing utility to parametric *and* radian mode. When choosing a viewing rectangle, you must not only set minimum and maximum values of x and y, you must also set minimum and maximum values of t.

A. Enter the parametric equations for x and y. The curve is shown in Figure 7.45(a). From the graph, you can see that y is *not* a function of x.

B. Enter the parametric equations for x and y. The curve is shown in Figure 7.45(b). From the graph, you can see that y *is* a function of x.

Technology Note

When you use a graphing utility to graph the curve represented by a set of parametric equations, be sure to set the utility to parametric mode. If the parameter involves an angle, you must also set the angle measure to the proper mode (usually radians).

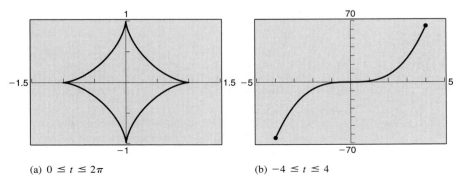

(a) $0 \le t \le 2\pi$ (b) $-4 \le t \le 4$

FIGURE 7.45

Eliminating the Parameter

Many curves that are represented by a set of parametric equations have graphs that can also be represented by a rectangular equation (in x and y). The process of finding the rectangular equation is called **eliminating the parameter.** Here is an example.

Parametric equations	$\rightarrow$	Solve for t in one equation	$\rightarrow$	Substitute into other equation	$\rightarrow$	Rectangular equation
$x = t^2 - 4$		$t = 2y$		$x = (2y)^2 - 4$		$x = 4y^2 - 4$
$y = \dfrac{t}{2}$						

After eliminating the parameter, you can recognize that the curve is a parabola with a horizontal axis and vertex at $(-4, 0)$.

Converting equations from parametric to rectangular form can change the ranges of x and y. In such cases, you should restrict x and y in the rectangular equation so that the graph of this equation matches the graph of the parametric equations.

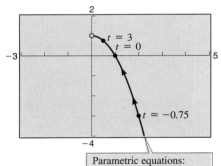

Parametric equations:
$$x = \frac{1}{\sqrt{t+1}}, \quad y = \frac{t}{t+1}$$

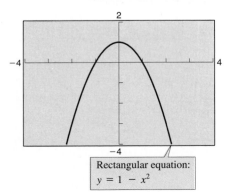

Rectangular equation:
$$y = 1 - x^2$$

FIGURE 7.46

EXAMPLE 3   **Adjusting the Domain After Eliminating the Parameter**

Identify the curve represented by the equations

$$x = \frac{1}{\sqrt{t+1}} \quad \text{and} \quad y = \frac{t}{t+1}.$$

SOLUTION

Solving for t in the equation for x produces

$$x = \frac{1}{\sqrt{t+1}} \quad \text{or} \quad x^2 = \frac{1}{t+1},$$

which implies that $t = (1 - x^2)/x^2$. Substituting into the equation for y, we obtain

$$y = \frac{t}{t+1} = \frac{\dfrac{1-x^2}{x^2}}{\dfrac{1-x^2}{x^2} + 1} = 1 - x^2.$$

From the rectangular equation, you can recognize the curve to be a parabola that opens downward and has its vertex at $(0, 1)$. The rectangular equation is defined for all values of x. From the parametric equation for x, however, you can see that the curve is defined only when $-1 < t$. Thus, you should restrict the domain of x to positive values, as shown in Figure 7.46.

It is not necessary for the parameter in a set of parametric equations to represent time. For instance, in the set of parametric equations in Example 4, *angle* is used as the parameter.

EXAMPLE 4   **Using a Trigonometric Identity to Eliminate a Parameter**

Identify the curve represented by

$$x = 3 \cos \theta \quad \text{and} \quad y = 4 \sin \theta, \qquad 0 \le \theta \le 2\pi.$$

SOLUTION

Begin by solving for $\cos \theta$ and $\sin \theta$ in the given equations.

$$\cos \theta = \frac{x}{3} \quad \text{and} \quad \sin \theta = \frac{y}{4} \qquad \textit{Solve for } \cos \theta \textit{ and } \sin \theta$$

You can use the identity $\cos^2 \theta + \sin^2 \theta = 1$ to form an equation involving only x and y.

$$\cos^2 \theta + \sin^2 \theta = 1 \qquad \textit{Trigonometric Identity}$$

$$\cos^2 \theta + \sin^2 \theta = \left(\frac{x}{3}\right)^2 + \left(\frac{y}{4}\right)^2 = 1 \qquad \textit{Substitute}$$

$$\frac{x^2}{9} + \frac{y^2}{16} = 1 \qquad \textit{Rectangular equation}$$

From the rectangular equation, you can see that the graph is an ellipse centered at $(0, 0)$, with vertices at $(0, 4)$ and $(0, -4)$, and minor axis of length $2b = 6$, as shown in Figure 7.47. Note that the ellipse is traced out *counterclockwise* as θ varies from 0 to 2π.

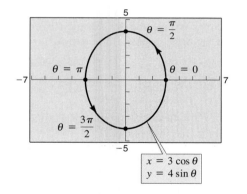

$$x = 3 \cos \theta$$
$$y = 4 \sin \theta$$

FIGURE 7.47

In Examples 3 and 4, it is important to realize that eliminating the parameter is primarily an aid to identifying the curve. If the parametric equations represent the path of a moving object, the graph alone is not sufficient to describe the object's motion. You still need the parametric equations to determine the *position, direction,* and *speed* at a given time.

Finding Parametric Equations for a Graph

How can we determine a set of parametric equations for a given graph or a given physical description? From the discussion following Example 1, we know that such a representation is not unique. This is further demonstrated in the following example, in which we find two different parametric representations for a graph.

EXAMPLE 5   Finding Parametric Equations for a Given Graph

Find a set of parametric equations to represent the graph of $y = 1 - x^2$, using the following parameters.

A. $t = x$ **B.** $t = 1 - x$

SOLUTION

A. Letting $t = x$, you obtain the parametric equations

$$x = t \quad \text{and} \quad y = 1 - x^2 = 1 - t^2.$$

B. Letting $t = 1 - x$, you obtain

$$x = 1 - t \quad \text{and} \quad y = 1 - (1 - t)^2 = 2t - t^2.$$

In Figure 7.48, note how the resulting curve is oriented by the increasing values of t. For part (A), the curve would have the opposite orientation.

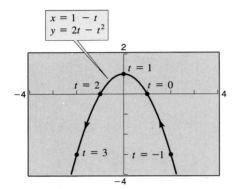

$x = 1 - t$
$y = 2t - t^2$

$t = 1$

$t = 2$ $t = 0$

$t = 3$ $t = -1$

Graph of $y = 1 - x^2$

FIGURE 7.48

EXAMPLE 6   Parametric Equations for a Cycloid

Determine the **cycloid** traced out by a point P on the circumference of a circle of radius a as the circle rolls along a straight line in a plane.

SOLUTION

Use the parameter θ, where θ is the measure of the circle's rotation. Assume that the point $P = (x, y)$ begins at the origin. When $\theta = 0$, P is at the origin; when $\theta = \pi$, P is at a maximum point $(\pi a, 2a)$; and when $\theta = 2\pi$, P is back on the x-axis at $(2\pi a, 0)$. In Figure 7.49, you can see that $\angle APC = 180° - \theta$.

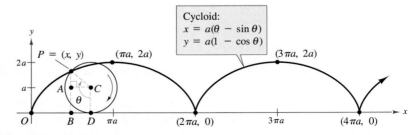

Cycloid:
$x = a(\theta - \sin \theta)$
$y = a(1 - \cos \theta)$

FIGURE 7.49

Hence, you have

$$\sin \theta = \sin(180° - \theta) = \sin(\angle APC) = \frac{AC}{a} = \frac{BD}{a}$$

$$\cos \theta = -\cos(180° - \theta) = -\cos(\angle APC) = \frac{AP}{-a},$$

which implies that $AP = -a \cos \theta$ and $BD = a \sin \theta$. Now, because the circle rolls along the x-axis, we know that $OD = \overset{\frown}{PD} = a\theta$. Furthermore, because $BA = DC = a$, we have

$$x = OD - BD = a\theta - a \sin \theta$$
$$y = BA + AP = a - a \cos \theta.$$

Therefore, the parametric equations are

$$x = a(\theta - \sin \theta) \quad \text{and} \quad y = a(1 - \cos \theta).$$

DISCUSSION PROBLEM

CHANGING THE ORIENTATION OF A CURVE

The **orientation** of a curve refers to the direction in which the curve is traced as the values of the parameter increase. For instance, as t increases, the circle given by

$$x = \cos t \quad \text{and} \quad y = \sin t$$

is traced out *counterclockwise*. Find a parametric representation for which the circle is traced out *clockwise*.

WARM-UP

The following warm-up exercises involve skills that were covered in earlier sections. You will use these skills in the exercise set for this section.

In Exercises 1–6, sketch the graph of the equation.

1. $y = -\frac{1}{4}x^2$ **2.** $y = 4 - \frac{1}{4}(x - 2)^2$

3. $16x^2 + y^2 = 16$ **4.** $-16x^2 + y^2 = 16$

5. $x + y = 4$ **6.** $x^2 + y^2 = 16$

In Exercises 7–10, simplify the expression.

7. $10 \sin^2 \theta + 10 \cos^2 \theta$ **8.** $5 \sec^2 \theta - 5$

9. $\sec^4 x - \tan^4 x$ **10.** $\dfrac{\sin 2\theta}{4 \cos \theta}$

SECTION 7.6 · EXERCISES

In Exercises 1–20, sketch the curve represented by the parametric equations (indicate the direction of the curve). Use a graphing utility to confirm your result. Then eliminate the parameter and write a rectangular equation whose graph represents the curve.

1. $x = t$
$y = -2t$

2. $x = t$
$y = \frac{1}{2}t$

3. $x = 3t - 1$
$y = 2t + 1$

4. $x = 3 - 2t$
$y = 2 + 3t$

5. $x = \frac{1}{4}t$
$y = t^2$

6. $x = t$
$y = t^3$

7. $x = t + 1$
$y = t^2$

8. $x = \sqrt{t}$
$y = 1 - t$

9. $x = t^3$
$y = \dfrac{t}{2}$

10. $x = t - 1$
$y = \dfrac{t}{t - 1}$

11. $x = 3 \cos \theta$
$y = 3 \sin \theta$

12. $x = 4 \sin 2\theta$
$y = 2 \cos 2\theta$

13. $x = \cos \theta$
$y = 2 \sin^2 \theta$

14. $x = \sec \theta$
$y = \cos \theta$

15. $x = 4 + 2 \cos \theta$
$y = -1 + 4 \sin \theta$

16. $x = 4 \sec \theta$
$y = 3 \tan \theta$

17. $x = e^t$
$y = e^{3t}$

18. $x = e^{2t}$
$y = e^t$

19. $x = t^3$
$y = 3 \ln t$

20. $x = \ln t$
$y = t^2$

In Exercises 21 and 22, describe how the curves differ from each other.

21. (a) $x = t$
$y = 2t + 1$

(b) $x = \cos \theta$
$y = 2 \cos \theta + 1$

(c) $x = e^{-t}$
$y = 2e^{-t} + 1$

(d) $x = e^t$
$y = 2e^t + 1$

22. (a) $x = t$
$y = t^2 - 1$

(b) $x = t^2$
$y = t^4 - 1$

(c) $x = \sin t$
$y = \sin^2 t - 1$

(d) $x = e^t$
$y = e^{2t} - 1$

In Exercises 23–26, eliminate the parameter and obtain the standard form of the rectangular equation.

23. Line through (x_1, y_1) and (x_2, y_2):
$x = x_1 + t(x_2 - x_1)$
$y = y_1 + t(y_2 - y_1)$

24. Circle:
$x = h + r \cos \theta$
$y = k + r \sin \theta$

25. Ellipse:
$x = h + a \cos \theta$
$y = k + b \sin \theta$

26. Hyperbola:
$x = h + a \sec \theta$
$y = k + b \tan \theta$

In Exercises 27–34, use the results of Exercises 23–26 to find a set of parametric equations for the line or conic.

27. Line: Passes through $(0, 0)$ and $(5, -2)$
28. Line: Passes through $(1, 4)$ and $(5, -2)$
29. Circle: Center: $(2, 1)$
       Radius: 4
30. Circle: Center: $(-3, 1)$
       Radius: 3
31. Ellipse: Vertices: $(\pm 5, 0)$
       Foci: $(\pm 4, 0)$
32. Ellipse: Vertices: $(4, 7), (4, -3)$
       Foci: $(4, 5), (4, -1)$
33. Hyperbola: Vertices: $(\pm 4, 0)$
       Foci: $(\pm 5, 0)$
34. Hyperbola: Vertices: $(0, \pm 1)$
       Foci: $(0, \pm 5)$

In Exercises 35 and 36, find two different sets of parametric equations for the rectangular equation.

35. $y = x^3$ **36.** $y = x^2$

In Exercises 37–42, use a graphing utility to graph the curve represented by the parametric equations.

37. Cycloid: $x = 2(\theta - \sin \theta)$
$y = 2(1 - \cos \theta)$

38. Cycloid: $x = \theta + \sin \theta$
$y = 1 - \cos \theta$

39. Prolate cycloid: $x = \theta - \frac{3}{2} \sin \theta$
$y = 1 - \frac{3}{2} \cos \theta$

40. Curtate cycloid: $x = 2\theta - \sin \theta$
$y = 2 - \cos \theta$

41. Witch of Agnesi: $x = 2 \cot \theta$
$y = 2 \sin^2 \theta$

42. Folium of Descartes: $x = \dfrac{3t}{1 + t^3}$
$y = \dfrac{3t^2}{1 + t^3}$

In Exercises 43–46, match the parametric equation with the correct graph and describe the viewing rectangle. [The graphs are labeled (a), (b), (c), and (d).]

43. Lissajous curve: $x = 4 \cos \theta$
$y = 2 \sin 2\theta$

44. Evolute of ellipse: $x = \cos^3 \theta$
$y = 2 \sin^3 \theta$

45. Involute of circle: $x = \cos \theta + \theta \sin \theta$
$y = \sin \theta - \theta \cos \theta$

46. Serpentine curve: $x = \cot \theta$
$y = 4 \sin \theta \cos \theta$

(a)
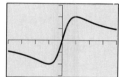

(b)
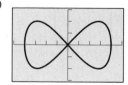

(c)
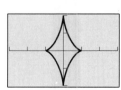

(d)
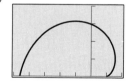

47. A wheel of radius a rolls along a straight line without slipping (see figure). Find the parametric equations for the curve generated by a point P that is b units from the center of the wheel. This curve is called a *curtate cycloid* when $b < a$.

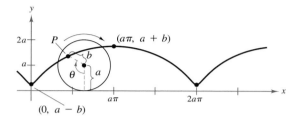

FIGURE FOR 47

48. A wheel of radius 1 rolls around the outside of a circle of radius 2 without slipping (see figure). Show that the parametric equations for the curve generated by a point on the rolling wheel are

$$x = 3 \cos \theta - \cos 3\theta \quad \text{and} \quad y = 3 \sin \theta - \sin 3\theta.$$

This curve is called an *epicycloid*.

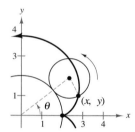

FIGURE FOR 48

7.7   POLAR COORDINATES

Introduction / Coordinate Conversion / Equation Conversion

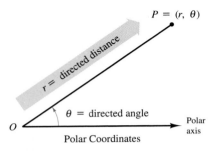

$P = (r, \theta)$

$r = $ directed distance

$\theta = $ directed angle

O Polar axis

Polar Coordinates

FIGURE 7.50

Introduction

To form the **polar coordinate system** in the plane, fix a point O, called the **pole** (or **origin**), and construct from O an initial ray called the **polar axis,** as shown in Figure 7.50. Each point P in the plane is assigned **polar coordinates** (r, θ) as follows.

1. $r = $ *directed distance* from O to P
2. $\theta = $ *directed angle*, counterclockwise from polar axis to segment $\overline{OP}$

In the polar coordinate system, it is convenient to locate points with respect to a grid of concentric circles intersected by **radial lines** through the pole. This procedure is shown in Example 1.

EXAMPLE 1   Plotting Points in the Polar Coordinate System

A. The point $(r, \theta) = (2, \pi/3)$ lies at the intersection of a circle of radius $r = 2$ and the terminal side of the angle $\theta = \pi/3$, as shown in Figure 7.51(a).

B. The point $(r, \theta) = (3, -\pi/6)$ lies in the fourth quadrant, 3 units from the pole. Note that negative angles are measured *clockwise,* as shown in Figure 7.51(b).

C. The point $(r, \theta) = (3, 11\pi/6)$ coincides with the point $(3, -\pi/6)$, as shown in Figure 7.51(c).

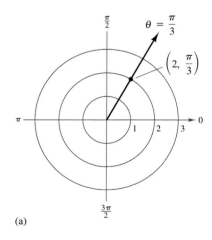

(a)

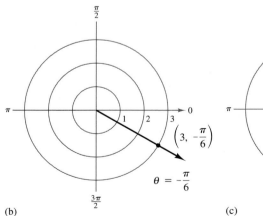

(b)

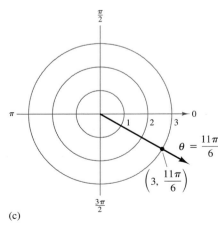

(c)

FIGURE 7.51

In rectangular coordinates, each point (x, y) has a unique representation. This is not true for polar coordinates. For instance, the coordinates (r, θ) and $(r, 2\pi + \theta)$ represent the same point. Another way to obtain multiple representations of a point is to use negative values of r. Because r is a *directed distance*, the coordinates (r, θ) and $(-r, \theta + \pi)$ represent the same point. In general, the point (r, θ) can be represented as

$$(r, \theta \pm 2n\pi) \quad \text{or} \quad (-r, \theta \pm (2n + 1)\pi),$$

where n is any integer. Moreover, the pole is represented by $(0, \theta)$, where θ is any angle.

EXAMPLE 2   Multiple Representation of Points

Plot the point $(3, -3\pi/4)$ and find three additional polar representations of this point, using $-2\pi < \theta < 2\pi$.

SOLUTION

The point is shown in Figure 7.52. Three other representations are as follows.

$$\left(3, \frac{-3\pi}{4} + 2\pi\right) = \left(3, \frac{5\pi}{4}\right) \qquad \text{\textit{Add } } 2\pi \text{ \textit{to} } \theta$$

$$\left(-3, \frac{-3\pi}{4} - \pi\right) = \left(-3, \frac{-7\pi}{4}\right) \qquad \begin{array}{l}\text{\textit{Replace r with } } -r; \\ \text{\textit{subtract } } \pi \text{ \textit{from} } \theta\end{array}$$

$$\left(-3, \frac{-3\pi}{4} + \pi\right) = \left(-3, \frac{\pi}{4}\right) \qquad \text{\textit{Replace r with } } -r; \text{ \textit{add} } \pi \text{ \textit{to} } \theta$$

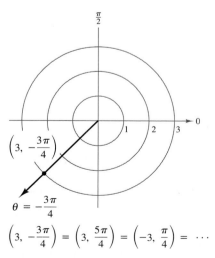

$$\left(3, -\frac{3\pi}{4}\right) = \left(3, \frac{5\pi}{4}\right) = \left(-3, \frac{\pi}{4}\right) = \cdots$$

FIGURE 7.52

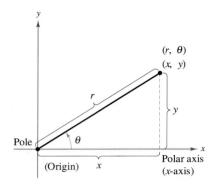

Relating Polar and Rectangular Coordinates

FIGURE 7.53

Coordinate Conversion

To establish the relationship between polar and rectangular coordinates, let the polar axis coincide with the positive x-axis and the pole with the origin, as shown in Figure 7.53. Because (x, y) lies on a circle of radius r, it follows that $r^2 = x^2 + y^2$. Moreover, for $r > 0$, the definitions of the trigonometric functions imply that

$$\tan \theta = \frac{y}{x}, \quad \cos \theta = \frac{x}{r}, \quad \text{and} \quad \sin \theta = \frac{y}{r}.$$

If $r < 0$, you can show that the same relationships hold. For example, consider the point (r, θ), where $r < 0$. Then, because $(-r, \theta + \pi)$ represents the same point and $-r > 0$, you have $-\sin \theta = \sin(\theta + \pi) = -y/r$, which implies that $\sin \theta = y/r$.

These relationships allow you to convert *coordinates* or *equations* from one system to the other, as indicated in the following rule.

COORDINATE CONVERSION

The polar coordinates (r, θ) are related to the rectangular coordinates (x, y) as follows.

$$x = r \cos \theta \quad \text{and} \quad \tan \theta = \frac{y}{x}$$

$$y = r \sin \theta \qquad\qquad r^2 = x^2 + y^2$$

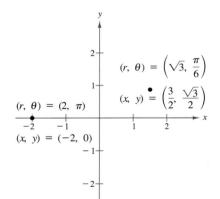

FIGURE 7.54

EXAMPLE 3 **Polar-to-Rectangular Conversion**

A. For the point $(r, \theta) = (2, \pi)$, you can write the following.

$$x = r \cos \theta = 2 \cos \pi = -2$$
$$y = r \sin \theta = 2 \sin \pi = 0$$

The rectangular coordinates are $(x, y) = (-2, 0)$, as shown in Figure 7.54.

*Technology Note*_____

You can use your graphing utility to convert polar coordinates to rectangular coordinates. For example, to convert the point $(r, \theta) =$ $(\sqrt{3}, \pi/6)$ to rectangular coordinates using the TI-81, go to the $P \triangleright R($ key in the MATH menu. Upon entering $(\sqrt{3}, \pi/6)$ as an ordered pair separated by a comma, and pressing ENTER, the calculator returns 1.5 as the x-coordinate. You can see the y-value ($\pi/6 = .8660254038$) by recalling Y, located above the number 1.

Similarly, you can convert from rectangular coordinates to polar coordinates using the $R \triangleright P($ key on the MATH menu. Try verifying the results of Example 4.

B. For the point $(r, \theta) = (\sqrt{3}, \pi/6)$, you can write the following.

$$x = \sqrt{3} \cos \frac{\pi}{6} = \sqrt{3}\left(\frac{\sqrt{3}}{2}\right) = \frac{3}{2}$$

$$y = \sqrt{3} \sin \frac{\pi}{6} = \sqrt{3}\left(\frac{1}{2}\right) = \frac{\sqrt{3}}{2}$$

The rectangular coordinates are $(x, y) = (3/2, \sqrt{3}/2)$, as shown in Figure 7.54.

EXAMPLE 4   **Rectangular-to-Polar Conversion**

A. For the second-quadrant point $(x, y) = (-1, 1)$, you can write

$$\tan \theta = \frac{y}{x} = -1 \quad \longrightarrow \quad \theta = \frac{3\pi}{4}.$$

Because θ lies in the same quadrant as (x, y), use positive r.

$$r = \sqrt{x^2 + y^2} = \sqrt{(-1)^2 + (1)^2} = \sqrt{2}$$

Thus, *one* set of polar coordinates is $(r, \theta) = (\sqrt{2}, 3\pi/4)$, as shown in Figure 7.55(a).

B. Because the point $(x, y) = (0, 2)$ lies on the positive y-axis, choose $\theta = \pi/2$ and $r = 2$. Thus, one set of polar coordinates is $(r, \theta) = (2, \pi/2)$, as shown in Figure 7.55(b).

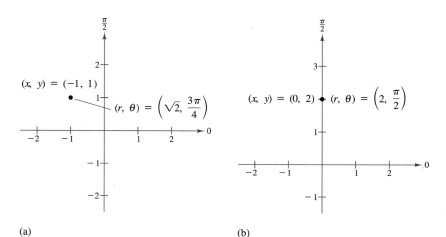

(a)   (b)

FIGURE 7.55

Equation Conversion

By comparing Examples 3 and 4, you can see that point conversion from the polar to the rectangular system is straightforward, whereas point conversion from the rectangular to the polar system is more involved. For equations, the opposite is true. To convert a rectangular equation to polar form, simply replace x by $r \cos \theta$ and y by $r \sin \theta$. For instance, the rectangular equation $y = x^2$ can be written in polar form as follows.

$$\underbrace{y = x^2}_{\substack{\text{Rectangular} \\ \text{equation}}} \longrightarrow \underbrace{r \sin \theta = (r \cos \theta)^2}_{\text{Polar equation}}$$

On the other hand, converting a polar equation to rectangular form can require considerable ingenuity.

EXAMPLE 5   Converting Polar Equations to Rectangular Form

Convert each polar equation to rectangular form. Then identify the graph.

A. $r = 2$ **B.** $\theta = \dfrac{\pi}{3}$ **C.** $r = \sec \theta$

SOLUTION

A. The graph of the polar equation $r = 2$ consists of all points that are two units from the pole. In other words, this graph is a circle centered at the origin and having a radius of 2, as shown in Figure 7.56(a). You can confirm this by converting to rectangular coordinates, using the relationship $r^2 = x^2 + y^2$.

$$\underbrace{r = 2}_{\text{Polar equation}} \longrightarrow r^2 = 2^2 \longrightarrow \underbrace{x^2 + y^2 = 2^2}_{\text{Rectangular equation}}$$

B. The graph of the polar equation $\theta = \pi/3$ consists of all points on the line that makes an angle of $\pi/3$ with the positive x-axis, as shown in Figure 7.56(b). To convert to rectangular form, you can make use of the relationship $\tan \theta = y/x$.

$$\underbrace{\theta = \frac{\pi}{3}}_{\text{Polar equation}} \longrightarrow \tan \theta = \sqrt{3} \longrightarrow \underbrace{y = \sqrt{3}x}_{\text{Rectangular equation}}$$

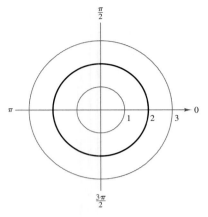

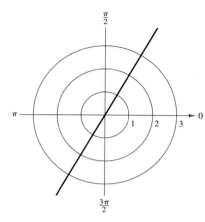

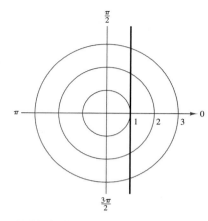

(a) Circle: $r = 2$

(b) Radial line: $\theta = \dfrac{\pi}{3}$

(c) Vertical line: $r = \sec \theta$

FIGURE 7.56

C. The graph of the polar equation $r = \sec \theta$ is not evident by simple inspection. You can convert to rectangular form by using the relationship $r \cos \theta = x$.

$$\underbrace{r = \sec \theta}_{\text{Polar equation}} \longrightarrow r \cos \theta = 1 \longrightarrow \underbrace{x = 1}_{\text{Rectangular equation}}$$

Now, you can see that the graph is a vertical line, as shown in Figure 7.56(c).

Curve sketching by converting to rectangular form is not always convenient. In the next section, you will learn other techniques for sketching polar equations.

DISCUSSION PROBLEM

SIMPLIFYING A POLAR EQUATION

In the discussion before Example 5, we showed how to convert the rectangular equation $y = x^2$ to the polar equation

$$r \sin \theta = (r \cos \theta)^2.$$

Simplifying this equation by dividing both sides by r produces

$$\sin \theta = r \cos^2 \theta \quad \text{or} \quad r = \frac{\sin \theta}{\cos^2 \theta}.$$

By simplifying in this way, however, you risk the possibility of losing the pole $(0, \theta)$ as a solution point. For the equation shown above, division by r does not change the set of solution points, because the pole is a solution point of the simplified equation. Find a polar equation for which division by r *would* change the set of solution points.

WARM-UP

The following warm-up exercises involve skills that were covered in earlier sections. You will use these skills in the exercise set for this section.

In Exercises 1 and 2, find a positive angle coterminal with the given angle.

1. $\dfrac{11\pi}{4}$ **2.** $-\dfrac{5\pi}{6}$

In Exercises 3 and 4, find the sine and cosine of the angle in standard position with terminal side passing through the given point.

3. $(2, 1)$ **4.** $(4, -3)$

In Exercises 5 and 6, find the magnitude (in radians) of an angle in standard position with terminal side passing through the given point.

5. $(-4, 4)$ **6.** $(3, 2)$

In Exercises 7 and 8, evaluate the trigonometric function without the aid of a calculator.

7. $\sin \dfrac{4\pi}{3}$ **8.** $\cos \dfrac{3\pi}{4}$

In Exercises 9 and 10, use a calculator to evaluate the trigonometric function.

9. $\cos \dfrac{3\pi}{5}$ **10.** $\sin 1.34$

SECTION 7.7 · EXERCISES

In Exercises 1–10, plot the point given in polar coordinates and find the corresponding rectangular coordinates for the point.

1. $\left(4, \dfrac{3\pi}{6}\right)$ **2.** $\left(4, \dfrac{3\pi}{2}\right)$

3. $\left(-1, \dfrac{5\pi}{4}\right)$ **4.** $(0, -\pi)$

5. $\left(4, -\dfrac{\pi}{3}\right)$ **6.** $\left(-1, -\dfrac{3\pi}{4}\right)$

7. $\left(0, -\dfrac{7\pi}{6}\right)$ **8.** $\left(\dfrac{3}{2}, \dfrac{5\pi}{2}\right)$

9. $(\sqrt{2}, 2.36)$ **10.** $(-3, -1.57)$

In Exercises 11–20, the rectangular coordinates of a point are given. Plot the point and find *two* sets of polar coordinates for the point for $0 \le \theta < 2\pi$.

11. $(1, 1)$
12. $(0, -5)$
13. $(-6, 0)$
14. $(-3, -3)$
15. $(-3, 4)$
16. $(3, -1)$
17. $(-\sqrt{3}, -\sqrt{3})$
18. $(-2, 0)$
19. $(4, 6)$
20. $(5, 12)$

In Exercises 21–34, convert the rectangular equation to polar form.

21. $x^2 + y^2 = 9$

22. $x^2 + y^2 = a^2$

23. $x^2 + y^2 - 2ax = 0$

24. $x^2 + y^2 - 2ay = 0$

25. $y = 4$

26. $y = b$

27. $x = 10$

28. $x = a$

29. $3x - y + 2 = 0$

30. $4x + 7y - 2 = 0$

31. $xy = 4$

32. $y = x$

33. $(x^2 + y^2)^2 - 9(x^2 - y^2) = 0$

34. $y^2 - 8x - 16 = 0$

In Exercises 35–44, convert the polar equation to rectangular form.

35. $r = 4 \sin \theta$

36. $r = 4 \cos \theta$

37. $\theta = \dfrac{\pi}{6}$

38. $r = 4$

39. $r = 2 \csc \theta$

40. $r^2 = \sin 2\theta$

41. $r = 2 \sin 3\theta$

42. $r = \dfrac{1}{1 - \cos \theta}$

43. $r = \dfrac{6}{2 - 3 \sin \theta}$

44. $r = \dfrac{6}{2 \cos \theta - 3 \sin \theta}$

In Exercises 45–50, convert the polar equation to rectangular form and sketch its graph.

45. $r = 3$

46. $r = 8$

47. $\theta = \dfrac{\pi}{4}$

48. $\theta = \dfrac{5\pi}{6}$

49. $r = 3 \sec \theta$

50. $r = 2 \csc \theta$

51. Show that the distance between the points (r_1, θ_1) and (r_2, θ_2) is given by

$$\sqrt{r_1^2 + r_2^2 - 2r_1 r_2 \cos(\theta_1 - \theta_2)}.$$

52. Choose two points in the polar coordinate system. Use the formula given in Exercise 51 to find the distance between the two points. Then choose different polar coordinate representations of the same two points and apply the distance formula again. Discuss your result.

53. Convert the polar equation

$$r = 2(h \cos \theta + k \sin \theta)$$

to rectangular form and verify that it is the equation of a circle. Find the radius and the rectangular coordinates of the center of the circle.

54. Convert the polar equation $r = \cos \theta + 3 \sin \theta$ to rectangular form and identify the graph.

7.8   GRAPHS OF POLAR EQUATIONS

Introduction / Using a Graphing Utility / Symmetry / Maximum
r-Values / Special Polar Graphs

Introduction

In previous chapters, you spent a lot of time learning how to sketch graphs in rectangular coordinates. You began with the basic point-plotting method. Then you used sketching aids such as a graphing utility, symmetry, intercepts, asymptotes, periods, and shifts to further investigate the nature of the graph. In this section, you will approach curve sketching in the polar coordinate system in a similar way.

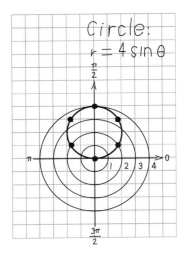

FIGURE 7.57

EXAMPLE 1   **Graphing a Polar Equation by Point Plotting**

Sketch the graph of the polar equation $r = 4 \sin \theta$.

SOLUTION

The sine function is periodic, so you can get a full range of *r*-values by considering values of θ in the interval $0 \le \theta \le 2\pi$, shown in the table.

θ	0	$\dfrac{\pi}{6}$	$\dfrac{\pi}{3}$	$\dfrac{\pi}{2}$	$\dfrac{2\pi}{3}$	$\dfrac{5\pi}{6}$	π	$\dfrac{7\pi}{6}$	$\dfrac{3\pi}{2}$	$\dfrac{11\pi}{6}$	2π
$4 \sin \theta$	0	2	$2\sqrt{3}$	4	$2\sqrt{3}$	2	0	-2	-4	-2	0

Plot these points as shown in Figure 7.57. In the figure, it appears that the graph is a circle of radius 2 whose center is at the point $(x, y) = (0, 2)$.

Using a Graphing Utility

Some graphing utilities have a polar-coordinate graphing mode. If your graphing utility *doesn't* have such a mode, but *does* have a parametric mode, you can use the following conversion to graph a polar equation.

POLAR EQUATIONS IN PARAMETRIC FORM

The graph of the polar equation $r = f(\theta)$ can be written in parametric form, using t as a parameter, as follows.

$$x = f(t) \cos t \quad \text{and} \quad y = f(t) \sin t$$

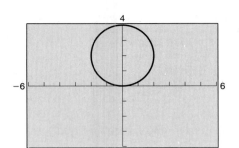

FIGURE 7.58

EXAMPLE 2   Graphing a Polar Equation with a Graphing Utility

Use a graphing utility to graph $r = 4 \sin \theta$.

SOLUTION

Begin by writing the equation in parametric form.

$$x = 4 \sin t \cos t$$
$$y = 4 \sin t \sin t$$

Set the graphing utility to parametric and radian mode. Use a viewing rectangle of $-6 \leq x \leq 6$ and $-4 \leq y \leq 4$, and let the parameter t vary from 0 to π. The graph is shown in Figure 7.58. In the figure, it appears that the graph is a circle, which reinforces the result obtained in Example 1.

REMARK   You can confirm that the graph given in Examples 1 and 2 is a circle by rewriting the polar equation in rectangular form. If you do this, you will find that the rectangular equation is $x^2 + y^2 = 4y$. By completing the square, this equation can be written as $x^2 + (y - 2)^2 = 2^2$, which you know is the equation of a circle.

Symmetry

In Figure 7.58, the entire graph of the polar equation was traced as t increased from 0 to π. Try resetting your graphing utility so that t varies from 0 to 2π. Then, use the trace feature to follow the curve through increasing values of t, and you will see that the curve is traced out *twice*.

In Figure 7.58, you should also notice that the graph is *symmetric with respect to the line $\theta = \pi/2$*. Symmetry with respect to the line $\theta = \pi/2$ is one of three important types of symmetry to consider in polar curve sketching. (See Figure 7.59.)

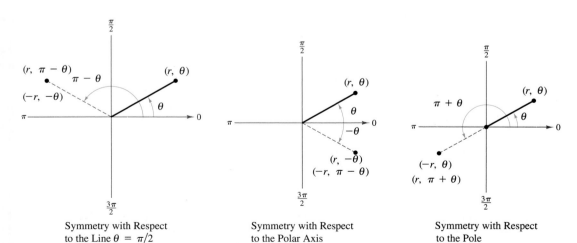

Symmetry with Respect to the Line $\theta = \pi/2$

Symmetry with Respect to the Polar Axis

Symmetry with Respect to the Pole

FIGURE 7.59

TEST FOR SYMMETRY IN POLAR COORDINATES

The graph of a polar equation is symmetric with respect to the following if the given substitution yields an equivalent equation.

1. The line $\theta = \pi/2$: replace (r, θ) by $(r, \pi - \theta)$ or $(-r, -\theta)$.
2. The polar axis: replace (r, θ) by $(r, -\theta)$ or $(-r, \pi - \theta)$.
3. The pole: replace (r, θ) by $(r, \pi + \theta)$ or $(-r, \theta)$.

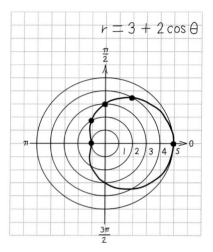

EXAMPLE 3   Finding Symmetry of the Graph of a Polar Equation

Sketch the graph of $r = 3 + 2 \cos \theta$, and discuss its symmetry.

SOLUTION

Replacing (r, θ) by $(r, -\theta)$ produces

$$r = 3 + 2 \cos(-\theta) = 3 + 2 \cos \theta.$$

Thus, you conclude that the curve is symmetric with respect to the polar axis. This conclusion can be confirmed by sketching the graph of the polar equation, as shown in Figure 7.60. A set of parameteric equations for the graph is

$$x = (3 + 2 \cos t)(\cos t)$$
$$y = (3 + 2 \cos t)(\sin t).$$

By letting t vary from 0 to 2π, you can obtain the graph shown in Figure 7.60.

The three tests given for symmetry in polar coordinates are sufficient to guarantee symmetry, but they are not necessary. For instance, Figure 7.61 shows the graph of $r = \theta + 2\pi$ to be symmetric with respect to the line $\theta = \pi/2$. Yet the test fails to indicate symmetry because neither of the following replacements yields an equivalent equation.

Original Equation	Replacement	New Equation
$r = \theta + 2\pi$	(r, θ) by $(-r, -\theta)$	$-r = -\theta + 2\pi$
$r = \theta + 2\pi$	(r, θ) by $(r, \pi - \theta)$	$r = -\theta + 3\pi$

The equations discussed in Examples 1–3 are of the form

$$r = 4 \sin \theta = f(\sin \theta) \quad \text{and} \quad r = 3 + 2 \cos \theta = g(\cos \theta).$$

The graph of the first equation is symmetric with respect to the line $\theta = \pi/2$, and the graph of the second equation is symmetric with respect to the polar axis. This observation can be generalized to yield the following *quick test for symmetry.*

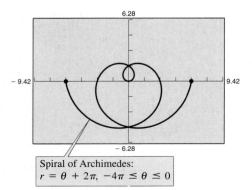

Spiral of Archimedes:
$r = \theta + 2\pi, \ -4\pi \le \theta \le 0$

FIGURE 7.61

1. The graph of $r = f(\sin \theta)$ is symmetric with respect to the line $\theta = \pi/2$.
2. The graph of $r = g(\cos \theta)$ is symmetric with respect to the polar axis.

Maximum *r*-Values

The graph of a polar equation has a maximum *r*-value if there is a point on the graph that is farthest from the pole. For instance, in Example 1, the maximum value of *r* for $r = 4 \sin \theta$ is 4. This value of *r* occurs when $\theta = \pi/2$.

EXAMPLE 4   Finding a Maximum *r*-Value of a Polar Graph

Find the maximum value of *r* for the graph of $r = 1 - 2 \cos \theta$.

SOLUTION

Because the polar equation is of the form $r = 1 - 2 \cos \theta = g(\cos \theta)$, you know that the graph is symmetric with respect to the polar axis. To sketch a graph of the equation, write it in parametric form.

$$x = (1 - 2 \cos t)(\cos t)$$
$$y = (1 - 2 \cos t)(\sin t)$$

Using a graphing utility and letting *t* vary from 0 to 2π, you can display the graph shown in Figure 7.62. Finally, using the trace feature of the graphing utility, you can conclude that the maximum value of *r* (the *r*-value of the point that is farthest from the pole) is 3. This value of *r* occurs when $t = \pi$ (or when $\theta = \pi$).

REMARK Note how the negative *r*-values determine the *inner loop* of the graph in Figure 7.62. This type of graph is called a **limaçon.**

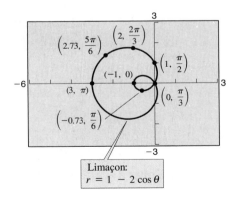

FIGURE 7.62

Some curves reach their maximum *r*-values at more than one point.

EXAMPLE 5 Finding a Maximum *r*-Value of a Polar Graph

Find the maximum *r*-value of the graph of $r = 2 \cos 3\theta$.

SOLUTION

Because $2 \cos 3\theta = 2 \cos(-3\theta)$, you can conclude that the graph is symmetric with respect to the polar axis. A set of parametric equations for the graph is given by

$$x = 2 \cos 3t \cos t$$
$$y = 2 \cos 3t \sin t.$$

REMARK   The graph shown in Figure 7.63 is called a **rose curve,** and each of the loops on the graph is called a *petal* of the rose curve.

By letting *t* vary from 0 to π, you can use a graphing utility to display the graph shown in Figure 7.63. The maximum *r*-value for this graph is 2. This value occurs at *three* different points on the graph: $(r, \theta) = (2, 0)$, $(-2, \pi/3)$, and $(2, 2\pi/3)$.

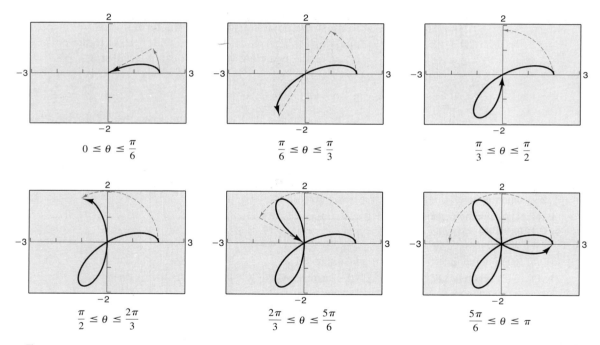

$$0 \le \theta \le \frac{\pi}{6} \qquad \frac{\pi}{6} \le \theta \le \frac{\pi}{3} \qquad \frac{\pi}{3} \le \theta \le \frac{\pi}{2}$$

$$\frac{\pi}{2} \le \theta \le \frac{2\pi}{3} \qquad \frac{2\pi}{3} \le \theta \le \frac{5\pi}{6} \qquad \frac{5\pi}{6} \le \theta \le \pi$$

FIGURE 7.63

Special Polar Graphs

Several important types of graphs have equations that are simpler in polar form than in rectangular form. For example, the circle $r = 4 \sin \theta$ in Example 1 has the more complicated rectangular equation $x^2 + (y - 2)^2 = 4$. The following list gives several other types of graphs that have simple polar equations.

Limaçons

$r = a \pm b \cos \theta$
$r = a \pm b \sin \theta$
$(0 < a, 0 < b)$

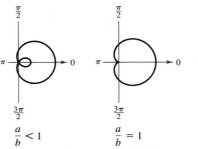
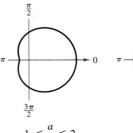
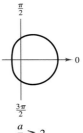

$\dfrac{a}{b} < 1$

Limaçon with
inner loop

$\dfrac{a}{b} = 1$

Cardioid
(heart-shaped)

$1 < \dfrac{a}{b} < 2$

Dimpled
limaçon

$\dfrac{a}{b} \geq 2$

Convex
limaçon

Rose Curves

n petals if n is odd
$2n$ petals if n is even
$(n \geq 2)$

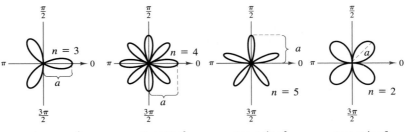

$r = a \cos n\theta$
Rose curve

$r = a \cos n\theta$
Rose curve

$r = a \sin n\theta$
Rose curve

$r = a \sin n\theta$
Rose curve

Circles and Lemniscates

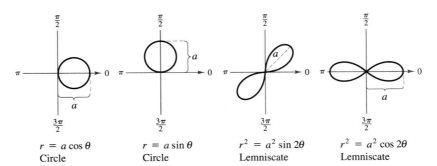

$r = a \cos \theta$
Circle

$r = a \sin \theta$
Circle

$r^2 = a^2 \sin 2\theta$
Lemniscate

$r^2 = a^2 \cos 2\theta$
Lemniscate

EXAMPLE 6 Sketching a Rose Curve

Sketch the graph of $r = 3 \cos 2\theta$.

SOLUTION

Begin with an analysis of the basic features of the graph.

Type of curve: Rose curve with $2n = 4$ petals

Symmetry: With respect to polar axis and the line $\theta = \dfrac{\pi}{2}$

Maximum value of r: $|r| = 3$ when $\theta = 0, \dfrac{\pi}{2}, \pi, \dfrac{3\pi}{2}$

Parametric equations: $x = 3 \cos 2t \cos t$
$y = 3 \cos 2t \sin t$

Using a graphing utility and letting t vary from 0 to 2π, you can graph the rose curve shown in Figure 7.64.

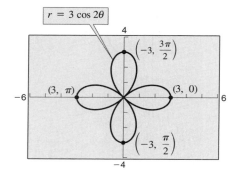

FIGURE 7.64

EXAMPLE 7 Sketching a Lemniscate

Sketch the graph of $r^2 = 9 \sin 2\theta$.

SOLUTION

Begin with an analysis of the basic features of the graph.

Type of curve: Lemniscate
Symmetry: With respect to the pole

Maximum value of r: $|r| = 3$ when $\theta = \dfrac{\pi}{4}$

Two sets of parametric equations: $x = 3\sqrt{\sin(2t)} \, (\cos t)$
$y = 3\sqrt{\sin(2t)} \, (\sin t)$
$x = -3\sqrt{\sin(2t)} \, (\cos t)$
$y = -3\sqrt{\sin(2t)} \, (\sin t)$

Using a graphing utility and letting t vary from 0 to $\pi/2$, you can graph the lemniscate shown in Figure 7.65.

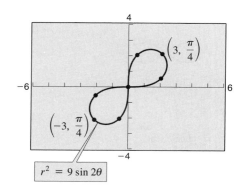

FIGURE 7.65

DISCUSSION PROBLEM

· · · · · · · · · · · · · · · ·

DIFFERENT TYPES OF LIMAÇONS

There are four different types of limaçons in the table showing special polar graphs. Explain the differences among the four types. Then classify each of the following as one of the special types and sketch its graph.

(a) $r = 3 + 3 \sin \theta$ (b) $r = 4 + 3 \sin \theta$

(c) $r = 2 + 3 \sin \theta$ (d) $r = 3 + \sin \theta$

WARM-UP

· · · · · · · · · · · · · · ·

The following warm-up exercises involve skills that were covered in earlier sections. You will use these skills in the exercise set for this section.

In Exercises 1–4, determine the amplitude and period of the function.

1. $y = 5 \sin 4x$ **2.** $y = 3 \cos 2\pi x$

3. $y = -5 \cos \dfrac{5x}{2}$ **4.** $y = -\dfrac{1}{2} \sin \dfrac{x}{2}$

In Exercises 5–8, sketch the graph of the function through two periods.

5. $y = 2 \sin x$ **6.** $y = 3 \cos x$

7. $y = 4 \cos 2x$ **8.** $y = 2 \sin \pi x$

In Exercises 9 and 10, use the sum and difference identities to simplify the trigonometric expressions.

9. $\sin\left(x - \dfrac{\pi}{6}\right)$ **10.** $\sin\left(x + \dfrac{\pi}{4}\right)$

SECTION 7.8 · EXERCISES

In Exercises 1–6, test for symmetry with respect to $\theta = \pi/2$, the polar axis, and the pole.

1. $r = 10 + 6 \cos \theta$ **2.** $r = 16 \cos 3\theta$

3. $r = \dfrac{2}{1 + \sin \theta}$ **4.** $r = 6 \sin \theta$

5. $r = 4 \sec \theta \csc \theta$ **6.** $r^2 = 25 \sin 2\theta$

In Exercises 7–10, find the maximum value of $|r|$.

7. $r = 5 \cos 3\theta$

8. $r = -2 \cos \theta$

9. $r = 10(1 - \sin \theta)$

10. $r = 6 + 12 \cos \theta$

In Exercises 11–30, sketch the graph of the polar equation.

11. $r = 5$ **12.** $r = 2$

13. $\theta = \dfrac{\pi}{6}$ **14.** $\theta = -\dfrac{\pi}{4}$

15. $r = 3 \sin \theta$ **16.** $r = 3(1 - \cos \theta)$
17. $r = 4(1 + \sin \theta)$ **18.** $r = 3 - 2 \cos \theta$
19. $r = 4 + 3 \cos \theta$ **20.** $r = 2 + 4 \sin \theta$
21. $r = 3 - 4 \cos \theta$ **22.** $r = 2 \cos 3\theta$
23. $r = 3 \sin 2\theta$ **24.** $r = 2 \sec \theta$

25. $r = \dfrac{3}{\sin \theta - 2 \cos \theta}$ **26.** $r = \dfrac{6}{2 \sin \theta - 3 \cos \theta}$

27. $r^2 = 4 \cos 2\theta$ **28.** $r^2 = 4 \sin \theta$

29. $r = \dfrac{\theta}{2}$ **30.** $r = \theta$

In Exercises 31–42, use a graphing utility to graph the polar equation.

31. $r = 6 \cos \theta$ **32.** $r = \dfrac{\theta}{2}$

33. $r = 3(2 - \sin \theta)$ **34.** $r = \cos 2\theta$
35. $r = 4 \sin \theta \cos^2 \theta$ **36.** $r = 3 \cos 2\theta \sec \theta$
37. $r = 2 \csc \theta + 5$ **38.** $r = 2 \cos(3\theta - 2)$
39. $r = 2 - \sec \theta$ **40.** $r = 2 + \csc \theta$

41. $r = \dfrac{2}{\theta}$ **42.** $r = 2 \cos 2\theta \sec \theta$

In Exercises 43–50, use a graphing utility to graph the polar equation. Find the interval for θ over which the graph is traced only once.

43. $r = 3 - 4 \cos \theta$ **44.** $r = 2(1 - 2 \sin \theta)$
45. $r = 2 + \sin \theta$ **46.** $r = 4 + 3 \cos \theta$

47. $r = 2 \cos\left(\dfrac{3\theta}{2}\right)$ **48.** $r = 3 \sin\left(\dfrac{5\theta}{2}\right)$

49. $r^2 = 4 \sin 2\theta$ **50.** $r^2 = 3 \cos 4\theta$

In Exercises 51 and 52, convert the polar equation to rectangular form and show that the indicated line is an asymptote of the graph.

Polar Equation	Asymptote
51. $r = 2 - \sec \theta$	$x = -1$
52. $r = 2 + \csc \theta$	$y = 1$

53. The graph of $r = f(\theta)$ is rotated about the pole through an angle ϕ. Show that the equation for the rotated graph is $r = f(\theta - \phi)$.

54. Consider the graph of $r = f(\sin \theta)$.
(a) Show that if the graph is rotated counterclockwise $\pi/2$ radians about the pole, then the equation for the rotated graph is $r = f(-\cos \theta)$.
(b) Show that if the graph is rotated counterclockwise π radians about the pole, then the equation for the rotated graph is $r = f(-\sin \theta)$.
(c) Show that if the graph is rotated counterclockwise $3\pi/2$ radians about the pole, then the equation for the rotated graph is $r = f(\cos \theta)$.

In Exercises 55–58, use the results of Exercises 53 and 54.

55. Write an equation for the rose curve $r = 2 - \sin \theta$ after it has been rotated by the given amount.

(a) $\dfrac{\pi}{4}$ (b) $\dfrac{\pi}{2}$ (c) π (d) $\dfrac{3\pi}{2}$

56. Write an equation for the rose curve $r = 2 \sin 2\theta$ after it has been rotated by the given amount.

(a) $\dfrac{\pi}{6}$ (b) $\dfrac{\pi}{2}$ (c) $\dfrac{2\pi}{3}$ (d) π

57. Sketch the graph of each equation.

(a) $r = 1 - \sin \theta$ (b) $r = 1 - \sin\left(\theta - \dfrac{\pi}{4}\right)$

58. Sketch the graph of each equation.

(a) $r = 3 \sec \theta$ (b) $r = 3 \sec\left(\theta - \dfrac{\pi}{4}\right)$

(c) $r = 3 \sec\left(\theta + \dfrac{\pi}{3}\right)$ (d) $r = 3 \sec\left(\theta - \dfrac{\pi}{2}\right)$

7.9   POLAR EQUATIONS OF CONICS

Alternative Definition of Conics / Polar Equations of Conics / Application

Alternative Definition of Conics

In Sections 7.3 and 7.4, you saw that the rectangular equations of ellipses and hyperbolas take simple forms when the origin lies at their *center*. There are, however, many important applications of conics in which it is more convenient to use one of the *foci* as the origin of the coordinate system. For example, the sun lies at the focus of the earth's orbit. Similarly, the light source of a parabolic reflector lies at its focus. In this section you will see that polar equations of conics take simple forms if one of the foci lies at the pole.

ALTERNATIVE DEFINITION OF CONIC

The locus of a point in the plane that moves so that its distance from a fixed point (focus) is in constant ratio to its distance from a fixed line (directrix) is a **conic.** The constant ratio is the **eccentricity** of the conic and is denoted by e. Moreover, the conic is an **ellipse** if $e < 1$, a **parabola** if $e = 1$, and a **hyperbola** if $e > 1$.

In Figure 7.66, note that for each type of conic, the pole corresponds to the fixed point (focus) given in the definition.

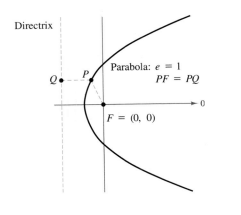

Parabola: $e = 1$
$PF = PQ$
$F = (0, 0)$

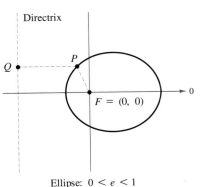

Ellipse: $0 < e < 1$
$$\frac{PF}{PQ} < 1$$
$F = (0, 0)$

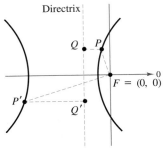

Hyperbola: $e > 1$
$$\frac{PF}{PQ} = \frac{P'F}{P'Q'} > 1$$
$F = (0, 0)$

FIGURE 7.66

Polar Equations of Conics

POLAR EQUATIONS OF CONICS

The graph of a polar equation of the form

1. $r = \dfrac{ep}{1 \pm e \cos \theta}$

2. $r = \dfrac{ep}{1 \pm e \sin \theta}$

is a conic, where $e > 0$ is the eccentricity and $|p|$ is the distance
between the focus (pole) and the directrix.

PROOF

We give a proof for $r = ep/(1 + e \cos \theta)$ with $p > 0$. In Figure 7.67 consider
a vertical directrix, p units to the right of the focus $F = (0, 0)$. If $P = (r, \theta)$
is a point on the graph of $r = ep/(1 + e \cos \theta)$, then the distance between P
and the directrix is

$$PQ = |p - x| = |p - r \cos \theta|$$

$$= \left| p - \left(\frac{ep}{1 + e \cos \theta} \right) \cos \theta \right|$$

$$= \left| p \left(1 - \frac{e \cos \theta}{1 + e \cos \theta} \right) \right|$$

$$= \left| \frac{p}{1 + e \cos \theta} \right|$$

$$= \left| \frac{r}{e} \right|.$$

Because the distance between P and the pole is simply $PF = |r|$, the ratio of
PF to PQ is

$$\frac{PF}{PQ} = \frac{|r|}{\left| \dfrac{r}{e} \right|} = |e| = e,$$

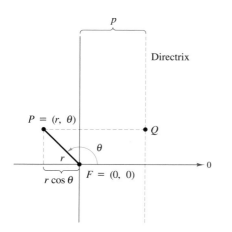

FIGURE 7.67

and, by definition, the graph of the equation must be a conic. ·········

By completing the proofs of the other three cases, you can see that the equations

$$r = \frac{ep}{1 \pm e \cos \theta}$$ *Vertical directrix*

correspond to conics with vertical directrices and the equations

$$r = \frac{ep}{1 \pm e \sin \theta}$$ *Horizontal directrix*

correspond to conics with horizontal directrices. The converse is also true. That is, any conic with a focus at the pole and having a horizontal or vertical directrix can be represented by one of the given equations.

EXAMPLE 1   Determining a Conic from Its Equation

Identify the conic given by

$$r = \frac{15}{3 - 2 \cos \theta}$$

and sketch its graph.

SOLUTION

To determine the type of conic, rewrite the equation as

$$r = \frac{15}{3 - 2 \cos \theta} = \frac{5}{1 - \frac{2}{3} \cos \theta}.$$

From this form you can conclude that the graph is an ellipse with $e = \frac{2}{3}$. To graph the ellipse, use a graphing utility and the following parametric equations.

$$x = \frac{15 \cos t}{3 - 2 \cos t} \quad \text{and} \quad y = \frac{15 \sin t}{3 - 2 \cos t}$$

Letting t vary from 0 to 2π produces the graph shown in Figure 7.68. Notice that the graph has symmetry with respect to the polar axis. ....

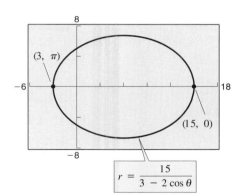

FIGURE 7.68

For the ellipse in Figure 7.68, the major axis is horizontal and the vertices lie at $(15, 0)$ and $(3, \pi)$. Thus, the length of the *major* axis is $2a = 18$. To find the length of the *minor* axis, you can use the equations $e = c/a$ and $b^2 = a^2 - c^2$ to conclude that

$$b^2 = a^2 - c^2 = a^2 - (ea)^2 = a^2(1 - e^2). \quad \textit{Ellipse}$$

Because $e = \frac{2}{3}$, it follows that $b^2 = 9^2[1 - (2/3)^2] = 45$, which implies that $b = \sqrt{45} = 3\sqrt{5}$. Thus, the length of the minor axis is $2b = 6\sqrt{5}$. A similar analysis for hyperbolas yields

$$b^2 = c^2 - a^2 = (ea)^2 - a^2 = a^2(e^2 - 1). \quad \textit{Hyperbola}$$

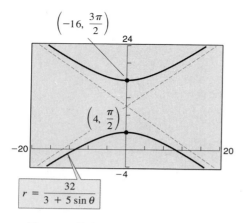

$\left(-16, \dfrac{3\pi}{2}\right)$

$\left(4, \dfrac{\pi}{2}\right)$

$r = \dfrac{32}{3 + 5 \sin \theta}$

FIGURE 7.69

EXAMPLE 2   **Sketching a Conic from Its Polar Equation**

Identify the conic given by

$$r = \frac{32}{3 + 5 \sin \theta}$$

and sketch its graph.

SOLUTION

To determine the type of conic, rewrite the equation as

$$r = \frac{\frac{32}{3}}{1 + \frac{5}{3} \sin \theta}.$$

Because $e = \frac{5}{3} > 1$, the graph is a hyperbola. The transverse axis of the hyperbola lies on the line $\theta = \pi/2$, and the vertices occur at $(4, \pi/2)$ and $(-16, 3\pi/2)$. Because the length of the transverse axis is 12, it follows that $a = 6$. To find b, you can write

$$b^2 = a^2(e^2 - 1) = 6^2\left[\left(\frac{5}{3}\right)^2 - 1\right] = 64.$$

Therefore, $b = 8$. Finally, use a and b to determine the asymptotes of the hyperbola and obtain the graph shown in Figure 7.69.

In the next example you are asked to find a polar equation for a specified conic. To do this, let p be the distance between the pole and the directrix. With this interpretation of p, we suggest the following guidelines for finding a polar equation for a conic.

1. Horizontal directrix above the pole: $r = \dfrac{ep}{1 + e \sin \theta}$

2. Horizontal directrix below the pole: $r = \dfrac{ep}{1 - e \sin \theta}$

3. Vertical directrix to the right of the pole: $r = \dfrac{ep}{1 + e \cos \theta}$

4. Vertical directrix to the left of the pole: $r = \dfrac{ep}{1 - e \cos \theta}$

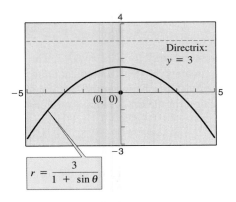

$$r = \frac{3}{1 + \sin \theta}$$

FIGURE 7.70

EXAMPLE 3 Finding the Polar Equation for a Conic

Find the polar equation for the parabola whose focus is the pole and whose directrix is the line $y = 3$.

SOLUTION

In Figure 7.70, you can see that the directrix is horizontal and above the pole. Thus, you should choose an equation of the form

$$r = \frac{ep}{1 + e \sin \theta}.$$

Moreover, because the eccentricity of a parabola is $e = 1$ and the distance between the pole and the directrix is $p = 3$, you obtain

$$r = \frac{3}{1 + \sin \theta}.$$

Application

Kepler's Laws (listed below), named after the German astronomer Johannes Kepler (1571–1630), can be used to describe the orbits of the planets about the sun.

1. Each planet moves in an elliptical orbit with the sun as a focus.
2. A ray from the sun to the planet sweeps out equal areas of the ellipse in equal times.
3. The square of the period is proportional* to the cube of the mean distance between the planet and the sun.

Although Kepler simply stated these laws on the basis of observation, they were later validated by Isaac Newton (1642–1727). In fact, Newton was able to show that each law can be deduced from a set of universal laws of motion and gravitation which govern the movement of all heavenly bodies, including comets and satellites. This is illustrated in the next example involving the comet named after the English mathematician and physicist Edmund Halley (1656–1742).

---

*Using Earth as a reference with a period of 1 year and a distance of 1 astronomical unit, the proportionality constant is 1. For example, because Mars has a mean distance to the sun of $d = 1.523$ AU, its period P is given by $d^3 = P^2$. Thus, the period for Mars is $P = 1.88$ years.

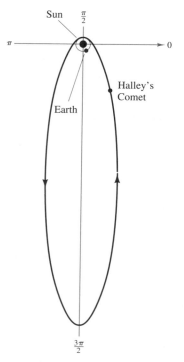

FIGURE 7.71

EXAMPLE 4   Halley's Comet

Halley's comet has an elliptical orbit with an eccentricity of $e \approx 0.97$. The length of the major axis of the orbit is approximately 36.18 astronomical units. (An astronomical unit is defined as the mean distance between the earth and the sun, 93 million miles.) Find a polar equation for the orbit. How close does Halley's comet come to the sun?

SOLUTION

Using a vertical axis, as shown in Figure 7.71, choose an equation of the form $r = ep/(1 + e \sin \theta)$. Because the vertices of the ellipse occur when $\theta = \pi/2$ and $\theta = 3\pi/2$, you can determine the length of the major axis to be the sum of the r-values of the vertices. That is,

$$2a = \frac{0.97p}{1 + 0.97} + \frac{0.97p}{1 - 0.97} \approx 32.83p \approx 36.18.$$

Thus, $p \approx 1.102$ and $ep \approx (0.97)(1.102) \approx 1.069$. Using this value in the equation, you have

$$r = \frac{1.069}{1 + 0.97 \sin \theta},$$

where r is measured in astronomical units. To find the closest point to the sun (the focus), you can write

$$c = ea \approx (0.97)(18.09) \approx 17.55.$$

Because c is the distance between the focus and the center, the closest point is

$$a - c \approx 18.09 - 17.55 \approx 0.54 \; AU \approx 50,000,000 \text{ miles.}$$

Halley's comet is not the only spectacular comet that is periodically visible to viewers on earth. Use your school library to find information about another comet, and write a paragraph describing some of its characteristics.

WARM-UP
. . . .

The following warm-up exercises involve skills that were covered in earlier sections. You will use these skills in the exercise set for this section.

In Exercises 1 and 2, plot the point given in polar coordinates and find the corresponding rectangular coordinates for the point.

1. $\left(-3, \dfrac{3\pi}{4}\right)$ **2.** $\left(4, -\dfrac{2\pi}{3}\right)$

In Exercises 3 and 4, plot the point given in rectangular coordinates and find two sets of polar coordinates for the point where $0 \le \theta < 2\pi$.

3. $(0, -3)$ **4.** $(-5, 12)$

In Exercises 5 and 6, convert the rectangular equation to polar form.

5. $x^2 + y^2 = 25$ **6.** $x^2 y = 4$

In Exercises 7 and 8, convert the polar equation to rectangular form.

7. $r \sin \theta = -4$ **8.** $r = 4 \cos \theta$

In Exercises 9 and 10, identify and sketch the graph of the polar equation.

9. $r = 1 - \sin \theta$ **10.** $r = 1 + 2 \cos \theta$

SECTION 7.9 · EXERCISES
. . . .

In Exercises 1–6, match the polar equation with the correct graph. [The graphs are labeled (a), (b), (c), (d), (e), and (f).]

1. $r = \dfrac{6}{1 - \cos \theta}$ **2.** $r = \dfrac{2}{2 - \cos \theta}$

3. $r = \dfrac{3}{1 - 2 \sin \theta}$ **4.** $r = \dfrac{2}{1 + \sin \theta}$

5. $r = \dfrac{6}{2 - \sin \theta}$ **6.** $r = \dfrac{2}{2 + 3 \cos \theta}$

(c)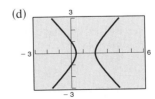

(d)

(a)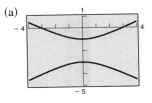

(b)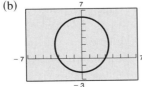

(e)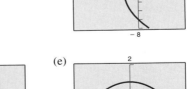

(f)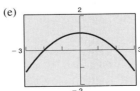

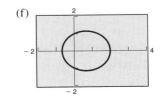

In Exercises 7–18, identify and sketch the graph of the polar equation. Use a graphing utility to confirm your sketch.

7. $r = \dfrac{2}{1 - \cos \theta}$

8. $r = \dfrac{4}{1 + \sin \theta}$

9. $r = \dfrac{5}{1 + \sin \theta}$

10. $r = \dfrac{6}{1 + \cos \theta}$

11. $r = \dfrac{2}{2 - \cos \theta}$

12. $r = \dfrac{3}{3 + \sin \theta}$

13. $r = \dfrac{4}{2 + \sin \theta}$

14. $r = \dfrac{6}{3 - 2 \cos \theta}$

15. $r = \dfrac{3}{2 + 4 \sin \theta}$

16. $r = \dfrac{5}{-1 + 2 \cos \theta}$

17. $r = \dfrac{3}{2 - 6 \cos \theta}$

18. $r = \dfrac{3}{2 + 6 \sin \theta}$

Conic	Vertex or Vertices
25. Parabola	$\left(1, -\dfrac{\pi}{2}\right)$
26. Parabola	$(4, 0)$
27. Parabola	$(5, \pi)$
28. Parabola	$\left(10, \dfrac{\pi}{2}\right)$
29. Ellipse	$(2, 0), (8, \pi)$
30. Ellipse	$\left(2, \dfrac{\pi}{2}\right), \left(4, \dfrac{3\pi}{2}\right)$
31. Ellipse	$(20, 0), (4, \pi)$
32. Hyperbola	$(2, 0), (10, 0)$
33. Hyperbola	$\left(1, \dfrac{3\pi}{2}\right), \left(9, \dfrac{3\pi}{2}\right)$
34. Hyperbola	$\left(4, \dfrac{\pi}{2}\right), \left(-1, \dfrac{3\pi}{2}\right)$

35. Show that the polar equation of the ellipse

$$\frac{x^2}{a^2} + \frac{y^2}{b^2} = 1 \quad \text{is} \quad r^2 = \frac{b^2}{1 - e^2 \cos^2 \theta}.$$

36. Show that the polar equation of the hyperbola

$$\frac{x^2}{a^2} - \frac{y^2}{b^2} = 1 \quad \text{is} \quad r^2 = \frac{-b^2}{1 - e^2 \cos^2 \theta}.$$

In Exercises 19–34, find a polar equation of the conic with its focus at the pole.

Conic	Eccentricity	Directrix
19. Parabola	$e = 1$	$x = -1$
20. Parabola	$e = 1$	$y = -2$
21. Ellipse	$e = \frac{1}{2}$	$y = 1$
22. Ellipse	$e = \frac{3}{4}$	$y = -2$
23. Hyperbola	$e = 2$	$x = 1$
24. Hyperbola	$e = \frac{3}{2}$	$x = -1$

In Exercises 37–42, use the results of Exercises 35 and 36 to write the polar form of the equation of the conic.

37. $\dfrac{x^2}{169} + \dfrac{y^2}{144} = 1$ **38.** $\dfrac{x^2}{25} + \dfrac{y^2}{16} = 1$

39. $\dfrac{x^2}{9} - \dfrac{y^2}{16} = 1$ **40.** $\dfrac{x^2}{36} - \dfrac{y^2}{4} = 1$

41. Hyperbola   One focus: $(5, 0)$
                   Vertices: $(4, 0), (4, \pi)$

42. Ellipse   One focus: $(4, 0)$
                   Vertices: $(5, 0), (5, \pi)$

In Exercises 43 and 44, sketch the graph of the rotated conic.

43. $r = \dfrac{2}{1 - \cos(\theta - \pi/4)}$ (See Exercise 7)

44. $r = \dfrac{4}{1 + \sin(\theta - \pi/3)}$ (See Exercise 8)

45. *Orbits of Planets*   The planets travel in elliptical orbits with the sun as a focus. Assume that the focus is at the pole, the major axis lies on the polar axis, and the length of the major axis is $2a$ (see figure). Show that the polar equation of the orbit is given by

$$r = \frac{(1 - e^2)a}{1 - e \cos \theta},$$

where e is the eccentricity.

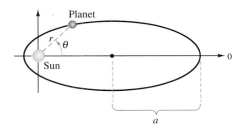

FIGURE FOR 45

46. *Orbits of Planets*   Use the result of Exercise 45 to show that the minimum distance (*perihelion distance*) from the sun to the planet is $r = a(1 - e)$ and the maximum distance (*aphelion distance*) is $r = a(1 + e)$.

Orbits of Planets   In Exercises 47 and 48, use the results of Exercises 45 and 46 to find the polar equation of the planet and the perihelion and aphelion distances.

47. Earth $a = 92.957 \times 10^6$ miles
        $e = 0.0167$

48. Pluto $a = 3.666 \times 10^9$ miles
        $e = 0.2481$

49. *Satellite Tracking*   A satellite in a 100-mile-high circular orbit around the earth has a velocity of approximately 17,500 miles per hour. If this velocity is multiplied by $\sqrt{2}$, then the satellite will have the minimum velocity necessary to escape the earth's gravity and it will follow a parabolic path with the center of the earth as the focus (see figure). Find a polar equation of the parabolic path of the satellite (assume the radius of the earth is 4000 miles).

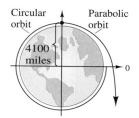

FIGURE FOR 49

50. *Explorer 18*   On November 26, 1963, the United States launched Explorer 18. Its low and high points over the surface of the earth were 119 miles and 122,000 miles, respectively (see figure). The center of the earth is the focus of the orbit. Find the polar equation for the orbit and find the distance between the surface of the earth and the satellite when $\theta = 60°$.

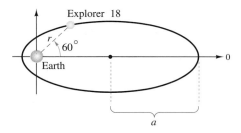

FIGURE FOR 50

CHAPTER 7 · REVIEW EXERCISES

In Exercises 1 and 2, find the slope of the line with the given inclinitation θ.

1. $\theta = 120°$

2. $\theta = 55.8°$

In Exercises 3 and 4, find the inclination θ of the given line.

3. $x + y - 10 = 0$

4. $3x - 2y - 4 = 0$

In Exercises 5 and 6, find the distance between the point and the line.

Point	Line
5. $(1, 2)$	$x - y - 3 = 0$
6. $(0, 4)$	$x + 2y - 2 = 0$

In Exercises 7–18, identify the conic and sketch its graph.

7. $4x - y^2 = 0$ **8.** $8y + x^2 = 0$

9. $x^2 - 6x + 2y + 9 = 0$

10. $y^2 - 12y - 8x + 20 = 0$

11. $x^2 + y^2 - 2x - 4y + 5 = 0$

12. $16x^2 + 16y^2 - 16x + 24y - 3 = 0$

13. $4x^2 + y^2 = 16$ **14.** $2x^2 + 6y^2 = 18$

15. $x^2 + 9y^2 + 10x - 18y + 25 = 0$

16. $4x^2 + y^2 - 16x + 15 = 0$

17. $5y^2 - 4x^2 = 20$

18. $x^2 - 9y^2 + 10x + 18y + 7 = 0$

In Exercises 19 and 20, consider the general form of a conic whose axes are not parallel to either the x-axis or the y-axis (note the xy term). Use the Quadratic Formula to solve for y, and use a graphing utility to graph the equation. Identify the conic.

19. $x^2 - 10xy + y^2 + 1 = 0$

20. $40x^2 + 36xy + 25y^2 - 52 = 0$

In Exercises 21–24, find an equation of the specified parabola.

21. Vertex: $(4, 2)$ **22.** Vertex: $(2, 0)$
    Focus: $(4, 0)$ Focus: $(0, 0)$

23. Vertex: $(0, 2)$ **24.** Vertex: $(2, 2)$
    Passes through $(-1, 0)$ Directrix: $y = 0$
    Horizontal axis

In Exercises 25–28, find an equation of the specified ellipse.

25. Vertices: $(-3, 0)$, $(7, 0)$ **26.** Vertices: $(2, 0)$, $(2, 4)$
    Foci: $(0, 0)$, $(4, 0)$ Foci: $(2, 1)$, $(2, 3)$

27. Vertices: $(0, ±6)$ **28.** Vertices: $(0, 1)$, $(4, 1)$
    Passes through $(2, 2)$ Minor axis endpoints:
                                  $(2, 0)$, $(2, 2)$

In Exercises 29–32, find an equation of the specified hyperbola.

29. Vertices: $(0, ±1)$
    Foci: $(0, ±3)$

30. Vertices: $(2, 2)$, $(-2, 2)$
    Foci: $(4, 2)$, $(-4, 2)$

31. Foci: $(0, 0)$, $(8, 0)$
    Asymptotes: $y = ±2(x - 4)$

32. Foci: $(3, ±2)$
    Asymptotes: $y = ±2(x - 3)$

33. *Satellite Antenna*   A cross section of a large parabolic antenna (see figure) is given by

$$y = \frac{x^2}{200}, \qquad 0 \le x \le 100.$$

The receiving and transmitting equipment is positioned at the focus. Find the coordinates of the focus.

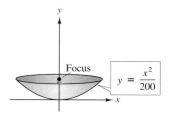

FIGURE FOR 33

34. *Semi-elliptical Archway*   A semi-elliptical archway is to be formed over the entrance to an estate. The arch is to be set on pillars that are 10 feet apart and is to have a height (atop the pillars) of 4 feet (see figure). Where should the foci be placed in order to sketch the semi-elliptical arch?

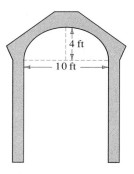

FIGURE FOR 34

In Exercises 35–38, find an equation of the specified tangent line. The tangent line at the point (x_0, y_0) on the conic $\dfrac{x^2}{a^2} \pm \dfrac{y^2}{b^2} = 1$ is given by $\dfrac{x_0 x}{a^2} \pm \dfrac{y_0 y}{b^2} = 1$.

Conic	Point
35. $\dfrac{x^2}{100} + \dfrac{y^2}{25} = 1$	$(-8, 3)$
36. $x^2 + 7y^2 = 16$	$(3, 1)$
37. $\dfrac{x^2}{9} - y^2 = 1$	$\left(6, \sqrt{3}\right)$
38. $\dfrac{x^2}{4} - \dfrac{y^2}{2} = 1$	$(6, 4)$

In Exercises 39–50, sketch the curve represented by the parametric equations. Use a graphing utility to confirm your sketch. If possible, write the corresponding rectangular equation by eliminating the parameter.

39. $x = 2t$
$\quad\,\,\, y = 4t$

40. $x = t^2$
$\quad\,\,\, y = \sqrt{t}$

41. $x = 1 + 4t$
$\quad\,\,\, y = 2 - 3t$

42. $x = t + 4$
$\quad\,\,\, y = t^2$

43. $x = \dfrac{1}{t}$
$\quad\,\,\, y = t^2$

44. $x = \dfrac{1}{t}$
$\quad\,\,\, y = 2t + 3$

45. $x = 6 \cos \theta$
$\quad\,\,\, y = 6 \sin \theta$

46. $x = 3 + 3 \cos \theta$
$\quad\,\,\, y = 2 + 5 \sin \theta$

47. $x = \cos^3 \theta$
$\quad\,\,\, y = 4 \sin^3 \theta$

48. $x = \sec \theta$
$\quad\,\,\, y = \tan \theta$

49. $x = e^t$
$\quad\,\,\, y = e^{-t}$

50. $x = 2\theta - \sin \theta$
$\quad\,\,\, y = 2 - \cos \theta$

In Exercises 51–66, identify and sketch the graph of the polar equation. Use a graphing utility to confirm your sketch.

51. $r = 4$

52. $\theta = \dfrac{\pi}{12}$

53. $r = 4 \sin 2\theta$

54. $r = 2\theta$

55. $r = -2(1 + \cos \theta)$

56. $r = 3 - 4 \cos \theta$

57. $r = 4 - 3 \cos \theta$

58. $r = \cos 5\theta$

59. $r = -3 \cos 3\theta$

60. $r^2 = \cos 2\theta$

61. $r^2 = 4 \sin 2\theta$

62. $r = 3 \csc \theta$

63. $r = \dfrac{3}{\cos\left(\theta - \dfrac{\pi}{4}\right)}$

64. $r = \dfrac{4}{5 - 3 \cos \theta}$

65. $r = \dfrac{2}{1 - \sin \theta}$

66. $r = \dfrac{1}{1 + 2 \sin \theta}$

In Exercises 67–72, convert the polar equation to rectangular form.

67. $r = 3 \cos \theta$

68. $r = 4 \sec\left(\theta - \dfrac{\pi}{3}\right)$

69. $r = \dfrac{2}{1 + \sin \theta}$

70. $r = \dfrac{1}{2 - \cos \theta}$

71. $r^2 = \cos 2\theta$

72. $r = 10$

In Exercises 73 and 74, convert the rectangular equation to polar form.

73. $(x^2 + y^2)^2 = ax^2 y$

74. $x^2 + y^2 - 4x = 0$

In Exercises 75–80, find a polar equation for the line or conic.

75. Circle $\quad$ Center: $\left(5, \dfrac{\pi}{2}\right)$
$\qquad\qquad\quad$ Solution point: $(0, 0)$

76. Line $\qquad$ Solution point: $(0, 0)$
$\qquad\qquad\quad$ Slope: $\quad\quad\quad \sqrt{3}$

77. Parabola $\quad$ Vertex: $\quad (2, \pi)$
$\qquad\qquad\quad$ Focus: $\quad\,\, (0, 0)$

78. Parabola $\quad$ Vertex: $\quad \left(2, \dfrac{\pi}{2}\right)$
$\qquad\qquad\quad$ Focus: $\quad\,\, (0, 0)$

79. Ellipse $\quad$ Vertices: $\quad (5, 0), (1, \pi)$
$\qquad\qquad\quad$ One focus: $(0, 0)$

80. Hyberbola $\quad$ Vertices: $\quad (1, 0), (7, 0)$
$\qquad\qquad\quad$ One focus: $(0, 0)$

81. Find a parametric representation of the ellipse with center at $(-3, 4)$, major axis horizontal and eight units in length, and minor axis six units in length.

82. Find a parametric representation of the hyperbola with vertices $(0, \pm 4)$ and foci at $(0, \pm 5)$.

83. Show that the Cartesian equation of a cycloid is

$$x = a \arccos \dfrac{a - y}{a} \pm \sqrt{2ay - y^2}.$$

84. The *involute of a circle* is described by the endpoint P of a string that is held taut as it is unwound from a spool (see figure). The spool does not rotate. Show that a parametric representation of the involute of a circle is given by

$$x = r(\cos \theta + \theta \sin \theta) \quad \text{and} \quad y = r(\sin \theta - \theta \cos \theta).$$

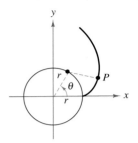

FIGURE FOR 84

85. Sketch the graph of the involute of a circle given by the parametric equations

$$x = 2(\cos \theta + \theta \sin \theta)$$

$$y = 2(\sin \theta - \theta \cos \theta).$$

CHAPTERS 6 AND 7 · CUMULATIVE TEST

Take this test as you would take a test in class. After you are done, check your work against the answers in the back of the book.

1. Sketch a graph of each of the following.

    (a) $f(x) = 6(2^{-x})$ (b) $g(x) = \log_3 x$

2. Write the exponential equation $4^{-2} = \frac{1}{16}$ in logarithmic form.

3. What is the relationship between the functions f and g if $f(x) = 3^x$ and $g(x) = \log_3 x$?

4. Use the properties of logarithms to write the expression

    $2 \ln x - \frac{1}{2} \ln(x + 5)$

    as the logarithm of a single quantity.

5. Solve each of the following, giving your answers accurate to two decimal places.

    (a) $6e^{2x} = 72$ (b) $\log_2 x + \log_2 5 = 6$

6. On the day a grandchild is born, a grandparent deposits $2500 in a fund earning 7.5% compounded continuously. Determine the balance in the account at the time of the grandchild's 25th birthday.

7. A certain truck that cost $25,000 new has a depreciated value of $18,000 after 2 years. Estimate the value of the truck after 4 years using the exponential model $y = Ce^{kt}$.

8. An object weighing W pounds is suspended from the ceiling by a steel spring. The weight is pulled downward from its equilibrium position and released. The resulting motion of the weight is described by the function

    $y = -\frac{2}{3}e^{-t/2} \cos 2t, \qquad t > 0,$

where y is the distance in feet and t is the time in seconds. Use a graphing utility to graph the function. Approximate the first two times that the weight passes the point of equilibrium ($y = 0$).

9. Find the inclination θ of the line $2x - 4y + 3 = 0$.

10. A road rises with an inclination of 6° from the horizontal. Find the slope of the road and the change in elevation after driving 2 miles.

11. Sketch the graph of each of the following equations.

    (a) $6x - y^2 = 0$ (b) $\dfrac{(x - 2)^2}{4} + \dfrac{(y + 1)^2}{9} = 1$

    (c) $y^2 - x^2 = 1$ (d) $xy - 4 = 0$

12. Find an equation of the hyperbola with foci $(0, 0)$ and $(0, 4)$ and asymptotes $y = \pm \frac{1}{2}x + 2$.

13. Find a set of parametric equations of the line passing through the points $(2, -3)$ and $(6, 4)$. (The answer is not unique.)

14. Sketch the curve represented by the parametric equations $x = 3(1 + \cos \theta)$ and $y = 2(1 + \sin \theta)$. Write the corresponding rectangular equation by eliminating the parameter.

15. Convert the polar equation $r = 6 \cos \theta$ to rectangular form and sketch its graph.

16. Use a graphing utility to graph the polar equation $r = 2 \cos 2\theta \sec \theta$. Identify any asymptotes of the graph and write the equation of any asymptotes in rectangular form.

17. Find a polar equation of the parabola with focus at the origin and vertex at $(1, \pi)$.

GRAPHING UTILITIES

Introduction

In Section 1.4, you studied the point-plotting method for sketching the graph of an equation. One of the disadvantages of the point-plotting method is that in order to get a good idea about the shape of a graph, you need to plot *many* points. With only a few points, you could badly misrepresent the graph. For instance, consider the equation

$$y = \frac{1}{30}x(39 - 10x^2 + x^4).$$

Suppose you plotted only five points: $(-3, -3)$, $(-1, -1)$, $(0, 0)$, $(1, 1)$, and $(3, 3)$, as shown in Figure A.1. From these five points, you might assume that the graph of the equation is a straight line. That, however, is not correct. By plotting several more points you can see that the actual graph is not straight at all! (See Figure A.2.)

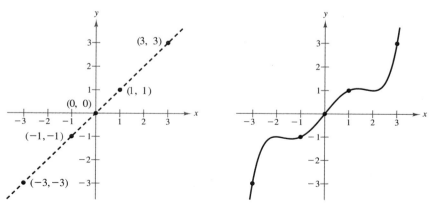

FIGURE A.1   FIGURE A.2

Thus, the point-plotting method leaves us with a dilemma. On the one hand, the method can be very inaccurate if only a few points are plotted. But, on the other hand, it is very time consuming to plot a dozen (or more) points.

Technology can help us solve this dilemma. Plotting several (even several hundred) points in a rectangular coordinate system is something that a graphing utility can do easily.

The point-plotting method is the method used by *all* graphing packages for computers and *all* graphing calculators. Each computer or calculator screen is made up of a grid of hundreds or thousands of small areas called **pixels.** Screens that have many pixels per inch are said to have a higher **resolution** than screens that don't have as many. For instance, the screen shown in Figure A.3(a) has a higher resolution than the screen shown in Figure A.3(b). Note that the "graph" of the line on the first screen looks more like a line than the "graph" on the second screen.

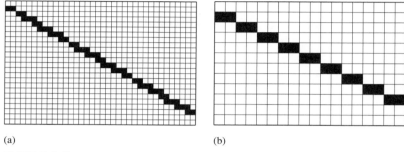

(a)   (b)

FIGURE A.3

Screens on most graphing calculators have 48 pixels per inch. Screens on computer monitors typically have between 32 and 100 pixels per inch.

EXAMPLE 1   **Using Pixels to Sketch a Graph**

Use the grid shown in Figure A.4 to sketch a graph of $y = \frac{1}{2}x^2$. Each pixel on the grid must be either on (shaded black) or off (unshaded).

SOLUTION

To shade the grid, we use the following rule. If a pixel contains a plotted point of the graph, then it will be "on"; otherwise, the pixel will be "off." Using this rule, the graph of the curve looks like that shown in Figure A.5.

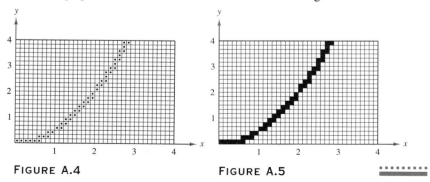

FIGURE A.4   **FIGURE A.5**

Basic Graphing

There are many different types of graphing utilities—graphing calculators and software packages for computers. The procedures used to draw a graph are similar with most of these utilities.

BASIC GRAPHING STEPS FOR A GRAPHING UTILITY

To draw the graph of an equation involving x and y with a graphing utility, use the following steps.

1. Rewrite the equation so that y is isolated on the left side of the equation.
2. Set the boundaries of the viewing rectangle by entering the minimum and maximum x-values and the minimum and maximum y-values.
3. Enter the equation in the form y = (expression involving x). Read the user's guide that accompanies your graphing utility to see how the equation should be entered.
4. Activate the graphing utility.

EXAMPLE 2   Sketching the Graph of an Equation

Sketch the graph of $2y + x^3 = 4x$.

SOLUTION

To begin, solve the given equation for y in terms of x.

$$2y + x^3 = 4x \qquad \text{\textit{Given equation}}$$

$$2y = -x^3 + 4x \qquad \text{\textit{Subtract }} x^3 \text{ \textit{from both sides}}$$

$$y = -\frac{1}{2}x^3 + 2x \qquad \text{\textit{Divide both sides by 2}}$$

Set the viewing rectangle so that $-10 \leq x \leq 10$ and $-10 \leq y \leq 10$. (On some graphing utilities, this is the default setting.) Next, enter the equation into the graphing utility.

$$Y = -X^\wedge 3/2 + 2 * X$$

Finally, activate the graphing utility. The display screen should look like that shown in Figure A.6.

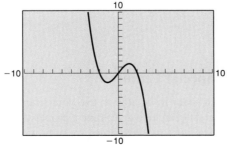

FIGURE A.6

In Figure A.6, notice that the calculator screen does not label the tick marks on the *x*-axis or the *y*-axis. To see what the tick marks represent, check the values in the utility's "range."

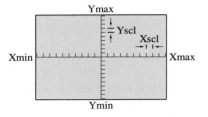

FIGURE A.7

Range

Xmin = −10	*The minimum x-value is −10.*
Xmax = 10	*The maximum x-value is 10.*
Xscl = 1	*The x-scale is 1 unit per tick mark.*
Ymin = −10	*The minimum y-value is −10.*
Ymax = 10	*The maximum y-value is 10.*
Yscl = 1	*The y-scale is 1 unit per tick mark.*
Xres = 1	*The x-resolution is 1 plotted point per 1 pixel.*

These settings are summarized visually in Figure A.7.

EXAMPLE 3   **Graphing an Equation Involving Absolute Value**

Sketch the graph of $y = |x - 3|$.

SOLUTION

This equation is already written so that *y* is isolated on the left side of the equation, so you can enter the equation as follows.

$$Y = \text{abs}(X - 3)$$

After activating the graphing utility, its screen should look like the one shown in Figure A.8.

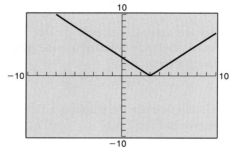

FIGURE A.8

Special Features

In order to be able to use your graphing calculator to its best advantage, you must be able to determine a proper viewing rectangle and use the zoom feature. The next two examples show how this is done.

EXAMPLE 4   Determining a Viewing Rectangle

Sketch the graph of $y = x^2 + 12$.

SOLUTION

Begin as usual by entering the equation.

$$Y = x^{\wedge}2 + 12$$

Activate the graphing utility. If you used a viewing rectangle in which $-10 \leq x \leq 10$ and $-10 \leq y \leq 10$, then no part of the graph will appear on the screen, as shown in Figure A.9(a). The reason for this is that the lowest point on the graph of $y = x^2 + 12$ occurs at the point (0, 12). With the viewing rectangle in Figure A.9(a), the largest y-value is 10. In other words, none of the graph is visible on a screen whose y-values range between -10 and 10.

To be able to see the graph, change Ymax $= 10$ to Ymax $= 30$, Yscl $= 1$ to Yscl $= 5$. Now activate the graphing utility and you will obtain the graph shown in Figure A.9(b). On this graph, note that each tick mark on the y-axis represents 5 units because you changed the y-scale to 5. Also note that the highest point on the y-axis is now 30 because you changed the maximum value of y to 30.

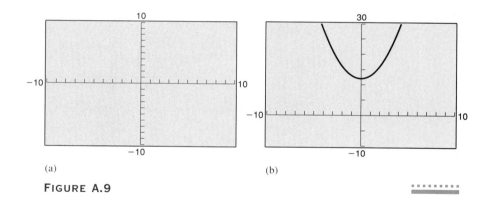

(a)   (b)

FIGURE A.9

EXAMPLE 5   Using the Zoom Feature

Sketch the graph of $y = x^3 - x^2 - x$. How many x-intercepts does this graph have?

SOLUTION

Begin by drawing the graph on a "standard" viewing rectangle as shown in Figure A.10(a). From the display screen, it is clear that the graph has at least one intercept (just to the left of $x = 2$), but it is difficult to determine whether

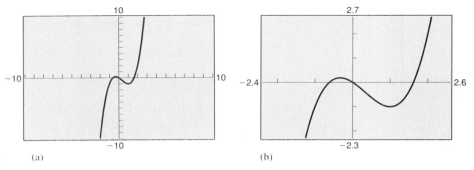

FIGURE A.10   (a)   (b)

the graph has other intercepts. To obtain a better view of the graph near $x = -1$, you can use the zoom feature of the graphing utility. The redrawn screen is shown in Figure A.10(b). From this screen you can tell that the graph has three x-intercepts whose x-coordinates are approximately -0.6, 0, and 1.6.

EXAMPLE 6   **Sketching More Than One Graph on the Same Screen**

Sketch the graphs of $y = -\sqrt{36 - x^2}$ and $y = \sqrt{36 - x^2}$ on the same screen.

SOLUTION

To begin, enter both equations in the graphing utility.

$$Y = \sqrt{36 - X^{\wedge}2)}$$
$$Y = -\sqrt{(36 - X^{\wedge}2)}$$

Then, activate the graphing utility to obtain the graph shown in Figure A.11(a). Notice that the graph should be the upper and lower parts of the circle given by $x^2 + y^2 = 6^2$. The reason it doesn't look like a circle is that, with the standard settings, the tick marks on the x-axis are farther apart than the tick marks on the y-axis. To correct this, change the viewing rectangle so that $-15 \leq x \leq 15$. The redrawn screen is shown in Figure A.11(b). Notice that in this screen the graph appears to be more circular.

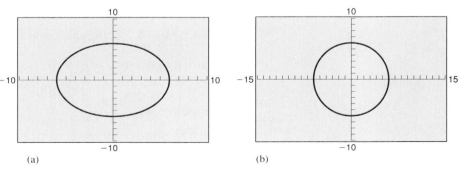

(a)   (b)

FIGURE A.11

DISCUSSION PROBLEM

A MISLEADING GRAPH

Sketch the graph of $y = x^2 - 12x$, using $-10 \le x \le 10$ and $-10 \le y \le 10$. The graph appears to be a straight line, as shown in the figure. However, this is misleading because the screen doesn't show an important portion of the graph. Can you find a range setting that reveals a better view of this graph?

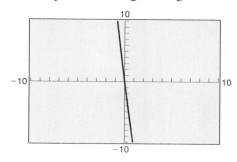

WARM-UP

In Exercises 1–10, solve for y in terms of x.

1. $3x + y = 4$ **2.** $x - y = 0$

3. $2x + 3y = 2$ **4.** $4x - 5y = -2$

5. $3x + 4y - 5 = 0$ **6.** $-2x - 3y + 6 = 0$

7. $x^2 + y - 4 = 0$ **8.** $-2x^2 + 3y + 2 = 0$

9. $x^2 + y^2 = 4$ **10.** $x^2 - y^2 = 9$

APPENDIX A · EXERCISES

In Exercises 1–20, use a graphing utility to sketch the graph of the equation. Use a setting on each graph of $-10 \le x \le 10$ and $-10 \le y \le 10$.

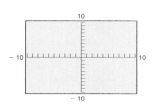

```
RANGE
Xmin=-10
Xmax=10
Xscl=1
Ymin=-10
Ymax=10
Yscl=1
Xres=1
```

1. $y = x - 5$ **2.** $y = -x + 4$

3. $y = -\frac{1}{2}x + 3$ **4.** $y = \frac{2}{3}x + 1$

5. $2x - 3y = 4$ **6.** $x + 2y = 3$

7. $y = \frac{1}{2}x^2 - 1$ **8.** $y = -x^2 + 6$

9. $y = x^2 - 4x - 5$ **10.** $y = x^2 - 3x + 2$

11. $y = -x^2 + 2x + 1$ **12.** $y = -x^2 + 4x - 1$

13. $2y = x^2 + 2x - 3$ **14.** $3y = -x^2 - 4x + 5$

15. $y = |x + 5|$ **16.** $y = \frac{1}{2}|x - 6|$

17. $y = \sqrt{x^2 + 1}$ **18.** $y = 2\sqrt{x^2 + 2} - 4$

19. $y = \frac{1}{5}(-x^3 + 16x)$ **20.** $y = \frac{1}{8}(x^3 + 8x^2)$

In Exercises 21–30, use a graphing utility to match the equation with its graph. [The graphs are labeled (a), (b), (c), (d), (e), (f), (g), (h), (i), and (j).]

21. $y = x$ **22.** $y = -x$

23. $y = x^2$ **24.** $y = -x^2$

25. $y = x^3$

26. $y = -x^3$

27. $y = |x|$

28. $y = -|x|$

29. $y = \sqrt{x}$

30. $y = -\sqrt{x}$

(a)
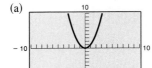

(b)
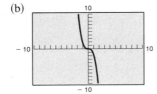

(c)
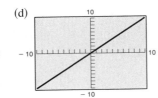

(d)
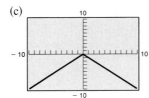

(e)
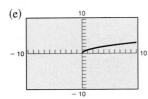

(f)
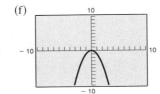

(g)
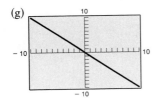

(h)
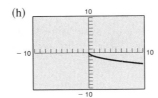

(i)
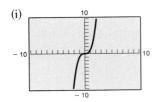

(j)
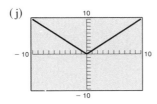

In Exercises 31–34, use a graphing utility to sketch the graph of the equation. Use the indicated setting.

31. $y = -2x^2 + 12x + 14$

32. $y = -x^2 + 5x + 6$

RANGE
Xmin=-5
Xmax=10
Xscl=1
Ymin=-5
Ymax=35
Yscl=5
Xres=1

RANGE
Xmin=-8
Xmax=4
Xscl=1
Ymin=-5
Ymax=15
Yscl=5
Xres=1

33. $y = x^3 + 6x^2$

34. $y = -x^3 + 16x$

RANGE
Xmin=-10
Xmax=5
Xscl=1
Ymin=-4
Ymax=36
Yscl=3
Xres=1

RANGE
Xmin=-6
Xmax=6
Xscl=1
Ymin=-25
Ymax=25
Yscl=5
Xres=1

In Exercises 35–38, find a setting on a graphing utility so that the graph of the equation agrees with the graph shown.

35. $y = -x^2 - 4x + 20$

36. $y = x^2 + 12x - 8$

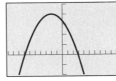

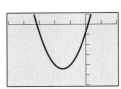

37. $y = -x^3 + x^2 + 2x$

38. $y = x^3 + 3x^2 - 2x$

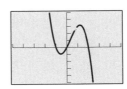

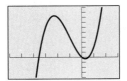

In Exercises 39–42, use a graphing utility to find the number of x-intercepts of the equation.

39. $y = \frac{1}{8}(4x^2 - 32x + 65)$

40. $y = \frac{1}{4}(-4x^2 + 16x - 15)$

41. $y = 4x^3 - 20x^2 - 4x + 61$

42. $y = \frac{1}{4}(2x^3 + 6x^2 - 4x + 1)$

In Exercises 43–46, use a graphing utility to sketch the graphs of the equations on the same screen. Using a "square setting," what geometrical shape is bounded by the graphs?

43. $y = |x| - 4$
$y = -|x| + 4$

44. $y = x + |x| - 4$
$y = x - |x| + 4$

45. $y = -\sqrt{25 - x^2}$
$y = \sqrt{25 - x^2}$

46. $y = 6$
$y = -\sqrt{3x} - 4$
$y = \sqrt{3x} - 4$

Ever Been Married?   In Exercises 47–50, use the following models, which relate ages to the percentages of American males and females who have never been married.

$$y = \frac{0.36 - 0.0056x}{1 - 0.0817x + 0.00226x^2}, \quad \begin{array}{l} \text{Males} \\ 20 \leq x \leq 50 \end{array}$$

$$y = \frac{100}{8.944 - 0.886x + 0.249x^2}, \quad \begin{array}{l} \text{Females} \\ 20 \leq x \leq 50 \end{array}$$

In these models, y is the percent of the population (in decimal form) who have never been married and x is the age of the person. (*Source:* U.S. Bureau of the Census)

47. Use a graphing utility to sketch the graph of both equations giving the percentages of American males and females who have never been married. Use the following range settings.

```
RANGE
Xmin=20
Xmax=50
Xscl=5
Ymin=0
Ymax=1
Yscl=0.1
Xres=1
```

48. Write a short paragraph describing the relationship between the two graphs that were plotted in Exercise 47.

49. Suppose an American male is chosen at random from the population. If the person is 25 years old, what is the probability that he has never been married?

50. Suppose an American female is chosen at random from the population. If the person is 25 years old, what is the probability that she has never been married?

Earnings and Dividends   In Exercises 51–54, use the following model, which approximates the relationship between dividends per share and earnings per share for the Pall Corporation between 1982 and 1989.

$$y = -0.166 + 0.502x - 0.0953x^2, \quad 0.25 \leq x \leq 2$$

In this model, y is the dividends per share (in dollars) and x is the earnings per share (in dollars). (*Source:* Standard ASE Stock Reports)

51. Use a graphing utility to sketch the graph of the model that gives the dividend per share in terms of the earnings per share. Use the following range settings.

```
RANGE
Xmin=0
Xmax=2
Xscl=0.25
Ymin=0
Ymax=0.5
Yscl=0.1
Xres=1
```

52. According to the given model, what size dividend would the Pall Corporation pay if the earnings per share were $1.30?

53. Use a trace feature on your graphing utility to estimate the earnings per share that would produce a dividend per share of $0.25. The choices are labeled (a), (b), (c), and (d). (Find the y-value that is as close to 0.25 as possible. The x-value that is displayed will then be the approximate earnings per share that would produce a dividend per share of $0.25.)
(a) $1.00   (b) $1.03   (c) $1.06   (d) 1.09

54. The **Payout Ratio** for a stock is the ratio of the dividend per share to earnings per share. Use the model to find the payout ratio for an earnings per share of (a) $0.75, (b) 1.00, and (c) $1.25.

APPENDIX B

PROGRAMS

Programs for the *TI-81* graphing calculator are given in several sections in the text. This appendix contains translations of these text programs for several other graphics calculators from Texas Instruments, Casio, Sharp, and Hewlett Packard®. Similar programs can be written for other brands and models of graphic calculators.

Enter a program in you calculator, then refer to the text discussion and apply the program as appropriate. Section references are provided to help you locate the text discussion of the programs and their use.

To illustrate the power and versatility of programmable calculators, a variety of types of programs are presented, including a simulation program (A Graph Reflecting Program) and a tutorial program (A Program for Practice).

Evaluating a Function (Section 1.6)

This program, shown in the Technology note on page 53, can be used to evaluate a function in one variable at several values of the variable.

Note: On the *TI-83* and the *TI-82*, the "Lbl" and "Goto" commands may be entered through the "CTL" menu accessed by pressing the PRGM key. The "Disp" and "Input" commands may be entered through the "I/O" menu accessed by pressing the PRGM key. On the *TI-83*, the symbol "Y1" may be entered through the "Y-VARS" and "Function" menus accessed by pressing the VARS key. On the *TI-82*, the symbol "Y1" may be entered through the "Function" menu accessed by pressing the Y-VARS key. Keystroke sequences required for similar commands on other calculators will vary. Consult the user manual for your calculator.

TI-80
```
PROGRAM:EVALUAT
:LBL A
:INPUT "ENTER X",X
:DISP Y1
:GOTO A
```

To use this program, enter an expression in Y1. Expressions may also be evaluated directly on the *TI-80*'s home screen.

TI-81
```
Prgm1:EVALUATE
:Lbl 1
:Disp "ENTER X"
:Input X
:Disp Y1
:Goto 1
```

To use this program, enter an expression in Y1. Expressions may also be evaluated directly on the *TI-81*'s home screen.

TI-83
TI-82
```
PROGRAM:EVALUATE
:Lbl A
:Input "ENTER X",X
:Disp Y1
:Goto A
```

To use this program, enter an expression in Y1. Expressions may also be evaluated directly on the *TI-83*'s or *TI-82*'s home screen.

TI-85
```
PROGRAM:EVALUATE
:Lbl A
:Input "Enter x",x
:Disp y1
:Goto A
```

To use this program, enter an expression in y1.

TI-92

evaluate( )
Prgm
Lbl one
Input "enter x",x
Disp y1(x)
Goto one
EndPrgm

To use this program, enter an expression in y1. Expressions may also be evaluated on the *TI-92*'s home screen.

Casio fx-7700G

EVALUATE
Lbl 1
"X="?→X
"F(X)=":f₁ ◣
Goto 1

To use this program, enter an expression in f₁.

Casio fx-7700GE
Casio fx-9700GE
Casio CFX-9800G

EVALUATE ↵
Lbl 1 ↵
"X="?→X ↵
"F(X)=":f₁ ◣
Goto 1

To use this program, enter an expression in f₁.

Sharp EL-9200C
Sharp EL-9300C

evaluate
————————REAL
Goto top
Label equation
Y=f(X)
Return
Label top
Input X
Gosub equation
Print Y
Goto top
End

To use this program, replace f(X) with your expression in X.

HP-38G

Use the Solve aplet to evaluate an expression.
1. Press ⌐LIB⌐. Highlight the Solve aplet. Press {{START}}.
2. Set your expression equal to *y*, enter the equation (*y = your expression*) in E1 and press {{OK}}. The equation should be checked.
3. Press ⌐NUM⌐.
4. Highlight the *x*-variable field. Enter a value for *x* and press {{OK}}.
5. Highlight the *y*-variable field and press {{SOLVE}}. The value of the expression will appear in the *y*-variable field.
6. Repeat steps 4 and 5 to evaluate the expression for other values of *x*.

Reflections and Shifts Program (Section 1.8)

This program, referenced in Example 5, A Program for Practice, on page 78, will sketch a graph of the function $y = R(x + H)^2 + V$, where $R = \pm 1$, H is an integer between -6 and 6, and V is an integer between -3 and 3. This program gives you practice working with reflections, horizontal shifts, and vertical shifts.

Note: On the *TI-83* and the *TI-82*, the "int" and "rand" commands may be entered through the "NUM" and "PRB" menus, respectively, accessed by pressing the MATH key. The "=" and "<" symbols may be entered through the "TEST" menu accessed by pressing the TEST key. Other commands, such as "If," "Then," "Else," and "End," may be entered through the "CTL" menu accessed by pressing the PRGM key. The commands "Xmin," "Xmax," "Xscl," "Ymin," "Ymax," and "Yscl" may be entered through the "Window" menu accessed by pressing the VARS key. The commands "DispGraph" and "Pause" may be entered through the "I/O" and "CTL" menus, respectively, accessed by pressing the PRGM key. For additional keystroke instructions, see previous programs in this appendix. Keystroke sequences for similar commands on other calculators will vary. Consult the user manual for your calculator.

TI-80

```
PROGRAM:PARABOL
:-6+INT (12RAND)→H
:-3+INT (6RAND)→V
:RAND→R
:IF R <.5
:THEN
:-1→R
:ELSE
:1→R
:END
:"R(X+H)²+V"→Y1
:-9→XMIN
:9→XMAX
:1→XSCL
:-6→YMIN
:6→YMAX
:1→YSCL
:DISPGRAPH
:PAUSE
:DISP "Y=R(X+H)²+V²"
:DISP "R=",R
:DISP "H=",H
:DISP "V=",V
:PAUSE
```

Press ENTER after viewing the graph to display the values of the integers.

TI-81

```
Prgm2:PARABOLA
:Rand→H
:-6+Int (12H)→H
:Rand→V
:-3+Int (6V)→V
:Rand→R
:If R <.5
:-1→R
:If R >.49
:1→R
:"R(X+H)²+V"→Y1
:-9→Xmin
:9→Xmax
:1→Xscl
:-6→Ymin
:6→Ymax
:1→Yscl
:DispGraph
:Pause
:Disp "Y=R(X+H)²+V"
:Disp "R="
:Disp R
:Disp "H="
:Disp H
:Disp "V="
:Disp V
:End
```

Press ENTER after viewing the graph to display the values of the integers.

TI-83
TI-82

PROGRAM:PARABOLA
:-6+int (12rand)→H
:-3+int (6rand)→V
:rand→R
:If R < .5
:Then
:-1→R
:Else
:1→R
:End
:"R(X+H)^2+V"→Y$_1$
:-9→Xmin
:9→Xmax
:1→Xscl
:-6→Ymin
:6→Ymax
:1→Yscl
:DispGraph
:Pause
:Disp "Y=R(X+H)^2+V"
:Disp "R=",R
:Disp "H=",H
:Disp "V=",V
:Pause

Press │ENTER│ after viewing the graph to display the values of the integers.

TI-85

PROGRAM:PARABOLA
:rand→H
:-6+int (12H)→H
:rand→V
:-3+int (6V)→V
:rand→R
:If R < .5
:-1→R
:If R > .49
:1→R
:y1=R(x+H)^2+V
:-9→xMin
:9→xMax
:1→xScl
:-6→yMin
:6→yMax
:1→yScl
:DispG
:Pause
:Disp "Y=R(X+H)^2+V"
:Disp "R=",R
:Disp "H=",H
:Disp "V=",V
:Pause

Press │ENTER│ after viewing the graph to display the values of the integers.

TI-92

```
Parabola( )
Prgm
ClrHome
ClrIO
setMode("Split Screen",
  "Left-Right")
setMode("Split 1 App","Home")
setMode("Split 2 App","Graph")
-6+int (12rand( ))→h
-3+int (6rand( ))→v
rand( )→r
If r < .5 Then
    -1→r
        Else
    1→r
EndIf
r*(x+h)^2+v→y1(x)
-9→xmin
9→xmax
1→xscl
-6→ymin
6→ymax
1→yscl
DispG
Disp "y1(x)=r(x+h)^2+v"
Output 20,1, "r=":Output 20,11,r
Output 40,1, "h=":Output 40,11,h
Output 60,1, "v=":Output 60,11,v
Pause
setMode("Split Screen","Full")
EndPrgm
```

Casio fx-7700G

```
PARABOLA
-6+INT (12Ran#)→H
-3+INT (6Ran#)→V
-1→R:Ran#<0.5 ⇒1→R
Range -9,9,1,-6,6,1
Graph Y=R(X+H)²+V ◢
"Y=R(X+H)²+V"
"R=":R ◢
"H=":H ◢
"V=":V
```

Press $\boxed{\text{EXE}}$ after viewing the graph to display the values of the integers.

Casio fx-7700GE
Casio fx-9700GE
Casio CFX-9800G

```
PARABOLA↵
-6+Int (12Ran#)→H↵
-3+Int (6Ran#)→V↵
Ran#→R↵
R< .5⇒-1→R↵
R≥ .5⇒1→R↵
Range -9,9,1,-6,6,1↵
Graph Y=R(X+H)²+V ◢
"Y=R(X+H)²+V"↵
"R=":R ◢
"H=":H ◢
"V=":V
```

Press $\boxed{\text{EXE}}$ after viewing the graph to display the values of the integers.

Sharp EL-9200C
Sharp EL-9300C

```
parabola
————————REAL
h=int (random*12) -6
v=int (random*6) -3
s=(random*2) -1
r=s/abs s
Range -9,9,1,-6,6,1
Graph r(X+h)²+v
Wait
Print "y=r(X+h)²+v
Print r
Print h
Print v
End
```

Press $\boxed{\text{ENTER}}$ after viewing the graph to display the values of the integers.

HP-38G

PARABOLA

PARABOLA PROGRAM
-6+INT(12RANDOM)▶H:
-3+INT(6RANDOM)▶V:
RANDOM ▶R:
IF R>.5
   THEN -1▶R:
   ELSE 1▶R:
END:
'R*(X+H)²+V'▶F1(X):
CHECK 1:

PARANS PROGRAM
ERASE:
DISP 2;"Y=R(X+H)²+V":
DISP 3;"R="R:
DISP 4;"H="H:
DISP 5;"V="V:
FREEZE:

PARABOLA.SV PROGRAM
SETVIEWS "RUN
PARABOLA";PARABOLA;1;
"ANSWER";PARANS;1;
" ";PARABOLA.SV;0:

1. Press ⃞LIB⃞. Highlight the
   Function aplet. Press
   {{SAVE}}. Enter the name
   PARABOLA for the new aplet
   and press {{OK}}.
2. Press ■ [SETUP-PLOT] and set
   XRNG: from −12 to 12,
   YRNG: from −6 to 6, and
   XTICK: and YTICK: to 1.
3. Enter the 3 programs
   PARABOLA, PARANS,
   PARABOLA.SV.
4. Run the program
   PARABOLA.SV.
5. Enter the PARABOLA aplet.
6. Press ■ [VIEWS]. Highlight
   RUN PARABOLA and press
   {{OK}}.

7. After viewing the graph press ■
   [VIEWS]. Highlight ANSWER
   and press {{OK}} to see the
   values of the integers.
8. Press {{OK}} to return to the
   graph.
9. Repeat steps 6, 7, and 8 for a
   new parabola.

Graph Reflection Program (Section 1.10)

This program, referenced in Example 5, A Graph Reflecting Program, on page 95, will graph a function f and its reflection in the line $y = x$.

Note: On the *TI-83* and the *TI-82*, the "While" command may be entered through the "CTL" menu accessed by pressing the PRGM key. The "Pt-On(" command may be entered through the "POINTS" menu accessed by pressing the DRAW key. For additional keystroke instructions, see previous programs in this appendix. Keystroke sequences required for similar commands on other calculators will vary. Consult the user manual for your calculator.

TI-80

PROGRAM:REFLECT
:47XMIN/63→YMIN
:47XMAX/63→YMAX
:XSCL→YSCL
:"X"→Y2
:DISPGRAPH
:(XMAX−XMIN)/62→I
:XMIN→X
:LBL A
:PT-ON(Y1,X)
:X+I→X
:If X>XMAX
:STOP
:GOTO A

To use this program, enter the function in Y1 and set a viewing rectangle.

TI-81

Prgm3:REFLECT
:2Xmin/3→Ymin
:2Xmax/3→Ymax
:Xscl→Yscl
:"X"→Y2
:DispGraph
:(Xmax−Xmin)/95→I
:Xmin→X
:Lbl 1
:Pt-On(Y1,X)
:X+I→X
:If X>Xmax
:End
:Goto 1

To use this program, enter the function in Y1 and set a viewing rectangle.

TI-83
TI-82

PROGRAM:REFLECT
:63Xmin/95→Ymin
:63Xmax/95→Ymax
:Xscl→Yscl
:"X"→Y2
:DispGraph
:(Xmax−Xmin)/94→I
:Xmin→X
:While X≤Xmax
:Pt-On(Y1,X)
:X+I→X
:End

To use this program, enter the function in Y1 and set a viewing rectangle.

TI-85

PROGRAM:REFLECT
:63*xMin/127→yMin
:63*xMax/127→yMax
:xScl→yScl
:y2=x
:DispG
:(xMax−xMin)/126→I
:xMin→x
:Lbl A
:PtOn(y1,x)
:x+I→x
:If x>xMax
:Stop
:Goto A

To use this program, enter the function in y1 and set a viewing rectangle.

TI-92

Prgm
103xmin/239→ymin
103xmax/239→ymax
xscl→yscl
x→y2(x)
DispG
(xmax−xmin)/238→n
xmin→x
While x<xmax
    PtOn y1(x),x
    x+n→x
EndWhile
EndPrgm

To use this program, enter a function in y1 and set an appropriate viewing window.

Casio fx-7700G

REFLECTION
"GRAPH -A TO A"
"A="?→A
Range -A,A,1,-2A÷3,2A÷3,1
Graph Y=f1
-A→B
Lbl 1
B→X
Plot f1,B
B+A÷32→B
B≤A⇒Goto1 :Graph Y=X

To use this program, enter the function in f1.

Casio fx-7700GE

REFLECTION
"GRAPH -A TO A"↵
"A="?→A↵
Range -A,A,1,-2A÷3,2A÷3,1↵
Graph Y=f₁↵
-A→B↵
Lbl 1↵
B→X↵
Plot f₁,B↵
B+A÷32→B↵
B≤A⇒Goto1:Graph Y=X

To use this program, enter the function in f₁.

Casio fx-9700GE

REFLECTION↵
63Xmin÷127→A↵
63Xmax÷127→B↵
Xscl→C↵
Range , , , A, B, C↵
(Xmax−Xmin)÷126→I↵
Xmax→M↵
Xmin→D↵
Graph Y=f₁↵
Lbl 1↵
D→X↵
Plot f₁,D↵
D+I→D↵
D≤M⇒Goto 1:Graph Y=X

To use this program, enter a function in f₁ and set a viewing rectangle.

Casio CFX-9800G

REFLECTION↵
63Xmin÷95→A↵
63Xmax÷95→B↵
Xscl→C↵
Range , , , A, B, C↵
(Xmax−Xmin)÷94→I↵
Xmax→M↵
Xmin→D↵
Graph Y=f₁↵

Casio CFX-9800G

(Continued)

Lbl 1↵
D→X↵
Plot f₁,D↵
D+I→D↵
D≤M⇒Goto 1:Graph Y=X

To use this program, enter a function in f₁ and set a viewing rectangle.

Sharp EL-9200C
Sharp EL-9300C

reflection
————REAL
Goto top
Label equation
Y=f(X)
Return
Label rng
xmin=-10
xmax=10
xstp=(xmax−xmin)/10
ymin=2xmin/3
ymax=2xmax/3
ystp=xstp
Range xmin,xmax,xstp,ymin,
    ymax,ystp
Return
Label top
Gosub rng
Graph X
step=(xmax−xmin)/(94*2)
X=xmin
Label 1
Gosub equation
Plot X,Y
Plot Y, X
X=X+step
If X<=xmax Goto 1
End

To use this program, replace f(X) with your expression in X.

Angles and Their Measures (Section 2.1)

This program, found in the Discussion Problem on page 114, An Angle-Drawing Program, can be used to draw several different angles in either radian or degree mode.

TI-80	TI-81
PROGRAM:ANGDRAW	PrgmD:ANGDRAW
:CLRDRAW	:ClrDraw
:CLRHOME	:ClrHome
:FNOFF	:All-Off
:RADIAN	:Rad
:PARAM	:Param
:DISP "ENTER MODE"	:Disp "ENTER MODE"
:DISP "0 RADIAN"	:Disp "0 RADIAN"
:DISP "1 DEGREE"	:Disp "1 DEGREE"
:INPUT M	:Input M
:INPUT "ENTER ANGLE",T	:Input "ENTER ANGLE"
:IF M=1	:Input T
:πT/180$\rightarrow$T	:If M=1
:-1.5$\rightarrow$XMIN	:πT/180$\rightarrow$T
:1.5$\rightarrow$XMAX	:-1.5$\rightarrow$Xmin
:1$\rightarrow$XSCL	:1.5$\rightarrow$Xmax
:-1$\rightarrow$YMIN	:1$\rightarrow$Xscl
:1$\rightarrow$YMAX	:-1$\rightarrow$Ymin
:1$\rightarrow$YSCL	:1$\rightarrow$Ymax
:0$\rightarrow$TMIN	:1$\rightarrow$Yscl
:ABS T$\rightarrow$TMAX	:0$\rightarrow$Tmin
:.15$\rightarrow$TSTEP	:abs T$\rightarrow$Tmax
:COS T$\rightarrow$A	:.15$\rightarrow$Tstep
:SIN T$\rightarrow$B	:cos T$\rightarrow$A
:1$\rightarrow$S	:sin T$\rightarrow$B
:IF T<0	:1$\rightarrow$S
:-1$\rightarrow$S	:If T<0
:"(.25+.04T)COS T"$\rightarrow$X1$_T$	:-1$\rightarrow$S
:"S(.25+.04T)SIN T"$\rightarrow$Y1$_T$	:"(.25+.04T)cos T"$\rightarrow$X1$_T$
:DISPGRAPH	:"S(.25+.04T)sin T"$\rightarrow$Y1$_T$
:LINE(0,0,A,B)	:DispGraph
:PAUSE	:Line(0,0,A,B)
:FUNC	:Pause
:STOP	:Function
	:End

TI-83
TI-82
PROGRAM:ANGDRAW
```
:ClrDraw
:ClrHome
:FnOff
:Radian
:Param
:Disp "ENTER MODE"
:Disp "0 RADIAN"
:Disp "1 DEGREE"
:Input M
:Input "ENTER ANGLE",T
:If M=1
:πT/180→T
:-1.5→Xmin
:1.5→Xmax
:1→Xscl
:-1→Ymin
:1→Ymax
:1→Yscl
:0→Tmin
:abs (T)→Tmax
:.15→Tstep
:cos (T)→A
:sin (T)→B
:1→S
:If T<0
:-1→S
:"(.25+.04T)cos T"→X₁ₜ
:"S(.25+.04T)sin T"→Y₁ₜ
:DispGraph
:Line(0,0,A,B)
:Pause
:Func
:Stop
```

TI-85
PROGRAM:ANGDRAW
```
:ClDrw
:ClLCD
:FnOff
:Radian
:Param
:Disp "Enter mode"
:Disp "0 radian"
:Disp "1 degree"
:Input M
:Input "enter angle",T
:If M==1
:πT/180→T
:-1.5→xMin
:1.5→xMax
:1→xScl
:-1→yMin
:1→yMax
:1→yScl
:0→tMin
:abs T→tMax
:.15→tStep
:cos T→A
:sin T→B
:1→S
:If T<0
:-1→S
:xt1=(.25+.04T)cos t
:yt1=S(.25+.04T)sin t
:DispG
:Line(0,0,A,B)
:Pause
:Func
```

TI-92

```
angdraw()
:Prgm
:ClrHome
:ClrGraph
:setMode("Angle","RADIAN")
:setMode("Graph",
    "PARAMETRIC")
:setMode("Split Screen",
    "LEFT-RIGHT")
:setMode("Split 1 App","Home")
:setMode("Split 2 App","Graph")
:Disp "enter mode"
:Disp "0 radian"
:Disp "1 degree"
:Input m
:Input "enter angle",t
:If m=1
:π*t/180→t
:-1.5→xmin
:1.5→xmax
:1→xscl
:-1→ymin
:1→ymax
:1→yscl
:0→tmin
:abs(t)→tmax
:.15→tstep
:cos (t)→a
:sin (t)→b
:1→s
:If t<0
:-1→s
:Graph(.25+.04*t)*cos (t),
    s*(.25+.04*t)*sin (t),t
:Line 0,0,a,b
:Pause
:setMode("Graph","FUNCTION")
:setMode("Split Screen","FULL")
:EndPrgm
```

Casio fx-7700G

```
ANGDRAW
"ENTER MODE"
"0 RADIAN"
"1 DEGREE"
?→M
"ENTER ANGLE"?→T
M=1⇒πT÷180→T
Rad
Cls
Range -1.5,1.5,1,-1,1,1,0,
    Abs T,.15
cos T→A
sin T→B
1→S
T<0⇒-1→S
Graph(X,Y)=((.25+.04T)cos T,
    S(.25+.04T)sin T)
Plot 0,0
Plot A,B
Line
```

Press MODE SHIFT × to change to
parametric mode before writing
this program.

Casio fx-7700GE
Casio fx-9700GE
Casio CFX-9800G

ANGDRAW↵
"ENTER MODE"↵
"0 RADIAN"↵
"1 DEGREE"↵
?→M ↵
"ENTER ANGLE"↵
?→T↵
M=1⟹πT÷180→T↵
Rad↵
Cls↵
Range -1.5,1.5,1,-1,1,1,0,
    Abs T,.15↵
cos T→A↵
sin T→B↵
1→S↵
T<0⟹-1→S↵
Graph(X,Y)=((.25+.04T)cos T,
    S(.25+.04T)sin T)↵
Plot 0,0↵
Plot A,B↵
Line

Use SET UP to change the GRAPH TYPE to parametric before writing this program.

SHARP EL-9200C
SHARP EL-9300C

angdraw
——————————REAL
Print "enter mode"
Print "0 radian"
Print "1 degree"
Input mode
Input angle
If mode=0 Goto 1
angle=angle∗π/180
Label 1
Range -2.25,2.25,1.5,-1.5,1.5,
    1.5
s=angle/abs angle
loops=ipart (s∗angle/(2π))+1
θ=0
ostep=angle/(30∗loops)
r=1/loops
rstep=(1−r)/(30∗loops)
xo=r
yo=0
n=0
Label top
r=r+rstep
θ=θ+ostep
x=r∗cos (θ)
y=r∗sin (θ)
Line xo,yo,x,y
xo=x
yo=y
n=n+1
If n<30∗loops Goto top
Line 0,0,1.5x,1.5y

Use SET UP to set the mode to radians before running the program.

HP-38G

ANGDRAW PROGRAM
INPUT M;"ENTER MODE";
   "0 OR 1";"0 RADIAN,
   1 DEGREE";1:
INPUT T;"ENTER ANGLE";
   "ANGLE?";" ";1:
IF M==1
   THEN T*π/180▶T:
END:
T/ABS(T)▶S:
0▶K:
INT(ST/(2π)+1▶L:
T/(30L)▶P:
1/L▶R:
(1-R/(30L)▶Q:
R▶A:
0▶B:
0▶N:
ERASE:
LINE -2.6;0;2.6;0:
LINE 0;-1.2;0;1.2:
FOR N=0 TO (29L)
   STEP 1;
   RUN "DRAWCURVE":
END:
LINE 0;0;1.5X;1.5Y:
FREEZE

DRAWURVE PROGRAM
R+Q▶R:
K+P▶K:
R*COS(K)▶X:
R*SIN(K)▶Y:
LINE A;B;X;Y:
X▶A:
Y▶B:

1. Enter the two programs ANGDRAW and DRAW-CURVE.
2. From the HOME screen set the mode to radians. Set the plot range in the Function aplet to $-2.6 \le x \le 2.6$ and $-1.2 \le y \le 1.2$. Set the angle measure to radians.
3. Run the ANGDRAW program.

Graphing a Sine Function (Section 2.5)

The program, shown in the Discussion Problem on page 158, will simultaneously draw a unit circle and the corresponding points on the sine curve. After the circle and sine curve are drawn, you can connect the points on the unit circle with their corresponding points on the sine curve by pressing ENTER or EXE .

TI-80

```
PROGRAM:SINESHO
:RADIAN
:CLRDRAW:FNOFF
:PARAM:SIMUL
:-2.25→XMIN
:π/2→XMAX
:3→XSCL
:-1.5→YMIN
:1.5→YMAX
:1→YSCL
:0→TMIN
:6.3→TMAX
:.15→TSTEP
:"-1.25+COS T"→X1T
:"SIN T"→Y1T
:"T/4"→X2T
:"SIN T"→Y2T
:DISPGRAPH
:FOR(N,1,12)
:Nπ/6.5→T
:"-1.25+COS T"→A
:SIN T→B
:T/4→C
:LINE(A,B,C,B)
:PAUSE
:END
:PAUSE:FUNC
:SEQUENTIAL:DISP
```

TI-81

```
PrgmA:SINESHOW
:Rad
:ClrDraw
:Param
:Simul
:-2.25→Xmin
:π/2→Xmax
:3→Xscl
:-1.19→Ymin
:1.19→Ymax
:1→Yscl
:0→Tmin
:6.3→Tmax
:.15→Tstep
:"-1.25+cos T"→X1T
:"sin T"→Y1T
:"T/4"→X2T
:"sin T"→Y2T
:DispGraph
:1→N
:Lbl 1
:IS>(N,12)
:Goto 2
:Pause
:Function
:Sequence
:Disp " "
:End
:Lbl 2
:Nπ/6.5→T
:-1.25+cos T→A
:sin T→B
:T/4→C
:Line(A,B,C,B)
:Pause
:Goto 1
```

TI-83
TI-82

PROGRAM:SINESHOW
:Radian
:ClrDraw:FnOff
:Param:Simul
:-2.25→Xmin
:π/2→Xmax
:3→Xscl
:-1.19→Ymin
:1.19→Ymax
:1→Yscl
:0→Tmin
:6.3→Tmax
:.15→Tstep
:"-1.25+cos (T)"→X1T
:"sin (T)"→Y1T
:"T/4"→X2T
:"sin (T)"→Y2T
:DispGraph
:For(N,1,12)
:Nπ/6.5→T
:-1.25+cos (T)→A
:sin(T)→B
:T/4→C
:Line(A,B,C,B)
:Pause
:End
:Pause :Func
:Sequential:Disp

TI-85

PROGRAM:SINESHOW
:Radian
:ClDrw:FnOff
:Param:SimulG
:-2.25→xMin
:π/2→xMax
:3→xScl
:-1.1→yMin
:1.1→yMax
:1→yScl
:0→tMin
:6.3→tMax
:.15→tStep
:xt1=-1.25+cos t
:yt1=sin t
:xt2=t/4
:yt2=sin t
:For(N,1,12)
:N*π/6.5→t
:-1.25+cos t→A
:sin t→B
:t/4→C
:Line(A,B,C,B)
:Pause
:End
:Pause :Func
:SeqG:Disp

TI-92

sineshow()
Prgm
Disp
ClrDraw:FnOff
setMode("Graph", "Parametric")
setGraph("Graph Order",
    "Simul")
-2.9→xmin
3π/4→xmax
3→xscl
-1.1→ymin
1.1→ymax
1→yscl
0→tmin
6.3→tmax
.15→tstep
-1.25+cos(t)→xt1(t)
sin(t)→yt1(t)
t/4→xt2(t)
sin(t)→yt2(t)
DispG
For N,1,12
N*π/6.5→t
-1.25+cos(t)→A
sin(t)→B
t/4→C
Line A,B,C,B
Pause
EndFor
Pause
setMode("Graph", "Function")
setGraph("Graph order",
    "Seq")
setMode("Split 1 App",
    "Home")
EndPrgm

Casio fx-7700G

SINESHOW
Rad
Range -2.25,π÷2,3,-1.19,1.19,
    10,6.3,.15
Graph(X,Y)=(-1.25+cos T,sinT)
Graph(X,Y)=(T÷4,sinT)
0→N
Lbl 1
N+1→N
Nπ÷6.5→T
-1.25+cos T→A
sin T→B
T÷4→C
Plot A,B
Plot C,B
Line ◢
N<12⇒Goto 1

Press MODE SHIFT × to change to
parametric mode when starting to
write this program.

Casio fx-7700GE
Casio fx-9700GE
Casio CFX-9800G

SINESHOW↵
Rad↵
Range -2.25,$\pi\div$2,3,-1.19,1.19,
    1,0,6.3,.15↵
Graph(X,Y)=(-1.25+cos T,sin T)↵
Graph(X,Y)=(T$\div$4,sin T)↵
0→N↵
Lbl 1↵
N+1→N↵
N$\pi\div$6.5→T↵
-1.25+cosT→A↵
sin T→B↵
T$\div$4→C↵
Plot A,B↵
Plot C,B↵
Line ◢
N<12⇒Goto 1↵
Cls

When starting to write this program press SHIFT SET UP and select PRM or PARM for the GRAPH TYPE to change to parametric mode.

Sharp EL-9200C
Sharp EL-9200C

sineshow
──────────REAL
m=sin^{-1} 1/(π/2)
Range -2.25,π/2,3,-1.19,1.19,1
step=π/15
θ=0
xco=-.25
xso=0
yo=0
Label 1
θ=θ+step
xc=cos(mθ)$-$1.25
xs=θ/4
y=sin (mθ)
Line xco,yo,xc,y
Line xso,yo,xs,y
xco=xc
xso=xs
yo=y
If θ<(2π) Goto 1
step=π/6
θ=0
Label 2
θ=θ+step
xc=cos (mθ)$-$1.25
xs=θ/4
y=sin (mθ)
Line xc,y,xs,y
Wait
If θ<2π Goto 2
End

HP-38G Programs

SINESHOW PROGRAM
ASIN(1)/(π/2)►M:
0►T:
-.25►A:
0►B:
0►C:
LINE -3;0;π/2;0:
LINE 0;-1.1;0;1.1:
FOR T=0 TO 31π/15
   STEP π/15;
   RUN "DRAW.SINE":
END:
0►T:
FOR T=0 TO 2π
   STEP π/6;
   RUN "DRAW.LINE":
END

DRAW.SINE PROGRAM
COS(MT)−1.25►D:
T/4►E:
SIN(MT)►F:
LINE A;C;D;F:
LINE B;C;E;F:
D►A:
E►B:
F►C:

DRAW.LINE PROGRAM
COS(MT)−1.25►D:
T/4►E:
SIN(MT)►F:
LINE D;F;E;F:
FREEZE

1. Enter the 3 programs SINESHOW, DRAW.SINE, and DRAW.LINE.
2. Set the plot range in the Function aplet to $-3 \le x \le \pi/2$ and $-1.1 \le y \le 1.1$. Set the angle measure to radians.
3. Run the SINESHOW program.

Vectors in the Plane (Section 4.3)

The program is presented in Exercise 73 on page 286. You are asked to explain what the program does.

TI-80

```
:PROGRAM:ADDVECT
:CLRDRAW
:DISP "ENTER(A,B)"
:INPUT "ENTER A",A
:INPUT "ENTER B",B
:DISP "ENTER (C,D)"
:INPUT "ENTER C",C
:INPUT "ENTER D",D
:LINE(0,0,A,B)
:LINE(0,0,C,D)
:A+C→E
:B+D→F
:LINE(0,0,E,F)
:LINE(A,B,E,F)
:LINE(C,D,E,F)
:PAUSE
```

TI-81

```
:PrgmC:ADDVECT
:ClrDraw
:Disp "ENTER(A,B)"
:Disp "ENTER A"
:Input A
:Disp "ENTER B"
:Input B
:Disp "ENTER (C,D)"
:Disp "ENTER C"
:Input C
:Disp "ENTER D"
:Input D
:Line(0,0,A,B)
:Line(0,0,C,D)
:A+C→E
:B+D→F
:Line(0,0,E,F)
:Line(A,B,E,F)
:Line(C,D,E,F)
:Pause
:End
```

TI-83
TI-82

```
:PROGRAM:ADDVECT
:ClrDraw
:Input "ENTER A",A
:Input "ENTER B",B
:Input "ENTER C",C
:Input "ENTER D",D
:Line(0,0,A,B)
:Line(0,0,C,D)
:A+C→E
:B+D→F
:Line(0,0,E,F)
:Line(A,B,E,F)
:Line(C,D,E,F)
:Pause
:Stop
```

TI-85

```
:PROGRAM:ADDVECT
:ClrDraw
:Input "enter A",A
:Input "enter B",B
:Input "enter C",C
:Input "enter D",D
:Line(0,0,A,B)
:Line(0,0,C,D)
:A+C→E
:B+D→F
:Line(0,0,E,F)
:Line(A,B,E,F)
:Line(C,D,E,F)
:Pause
:Disp
```

TI-92

```
addvect( )
Prgm
ClrIO
Input "ENTER a ",a
Input "ENTER b ",b
Input "ENTER c ",c
Input "ENTER d ",d
ClrDraw
Line(0,0,a,b)
Line(0,0,c,d)
a+c→e
b+d→f
Line 0,0,e,f
Line a,b,e,f
Line c,d,e,f
Pause
setMode("Split 1 App","Home")
Stop
EndPrgm
```

Casio fx-7700G

```
ADDVECT
Cls
"A="?→A
"B="?→B
"C="?→C
"D="?→D
Plot 0,0
Plot A,B
Line
Plot 0,0
Plot C,D
Line ◢
A+C→E
B+D→F
Plot 0,0
Plot E,F
Line
Plot A,B
Plot E,F
Line
Plot C,D
Plot E,F
Line ◢
```

Casio fx-7700GE
Casio fx-9700GE
Casio CFX-9800G

```
ADDVECT↵
Cls↵
"A="?→A↵
"B="?→B↵
"C="?→C↵
"D="?→D↵
Plot 0,0↵
Plot A,B↵
Line↵
Plot 0,0↵
Plot C,D↵
Line▲
A+C→E↵
B+D→F↵
Plot 0,0↵
Plot E,F↵
Line↵
Plot A,B↵
Plot E,F↵
Line↵
Plot C,D↵
Plot E,F↵
Line▲
```

Sharp EL-9200C
Sharp EL-9300C

addvect
————————REAL

```
ClrG
Input a
Input b
Input c
Input d
Line 0,0,a,b
Line 0,0,c,d
e=a+c
f=b+d
Line 0,0,e,f
Line a,b,e,f
Line c,d,e,f
Wait
End
```

HP-38G PROGRAMS

ADDVECT PROGRAM
```
INPUT A;; "ENTER A";;1:
INPUT B;; "ENTER B";;1:
INPUT C;; "ENTER C";;1:
INPUT D;; "ENTER D";;1:
ERASE:
LINE−10;0;10;0:
LINE 0;−10;0;10:
LINE 0;0;A;B:
LINE 0;0;C;D:
FREEZE:
A+C▶ E
B+D▶ F
LINE 0;0;E;F:
LINE A;B;E;F:
LINE C;D;E;F:
FREEZE
```

The Function aplet should have a plot range of $-10 \le x \le 10$ and $-10 \le y \le 10$.

Finding the Angle between Two Vectors (Section 4.4)

The program, shown in the Discovery note on page 288, will sketch two vectors and calculate the measure of the angle between the vectors. Be sure to set an appropriate viewing rectangle.

TI-80

```
:PROGRAM:VECANGL
:CLRHOME
:DEGREE
:DISP "ENTER (A,B)"
:INPUT "ENTER A",A
:INPUT "ENTER B",B
:CLRHOME
:DISP "ENTER (C,D)"
:INPUT "ENTER C",C
:INPUT "ENTER D",D
:LINE(0,0,A,B)
:LINE(0,0,C,D)
:PAUSE
:AC+BD→E
:√ (A²+B²)→U
:√ (C²+D²)→V
:COS⁻¹(E/(UV))→θ
:DISP "θ=", θ
:CLRDRAW
```

TI-81

```
:PrgmB:VECANGL
:ClrHome
:Deg
:Disp "ENTER (A,B)"
:Disp "ENTER A"
:Input A
:Disp "ENTER B"
:Input B
:ClrHome
:Disp "ENTER (C,D)"
:Disp "ENTER C"
:Input C
:Disp "ENTER D"
:Input D
:Line(0,0,A,B)
:Line(0,0,C,D)
:Pause
:AC+BD→E
:√ (A²+B²)→U
:√ (C²+D²)→V
:cos⁻¹(E/(UV))→θ
:Disp "θ="
:Disp θ
:ClrDraw
:End
```

TI-83
TI-82

:PROGRAM:VECANGL
:ClrHome
:Degree
:Disp "ENTER (A,B)"
:Input "ENTER A",A
:Input "ENTER B",B
:ClrHome
:Disp "ENTER (C,D)"
:Input "ENTER C",C
:Input "ENTER D",D
:Line(0,0,A,B)
:Line(0,0,C,D)
:Pause
:AC+BD→E
:$\sqrt{\ }$ (A^2+B^2)→U
:$\sqrt{\ }$ (C^2+D^2)→V
:cos^{-1}(E/(UV))→θ
:ClrDraw:ClrHome
:Disp "θ=",θ
:Stop

TI-85

:PROGRAM:VECANGL
:CILCD
:Radian
:Disp "enter (A,B)"
:Input "enter A",A
:Input "enter B",B
:CILCD
:Disp "enter (C,D)"
:Input "enter C",C
:Input "enter D",D
:Line(0,0,A,B)
:Line(0,0,C,D)
:Pause
:A*C+B*D→E
:$\sqrt{\ }$ (A^2+B^2)→U
:$\sqrt{\ }$ (C^2+D^2)→V
:cos^{-1}(E/(U*V))→T
:T*180/π→T
:Disp "T=",T
:CIDrw

TI-92

vecangl( )
Prgm
FnOff
ClrHome:ClrDraw
SetMode("Split Screen",
    "Left-Right")
SetMode("Split 1 App", "Home")
SetMode("Split 2 App", "Graph")
SetMode("Exact/Approx",
    "Approximate")
ClrIO
Disp "ENTER (A,B)"
Input "ENTER A", A
Input "ENTER B", B
Line(0,0,A,B)
Pause
ClrIO
Disp "ENTER (C,D)"
Input "ENTER C",C
Input "ENTER D",D
Line(0,0,C,D)
Pause
ClrIO
A*C+B*D→E
$\sqrt{\ }$ (A^2+B^2)→U
$\sqrt{\ }$ (C^2+D^2)→V
$\cos^{-1}$ (E/(U*V))→θ
Disp "θ=",θ
Pause
SetMode("Exact/Approx", "Auto")
SetMode("Split Screen", "Full")
SetMode("Split 1 App", "Home")
Stop
EndPrgm

Casio fx-7700G

VECANGL
Cls
Deg
"ENTER (A,B)"
"A="?→A
"B="?→B
"ENTER (C,D)"
"C="?→C
"D="?→D
Plot 0,0
Plot A,B
Line
Plot 0,0
Plot C,D
Line◢
AC+BD→E
$\sqrt{\ }$ (A^2+B^2)→U
$\sqrt{\ }$ (C^2+D^2)→V
$\cos^{-1}$(E÷UV)→θ
"θ="
θ

Casio fx-7700GE
Casio fx-9700GE
Casio CFX-9800G

VECANGL⏎
Cls⏎
Deg⏎
"ENTER (A,B)"⏎
"A="?→A⏎
"B="?→B⏎
"ENTER (C,D)"⏎
"C="?→C⏎
"D="?→D⏎
Plot 0,0⏎
Plot A,B⏎
Line⏎
Plot 0,0⏎
Plot C,D⏎
Line ◢
AC+BD→E⏎
$\sqrt{\ }$ (A²+B²)→U⏎
$\sqrt{\ }$ (C²+D²)→V⏎
cos⁻¹(E÷UV)→θ⏎
"θ="⏎
θ

Sharp EL-9200C
Sharp EL-9300C

vecangl
────────REAL
ClrG
ClrT
Print"enter (a,b)"
Input a
Input b
ClrT
Print"enter (c,d)"
Input c
Input d
Line 0,0,a,b
Line 0,0,c,d
Wait
e=a*c+b*d
u=$\sqrt{\ }$ (a²+b²)
v=$\sqrt{\ }$ (c²+d²)
t=cos⁻¹(e/(u*v))
Print t
End

Set the calculator to degree mode before running the program.

HP-38G

VECANGL PROGRAM
INPUT A; "ENTER (A,B)";
    "ENTER A";;1:
INPUT B; "ENTER (A,B)";
    "ENTER B";;1:
INPUT C; "ENTER (C,D)";
    "ENTER C";;1:
INPUT D; "ENTER (C,D)";
    "ENTER D";;1:
ERASE:
LINE −10;0;10;0:
LINE 0;−10;0;10:
LINE 0;0;A;B:
LINE 0;0;C;D:
FREEZE:
AC+BD▶ E
$\sqrt{}$ (A²+B²)▶U:
$\sqrt{}$ (C²+D²)▶V:
ACOS(E/(UV))▶T:
ERASE:
DISP 3; "ANGLE= "T:
FREEZE

The Function aplet should have a plot range of $-10 \leq x \leq 10$ and $-10 \leq y \leq 10$. Set the MODE to degrees before running the program.

....

SERIES AND TRIGONOMETRIC FUNCTIONS

In this section you will learn how to represent trigonometric functions in terms of "infinite polynomials" called **power series.** These representations are a practical method to approximate trigonometric values using simple polynomials. The development of these ideas begins with another important mathematical topic, sequences and series.

Sequences

When we say that a collection of objects is listed "in sequence," we usually mean that the collection is ordered so that it has a first member, a second member, a third member, and so on. In mathematics, the word "sequence" is used in much the same way. Here are two examples.

$$1, 3, 5, 7, \ldots \quad \text{and} \quad \frac{1}{2}, \frac{1}{4}, \frac{1}{8}, \frac{1}{16}, \ldots$$

DEFINITION OF A SEQUENCE

A **sequence** is a function whose domain is the set of positive integers. The function values

$$a_1, a_2, a_3, a_4, \ldots, a_n, \ldots$$

are the **terms** of the sequence. If the domain consists of the first n positive integers only, then the sequence is a **finite sequence.**

Sometimes it is convenient to include zero in the domain of a sequence and list the terms as

$$a_0, a_1, a_2, a_3, a_4, \ldots$$

Unless otherwise stated, however, you may assume a sequence begins with a_1.

When you are given a formula for the nth term of a sequence, you can find the terms by substituting 1, 2, 3, and so on, for n.

A. Here are the first four terms of the sequence whose nth term is $a_n = 4n - 3$.

$$a_1 = 4(1) - 3 = 1$$
$$a_2 = 4(2) - 3 = 5$$
$$a_3 = 4(3) - 3 = 9$$
$$a_4 = 4(4) - 3 = 13$$

B. Here are the first four terms of the sequence whose nth term is $a_n = 2 + (-1)^n$.

$$a_1 = 2 + (-1)^1 = 2 - 1 = 1$$
$$a_2 = 2 + (-1)^2 = 2 + 1 = 3$$
$$a_3 = 2 + (-1)^3 = 2 - 1 = 1$$
$$a_4 = 2 + (-1)^4 = 2 + 1 = 3$$

Note that simply listing the first few terms is not sufficient to define a sequence—the nth term *must be given*. To see this, consider the following sequences, both of which have the same first three terms.

$$\frac{1}{2}, \frac{1}{4}, \frac{1}{8}, \frac{1}{16}, \cdots, \frac{1}{2^n}, \cdots$$

$$\frac{1}{2}, \frac{1}{4}, \frac{1}{8}, \frac{1}{15}, \cdots, \frac{6}{(n+1)(n^2 - n + 6)}, \cdots$$

Some very important sequences in mathematics, and particularly those that arise in trigonometry, involve terms that are defined with special types of products called **factorials.**

DEFINITION OF FACTORIAL

If n is a positive integer, then n **factorial** is defined by

$$n! = 1 \cdot 2 \cdot 3 \cdot 4 \cdots (n - 1) \cdot n.$$

As a special case, zero factorial is defined to be $0! = 1$.

Here are some values of $n!$ for the first several nonnegative integers

$0! = 1$

$1! = 1$

$2! = 1 \cdot 2 = 2$

$3! = 1 \cdot 2 \cdot 3 = 6$

$4! = 1 \cdot 2 \cdot 3 \cdot 4 = 24$

$5! = 1 \cdot 2 \cdot 3 \cdot 4 \cdot 5 = 120$

The value of n does not have to be very large before $n!$ becomes huge. For instance, $10! = 3,628,800$. Many calculators have a factorial key, denoted by $\boxed{\times!}$.

Factorials follow the same conventions for order of operations as do exponents. For instance,

$$2n! = 2(n!) = 2(1 \cdot 2 \cdot 3 \cdot 4 \cdots n),$$

whereas $(2n)! = 1 \cdot 2 \cdot 3 \cdot 4 \cdots n \cdots (2n - 1)(2n)$.

Write the first six terms of the sequence whose nth term is $a_n = \dfrac{30}{n!}$. Begin with $n = 0$.

$$a_0 = \frac{30}{0!} = \frac{30}{1} = 30$$

$$a_1 = \frac{30}{1!} = \frac{30}{1} = 30$$

$$a_2 = \frac{30}{2!} = \frac{30}{2} = 15$$

$$a_3 = \frac{30}{3!} = \frac{30}{6} = 5$$

$$a_4 = \frac{30}{4!} = \frac{30}{24} = \frac{5}{4}$$

$$a_5 = \frac{30}{5!} = \frac{30}{120} = \frac{1}{4}$$

Summation Notation

There is a convenient notation for the sum of the terms of a *finite sequence*. It is called **summation notation** or **sigma notation** because it uses the upper-case Greek letter *sigma*, written as Σ.

DEFINITION OF SUMMATION NOTATION

The expression formed by adding the first n terms of a sequence is called a **series** and is represented by

$$\sum_{i=1}^{n} a_i = a_1 + a_2 + a_3 + a_4 + \cdots + a_n$$

where i is called the **index of summation,** n is the **upper limit of summation,** and 1 is the **lower limit of summation.** It is read as "the sum from $i = 1$ to n of a_i."

The index of summation does not have to be i—any letter can be used. Also, the index does not have to begin at 1. For instance, in part b of the next example, the index begins with 3. For sums that arise in trigonometry, we will see that it is often convenient to begin with $i = 0$.

A. Let i take on values from 1 to 5 to evaluate the sum.

$$\sum_{i=1}^{5} 3i = 3(1) + 3(2) + 3(3) + 3(4) + 3(5)$$
$$= 3(1 + 2 + 3 + 4 + 5)$$
$$= 3(15)$$
$$= 45$$

B. Let k take on values from 3 to 6 to evaluate the sum.

$$\sum_{k=3}^{6} (1 + k^2) = (1 + 3^2) + (1 + 4^2) + (1 + 5^2) + (1 + 6^2)$$
$$= 10 + 17 + 26 + 37$$
$$= 90$$

C. Let j take on values from 0 to 4 to evaluate the sum.

$$\sum_{j=0}^{4} (-1)^j \frac{1}{(2j)!} = (-1)^0 \frac{1}{0!} + (-1)^1 \frac{1}{2!} + (-1)^2 \frac{1}{4!} + (-1)^3 \frac{1}{6!} + (-1)^4 \frac{1}{8!}$$
$$= 1 - \frac{1}{2} + \frac{1}{24} - \frac{1}{720} + \frac{1}{40,320}$$
$$\approx 0.5403026$$

It is interesting to note that this answer is very close to the exact value of cos 1. You will see later how to use series to approximate trigonometric functions.

The sum of the terms of a finite sequence can be found by simply adding the terms. For sequences with many terms, however, adding the terms can be tedious. It is often more efficient to use a formula for the sum of special types of sequences.

FORMULAS FOR SPECIAL SERIES

1. $\displaystyle\sum_{i=1}^{n} 1 = n$

2. $\displaystyle\sum_{i=1}^{n} i = \frac{1}{2}(n)(n+1)$

3. $\displaystyle\sum_{i=1}^{n} i^2 = \frac{1}{6}(n)(n+1)(2n+1)$

You are working in the produce department of a grocery store. The manager of the produce department asks you to create a display in which oranges are stacked in a pyramid as shown below. The pyramid is to have 16 layers. How many oranges do you need?

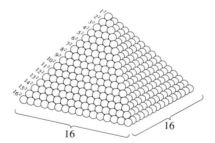

FIGURE A.12

To solve the problem, you first need to recognize that the bottom layer has 16^2 oranges, the next layer 15^2, and so on. This means that the total number of oranges is

$$\sum_{i=1}^{16} i^2 = 1^2 + 2^2 + \cdots + 16^2.$$

One way to find the sum is to add the 16 terms. Another way is to use formula 3 shown above.

$$\sum_{i=1}^{16} i^2 = \frac{1}{6}(16)(16+1)(2 \cdot 16 + 1) = \frac{1}{6}(16)(17)(33) = 1496$$

Hence, there are 1496 oranges in the pyramid.

Geometric Sequences and Series

A sequence whose consecutive terms have a common ratio is a **geometric sequence.**

DEFINITION OF A GEOMETRIC SEQUENCE

The sequence $a_1, a_2, a_3, a_4, \ldots, a_n, \ldots$ is **geometric** if there is a nonzero number r such that

$$\frac{a_2}{a_1} = r, \quad \frac{a_3}{a_2} = r, \quad \frac{a_4}{a_3} = r$$

and so on. The number r is the *common ratio* of the sequence.

A. The first 4 terms of the sequence whose nth term is 2^n are 2, 4, 8, and 16. The common ratio between consecutive terms is 2. Therefore, the sequence is geometric.

$$2, 4, 8, 16, \ldots, 2^n, \ldots$$
$$\frac{4}{2} = 2$$

B. The first 4 terms of the sequence whose nth term is $4(3^n)$ are 12, 36, 108, and 324. The common ratio between consecutive terms is 3. Therefore, the sequence is geometric.

$$12, 36, 108, 324, \ldots, 4(3^n), \ldots$$
$$\frac{36}{12} = 3$$

C. The first 4 terms of the sequence whose nth term is $9(-\frac{1}{3})^n$ are $-3, 1, -\frac{1}{3}$ and $\frac{1}{9}$. The common ratio between consecutive terms is $-\frac{1}{3}$. Therefore, the sequence is geometric.

$$-3, 1, -\frac{1}{3}, \frac{1}{9}, \ldots, 9\left(-\frac{1}{3}\right)^n, \ldots$$
$$\frac{1}{-3} = -\frac{1}{3}$$

If r is the common ratio of a geometric sequence, then the formula for the nth term of the sequence is

$$a_n = a_1 r^{n-1}.$$

Find the twelfth term of the geometric sequence whose first three terms are 5, 15, and 45.

The common ratio of this sequence is $r = \frac{15}{5} = 3$. Because the first term is $a_1 = 5$, the formula for the nth term is as follows.

$$a_n = a_1 r^{n-1}$$
$$= 5(3)^{n-1}$$

The twelfth term is $a_{12} = 5(3)^{11} = 885{,}735$.

The expression formed by adding the terms of a geometric sequence is called a **geometric series.** You can develop a formula for the sum, S, of a (finite) geometric series as follows.

$$S = a_1 + a_1 r + a_1 r^2 + a_1 r^3 + \cdots + a_1 r^{n-1}$$
$$-rS = \qquad - a_1 r - a_1 r^2 - a_1 r^3 - \cdots - a_1 r^{n-1} - a_1 r^n$$

Adding these two equations together gives

$$S(1 - r) = a_1 - a_1 r^n = a_1(1 - r^n).$$

By dividing both sides of this equation by $(1 - r)$, you can obtain the following formula for S.

THE SUM OF A FINITE GEOMETRIC SERIES

The sum of the geometric series $a_1 + a_1 r + a_1 r^2 + \cdots + a_1 r^{n-1}$ with common ratio $r \neq 1$ is

$$\sum_{i=1}^{n} a_1 r^{i-1} = a_1 \left(\frac{1 - r^n}{1 - r} \right).$$

Evaluate the sum

$$\sum_{i=1}^{12} 4(0.3)^{i-1} = 4 + 4(0.3) + 4(0.3)^2 + 4(0.3)^3 + \cdots + 4(0.3)^{11}.$$

Because $a_1 = 4$, $r = 0.3$, and $n = 12$, you can apply the formula for the sum of a geometric series.

$$\sum_{i=1}^{12} 4(0.3)^{i-1} = a_1 \left(\frac{1 - r^n}{1 - r} \right)$$

$$= 4 \left(\frac{1 - (0.3)^{12}}{1 - 0.3} \right)$$

$$\approx 5.7143$$

Infinite Geometric Series

It should not surprise you that all finite geometric series have a finite sum. You may, however, be surprised that some *infinite* geometric series have a finite sum. Here is an example.

$$\sum_{i=0}^{\infty} 0.3 \left(\frac{1}{10} \right)^i = 0.3 + 0.03 + 0.003 + 0.0003 + \cdots$$

$$= 0.3333 \ldots$$

$$= \frac{1}{3}$$

To indicate that an **infinite series** has no upper limit of summation, we write the symbol for *infinity*, ∞, in place of the upper limit. This symbol is never used to represent a number. Instead, it is used to indicate a type of unbounded or unending situation. For an infinite series, the infinity symbol is used to indicate that the series is composed of infinitely many terms.

SUM OF AN INFINITE GEOMETRIC SERIES

If $|r| < 1$, then the infinite geometric series

$$a_1 + a_1 r + a_1 r^2 + a_1 r^3 + \cdots + a_1 r^{n-1} + \cdots$$

has the sum

$$\sum_{n=1}^{\infty} a_1 r^{n-1} = \frac{a_1}{1 - r}.$$

If $|r| \geq 1$, then the series has no sum.

Find the sum.

$$\sum_{n=1}^{\infty} 4(0.6)^{n-1} = 4 + 4(0.6) + 4(0.6)^2 + 4(0.6)^3 + \cdots$$

For this series, $a_1 = 4$ and $r = 0.6$. Because $|r| < 1$, you can apply the formula for an infinite geometric series.

$$\sum_{n=1}^{\infty} 4(0.6)^{n-1} = \frac{a_1}{1-r} = \frac{4}{1-0.6} = 10$$

Power Series for Trigonometric Functions

The geometric series discussed above are a special case of a **power series.** Power series can be thought of as "infinite polynomials," as indicated in the following definition.

DEFINITION OF POWER SERIES

If x is a variable, then a series of the form

$$\sum_{n=0}^{\infty} a_n x^n = a_0 + a_1 x + a_2 x^2 + a_3 x^3 + \cdots + a_n x^n + \cdots$$

is called a **power series.**

Most of the common functions in mathematics can be represented by power series. A detailed study of power series is beyond the scope of this text, and better left for a course in calculus. However, the trigonometric functions $\sin x$ and $\cos x$ have the following power series representations.

$$\sin x = x - \frac{x^3}{3!} + \frac{x^5}{5!} - \frac{x^7}{7!} + \cdots + (-1)^n \frac{x^{2n+1}}{(2n+1)!} + \cdots = \sum_{n=0}^{\infty} (-1)^n \frac{x^{2n+1}}{(2n+1)!}$$

$$\cos x = 1 - \frac{x^2}{2!} + \frac{x^4}{4!} - \frac{x^6}{6!} + \cdots + (-1)^n \frac{x^{2n}}{(2n)!} + \cdots = \sum_{n=0}^{\infty} (-1)^n \frac{x^{2n}}{(2n)!}$$

Notice that the *partial sums* of a power series are simply polynomials in the variable x. For instance, you can approximate $\sin x$ and $\cos x$ by using a finite number of terms of the appropriate power series, as illustrated in the following example.

Use a graphing utility to graph the following polynomials together with the function $f(x) = \sin x$. What do you observe as the degree of the polynomials increases?

$$p(x) = x$$

$$q(x) = x - \frac{x^3}{3!}$$

$$r(x) = x - \frac{x^3}{3!} + \frac{x^5}{5!}$$

$$s(x) = x - \frac{x^3}{3!} + \frac{x^5}{5!} - \frac{x^7}{7!}$$

The graphs are shown in Figure A.13. As the degree of the polynomial increases, the graph of the polynomial more closely resembles that of the sine function.

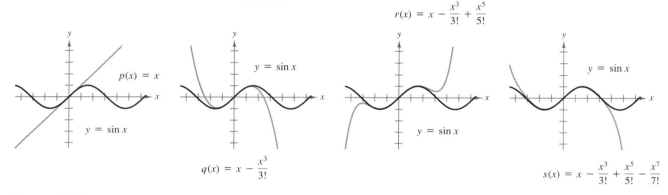

FIGURE A.13

The power series for $\sin x$ and $\cos x$ can be used to approximate these trigonometric functions for any value of x. The approximations are better if x is near 0, as shown in the following example.

Use the first 4 terms of the power series for $\cos x$ to approximate $\cos 1$ and $\cos 2$. Which approximation is better?

The first 4 terms of the power series for $\cos x$ are

$$\cos x \approx 1 - \frac{x^2}{2!} + \frac{x^4}{4!} - \frac{x^6}{6!}.$$

Letting $x = 1$ and $x = 2$, you obtain the following approximations.

$$\cos 1 \approx 1 - \frac{1^2}{2!} + \frac{1^4}{4!} - \frac{1^6}{6!} \approx 0.540278$$

$$\cos 2 \approx 1 - \frac{2^2}{2!} + \frac{2^4}{4!} - \frac{2^6}{6!} \approx -0.42222$$

Since $\cos 1 = 0.5403023 \ldots$ and $\cos 2 = -0.4161468 \ldots$, you see that the first approximation is better.

The approximations in Example 10 could be improved if more terms of the cosine series were used. For instance, if the first 10 terms of the series for $\cos 1$ are summed, the error of the approximation is about 10^{-11}. In fact, some computers use the method of truncated power series for approximating trigonometric functions (and other functions as well).

There are power series representations of the other four trigonometric functions and their inverses, but the series are generally more complicated. Furthermore, these series are only valid for certain restricted values of x, unlike the sine and cosine series which are valid for all x. The final example of this section uses the power series for $f(x) = \arctan x$ to obtain an approximation of π.

The power series for $f(x) = \arctan x$ (valid for $-1 \le x \le 1$) is

$$\arctan x = x - \frac{x^3}{3} + \frac{x^5}{5} - \frac{x^7}{7} + \cdots$$

$$= \sum_{n=0}^{\infty} (-1)^n \frac{x^{2n+1}}{(2n + 1)!}, \quad -1 \le x \le 1$$

Use the first 10 terms of this series to approximate $\arctan 1$. Then use your answer to approximate π.

Substituting 1 for x and summing the first 10 terms of the series you obtain

$$\arctan 1 \approx 1 - \frac{1^3}{3} + \frac{1^5}{5} - \frac{1^7}{7} + \frac{1^9}{9} - \frac{1^{11}}{11} + \frac{1^{13}}{13} - \frac{1^{15}}{15} + \frac{1^{17}}{17} - \frac{1^{19}}{19}$$

$$= 1 - \frac{1}{3} + \frac{1}{5} - \frac{1}{7} + \frac{1}{9} - \frac{1}{11} + \frac{1}{13} - \frac{1}{15} + \frac{1}{17} - \frac{1}{19}$$

$$\approx 0.76046.$$

Since $\arctan 1 = \pi/4$, you can multiply this answer by 4 to obtain a relatively poor approximation of π: $4(0.76046) = 3.04184$. If you were to use 1000 terms, the approximation would improve to $\pi \approx 3.14059$.

In Exercises 1–24, write the first five terms of the sequence. Begin with $n = 1$.

1. $a_n = 3n - 5$

2. $a_n = 4 - n$

3. $a_n = n^2 + 2$

4. $a_n = (n + 1)^2$

5. $a_n = 2^n$

6. $a_n = \left(\dfrac{1}{2}\right)^n$

7. $a_n = \left(-\dfrac{1}{2}\right)^n$

8. $a_n = (-1)^{n-1}$

9. $a_n = \dfrac{n}{n + 1}$

10. $a_n = \dfrac{n^2}{2n}$

11. $a_n = (-2)^n$

12. $a_n = (-5)^n$

13. $a_n = n! - 2$

14. $a_n = \dfrac{n!}{2}$

15. $a_n = \dfrac{3^n}{n!}$

16. $a_n = (2n + 1)!$

17. $a_n = 2(n!)$

18. $a_n = 2(n + 2)!$

19. $a_n = (n - 1)!$

20. $a_n = \dfrac{n^2}{n!}$

21. $a_n = 4 + (-1)^n$

22. $a_n = 2(-1)^n$

23. $a_n = 4 - (-1)^{n+1}$

24. $a_n = 2(-1)^{n+1}$

25. *Compound Interest* A deposit of $100 is made *each* month in an account that earns 12% interest compounded monthly. The balance in the account after n months is given by
$$A_n = 100(101)[(1.01)^n - 1], \quad n = 1, 2, 3, \ldots .$$
(a) Compute the first six terms of this sequence.
(b) Find the balance after five years by computing the 60th term of the sequence.
(c) Find the balance after 20 years by computing the 240th term of the sequence.

26. *Federal Debt* It took more than 200 years for the U.S. to accumulate a $1 trillion debt. Then it took just 8 years to get to $3 trillion. (*Source:* Treasury Department) The federal debt during the decade of the 1980s is approximated by the model
$$a_n = 0.1\sqrt{82 + 9n^2}, \quad n = 0, 1, 2, \ldots, 10,$$
where a_n is the debt in trillions and n is the year, with $n = 0$ corresponding to 1980. Find the terms of the finite sequence and construct a bar graph that represents the sequence.

In Exercises 27–38, write the series represented by the summation notation. Then evaluate the sum.

27. $\displaystyle\sum_{n=0}^{4} n^2$

28. $\displaystyle\sum_{i=0}^{6} (2i + 5)$

29. $\displaystyle\sum_{k=1}^{5} (k^2 + 1)$

30. $\displaystyle\sum_{j=3}^{7} (6j - 10)$

31. $\displaystyle\sum_{n=1}^{4} n(n + 1)$

32. $\displaystyle\sum_{n=0}^{5} 2n^2$

33. $\displaystyle\sum_{k=2}^{6} (k! - k)$

34. $\displaystyle\sum_{j=0}^{4} \dfrac{6}{j!}$

35. $\displaystyle\sum_{m=2}^{6} \dfrac{2m}{2(m - 1)}$

36. $\displaystyle\sum_{i=2}^{5} -2i!$

37. $\displaystyle\sum_{n=0}^{6} (n! + 10)$

38. $\displaystyle\sum_{k=1}^{5} \dfrac{10k}{k + 2}$

In Exercises 39–44, use the formulas for special series to evaluate the sum.

39. $\displaystyle\sum_{i=1}^{4} 10$

40. $\displaystyle\sum_{i=1}^{7} 6$

41. $\displaystyle\sum_{i=1}^{12} i$

42. $\displaystyle\sum_{k=1}^{100} k$

43. $\displaystyle\sum_{n=1}^{5} n^2$

44. $\displaystyle\sum_{i=1}^{10} i^2$

In Exercises 45 and 46, use a graphing utility to find the sum.

45. $\displaystyle\sum_{j=1}^{20} (j^2 - 1)$

46. $\displaystyle\sum_{n=1}^{12} \dfrac{(-1)^{n-1}}{n}$

47. *Corporate Dividends* The dividends declared per share of common stock of Ameritech Corporation for the years 1985 through 1990 are shown in the figure. (*Source:* Ameritech 1990 Annual Report) These dividends can be approximated by the model
$$a_n = 0.20n + 1.17, \quad n = 5, 6, 7, 8, 9, 10,$$
where a_n is the dividend in dollars and n is the year, with $n = 5$ corresponding to 1985. Approximate the sum of the dividends per share of common stock for the years 1985 through 1990 by evaluating
$$\sum_{n=5}^{10} (0.20n + 1.17).$$

Compare this sum with the result of adding the dividends as shown in the figure.

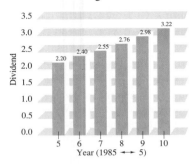

FIGURE FOR 47

48. *Total Revenue*   From 1980 through 1989, the total annual sales for MCI Communications can be approximated by the model

$$a_n = 48.217n^2 + 228.1n + 311.28, \quad n = 0, 1, \ldots, 9,$$

where a_n is the annual sales (in millions of dollars) and n is the year, with $n = 0$ corresponding to 1980. (*Source: MCI Communications*) Find the total revenue from 1980 through 1989 by evaluating the sum

$$\sum_{n=0}^{9} (48.217n^2 + 228.1n + 311.28).$$

MCI Communications

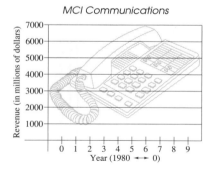

FIGURE FOR 48

In Exercises 49–56, write the first five terms of the geometric sequence.

49. $a_1 = 2, r = 3$ **50.** $a_1 = 6, r = 2$

51. $a_1 = 1, r = \frac{1}{2}$ **52.** $a_1 = 1, r = \frac{1}{3}$

53. $a_1 = 5, r = -\frac{1}{10}$ **54.** $a_1 = 6, r = -\frac{1}{4}$

55. $a_1 = 1, r = e$ **56.** $a_1 = 2, r = \sqrt{3}$

In Exercises 57–62, find the nth term of the geometric sequence.

57. $a_1 = 4, r = \frac{1}{2}, n = 10$

58. $a_1 = 5, r = \frac{3}{2}, n = 8$

59. $a_1 = 6, r = -\frac{1}{3}, n = 12$

60. $a_1 = 8, r = \sqrt{5}, n = 9$

61. $a_1 = 500, r = 1.02, n = 40$

62. $a_1 = 1000, r = 1.005, n = 60$

In Exercises 63–72, find the indicated sum.

63. $\displaystyle\sum_{n=1}^{9} 2^{n-1}$ **64.** $\displaystyle\sum_{n=1}^{9} (-2)^{n-1}$ **65.** $\displaystyle\sum_{i=1}^{7} 64\left(-\frac{1}{2}\right)^{i-1}$

66. $\displaystyle\sum_{i=1}^{6} 32\left(\frac{1}{4}\right)^{i-1}$ **67.** $\displaystyle\sum_{i=1}^{10} 8\left(\frac{-1}{4}\right)^{i-1}$ **68.** $\displaystyle\sum_{i=1}^{10} 5\left(\frac{-1}{3}\right)^{i-1}$

69. $\displaystyle\sum_{n=0}^{20} 3\left(\frac{3}{2}\right)^{n}$ **70.** $\displaystyle\sum_{n=0}^{15} 2\left(\frac{4}{3}\right)^{n}$

71. $\displaystyle\sum_{n=0}^{5} 300(1.06)^{n}$ **72.** $\displaystyle\sum_{n=0}^{6} 500(1.04)^{n}$

73. *Salary*   You accept a job with a salary of $30,000 for the first year. Suppose that during the next 39 years, you receive a 5% raise each year. What would be your total compensation over the 40-year period?

74. *Area*   The sides of a square are 16 inches in length. The square is divided into nine smaller squares and the center square is shaded (see figure). Each of the eight unshaded squares is then divided into nine smaller squares and the center square of each is shaded. If this process is repeated four more times, determine the area of the shaded region.

FIGURE FOR 74

In Exercises 75–80, find the sum of the infinite geometric series.

75. $\displaystyle\sum_{n=0}^{\infty} \left(\tfrac{1}{2}\right)^{n} = 1 + \frac{1}{2} + \frac{1}{4} + \frac{1}{8} + \cdots$

76. $\displaystyle\sum_{n=0}^{\infty} 2\left(\tfrac{2}{3}\right)^{n} = 2 + \frac{4}{3} + \frac{8}{9} + \frac{16}{27} + \cdots$

77. $\displaystyle\sum_{n=0}^{\infty} \left(-\tfrac{1}{2}\right)^{n} = 1 - \frac{1}{2} + \frac{1}{4} - \frac{1}{8} + \cdots$

78. $\displaystyle\sum_{n=0}^{\infty} 2\left(-\frac{2}{3}\right)^n = 2 - \frac{4}{3} + \frac{8}{9} - \frac{16}{27} + \cdots$

79. $\displaystyle\sum_{n=0}^{\infty} 4\left(\frac{1}{4}\right)^n = 4 + 1 + \frac{1}{4} + \frac{1}{16} + \cdots$

80. $\displaystyle\sum_{n=0}^{\infty} \left(\frac{1}{10}\right)^n = 1 + 0.1 + 0.01 + 0.001 + \cdots$

In Exercises 81–90, use the first k terms of the power series for sine or cosine to approximate the value of the function. Compare the result with the value of the function given by a graphing utility.

81. $\sin 0.25, \ k = 1$

82. $\cos 0.25, \ k = 1$

83. $\sin 0.25, \ k = 3$

84. $\cos 0.25, \ k = 3$

85. $\cos 0.5, \ k = 2$

86. $\sin 0.5, \ k = 2$

87. $\cos 0.5, \ k = 6$

88. $\sin 0.5, \ k = 6$

89. $\sin 1, \ k = 2$

90. $\sin 1, \ k = 4$

91. Use a graphing utility to graph the following polynomials together with the function $f(x) = \arctan x$. What do you observe as the degree of the polynomial increases?

$$p(x) = x \qquad q(x) = x - \frac{x^3}{3}$$

$$r(x) = x - \frac{x^3}{3} + \frac{x^5}{5} \qquad s(x) = x - \frac{x^3}{3} + \frac{x^5}{5} - \frac{x^7}{7}$$

92. Use a graphing utility to graph the following polynomials together with the function $f(x) = \sec x$. What do you observe as the degree of the polynomial increases?

$$p(x) = 1 + \frac{x^2}{2} \qquad q(x) = 1 + \frac{x^2}{2} + \frac{5x^4}{24}$$

$$r(x) = 1 + \frac{x^2}{2} + \frac{5x^4}{24} + \frac{61x^6}{720}$$

ADDITIONAL PROBLEM SOLVING WITH TECHNOLOGY

1.4 · EXERCISES

In Exercises 1–4, explain how to use a graphing utility to verify that $y_1 = y_2$. Identify the rule of algebra that is illustrated.

1. $y_1 = \frac{1}{4}(x^2 - 8)$

$y_2 = \frac{1}{4}x^2 - 2$

2. $y_1 = \frac{1}{2}x + (x + 1)$

$y_2 = \frac{3}{2}x + 1$

3. $y_1 = \frac{1}{5}[10(x^2 - 1)]$

$y_2 = 2(x^2 - 1)$

4. $y_1 = (x - 3) \cdot \dfrac{1}{x - 3}$

$y_2 = 1$

1.5 · EXERCISES

Graphical Analysis In Exercises 5–8, use a graphing utility to graph the three equations in the same viewing rectangle. Adjust the viewing rectangle so that the slope appears visually correct. Identify any lines that are parallel or perpendicular.

5. $L_1: y = 2x$

$L_2: y = -2x$

$L_3: y = \frac{1}{2}x$

6. $L_1: y = \frac{2}{3}x$

$L_2: y = -\frac{3}{2}x$

$L_3: y = \frac{2}{3}x + 2$

7. $L_1: y = -\frac{1}{2}x$

$L_2: y = -\frac{1}{2}x + 3$

$L_3: y = 2x - 4$

8. $L_1: y = x - 8$

$L_2: y = x + 1$

$L_3: y = -x + 3$

9. *Data Analysis* The average annual salaries y of major league baseball players (in thousands of dollars) from 1983 to 1992 are given as ordered pairs (t, y), where t is the time in years, with $t = 0$ corresponding to 1980. The ordered pairs are $(3, 289)$, $(4, 329)$, $(5, 371)$, $(6, 413)$, $(7, 412)$, $(8, 439)$, $(9, 497)$, $(10, 598)$, $(11, 851)$, and $(12, 1029)$. (Source: Major League Baseball Players Association)

(a) Use the regression capabilities of a graphing utility to find the least squares regression line.

(b) Use a graphing utility to plot the points and graph the regression line in the same viewing rectangle.

(c) Use the regression line to predict the average salary in the year 2000.

(d) Interpret the meaning of the slope of the regression line.

10. *Data Analysis* An instructor gives regular 20-point quizzes and 100-point exams in a mathematics course. Average scores for six students, given as ordered pairs (x, y), where x is the average quiz score and y is the average test score, are $(18, 87)$, $(10, 55)$, $(19, 96)$, $(16, 79)$, $(13, 76)$, and $(15, 82)$.

(a) Use the regression capabilities of a graphing utility to find the least squares regression line.

(b) Use a graphing utility to plot the points and graph the regression line in the same viewing rectangle.

(c) Use the regression line to predict the average exam score for a student whose average quiz score is 17.

(d) Interpret the meaning of the slope of the regression line.

(e) If the instructor added 4 points to the average exam score of everyone in the class, describe the changes in the positions of the plotted points and the change in the equation of the line.

1.7 · EXERCISES

In Exercises 11 and 12, use a graphing utility to graph the function. State the domain and range of the function. Describe the pattern of the graph.

11. $s(x) = 2\left(\frac{1}{4}x - \left[\!\left[\frac{1}{4}x\right]\!\right]\right)$ 12. $g(x) = 2\left(\frac{1}{4}x - \left[\!\left[\frac{1}{4}x\right]\!\right]\right)^2$

13. *Cost of a Telephone Call* The cost of a telephone call between two cities is \$0.65 for the first minute and \$0.40 for each additional minute.

(a) It is required that a model be created for the cost C of a telephone call between the two cities lasting t minutes. Which of the following is the appropriate model? Explain.

$C_1(t) = 0.65 + 0.4[\!\![t - 1]\!\!]$

$C_2(t) = 0.65 - 0.4[\!\![-(t - 1)]\!\!]$

(b) Use a graphing utility to graph the appropriate model and use the zoom and trace features to determine the cost of an 18-minute, 45-second call.

14. *Data Analysis* The amounts y of the merchandise trade balance of the United States in billions of dollars for the years 1986 through 1993 are given in the table. (Source: U.S. Bureau of the Census)

Year	1986	1987	1988	1989
y	-152.7	-152.1	-118.6	-109.6

Year	1990	1991	1992	1993
y	-101.7	-65.4	-84.5	-115.8

(a) Let t be the time in years, with $t = 0$ corresponding to 1990. Use the regression capabilities of a graphing utility to find a cubic model for the data. What is the domain of the model?

(b) Use a graphing utility to graph the data and the model.

(c) For which year does the model most accurately estimate the actual data? During which year is it least accurate?

(d) Why would economists be concerned if this model remained valid in the future?

1.9 · EXERCISES

Data Analysis In Exercises 15 and 16, use the data in the table, which gives the variable costs for operating an automobile in the United States for the years 1985 through 1991. The variables y_1, y_2, and y_3 represent the costs in cents per mile for gas and oil, maintenance, and tires, respectively. (Source: American Automobile Manufacturers Association)

Year	1985	1986	1987	1988	1989	1990	1991
y_1	6.16	4.48	4.80	5.20	5.20	5.40	6.70
y_2	1.23	1.37	1.60	1.60	1.90	2.10	2.20
y_3	0.65	0.67	0.80	0.80	0.80	0.90	0.90

15. Create a stacked bar graph for the data.

16. Mathematical models for the data are given by

$$y_1 = 0.16t^2 - 2.43t + 13.96$$
$$y_2 = 0.17t + 0.38$$
$$y_3 = 0.04t + 0.44$$

where $t = 5$ represents 1985. Use a graphing utility to graph y_1, y_2, y_3, and $y_1 + y_2 + y_3$ in the same viewing rectangle. Use the model to estimate the total variable cost per mile in 1995.

1.10 · EXERCISES

Graphical Reasoning   In Exercises 17–20, (a) use a graphing utility to graph the function, (b) use the DrawInv feature of the graphing utility to draw the inverse of the function, and (c) determine if the graph of the inverse relation is an inverse function (explain).

17. $f(x) = x^3 + x + 1$ **18.** $h(x) = x\sqrt{4 - x^2}$

19. $g(x) = \dfrac{3x^2}{x^2 + 1}$ **20.** $f(x) = \dfrac{4x}{\sqrt{x^2 + 15}}$

2.1 · EXERCISES

In Exercises 21–24, use the angle-conversion capabilities of a graphing utility to convert the angle measure to decimal degree form.

21. (a) $54° 45'$ (b) $-128° 30'$
22. (a) $245° 10'$ (b) $2° 12'$
23. (a) $85° 18' 30''$ (b) $330° 25''$
24. (a) $-135° 36''$ (b) $-408° 16' 25''$

In Exercises 25–28, use the angle-conversion capabilities of a graphing utility to convert the angle measure to $D° M' S''$ form.

25. (a) $240.6°$ (b) $-145.8°$
26. (a) $-345.12°$ (b) 0.45
27. (a) 2.5 (b) -3.58
28. (a) -0.355 (b) 0.7865

2.3 · EXERCISES

29. *Exploration*

(a) Use a graphing utility to complete the table.

θ	0	0.1	0.2	0.3	0.4	0.5
$\sin \theta$						

(b) Is θ or $\sin \theta$ greater for θ in the interval $(0, 0.5]$?

(c) As θ approaches 0, how do θ and $\sin \theta$ compare? Explain.

30. *Exploration*   Use a graphing utility to complete the table and make a conjecture about the relationship between $\cos \theta$ and $\sin(90° - \theta)$. What are the angles θ and $90° - \theta$ called?

θ	0°	20°	40°	60°	80°
$\cos \theta$					
$\sin(90° - \theta)$					

31. *Numerical Analysis*   The range R of a projectile fired at an angle of θ degrees with the horizontal with an initial velocity of 50 meters per second is given by $R = 530 \sin \theta \cos \theta$.

(a) Use a graphing utility to complete the table.

θ	20°	30°	40°	50°	60°	70°
R						

(b) Identify a pattern in the range of the projectile versus the angle θ.

(c) Based on the table, make a conjecture about the angle θ that will yield a maximum range.

(d) Test the conjecture by creating a new table with values of θ near the value of θ that you believe yields the maximum range.

32. *Numerical Analysis* A 3000-pound automobile is negotiating a circular interchange of radius 300 feet at a speed of s miles per hour (see figure). The relationship between the speed and the angle θ (in degrees) at which the roadway should be banked so that no lateral frictional force is exerted on the tires is given by

$$\tan \theta = \frac{0.672s^2}{3000}.$$

(a) Use a graphing utility to complete the table.

s	10	20	30	40	50	60
θ						

(b) The speeds are incremented by 10 miles per hour in the table, but θ does not increase by equal increments. Explain.

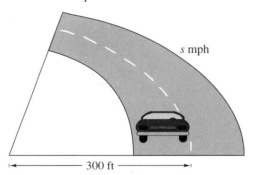

s mph

300 ft

2.4 · EXERCISES

33. *Exploration* Use a graphing utility to complete the table to demonstrate the properties of even and odd functions. Identify each function as being either even or odd.

θ	0°	20°	40°	60°	80°
$\sec \theta$					
$\sec(-\theta)$					
$\tan \theta$					
$\tan(-\theta)$					

34. *Conjecture*
(a) Use a graphing utility to complete the table.

θ	0°	20°	40°	60°	80°
$\sin \theta$					
$\sin(180° - \theta)$					

(b) Make a conjecture about the relationship between $\sin \theta$ and $\sin(180° - \theta)$.

35. *Conjecture*
(a) Use a graphing utility to complete the table.

θ	0	0.3	0.6	0.9	1.2	1.5
$\cos\left(\frac{3\pi}{2} - \theta\right)$						
$-\sin \theta$						

(b) Make a conjecture about the relationship between $\cos\left(\frac{3\pi}{2} - \theta\right)$ and $-\sin \theta$.

36. *Conjecture*
(a) Use a graphing utility to complete the table.

θ	0.1	0.4	0.7	1.0	1.3
$\tan\left(\theta - \frac{\pi}{2}\right)$					
$-\cot \theta$					

(b) Make a conjecture about the relationship between $\tan\left(\theta - \frac{\pi}{2}\right)$ and $-\cot \theta$.

2.5 · EXERCISES

37. *Essay* Use a graphing utility to graph the function $y = a \sin x$ for $a = \frac{1}{2}, a = \frac{3}{2}$, and $a = -3$. Write a paragraph describing the changes in the graph for the specified changes in a.

38. *Essay* Use a graphing utility to graph the function $y = d + \sin x$ for $d = 2, d = 3.5$, and $d = -2$. Write a paragraph describing the changes in the graph for the specified changes in d.

39. *Essay* Use a graphing utility to graph the function $y = \sin bx$ for $b = \frac{1}{2}$, $b = \frac{3}{2}$, and $b = 4$. Write a paragraph describing the changes in the graph for the specified changes in b.

40. *Essay* Use a graphing utility to graph the function $y = \sin(x - c)$ for $c = 1$, $c = 3$, and $c = -2$. Write a paragraph describing the change in the graph for the specified changes in c.

41. *Data Analysis* The motion of an oscillating weight suspended by a spring was measured by a motion detector. The data was collected and the approximate maximum displacements from equilibrium ($y = 2$) are labeled in the figure. The displacement y from the motion detector is measured in centimeters and the time t is measured in seconds.

(a) Is y a function of t? Explain.

(b) Approximate the amplitude and period.

(c) Find a model for the data.

(d) Use a graphing utility to graph the model of part (c). Compare the result with the data in the figure.

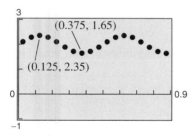

42. *Data Analysis* The table gives the normal daily high temperatures for Honolulu H and Chicago C (in degrees Fahrenheit) for month t, with $t = 1$ corresponding to January. (Source: NOAA)

t	1	2	3	4	5	6
H	80.1	80.5	81.6	82.8	84.7	86.5
C	29.0	33.5	45.8	58.6	70.1	79.6

t	7	8	9	10	11	12
H	87.5	88.7	88.5	86.9	84.1	81.2
C	83.7	81.8	74.8	63.3	48.4	34.0

(a) A model for Honolulu is given by
$$H(t) = 84.40 + 4.28 \sin\left(\frac{\pi t}{6} + 3.86\right).$$
Find a trigonometric model for Chicago.

(b) Use a graphing utility to graph the data points and the model for the temperatures in Honolulu. How well does the model fit?

(c) Use a graphing utility to graph the data points and the model for the temperatures in Chicago. How well does the model fit?

(d) Use the models to estimate the average annual temperature in each city. Which term of the models did you use? Explain.

(e) What are the periods of the two models? Are they what you expected? Explain.

(f) Which city has a greater variability in temperature throughout the year? Which factor of the models determines this variability? Explain.

2.6 · EXERCISES

In Exercises 43–46, use a graphing utility to graph the two equations in the same viewing rectangle. Determine analytically whether the expressions are equivalent.

43. $y_1 = \sin x \csc x$, $y_2 = 1$

44. $y_1 = \sin x \sec x$, $y_2 = \tan x$

45. $y_1 = \dfrac{\cos x}{\sin x}$, $y_2 = \cot x$

46. $y_1 = \sec^2 x - 1$, $y_2 = \tan^2 x$

47. *Approximation* Using calculus, it can be shown that the tangent function can be approximated by the polynomial
$$\tan x \approx x + \frac{2x^3}{3!} + \frac{16x^5}{5!}$$
where x is in radians. Use a graphing utility to graph the tangent function and its polynomial approximation in the same viewing rectangle. How do the graphs compare?

48. *Approximation*   Using calculus, it can be shown that the secant function can be approximated by the polynomial

$$\sec x \approx 1 + \frac{x^2}{2!} + \frac{5x^4}{4!}$$

where x is in radians. Use a graphing utility to graph the secant function and its polynomial approximation in the same viewing rectangle. How do the graphs compare?

49. *Numerical and Graphical Reasoning*   A crossed belt connects a 10-centimeter pulley on an electric motor with a 20-centimeter pulley on a saw arbor (see figure). The electric motor runs at 1700 revolutions per minute.

(a) Determine the number of revolutions per minute of the saw.

(b) How does crossing the belt affect the saw in relation to the motor?

(c) Let L be the total length of the belt. Write L as a function of ϕ, where ϕ is measured in radians. What is the domain of the function? (*Hint:* Add the lengths of the straight sections of the belt and the length of the belt around each pulley.)

(d) Use a graphing utility to complete the table.

ϕ	0.3	0.6	0.9	1.2	1.5
L					

(e) As ϕ increases, do the lengths of the straight sections of the belt change faster or slower than the lengths of the belt around each pulley?

(f) Use a graphing utility to graph the function over the appropriate domain.

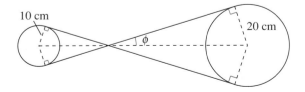
10 cm   20 cm ϕ

2.7 · EXERCISES

50. *Numerical and Graphical Analysis*   Consider the function $y = \arcsin x$.

(a) Use a graphing utility to complete the table.

x	-1	-0.8	-0.6	-0.4	-0.2
y					

x	0	0.2	0.4	0.6	0.8	1
y						

(b) Plot the points from the table in part (a) and graph the function. (Do not use a graphing utility.)

(c) Use a graphing utility to graph the inverse sine function and compare the result with your hand drawn graph in part (b).

(d) Determine any intercepts and symmetry of the graph.

51. *Numerical and Graphical Analysis*   Consider the function $y = \arccos x$.

(a) Use a graphing utility to complete the table.

x	-1	-0.8	-0.6	-0.4	-0.2
y					

x	0	0.2	0.4	0.6	0.8	1
y						

(b) Plot the points from the table in part (a) and graph the function. (Do not use a graphing utility.)

(c) Use a graphing utility to graph the inverse cosine function and compare the result with your hand drawn graph in part (b).

(d) Determine any intercepts and symmetry of the graph.

52. *Numerical and Graphical Analysis* Consider the function $y = \arctan x$.

(a) Use a graphing utility to complete the table.

x	-10	-8	-6	-4	-2
y					

x	0	2	4	6	8	10
y						

(b) Plot the points from the table in part (a) and graph the function. (Do not use a graphing utility.)

(c) Use a graphing utility to graph the inverse tangent function and compare the result with your hand drawn graph in part (b).

(d) Determine the horizontal asymptotes of the graph.

53. *Think About It* Consider the functions $f(x) = \sin x$ and $f^{-1}(x) = \arcsin x$.

(a) Use a graphing utility to graph the composite functions $f \circ f^{-1}$ and $f^{-1} \circ f$.

(b) Explain why the graphs in part (a) are not the graph of the line $y = x$. Why do the graphs of $f \circ f^{-1}$ and $f^{-1} \circ f$ differ?

54. *Think About It* Use a graphing utility to graph the functions $f(x) = \sqrt{x}$ and $g(x) = 6 \arctan x$. For $x > 0$, it appears that $g > f$. Explain why you know that there exists a positive real number a such that $g < f$ for $x > a$. Approximate the number a.

2.8 · EXERCISES

55. *Length of a Shadow* A shadow of length L is created by a 60-foot silo when the sun is $\theta°$ above the horizon.

(a) Write L as a function of θ.

(b) Use a graphing utility to complete the table.

θ	$10°$	$20°$	$30°$	$40°$	$50°$
L					

(c) The angle measure increases in equal increments in the table. Does the length of the shadow change in equal increments? Explain.

56. *Length of a Shadow* A shadow of length L is created by a 600-foot building when the sun is $\theta°$ above the horizon.

(a) Write L as a function of θ.

(b) Use a graphing utility to complete the table.

θ	$10°$	$20°$	$30°$	$40°$	$50°$
L					

(c) The angle measure increases in equal increments in the table. Does the length of the shadow change in equal increments? Explain.

57. *Height* A ladder of length 20 feet leans against the side of a house. The angle of elevation of the ladder is $\theta°$.

(a) Write the height h of the ladder as a function of θ.

(b) Use a graphing utility to complete the table.

θ	$60°$	$65°$	$70°$	$75°$	$80°$
h					

58. *Height* The length of a shadow of a tree is 110 feet when the angle of elevation of the sun is $\theta°$.

(a) Write the height h of the tree as a function of θ.

(b) Use a graphing utility to complete the table.

θ	$10°$	$15°$	$20°$	$25°$	$30°$
h					

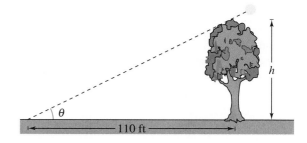

59. *Numerical and Graphical Analysis*   A 2-meter-high fence is 3 meters from the side of a grain storage bin. A grain elevator must reach from ground level outside the fence to the storage bin (see figure). The objective is to determine the shortest elevator meeting the constraints.

(a) Complete four rows of the table.

θ	L_1	L_2	$L_1 + L_2$
0.1	$\dfrac{2}{\sin 0.1}$	$\dfrac{3}{\cos 0.1}$	23.0
0.2	$\dfrac{2}{\sin 0.2}$	$\dfrac{3}{\cos 0.2}$	13.1

(b) Use a graphing utility to generate additional rows of the table. Use the table to estimate the minimum length.

(c) Write the length L as a function of θ.

(d) Use a graphing utility to graph the function. Use the graph to estimate the minimum length. How does your estimate compare with that of part (b)?

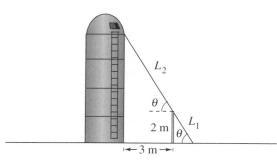

60. *Numerical and Graphical Analysis*   The cross sections of an irrigation canal are isosceles trapezoids where the length of three of the sides is 8 feet (see figure). The objective is to find the angle θ that maximizes the area of the cross sections.

(a) Complete six rows of the table.

Base 1	Base 2	Altitude	Area
8	$8 + 16\cos 10°$	$8\sin 10°$	22.1
8	$8 + 16\cos 20°$	$8\sin 20°$	42.5

(b) Use a graphing utility to generate additional rows of the table. Use the table to estimate the maximum cross-sectional area.

(c) Write the area A as a function of θ.

(d) Use a graphing utility to graph the function. Use the graph to estimate the maximum cross-sectional area. How does your estimate compare with that of part (b)?

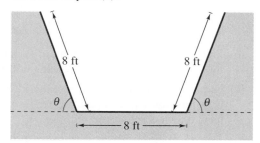

61. *Data Analysis*   The table gives the time S of sunset (Greenwich Mean Time) at 40° north latitude on the 15th of each month. The month is represented by t, with $t = 1$ corresponding to January.

t	1	2	3	4	5	6
H	16:59	17:35	18:06	18:38	19:08	19:30

t	7	8	9	10	11	12
H	19:28	18:57	18:09	17:21	16:44	16:36

A model (where minutes have been converted to the decimal part of an hour) for this data is

$$S(t) = 18.09 + 1.41 \sin\left(\frac{\pi t}{6} + 4.60\right).$$

(a) Use a graphing utility to graph the data points and the model in the same viewing rectangle.

(b) What is the period of the model? Is it what you expected? Explain.

(c) What is the amplitude of the function? What does it represent in the model? Explain.

62. *Numerical and Graphical Analysis* Consider the shaded region outside the sector of the circle of radius 10 meters and inside the right triangle (see figure).

(a) Write the area A of the region as a function of θ. Determine the domain of the function.

(b) Use a graphing utility to complete the table.

θ	0	0.3	0.6	0.9	1.2	1.5
A						

(c) Use a graphing utility to graph the function over the appropriate domain.

(d) What does the area approach as θ approaches $\pi/2$?

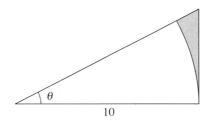

3.1 · EXERCISES

Numerical and Graphical Analysis In Exercises 63–66, use a graphing utility to complete the table and graph the functions. Make a conjecture about y_1 and y_2.

x	0.2	0.4	0.6	0.8	1.0	1.2	1.4
y_1							
y_2							

63. $y_1 = \cos\left(\dfrac{\pi}{2} - x\right), \quad y_2 = \sin x$

64. $y_1 = \cos x + \sin x \tan x, \quad y_2 = \sec x$

65. $y_1 = \dfrac{\cos x}{1 - \sin x}, \quad y_2 = \dfrac{1 + \sin x}{\cos x}$

66. $y_1 = \sec^4 x - \sec^2 x, \quad y_2 = \tan^2 x + \tan^4 x$

In Exercises 67 and 68, use a graphing utility to determine which of the six trigonometric functions is equal to the expression.

67. $\cos x \cot x + \sin x$

68. $\dfrac{1}{2}\left(\dfrac{1 + \sin\theta}{\cos\theta} + \dfrac{\cos\theta}{1 + \sin\theta}\right)$

3.2 · EXERCISES

Conjecture In Exercises 69–72, use a graphing utility to graph the trigonometric function. Use the graph to make a conjecture about a simplification of the expression. Verify the resulting identity analytically.

69. $\dfrac{1}{\cot x + 1} + \dfrac{1}{\tan x + 1}$

70. $\dfrac{\cos x}{1 - \tan x} + \dfrac{\sin x \cos x}{\sin x - \cos x}$

71. $\dfrac{1}{\sin x} - \dfrac{\cos^2 x}{\sin x}$

72. $\sin t + \dfrac{\cot^2 t}{\csc t}$

3.3 · EXERCISES

In Exercises 73–76, consider the function in the interval $[0, 2\pi)$. (a) Use a graphing utility to complete the table. Then make a conjecture about any unit intervals containing a zero of the function. Give a reason for your answer. (b) Use a graphing utility to graph the function and determine any unit intervals containing a zero of the function. Do the results agree with those of part (a)? Explain. (c) Use a graphing utility to approximate the zero of the function.

x	0	1	2	3	4	5	6
$f(x)$							

73. $f(x) = \cot\dfrac{x}{2} - x$ 　　　**74.** $f(x) = x \sec x - 1$

75. $f(x) = \sin^2 x + 2\sin x - 1$

76. $f(x) = 4\cos^2 x - 4\cos x - 1$

3.4 · EXERCISES

77. Verify the identity used in calculus.

$$\frac{\cos(x + h) - \cos x}{h} = \cos x\left(\frac{\cos h - 1}{h}\right) - \sin x\left(\frac{\sin h}{h}\right)$$

78. *Exploration* Let $x = \frac{\pi}{6}$ in the identity in Exercise 77. Define the functions f and g as follows.

$$f(h) = \frac{\cos\left(\frac{\pi}{6} + h\right) - \cos\frac{\pi}{6}}{h}$$

$$g(h) = \cos\frac{\pi}{6}\left(\frac{\cos h - 1}{h}\right) - \sin\frac{\pi}{6}\left(\frac{\sin h}{h}\right)$$

(a) What are the domains of the functions?

(b) Use a graphing utility to complete the table.

h	0.01	0.02	0.05	0.10	0.20	0.50
$f(h)$						
$g(h)$						

(c) Use a graphing utility to graph the functions.

(d) Use the table and graph to make a conjecture about the values of the functions as $h \to 0$.

3.5 · EXERCISES

In Exercises 79 and 80, (a) use a graphing utility to graph the function and approximate the maximum and minimum points on the graph in the interval $[0, 2\pi)$, and (b) solve the trigonometric equation and verify that its solutions are the x-coordinates of the maximum and minimum points of f (calculus is required to find the trigonometric equation).

Function	*Trigonometric Equation*
79. $f(x) = 4\sin\dfrac{x}{2} + \cos x$	$2\cos\dfrac{x}{2} - \sin x = 0$
80. $f(x) = \cos 2x - 2\sin x$	$-2\cos x(2\sin x + 1) = 0$

81. *Exploration* Consider the function

$$f(x) = \sin^4 x + \cos^4 x.$$

(a) Use the power-reducing formulas to write the expression in terms of the first power of the cosine.

(b) Determine another way of rewriting the function. Use a graphing utility to rule out incorrectly rewritten functions.

(c) Determine a trigonometric expression such that the sum of the expression and the function is a perfect square trinomial. Rewrite the function as a perfect square trinomial minus the term that you added. Use a graphing utility to rule out incorrectly rewritten functions.

(d) Rewrite the result of part (c) in terms of the sine of a double angle. Use a graphing utility to rule out incorrectly rewritten functions.

(e) In how many ways have you rewritten the trigonometric function? When rewriting a trigonometric expression, your results may not be the same as a friend's results. Does this necessarily mean that one of you is wrong? Explain.

4.1 · EXERCISES

82. *Graphical and Numerical Analysis* In the figure, α and β are positive angles.

(a) Write α as a function of β.

(b) Use a graphing utility to graph the function. Determine its domain and range.

(c) Write c as a function of β.

(d) Use a graphing utility to graph the function in part (c). Determine its domain and range.

(e) Use a graphing utility to complete the table. What can you infer?

β	0	0.4	0.8	1.2	1.6	2.0	2.4	2.8
α								
c								

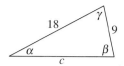

FIGURE FOR 82

83. *Graphical Analysis*

(a) Write the area A of the shaded region in the figure as a function of θ.

(b) Use a graphing utility to graph the area function.

(c) Determine the domain of the area function. Explain how the area of the region and the domain of the function would change if the 8-centimeter line segment were decreased in length.

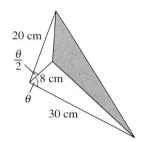

4.2 · EXERCISES

84. *Engine Design*   An engine has a 7-inch connecting rod fastened to a crank (see figure).

(a) Use the Law of Cosines to write an equation giving the relationship between x and θ.

(b) Write x as a function of θ. (Select the sign that yields positive values of x.)

(c) Use a graphing utility to graph the function in part (b).

(d) Use the graph in part (c) to determine the maximum distance the piston moves in one cycle.

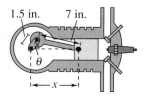

4.3 · EXERCISES

85. *Graphical Reasoning*   Consider two forces $\mathbf{F}_1 = \langle 10, 0 \rangle$ and $\mathbf{F}_2 = 5\langle \cos\theta, \sin\theta \rangle$.

(a) Find $\|\mathbf{F}_1 + \mathbf{F}_2\|$.

(b) Determine the magnitude of the resultant as a function of θ. Use a graphing utility to graph the function for $0 \le \theta < 2\pi$.

(c) Use the graph in part (b) to determine the range of the function. What is its maximum and for what value of θ does it occur? What is its minimum and for what value of θ does it occur?

(d) Explain why the magnitude of the resultant is never 0.

86. *Numerical and Graphical Analysis*   Forces with magnitudes of 150 newtons and 220 newtons act on a hook (see figure).

(a) If $\theta = 30°$, find the direction and magnitude of the resultant of these forces.

(b) Express the magnitude M of the resultant and the direction α of the resultant as functions of θ, where $0° \le \theta \le 180°$.

(c) Use a graphing utility to complete the table.

θ	0°	30°	60°	90°	120°	150°	180°
M							
α							

(d) Use a graphing utility to graph the two functions.

(e) Explain why one function decreases for increasing θ, whereas the other doesn't.

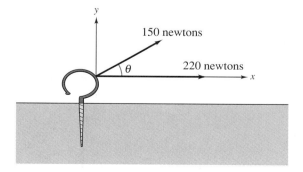

5.2 · EXERCISES

In Exercises 87–89, use a graphing utility to (a) graph the function, (b) determine the number of real zeros of the function, and (c) approximate the real zeros of the function to the nearest hundredth.

87. $f(x) = x^4 + 2x + 1$

88. $g(x) = x^3 - 3x^2 + 3x + 2$

89. $h(x) = x^3 - 6x^2 + 12x - 10$

90. *Exploration* Use a graphing utility to graph the function $f(x) = x^4 - 4x^2 + k$ for different values of k. Find values of k such that the zeros of f satisfy the specified characteristics. (Some parts do not have unique answers.)

(a) Four real zeros

(b) Two real zeros each of multiplicity 2

(c) Two real zeros and two complex zeros

(d) Four complex zeros

5.3 · EXERCISES

In Exercises 91–94, use a graphing utility to represent the complex number in trigonometric form.

91. $5 + 2i$ **92.** $-3 + i$

93. $3\sqrt{2} - 7i$ **94.** $-8 - 5\sqrt{3}i$

In Exercises 95–98, use a graphing utility to represent the complex number in standard form.

95. $5\left(\cos \dfrac{\pi}{9} + i \sin \dfrac{\pi}{9}\right)$ **96.** $12\left(\cos \dfrac{3\pi}{5} + i \sin \dfrac{3\pi}{5}\right)$

97. $4(\cos 216.5° + i \sin 216.5°)$

98. $9(\cos 58° + i \sin 58°)$

5.4 · EXERCISES

In Exercises 99–102, use a graphing utility and DeMoivre's Theorem to find the indicated power of the complex number. Express the result in standard form.

99. $(3 - 2i)^5$ **100.** $\left(\sqrt{5} - 4i\right)^3$

101. $[3(\cos 15° + i \sin 15°)]^4$

102. $\left[2\left(\cos \dfrac{\pi}{10} + i \sin \dfrac{\pi}{10}\right)\right]^5$

6.1 · EXERCISES

In Exercises 103–106, use a graphing utility to graph the exponential function. Identify any asymptotes of the graph.

103. $y = 2^{-x^2}$ **104.** $y = 3^{-|x|}$

105. $y = 3^{x-2} + 1$ **106.** $y = 4^{x+1} - 2$

107. *Data Analysis* A meteorologist measures the atmospheric pressure P (in kilograms per square meter) at altitude h (in kilometers). The data is shown in the table.

h	0	5	10	15	20
P	10,332	5583	2376	1240	517

(a) Use a graphing utility to plot the data points.

(b) A model for the data is given by
$$P = 10,958e^{-0.15h}.$$
Use a graphing utility to graph the model in the same viewing rectangle used in part (a). How well does the model fit the data?

(c) Use a graphing utility to create a table comparing the model with the sample data.

(d) Estimate the atmospheric pressure at a height of 8 kilometers.

(e) Estimate the altitude at which the atmospheric pressure is 2000 kilograms per square meter.

108. *Data Analysis*   A cup of water at an initial temperature of 78°C is placed in a room at a constant temperature of 21°C. The temperature of the water is measured every 5 minutes for a period of $\frac{1}{2}$ hour. The results are recorded in the table, where t is time in minutes and T is the temperature in degrees Celsius.

t	0	5	10	15
T	78.0°	66.0°	57.5°	51.2°

t	20	25	30
T	46.3°	42.5°	39.6°

(a) Use the regression capabilities of a graphing utility to fit a line to the data. Use the graphing utility to plot the data points and the regression line. Does the data appear linear? Explain.

(b) Use the regression capabilities of a graphing utility to fit a parabola to the data. Use the graphing utility to plot the data points and the regression parabola. Does the data appear quadratic? Even though the quadratic model appears to be a "good" fit, explain why it may not be a good model for predicting the temperature of the water when $t = 60$.

(c) The graph of the model should be asymptotic with the temperature of the room. Subtract the room temperature from each of the temperatures in the table. Use a graphing utility to fit an exponential model to the revised data. Add the room temperature to this regression model. Use a graphing utility to plot the original data points and the model.

(d) Explain why the procedure in part (c) was necessary for finding the exponential model.

109. *Finding a Pattern*   Use a graphing utility to compare the graph of the function $y = e^x$ with the graphs of the following functions.

(a) $y_1 = 1 + \dfrac{x}{1!}$

(b) $y_2 = 1 + \dfrac{x}{1!} + \dfrac{x^2}{2!}$

(c) $y_3 = 1 + \dfrac{x}{1!} + \dfrac{x^2}{2!} + \dfrac{x^3}{3!}$

110. *Finding A Pattern*   Identify the pattern of successive polynomials given in Exercise 109. Extend the pattern one more term and compare the graph of the resulting polynomial function with the graph of $y = e^x$. What do you think this pattern implies?

6.2 · EXERCISES

In Exercises 111–116, use a graphing utility to graph the logarithmic function. Determine the domain and identify any vertical asymptote and x-intercept.

111. $y = \log_{10}\left(\dfrac{x}{5}\right)$ **112.** $y = \log_{10}(-x)$

113. $f(x) = \ln(x - 2)$ **114.** $h(x) = \ln(x + 1)$

115. $g(x) = \ln(-x)$ **116.** $f(x) = \ln(3 - x)$

In Exercises 117 and 118, use a graphing utility to graph the function. What is the domain? Use the graph to determine the intervals in which the function is increasing and decreasing and approximate any relative maximum or minimum values of the function.

117. $h(x) = 4x \ln x$ **118.** $f(x) = \dfrac{x}{\ln x}$

119. *Exploration*   Use a graphing utility to compare the graph of the function $y = \ln x$ with the graphs of the following functions.

(a) $y = x - 1$

(b) $y = (x - 1) - \frac{1}{2}(x - 1)^2$

(c) $y = (x - 1) - \frac{1}{2}(x - 1)^2 + \frac{1}{3}(x - 1)^3$

120. *Finding a Pattern*   Identify the pattern of successive polynomials given in Exercise 119. Extend the pattern one more term and compare the graph of the resulting polynomial function with the graph of $y = \ln x$. What do you think the pattern implies?

121. *Data Analysis* A meteorologist measures the atmospheric pressure P (in kilograms per square meter) at altitude h (in kilometers). The data is shown in the table.

h	0	5	10	15	20
P	10,332	5583	2376	1240	517

A model for the data is given by

$$P = 10{,}957e^{-0.15h}.$$

(a) Use a graphing utility to plot the data points and graph the model in the same viewing rectangle.

(b) Use a graphing utility to plot the following points: $(h, \ln P)$. Use the regression capabilities of the graphing utility to fit a regression line to the revised data points.

(c) The line in part (b) has the form

$$\ln P = ah + b.$$

Write the line in exponential form.

(d) Verify, graphically and analytically, that the result of part (c) is equivalent to the given exponential model for the data.

122. *Tractrix* A person walking along a dock (the y-axis) drags a boat by a 10-foot rope. The boat travels along a path known as a *tractrix*. The equation of this path is

$$y = 10 \ln\left(\frac{10 + \sqrt{100 - x^2}}{x}\right) - \sqrt{100 - x^2}.$$

(a) Use a graphing utility to obtain a graph of the function. What is the domain of the function?

(b) Identify any asymptotes of the graph.

(c) Determine the position of the person when the x-coordinate of the position of the boat is $x = 2$.

(d) Let $(0, p)$ be the position of the person. Determine p as a function of x, the x-coordinate of the position of the boat.

(e) Use a graphing utility to graph the function p. When does the position of the person change most for a small change in the position of the boat? Explain.

6.3 · EXERCISES

Graphical Analysis In Exercises 123 and 124, use a graphing utility to graph the two equations in the same viewing rectangle. What do the graphs suggest? Verify your conclusion algebraically.

123. $y_1 = 2[\ln 8 - \ln(x^2 + 1)]$, $\quad y_2 = \ln\left[\dfrac{64}{(x^2 + 1)^2}\right]$

124. $y_1 = \ln x + \frac{1}{3}\ln(x + 1)$, $\quad y_2 = \ln(x\sqrt[3]{x + 1})$

Think About It In Exercises 125 and 126, use a graphing utility to graph the two equations in the same viewing rectangle. Are the expressions equivalent? Explain.

125. $y_1 = \ln x^2$, $\quad y_2 = 2\ln x$

126. $y_1 = \frac{1}{4}\ln[x^4(x^2 + 1)]$, $\quad y_2 = \ln x + \frac{1}{4}\ln(x^2 + 1)$

6.4 · EXERCISES

In Exercises 127–130, use a graphing utility to approximate the point of intersection of the graphs. Round the result to three decimal places.

127. $y_1 = 7$ 　　　　　**128.** $y_1 = 500$
$\quad\;\; y_2 = 2^x$ 　　　　　$\quad\;\;\; y_2 = 1500e^{-x/2}$

129. $y_1 = 3$ 　　　　　**130.** $y_1 = 10$
$\quad\;\; y_2 = \ln x$ 　　　　$\quad\;\;\; y_2 = 4\ln(x - 2)$

6.5 · EXERCISES

131. *Data Analysis* The time t (in seconds) required to attain a speed of s miles per hour from a standing start for a 1995 Dodge Avenger is given in the table. (Source: *Road & Track*, March 1995)

s	30	40	50	60	70	80	90
t	3.4	5.0	7.0	9.3	12.0	15.8	20.0

Two models for this data are

$$t_1 = 40.757 + 0.556s - 15.817 \ln s$$

and

$$t_2 = 1.2259 + 0.0023s^2.$$

(a) Use a graphing utility to fit a linear model t_3 and an exponential model t_4 to the data.

(b) Use a graphing utility to graph the data points and each of the four models.

(c) Use a graphing utility to create a table comparing the given data with the estimates obtained from each model.

(d) Use the results of part (c) to find the sum of the absolute values of the differences between the data and the estimated values given by each model. Based on the four sums, which model do you think best fits the data? Explain.

6.6 · EXERCISES

132. *Foreign Travel*   The numbers of United States citizens y (in millions) who traveled to foreign countries in the years 1984 through 1993 are given in the table, where $t = 4$ represents the year 1984. (Source: U.S. Bureau of Economic Analysis)

t	4	5	6	7	8
y	34.0	34.7	37.2	39.4	40.7

t	9	10	11	12	13
y	41.1	44.6	41.6	43.9	45.5

(a) Use the regression capabilities of a graphing utility to fit an appropriate model to the data.

(b) Use the graphing utility to plot the data and graph the model.

(c) Use the model to predict the number of foreign travelers in the year 2001.

7.2 · EXERCISES

133. *Exploration*   Consider the parabola $x^2 = 4py$.

(a) Use a graphing utility to graph the parabola for $p = 1, p = 2, p = 3$, and $p = 4$. Describe the effect on the graph when p increases.

(b) Locate the focus for each parabola in part (a).

(c) For each parabola in part (a), find the length of the chord passing through the focus and parallel to the directrix. How can the length of this chord be determined directly from the standard form of the equation of the parabola?

(d) Explain how the result of part (c) can be used as a sketching aid when graphing parabolas.

7.3 · EXERCISES

134. *Exploration*   The area A of the ellipse

$$\frac{x^2}{a^2} + \frac{y^2}{b^2} = 1$$

is $A = \pi ab$. For a particular application, $a + b = 20$.

(a) Write the area of the ellipse as a function of a.

(b) Find the equation of an ellipse with an area of 264 square centimeters.

(c) Complete the table and make a conjecture about the shape of the ellipse with a maximum area.

a	8	9	10	11	12	13
A						

(d) Use a graphing utility to graph the area function, and use the graph to make a conjecture about the shape of the ellipse that yields a maximum area.

7.6 · EXERCISES

Projectile Motion   In Exercises 135 and 136, consider a projectile launched at a height h feet above the ground at an angle θ with the horizontal. If the initial velocity is v_0 feet per second, the path of the projectile is modeled by the parametric equations

$$x = (v_0 \cos \theta)t \quad \text{and} \quad y = h + (v_0 \sin \theta)t - 16t^2.$$

135. *Baseball*   The centerfield fence in a ballpark is 10 feet high and 400 feet from home plate. The ball is hit 3 feet above the ground. It leaves the bat at an angle of θ degrees with the horizontal and at a speed of 100 miles per hour (see figure).

  (a) Write a set of parametric equations for the path of the baseball.

  (b) Use a graphing utility to sketch the path of the baseball for $\theta = 15°$. Is the hit a home run?

  (c) Use a graphing utility to sketch the path of the baseball if $\theta = 23°$. Is the hit a home run?

  (d) Find the minimum angle required for the hit to be a home run.

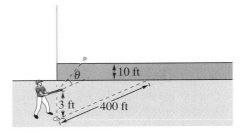

136. *Football*   The quarterback of a football team releases a pass at a height of 7 feet above the playing field, and the football is caught by a receiver at a height of 4 feet, 30 yards directly downfield. The pass is released at an angle of $35°$ with the horizontal.

  (a) Write a set of parametric equations for the path of the football.

  (b) Find the speed of the football when it is released.

  (c) Use a graphing utility to graph the path of the football and approximate its maximum height.

  (d) Find the time the receiver has to position himself after the quarterback releases the ball.

7.7 · EXERCISES

In Exercises 137–140, use a graphing utility to find the rectangular coordinates for the point given in polar coordinates.

137. $\left(2, \dfrac{3\pi}{4}\right)$ **138.** $\left(-2, \dfrac{7\pi}{6}\right)$

139. $(-4.5, 1.3)$ **140.** $(8.25, 3.5)$

In Exercises 141–146, use a graphing utility to find one set of polar coordinates for the point given in rectangular coordinates.

141. $(3, -2)$ **142.** $(-4, 1)$

143. $\left(\sqrt{3}, 2\right)$ **144.** $\left(3\sqrt{2}, 3\sqrt{2}\right)$

145. $\left(\dfrac{5}{2}, \dfrac{4}{3}\right)$ **146.** $(0, -5)$

7.8 · EXERCISES

In Exercises 147–150, use a graphing utility to graph the polar equation and show that the indicated line is an asymptote of the graph.

	Name of Graph	*Polar Equation*	*Asymptote*
147.	Conchoid	$r = 2 - \sec\theta$	$x = -1$
148.	Conchoid	$r = 2 + \csc\theta$	$y = 1$
149.	Hyperbolic spiral	$r = \dfrac{2}{\theta}$	$y = 2$
150.	Strophoid	$r = 2\cos 2\theta \sec\theta$	$x = -2$

7.9 · EXERCISES

In Exercises 151–156, use a graphing utility to graph the polar equation. Identify the graph.

151. $r = \dfrac{-1}{1 - \sin\theta}$ **152.** $r = \dfrac{-3}{2 + 4\sin\theta}$

153. $r = \dfrac{3}{-4 + 2\cos\theta}$ **154.** $r = \dfrac{4}{1 - 2\cos\theta}$

155. $r = \dfrac{6}{2 - \cos\theta}$ **156.** $r = \dfrac{2}{2 + 3\sin\theta}$

In Exercises 157 and 158, use a graphing utility to graph the rotated conic.

157. $r = \dfrac{4}{2 + \sin(\theta + \pi/6)}$ *(See page 494, Exercise 13.)*

158. $r = \dfrac{5}{-1 + 2\cos(\theta + 2\pi/3)}$ *(See page 494, Exercise 16.)*

ANSWERS

Warm-Ups, Odd-Numbered Exercises, and Cumulative Tests

CHAPTER 1

Section 1.1 (*page 7*)

1. (a) 5, 1   (b) $-9, 5, 0, 1$ (c) $-9, -\frac{7}{2}, 5, \frac{2}{3}, 0, 1$
(d) $\sqrt{2}$

3. (a) 12, 1, $\sqrt{4}$ (b) 12, $-13, 1, \sqrt{4}$
(c) 12, $-13, 1, \sqrt{4}, \frac{3}{2}$ (d) $\sqrt{6}$

5. (a) $\frac{6}{3}$ (b) $\frac{6}{3}$ (c) $-\frac{1}{3}, \frac{6}{3}, -7.5$ (d) $-\pi, \frac{1}{2}\sqrt{2}$

7. $\frac{3}{2} < 7$ **9.** $-4 > -8$

11. $\frac{5}{6} > \frac{2}{3}$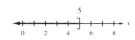

13. $-1 \le x \le 3$, bounded   **15.** $10 < x$, unbounded
17. $(-\infty, 5]$ **19.** $(-\infty, 0)$

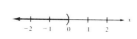

21. $[4, \infty)$ **23.** $(-2, 2)$

25. $[-1, 0)$

27. $(-\infty, 0)$ **29.** $(-\infty, 25]$
31. $[30, \infty)$ **33.** $[0.035, 0.06]$ **35.** -1 **37.** -9
39. 3.75   **41.** $|-3| > -|-3|$ **43.** $-5 = -|5|$
45. $-|-2| = -|2|$ **47.** $\frac{5}{2}$ **49.** 51   **51.** 14.99
53. $|x - 5| \le 3$ **55.** $|z - \frac{3}{2}| > 1$ **57.** $|y| \ge 6$
59. $\frac{127}{90}, \frac{584}{413}, \frac{7071}{5000}, \sqrt{2}, \frac{47}{33}$ **61.** 0.625   **63.** 0.123123 . . .
65. False. The reciprocal of 2 is $\frac{1}{2}$, which is not an integer.
67. True

Section 1.2 (*page 16*)

WARM-UP   **1.** $-6x + 30$ **2.** $-40x + 24$ **3.** x
4. $x + 26$ **5.** $4x^2 - 25$ **6.** $9x^2 - 24x + 16$
7. $\frac{8x}{15}$ **8.** $\frac{3x}{4}$ **9.** $-\dfrac{1}{x(x + 1)}$ **10.** $\dfrac{5}{x}$

1. Identity   **3.** Conditional   **5.** Conditional
7. (a) No   (b) No   (c) Yes   (d) No
9. (a) Yes   (b) Yes   (c) No   (d) No
11. (a) Yes   (b) No   (c) No   (d) No
13. 9   **15.** -4 **17.** No solution   **19.** 10   **21.** 4
23. 5   **25.** $\frac{11}{6}$ **27.** 0   **29.** $\dfrac{1 + 4b}{2 + a}, a \ne -2$
31. $0, -\frac{1}{2}$ **33.** $4, -2$ **35.** $-\frac{7}{4}$ **37.** $-\frac{3}{2}, 11$
39. $\frac{1}{2}, -1$ **41.** $\frac{1}{4}, -\frac{3}{4}$ **43.** $1 \pm \sqrt{3}$ **45.** $\dfrac{2}{3} \pm \dfrac{\sqrt{7}}{3}$
47. $-\frac{1}{2} \pm \sqrt{2}$ **49.** $6 \pm \sqrt{11}$ **51.** $-\dfrac{1}{2} \pm \dfrac{\sqrt{21}}{6}$
53. $-0.643, 0.976$ **55.** 0.126, 0.561   **57.** $0, \pm\frac{3}{2}$
59. $3, -1, 0$ **61.** ± 3 **63.** $-3, 0$ **65.** $1, -2$
67. $7600
69. $x = 35$ ft, $y = 20$ ft or $x = 15$ ft, $y = \frac{140}{3}$ ft   **71.** 34

Section 1.3 (page 24)

WARM-UP   **1.** 5   **2.** $3\sqrt{2}$ **3.** 1   **4.** -2
5. $3\left(\sqrt{2} + \sqrt{5}\right)$ **6.** $2\left(\sqrt{3} + \sqrt{11}\right)$ **7.** $-3, 11$
8. 9, 1   **9.** 11   **10.** 4

1.
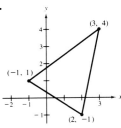

3.
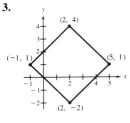

5. 8   **7.** 5   **9.** $a = 4, b = 3, c = 5$
11. $a = 10, b = 3, c = \sqrt{109}$
13. (a)   (b) 10   (c) (5, 4)
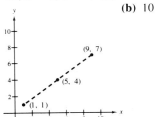

15. (a)   (b) 17   (c) $\left(0, \frac{5}{2}\right)$
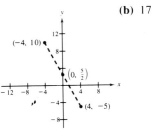

17. (a)   (b) $2\sqrt{10}$ (c) (2, 3)
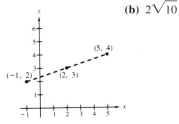

19. (a)   (b) $\dfrac{\sqrt{82}}{3}$ (c) $\left(-1, \dfrac{7}{8}\right)$
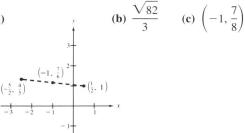

21. (a)   (b) $\sqrt{110.97}$ (c) (1.25, 3.6)
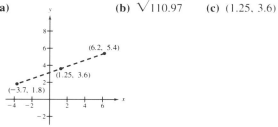

23. (a)   (b) $6\sqrt{277}$ (c) $(6, -45)$
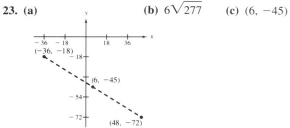

25. \$630,000   **27.** $\left(\sqrt{5}\right)^2 + \left(\sqrt{45}\right)^2 = \left(\sqrt{50}\right)^2$
29. All sides have a length of $\sqrt{5}$.   **31.** $x = 6, -4$
33. $y = \pm 15$ **35.** $3x - 2y - 1 = 0$
37. Quadrant IV   **39.** Quadrant I   **41.** Quadrant II
43. Quadrant III or IV   **45.** Quadrant I or III
47. Quadrant III
49.
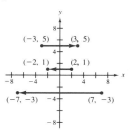

The points are reflected through the y-axis.

51. $x^2 + y^2 = 9$ **53.** $(x - 2)^2 + (y + 1)^2 = 16$
55. $(x + 1)^2 + (y - 2)^2 = 5$
57. $(x - 3)^2 + (y - 4)^2 = 25$

Section 1.4 (*page 33*)

WARM-UP   **1.** $14x - 42$ **2.** $-17s$ **3.** $-24y^7$
4. $\dfrac{2a}{3b^5}$ **5.** $5x^2\sqrt{6}$ **6.** $\dfrac{1}{2}$ **7.** $2x^2(x - 3)$
8. $2(t + 3)(t^2 + 3t - 2)$ **9.** $(3z + 10)(2z - 5)$
10. $(2s + 5)(2s - 5)$

1. (a) Yes   **(b)** Yes   **3. (a)** No   **(b)** Yes
5. (a) Yes   **(b)** Yes   **7.** 2   **9.** 4

11.

x	-4	-2	0	2	4
y	11	7	3	-1	-5
(x, y)	$(-4, 11)$	$(-2, 7)$	$(0, 3)$	$(2, -1)$	$(4, -5)$

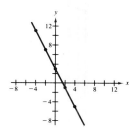

13. c

```
Xmin=-2
Xmax=6
Xscl=1
Ymin=-1
Ymax=5
Yscl=1
```

14. f

```
Xmin=-4
Xmax=2
Xscl=1
Ymin=-2
Ymax=4
Yscl=1
```

15. d

```
Xmin=-3
Xmax=3
Xscl=1
Ymin=-1
Ymax=3
Yscl=1
```

16. a

```
Xmin=-1
Xmax=8
Xscl=1
Ymin=-1
Ymax=4
Yscl=1
```

17. e

```
Xmin=-2
Xmax=2
Xscl=1
Ymin=-2
Ymax=2
Yscl=1
```

18. b

```
Xmin=-4
Xmax=4
Xscl=1
Ymin=-3
Ymax=3
Yscl=1
```

19.

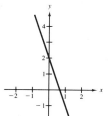

21.

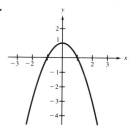

23.

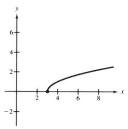

25.

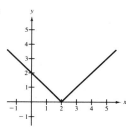

27.

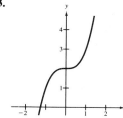

29.

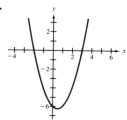

31.

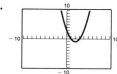

The graph intersects the x-axis twice.
The graph intersects the y-axis once.

33.

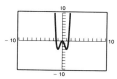

The graph intersects the x-axis three times.
The graph intersects the y-axis once.

35.

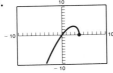

The graph intersects the x-axis twice.
The graph intersects the y-axis once.

37.

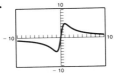

The graph intersects the x-axis once.
The graph intersects the y-axis once.

39.

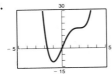

The graph intersects the x-axis twice.
The graph intersects the y-axis once.

41.

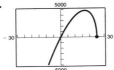

The graph intersects the x-axis twice.
The graph intersects the y-axis once.

43.

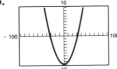

The graph intersects the x-axis twice.
The graph intersects the y-axis once.

45. $y = \sqrt{64 - x^2}, y = -\sqrt{64 - x^2}$

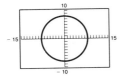

47. $y = \sqrt{6(12 - x^2)}, y = -\sqrt{6(12 - x^2)}$

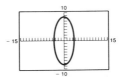

49.

```
Xmin=-1
Xmax=100
Xscl=5
Ymin=-1
Ymax=40
Yscl=2
```

51.

```
Xmin=-1
Xmax=20
Xscl=1
Ymin=-25
Ymax=150
Yscl=10
```

53.

The person would select the first viewing rectangle.

55. (a)

x	-1	0	1
y	-1	0	1

(b)

x	-1	$-\frac{3}{4}$	$-\frac{1}{2}$	$-\frac{1}{4}$	0	$\frac{1}{4}$	$\frac{1}{2}$	$\frac{3}{4}$	1
y	-1	-0.91	-0.79	-0.63	0	0.63	0.79	0.91	1

(c)

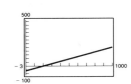

57.

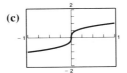

(a) $y = 1.73$
(b) $x = -4$

59.

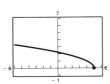

(a) $y = 2.47$
(b) $x = -1.65, 1$

61. (b)

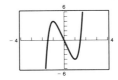

63.

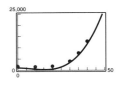

$\$16,890$

65.

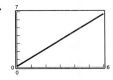

67. There are an unlimited number of correct answers, one of which is $a = 1$ and $b = -5$.

Section 1.5   (page 47)

WARM-UP   **1.** $-\frac{9}{2}$ **2.** $-\frac{13}{3}$ **3.** $-\frac{5}{4}$ **4.** $\frac{1}{2}$
5. $y = \frac{2}{3}x - \frac{5}{3}$ **6.** $y = -2x$ **7.** $y = 3x - 1$
8. $y = \frac{2}{3}x + 5$ **9.** $y = -2x + 7$ **10.** $y = x + 3$

1. $\frac{6}{5}$ **3.** 0 **5.** -3

7. $m = -\frac{3}{5}$ $m = 1$

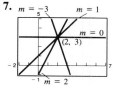

9. $m = 2$

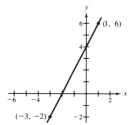

11. m is undefined.

13. $m = \frac{4}{3}$

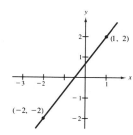

15. $(0, 1), (3, 1), (-1, 1)$ **17.** $(6, -5), (7, -4), (8, -3)$
19. Perpendicular   **21.** Parallel   **23.** Collinear
25. $16,667$ ft ≈ 3.16 mi

27. $m = 5$; Intercept: $(0, 3)$

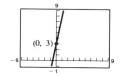

29. m is undefined. There is no y-intercept.

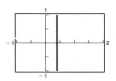

31. $3x + 5y - 10 = 0$ **33.** $x + 2y - 3 = 0$
35. $x + 8 = 0$ **37.** $3x - y - 2 = 0$
39. $8x + 6y - 47 = 0$
41. $x - 6 = 0$
43.

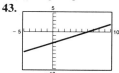

45. $y = -2x$ $y = 2x$ **47.** $y = -\frac{1}{2}x + 3$ $y = 2x - 4$
$y = \frac{1}{2}x$ $y = -\frac{1}{2}x$

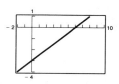

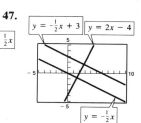

49.

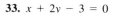

a and b are the x- and y-intercepts of the graph of the line.
51. $3x + 2y - 6 = 0$
53. **(a)** $2x - y - 3 = 0$ **(b)** $x + 2y - 4 = 0$
55. **(a)** $y = 0$ **(b)** $x + 1 = 0$ **57.** $V = 125t + 2540$
59. $V = 2000t + 20,400$ **61.** b   **62.** c   **63.** a
64. d   **65.** $F = \frac{9}{5}C + 32$
67. **(a)** $S = 2200t + 28,500$ **(b)** \$39,500

69. (a) $V = -175t + 875$ **(b)**
   (c) \$525   **(d)** 3.86 yr

71. $W = 0.07S + 2500$

73. (a) $C = 16.75t + 36{,}500$ **(b)** $R = 37t$
   (c) $t \approx 1802$ hr

75. The model approximates the daily cost y of producing x units of a product.

Section 1.6   *(page 59)*

WARM-UP   **1.** -73 **2.** 13   **3.** $2(x + 2)$
4. $-8(x - 2)$ **5.** $y = \frac{7}{5} - \frac{2}{5}x$ **6.** $y = \pm x$
7. $5xy\sqrt{2x}$ **8.** $3z\sqrt{2z^2 - 1}$ **9.** $\dfrac{x - y}{x + y}$
10. $\dfrac{x(x - 3)}{2}$

1. (a) Function
   (b) Not a function because the element 1 in A is matched with two elements, -2 and 1, in B.
   (c) Function
   (d) Not a function because not all elements of A are matched with an element in B.
3. Not a function   **5.** Function   **7.** Function
9. Not a function
11. (a) $6 - 4(3) = -6$ **(b)** $6 - 4(-7) = 34$
   (c) $6 - 4(t) = 6 - 4t$
   (d) $6 - 4(c + 1) = 2(1 - 2c)$
13. (a) $\dfrac{1}{4 + 1} = \dfrac{1}{5}$ **(b)** $\dfrac{1}{0 + 1} = 1$ **(c)** $\dfrac{1}{4x + 1}$
   (d) $\dfrac{1}{(x + h) + 1}$
15. (a) -1 **(b)** -9 **(c)** $2x - 5$ **(d)** $-\frac{5}{2}$
17. (a) 0   **(b)** 3   **(c)** $x^2 + 2x$ **(d)** -0.75
19. (a) 1   **(b)** -1 **(c)** 1   **(d)** $\dfrac{|x - 1|}{x - 1}$
21. (a) -1 **(b)** 2   **(c)** 4   **(d)** 6
23. All real numbers   **25.** $y \geq 10$
27. All real numbers except $t = 0$
29. All real numbers except $x = 0, -2$
31. $(-2, 4), (-1, 1), (0, 0), (1, 1), (2, 4)$
33. $(-2, 0), (-1, 1), \left(0, \sqrt{2}\right), \left(1, \sqrt{3}\right), (2, 2)$
35. (a) 1056.250   **(b)** 1470.084
37. (a) 0.118   **(b)** 21.277   **39.** $g(x) = cx^2, c = -2$

41. $r(x) = \dfrac{c}{x}, c = 32$ **43.** 2, $h \neq 0$

45. $3xh + 3x^2 + h^2, h \neq 0$ **47.** $A = \dfrac{C^2}{4\pi}$

49. $A = \dfrac{x^2}{x - 1}, x > 1$

51. $V = 4x^2(27 - x), 0 < x < 27$

53. $h = \sqrt{d^2 - 2000^2}, 0 \leq d$

55. (a) $C = 12.30x + 98{,}000$ **(b)** $R = 17.98x$
   (c) $P = 5.68x - 98{,}000$

Section 1.7   *(page 71)*

WARM-UP   **1.** 2   **2.** 0   **3.** $-\dfrac{3}{x}$ **4.** $x^2 + 3$
5. x-intercept: $\left(-\frac{12}{5}, 0\right)$
   y-intercept: $(0, 6)$
6. x-intercept: $(5, 0)$
   y-intercept: $(0, 3)$
7. All real numbers except $x = 4$
8. All real numbers except $x = 4, 5$ **9.** $t \leq \frac{5}{3}$
10. All real numbers

1. 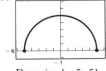   **3.**
Domain: $[1, \infty)$ Domain: $(-\infty, -2], [2, \infty)$
Range: $[0, \infty)$ Range: $[0, \infty)$

5.
Domain: $[-5, 5]$
Range: $[0, 5]$

7. Function
```
Xmin=-3
Xmax=3
Xscl=1
Ymin=-1
Ymax=4
Yscl=1
```

9. Not a function   **11.** Function
```
Xmin=-1
Xmax=5
Xscl=1
Ymin=-3
Ymax=3
Yscl=1
```
```
Xmin=-5
Xmax=5
Xscl=1
Ymin=-4
Ymax=4
Yscl=1
```

13. b   **15.** a

17.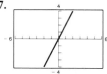

Increasing on $(-\infty, \infty)$

19.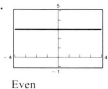

Increasing on $(-\infty, 0)$, $(2, \infty)$
Decreasing on $(0, 2)$

21.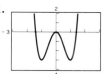

Increasing on $(-1, 0)$, $(1, \infty)$
Decreasing on $(-\infty, -1)$, $(0, 1)$

23.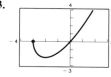

Increasing on $(-2, \infty)$
Decreasing on $(-3, -2)$

25. Relative minimum: $(3, -9)$
27. Relative minimum: $(1, -7)$
    Relative maximum: $(-2, 20)$
29. Relative minimum: $(0.33, -0.38)$
31. **(b)**

33.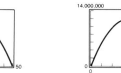

(c) Maximum area:
625 sq ft, 25 ft $\times$ 25 ft

350,000 units

35.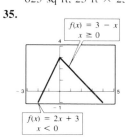

$f(x) = 3 - x$
$x \geq 0$
$f(x) = 2x + 3$
$x < 0$

37.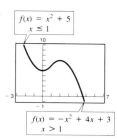

$f(x) = x^2 + 5$
$x \leq 1$
$f(x) = -x^2 + 4x + 3$
$x > 1$

39.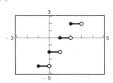

41. **(a)** $C = 0.65 + 0.42[\![t]\!]$
(b)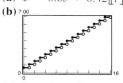

(c) Less than 13 minutes

43. **(a)** $(-5, 6)$ **(b)** $(-5, -6)$
45. **(a)** $\left(\frac{3}{2}, -2\right)$ **(b)** $\left(\frac{3}{2}, 2\right)$

47.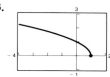

Even

49.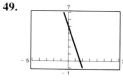

Neither even nor odd

51.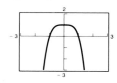

Odd

53.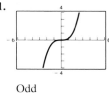

Neither even nor odd

55.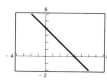

Odd

57. Even

59. Odd
61. Neither even nor odd
63. $(-\infty, 4]$
65. $[-1, 1]$

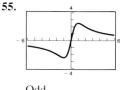 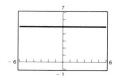

67. $f(x) > 0$ for all x
69. $h = (4x - x^2) - 3$

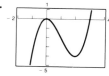

71. $h = 4x - 2x^2$ **73.** $L = (4 - y^2) - (y + 2)$

75.

Interval	Intake Pipe	Drain Pipe 1	Drain Pipe 2
[0, 5]	Open	Closed	Closed
[5, 10]	Open	Open	Closed
[10, 20]	Closed	Closed	Closed
[20, 30]	Closed	Closed	Open
[30, 40]	Open	Open	Open
[40, 45]	Open	Closed	Open
[45, 50]	Open	Open	Open
[50, 60]	Open	Open	Closed

Section 1.8   (page 81)

WARM-UP

1.

Domain: $[-4, 4]$
Range: $[0, 4]$

2.

Domain: $(-\infty, \infty)$
Range: $[0, \infty)$

3.

Domain: $(-\infty, \infty)$
Range: $[-1, \infty)$

4.

Domain: $(-\infty, \infty)$
Range: $(-\infty, \infty)$

5.

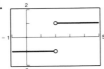

Domain: all real numbers
except $x = 2$; Range: $\{-1, 1\}$

6.

Domain: $[2, \infty)$
Range: $[0, \infty)$

7. Odd   **8.** Even   **9.** Even
10. Neither even nor odd

1.

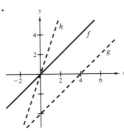

3.

5.

7.

9. $g(x) = (x - 1)^2 + 1$
$h(x) = -(x - 2)^2 + 4$

11. $g(x) = (x - 3)^2 - 2$
$h(x) = -x^2 + 3$

13.

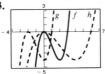

$g(x)$ is shifted two units to the left of $f(x)$.
$h(x)$ is a vertical shrink of $f(x)$.

15.

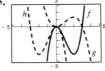

$g(x)$ is a vertical shrink of $f(x)$ and a reflection across
the x-axis.
$h(x)$ is a reflection across the y-axis.

17. $y = -(x^3 - 3x^2) + 1$
19. y is $f(x)$ shifted up two units.
21. y is $f(x)$ shifted right two units.
23. y is a vertical stretch of $f(x)$ by $\sqrt{2}$.
25. y is $f(x)$ shifted down one unit.
27. y is $f(x)$ shifted right one unit.
29. y is $f(x)$ reflected in the y-axis.

31.

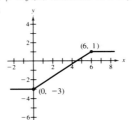

33.

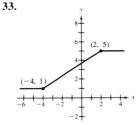

35.

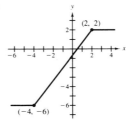

37. A reflection in the x-axis followed by a vertical shift of
four units upward.
39. A horizontal shift of two units to the left and a vertical
shrink.
41. A horizontal stretch and a vertical shift of two units
upward.

43. (a)

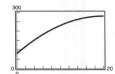

(b) $P(x) = -2420 + 20x - 0.5x^2$
Vertical shift
(c) $P(x) = 80 + 0.2x - 0.00005x^2$
Horizontal stretch

45.

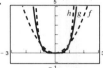

The three even functions are nonnegative. As the exponents increase, the graphs become flatter in the interval $(-1, 1)$.

47.

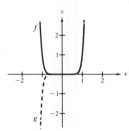

49.

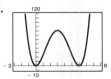

51.

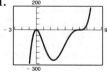

Section 1.9 (page 89)

WARM-UP

1. $\dfrac{1}{x(1 - x)}$ **2.** $-\dfrac{12}{(x + 3)(x - 3)}$

3. $\dfrac{3x - 2}{x(x - 2)}$ **4.** $\dfrac{4x - 5}{3(x - 5)}$ **5.** $\sqrt{\dfrac{x - 1}{x + 1}}$

6. $\dfrac{x + 1}{x(x + 2)}$ **7.** $5(x - 2)$ **8.** $\dfrac{x + 1}{(x - 2)(x + 3)}$

9. $\dfrac{1 + 5x}{3x - 1}$ **10.** $\dfrac{x + 4}{4x}$

1. (a) $2x$ **(b)** 2 **(c)** $x^2 - 1$ **(d)** $\dfrac{x + 1}{x - 1}, x \neq 1$

3. (a) $x^2 + 5 + \sqrt{1 - x}$ **(b)** $x^2 + 5 - \sqrt{1 - x}$
(c) $(x^2 + 5)\sqrt{1 - x}$ **(d)** $\dfrac{x^2 + 5}{\sqrt{1 - x}}, x < 1$

5. (a) $\dfrac{x + 1}{x^2}$ **(b)** $\dfrac{x - 1}{x^2}$ **(c)** $\dfrac{1}{x^3}$ **(d)** $x, x \neq 0$

7. 9 **9.** $4t^2 - 2t + 5$ **11.** 0 **13.** 26 **15.** $\frac{3}{5}$

17.

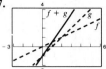

19.

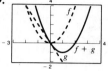

21.

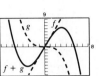

23. $T = \frac{3}{4}x + \frac{1}{15}x^2$

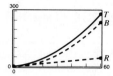

$0 \leq x \leq 2, f$
$x > 5, g$

25. (a) $(x - 1)^2$ **(b)** $x^2 - 1$ **(c)** x^4
27. (a) $20 - 3x$ **(b)** $-3x$ **(c)** $9x + 20$
29. (a) $\sqrt{x^2 + 4}$ **(b)** $x + 4$
31. (a) $x - \frac{8}{3}$ **(b)** $x - 8$
33. (a) $|x + 6|$ **(b)** $|x| + 6$ **35. (a)** 3 **(b)** 0
37. (a) 0 **(b)** 4 **39.** $f(x) = x^2, g(x) = 2x + 1$
41. $f(x) = \sqrt[3]{x}, g(x) = x^2 - 4$
43. $f(x) = \dfrac{1}{x}, g(x) = x + 2$
45. $f(x) = x^2 + 2x, g(x) = x + 4$
47. (a) $x \geq 0$ **(b)** All real numbers
(c) All real numbers
49. (a) All real numbers except $x = \pm 1$
(b) All real numbers
(c) All real numbers except $x = -2, 0$
51. (a) $(A \circ r)(t) = 0.36\pi t^2$
$A \circ r$ represents the area of the circle at time t.
(b)

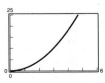

4.2 sec

53. (a) $(C \circ x)(t) = 3000t + 750$

$C \circ x$ represents the cost after t production hours.

(b)
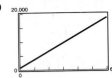

4.75 hr

55. $R(x) = x - 1200$, $D(x) = 0.85x$

$(D \circ R)(x) = 0.85(x - 1200)$

$(D \circ R)(18,400) = \$14,620$

$(R \circ D)(x) = 0.85x - 1200$

$(R \circ D)(18,400) = \$14,440$

57. Odd

Section 1.10 *(page 100)*

WARM-UP   **1.** All real numbers   **2.** $[-1, \infty)$

3. All real numbers except $x = 0, 2$

4. All real numbers except $x = -\frac{5}{3}$ **5.** x **6.** x

7. x **8.** x **9.** $x = \frac{3}{2}y + 3$ **10.** $x = \frac{y^3}{2} + 2$

1. $f^{-1}(x) = \frac{1}{8}x$ **3.** $f^{-1}(x) = x - 10$

5. $f^{-1}(x) = x^3$

7. (a) $(f \circ g)(x) = f\left(\frac{x}{2}\right) = 2\left(\frac{x}{2}\right) = x$

$(g \circ f)(x) = g(2x) = \frac{2x}{2} = x$

(b)
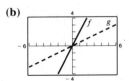

9. (a) $(f \circ g)(x) = f\left(\frac{x - 1}{5}\right) = \frac{5x + 1 - 1}{5} = x$

$(g \circ f)(x) = g(5x + 1) = 5\left(\frac{x - 1}{5}\right) + 1 = x$

(b)
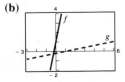

11. (a) $(f \circ g)(x) = f\left(\sqrt[3]{x}\right) = \left(\sqrt[3]{x}\right)^3 = x$

$(g \circ f)(x) = g(x^3) = \sqrt[3]{x^3} = x$

(b)
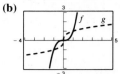

13. (a) $(f \circ g)(x) = f(x^2 + 4)$, $x \geq 0$

$= \sqrt{(x^2 + 4) - 4} = x$

$(g \circ f)(x) = g\left(\sqrt{x - 4}\right)$

$= \left(\sqrt{x - 4}\right)^2 + 4 = x$

(b)
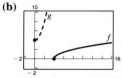

15. (a) $(f \circ g)(x) = f\left(\sqrt{9 - x}\right)$, $x \leq 9$

$= 9 - \left(\sqrt{9 - x}\right)^2 = x$

$(g \circ f)(x) = g(9 - x^2)$

$= \sqrt{9 - (9 - x)^2} = x$

(b)
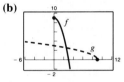

17. One-to-one   **19.** Not one-to-one

21. Not one-to-one   **23.** $f^{-1}(x) = \dfrac{x + 3}{2}$

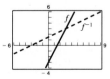

25. $f^{-1}(x) = \sqrt[5]{x}$ **27.** $f^{-1}(x) = x^2$, $x \geq 0$

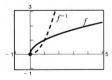

29. $f^{-1}(x) = \sqrt{4 - x^2}, 0 \le x \le 2$ **31.** $f^{-1}(x) = x^3 + 1$

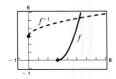

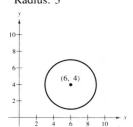

33. Not one-to-one   **35.** $g^{-1}(x) = 8x$

37. $f^{-1}(x) = \sqrt{x} - 3, x \ge 0$ **39.** $h^{-1}(x) = \dfrac{1}{x}$

41. $f^{-1}(x) = \dfrac{x^2 - 3}{2}, x \ge 0$

43. Not one-to-one   **45.** $f^{-1}(x) = -\sqrt{25 - x}$

47. $f^{-1}(x) = \sqrt{x} + 3, x \ge 0$ **49.** $f^{-1}(x) = x - 3, x \ge 0$

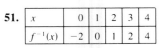

51.

x	0	1	2	3	4
$f^{-1}(x)$	-2	0	1	2	4

53. 32   **55.** 600   **57.** $\dfrac{x + 1}{2}$ **59.** $\dfrac{x + 1}{2}$

61. (a) $y = \sqrt{\dfrac{x - 254.50}{0.03}}$

   y: percentage load; x: exhaust temperature

(b)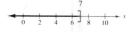

63. False   **65.** True

Chapter 1 Review Exercises   (page 101)

1. (a) 11   (b) 11, -14 (c) 11, $-14, -\frac{8}{9}, \frac{5}{2}, 0.4$
   (d) $\sqrt{6}$

3. $-4 < -3$

5. The set consists of all real numbers less than or equal to 7.

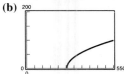

7. -14 **9.** (a) No   (b) Yes   (c) Yes   (d) No
11. 20   **13.** $\frac{1}{5}$ **15.** 0, 2   **17.** $6 \pm \sqrt{6}$
19. $0, \frac{12}{5}$ **21.** 0, 1, 2   **23.** (a) 10   (b) (0, 5)
(c) $x^2 + (y - 5)^2 = 25$ **25.** (a) 13   (b) $(8, \frac{7}{2})$
(c) $(x - 8)^2 + (y - \frac{7}{2})^2 = \frac{169}{4}$ **27.** 725,000
29. Verification   **31.** Intercept: (0, 0), Symmetry: x-axis
33. Intercepts: (0, 0), $(-2\sqrt{2}, 0), (2\sqrt{2}, 0)$,
   Symmetry: y-axis
35. Intercepts: (0, 0), $(-2, 0)$, (2, 0), Symmetry: Origin
37. Center: (6, 4)   **39.** Center: $(\frac{1}{2}, 5)$
   Radius: 3   Radius: $\frac{3}{2}$

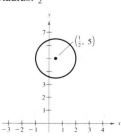

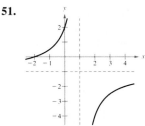

41. 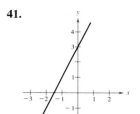   **43.**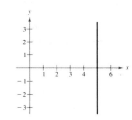

45. 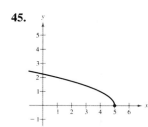   **47.**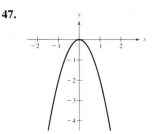

49. 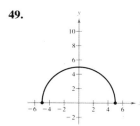   **51.**

53.

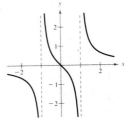

55.

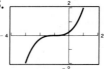

The graph intersects the x-axis once.
The graph intersects the y-axis once.

57.

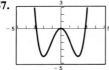

The graph intersects the x-axis three times.
The graph intersects the y-axis once.

59.

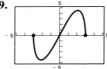

The graph intersects the x-axis three times.
The graph intersects the y-axis once.

61. 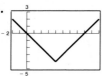   **63.**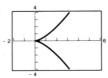

The graph intersects the x-axis twice.
The graph intersects the y-axis once.

65. $t = \frac{7}{3}$ **67.** $t = 3$ **69.** $x = 0$
71. $5x - 12y + 2 = 0$ **73.** $2x - 7y + 2 = 0$
75. $3x - 2y - 10 = 0$ **77.** $2x + 3y - 6 = 0$

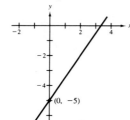   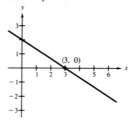

79. (a) 5   **(b)** 17   **(c)** $t^4 + 1$ **(d)** $-x^2 - 1$
81. \$210,000
83. $[-5, 5]$ **85.** All real numbers except $s = 3$
87. All real numbers except $x = -2, 3$
89. (a)

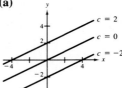

(b)

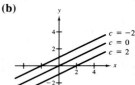

(c)

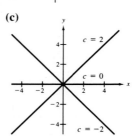

91. (a) $f^{-1}(x) = 2x + 6$
(b)

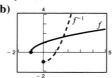

(c) $f^{-1}(f(x)) = f^{-1}(\frac{1}{2}x - 3)$
$= 2(\frac{1}{2}x - 3) + 6 = x$
$f(f^{-1}(x)) = f(2x + 6)$
$= \frac{1}{2}(2x + 6) - 3 = x$

93. (a) $f^{-1}(x) = x^2 - 1,\ x \geq 0$
(b)

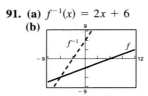

(c) $f^{-1}(f(x)) = f^{-1}\left(\sqrt{x + 1}\right),\ x \geq 0$
$= \left(\sqrt{x + 1}\right)^2 - 1 = x$
$f(f^{-1}(x)) = f(x^2 - 1)$
$= \sqrt{(x^2 - 1) + 1} = x$

95. (a) $f^{-1}(x) = \sqrt{x + 5},\ x \geq -5$
(b)

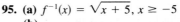
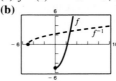

(c) $f^{-1}(f(x)) = f^{-1}(x^2 - 5), x \geq -5$
$= \sqrt{(x^2 - 5) + 5} = x$
$f(f^{-1}(x)) = f\left(\sqrt{x + 5}\right), x \geq 0$
$= \left(\sqrt{x + 5}\right)^2 - 5 = x$

97. $x \geq 4, f^{-1}(x) = \sqrt{\dfrac{x}{2}} + 4$

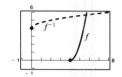

99. $x \geq 2, f^{-1}(x) = \sqrt{x^2 + 4}, x \geq 0$

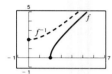

101. -7 **103.** 5   **105.** 23   **107.** 9
109. September: \$325,000, October: \$364,000
111. 4 farmers   **113. (a)** 16 ft/sec   **(b)** 1.5 sec
(c) -16 ft/sec
115. $A = x(12 - x), (0, 6]$

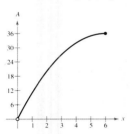

CHAPTER 2

Section 2.1 *(page 115)*

WARM-UP   **1.** 45   **2.** 70   **3.** $\dfrac{\pi}{6}$ **4.** $\dfrac{\pi}{3}$ **5.** $\dfrac{\pi}{4}$

6. $\dfrac{4\pi}{3}$ **7.** $\dfrac{\pi}{9}$ **8.** $\dfrac{11\pi}{6}$ **9.** 45   **10.** 45

---

1. (a) Quadrant I   **3. (a)** Quadrant IV
(b) Quadrant III   **(b)** Quadrant III
5. (a) Quadrant II
(b) Quadrant IV

7. (a)   **(b)**

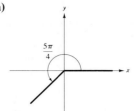

9. (a)   **(b)**

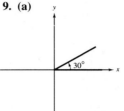

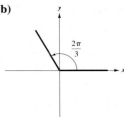

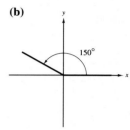

11. (a) $\dfrac{19\pi}{9}, -\dfrac{17\pi}{9}$ **(b)** $\dfrac{10\pi}{3}, -\dfrac{2\pi}{3}$

13. (a) $396°, -324°$ **15. (a)** $660°, -60°$
(b) $315°, -405°$ **(b)** $20°, -340°$

17. (a) Complement: $\dfrac{\pi}{6}$; Supplement: $\dfrac{2\pi}{3}$

(b) Complement: none; Supplement: $\dfrac{\pi}{4}$

19. (a) Complement: $72°$; Supplement: $162°$
(b) Complement: none; Supplement: $65°$
21. (a) $270°$ **23. (a)** $420°$
(b) $210°$ **(b)** $-66°$

25. (a) $\dfrac{\pi}{6}$ **27. (a)** $-\dfrac{\pi}{9}$

(b) $\dfrac{5\pi}{6}$ **(b)** $-\dfrac{4\pi}{3}$

29. (a) 2.007   **31. (a)** 9.285
(b) 1.525   **(b)** 0.009
33. (a) $25.714°$ **35. (a)** $-756°$
(b) $81.818°$ **(b)** $275.020°$
37. (a) $245.167°$
(b) $2.2°$
39. (a) $240° \, 36'$ **41. (a)** $143° \, 14' \, 22''$
(b) $-145° \, 48'$ **(b)** $-205° \, 7' \, 8''$
43. $\frac{4}{15}$ rad   **45.** 1.724 rad   **47.** 15π in. ≈ 47.12 in.
49. 12 m   **51.** 591.72 mi   **53.** 1141 mi   **55.** $4.655°$
57. $\frac{1}{4}$ rad $\approx 14.32°$
59. (a) 560.2 rev/min
(b) 3520 rad/min

61. (a) Angular speed of motor pulley $= 3400\pi$ rad/min
Angular speed of saw arbor $= 1700\pi$ rad/min
(b) 850 rev/min
63. (a) 80π rad/sec
(b) 78.54 ft/sec

Section 2.2 (page 125)

WARM-UP **1.** $-\dfrac{\sqrt{3}}{3}$ **2.** -1 **3.** $\dfrac{2\pi}{3}$ **4.** $\dfrac{7\pi}{4}$

5. $\dfrac{\pi}{6}$ **6.** $\dfrac{3\pi}{4}$ **7.** $60°$ **8.** $-270°$ **9.** 2π

10. π

1. $\left(\dfrac{\sqrt{2}}{2}, \dfrac{\sqrt{2}}{2}\right)$ **3.** $\left(-\dfrac{\sqrt{3}}{2}, \dfrac{1}{2}\right)$ **5.** $\left(-\dfrac{1}{2}, -\dfrac{\sqrt{3}}{2}\right)$

7. $(0, -1)$

9. $\sin\dfrac{\pi}{4} = \dfrac{\sqrt{2}}{2}$ **11.** $\sin -\dfrac{5\pi}{4} = \dfrac{\sqrt{2}}{2}$

$\cos\dfrac{\pi}{4} = \dfrac{\sqrt{2}}{2}$ $\cos -\dfrac{5\pi}{4} = -\dfrac{\sqrt{2}}{2}$

$\tan\dfrac{\pi}{4} = 1$ $\tan -\dfrac{5\pi}{4} = -1$

13. $\sin\dfrac{11\pi}{6} = -\dfrac{1}{2}$ **15.** $\sin\dfrac{4\pi}{3} = -\dfrac{\sqrt{3}}{2}$

$\cos\dfrac{11\pi}{6} = \dfrac{\sqrt{3}}{2}$ $\cos\dfrac{4\pi}{3} = -\dfrac{1}{2}$

$\tan\dfrac{11\pi}{6} = -\dfrac{\sqrt{3}}{3}$ $\tan\dfrac{4\pi}{3} = \sqrt{3}$

17. $\sin\dfrac{3\pi}{4} = \dfrac{\sqrt{2}}{2}$ **19.** $\sin\dfrac{\pi}{2} = 1$

$\cos\dfrac{3\pi}{4} = -\dfrac{\sqrt{2}}{2}$ $\cos\dfrac{\pi}{2} = 0$

$\tan\dfrac{3\pi}{4} = -1$ $\tan\dfrac{\pi}{2}$ is undefined

$\csc\dfrac{3\pi}{4} = \sqrt{2}$ $\csc\dfrac{\pi}{2} = 1$

$\sec\dfrac{3\pi}{4} = -\sqrt{2}$ $\sec\dfrac{\pi}{2}$ is undefined

$\cot\dfrac{3\pi}{4} = -1$ $\cot\dfrac{\pi}{2} = 0$

21. $\sin\left(-\dfrac{4\pi}{3}\right) = \dfrac{\sqrt{3}}{2}$

$\cos\left(-\dfrac{4\pi}{3}\right) = -\dfrac{1}{2}$

$\tan\left(-\dfrac{4\pi}{3}\right) = -\sqrt{3}$

$\csc\left(-\dfrac{4\pi}{3}\right) = \dfrac{2\sqrt{3}}{3}$

$\sec\left(-\dfrac{4\pi}{3}\right) = -2$

$\sin\left(-\dfrac{4\pi}{3}\right) = -\dfrac{\sqrt{3}}{3}$

23. $\sin\pi = 0$ **25.** $\cos\dfrac{2\pi}{3} = -\dfrac{1}{2}$

27. $\cos\dfrac{7\pi}{6} = -\dfrac{\sqrt{3}}{2}$ **29.** $\sin\dfrac{7\pi}{4} = -\dfrac{\sqrt{2}}{2}$

31. (a) $-\frac{1}{3}$ **(b)** -3 **33. (a)** $-\frac{7}{8}$ **(b)** $-\frac{8}{7}$

35. (a) $\frac{4}{5}$ **(b)** $-\frac{4}{5}$ **37.** 0.7071 **39.** 0.8271

41. -2.6746 **43.** 1.3940 **45. (a)** -1 **(b)** -0.4

47. (a) 0.25, 2.9 **(b)** 1.8, 4.5

49. (a) 0.2500 ft **(b)** 0.0177 ft **(c)** -0.2475 ft

51. 0.794

Section 2.3 (page 134)

WARM-UP **1.** $2\sqrt{5}$ **2.** $3\sqrt{10}$ **3.** 10
4. $3\sqrt{2}$ **5.** 1.24 **6.** 317.55 **7.** 63.13
8. 133.57 **9.** 2,785,714.29 **10.** 28.80

1. $\sin\theta = \dfrac{1}{2}$ **3.** $\sin\theta = \dfrac{3}{5}$

$\cos\theta = \dfrac{\sqrt{3}}{2}$ $\cos\theta = \dfrac{4}{5}$

$\tan\theta = \dfrac{\sqrt{3}}{3}$ $\tan\theta = \dfrac{3}{4}$

$\csc\theta = 2$ $\csc\theta = \dfrac{5}{3}$

$\sec\theta = \dfrac{2\sqrt{3}}{3}$ $\sec\theta = \dfrac{5}{4}$

$\cot\theta = \sqrt{3}$ $\cot\theta = \dfrac{4}{3}$

5. $\sin \theta = \dfrac{\sqrt{161}}{15}$

$\cos \theta = \dfrac{8}{15}$

$\tan \theta = \dfrac{\sqrt{161}}{8}$

$\csc \theta = \dfrac{15\sqrt{161}}{161}$

$\sec \theta = \dfrac{15}{8}$

$\cot \theta = \dfrac{8\sqrt{161}}{161}$

9. $\cos \theta = \dfrac{\sqrt{5}}{3}$

$\tan \theta = \dfrac{2\sqrt{5}}{5}$

$\csc \theta = \dfrac{3}{2}$

$\sec \theta = \dfrac{3\sqrt{5}}{5}$

$\cot \theta = \dfrac{\sqrt{5}}{2}$

11. $\sin \theta = \dfrac{\sqrt{3}}{2}$

$\cos \theta = \dfrac{1}{2}$

$\tan \theta = \sqrt{3}$

$\csc \theta = \dfrac{2\sqrt{3}}{3}$

$\cot \theta = \dfrac{\sqrt{3}}{3}$

13. $\sin \theta = \dfrac{3\sqrt{10}}{10}$

$\cos \theta = \dfrac{\sqrt{10}}{10}$

$\csc \theta = \dfrac{\sqrt{10}}{3}$

$\sec \theta = \sqrt{10}$

$\cot \theta = \dfrac{1}{3}$

7. $\sin \theta = \dfrac{15}{17}$

$\cos \theta = \dfrac{8}{17}$

$\tan \theta = \dfrac{15}{8}$

$\csc \theta = \dfrac{17}{15}$

$\sec \theta = \dfrac{17}{8}$

$\cot \theta = \dfrac{8}{15}$

15. $\sin \theta = \dfrac{2\sqrt{13}}{13}$

$\cos \theta = \dfrac{3\sqrt{13}}{13}$

$\tan \theta = \dfrac{2}{3}$

$\csc \theta = \dfrac{\sqrt{13}}{2}$

$\sec \theta = \dfrac{\sqrt{13}}{3}$

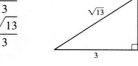

17. (a) $\sqrt{3}$ **19. (a)** $\dfrac{1}{3}$

(b) $\dfrac{1}{2}$ **(b)** $\dfrac{2\sqrt{2}}{3}$

(c) $\dfrac{\sqrt{3}}{2}$ **(c)** $\dfrac{\sqrt{2}}{4}$

(d) $\dfrac{\sqrt{3}}{3}$ **(d)** 3

21. (a) $\dfrac{1}{2}$ **23. (a)** 1

(b) 1 **(b)** $\dfrac{\sqrt{2}}{2}$

25. (a) 0.1736   **27. (a)** 0.2815
(b) 0.1736   **(b)** 3.5523

29. (a) 1.3499   **31. (a)** 5.0273
(b) 1.3432   **(b)** 0.1989

33. (a) 1.1884   **35. (a)** $30° = \dfrac{\pi}{6}$

(b) 1.1884   **(b)** $30° = \dfrac{\pi}{6}$

37. (a) $60° = \dfrac{\pi}{3}$ **39. (a)** $60° = \dfrac{\pi}{3}$

(b) $45° = \dfrac{\pi}{4}$ **(b)** $45° = \dfrac{\pi}{4}$

41. (a) $55° \approx 0.96$ **43. (a)** $50° \approx 0.873$
(b) $89° \approx 1.55$ **(b)** $25° \approx 0.436$

45. 57.74   **47.** 14.43   **49.** 15.56
51. 9.19   **53.** 15 ft   **55.** 19.32 ft   **57.** 2145.10 ft
59. $(x_1, y_1) = (10\sqrt{3}, 10), (x_2, y_2) = (10, 10\sqrt{3})$
61. $\sin 25° \approx 0.42$, $\cos 25° \approx 0.91$, $\tan 25° \approx 0.47$,
$\csc 25° \approx 2.37$, $\sec 25° \approx 1.10$, $\cot 25° \approx 2.14$

63. True, $\csc x = \dfrac{1}{\sin x}$ **65.** False, $\dfrac{\sqrt{2}}{2} + \dfrac{\sqrt{2}}{2} \neq 1$

67. False, $1.7321 \neq \sin 2°$

69. (a) 29.73 ft   **(d)** The difference is
(b) 82.34°   due to rounding.
(c) 29.74 ft

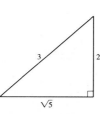

Section 2.4   *(page 147)*

WARM-UP   **1.** $\sin 30° = \dfrac{1}{2}$ **2.** $\tan 45° = 1$

3. $\cos \dfrac{\pi}{4} = \dfrac{\sqrt{2}}{2}$ **4.** $\cot \dfrac{\pi}{3} = \dfrac{\sqrt{3}}{3}$

5. $\sec \dfrac{\pi}{6} = \dfrac{2\sqrt{3}}{3}$ **6.** $\csc \dfrac{\pi}{4} = \sqrt{2}$

7. $\sin \theta = \dfrac{3\sqrt{13}}{13}$

$\cos \theta = \dfrac{2\sqrt{13}}{13}$

$\csc \theta = \dfrac{\sqrt{13}}{3}$

$\sec \theta = \dfrac{\sqrt{13}}{2}$

$\cot \theta = \dfrac{2}{3}$

8. $\sin \theta = \dfrac{\sqrt{5}}{3}$

$\tan \theta = \dfrac{\sqrt{5}}{2}$

$\csc \theta = \dfrac{3\sqrt{5}}{5}$

$\sec \theta = \dfrac{3}{2}$

$\cot \theta = \dfrac{2\sqrt{5}}{5}$

9. $\cos \theta = \dfrac{2\sqrt{6}}{5}$

$\tan \theta = \dfrac{\sqrt{6}}{12}$

$\csc \theta = 5$

$\sec \theta = \dfrac{5\sqrt{6}}{12}$

$\cot \theta = 2\sqrt{6}$

10. $\sin \theta = \dfrac{2\sqrt{2}}{3}$

$\cos \theta = \dfrac{1}{3}$

$\tan \theta = 2\sqrt{2}$

$\csc \theta = \dfrac{3\sqrt{2}}{4}$

$\cot \theta = \dfrac{\sqrt{2}}{4}$

1. (a) $\sin \theta = \dfrac{4}{5}$

$\cos \theta = \dfrac{3}{5}$

$\tan \theta = \dfrac{4}{3}$

$\csc \theta = \dfrac{5}{4}$

$\sec \theta = \dfrac{5}{3}$

$\cot \theta = \dfrac{3}{4}$

(b) $\sin \theta = -\dfrac{15}{17}$

$\cos \theta = \dfrac{8}{17}$

$\tan \theta = -\dfrac{15}{8}$

$\csc \theta = -\dfrac{17}{15}$

$\sec \theta = \dfrac{17}{8}$

$\cot \theta = -\dfrac{8}{15}$

3. (a) $\sin \theta = \dfrac{1}{2}$

$\cos \theta = -\dfrac{\sqrt{3}}{2}$

$\tan \theta = -\dfrac{\sqrt{3}}{3}$

$\csc \theta = 2$

$\sec \theta = -\dfrac{2\sqrt{3}}{3}$

$\cot \theta = -\sqrt{3}$

(b) $\sin \theta = -\dfrac{\sqrt{2}}{2}$

$\cos \theta = -\dfrac{\sqrt{2}}{2}$

$\tan \theta = 1$

$\csc \theta = -\sqrt{2}$

$\sec \theta = -\sqrt{2}$

$\cot \theta = 1$

5. (a) $\sin \theta = \dfrac{24}{25}$

$\cos \theta = \dfrac{7}{25}$

$\tan \theta = \dfrac{24}{7}$

$\csc \theta = \dfrac{25}{24}$

$\sec \theta = \dfrac{25}{7}$

$\cot \theta = \dfrac{7}{24}$

(b) $\sin \theta = -\dfrac{24}{25}$

$\cos \theta = \dfrac{7}{25}$

$\tan \theta = -\dfrac{24}{7}$

$\csc \theta = -\dfrac{25}{24}$

$\sec \theta = \dfrac{25}{7}$

$\cot \theta = -\dfrac{7}{24}$

7. (a) $\sin \theta = \dfrac{5\sqrt{29}}{29}$

$\cos \theta = -\dfrac{2\sqrt{29}}{29}$

$\tan \theta = -\dfrac{5}{2}$

$\csc \theta = \dfrac{\sqrt{29}}{5}$

$\sec \theta = -\dfrac{\sqrt{29}}{2}$

$\cot \theta = -\dfrac{2}{5}$

(b) $\sin \theta = -\dfrac{5\sqrt{34}}{34}$

$\cos \theta = \dfrac{3\sqrt{34}}{34}$

$\tan \theta = -\dfrac{5}{3}$

$\csc \theta = -\dfrac{\sqrt{34}}{5}$

$\sec \theta = \dfrac{\sqrt{34}}{3}$

$\cot \theta = -\dfrac{3}{5}$

9. (a) $c_1 = 5$

$b_2 = 12$

$c_2 = 15$

(b) $\sin \alpha_1 = \dfrac{a_1}{c_1} = \dfrac{3}{5} = \dfrac{a_2}{c_2} = \sin \alpha_2$

$\cos \alpha_1 = \dfrac{b_1}{c_1} = \dfrac{4}{5} = \dfrac{b_2}{c_2} = \cos \alpha_2$

$\tan \alpha_1 = \dfrac{a_1}{b_1} = \dfrac{3}{4} = \dfrac{a_2}{b_2} = \tan \alpha_2$

$\csc \alpha_1 = \dfrac{c_1}{a_1} = \dfrac{5}{3} = \dfrac{c_2}{a_2} = \csc \alpha_2$

$\sec \alpha_1 = \dfrac{c_1}{b_1} = \dfrac{5}{4} = \dfrac{c_2}{b_2} = \sec \alpha_2$

$\cot \alpha_1 = \dfrac{b_1}{a_1} = \dfrac{4}{3} = \dfrac{b_2}{a_2} = \cot \alpha_2$

11. (a) $b_1 = \sqrt{3}$

$a_2 = \dfrac{5\sqrt{3}}{3}$

$c_2 = \dfrac{10\sqrt{3}}{3}$

(b) $\sin \alpha_1 = \dfrac{a_1}{c_2} = \dfrac{1}{2} = \dfrac{a_2}{c_2} = \sin \alpha_2$

$\cos \alpha_1 = \dfrac{b_1}{c_1} = \dfrac{\sqrt{3}}{2} = \dfrac{b_2}{c_2} = \cos \alpha_2$

$\tan \alpha_1 = \dfrac{a_1}{b_1} = \dfrac{\sqrt{3}}{3} = \dfrac{a_2}{b_2} = \tan \alpha_2$

$\csc \alpha_1 = \dfrac{c_1}{a_1} = 2 = \dfrac{c_2}{a_2} = \csc \alpha_2$

$\sec \alpha_1 = \dfrac{c_1}{b_1} = \dfrac{2\sqrt{3}}{3} = \dfrac{c_2}{b_2} = \sec \alpha_2$

$\cot \alpha_1 = \dfrac{b_1}{a_1} = \sqrt{3} = \dfrac{b_2}{a_2} = \cot \alpha_2$

13. (a) Quadrant III
   (b) Quadrant II

15. (a) Quadrant II
   (b) Quadrant IV

17. $\sin \theta = \frac{3}{5}$
   $\cos \theta = -\frac{4}{5}$
   $\tan \theta = -\frac{3}{4}$
   $\csc \theta = \frac{5}{3}$
   $\sec \theta = -\frac{5}{4}$
   $\cot \theta = -\frac{4}{3}$

19. $\sin \theta = -\frac{15}{17}$
   $\cos \theta = \frac{8}{17}$
   $\tan \theta = -\frac{15}{8}$
   $\csc \theta = -\frac{17}{15}$
   $\sec \theta = \frac{17}{8}$
   $\cot \theta = -\frac{8}{15}$

21. $\sin \theta = \dfrac{\sqrt{3}}{2}$

   $\cos \theta = -\dfrac{1}{2}$

   $\tan \theta = -\sqrt{3}$

   $\csc \vartheta = \dfrac{2\sqrt{3}}{3}$

   $\sec \theta = -2$

   $\cot \theta = -\dfrac{\sqrt{3}}{3}$

23. $\sin \theta = 0$
   $\cos \theta = -1$
   $\tan \theta = 0$
   $\csc \theta$ is undefined.
   $\sec \theta = -1$
   $\cot \theta$ is undefined.

25. $\sin \theta = -\dfrac{2\sqrt{5}}{5}$

   $\cos \theta = -\dfrac{\sqrt{5}}{5}$

   $\tan \theta = 2$

   $\csc \theta = \dfrac{-\sqrt{5}}{2}$

   $\sec \theta = -\sqrt{5}$

   $\cot \theta = \dfrac{1}{2}$

27. (a) $\theta' = 23°$ **(b)** $\theta' = 53°$

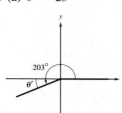
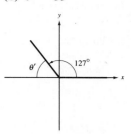

29. (a) $\theta' = 65°$ **(b)** $\theta' = 72°$

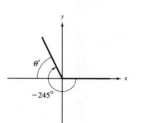
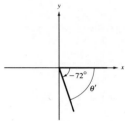

31. (a) $\theta' = \dfrac{\pi}{3}$ **(b)** $\theta' = \dfrac{\pi}{6}$

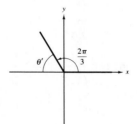
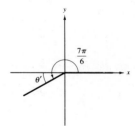

33. (a) $\theta' = 3.5 - \pi$ **(b)** $\theta' = 2\pi - 5.8$

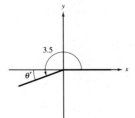
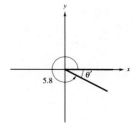

35. (a) $\sin(225°) = -\dfrac{\sqrt{2}}{2}$ **(b)** $\sin(-225°) = \dfrac{\sqrt{2}}{2}$

$\cos(225°) = -\dfrac{\sqrt{2}}{2}$ $\cos(-225°) = -\dfrac{\sqrt{2}}{2}$

$\tan(225°) = 1$ $\tan(-225°) = -1$

37. (a) $\sin(750°) = \dfrac{1}{2}$ **(b)** $\sin(510°) = \dfrac{1}{2}$

$\cos(750°) = \dfrac{\sqrt{3}}{2}$ $\cos(510°) = -\dfrac{\sqrt{3}}{2}$

$\tan(750°) = \dfrac{\sqrt{3}}{3}$ $\tan(510°) = -\dfrac{\sqrt{3}}{3}$

39. (a) $\sin\dfrac{4\pi}{3} = -\dfrac{\sqrt{3}}{2}$ **(b)** $\sin\dfrac{2\pi}{3} = \dfrac{\sqrt{3}}{2}$

$\cos\dfrac{4\pi}{3} = -\dfrac{1}{2}$ $\cos\dfrac{2\pi}{3} = -\dfrac{1}{2}$

$\tan\dfrac{4\pi}{3} = \sqrt{3}$ $\tan\dfrac{2\pi}{3} = -\sqrt{3}$

41. (a) $\sin\left(-\dfrac{\pi}{6}\right) = -\dfrac{1}{2}$ **(b)** $\sin\dfrac{5\pi}{6} = \dfrac{1}{2}$

$\cos\left(-\dfrac{\pi}{6}\right) = \dfrac{\sqrt{3}}{2}$ $\cos\dfrac{5\pi}{6} = -\dfrac{\sqrt{3}}{2}$

$\tan\left(-\dfrac{\pi}{6}\right) = -\dfrac{\sqrt{3}}{3}$ $\tan\dfrac{5\pi}{6} = -\dfrac{\sqrt{3}}{3}$

43. (a) $\sin\dfrac{11\pi}{4} = \dfrac{\sqrt{2}}{2}$ **(b)** $\sin\left(-\dfrac{13\pi}{6}\right) = -\dfrac{1}{2}$

$\cos\dfrac{11\pi}{4} = -\dfrac{\sqrt{2}}{2}$ $\cos\left(-\dfrac{13\pi}{6}\right) = \dfrac{\sqrt{3}}{2}$

$\tan\dfrac{11\pi}{4} = -1$ $\tan\left(-\dfrac{13\pi}{6}\right) = -\dfrac{\sqrt{3}}{3}$

45. (a) 0.1736   **47. (a)** -0.3420
(b) 5.7588   **(b)** -0.3420

49. (a) 1.7321   **51. (a)** 0.3640
(b) 1.7321   **(b)** 0.3640

53. (a) $30° = \dfrac{\pi}{6},\ 150° = \dfrac{5\pi}{6}$

(b) $210° = \dfrac{7\pi}{6},\ 330° = \dfrac{11\pi}{6}$

55. (a) $60° = \dfrac{\pi}{3},\ 120° = \dfrac{2\pi}{3}$

(b) $135° = \dfrac{3\pi}{4},\ 315° = \dfrac{7\pi}{4}$

57. (a) $45° = \dfrac{\pi}{4},\ 225° = \dfrac{5\pi}{4}$ **59. (a)** $54.99°,\ 125.01°$

(b) $150° = \dfrac{5\pi}{6},\ 330° = \dfrac{11\pi}{6}$ **(b)** $195.00°,\ 345.00°$

61. (a) 0.175, 6.109   **63. (a)** 0.873, 4.014
(b) 2.201, 4.083   **(b)** 1.693, 4.835

65. $\dfrac{4}{5}$ **67.** $-\sqrt{3}$

69. $\sin^2\theta + \cos^2\theta = 1$
$\sin^2 2 + \cos^2 2 = 1$

71. (a) 25.2°F   **73. (a)** 10 mi
(b) 65.1°F   **(b)** 5.18 mi
(c) 50.8°F   **(c)** 5 mi

Section 2.5 *(page 159)*

1. Period: π
Amplitude: 2

```
Xmin=0
Xmax=3π
Xscl=π/2
Ymin=-3
Ymax=3
Yscl=1
```

3. Period: 4π
Amplitude: $\frac{3}{2}$

```
Xmin=0
Xmax=4π
Xscl=π
Ymin=-2
Ymax=2
Yscl=1
```

5. Period: 2
Amplitude: $\frac{1}{2}$

```
Xmin=0
Xmax=6
Xscl=1
Ymin=-2
Ymax=2
Yscl=0.5
```

7. Period: $\dfrac{\pi}{5}$; Amplitude: 3   **9.** Period: $\dfrac{1}{2}$; Amplitude: 3

11. *Shift* the graph of f π units to the right to obtain the graph of g.   **13.** *Reflect* the graph of f about the x-axis to obtain the graph of g.   **15.** *Shift* the graph of f two units up to obtain the graph of g.

17.   **19.**

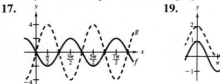

21.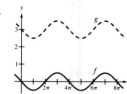

23.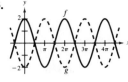

45.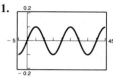

47.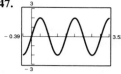

25.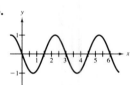

27.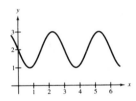

49.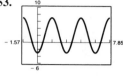

51.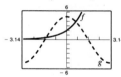

53.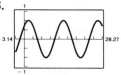

29.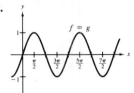

31.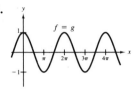

55. $-\dfrac{5\pi}{6}, -\dfrac{\pi}{6}, \dfrac{7\pi}{6}, \dfrac{11\pi}{6}$ **57.** $-\dfrac{7\pi}{4}, -\dfrac{\pi}{4}, \dfrac{\pi}{4}, \dfrac{7\pi}{4}$

59. $f(x) = 2\cos x + 2$ **61.** $y = 2\sin 4x$

63. $y = \cos\left(2x + \dfrac{\pi}{2}\right)$

33.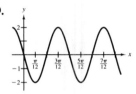

35.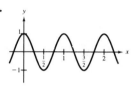

65.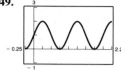

$(-1.480, 0.455), (0.672, 3.914)$

67. (a) 6
(b) 10 cycles/min
(c)

69. (a) $\frac{1}{440}$
(b) 440
(c)

37.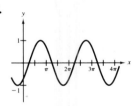

39.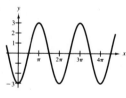

41.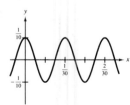

43.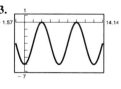

71.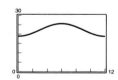

Maximum sales: June
Minimum sales: December

73. $g(x) = 2f(x)$ **75.** $g(x) = f(x - \pi)$

Section 2.6   (page 171)

WARM-UP

1. $f(x) = -1: \dfrac{3\pi}{2}$

$f(x) = 0: 0, \pi, 2\pi$

$f(x) = 1: \dfrac{\pi}{2}$

2. $f(x) = -1: \pi$

$f(x) = 0: \dfrac{\pi}{2}, \dfrac{3\pi}{2}$

$f(x) = 1: 0, 2\pi$

3. $f(x) = -1: \dfrac{3\pi}{4}, \dfrac{7\pi}{4}$

$f(x) = 0: 0, \dfrac{\pi}{2}, \pi, \dfrac{3\pi}{2}, 2\pi$

$f(x) = 1: \dfrac{\pi}{4}, \dfrac{5\pi}{4}$

4. $f(x) = -1: 2\pi$

$f(x) = 0: \pi$

5.

6.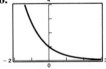

7.

8.

9. $0, \dfrac{\sqrt{3}\pi}{12}, \dfrac{\sqrt{2}\pi}{8}, \dfrac{\pi}{6}, 0$

10. $0, \dfrac{3+\pi}{6}, \dfrac{2\sqrt{2}+\pi}{4}, \dfrac{3\sqrt{3}+2\pi}{6}, \dfrac{\pi+2}{2}$

1. c

```
Xmin=-π/2
Xmax=3π/2
Xscl=π/4
Ymin=-3
Ymax=3
Yscl=1
```

2. g

```
Xmin=-π/3
Xmax=π
Xscl=π/6
Ymin=-2
Ymax=2
Yscl=1
```

3. e

```
Xmin=-2π
Xmax=2π
Xscl=π/2
Ymin=-3
Ymax=3
Yscl=1
```

4. a

```
Xmin=-π
Xmax=5π
Xscl=π
Ymin=-5
Ymax=5
Yscl=1
```

5. d

```
Xmin=-1
Xmax=4
Xscl=1
Ymin=-2
Ymax=2
Yscl=1
```

6. h

```
Xmin=-1
Xmax=3
Xscl=0.5
Ymin=-2
Ymax=2
Yscl=1
```

7. b

```
Xmin=-π
Xmax=2π
Xscl=π/2
Ymin=-3
Ymax=3
Yscl=1
```

8. f

```
Xmin=-0.5
Xmax=2
Xscl=0.25
Ymin=-4
Ymax=4
Yscl=1
```

9.

11.

13.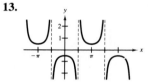

15.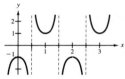

17.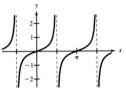

19.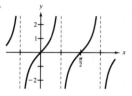

21.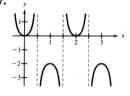

23.

25.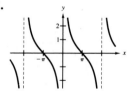

27.

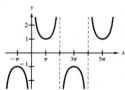

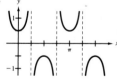

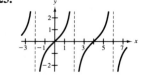

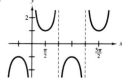

29.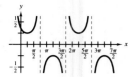

31.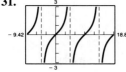

57.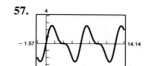

Period: 2π
Relative maximum: $(1.047, 2.598)$
Relative minimum: $(5.236, -2.598)$

33.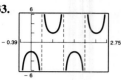

35.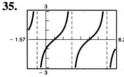

59.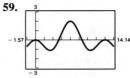

Period: 4π
Relative maxima: $(0, 0)$, $(6.2831, 2)$
Relative minima: $(2.636, -1.125)$, $(9.930, -1.125)$

37.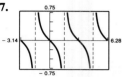

39.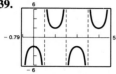

41. $-\dfrac{7\pi}{4}, -\dfrac{3\pi}{4}, \dfrac{\pi}{4}, \dfrac{5\pi}{4}$

43. $-\dfrac{4\pi}{3}, -\dfrac{2\pi}{3}, \dfrac{2\pi}{3}, \dfrac{4\pi}{3}$

45. Even

47. **(a)**

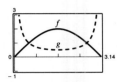

(b) $(0.524, 2.618)$
(c) As x approaches π, f approaches 0 and g increases without bound. g is the reciprocal of f.

61.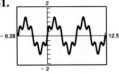

63.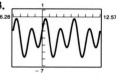

65.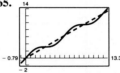

67.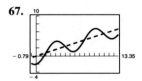

49.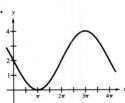

51.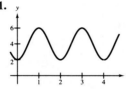

69.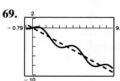

71.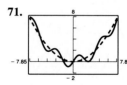

53.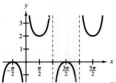

55.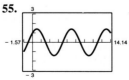

Period: 2π
Relative maximum: $(0.785, 1.414)$
Relative minimum: $(3.927, -1.414)$

73.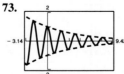

The functional values approach 0 as x increases without bound.

75.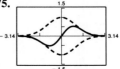

The functional values approach 0 as x increases without bound.

77.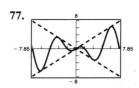

The functional values approach 0 as x approaches 0.

79.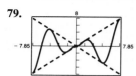

The functional values approach 0 as x approaches 0.

81.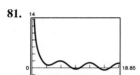

y increases without bound.

83.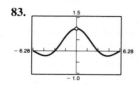

$g(x)$ approaches the value of 1.

85.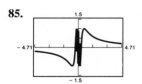

$f(x)$ oscillates between -1 and 1.

87.

$f = g$

89.

$f = g$

91.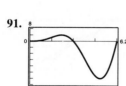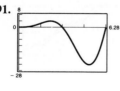

93.

95. $d = 6 \cot x$

97.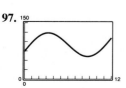

99.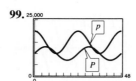

We can explain the cycles of this predator-prey population by noting the cause and effect pattern.

101.

Section 2.7 *(page 183)*

WARM-UP   **1.** -1 **2.** -1 **3.** -1 **4.** $\dfrac{\sqrt{2}}{2}$

5. 0   **6.** $\dfrac{\pi}{6}$ **7.** π **8.** $\dfrac{\pi}{4}$ **9.** 0   **10.** $-\dfrac{\pi}{4}$

---

1. $\dfrac{\pi}{6}$ **3.** $\dfrac{\pi}{3}$ **5.** $\dfrac{\pi}{6}$ **7.** $\dfrac{5\pi}{6}$ **9.** $-\dfrac{\pi}{3}$ **11.** $\dfrac{2\pi}{3}$

13. $\dfrac{\pi}{3}$ **15.** 0   **17.** 1.29   **19.** -0.85 **21.** -1.11

23. 0.32   **25.** 1.99   **27.** 0.74   **29.**

31. 0.3   **33.** -0.1 **35.** 0

37. $\dfrac{3}{5}$ **39.** $\dfrac{\sqrt{5}}{5}$ **41.** $\dfrac{12}{13}$

43. $\dfrac{\sqrt{34}}{5}$

45. $y = \pm 1$ **47.** $\dfrac{1}{x}$

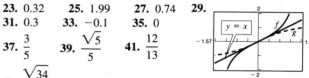

49. $\sqrt{1 - 4x^2}$ **51.** $\sqrt{1 - x^2}$ **53.** $\dfrac{\sqrt{9 - x^2}}{x}$

55. $\dfrac{\sqrt{2}}{\sqrt{2 - x^2}}$ **57.** $\dfrac{9}{\sqrt{x^2 + 81}}$ **59.** $\dfrac{|x - 1|}{\sqrt{x^2 - 2x + 10}}$

61. 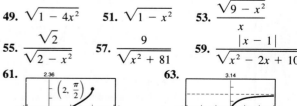   **63.**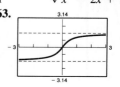

65. $f(t) = 3\sqrt{2} \sin\left(2t + \dfrac{\pi}{4}\right)$

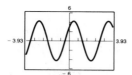

67. (a)

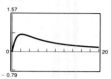

   (b) 2.236 ft

   (c) $\beta = 0$; As the distance from the picture increases without bound, the angle β approaches 0.

69. (a) 14.5°

   (b) 30.0°

71. $y = \operatorname{arccot} x$ if and only if $\cot y = x$, where $0 < y < \pi$.

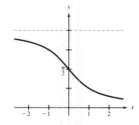

73. $y = \operatorname{arccsc} x$ if and only if $\csc y = x$, where $-\dfrac{\pi}{2} \le y < 0$ or $0 < y \le \dfrac{\pi}{2}$.

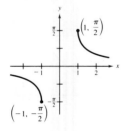

Section 2.8 (page 194)

WARM-UP **1.** 8.45 **2.** 78.99 **3.** 1.06 **4.** 1.24
5. 4.88 **6.** 34.14 **7.** 4, π **8.** $\frac{1}{2}$, 2 **9.** 3, $\frac{2}{3}$
10. 0.2, 8π

1. $a \approx 3.64$ **3.** $a \approx 8.26$
    $c \approx 10.64$ $c \approx 25.38$
    $B = 70°$ $A = 19°$
5. $a \approx 91.34$ **7.** $c \approx 11.66$
    $b \approx 420.70$ $A \approx 30.96°$
    $B = 77° \, 45'$ $B \approx 59.04°$

9. $a \approx 49.48$
   $A \approx 72.08°$
   $B \approx 17.92°$
11. 2.56 in.   **13.** 121.2 ft   **15.** 15.4 ft   **17.** 56.3°
19. 12.68°   **21.** 5099 ft   **23.** 19.9 ft
25. 508 mi north; 650 mi east   **27.** N 56.3° W
29. (a) N 58° E   **(b)** 68.8 yd   **31.** 1657 ft
33. 17,054 ft $\approx$ 3.23 mi   **35.** 29.389 in.
37. $y = \sqrt{3}r$ **39.** $a \approx 7$, $c \approx 12.2$
41. (a) 4   **43. (a)** $\frac{1}{16}$
   (b) 4   **(b)** 60
   (c) $\frac{1}{16}$ **(c)** $\frac{1}{120}$
45. $\omega = 528\pi$

Chapter 2 Review Exercises (page 199)

1. $\dfrac{3\pi}{4}, -\dfrac{5\pi}{4}$ **3.** 250°, $-470°$

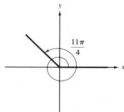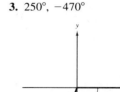

5. 135.28°   **7.** 5.38°   **9.** 135° 16′ 12″   **11.** $-85°\,9'$
13. 128.57°   **15.** $-200.54°$ **17.** 8.3776

19. -0.5890 **21.** 72°   **23.** $\dfrac{\pi}{5}$

25. $\sin \theta = \dfrac{4}{5}$ **27.** $\sin \theta = \dfrac{2\sqrt{53}}{53}$

    $\cos \theta = \dfrac{3}{5}$ $\cos \theta = -\dfrac{7\sqrt{53}}{53}$

    $\tan \theta = \dfrac{4}{3}$ $\tan \theta = -\dfrac{2}{7}$

    $\csc \theta = \dfrac{5}{4}$ $\csc \theta = \dfrac{\sqrt{53}}{2}$

    $\sec \theta = \dfrac{5}{3}$ $\sec \theta = -\dfrac{\sqrt{53}}{7}$

    $\cot \theta = \dfrac{3}{4}$ $\cot \theta = -\dfrac{7}{2}$

29. $\sin \theta = -\dfrac{3\sqrt{13}}{13}$

$\cos \theta = -\dfrac{2\sqrt{13}}{13}$

$\tan \theta = \dfrac{3}{2}$

$\csc \theta = -\dfrac{\sqrt{13}}{3}$

$\sec \theta = -\dfrac{\sqrt{13}}{2}$

$\cot \theta = \dfrac{2}{3}$

31. $\sin \theta = -\dfrac{\sqrt{11}}{6}$

$\cos \theta = \dfrac{5}{6}$

$\tan \theta = -\dfrac{\sqrt{11}}{5}$

$\csc \theta = -\dfrac{6\sqrt{11}}{11}$

$\cot \theta = -\dfrac{5\sqrt{11}}{11}$

33. $\cos \theta = -\dfrac{\sqrt{55}}{8}$

$\tan \theta = -\dfrac{3\sqrt{55}}{55}$

$\csc \theta = \dfrac{8}{3}$

$\sec \theta = -\dfrac{8\sqrt{55}}{55}$

$\cot \theta = -\dfrac{\sqrt{55}}{3}$

35. $\sqrt{3}$ **37.** $-\dfrac{\sqrt{3}}{2}$ **39.** $-\dfrac{\sqrt{2}}{2}$

41. 0.65   **43.** 3.24   **45.** $135° = \dfrac{3\pi}{4}, 225° = \dfrac{5\pi}{4}$

47. $210° = \dfrac{7\pi}{6}, 330° = \dfrac{11\pi}{6}$

49. $57° \approx 0.9948, 123° \approx 2.1467$

51. $165° \approx 2.8798, 195° \approx 3.4034$

53. $\dfrac{\sqrt{-x^2 + 2x}}{-x^2 + 2x}$ **55.** $\dfrac{2\sqrt{4 - 2x^2}}{4 - x^2}$

57.

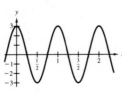

59.

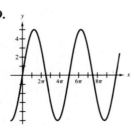

61.

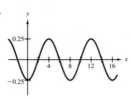

63.

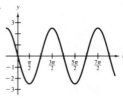

65.

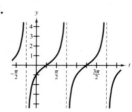

67.

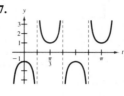

69.

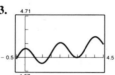

71.

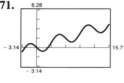

73.

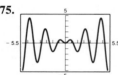

75.

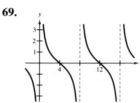

77.

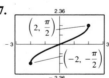

79.

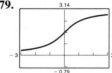

81.

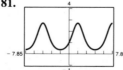

Periodic
Relative maximum: (1.571, 2.718)
Relative minimum: (4.712, 0.368)

83.

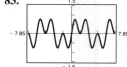

Periodic
Relative maxima: (0.615, 0.770), (2.526, 0.770), (4.712, 0)
Relative minima: (1.571, 0), (3.757, −0.770),
(5.668, −0.770)

85. 7.66 m   **87.** 1.33 mi   **89.** 268.8 ft

91. (a)

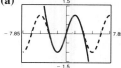

(b)

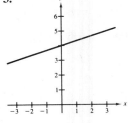

(c) $\sin x = x - \dfrac{x^3}{3!} + \dfrac{x^5}{5!} - \dfrac{x^7}{7!} + \dfrac{x^9}{9!}$

$\cos x = 1 - \dfrac{x^2}{2!} + \dfrac{x^4}{4!} - \dfrac{x^6}{6!} + \dfrac{x^8}{8!}$

The accuracy increases with additional terms.

CUMULATIVE TEST for Chapters 1 and 2   (page 202)

1. $\dfrac{7}{2}$ **2.** $-2, \dfrac{5}{3}$

3.

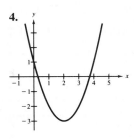

4.

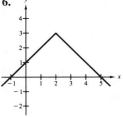

5.

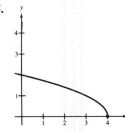

6.

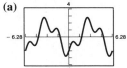

7. (1, 11), (7, 7), (10, 5)   **8.** $2x - y + 2 = 0$

9. (a) $\dfrac{3}{2}$ **(b)** Undefined   **(c)** $\dfrac{t}{t-1}$ **(d)** $\dfrac{s+2}{s}$

10. $A = \dfrac{\sqrt{3}}{4} s^2$

11. 80°   **12.** $-\dfrac{2\pi}{3}$

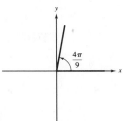

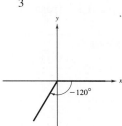

13. $\sin \theta = \frac{5}{13}$ $\csc \theta = \frac{13}{5}$
$\cos \theta = \frac{12}{13}$ $\sec \theta = \frac{13}{12}$
$\tan \theta = \frac{5}{12}$ $\cot \theta = \frac{12}{5}$

14. $\sin t = \dfrac{\sqrt{5}}{3}$, $\tan t = -\dfrac{\sqrt{5}}{2}$

15. 61.93°, 298.07°

16. (a)

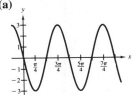

(b)

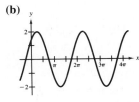

(c)

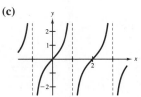

(d)

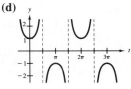

17. (a) The graph of g is 10 units above the graph of f.
(b) The period of f is 2π and the period of g is 4.
(c) The graph of g is $\pi/4$ units to the left of the graph
of f.
(d) The graph of g is a reflection in the x-axis of the
graph of f.

18. (a)

(b) $x \approx 1.9$
(c) $f(2.7) \approx 2.8$

19. (a) $\dfrac{\pi}{6}$ **(b)** $\dfrac{\pi}{3}$ **20.** $\sqrt{1 - 4x^2}$

21. 382 ft

CHAPTER 3

Section 3.1 (*page 210*)

WARM-UP

1. $\sin \theta = \dfrac{3\sqrt{13}}{13}$

$\cos \theta = \dfrac{2\sqrt{13}}{13}$

$\csc \theta = \dfrac{\sqrt{13}}{3}$

$\sec \theta = \dfrac{\sqrt{13}}{2}$

$\cot \theta = \dfrac{2}{3}$

2. $\sin \theta = \dfrac{2\sqrt{2}}{3}$

$\cos \theta = \dfrac{1}{3}$

$\tan \theta = 2\sqrt{2}$

$\csc \theta = \dfrac{3\sqrt{2}}{4}$

$\cot \theta = \dfrac{\sqrt{2}}{4}$

3. $\sin \theta = -\dfrac{3\sqrt{58}}{58}$

$\cos \theta = \dfrac{7\sqrt{58}}{58}$

$\tan \theta = -\dfrac{3}{7}$

$\csc \theta = -\dfrac{\sqrt{58}}{3}$

$\sec \theta = \dfrac{\sqrt{58}}{7}$

$\cot \theta = -\dfrac{7}{3}$

4. $\sin \theta = \dfrac{\sqrt{5}}{5}$

$\cos \theta = -\dfrac{2\sqrt{5}}{5}$

$\tan \theta = -\dfrac{1}{2}$

$\csc \theta = \sqrt{5}$

$\sec \theta = -\dfrac{\sqrt{5}}{2}$

$\cot \theta = -2$

5. $\dfrac{1}{2}$ **6.** $\dfrac{5}{4}$ **7.** $\dfrac{\sqrt{73}}{8}$ **8.** $\dfrac{2}{3}$ **9.** $\dfrac{x^2 + x + 16}{4(x + 1)}$

10. $\dfrac{8x - 2}{1 - x^2}$

1. $\tan x = \dfrac{\sqrt{3}}{3}$

$\csc x = 2$

$\sec x = \dfrac{2\sqrt{3}}{3}$

$\cot x = \sqrt{3}$

3. $\cos \theta = \dfrac{\sqrt{2}}{2}$

$\tan \theta = -1$

$\csc \theta = -\sqrt{2}$

$\cot \theta = -1$

5. $\sin x = \dfrac{2}{3}$

$\cos x = -\dfrac{\sqrt{5}}{3}$

$\csc x = \dfrac{3}{2}$

$\sec x = -\dfrac{3\sqrt{5}}{5}$

$\cot x = -\dfrac{\sqrt{5}}{2}$

7. $\sin \theta = -\dfrac{2\sqrt{5}}{5}$

$\cos \theta = -\dfrac{\sqrt{5}}{5}$

$\csc \theta = -\dfrac{\sqrt{5}}{2}$

$\sec \theta = -\sqrt{5}$

$\cot \theta = \dfrac{1}{2}$

9. $\cos \theta = 0$; $\tan \theta$ is undefined. $\csc \theta = -1$; $\sec \theta$ is undefined.

11. d **12.** e **13.** a **14.** f **15.** b **16.** c
17. b **18.** c **19.** e **20.** a **21.** f **22.** d
23. $\sec \phi$ **25.** $\sin \beta$ **27.** $\cos x$ **29.** 1
31. $-\tan x$ **33.** $\tan x$ **35.** $1 + \sin y$ **37.** $\sin^2 x$
39. $\sin^2 x \tan^2 x$ **41.** $\sec^4 x$ **43.** $\sin^2 x - \cos^2 x$
45. $1 + 2 \sin x \cos x$ **47.** $\tan^2 x$ **49.** $2 \csc^2 x$
51. $2 \sec x$ **53.** $1 + \cos y$ **55.** $3(\sec x + \tan x)$
57. Identity **59.** Not an identity

61. (a) $0 \le x \le \dfrac{\pi}{2}, \dfrac{3\pi}{2} \le x \le 2\pi$ **(b)** $\dfrac{\pi}{2} \le x \le \dfrac{3\pi}{2}$

63. $5 \cos \theta$ **65.** $3 \tan \theta$ **67.** $5 \sec \theta$ **69.** $\cos \theta$

71. $27 \sec^3 \theta$ **73.** $0 \le \theta < \dfrac{\pi}{2}, \dfrac{3\pi}{2} < \theta \le 2\pi$

75. $\ln |\cot \theta|$ **77.** False, $\dfrac{\sin k\theta}{\cos k\theta} = \tan k\theta$ **79.** True

81. (a) $\csc^2 132° - \cot^2 132° \approx 1.8107 - 0.8107 = 1$

(b) $\csc^2 \dfrac{2\pi}{7} - \cot^2 \dfrac{2\pi}{7} \approx 1.6360 - 0.6360 = 1$

83. (a) $\cos(90° - 80°) = \sin 80° \approx 0.9848$

(b) $\cos\left(\dfrac{\pi}{2} - 0.8\right) = \sin 0.8 \approx 0.7174$

85. $\cos \theta = \pm\sqrt{1 - \sin^2 \theta}$

$\tan \theta = \pm\dfrac{\sin \theta}{\sqrt{1 - \sin^2 \theta}}$

$\csc \theta = \dfrac{1}{\sin \theta}$

$\sec \theta = \pm\dfrac{1}{\sqrt{1 - \sin^2 \theta}}$

$\cot \theta = \pm\dfrac{\sqrt{1 - \sin^2 \theta}}{\sin \theta}$

Section 3.2   (page 217)

WARM-UP
1. (a) $x^2(1 - y^2)$ **2. (a)** $x^2(1 + y^2)$
   (b) $\sin^4 x$ **(b)** 1
3. (a) $(x^2 + 1)(x^2 - 1)$
   (b) $\sec^2 x(\tan^2 x - 1)$
4. (a) $(z + 1)(z^2 - z + 1)$
   (b) $(\tan x + 1)(\tan^2 x - \tan x + 1)$
5. (a) $(x - 1)(x^2 + 1)$ **6. (a)** $(x^2 - 1)^2$
   (b) $(\cot x - 1) \csc^2 x$ **(b)** $\cos^4 x$
7. (a) $\dfrac{y^2 - x^2}{x}$ **8. (a)** $\dfrac{x^2 - 1}{x^2}$
   (b) $\tan x$ **(b)** $\sin^2 x$
9. (a) $\dfrac{y^2 + (1 + z)^2}{y(1 + z)}$ **10. (a)** $\dfrac{y(1 + y) - z^2}{z(1 + y)}$
   (b) $2 \csc x$ **(b)** $\dfrac{\tan x - 1}{\sec x(1 + \tan x)}$

Exercises 1–50 are proofs.

51. $\sin \theta = \pm\sqrt{1 - \cos^2 \theta}$; $\dfrac{7\pi}{4}$

53. $\sqrt{\tan^2 x} = |\tan x|$; $\dfrac{3\pi}{4}$ **55.** $\mu = \tan \theta$

Section 3.3   (page 228)

WARM-UP
1. $\dfrac{2\pi}{3}, \dfrac{4\pi}{3}$ **2.** $\dfrac{\pi}{3}, \dfrac{2\pi}{3}$ **3.** $\dfrac{\pi}{4}, \dfrac{7\pi}{4}$ **4.** $\dfrac{7\pi}{4}, \dfrac{5\pi}{4}$
5. $\dfrac{\pi}{3}, \dfrac{4\pi}{3}$ **6.** $\dfrac{3\pi}{4}, \dfrac{7\pi}{4}$ **7.** $\dfrac{15}{8}$ **8.** $-3, \dfrac{5}{2}$
9. $\dfrac{2 \pm \sqrt{14}}{2}$ **10.** $-1, 3$

Exercises 1–6 are proofs.

7. $\dfrac{2\pi}{3} + 2n\pi, \dfrac{4\pi}{3} + 2n\pi$ **9.** $\dfrac{\pi}{3} + 2n\pi, \dfrac{2\pi}{3} + 2n\pi$

11. $\dfrac{\pi}{4} + \dfrac{n\pi}{2}$ **13.** $\dfrac{\pi}{6} + n\pi, \dfrac{5\pi}{6} + n\pi$ **15.** $n\pi, \dfrac{\pi}{4} + n\pi$

17. $n\pi, \dfrac{3\pi}{2} + 2n\pi$ **19.** $\dfrac{\pi}{3} + n\pi, \dfrac{2\pi}{3} + n\pi$

21. $\dfrac{\pi}{3}, \dfrac{5\pi}{3}$ **23.** $\dfrac{7\pi}{6}, \dfrac{3\pi}{2}, \dfrac{11\pi}{6}$ **25.** $\dfrac{\pi}{6}, \dfrac{5\pi}{6}, \dfrac{7\pi}{6}, \dfrac{11\pi}{6}$
27. No solution   **29.** $\dfrac{2\pi}{3}, \dfrac{5\pi}{6}, \dfrac{5\pi}{3}, \dfrac{11\pi}{6}$

31. $\dfrac{\pi}{2}$ **33.** $\dfrac{2}{3}, \dfrac{3}{2}$; 0.8411, 5.4421
35. $4 \pm \sqrt{3}$; 1.1555, 1.3981, 4.2971, 4.5397
37. 1.107, 4.249   **39.** 1.0472, 5.2360
41. 0, 1.895   **43.** 0, 2.678, 3.142, 5.820, 6.283
45. 0.9828, 1.7682, 4.1244, 4.9098
47. 0.3398, 0.8481, 2.2935, 2.8018   **49.** 0.4271, 2.7145
51.   **53.** 0.7391, approximate the zeros of $f(x) = \cos x - x$.

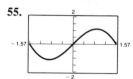

Maximum: (0.785, 1.414)
Minimum: (3.927, −1.414)

55.

1.122 square units
57. 0.04, 0.43, 0.83   **59.** 37°, 53°   **61.** 1   **63.** $\sqrt{2}$

Section 3.4   (page 237)

WARM-UP
1. $\dfrac{\sqrt{10}}{10}$ **2.** $\dfrac{-5\sqrt{34}}{34}$ **3.** $-\dfrac{\sqrt{7}}{4}$ **4.** $\dfrac{2\sqrt{2}}{3}$
5. $\dfrac{\pi}{4}, \dfrac{3\pi}{4}$ **6.** $\dfrac{\pi}{2}, \dfrac{3\pi}{2}$ **7.** $\tan^3 x$ **8.** $\cot^2 x$
9. $\sec x$ **10.** $1 - \tan^2 x$

1. $\sin 75° = \dfrac{\sqrt{2}}{4}\left(1 + \sqrt{3}\right)$
   $\cos 75° = \dfrac{\sqrt{2}}{4}\left(\sqrt{3} - 1\right)$
   $\tan 75° = \sqrt{3} + 2$

3. $\sin 105° = \dfrac{\sqrt{2}}{4}(\sqrt{3} + 1)$

$\cos 105° = \dfrac{\sqrt{2}}{4}(1 - \sqrt{3})$

$\tan 105° = -2 - \sqrt{3}$

5. $\sin 195° = \dfrac{\sqrt{2}}{4}(1 - \sqrt{3})$

$\cos 195° = -\dfrac{\sqrt{2}}{4}(\sqrt{3} + 1)$

$\tan 195° = 2 - \sqrt{3}$

7. $\sin \dfrac{11\pi}{12} = \dfrac{\sqrt{2}}{4}(\sqrt{3} - 1)$

$\cos \dfrac{11\pi}{12} = -\dfrac{\sqrt{2}}{4}(\sqrt{3} + 1)$

$\tan \dfrac{11\pi}{12} = -2 + \sqrt{3}$

9. $\sin \dfrac{17\pi}{12} = -\dfrac{\sqrt{2}}{4}(\sqrt{3} + 1)$ **11.** $\cos 40°$

$\cos \dfrac{17\pi}{12} = \dfrac{\sqrt{2}}{4}(1 - \sqrt{3})$

$\tan \dfrac{17\pi}{12} = 2 + \sqrt{3}$

13. $\sin 200°$ **15.** $\tan 239°$ **17.** $\sin 1.8$ **19.** $\tan 3x$

21. $\frac{33}{65}$ **23.** $-\frac{56}{65}$ **25.** $-\frac{3}{5}$ **27.** $\frac{44}{125}$

Exercises 29–46 are proofs.

47. (a) $\sqrt{2} \sin\left(\theta + \dfrac{\pi}{4}\right)$ **49. (a)** $13 \sin(3\theta + 0.3948)$

(b) $\sqrt{2} \cos\left(\theta - \dfrac{\pi}{4}\right)$ **(b)** $13 \cos(3\theta - 1.1760)$

51. $\sqrt{2} \sin \theta + \sqrt{2} \cos \theta$ **53.** 1

55. $\dfrac{\pi}{2}$ **57.** $\dfrac{\pi}{4}, \dfrac{7\pi}{4}$ **59.** 3.927, 5.498

Section 3.5 (page 247)

WARM-UP   **1.** $\sin x(2 + \cos x)$

2. $(\cos x - 2)(\cos x + 1)$ **3.** $0, \dfrac{\pi}{2}, \pi, \dfrac{3\pi}{2}$

4. $\dfrac{\pi}{4}, \dfrac{3\pi}{4}, \dfrac{5\pi}{4}, \dfrac{7\pi}{4}$ **5.** π **6.** 0   **7.** $\dfrac{2 - \sqrt{2}}{4}$

8. $\frac{3}{4}$ **9.** $\tan 3x$ **10.** $\cos x(1 - 4 \sin^2 x)$

1.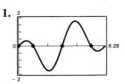

$0, \dfrac{\pi}{3}, \pi, \dfrac{5\pi}{3}$

3.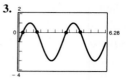

$\dfrac{\pi}{12}, \dfrac{5\pi}{12}, \dfrac{13\pi}{12}, \dfrac{17\pi}{12}$

5.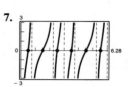

$0, \dfrac{2\pi}{3}, \dfrac{4\pi}{3}$

7.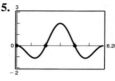

$\dfrac{\pi}{2}, \dfrac{\pi}{6}, \dfrac{5\pi}{6}, \dfrac{7\pi}{6}, \dfrac{3\pi}{2}, \dfrac{11\pi}{6}$

9.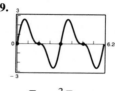

$0, \dfrac{\pi}{2}, \pi, \dfrac{3\pi}{2}$

11. $f(x) = 3 \sin 2x$

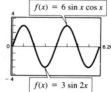

Relative maxima: (0.785, 3), (3.927, 3)
Relative minima: (2.356, −3), (5.498, −3)

13. $g(x) = 4 \cos 2x$ **15.** $\sin 2u = \frac{24}{25}$

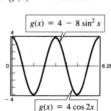

$\cos 2u = \frac{7}{25}$

$\tan 2u = \frac{24}{7}$

Relative maxima: (0, 4), (3.142, 4)
Relative minima: (1.571, −4), (4.712, −4)

17. $\sin 2u = \dfrac{4}{5}$ **19.** $\sin 2u = -\dfrac{4\sqrt{21}}{25}$

$\cos 2u = \dfrac{3}{5}$ $\cos 2u = -\dfrac{17}{25}$

$\tan 2u = \dfrac{4}{3}$ $\tan 2u = \dfrac{4\sqrt{21}}{17}$

21. $\frac{1}{8}(3 + 4 \cos 2x + \cos 4x)$ **23.** $\frac{1}{8}(1 - \cos 4x)$

25. $\frac{1}{16}(1 + \cos 2x)(1 - \cos 4x)$

27. $\sin 105° = \frac{1}{2}\sqrt{2 + \sqrt{3}}$
$\cos 105° = -\frac{1}{2}\sqrt{2 - \sqrt{3}}$
$\tan 105° = -2 - \sqrt{3}$

29. $\sin 112° \, 30' = \frac{1}{2}\sqrt{2 + \sqrt{2}}$
$\cos 112° \, 30' = -\frac{1}{2}\sqrt{2 - \sqrt{2}}$
$\tan 112° \, 30' = -1 - \sqrt{2}$

31. $\sin \frac{\pi}{8} = \frac{1}{2}\sqrt{2 - \sqrt{2}}$

$\cos \frac{\pi}{8} = \frac{1}{2}\sqrt{2 + \sqrt{2}}$

$\tan \frac{\pi}{8} = \sqrt{2} - 1$

33. $\sin \frac{u}{2} = \frac{5\sqrt{26}}{26}$

$\cos \frac{u}{2} = \frac{\sqrt{26}}{26}$

$\tan \frac{u}{2} = 5$

35. $\sin \frac{u}{2} = \sqrt{\frac{89 - 8\sqrt{89}}{178}}$

$\cos \frac{u}{2} = -\sqrt{\frac{89 + 8\sqrt{89}}{178}}$

$\tan \frac{u}{2} = \frac{8 - \sqrt{89}}{5}$

37. $\sin \frac{u}{2} = \frac{3\sqrt{10}}{10}$

$\cos \frac{u}{2} = -\frac{\sqrt{10}}{10}$

$\tan \frac{u}{2} = -3$

39. $|\sin 3x|$ **41.** $-|\tan 4x|$

43. π

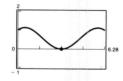

45. $\frac{\pi}{3}, \pi, \frac{5\pi}{3}$

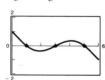

47. $3\left(\sin \frac{\pi}{2} + \sin 0\right)$ **49.** $\frac{1}{2}(\sin 8\theta + \sin 2\theta)$

51. $\frac{5}{2}(\cos 8\beta + \cos 2\beta)$ **53.** $\frac{1}{2}(\cos 2y - \cos 2x)$

55. $\frac{1}{2}(\sin 2\theta + \sin 2\pi)$ **57.** $2 \sin 45° \cos 15°$

59. $-2 \sin \frac{\pi}{2} \sin \frac{\pi}{4}$ **61.** $2 \cos 4x \cos 2x$

63. $2 \cos \alpha \sin \beta$ **65.** $2 \cos(\phi + \pi) \cos \pi$

67. $0, \frac{\pi}{4}, \frac{\pi}{2}, \frac{3\pi}{4}, \pi, \frac{5\pi}{4}, \frac{3\pi}{2}, \frac{7\pi}{4}$ **69.** $\frac{\pi}{6}, \frac{5\pi}{6}$

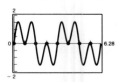   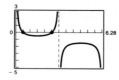

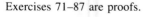
Exercises 71–87 are proofs.

89.

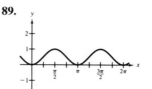

91. $2x\sqrt{1 - x^2}$ **93.** Proof

95. **(a)** $A = 100 \sin \frac{\theta}{2} \cos \frac{\theta}{2}$

(b) $A = 50 \sin \theta$; The area is maximum when $\theta = \frac{\pi}{2}$.

Chapter 3 Review Exercises (*page 250*)

1. $\sin^2 x$ **3.** $1 + \cot \alpha$ **5.** 1   **7.** 1
9. $\cos^2 2x$

Exercises 11–35 are proofs.

37. $\frac{\sqrt{2}}{4}\left(\sqrt{3} + 1\right)$ **39.** $-\frac{1}{2}\sqrt{2 + \sqrt{2}}$

41. $-\frac{3}{52}\left(5 + 4\sqrt{7}\right)$ **43.** $\frac{1}{52}\left(36 + 5\sqrt{7}\right)$

45. $\frac{1}{4}\sqrt{2\left(4 - \sqrt{7}\right)}$

47. False; If $\frac{\pi}{2} < \theta < \pi$, then $\cos \frac{\theta}{2} > 0$.   **49.** True

51. $0, \pi$

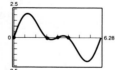

53. $0, \frac{3\pi}{4}, \pi, \frac{5\pi}{4}$

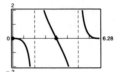

55. $0, \frac{\pi}{2}, \pi$

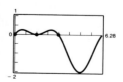

57. $\dfrac{\pi}{3}, \dfrac{5\pi}{3}$

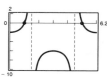

59. $\dfrac{\pi}{4}, \dfrac{5\pi}{4}$

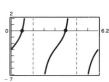

61. $2 \cos \dfrac{5\theta}{2} \cos \dfrac{\theta}{2}$

63. $\frac{1}{2}(\cos \alpha - \cos 5\alpha)$ **65.** $8x^2 - 1$ **67.** Proof

69. (a) $y = \frac{1}{2}\sqrt{10} \sin(8t - \arctan \frac{1}{3})$

(b)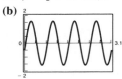

(c) $\frac{1}{2}\sqrt{10}$

(d) $\dfrac{4}{\pi}$

CHAPTER 4

Section 4.1   (page 261)

WARM-UP   **1.** $b = 3\sqrt{3}, A = 30°, B = 60°$
2. $c = 5\sqrt{2}, A = 45°, B = 45°$
3. $a = 8, A \approx 28.07°, B \approx 61.93°$
4. $b \approx 8.33, c \approx 11.21, B = 48°$
5. $a \approx 22.69, c \approx 23.04, A = 80°$
6. $a \approx 45.73, b \approx 142.86, A = 17° 45'$
7. 8.48   **8.** 12.94   **9.** 2.25   **10.** 91.06

1. $C = 105°, b \approx 14.14, c \approx 19.32$
3. $C = 110°, b \approx 22.44, c \approx 24.35$
5. $B \approx 21.55°, C \approx 122.45°, c \approx 11.49$
7. $B = 10°, b \approx 69.46, c \approx 136.81$
9. $B = 42° 4', a \approx 22.05, b \approx 14.88$
11. $A \approx 10° 11', C \approx 154° 19', c \approx 11.03$
13. $A \approx 25.57°, B \approx 9.43°, a \approx 10.5$

15. $B \approx 18° 13', C \approx 51° 32', c \approx 40.06$
17. No solution
19. Two solutions: $B \approx 70.4°, C \approx 51.6°, c \approx 4.16$;
$B \approx 109.6°, C \approx 12.4°, c \approx 1.14$
21. No solution
23. (a) $b \leq 5, b = \dfrac{5}{\sin 36°}$

(b) $5 < b < \dfrac{5}{\sin 36°}$

(c) $b > \dfrac{5}{\sin 36°}$

25. 10.4   **27.** 1675.2   **29.** 474.9   **31.** 6 ft
33. 77 yd   **35.** 5 mi   **37.** 26.1 mi, 15.9 mi
39. 4.55 mi   **41.** No solution

Section 4.2   (page 269)

WARM-UP   **1.** $2\sqrt{13}$ **2.** $3\sqrt{5}$ **3.** $4\sqrt{10}$
4. $3\sqrt{13}$ **5.** 20   **6.** 48
7. $a \approx 4.62, c \approx 26.20, B = 70°$
8. $a \approx 34.20, b \approx 93.97, B = 70°$ **9.** No solution
10. $a \approx 15.09, B \approx 18.97°, C \approx 131.03°$

1. $A \approx 27.7°, B \approx 40.5°, C \approx 111.8°$
3. $B \approx 23.8°, C \approx 126.2°, a \approx 12.4$
5. $A \approx 36.9°, B \approx 53.1°, C \approx 90°$
7. $A \approx 92.9°, B \approx 43.55°, C \approx 43.55°$
9. $a \approx 11.79, B \approx 12.7°, C \approx 47.3°$
11. $A \approx 158° 36', C \approx 12° 38', b \approx 10.4$
13. $A = 27° 10', B = 27° 10', c \approx 56.9$

	a	b	c	d	θ	ϕ
15.	4	6	9.67	3.23	30°	150°
17.	10	14	20	13.86	68.2°	111.8°
19.	10	11.57	18	12	67.1°	112.9°

21. 16.25   **23.** 54   **25.** 96.82   **27.** 97,979.6 ft^2
29. N 52° 37' E, S 64° 40' E   **31.** 43.3 mi
33. 1344 ft   **35.** 114.95°
37. $\overline{PQ} \approx 9.4$ ft, $\overline{QS} \approx 5.0$ ft, $\overline{RS} \approx 12.8$ ft
39. (a) N 58.3° W   **41.** (a) 63.7 ft
(b) S 81.6° W   (b) 47.6 ft
43. 3.26 ft
45. (a) 570.60   (b) 5910.68   (c) 177.09

Section 4.3   *(page 284)*

WARM-UP   **1.** $7\sqrt{10}$ **2.** $\sqrt{58}$
3. $3x + 5y - 14 = 0$ **4.** $4x - 3y - 1 = 0$
5. $111.8°$ **6.** $323.1°$ **7.** $\dfrac{1}{2}, \dfrac{\sqrt{3}}{2}$ **8.** $\dfrac{\sqrt{3}}{2}, -\dfrac{1}{2}$
9. $-\dfrac{\sqrt{3}}{2}, \dfrac{1}{2}$ **10.** $-\dfrac{1}{2}, -\dfrac{\sqrt{3}}{2}$

1.
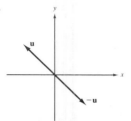

3.
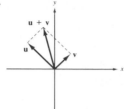

5.
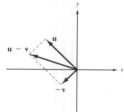

7. $\mathbf{v} = \langle 3, 4 \rangle$, $\|\mathbf{v}\| = 5$ **9.** $\mathbf{v} = \langle -3, 2 \rangle$, $\|\mathbf{v}\| = \sqrt{13}$
11. $\mathbf{v} = \langle 0, 5 \rangle$, $\|\mathbf{v}\| = 5$
13. $\mathbf{v} = \langle 16, -3 \rangle$, $\|\mathbf{v}\| = \sqrt{265}$
15. $\mathbf{v} = \langle 8, 4 \rangle$, $\|\mathbf{v}\| = 4\sqrt{5}$
17. (a) $\langle 4, 3 \rangle$ **19.** (a) $\langle -4, 4 \rangle$
　　(b) $\langle -2, 1 \rangle$　　　(b) $\langle 0, 2 \rangle$
　　(c) $\langle -7, 1 \rangle$　　　(c) $\langle 2, 3 \rangle$
21. (a) $\langle 4, -2 \rangle$ **23.** (a) $3\mathbf{i} - 2\mathbf{j}$
　　(b) $\langle 4, -2 \rangle$　　　(b) $-\mathbf{i} + 4\mathbf{j}$
　　(c) $\langle 8, -4 \rangle$　　　(c) $-4\mathbf{i} + 11\mathbf{j}$
25. (a) $2\mathbf{i} + \mathbf{j}$ **27.** $\|\mathbf{v}\| = 5$, $\theta = 30°$
　　(b) $2\mathbf{i} - \mathbf{j}$
　　(c) $4\mathbf{i} - 3\mathbf{j}$
29. $\|\mathbf{v}\| = 6\sqrt{2}$, $\theta = 315°$

31. $\mathbf{v} = \langle 3, 0 \rangle$
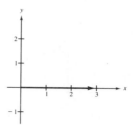

33. $\mathbf{v} = \left\langle -\dfrac{\sqrt{3}}{2}, \dfrac{1}{2} \right\rangle$
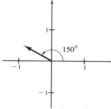

35. $\mathbf{v} = \left\langle -\dfrac{3\sqrt{6}}{2}, \dfrac{3\sqrt{2}}{2} \right\rangle$
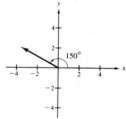

37. $\mathbf{v} = \left\langle \dfrac{\sqrt{10}}{5}, \dfrac{3\sqrt{10}}{5} \right\rangle$

39. $\mathbf{v} = \left\langle 3, -\dfrac{3}{2} \right\rangle$
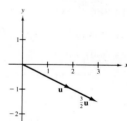

41. $\mathbf{v} = \langle 4, 3 \rangle$
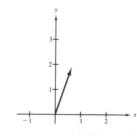

43. $\mathbf{v} = \left\langle \dfrac{7}{2}, -\dfrac{1}{2} \right\rangle$
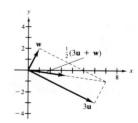

45. $\langle 5, 5 \rangle$ **47.** $\left\langle \left(10\sqrt{2} - 50\right), 10\sqrt{2} \right\rangle$ **49.** $\dfrac{4}{5}\mathbf{i} - \dfrac{3}{5}\mathbf{j}$
51. $\mathbf{j}$ **53.** $90°$ **55.** $63.4°$ **57.** $62.7°$
59. $12.3°$, 82.2 lb   **61.** $71.3°$, 228.5 lb

63. Horizontal component: 80 cos 50° ≈ 51.42 ft/sec;
Vertical component: 80 sin 50° ≈ 61.28 ft/sec
65. $T_{AC} \approx 879.4$ lb, $T_{BC} \approx 652.7$ lb
67. 3192.5 lb   **69.** N 25.2° E, 82.8 mph
71. 425 ft-lb

Section 4.4 (page 294)

WARM-UP   **1.** (a) $\langle -14, -5 \rangle$ (b) $3\sqrt{5}$
2. (a) $\langle \frac{43}{8}, \frac{3}{5} \rangle$ (b) $\frac{1}{40}\sqrt{1249} \approx 0.88$
3. (a) $-6\mathbf{i} + 4\mathbf{j} - 18\mathbf{k}$ (b) $4\sqrt{26} \approx 20.4$
4. (a) $8.7\mathbf{i} - 2.2\mathbf{j}$ (b) $\sqrt{12.45} \approx 3.53$
5. $\dfrac{2\pi}{5}, \dfrac{4\pi}{3}$ **6.** $\dfrac{\pi}{2}, \dfrac{3\pi}{2}$ **7.** 1.00, 5.28
8. 2.89, 3.39
9. $\dfrac{1}{\sqrt{569}}\langle 12, -5, 20 \rangle$, $-\dfrac{1}{\sqrt{569}}\langle 12, -5, 20 \rangle$
10. $\dfrac{1}{\sqrt{394}}\langle 12, 5, -15 \rangle$, $-\dfrac{1}{\sqrt{394}}\langle 12, 5, -15 \rangle$

1. 6   **3.** -15 **5.** -2
7. 8, scalar   **9.** $\langle -6, 8 \rangle$, vector   **11.** -20
13. $37,289; It is the total revenue if all the units are sold at the given price.
15. 90°   **17.** ≈22.17°   **19.** ≈78.69°
21. $\dfrac{5\pi}{12}$ or 75°
23. ≈91.33°   **25.** 90°

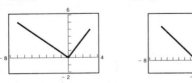

27. Orthogonal   **29.** Parallel
31. Neither   **33.** $\mathbf{w}_1 = \langle \frac{64}{17}, \frac{16}{17} \rangle$; $\mathbf{w}_2 = \langle -\frac{13}{17}, \frac{52}{17} \rangle$
35. $\mathbf{w}_1 = \langle 0 \rangle$; $\mathbf{w}_2 = \langle 4, 2 \rangle$ **37.** $\mathbf{w}_1 = \langle 0, 1 \rangle$; $\mathbf{w}_2 = \langle 2, 0 \rangle$
39. (a) ≈4514.9 lb   (b) ≈25,605 lb
41. ≈7.37 × 10⁶ ft-lb

Section 4.5 (page 302)

WARM-UP   **1.** 13, $(\frac{5}{2}, 6)$ **2.** $7\sqrt{2}, (-\frac{1}{2}, \frac{9}{2})$
3. $\sqrt{74}, (\frac{7}{2}, \frac{5}{2})$ **4.** $\sqrt{13.13} \approx 3.62, (-0.35, -0.8)$
5. $\frac{1}{2}\sqrt{17} \approx 2.06, (\frac{3}{2}, \frac{3}{4})$ **6.** $\frac{5}{6}, (1, \frac{5}{4})$
7. $(x - 4)^2 + (y + 5)^2 = 16$

8. $(x + 1)^2 + (y - 3)^2 = 1$
9. $(x - 3)^2 + (y - 3)^2 = 5$
10. $(x + \frac{1}{2})^2 + (y - 3)^2 = \frac{37}{4}$

1.   **3.**

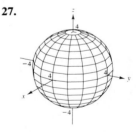

5. $AB = \sqrt{22}$ **7.** $AB = \sqrt{26}$
$BC = 9$ $BC = \sqrt{40}$
$AC = 7$ $AC = \sqrt{22}$
Neither   Neither
9. $(0, -2, 6)$ **11.** $(0, 0, \frac{13}{2})$
13. $x^2 + (y - 4)^2 + (z - 3)^2 = 16$
15. $(x + 3)^2 + (y - 7)^2 + (z - 5)^2 = 25$
17. $(x - \frac{3}{2})^2 + y^2 + (z - 3)^2 = \frac{45}{4}$
19. Center: $(2, -1, 3)$; Radius: 2
21. Center: $(-2, 0, 4)$; Radius: 1
23. Center: $(1, \frac{1}{3}, 4)$; Radius: 3
25.   **27.**

29.

Section 4.6   (*page 308*)

WARM-UP

1. $\vec{AB} = \langle 1, -2 \rangle$
$\vec{AC} = \langle 2, 4 \rangle$
$\vec{AB} + \vec{AC} = \langle 3, 2 \rangle$

2. $\vec{AB} = \langle -4, -2 \rangle$
$\vec{AC} = \langle -2, 3 \rangle$
$\vec{AB} + \vec{AC} = \langle -6, 1 \rangle$

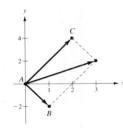

3. $\vec{AB} = \langle -6, -5 \rangle$
$\vec{AC} = \langle -8, 0 \rangle$
$\vec{AB} + \vec{AC} = \langle -14, -5 \rangle$

4. $\vec{AB} = \langle 7, 4 \rangle$
$\vec{AC} = \langle 4, 8 \rangle$
$\vec{AB} + \vec{AC} = \langle 11, 12 \rangle$

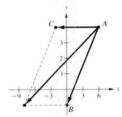

5. $\langle \frac{5}{13}, -\frac{12}{13} \rangle$

6. $\langle \cos 38°, \sin 38° \rangle$

7. $8, 45°$

8. $7, \approx 73.74°$

9. $0, 90°$

10. $-102, 180°$

1. $\mathbf{v} = \langle -2, 3, 1 \rangle$

3. $\mathbf{v} = \langle 4, 4, 2 \rangle$

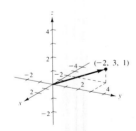

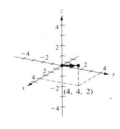

5. $\mathbf{v} = \langle 4, 4, 4 \rangle; c\langle 1, 1, 1 \rangle, c > 0; c\langle 1, 1, 1 \rangle, c < 0$

7. $\mathbf{z} = \langle -3, 7, 6 \rangle$ **9.** $\mathbf{z} = \langle 1, -10, -7 \rangle$ **11.** $\mathbf{z} = \langle \frac{1}{2}, 6, \frac{3}{2} \rangle$

13. $\sqrt{33}$ **15.** $\dfrac{1}{\sqrt{74}}(8\mathbf{i} + 3\mathbf{j} - \mathbf{k})$ **17.** 0

19. -3 **21.** $\approx 124.45°$ **23.** $\approx 109.92°$

25. Parallel   **27.** Orthogonal

29. $\mathbf{w}_1 = \langle 0, \frac{64}{73}, \frac{24}{73} \rangle$; $\mathbf{w}_2 = \langle 2, \frac{9}{73}, -\frac{24}{73} \rangle$

31. Collinear   **33.** Not collinear

35. $\langle 3, 1, 7 \rangle$ **37.** $\langle 0, 2\sqrt{2}, 2\sqrt{2} \rangle$

39. Sphere: $(x - x_1)^2 + (y - y_1)^2 + (z - z_1)^2 = 81$

41. 68 (work units)

Section 4.7   (*page 315*)

WARM-UP

1. (a) 64   (b) 64   (c) $\langle 384, 512 \rangle$

2. (a) 0   (b) 850   (c) $\langle 0, 0 \rangle$

3. (a) 327   (b) 436   (c) $\langle 2943, -\frac{2943}{2}, 3924 \rangle$

4. (a) -8 (b) 25   (c) $16\mathbf{i} - 48\mathbf{j}$

5. $k = -\frac{15}{2}$ **6.** $k = 18$ **7.** $k = \pm 3$

8. 44   **9.** -10 **10.** -72

1. $\mathbf{i} \times \mathbf{j} = \mathbf{k}$ **3.** $\mathbf{i} \times \mathbf{k} = -\mathbf{j}$

5. $\langle 0, 0, 14 \rangle$ **7.** $\langle -3, 5, 23 \rangle$ **9.** $\langle 10, -2, -4 \rangle$

11. $-7\mathbf{i} + 13\mathbf{j} + 16\mathbf{k}$ **13.** $-\frac{3}{2}\mathbf{i} - \frac{3}{2}\mathbf{j} - \frac{3}{2}\mathbf{k}$

15. $-3\mathbf{i} - \frac{11}{3}\mathbf{j} - \frac{1}{3}\mathbf{k}$

17. $\dfrac{1}{\sqrt{19}}(\mathbf{i} - 3\mathbf{j} + 3\mathbf{k})$

19. $\dfrac{1}{\sqrt{342}}(-6\mathbf{i} + 15\mathbf{j} - 9\mathbf{k})$

21. $\dfrac{1}{2\sqrt{2}}(2\mathbf{i} - 2\mathbf{j})$ **23.** 1   **25.** $\sqrt{806}$ **27.** 14

29. $\vec{AB} = \langle 1, 2, -2 \rangle$ is parallel to $\vec{DC} = \langle 1, 2, -2 \rangle$.
$\vec{AD} = \langle -3, 4, 4 \rangle$ is parallel to $\vec{BC} = \langle -3, 4, 4 \rangle$. Area is
$\| \vec{AB} \times \vec{AD} \| = 6\sqrt{10}$.

31. $\sqrt{349}$ sq. units   **33.** $\frac{1}{2}\sqrt{4290}$ sq. units

35. -16 **37.** 4 cu. units   **39.** 84 cu. units

41. $\begin{vmatrix} \mathbf{i} & \mathbf{j} & \mathbf{k} \\ u_1 & u_2 & u_3 \\ u_1 & u_2 & u_3 \end{vmatrix} = 0\mathbf{i} - 0\mathbf{j} + 0\mathbf{k} = \mathbf{0}$

43. $\begin{vmatrix} \mathbf{i} & \mathbf{j} & \mathbf{k} \\ \cos \beta & \sin \beta & 0 \\ \cos \alpha & \sin \alpha & 0 \end{vmatrix} = (\sin \alpha \cos \beta - \cos \alpha \sin \beta)\mathbf{k}$

Chapter 4 Review Exercises   (*page 317*)

1. $A \approx 29.7°, B \approx 52.4°, C \approx 97.9°$
3. $C \approx 110°, b \approx 20.4, c \approx 22.6$
5. $A \approx 35°, C \approx 35°, b \approx 6.6$ **7.** No solution
9. $A \approx 25.9°, C \approx 39.1°, c \approx 10.1$
11. $B \approx 31.2°, C \approx 133.8°, c \approx 13.9$
$B \approx 148.8°, C \approx 16.2°, c \approx 5.39$
13. $A \approx 9.9°, C \approx 20.1°, b \approx 29.1$
15. $A \approx 40.9°, C \approx 114.1°, c \approx 8.6$
$A \approx 139.1°, C \approx 15.9°, c \approx 2.6$
17. 9.798   **19.** 9.08   **21.** 31 ft   **23.** 31.1 m
25. 1135 mi
27. $\overrightarrow{AB} = \langle 3, -1 \rangle$
$\overrightarrow{AC} = \langle 2, 2 \rangle$
$\overrightarrow{AB} + \overrightarrow{AC} = \langle 5, 1 \rangle$

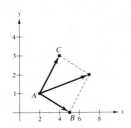

29. $\langle 7, -7 \rangle$ **31.** $\langle 8 \cos 120°, 8 \sin 120° \rangle$
33. $\mathbf{v} = 3\langle \cos 135°, \sin 135° \rangle$

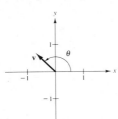

35. $\left\langle \dfrac{6}{\sqrt{61}}, -\dfrac{5}{\sqrt{61}} \right\rangle$ **37.** $\langle -26, -35 \rangle$

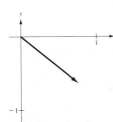

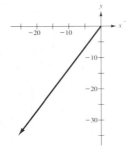

39. (a) $\frac{5}{2}\mathbf{i} + \mathbf{j}$ **(b)** $\frac{9}{2}\mathbf{i} - 3\mathbf{j}$ **(c)** $10\mathbf{i} - 8\mathbf{j}$
41. $\|\mathbf{v}\| = 4, \theta = 300°$

43. 92.3°, 117.0   **45.** 100 lb
47. 460.3 mph; N 32.2° E
49. $(x - 2)^2 + (y - 3)^2 + (z - 2)^2 = 17$
51. $\dfrac{1}{\sqrt{185}} \langle -10, 6, 7 \rangle$ **53.** Parallel   **55.** $\dfrac{11\pi}{12}$ or 165°
57. 90°   **59.** $-\frac{13}{17}\langle 4, 1 \rangle$ **61.** $-\frac{5}{2}\langle 1, -1, 0 \rangle$
63. $\langle -10, 0, -10 \rangle$ **65.** $\mathbf{u} \cdot \mathbf{u} = 6 = \|\mathbf{u}\|^2$
67. $\mathbf{u} \times \mathbf{v} = \langle 4, -2, 8 \rangle = -(\mathbf{v} \times \mathbf{u})$ **69.** .44 cu. units
71. $\approx 78.46°$

CHAPTER 5

Section 5.1   (*page 329*)

WARM-UP   **1.** $2\sqrt{3}$ **2.** $10\sqrt{5}$ **3.** $\sqrt{5}$
4. $-6\sqrt{3}$ **5.** 12   **6.** 48   **7.** $\dfrac{\sqrt{3}}{3}$ **8.** $\sqrt{2}$
9. $-\dfrac{1}{2} \pm \dfrac{\sqrt{5}}{2}$ **10.** $-1 \pm \sqrt{2}$

1. $i, -1, -i, 1, i, -1, -i, 1, i, -1, -i, 1, i, -1, -i, 1$
3. $a = -10, b = 6$ **5.** $a = 6, b = 5$ **7.** $4 + 3i$
9. $2 - 3\sqrt{3}i$ **11.** $5\sqrt{3}i$ **13.** $-1 - 6i$ **15.** $-5i$
17. 8   **19.** $11 - i$ **21.** 4   **23.** $3 - 3\sqrt{2}i$
25. $\frac{1}{6} + \frac{7}{6}i$ **27.** $5 - 3i, 34$ **29.** $-2 + \sqrt{5}i, 9$
31. $-2\sqrt{3}$ **33.** -10 **35.** $5 + i$ **37.** $12 + 30i$
39. 24   **41.** $-9 + 40i$ **43.** -10 **45.** $\frac{16}{41} + \frac{20}{41}i$
47. $\frac{3}{5} + \frac{4}{5}i$ **49.** $-7 - 6i$ **51.** $-\frac{5}{4} - \frac{5}{4}i$
53. $\frac{35}{29} + \frac{595}{29}i$

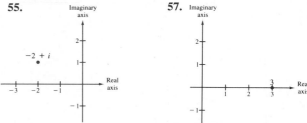

61. $0, 0, 0, 0, 0, 0$

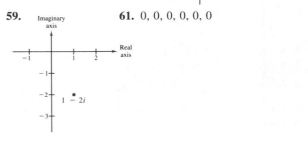

63. $0.5i$, $-0.25 + 0.5i$, $-0.1875 + 0.25i$,
$-0.0273 + 0.4063i$, $-0.1643 + 0.4778i$,
$-0.2013 + 0.3430i$

65. $1, 2, 5, 26, 677, 458, 330$ **67.** $8, 8, 8$

Section 5.2   (*page 335*)

WARM-UP   **1.** $4 - \sqrt{29}i$, $4 + \sqrt{29}i$
2. $-5 - 12i$, $-5 + 12i$ **3.** $-1 + 4\sqrt{2}i$, $-1 - 4\sqrt{2}i$
4. $6 + \frac{1}{2}i$, $6 - \frac{1}{2}i$ **5.** $-13 + 9i$ **6.** $12 + 16i$
7. $26 + 22i$ **8.** 29 **9.** i **10.** $-9 + 46i$

---

1. Three solutions   **3.** Four solutions
5. One real solution   **7.** Two real solutions
9. No real solutions   **11.** Two real solutions
13. $\pm\sqrt{5}$ **15.** $-5 \pm \sqrt{6}$ **17.** 4 **19.** $-1 \pm 2i$
21. $\frac{1}{2} \pm i$ **23.** $20 \pm 2\sqrt{215}$
25. $1 \pm i$
$f(z) = (z - 1 + i)(z - 1 - i)$
27. $2 \pm \sqrt{3}$
$h(x) = \left(x - 2 + \sqrt{3}\right)\left(x - 2 - \sqrt{3}\right)$
29. ±3, $\pm3i$
$f(x) = (x + 3)(x - 3)(x + 3i)(x - 3i)$
31. $2, 2 \pm i$
$g(x) = (x - 2)(x - 2 + i)(x - 2 - i)$
33. $-5, 4 \pm 3i$
$f(t) = (t + 5)(t - 4 + 3i)(t - 4 - 3i)$
35. $-10, -7 \pm 5i$
$f(x) = (x + 10)(x + 7 + 5i)(x + 7 - 5i)$
37. $-\frac{3}{4}, 1 \pm \frac{1}{2}i$
$f(x) = (4x + 3)(2x - 2 + i)(2x - 2 - i)$
39. $1 \pm \sqrt{2}i, -2$
$h(x) = (x + 2)\left(x - 1 + \sqrt{2}i\right)\left(x - 1 - \sqrt{2}i\right)$
41. $-\frac{1}{5}, 1 \pm \sqrt{5}i$
$f(x) = (5x + 1)\left(x - 1 + \sqrt{5}i\right)\left(x - 1 - \sqrt{5}i\right)$
43. $2, \pm2i$
$g(x) = (x - 2)^2(x + 2i)(x - 2i)$
45. $\pm i, \pm3i$
$f(x) = (x + i)(x - i)(x + 3i)(x - 3i)$
47. $-\frac{3}{2}, \pm5i$
$f(x) = (2x + 3)(x - 5i)(x + 5i)$
49. $-\frac{1}{2}, 1, \pm2i$
$f(x) = (x - 1)(2x + 1)(x - 2i)(x + 2i)$
51. $\frac{1}{4}, -3 \pm i$
$g(x) = (4x - 1)(x + 3 - i)(x + 3 + i)$
53. $1, 2, -3 \pm \sqrt{2}i$
$f(x) = (x - 1)(x - 2)\left(x + 3 - \sqrt{2}i\right)\left(x + 3 + \sqrt{2}i\right)$

55. $\dfrac{3}{4}, \dfrac{1 \pm \sqrt{5}i}{2}$
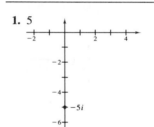
$h(x) = (8x - 6)\left(x - \dfrac{1 + \sqrt{5}i}{2}\right)\left(x - \dfrac{1 - \sqrt{5}i}{2}\right)$

57. $f(x) = (x + 3i)(x - 3i)\left(x + \sqrt{3}\right)\left(x - \sqrt{3}\right)$
59. $f(x) = \left(x - 1 - \sqrt{3}\right)\left(x - 1 + \sqrt{3}\right)\left(x - 1 - \sqrt{2}i\right)$
$\left(x - 1 + \sqrt{2}i\right)$
61. $x^3 - x^2 + 25x - 25$ **63.** $x^3 - 10x^2 + 33x - 34$
65. $x^4 + 37x^2 + 36$ **67.** $x^4 + 8x^3 + 9x^2 - 10x + 100$
69. $16x^4 + 36x^3 + 16x^2 + x - 30$
71. Setting $h = 64$ and solving the resulting equation yields
imaginary roots.   **73.** $x^2 + b$

Section 5.3   (*page 343*)

WARM-UP   **1.** $-5 - 10i$ **2.** $7 + 3\sqrt{6}i$
3. $-1 - 4i$ **4.** $-3i$ **5.** $6 - 14i$
6. $6 + 4\sqrt{2}i$ **7.** $-22 + 16i$ **8.** 13
9. $-\frac{3}{2} + \frac{5}{2}i$ **10.** $-\frac{5}{2} - \frac{3}{2}i$

---

1. 5
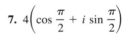

3. $4\sqrt{2}$
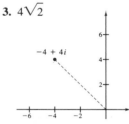

5. $\sqrt{85}$

7. $4\left(\cos\dfrac{\pi}{2} + i\sin\dfrac{\pi}{2}\right)$

9. $3\sqrt{2}\left(\cos\dfrac{5\pi}{4} + i\sin\dfrac{5\pi}{4}\right)$

11. $3\sqrt{2}\left(\cos\dfrac{7\pi}{4} + i\sin\dfrac{7\pi}{4}\right)$ **13.** $2\left(\cos\dfrac{\pi}{6} + i\sin\dfrac{\pi}{6}\right)$
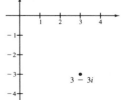
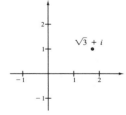

15. $4\left(\cos\dfrac{4\pi}{3} + i\sin\dfrac{4\pi}{3}\right)$

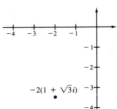

17. $6\left(\cos\dfrac{\pi}{2} + i\sin\dfrac{\pi}{2}\right)$

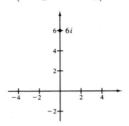

31. $\dfrac{-15\sqrt{2}}{8} + \dfrac{15\sqrt{2}}{8}i$

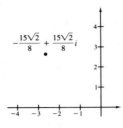

33. $-4i$

19. $\sqrt{65}(\cos 2.62 + i\sin 2.62)$

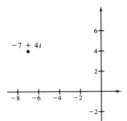

21. $7(\cos 0 + i\sin 0)$

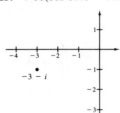

35. $2.8408 + 0.9643i$

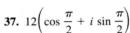

37. $12\left(\cos\dfrac{\pi}{2} + i\sin\dfrac{\pi}{2}\right)$

39. $\dfrac{10}{9}(\cos 200° + i\sin 200°)$

41. $0.27(\cos 150° + i\sin 150°)$

43. $\dfrac{1}{2}(\cos 80° + i\sin 80°)$

45. $\cos\dfrac{2\pi}{3} + i\sin\dfrac{2\pi}{3}$

47. $4[\cos(-58°) + i\sin(-58°)]$

23. $\sqrt{37}(\cos 1.41 + i\sin 1.41)$

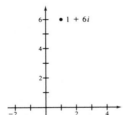

25. $\sqrt{10}(\cos 3.46 + i\sin 3.46)$

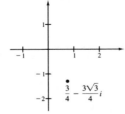

49. (a) $2\sqrt{2}(\cos 45° + i\sin 45°)$,
$\sqrt{2}[\cos(-45°) + i\sin(-45°)]$
(b) $4(\cos 0° + i\sin 0°) = 4$
(c) 4

51. (a) $2[\cos(-90°) + i\sin(-90°)]$,
$\sqrt{2}(\cos 45° + i\sin 45°)$
(b) $2\sqrt{2}[\cos(-45°) + i\sin(-45°)] = 2 - 2i$
(c) $-2i - 2i^2 = -2i + 2 = 2 - 2i$

53. (a) $5(\cos 0° + i\sin 0°), \sqrt{13}(\cos 56.31° + i\sin 56.31°)$
(b) $\dfrac{5}{\sqrt{13}}[\cos(-56.31°) + i\sin(-56.31°)]$
$\approx 0.7692 - 1.154i$
(c) $\dfrac{10}{13} - \dfrac{15}{13}i \approx 0.7692 - 1.154i$

57. (a) r^2
(b) $\cos 2\theta + i\sin 2\theta$

27. $-\sqrt{3} + i$

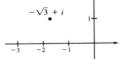

29. $\dfrac{3}{4} - \dfrac{3\sqrt{3}}{4}i$

59.

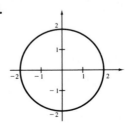

Section 5.4   (page 350)

WARM-UP   **1.** $3\sqrt[3]{2}$ **2.** $2\sqrt{2}$

3. $5\sqrt{2}(\cos 135° + i \sin 135°)$

4. $3(\cos 270° + i \sin 270°)$

5. $12(\cos 180° + i \sin 180°)$ **6.** $12(\cos 0° + i \sin 0°)$

7. $\cos \dfrac{3\pi}{4} + i \sin \dfrac{3\pi}{4}$ **8.** $\cos \dfrac{11\pi}{12} + i \sin \dfrac{11\pi}{12}$

9. $2\left(\cos \dfrac{\pi}{2} + i \sin \dfrac{\pi}{2}\right)$ **10.** $\dfrac{2}{3}(\cos 45° + i \sin 45°)$

---

1. $-4 - 4i$ **3.** $-32i$ **5.** $-128\sqrt{3} - 128i$

7. $\dfrac{125}{2} + \dfrac{125\sqrt{3}}{2}i$ **9.** i **11.** $608.02 + 144.69i$

13. (a) $3(\cos 60° + i \sin 60°)$
$3(\cos 240° + i \sin 240°)$

(b)

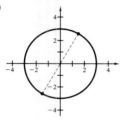

(c) $\dfrac{3}{2} + \dfrac{3\sqrt{3}}{2}i, \ -\dfrac{3}{2} - \dfrac{3\sqrt{3}}{2}i$

15. (a) $2\left(\cos \dfrac{\pi}{3} + i \sin \dfrac{\pi}{3}\right)$
$2\left(\cos \dfrac{5\pi}{6} + i \sin \dfrac{5\pi}{6}\right)$
$2\left(\cos \dfrac{4\pi}{3} + i \sin \dfrac{4\pi}{3}\right)$
$2\left(\cos \dfrac{11\pi}{6} + i \sin \dfrac{11\pi}{6}\right)$

(b)

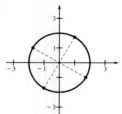

(c) $1 + \sqrt{3}i, \ -\sqrt{3} + i, \ -1 - \sqrt{3}i, \ \sqrt{3} - i$

17. (a) $5\left(\cos \dfrac{3\pi}{4} + i \sin \dfrac{3\pi}{4}\right)$
$5\left(\cos \dfrac{7\pi}{4} + i \sin \dfrac{7\pi}{4}\right)$

(b)

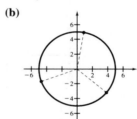

(c) $-\dfrac{5\sqrt{2}}{2} + \dfrac{5\sqrt{2}}{2}i, \ \dfrac{5\sqrt{2}}{2} - \dfrac{5\sqrt{2}}{2}i$

19. (a) $5\left(\cos \dfrac{4\pi}{9} + i \sin \dfrac{4\pi}{9}\right)$
$5\left(\cos \dfrac{10\pi}{9} + i \sin \dfrac{10\pi}{9}\right)$
$5\left(\cos \dfrac{16\pi}{9} + i \sin \dfrac{16\pi}{9}\right)$

(b)

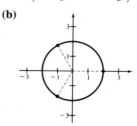

(c) $0.8682 + 4.924i, \ -4.698 - 1.710i, \ 3.830 - 3.214i$

21. (a) $2(\cos 0 + i \sin 0)$
$2\left(\cos \dfrac{2\pi}{3} + i \sin \dfrac{2\pi}{3}\right)$
$2\left(\cos \dfrac{4\pi}{3} + i \sin \dfrac{4\pi}{3}\right)$

(b)

(c) $2, \ -1 + \sqrt{3}i, \ -1 - \sqrt{3}i$

23. (a) $\cos 0 + i \sin 0$

$\cos \dfrac{2\pi}{5} + i \sin \dfrac{2\pi}{5}$

$\cos \dfrac{4\pi}{5} + i \sin \dfrac{4\pi}{5}$

$\cos \dfrac{6\pi}{5} + i \sin \dfrac{6\pi}{5}$

$\cos \dfrac{8\pi}{5} + i \sin \dfrac{8\pi}{5}$

(b)

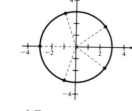

(c) 1, $0.3090 + 0.9511i$, $-0.8090 + 0.5878i$,
$-0.8090 - 0.5878i$, $0.3090 - 0.9511i$

27. $3\left(\cos \dfrac{\pi}{5} + i \sin \dfrac{\pi}{5}\right)$

$3\left(\cos \dfrac{3\pi}{5} + i \sin \dfrac{3\pi}{5}\right)$

$3(\cos \pi + i \sin \pi)$

$3\left(\cos \dfrac{7\pi}{5} + i \sin \dfrac{7\pi}{5}\right)$

$3\left(\cos \dfrac{9\pi}{5} + i \sin \dfrac{9\pi}{5}\right)$

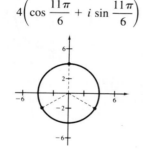

31. $\sqrt[6]{2}(\cos 105° + i \sin 105°)$
$\sqrt[6]{2}(\cos 225° + i \sin 225°)$
$\sqrt[6]{2}(\cos 345° + i \sin 345°)$

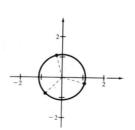

25. $\cos \dfrac{\pi}{8} + i \sin \dfrac{\pi}{8}$

$\cos \dfrac{5\pi}{8} + i \sin \dfrac{5\pi}{8}$

$\cos \dfrac{9\pi}{8} + i \sin \dfrac{9\pi}{8}$

$\cos \dfrac{13\pi}{8} + i \sin \dfrac{13\pi}{8}$

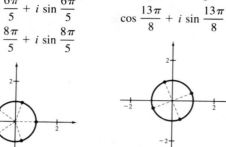

29. $4\left(\cos \dfrac{\pi}{2} + i \sin \dfrac{\pi}{2}\right)$

$4\left(\cos \dfrac{7\pi}{6} + i \sin \dfrac{7\pi}{6}\right)$

$4\left(\cos \dfrac{11\pi}{6} + i \sin \dfrac{11\pi}{6}\right)$

Chapter 5 Review Exercises   (*page 350*)

1. $3 + 7i$ **3.** $-\sqrt{2}i$ **5.** $40 + 65i$
7. $-4 - 46i$ **9.** $1 - 6i$ **11.** $\frac{4}{3}i$
13. Two real solutions   **15.** One real solution
17. No real solutions   **19.** Two real solutions

21. $0, 2$ **23.** $-4 \pm \sqrt{6}$ **25.** $-\dfrac{1}{2} \pm \dfrac{\sqrt{5}}{2}i$

27. $\frac{3}{4}, 1$
$f(x) = (4x - 3)$
$(x - 1)^2$

29. $1 \pm 2i, -5$
$f(x) = (x + 5)$
$(x - 1 - 2i)$
$(x - 1 + 2i)$

31. $7 \pm i, 4$
$h(x) = (x - 4)(x - 7 - i)(x - 7 + i)$

33. $-3 \pm \sqrt{5}i, 3, -2$
$f(x) = (x + 2)(x - 3)\left(x + 3 - \sqrt{5}i\right)\left(x + 3 + \sqrt{5}i\right)$
35. $6x^4 + 13x^3 + 7x^2 - x - 1$
37. $3x^4 - 14x^3 + 17x^2 - 42x + 24$
39. $5\sqrt{2}(\cos 315° + i \sin 315°)$
41. $13(\cos 67.38° + i \sin 67.38°)$ **43.** $-50 - 50\sqrt{3}i$
45. 13

47. (a) $z_1 = -6 = 6(\cos \pi + i \sin \pi)$

$z_2 = 5i = 5\left(\cos \dfrac{\pi}{2} + i \sin \dfrac{\pi}{2}\right)$

(b) $z_1 z_2 = 30\left(\cos \dfrac{3\pi}{2} + i \sin \dfrac{3\pi}{2}\right)$

$\dfrac{z_1}{z_2} = \dfrac{6}{5}\left(\cos \dfrac{\pi}{2} + i \sin \dfrac{\pi}{2}\right)$

49. (a) $z_1 = -3(1 + i) = 3\sqrt{2}(\cos 225° + i \sin 225°)$
$z_2 = 2\left(\sqrt{3} + i\right) = 4(\cos 30° + i \sin 30°)$
(b) $z_1 z_2 = -4.392 - 16.39i$

$\dfrac{z_1}{z_2} = -1.025 - 0.2745i$

51. $\dfrac{625}{2} + \dfrac{625\sqrt{3}}{2}i$ **53.** $2035 - 828i$

55. $3\left(\cos \dfrac{\pi}{4} + i \sin \dfrac{\pi}{4}\right)$

$3\left(\cos \dfrac{7\pi}{12} + i \sin \dfrac{7\pi}{12}\right)$

$3\left(\cos \dfrac{11\pi}{12} + i \sin \dfrac{11\pi}{12}\right)$

$3\left(\cos \dfrac{5\pi}{4} + i \sin \dfrac{5\pi}{4}\right)$

$3\left(\cos \dfrac{19\pi}{12} + i \sin \dfrac{19\pi}{12}\right)$

$3\left(\cos \dfrac{23\pi}{12} + i \sin \dfrac{23\pi}{12}\right)$

57. $\cos \dfrac{\pi}{3} + i \sin \dfrac{\pi}{3} = \dfrac{1}{2} + \dfrac{\sqrt{3}}{2}i$

$\cos \pi + i \sin \pi = -1$

$\cos \dfrac{5\pi}{3} + i \sin \dfrac{5\pi}{3} = \dfrac{1}{2} - \dfrac{\sqrt{3}}{2}i$

59. $3\left(\cos \dfrac{\pi}{4} + i \sin \dfrac{\pi}{4}\right) = \dfrac{3\sqrt{2}}{2} + \dfrac{3\sqrt{2}}{2}i$

$3\left(\cos \dfrac{3\pi}{4} + i \sin \dfrac{3\pi}{4}\right) = -\dfrac{3\sqrt{2}}{2} + \dfrac{3\sqrt{2}}{2}i$

$3\left(\cos \dfrac{5\pi}{4} + i \sin \dfrac{5\pi}{4}\right) = -\dfrac{3\sqrt{2}}{2} - \dfrac{3\sqrt{2}}{2}i$

$3\left(\cos \dfrac{7\pi}{4} + i \sin \dfrac{7\pi}{4}\right) = \dfrac{3\sqrt{2}}{2} - \dfrac{3\sqrt{2}}{2}i$

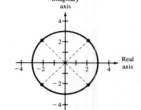

61. $\cos 0 + i \sin 0 = 1$

$\cos \dfrac{\pi}{2} + i \sin \dfrac{\pi}{2} = i$

$\cos \dfrac{2\pi}{3} + i \sin \dfrac{2\pi}{3} = -\dfrac{1}{2} + \dfrac{\sqrt{3}}{2}i$

$\cos \dfrac{4\pi}{3} + i \sin \dfrac{4\pi}{3} = -\dfrac{1}{2} - \dfrac{\sqrt{3}}{2}i$

$\cos \dfrac{3\pi}{2} + i \sin \dfrac{3\pi}{2} = -i$

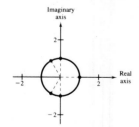

CUMULATIVE TEST for Chapters 3–5 *(page 351)*

1. $2 \csc \beta$ **2.** $5 \tan \theta$

3. $\dfrac{\pi}{3}, \dfrac{\pi}{2}, \dfrac{3\pi}{2}, \dfrac{5\pi}{3}$ **4.** $\dfrac{4\sqrt{5}}{9}$ **5.** $\dfrac{\sqrt{2}}{3}\left(1 - \sqrt{3}\right)$

6. (a) $c \approx 57.57,\ B \approx 22.69°,\ C \approx 112.31°$
   (b) $a \approx 9.91,\ B \approx 23.79°,\ C \approx 126.21°$

7. 42.5 ft   **8.** $\left\langle \dfrac{3\sqrt{3}}{3}, \dfrac{3}{2} \right\rangle$

9. Horizontal component: 115.91 ft/sec
   Vertical component: 31.06 ft/sec

10. $\mathbf{u} \cdot \mathbf{v} = -13;\ \mathbf{u} \times \mathbf{v} = 8\mathbf{i} + 11\mathbf{j} - 20\mathbf{k}$

11. $\arccos \left(-\dfrac{1}{\sqrt{10}}\right) \approx 108.4°$

12. $x^2 + y^2 + z^2 - 4y - 3z = 0$
13. (a) $-3\sqrt{6}$ **(b)** $0.8 + 3i$

   (c) $7 + 22i$ **(d)** $\dfrac{20}{17} + \dfrac{5}{17}i$

14. $\dfrac{3}{2},\ 2 \pm 3i$

15. $z = 2\sqrt{2}(\cos 315° + i \sin 315°)$
   $z^3 = 16\sqrt{2}(\cos 225° + i \sin 225°) = -16 - 16i$
16. $-2\sqrt{3} - 2i$

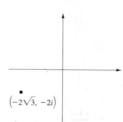

$(-2\sqrt{3},\ -2i)$

17. $3,\ \dfrac{-3}{2} \pm \dfrac{3\sqrt{3}}{2}i$

CHAPTER 6

Section 6.1   *(page 363)*

WARM-UP	**1.** 5^x	**2.** 3^{2x}	**3.** 4^{3x}	**4.** 10^x
5. 4^{2x}	**6.** 4^{10x}	**7.** $\left(\dfrac{3}{2}\right)^x$	**8.** 4^{3x}	
9. 2^{-x}	**10.** 2^x			

1. 946.852   **3.** 747.258   **5.** 0.006   **7.** 673.639
9. 0.472
11. g   **12.** e   **13.** b

Xmin=-3	Xmin=-3	Xmin=-3
Xmax=3	Xmax=3	Xmax=3
Xscl=1	Xscl=1	Xscl=1
Ymin=-1	Ymin=-5	Ymin=-1
Ymax=3	Ymax=1	Ymax=3
Yscl=1	Yscl=1	Yscl=1

14. h

```
Xmin=-3
Xmax=3
Xscl=1
Ymin=-5
Ymax=1
Yscl=1
```

15. d

```
Xmin=-3
Xmax=3
Xscl=1
Ymin=-5
Ymax=1
Yscl=1
```

16. a

```
Xmin=-3
Xmax=3
Xscl=1
Ymin=0
Ymax=4
Yscl=1
```

17. f

```
Xmin=-1
Xmax=5
Xscl=1
Ymin=-5
Ymax=1
Yscl=1
```

18. c

```
Xmin=-2
Xmax=4
Xscl=1
Ymin=-1
Ymax=3
Yscl=1
```

19.

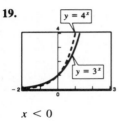

$y = 4^x$

$y = 3^x$

$x < 0$

21.

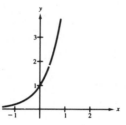

23.

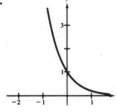

25.

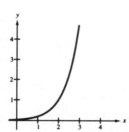

27.

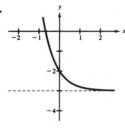

29.

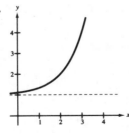

31.

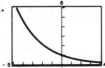

33.

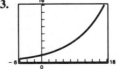

35.

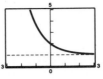

37.

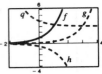

q f g

h

39. (a)

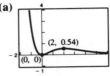

(2, 0.54)

(0, 0)

Increasing: (0, 2)
Decreasing: $(-\infty, 0)$, $(2, \infty)$
Relative minimum: (0, 0)
Relative maximum: (2, 0.54)

(b)

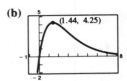

(1.44, 4.25)

Increasing: $(-\infty, 1.44)$
Decreasing: $(1.44, \infty)$
Relative maximum:
(1.44, 4.25)

41.

n	1	2	4
A	\$7764.62	\$8017.84	\$8155.09

n	12	365	Continuous compounding
A	\$8250.97	\$8298.66	\$8300.29

43.

n	1	2	4
A	\$24,115.73	\$25,714.29	\$26,602.23

n	12	365	Continuous compounding
A	\$27,231.38	\$27,547.07	\$27,557.94

45.

t	1	10	20
P	\$91,393.12	\$40,656.97	\$16,529.89

t	30	40	50
P	\$6720.55	\$2732.37	\$1110.90

47.

t	1	10	20
P	\$90,521.24	\$36,940.70	\$13,646.15

t	30	40	50
P	\$5040.98	\$1862.17	\$687.90

49. \$222,822.57

51. (a) \$472.70

   (b) \$298.29; The graph of p is a decreasing function with domain (0, 1725) and range (0, 499.50).

53. (a) 100
(b) 300
(c) 900

55. (a) 25 units
(b) 16.30 units
(c)

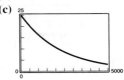

57. (a)

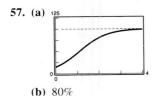

(b) 80%
(c) 1550

59.

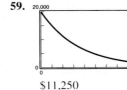

$11,250

34. c

```
Xmin=-1
Xmax=4
Xscl=1
Ymin=-2
Ymax=2
Yscl=1
```

35. f

```
Xmin=-3
Xmax=2
Xscl=1
Ymin=-2
Ymax=2
Yscl=1
```

36. b

```
Xmin=-4
Xmax=1
Xscl=1
Ymin=-2
Ymax=2
Yscl=1
```

37. Domain: $(0, \infty)$
Vertical asymptote: $x = 0$
Intercept: $(1, 0)$

39. Domain: $(-2, \infty)$
Vertical asymptote: $x = -2$
Intercepts: $(-1, 0)$, $(0, -\log_3 2)$

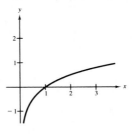

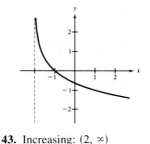

Section 6.2 (*page 372*)

WARM-UP   **1.** 3   **2.** 0   **3.** -1 **4.** 1   **5.** 7.389
6. 0.368   **7.** Graph is shifted two units to the left.
8. Graph is reflected about the x-axis.
9. Graph is shifted down one unit.
10. Graph is reflected about the y-axis.

1. 4   **3.** $\frac{1}{2}$ **5.** 0   **7.** -2 **9.** 3
11. 2   **13.** $\log_5 125 = 3$ **15.** $\log_{81} 3 = \frac{1}{4}$
17. $\log_6 \frac{1}{36} = -2$ **19.** $\ln 20.0855 \approx 3$ **21.** 2.538
23. -0.319 **25.** 2.913

27.

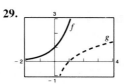

29.

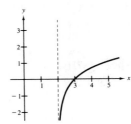

31. d

```
Xmin=-1
Xmax=4
Xscl=1
Ymin=-1
Ymax=3
Yscl=1
```

32. e

```
Xmin=-1
Xmax=4
Xscl=1
Ymin=-2
Ymax=2
Yscl=1
```

33. a

```
Xmin=-3
Xmax=3
Xscl=1
Ymin=-2
Ymax=2
Yscl=1
```

41. Domain: $(2, \infty)$
Vertical asymptote: $x = 2$
Intercept: $(3, 0)$

43. Increasing: $(2, \infty)$
Decreasing: $(0, 2)$
Relative minimum: $(2, 1.693)$

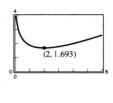

(2, 1.693)

45. (a)

$\sqrt{x}$ is increasing more rapidly than $\ln x$.

(b)

$\sqrt[4]{x}$ is increasing more rapidly than $\ln x$.

47. (a) 80   **(b)** 68.1   **(c)** 62.3

49.

r	0.005	0.010	0.015	0.020	0.025	0.030
t	138.6 yr	69.3 yr	46.2 yr	34.7 yr	27.7 yr	23.1 yr

51.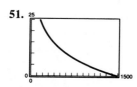

17.66 ft³/min

53. 20 yr

55. $280,178.40   **57.** 21,357 ft·lb

59. (a)

x	1	5	10	10²	10⁴	10⁶
f(x)	0	0.322	0.230	0.046	0.00092	0.0000138

(b) 0

Section 6.3   (*page 381*)

WARM-UP   **1.** 2   **2.** −5   **3.** −2   **4.** −3
5. e^5 **6.** $\dfrac{1}{e}$ **7.** e^6 **8.** 1   **9.** x^{-2}
10. $x^{1/2}$

1. $\dfrac{\log_{10} 5}{\log_{10} 3}$ **3.** $\dfrac{\log_{10} x}{\log_{10} 2}$ **5.** $\dfrac{\ln 5}{\ln 3}$ **7.** $\dfrac{\ln x}{\ln 2}$
9. 1.771   **11.** −2.000   **13.** −0.417   **15.** 2.633
17. $\log_{10} 5 + \log_{10} x$ **19.** $\log_{10} 5 - \log_{10} x$
21. $4 \log_8 x$ **23.** $\frac{1}{2} \ln z$ **25.** $\ln x + \ln y + \ln z$
27. $\frac{1}{2} \ln(a - 1)$
29. $\ln z + 2 \ln(z - 1)$ **31.** $\frac{1}{3} \ln x - \frac{1}{3} \ln y$
33. $4 \ln x + \frac{1}{2} \ln y - 5 \ln z$
35. $2 \log_b x - 2 \log_b y - 3 \log_b z$ **37.** $\ln 2x$
39. $\log_4 \dfrac{z}{y}$ **41.** $\log_2(x + 4)^2$ **43.** $\log_3 \sqrt[3]{5x}$
45. $\ln \dfrac{x}{(x + 1)^3}$ **47.** $\log_3 \dfrac{x - 2}{x + 2}$
49. $\ln \dfrac{x}{(x^2 - 4)^2}$ **51.** $\ln \sqrt[3]{\dfrac{x(x + 3)^2}{x^2 - 1}}$
53. $\ln \dfrac{\sqrt[3]{y(y + 4)^2}}{y - 1}$ **55.** $\ln \dfrac{9}{\sqrt{x^2 + 1}}$ **57.** 0.9208
59. 1.8957   **61.** −0.1781   **63.** 0.9136   **65.** 2.0365
67. 2   **69.** 2.4   **71.** 4.5   **73.** $\frac{3}{2}$
75. $\frac{1}{2} + \frac{1}{2} \log_7 10$ **77.** $-3 - \log_5 2$ **79.** $6 + \ln 5$
81. $\beta = 10(\log_{10} I + 16)$, 60 db
83.

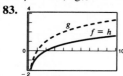

f and h have the same graphs.

Section 6.4   (*page 390*)

WARM-UP   **1.** $\dfrac{\ln 3}{\ln 2}$ **2.** $1 + \dfrac{2}{\ln 4}$ **3.** $\dfrac{e}{2}$ **4.** $2e$
5. $2 \pm i$ **6.** $\frac{1}{2}$, 1   **7.** $2x$ **8.** $3x$ **9.** $2x$
10. $-x^2$

1. 2   **3.** −2   **5.** 3   **7.** 64   **9.** $\frac{1}{10}$ **11.** x^2
13. $5x + 2$ **15.** x^2 **17.** $\ln 10 \approx 2.303$ **19.** 0
21. $\dfrac{\ln 12}{3} \approx 0.828$ **23.** $\ln \dfrac{5}{3} \approx 0.511$ **25.** $\ln 5 \approx 1.609$
27. $2 \ln 75 \approx 8.6350$ **29.** $\log_{10} 42 \approx 1.623$
31. $\dfrac{\ln 80}{2 \ln 3} \approx 1.994$ **33.** 2
35. $\dfrac{\ln 2}{12 \ln(1 + 0.10/12)} \approx 6.960$
37. 3.847   **39.** 12.207
41. 0.059   **43.** 21.330   **45.** $e^{-3} \approx 0.050$
47. $\dfrac{e^{2.4}}{2} \approx 5.512$ **49.** $e^2 - 2 \approx 5.389$
51. $1 + \sqrt{1 + e} \approx 2.928$ **53.** 103
55. $\dfrac{-1 + \sqrt{17}}{2} \approx 1.562$ **57.** 4   **59.** No solution
61. 33.115   **63.** 14.369   **65.** 8.2 yr   **67.** 12.9 yr
69. (a) 1426 units   **(b)** 1498 units
71. (a)

(b) $V = 6.7$
The asymptote represents the limiting yield per acre.
(c) 29.33 yr
73. (a)

(b) $y = 0$, $y = 100$
(c) Males: 69.71 in.; Females: 64.51 in.

Section 6.5 (*page 400*)

WARM-UP

1.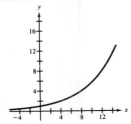

2.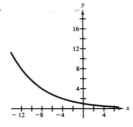

3.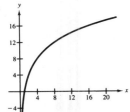

4.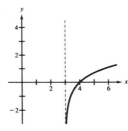

5.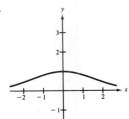

6.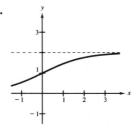

7. $\frac{1}{2} \ln \frac{7}{3} \approx 0.424$ **8.** $\frac{\ln(0.001)}{-0.2} \approx 34.539$

9. $\frac{1}{5}e^{7/2} \approx 6.623$ **10.** $\frac{1}{2}e^2 \approx 3.695$

Initial Investment	Annual % Rate	Effective Yield	Time to Double	Amount After 10 Years
1. $1000	12%	12.75%	5.78 yr	$3320.12
3. $750	8.94%	9.35%	$7\frac{3}{4}$ yr	$1833.67
5. $500	9.5%	9.97%	7.30 yr	$1292.85
7. $6392.79	11%	11.63%	6.30 yr	$19,205.00
9. $5000	8%	8.33%	8.66 yr	$11,127.70

11. $112,087.09

13. (a) 6.642 yr   (b) 6.330 yr   (c) 6.302 yr
(d) 6.301 yr

15.

r	2%	4%	6%	8%	10%	12%
t	54.93	27.47	18.31	13.73	10.99	9.16

17.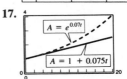

Continuous compounding at 7% grows faster.

Isotope	Half-life (Years)	Initial Quantity	Amount After 1000 Years	Amount After 10,000 Years
19. ^{226}Ra	1620	10 g	6.52 g	0.14 g
21. ^{14}C	5730	6.70 g	5.95 g	2 g
23. ^{230}Pu	24,360	2.16 g	2.1 g	1.63 g

25. $\frac{1}{4}\ln 10 \approx 0.5756$ **27.** $\frac{1}{4}\ln\frac{1}{4} \approx -0.3466$

29. 105,300, 2013   **31.** $k = 0.0137$, 3288

33. $y = 4.22e^{0.0430t}$, 9.97 million   **35.** 3.15 hr

37. 95.8%   **39.** $9281

41. (a) $S(t) = 100(1 - e^{-0.1625t})$ (b) 55,625

43.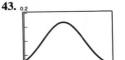

64.5 in

45. (a)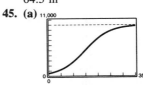

$p = 10,000$. This represents the carrying capacity of the lake for this species of fish.

(b) 1252 fish   (c) 7.8 months

47. (a) $S = 10(1 - e^{-0.0575x})$ (b) 3314

49. (a) 7.91   (b) 7.68

51. (a) 20   (b) 70   (c) 95   (d) 120

53. 95%   **55.** 4.64   **57.** 1.58×10^{-6} moles per liter

59. 10^7

61. (a) 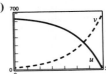   (b) Interest payments, 27.7 yr

63. 7:30 A.M.

Section 6.6   (*page 411*)

WARM-UP

1.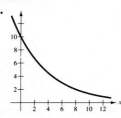

2.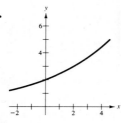

3.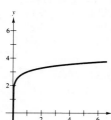

4.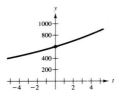

5.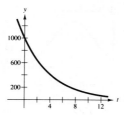

6.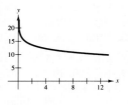

7. $y = -0.7x + 2.6$

8. $y = \frac{2}{3}x + 4.75$

9. $y = 3.9x + 9.9$

10. $y = -x + 8.4$

1. $y = 2.152 + 2.704 \ln x$ **3.** $y = 3.114e^{0.2935x}$

5. $y = -0.788x + 8.257$

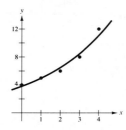

7. $y = 3.807e^{0.2667x}$ **9.** $y = 8.463e^{-0.2517x}$

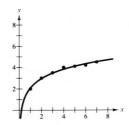 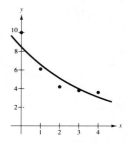

11. $y = 2.083 + 1.257 \ln x$ **13.** $y = 9.826 - 4.097 \ln x$

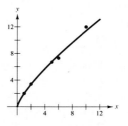 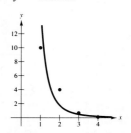

15. $y = 1.985x^{0.760}$ **17.** $y = 16.103x^{-3.174}$

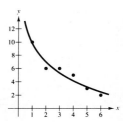

19. $y = 4.754(6.774)^x$ **21.** $y = 19.752 + 3.541 \ln t$

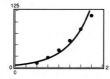 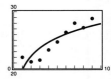

23. $y = 5.088x^{0.645}$ **25.** $y = 28.5 \cos(0.491t) + 31.5$

27. $y = 1.56 \cos \dfrac{2\pi}{365}(t - 172) + 18.44$

Chapter 6 Review Exercises   (*page 413*)

1. d

```
Xmin=-4
Xmax=3
Xscl=1
Ymin=-1
Ymax=4
Yscl=1
```

2. f

```
Xmin=-3
Xmax=4
Xscl=1
Ymin=-1
Ymax=4
Yscl=1
```

3. a

```
Xmin=-4
Xmax=4
Xscl=1
Ymin=-5
Ymax=1
Yscl=1
```

4. b

```
Xmin=-4
Xmax=4
Xscl=1
Ymin=0
Ymax=6
Yscl=1
```

5. c

```
Xmin=-3
Xmax=6
Xscl=1
Ymin=-3
Ymax=3
Yscl=1
```

6. e

```
Xmin=-1
Xmax=7
Xscl=1
Ymin=-4
Ymax=3
Yscl=1
```

7.

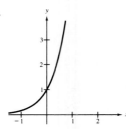

9.

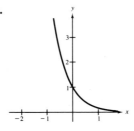

11.

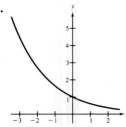

13.

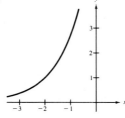

15.

n	1	2	4
A	\$9499.28	\$9738.91	\$9867.22

n	12	365	Continuous compounding
A	\$9956.20	\$10,000.27	\$10,001.78

17.

t	1	10	20
P	\$184,623.27	\$89,865.79	\$40,379.30

t	30	40	50
P	\$18,143.59	\$8152.44	\$3663.13

19. \$1,069,047.14   **21.** 229.2 units per milliliter

23.

Speed	50	55	60	65	70
Miles per gallon	28	26.4	24.8	23.4	22.0

25.

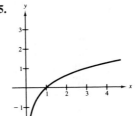

27.

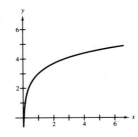

29.

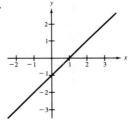

31. $\log_4 64 = 3$

33. 3   **35.** -2 **37.** 7   **39.** 0   **41.** 1.585
43. 2.132   **45.** $1 + 2 \log_5 x$
47. $\log_{10} 5 + \frac{1}{2} \log_{10} y - 2 \log_{10} x$
49. $\ln(x^2 + 1) + \ln(x - 1)$
51. $\log_2 5x$ **53.** $\ln \dfrac{\sqrt{|2x - 1|}}{(x + 1)^2}$ **55.** $\ln \dfrac{3\sqrt[3]{4 - x^2}}{x}$
57. False   **59.** False   **61.** 1.6542   **63.** 0.2823
65. 27.16 mi   **67.** $\ln 12 \approx 2.485$
69. $-\dfrac{\ln 44}{5} \approx -0.757$ **71.** $\ln 2 \approx 0.693, \ln 5 \approx 1.609$
73. $\frac{1}{3} e^{8.2} \approx 1213.650$ **75.** $3e^2 \approx 22.167$
77. $y = 2e^{0.1014t}$ **79.** $y = 4e^{-0.4159t}$
81. **(a)** 1151 units   **(b)** 1325 units
83. **(a)** 8.94%   **(b)** \$1834.37   **(c)** 9.36%
85. $10^{-3.5}$

87. $y = 234.684e^{-0.134x}$

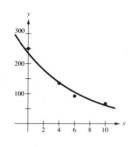

CHAPTER 7

Section 7.1   (page 423)

WARM-UP   **1.** $\sqrt{89}$ **2.** $2\sqrt{13}$ **3.** 1   **4.** $\frac{6}{7}$
5. $-\frac{3}{2}$ **6.** $-\frac{1}{2}$ **7.** $2x - 3y + 9 = 0$
8. $7x - 2y - 24 = 0$ **9.** $4x + y - 32 = 0$
10. $3x + 4y + 2 = 0$

---

1. $\dfrac{\sqrt{3}}{3}$ **3.** -1 **5.** 0.7869   **7.** -2.7475
9. $135°$ **11.** $36.9°$ **13.** $60.3°$ **15.** $121.0°$
17. $78.7°$ **19.** $121.0°$ **21.** $71.6°$ **23.** $11.3°$
25. $81.9°$ **27.** $53.1°$ **29.** $46.9°$ **31.** 2
33. 1.4   **35.** 7   **37.** $\dfrac{8\sqrt{37}}{37} \approx 1.3152$
39. $2\sqrt{2} \approx 2.8284$ **41.** 0.1139, 1195 ft
43. (a) $69°$ (b) 5970 ft, 2415 ft   **45.** $33.7°$
47. (a) $\dfrac{7\sqrt{10}}{5}$ (b) 7   **49.** (a) $\dfrac{35\sqrt{37}}{74}$ (b) $\dfrac{35}{8}$
51. $(2, 1)$, $42.3°$; $(4, 4)$, $78.7°$; $(6, 2)$, $59.0°$
53. $(-4, -1)$, $11.9°$; $(3, 2)$, $21.8°$; $(1, 0)$, $146.3°$

Section 7.2   (page 431)

WARM-UP   **1.** $x^2 - 10x + 5$ **2.** $x^2 + 6x + 8$
3. $-x^2 - 8x - 6$ **4.** $-x^2 + 4x$ **5.** $(x + 3)^2 - 1$
6. $(x - 5)^2 - 4$ **7.** $2 - (x - 1)^2$ **8.** $-2(x - 1)^2$
9. $2x + 3y - 20 = 0$ **10.** $3x - 4y - 17 = 0$

---

1. e   **2.** f   **3.** a   **4.** c   **5.** d   **6.** b
7. Vertex: $(0, 0)$; Focus: $\left(0, \frac{1}{16}\right)$; Directrix: $y = -\frac{1}{16}$

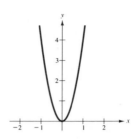

9. Vertex: $(0, 0)$; Focus: $\left(-\frac{3}{2}, 0\right)$; Directrix: $x = \frac{3}{2}$

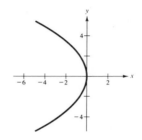

11. Vertex: $(0, 0)$; Focus: $(0, -2)$; Directrix: $y = 2$

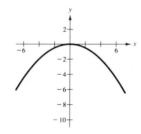

13. Vertex: $(1, -2)$; Focus: $(1, -4)$; Directrix: $y = 0$

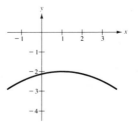

15. Vertex: $\left(5, -\frac{1}{2}\right)$; Focus: $\left(\frac{11}{2}, -\frac{1}{2}\right)$; Directrix: $x = \frac{9}{2}$

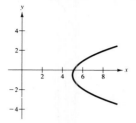

17. Vertex: $(1, 1)$; Focus: $(1, 2)$; Directrix: $y = 0$

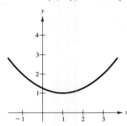

19. Vertex: $(-1, 2)$; Focus: $(0, 2)$; Directrix: $x = -2$

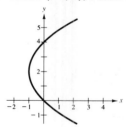

21. Vertex: $(2, -2)$; Focus: $(0, -2)$; Directrix: $x = 4$

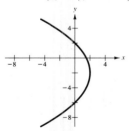

23. Vertex: $(-2, 1)$; Focus: $\left(-2, -\frac{1}{2}\right)$; Directrix: $y = \frac{5}{2}$

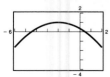

25. Vertex: $\left(\frac{1}{4}, -\frac{1}{2}\right)$; Focus: $\left(0, -\frac{1}{2}\right)$; Directrix: $x = \frac{1}{2}$

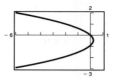

27. $x^2 = -6y$ **29.** $y^2 = -12x$
31. $x^2 - 8y + 32 = 0$ **33.** $5x^2 - 14x - 3y + 9 = 0$
35. $x^2 + y - 4 = 0$ **37.** $x^2 = 12y$
39. (a) $y = \frac{1}{800}x^2$ **(b)** $\frac{25}{2}$ ft
41. (a) 24,749 mph   **(b)** $x^2 = -16,400(y - 4100)$
43. (a) $y = -\frac{1}{64}x^2 + 75$ **(b)** 69.3 ft   **45.** 41,329 ft
47. $y = 4x - 8$; $(2, 0)$ **49.** $y = 4x + 2$; $\left(-\frac{1}{2}, 0\right)$

Section 7.3   (*page 440*)

WARM-UP
1.

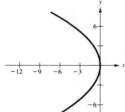

2.

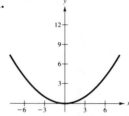

3.

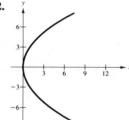

4.

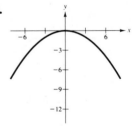

5. $c = 12$ **6.** $b = 1$ **7.** $a = 10$ **8.** $c = 2\sqrt{6}$
9. $4x^2 + 3y^2$ **10.** $\dfrac{9(x - 1)^2}{4} + 9(y + 2)^2$

1. e   **3.** c   **5.** f
7. Center: $(0, 0)$; Foci: $(\pm 3, 0)$;
    Vertices: $(\pm 5, 0)$: $e = \frac{3}{5}$

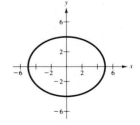

9. Center: $(0, 0)$; Foci: $(0, \pm3)$; Vertices: $(0, \pm5)$: $e = \frac{3}{5}$

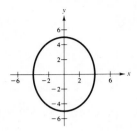

11. Center: $(0, 0)$; Foci: $(\pm2, 0)$; Vertices: $(\pm3, 0)$: $e = \frac{2}{3}$

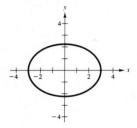

13. Center: $(0, 0)$; Foci: $\left(\pm\sqrt{3}, 0\right)$; Vertices: $(\pm2, 0)$:
$e = \dfrac{\sqrt{3}}{2}$

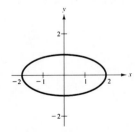

15. Center: $(0, 0)$; Foci: $(0, \pm1)$; Vertices: $\left(0, \pm\sqrt{3}\right)$;
$e = \dfrac{1}{\sqrt{3}} = \dfrac{\sqrt{3}}{3}$

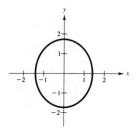

17. Center: $(1, 5)$; Foci: $(1, 9)$, $(1, 1)$; Vertices: $(1, 10)$; $(1, 0)$; $e = \frac{4}{5}$

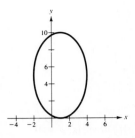

19. Center: $(-2, 3)$; Foci: $\left(-2, 3 \pm \sqrt{5}\right)$;
Vertices: $(-2, 6)$, $(-2, 0)$; $e = \dfrac{\sqrt{5}}{3}$

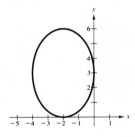

21. Center: $(1, -1)$; Foci: $\left(\frac{7}{4}, -1\right)$, $\left(\frac{1}{4}, -1\right)$;
Vertices: $\left(\frac{9}{4}, -1\right)$, $\left(-\frac{1}{4}, -1\right)$; $e = \frac{3}{5}$

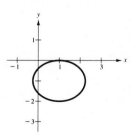

23. Center: $(0, 0)$; Foci: $\left(0, \dfrac{\pm\sqrt{3}}{2}\right)$; Vertices: $(0, \pm1)$

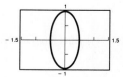

25. Center: $\left(\frac{1}{2}, -1\right)$; Foci: $\left(\frac{1}{2} \pm \sqrt{2}, -1\right)$;
Vertices: $\left(\frac{1}{2} \pm \sqrt{5}, -1\right)$

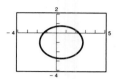

27. Center: $\left(\frac{3}{2}, -1\right)$; Foci: $\left(\frac{3}{2} \pm \sqrt{2}, -1\right)$;
Vertices: $\left(-\frac{1}{2}, -1\right)$, $\left(\frac{7}{2}, -1\right)$

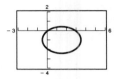

29. $\dfrac{x^2}{36} + \dfrac{y^2}{11} = 1$ **31.** $\dfrac{y^2}{4} + x^2 = 1$

33. $\dfrac{(y-4)^2}{64} + \dfrac{x^2}{48} = 1$ **35.** $\dfrac{(x-3)^2}{36} + \dfrac{(y-2)^2}{32} = 1$

37. Place tacks 1.5 ft from center. Length of string:
$2a = 5$ ft

39.

41.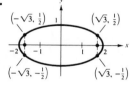

43.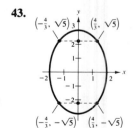

45. Least distance: $a - c \approx 91.377$ million miles; Greatest
distance: $a + c \approx 94.537$ million miles
47. $e = 0.0543$ **49.** $e = 0.052$

Section 7.4   (page 450)

WARM-UP   **1.** $\sqrt{61}$ **2.** $\sqrt{65}$

3. 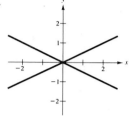   **4.**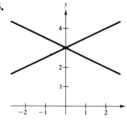

5. 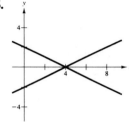   **6.**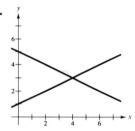

7. Parabola   **8.** Ellipse   **9.** Circle   **10.** Parabola

1. e   **3.** f   **5.** d
7. Center: $(0, 0)$; Vertices: $(\pm 1, 0)$; Foci: $\left(\pm\sqrt{2}, 0\right)$;
Asymptotes: $y = \pm x$

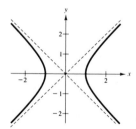

9. Center: $(0, 0)$; Vertices: $(0, \pm 1)$; Foci: $\left(0, \pm\sqrt{5}\right)$;
Asymptotes: $y = \pm\frac{1}{2}x$

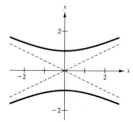

11. Center: $(0, 0)$; Vertices: $(0, \pm 5)$; Foci: $(0, \pm 13)$; Asymptotes: $y = \pm \frac{5}{12}x$

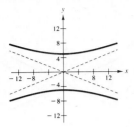

13. Center: $(0, 0)$; Vertices: $(0, \pm 2)$; Foci: $(0, \pm 3)$; Asymptotes: $y = \pm \dfrac{2}{\sqrt{5}}x$

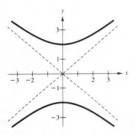

15. Center: $(1, -2)$; Vertices: $(-1, -2)$, $(3, -2)$; Foci: $\left(1 \pm \sqrt{5}, -2\right)$; Asymptotes: $y = -2 \pm \frac{1}{2}(x - 1)$

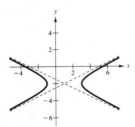

17. Center: $(2, -6)$; Vertices: $(2, -5)$, $(2, -7)$; Foci: $\left(2, -6 \pm \sqrt{2}\right)$; Asymptotes: $y = -6 \pm (x - 2)$

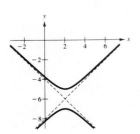

19. Center: $(2, -3)$; Vertices: $(1, -3)$, $(3, -3)$; Foci: $\left(2 \pm \sqrt{10}, -3\right)$; Asymptotes: $y = -3 \pm 3(x - 2)$

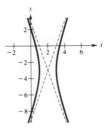

21. Degenerate hyperbola is two intersecting lines

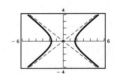

23. Center: $(0, 0)$; Vertices: $\left(\pm\sqrt{3}, 0\right)$; Foci: $\left(\pm\sqrt{5}, 0\right)$; Asymptotes: $y = \pm\sqrt{\frac{2}{3}}x$

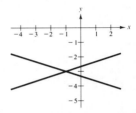

25. Center: $(1, -3)$; Vertices: $\left(1, -3 \pm \sqrt{2}\right)$; Foci: $\left(1, -3 \pm 2\sqrt{5}\right)$; Asymptotes: $y = -3 \pm \frac{1}{3}(x - 1)$

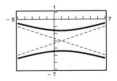

27. Center: $(1, -3)$; Vertices: $(1 \pm 2, -3)$; Foci: $\left(1 \pm \sqrt{10}, -3\right)$; Asymptotes: $y = -3 \pm \dfrac{\sqrt{6}}{2}(x - 1)$

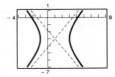

29. $\dfrac{y^2}{4} - \dfrac{x^2}{12} = 1$ **31.** $x^2 - \dfrac{y^2}{9} = 1$

33. $\dfrac{(x-4)^2}{4} - \dfrac{y^2}{12} = 1$ **35.** $\dfrac{y^2}{9} - \dfrac{(x-2)^2}{9/4} = 1$

37. $\dfrac{(x-3)^2}{9} - \dfrac{(y-2)^2}{4} = 1$ **39.** $(4400, -4290)$

41. $\left(\sqrt{216 - 72\sqrt{5}},\ 0\right) \approx (7.42, 0)$ **43.** Circle

45. Hyperbola   **47.** Ellipse   **49.** Parabola

Section 7.5 *(page 460)*

WARM-UP   **1.** b   **2.** e   **3.** d   **4.** a   **5.** f

6. c   **7.** $\dfrac{1}{2}x - \dfrac{\sqrt{3}}{2}y$ **8.** $-\dfrac{1}{2}x + \dfrac{\sqrt{3}}{2}y$

9. $\dfrac{4x^2 - 12xy + 9y^2}{13}$ **10.** $\dfrac{x^2 - 2\sqrt{2}xy + 2y^2}{3}$

1. $\dfrac{(y')^2}{2} - \dfrac{(x')^2}{2} = 1$ **3.** $\dfrac{(x')^2}{1/4} - \dfrac{(y')^2}{1/6} = 1$

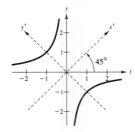

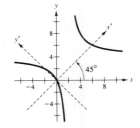

5. $\dfrac{(x' - 3\sqrt{2})^2}{16} - \dfrac{(y' - \sqrt{2})^2}{16} = 1$

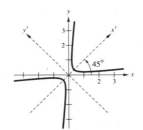

7. $\dfrac{(x')^2}{3} + \dfrac{(y')^2}{2} = 1$ **9.** $4(y')^2 + 4x' = 0,\ x' = -(y')^2$

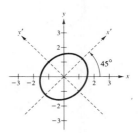

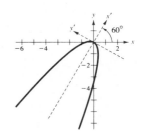

11. $(x' - 1)^2 = 4\left(\dfrac{3}{2}\right)\left(y' + \dfrac{1}{6}\right)$

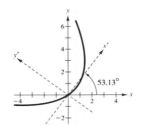

13.

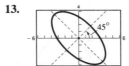

15.

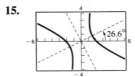

17.

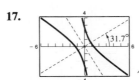

19. Parabola   **21.** Ellipse or circle
23. Hyperbola   **25.** Parabola
27.

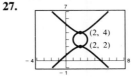

29.
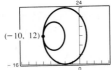
$(-10, 12)$

31.
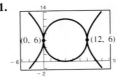
$(0, 6)$ $(12, 6)$

5.
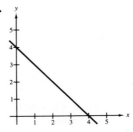

6.
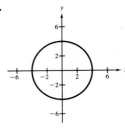

33.
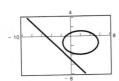
$(0, 4)$

35.
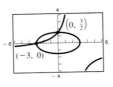
$(4, 3)$
$(4, -3)$

7. 10   **8.** $5 \tan^2 \theta$
9. $\sec^2 x + \tan^2 x$ **10.** $\frac{1}{2} \sin \theta$

37. No points of intersection

39.
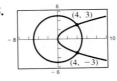
$\left(0, \frac{3}{2}\right)$
$(-3, 0)$

1.
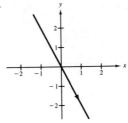
$y = -2x$

3.
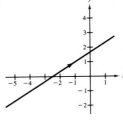
$2x - 3y + 5 = 0$

Section 7.6   (*page 468*)

WARM-UP

1.
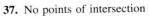

2.
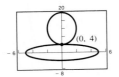

5.
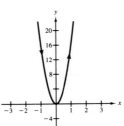
$y = 16x^2$

7.
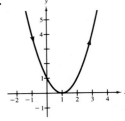
$y = (x - 1)^2$

3.
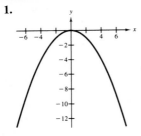

4.
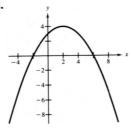

9.

$y = \frac{1}{2}\sqrt[3]{x}$

11.
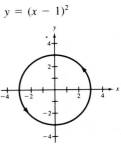
$x^2 + y^2 = 9$

13.

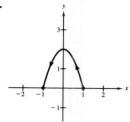

$$y = 2 - 2x^2, \ -1 \le x \le 1$$

15.

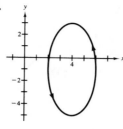

$$\frac{(x-4)^2}{4} + \frac{(y+1)^2}{16} = 1$$

17.

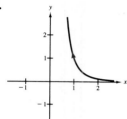

$$\sqrt[3]{y} = \frac{1}{x}, \ y = \frac{1}{x^3}, \ x > 0, \ y > 0$$

19.

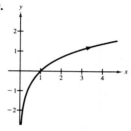

$$y = \ln x$$

21. Each curve represents a portion of the line $y = 2x + 1$.

Domain	Orientation
(a) $-\infty < x < \infty$	Up
(b) $-1 \le x \le 1$	Oscillates
(c) $0 < x < \infty$	Down
(d) $0 < x < \infty$	Up

23. $y - y_1 = \dfrac{y_2 - y_1}{x_2 - x_1}(x - x_1)$

25. $\dfrac{(x-h)^2}{a^2} + \dfrac{(y-k)^2}{b^2} = 1$

27. $x = 5t$
$y = -2t$
Solution not unique

29. $x = 2 + 4 \cos \theta$
$y = 1 + 4 \sin \theta$
Solution not unique

31. $x = 5 \cos \theta$
$y = 3 \sin \theta$
Solution not unique

33. $x = 4 \sec \theta$
$y = 3 \tan \theta$
Solution not unique

35. *Examples*

$x = t, \ y = t^3$
$x = \sqrt[3]{t}, \ y = t$
$x = \tan t, \ y = \tan^3 t$

37.

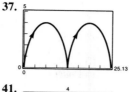

39.

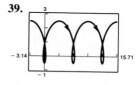

41.

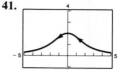

43. b
45. d
47. $x = a\theta - b \sin \theta$ and
$y = a - b \cos \theta$.

Section 7.7 (page 476)

WARM-UP

1. $\dfrac{3\pi}{4}$ **2.** $\dfrac{7\pi}{6}$ **3.** $\sin \theta = \dfrac{\sqrt{5}}{5}, \cos \theta = \dfrac{2\sqrt{5}}{5}$

4. $\sin \theta = -\dfrac{3}{5}, \cos \theta = \dfrac{4}{5}$ **5.** $\dfrac{3\pi}{4}$ **6.** 0.5880

7. $-\dfrac{\sqrt{3}}{2}$ **8.** $-\dfrac{\sqrt{2}}{2}$ **9.** -0.3090 **10.** 0.9735

1.

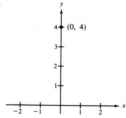

$(0, 4)$

3.

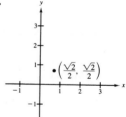

$\left(\dfrac{\sqrt{2}}{2}, \dfrac{\sqrt{2}}{2}\right)$

5.

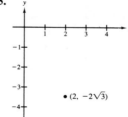

$\left(2, -2\sqrt{3}\right)$

7.

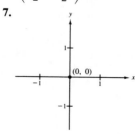

$(0, 0)$

9.

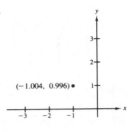

$(-1.004, 0.996)$

11.

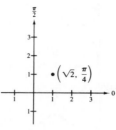

$\left(\sqrt{2}, \dfrac{\pi}{4}\right), \left(-\sqrt{2}, \dfrac{5\pi}{4}\right)$

13.

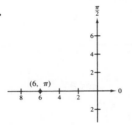

$(6, \pi), (-6, 0)$

15.

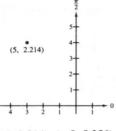

$(5, 2.214), (-5, 5.356)$

17.

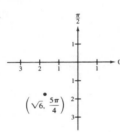

$\left(\sqrt{6}, \dfrac{5\pi}{4}\right), \left(-\sqrt{6}, \dfrac{\pi}{4}\right)$

19.

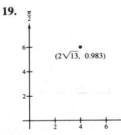

$\left(2\sqrt{13}, 0.983\right), \left(-2\sqrt{13}, 4.124\right)$

21. $r = 3$ **23.** $r = 2a \cos\theta$ **25.** $r = 4 \csc\theta$

27. $r = 10 \sec\theta$ **29.** $r = \dfrac{-2}{3\cos\theta - \sin\theta}$

31. $r^2 = 4 \sec\theta \csc\theta = 8 \csc 2\theta$

33. $r^2 = 9 \cos 2\theta$ **35.** $x^2 + y^2 - 4y = 0$

37. $\sqrt{3}x - 3y = 0$ **39.** $y = 2$

41. $(x^2 + y^2)^2 = 6x^2y - 2y^3$

43. $4x^2 - 5y^2 - 36y - 36 = 0$

45. $x^2 + y^2 = 9$

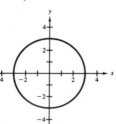

49. $x - 3 = 0$

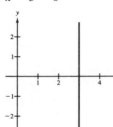

47. $x - y = 0$

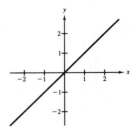

51. Proof

53. $(x - h)^2 + (y - k)^2 = h^2 + k^2$

Center: (h, k)

Radius: $\sqrt{h^2 + k^2}$

Section 7.8   (*page 485*)

WARM-UP   **1.** Amplitude: 5; Period: $\pi/2$

2. Amplitude: 3; Period: 1

3. Amplitude: 5; Period: $\frac{4}{5}$ **4.** Amplitude: $\frac{1}{2}$; Period: 4π

5.

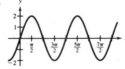

6.

7.

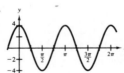

8.

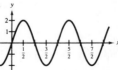

9. $\dfrac{1}{2}\left(\sqrt{3}\sin x - \cos x\right)$ **10.** $\dfrac{\sqrt{2}}{2}(\cos x + \sin x)$

1. Polar axis   **3.** $\theta = \dfrac{\pi}{2}$ **5.** $\theta = \dfrac{\pi}{2}$, polar axis, pole

7. Maximum: $|r| = 5$ when $\theta = 0, \dfrac{\pi}{3}, \dfrac{2\pi}{3}$

9. Maximum: $|r| = 20$ when $\theta = \dfrac{3\pi}{2}$

Zero: $r = 0$ when $\theta = \dfrac{\pi}{2}$

11.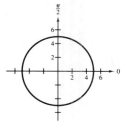

13.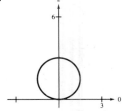

15.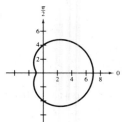

17.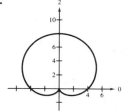

19.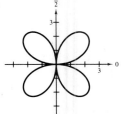

21.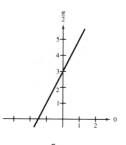

23.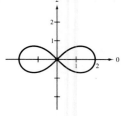

25.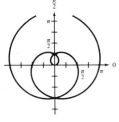

27.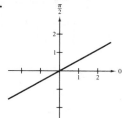

29.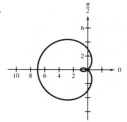

31.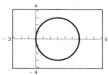

33.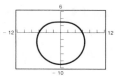

35.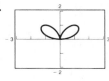

37.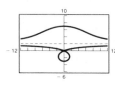

39.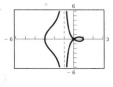

41.

43.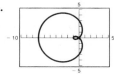

45.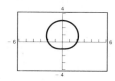

47.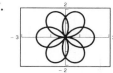

49.

51. $y = \pm\left|\dfrac{x}{x+1}\right|\sqrt{3 - 2x - x^2}$ **53.** Proof

55. (a) $r = 2 - \dfrac{\sqrt{2}}{2}(\sin\theta - \cos\theta)$
(b) $r = 2 + \cos\theta$
(c) $r = 2 + \sin\theta$
(d) $r = 2 - \cos\theta$

57. (a) 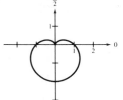   (b)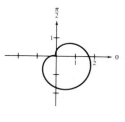

Section 7.9 (page 493)

WARM-UP

1. $\left(\dfrac{3\sqrt{2}}{2}, -\dfrac{3\sqrt{2}}{2}\right)$ **2.** $(-2, -2\sqrt{3})$

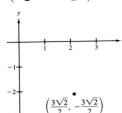

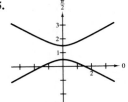

3. $\left(3, \dfrac{3\pi}{2}\right), \left(-3, \dfrac{\pi}{2}\right)$ **4.** $(13, 1.9656), (-13, 5.1072)$

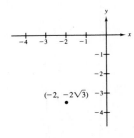

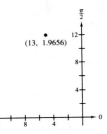

5. $r = 5$ **6.** $r^3 = 4\sec^2\theta\csc\theta$
7. $y = -4$ **8.** $x^2 + y^2 - 4x = 0$

9. 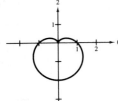 **10.**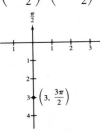

11. 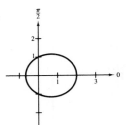 **13.**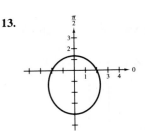

15. 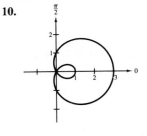 **17.**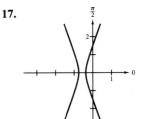

19. $r = \dfrac{1}{1 - \cos\theta}$ **21.** $r = \dfrac{1}{2 + \sin\theta}$

23. $r = \dfrac{2}{1 + 2\cos\theta}$ **25.** $r = \dfrac{2}{1 - \sin\theta}$

27. $r = \dfrac{10}{1 - \cos\theta}$ **29.** $r = \dfrac{16}{5 + 3\cos\theta}$

31. $r = \dfrac{20}{3 - 2\cos\theta}$ **33.** $r = \dfrac{9}{4 - 5\sin\theta}$

35. Proof **37.** $r^2 = \dfrac{24{,}336}{169 - 25\cos^2\theta}$

39. $r^2 = \dfrac{144}{25\cos^2\theta - 9}$ **41.** $r^2 = \dfrac{144}{25\cos^2\theta - 16}$

43. 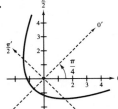 **45.** Proof

47. $r = \dfrac{9.2931 \times 10^7}{1 - 0.0167\cos\theta}$, 9.1405×10^7, 9.4509×10^7

49. $r = \dfrac{8200}{1 + \sin\theta}$

1. c **3.** a **5.** b
7. 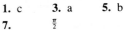 **9.**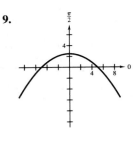

Chapter 7 Review Exercises (*page 496*)

1. $-\sqrt{3}$ **3.** $135°$ **5.** $2\sqrt{2}$

7. Parabola

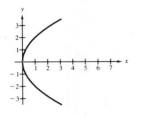

9. Parabola

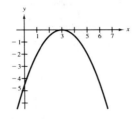

11. Degenerate circle

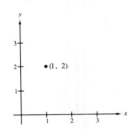

13. Ellipse

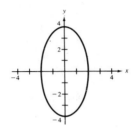

15. Ellipse

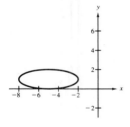

17. Hyperbola

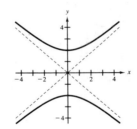

19. Hyperbola
$$y = 5x + \sqrt{24x^2 - 1}$$
$$y = 5x - \sqrt{24x^2 - 1}$$

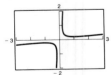

21. $(x - 4)^2 = -8(y - 2)$ **23.** $(y - 2)^2 = -4x$

25. $\dfrac{(x - 2)^2}{25} + \dfrac{y^2}{21} = 1$ **27.** $\dfrac{2x^2}{9} + \dfrac{y^2}{36} = 1$

29. $\dfrac{y^2}{1} - \dfrac{x^2}{8} = 1$ **31.** $\dfrac{5(x - 4)^2}{16} - \dfrac{5y^2}{64} = 1$

33. $(0, 50)$ **35.** $-2x + 3y = 25$

37. $\dfrac{2}{3}x - \sqrt{3}y = 1$

39. $2x - y = 0$

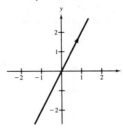

41. $3x + 4y - 11 = 0$

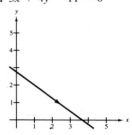

43. $y = \dfrac{1}{x^2}$

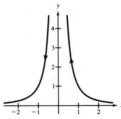

45. $x^2 + y^2 = 36$

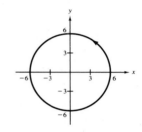

47. $x^{2/3} + \left(\dfrac{y}{4}\right)^{2/3} = 1$

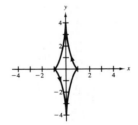

49. $xy = 1, x > 0, y > 0$

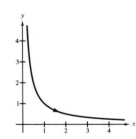

51. Circle

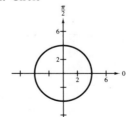

53. Rose curve

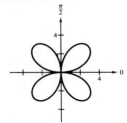

55. Cardioid

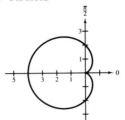

57. Limaçon

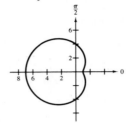

59. Rose curve

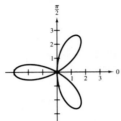

61. Lemniscate

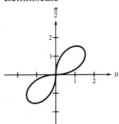

63. Line

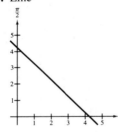

65. Parabola

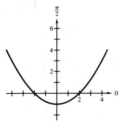

67. $x^2 + y^2 - 3x = 0$ **69.** $x^2 + 4y - 4 = 0$

71. $(x^2 + y^2)^2 - x^2 + y^2 = 0$ **73.** $r = a \cos^2 \theta \sin \theta$

75. $r = 10 \sin \theta$ **77.** $r = \dfrac{4}{1 - \cos \theta}$

79. $r = \dfrac{5}{3 - 2 \cos \theta}$

81. $x = -3 + 4 \cos \theta, y = 4 + 3 \sin \theta$

83. Proof

85.

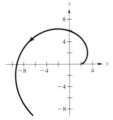

CUMULATIVE TEST for Chapters 6 and 7   *(page 498)*

1. (a)

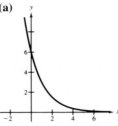

(b)

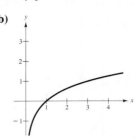

2. $\log_4 \dfrac{1}{16} = -2$ **3.** Inverses   **4.** $\ln \dfrac{x^2}{\sqrt{x + 5}}$

5. (a) $\dfrac{\ln 12}{2} \approx 1.2425$ **(b)** $\dfrac{64}{5}$

6. $16,302.05   **7.** $12,960

8.

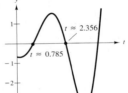

9. $\theta = \arctan \dfrac{1}{2} \approx 26.6°$

10. $m = \tan 6° \approx 0.105$
1104 feet

11. (a)

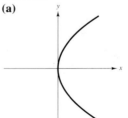

(b)

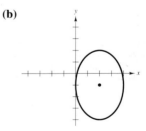

(c) **(d)**

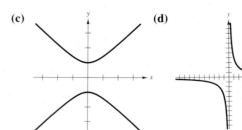

12. $\dfrac{(y-2)^2}{4/5} - \dfrac{x^2}{16/5} = 1$

13. $x = t$

$y = \dfrac{7}{4}t - \dfrac{13}{2}$

14. $\dfrac{(x-3)^2}{9} + \dfrac{(y-2)^2}{4} = 1$ **15.** $x^2 + y^2 - 6x = 0$

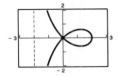

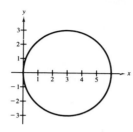

16. Vertical asymptote: $x = -2$ **17.** $r = \dfrac{2}{1 - \cos\theta}$

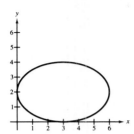

APPENDIX A   (*page A7*)

WARM-UP   **1.** $y = 4 - 3x$ **2.** $y = x$

3. $y = \frac{2}{3}(1 - x)$ **4.** $y = \frac{2}{5}(2x + 1)$

5. $y = \frac{1}{4}(5 - 3x)$ **6.** $y = \frac{2}{3}(-x + 3)$

7. $y = 4 - x^2$ **8.** $y = \frac{2}{3}(x^2 - 1)$

9. $y = \pm\sqrt{4 - x^2}$ **10.** $y = \pm\sqrt{x^2 - 9}$

1.

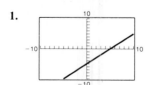

3.

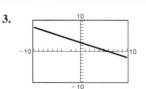

5.

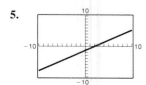

7.

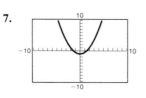

9.

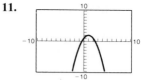

11.

13.

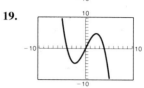

15.

17.

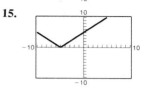

19.

21. (d)   **23.** (a)   **25.** (i)   **27.** (j)   **29.** (e)

31.

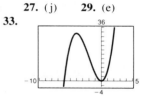

33.

35.

```
Xmin = -10
Xmax = 10
Xscl = 1
Ymin = -12
Ymax = 30
Yscl = 6
Xres = 1
```

37.

```
Xmin = -10
Xmax = 10
Xscl = 1
Ymin = -10
Ymax = 10
Yscl = 1
Xres = 1
```

39. No intercepts   **41.** Three x-intercepts

43. Square   **45.** Circle

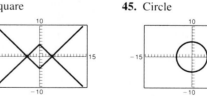

47.

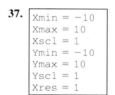

49. 0.59

51.

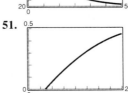

53. (b)

APPENDIX C   (*page A39*)

1. $-2, 1, 4, 7, 10$

3. $3, 6, 11, 18, 27$

5. $2, 4, 8, 16, 32$

7. $-\frac{1}{2}, \frac{1}{4}, -\frac{1}{8}, \frac{1}{16}, -\frac{1}{32}$

9. $\frac{1}{2}, \frac{2}{3}, \frac{3}{4}, \frac{4}{5}, \frac{5}{6}$

11. $-2, 4, -8, 16, -32$

13. $-1, 0, 4, 22, 118$

15. $3, \frac{9}{2}, \frac{9}{2}, \frac{27}{8}, \frac{81}{40}$

17. $2, 4, 12, 48, 240$

19. $1, 1, 2, 6, 24$

21. $3, 5, 3, 5, 3$

23. $3, 5, 3, 5, 3$

25. (a) \$101.00, \$203.01, \$306.04, \$410.10, \$515.20, \$621.35

(b) \$8248.64

(c) \$99,914.79

27. $0 + 1 + 4 + 9 + 16 = 30$

29. $2 + 5 + 10 + 17 + 26 = 60$

31. $2 + 6 + 12 + 20 = 40$

33. $0 + 3 + 20 + 115 + 714 = 852$

35. $2 + \frac{3}{2} + \frac{4}{3} + \frac{5}{4} + \frac{6}{5} = \frac{437}{60}$

37. $11 + 11 + 12 + 16 + 34 + 130 + 730 = 944$

39. 40 **41.** 78 **43.** 55 **45.** 2850

47. 16.02 **49.** $2, 6, 18, 54, 162$ **51.** $1, \frac{1}{2}, \frac{1}{4}, \frac{1}{8}, \frac{1}{16}$

53. $5, -\frac{1}{2}, \frac{1}{20}, -\frac{1}{200}, \frac{1}{2,000}$ **55.** $1, e, e^2, e^3, e^4$

57. $\left(\frac{1}{2}\right)^7$ **59.** $-\frac{2}{3^{10}}$ **61.** $500(1.02)^{39}$ **63.** 511

65. 43 **67.** 6.4 **69.** $29,921.31$ **71.** $2,092.60$

73. \$3,623,993.23   **75.** 2 **77.** $\frac{2}{3}$ **79.** $\frac{16}{3}$

81. Polynomial: 0.25000
Graphing utility: 0.24740

83. Polynomial: 0.24740
Graphing utility: 0.24740

85. Polynomial: 0.87500
Graphing utility: 0.87758

87. Polynomial: 0.87758
Graphing utility: 0.87758

89. Polynomial: 0.83333
Graphing utility: 0.84147

91.

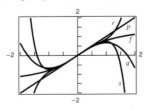

APPENDIX D

Section 1.4   (*page A53*)

1. The graphs are identical. Distributive Property

3. The graphs are identical. Associative Property

Section 1.5   (*page A53*)

5.

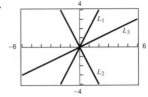

L_2 is perpendicular to L_3.

7.

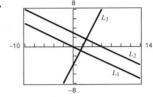

L_1 is parallel to L_2.

L_3 is perpendicular to L_1 and L_2.

9. (a) $y = 71.08t - 10.29$

(b)

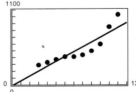

(c) \$1,411,000

(d) Average increase per year in thousands of dollars

Section 1.7   (page A54)

11.

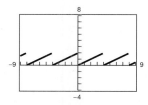

Domain: $(-\infty, \infty)$

Range: $[0, 2)$

Sawtooth pattern

13. (a) C_2 is the appropriate model because the cost does not increase until after the next minute of conversation has started.

(b) $7.85

Section 1.9   (page A54)

15.

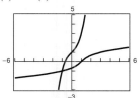

Section 1.10   (page A55)

17. (a) and (b)

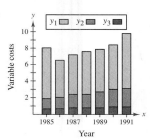

(c) Inverse function because it satisfies the Vertical Line Test

19. (a) and (b)

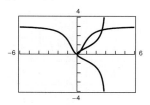

(c) Not an inverse function because it does not satisfy the Vertical Line Test

Section 2.1   (page A55)

21. (a) $54.75°$ (b) $-128.5°$

23. (a) $85.308°$ (b) $330.007°$

25. (a) $240°36'$ (b) $-145°48'$

27. (a) $143°14'22''$ (b) $-205°7'8''$

Section 2.3   (page A55)

29. (a)

θ	0	0.1	0.2	0.3	0.4	0.5
$\sin \theta$	0	0.0998	0.1987	0.2955	0.3894	0.4794

(b) θ

(c) $\sin \theta$ approaches θ as θ approaches 0.

31. (a)

θ	20°	30°	40°	50°	60°	70°
R	170.3	229.5	261.0	261.0	229.5	170.3

(b) The range increases as θ increases from 20° through 40° and decreases as θ increases from 50° through 70°.

(c) 45°

(d)

θ	43°	44°	45°	46°	47°
R	264.4	264.8	265.0	264.8	264.4

Section 2.4 (page A56)

33.

θ	0°	20°	40°	60°	80°
$\sec\theta$	1	1.0642	1.3054	2	5.7588
$\sec(-\theta)$	1	1.0642	1.3054	2	5.7588
$\tan\theta$	0	0.3640	0.8391	1.7321	5.6713
$\tan(-\theta)$	0	−0.3640	−0.8391	−1.7321	5.6713

The secant is an even function and the tangent is an odd function.

35. (a)

θ	0	0.3	0.6	0.9
$\cos\left(\dfrac{3\pi}{2}-\theta\right)$	0	−0.2955	−0.5646	−0.7833
$-\sin\theta$	0	−0.2955	−0.5646	−0.7833

θ	1.2	1.5
$\cos\left(\dfrac{3\pi}{2}-\theta\right)$	−0.9320	−0.9975
$-\sin\theta$	−0.9320	−0.9975

(b) $\cos\left(\dfrac{3\pi}{2}-\theta\right) = -\sin\theta$

Section 2.5 (page A56)

37.

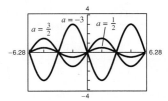

Amplitude changes

39.

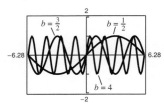

Period changes

41. (a) Yes. For each value of t there corresponds one and only one value of y.

(b) Period: 0.5 second; Amplitude: 0.35 centimeter

(c) $y = 0.35 \sin 4\pi t + 2$

(d)

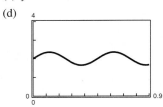

Section 2.6 (page A57)

43.

Not equivalent

45.

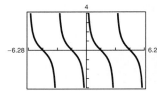

Equivalent

47.

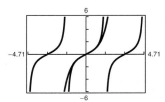

49. (a) 850 revolutions per minute

(b) The direction of the saw is reversed.

(c) $L = 60\left[\left(\dfrac{\pi}{2} + \phi\right) + \cot\phi\right], \quad 0 < \phi < \dfrac{\pi}{2}$

(d)

ϕ	0.3	0.6	0.9	1.2	1.5
L	306.2	217.9	195.9	189.6	188.5

(e) faster

(f)

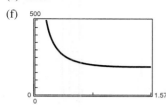

Section 2.7 (page A58)

51. (a)

x	−1	−0.8	−0.6	−0.4	−0.2
y	3.1416	2.4981	2.2143	1.9823	1.7722

x	0	0.2	0.4	0.6	0.8	1
y	1.5708	1.3694	1.1593	0.9273	0.6435	0

(b)

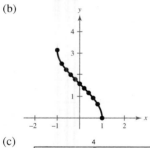

(c)

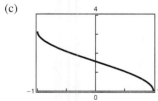

(d) $\left(0, \dfrac{\pi}{2}\right)$, $(1, 0)$

53. (a) $f \circ f^{-1}$ $f^{-1} \circ f$

(b) Differ because of the domain and range of f and f^{-1}.

Section 2.8 (page A59)

55. (a) $L = 60 \cot \theta$

(b)

θ	10°	20°	30°	40°	50°
L	3403	1648	1039	715	503

(c) No. Cotangent is not a linear function.

57. (a) $h = 20 \sin \theta$

(b)

θ	60°	65°	70°	75°	80°
h	17.3	18.1	18.8	19.3	19.7

59. (a)

θ	L_1	L_2	$L_1 + L_2$
0.1	$\dfrac{2}{\sin 0.1}$	$\dfrac{3}{\cos 0.1}$	23.0
0.2	$\dfrac{2}{\sin 0.2}$	$\dfrac{3}{\cos 0.2}$	13.1
0.3	$\dfrac{2}{\sin 0.3}$	$\dfrac{3}{\cos 0.3}$	9.9
0.4	$\dfrac{2}{\sin 0.4}$	$\dfrac{3}{\cos 0.4}$	8.4

(b)

θ	L_1	L_2	$L_1 + L_2$
0.5	$\dfrac{2}{\sin 0.5}$	$\dfrac{3}{\cos 0.5}$	7.6
0.6	$\dfrac{2}{\sin 0.6}$	$\dfrac{3}{\cos 0.6}$	7.2
0.7	$\dfrac{2}{\sin 0.7}$	$\dfrac{3}{\cos 0.7}$	7.0
0.8	$\dfrac{2}{\sin 0.8}$	$\dfrac{3}{\cos 0.8}$	7.1

7.0 meters

(c) $L = \dfrac{2}{\sin \theta} + \dfrac{3}{\cos \theta}$

(d)

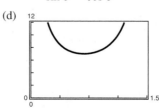

7.0 meters

61. (a)

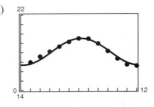

(b) 12 months. Yes, 1 period is 1 year.

(c) 1.41 hours. 1.41 represents the maximum change in time from the average time ($d = 18.09$) of sunset.

Section 3.1 (page A61)

63.

x	0.2	0.4	0.6	0.8	1.0
y_1	0.1987	0.3894	0.5646	0.7174	0.8415
y_2	0.1987	0.3894	0.5646	0.7174	0.8415

x	1.2	1.4
y_1	0.9320	0.9854
y_2	0.9320	0.9854

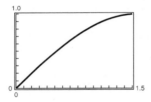

$y_1 = y_2$

65.

x	0.2	0.4	0.6	0.8	1.0
y_1	1.2230	1.5085	1.8958	2.4650	3.4082
y_2	1.2230	1.5085	1.8958	2.4650	3.4082

x	1.2	1.4
y_1	5.3319	11.6814
y_2	5.3319	11.6814

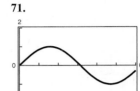

$y_1 = y_2$

67. $\csc x$

Section 3.2 (page A61)

69.

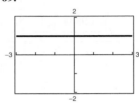

1

71.

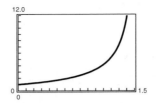

$\sin x$

Section 3.3 (page A61)

73. (a)

x	0	1	2	3	4
$f(x)$	Undef.	0.83	-1.36	-2.93	-4.46

x	5	6
$f(x)$	-6.34	-13.02

The zero is in the interval $(1, 2)$ because f changes signs in the interval.

(b)

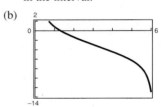

The interval is the same as in part (a).

(c) 1.3065

75. (a)

x	0	1	2	3	4
$f(x)$	−1	1.39	1.65	−0.70	−1.94

x	5	6
$f(x)$	−2.00	−1.48

The zeros are in the intervals (0, 1) and (2, 3) because f changes signs in these intervals.

(b)

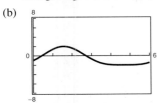

The intervals are the same as in part (a).

(c) 0.4271, 2.7145

Section 3.4   (page A62)

77. Answers will vary.

Section 3.5   (page A62)

79.

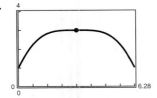

Maximum: $(\pi, 3)$

81. (a) $f(x) = \frac{1}{4}(3 + \cos 4x)$

(b) $f(x) = 2 \cos^4 x - 2 \cos^2 x + 1$

(c) $f(x) = 1 - 2 \sin^2 x \cos^2 x$

(d) $f(x) = 1 - \frac{1}{2} \sin^2 2x$

(e) There is often more than one way to rewrite a trigonometric expression.

Section 4.1   (page A62)

83. (a) $20\left[15 \sin \dfrac{3\theta}{2} - 4 \sin \dfrac{\theta}{2} - 6 \sin \theta\right]$

(b)

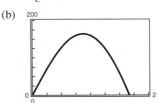

(c) Domain: $0 \le \theta \le 1.6690$

The domain would increase in length and the area would increase.

Section 4.3   (page A63)

85. (a) $5\sqrt{5 + 4 \cos \theta}$

(b)

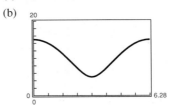

(c) Range is $[5, 15]$. Maximum is 15 when $\theta = 0$.

Minimum is 5 when $\theta = \pi$.

(d) The magnitudes of $\mathbf{F}_1$ and $\mathbf{F}_2$ are not the same.

Section 5.2   (page A64)

87. (a)

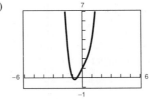

(b) 2

(c) $-1, -0.54$

89. (a)

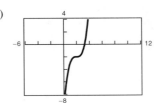

(b) 1

(c) 3.26

Section 5.3   (page A64)

91. $5.39(\cos 0.38 + i \sin 0.38)$

93. $8.19(\cos 5.26 + i \sin 5.26)$

95. $4.70 + 1.71i$

97. $-3.22 - 2.38i$

Section 5.4   (page A64)

99. $-597 - 122i$

101. $\dfrac{81}{2} + \dfrac{81\sqrt{3}}{2}i$

Section 6.1   (page A64)

103. Asymptote: $y = 0$

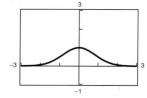

105. Asymptote: $y = 1$

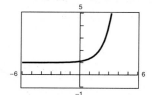

107. (a) and (b)

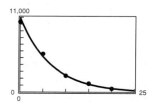

(c)

h	0	5	10	15	20
P	10,958	5176	2445	1155	546

(d) 3300 kilograms per square meter

(e) 11.3 kilometers

109.

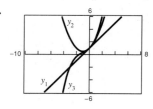

Section 6.2   (page A65)

111. Domain: $(0, \infty)$

Vertical asymptote: $x = 0$

Intercept: $(5, 0)$

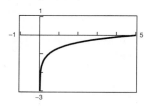

113. Domain: $(2, \infty)$

Vertical asymptote: $x = 2$

Intercept: $(3, 0)$

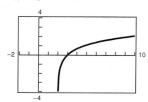

115. Domain: $(-\infty, 0)$

Vertical asymptote: $x = 0$

Intercept: $(-1, 0)$

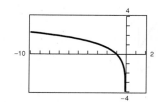

117. Domain: $(0, \infty)$

Decreasing: $(0, 0.37)$

Increasing: $(0.37, \infty)$

Relative minimum: $(0.37, -1.47)$

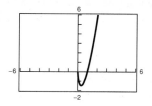

119.

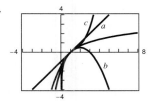

121. (a)

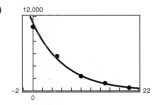

(b) $\ln P = -0.1499h + 9.3018$

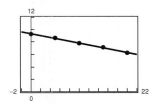

(c) $P = 10{,}957.7e^{-0.1499h}$

Section 6.3 (page A66)

123.

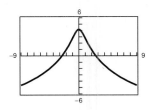

$y_1 = y_2$

125.

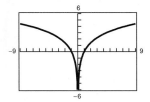

No. The domains differ.

Section 6.4 (page A66)

127.

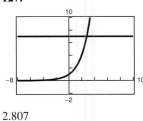

2.807

129.

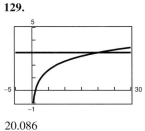

20.086

Section 6.5 (page A66)

131. (a) $t_3 = 0.2729s - 6.0143$

$t_4 = 1.5385e^{0.0291s}$

(b)

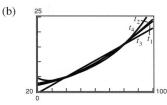

(c)

s	30	40	50	60	70	80	90
t_1	3.6	4.7	6.7	9.4	12.5	15.9	19.6
t_2	3.3	4.9	7.0	9.5	12.5	15.9	19.9
t_3	2.2	4.9	7.6	10.4	13.1	15.8	18.5
t_4	3.7	4.9	6.6	8.8	11.8	15.8	21.1

(d) Model: t_1; Sum ≈ 1.9

Model: t_2; Sum ≈ 1.2

Model: t_3; Sum ≈ 5.6

Model: t_4; Sum ≈ 2.6

Quadratic model fits best.

Section 7.2 (page A67)

133. (a)

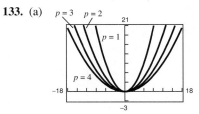

As p increases the parabola becomes wider.

(b) $(0, 1), (0, 2), (0, 3), (0, 4)$

(c) $4, 8, 12, 16; 4p$

(d) Easy way to determine two additional points on the graph

Section 7.6   (page A68)

135. (a) $x = (146.67 \cos \theta)t$

$y = 3 + (146.67 \sin \theta)t - 16t^2$

(b) $x = 141.7t$

$y = 3 + 38.0t - 16t^2$

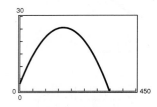

No

(c) $x = 135.0t$

$y = 3 + 57.3t - 16t^2$

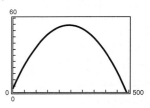

Yes

(d) About $19.4°$

Section 7.7   (page A68)

137. $\left(-\sqrt{2}, \sqrt{2}\right)$ **139.** $(-1.204, -4.336)$

141. $\left(\sqrt{13}, -0.588\right)$ **143.** $\left(\sqrt{7}, 0.857\right)$

145. $\left(\dfrac{17}{6}, 0.490\right)$

Section 7.8   (page A68)

147.

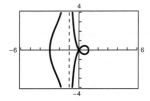

149.

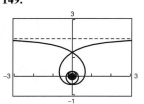

Section 7.9   (page A68)

151.

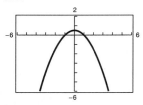

Parabola

153.

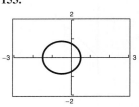

Ellipse

155.

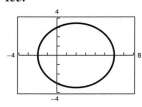

Ellipse

157.

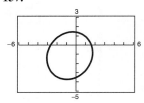

INDEX OF APPLICATIONS

U.S. Demographics Applications

INDEX